A   Courtesy Champion Petfoods, Inc.

B   Courtesy Dr. Michael J. Adkesson.

C   Courtesy Tom Schaefges.

D   Curtesy Paul E. Miller, Sullivan, MO.

E   Courtesy Paul E. Miller, Sullivan, MO.

F   Courtesy Dr. Michael J. Adkesson.

G   Courtesy Champion Petfoods, Inc.

H   Courtesy Dr. Michael J. Adkesson.

I   Courtesy Questhavenpets.com.

J   Courtesy Questhavenpets.com.

K   Courtesy Champion Petfoods, Inc.

L   Courtesy anonymous contributor.

M   Courtesy Melissa Maitland.

N   Courtesy anonymous contributor.

# Companion Animals

## Their Biology, Care, Health, and Management

**Karen L. Campbell, DVM, MS**
*Diplomate, American College of Veterinary Internal Medicine*
*Diplomate, American College of Veterinary Dermatology*
*Professor and Section Head, Specialty Medicine*
*University of Illinois*

**James E. Corbin, Ph.D.**
*Professor Emeritus of Animal Sciences, University of Illinois*

**John R. Campbell, Ph.D.**
*President Emeritus*
*Professor of Animal Science*
*Oklahoma State University*
*Formerly Dean, College of Agriculture, University of Illinois*
*Formerly Professor of Dairy Science, University of Missouri*

PEARSON
Prentice
Hall

Upper Saddle River, New Jersey 07458

**Library of Congress Cataloging-in-Publication Data**
Campbell, Karen L.
  Companion animals : their biology, care, health, and management/Karen L. Campbell,
James E. Corbin, John R. Campbell.
    p. cm.
  Includes bibliographical references and index.
  ISBN 0–13–113610–0
    1. Pets. I. Campbell, John R. II. Corbin, James E. (James Edward). III. Title.

SF413.C36 2004
636.088'7—dc22

2004044659

**Executive Editor:** Debbie Yarnell
**Associate Editor:** Kimberly Yehle
**Production Editor:** Trish Finley, Carlisle Publishers Services
**Production Liaison:** Janice Stangel
**Director of Production and Manufacturing:** Bruce Johnson
**Managing Editor:** Mary Carnis
**Manufacturing Buyer:** Cathleen Petersen
**Creative Director:** Cheryl Asherman
**Cover Design Coordinator:** Christopher Weigand
**Cover Design:** Christopher Weigand
**Cover Images:** top left: Bearded Dragon Lizard, photographed by G. K. and Vikki Hart, courtesy of Getty Images, Inc.–Photodisc;
top middle: Rabbit, photographed by G. K. and Vikki Hart, courtesy of Getty Images, Inc.–Photodisc; top right: Macaw,
photographed by Siede Preis, courtesy of Getty Images, Inc.–Photodisc; bottom left: German Shephed Dog, photographed by
G. K. and Vikki Hart, courtesy of Getty Images, Inc.–Photodisc; middle bottom: Ferret, photographed by G. K. and Vikki Hart,
courtesy of Getty Images, Inc.–Photodisc; bottom right: Domestic Cat, photographed by G. K. and Vikki Hart, courtesy of Getty
Images, Inc.–Photodisc.
**Marketing Manager:** Jimmy Stephens
**Composition:** Carlisle Communications, Ltd.
**Printing and Binding:** Courier Westford

Pearson Education Ltd.
Pearson Education Singapore, Pte. Ltd.
Pearson Education Canada, Ltd.
Pearson Education—Japan

Pearson Education Australia PTY, Limited
Pearson Education North Asia Ltd.
Pearson Educación de Mexico, S.A. de C.V.
Pearson Education Malaysia, Pte. Ltd.

10 9 8 7 6 5 4 3 2 1
ISBN: 0-13-113610-0

# BRIEF CONTENTS

# Contents

        7.3.1    Introduction      126
        7.3.2    Biology and Behavior      127
        7.3.3    Husbandry      128
        7.3.4    Breeding      128
        7.3.5    Common Diseases      129
        7.3.6    Miscellaneous Diseases      130
        7.3.7    Zoonoses      130
7.4      Hamsters      131
        7.4.1    Introduction      131
        7.4.2    Anatomy and Selected Characteristics      132
        7.4.3    Biology and Behavior      133
        7.4.4    Husbandry      133
        7.4.5    Breeding      134
        7.4.6    Common Diseases      134
        7.4.7    Parasites      135
        7.4.8    Kidney Failure      135
        7.4.9    Atrial Thrombosis      135
        7.4.10   Neoplasia/Tumors      135
        7.4.11   Zoonoses      135
7.5      Ferrets      136
        7.5.1    Introduction      136
        7.5.2    Biology      136
        7.5.3    Husbandry      138
        7.5.4    Diseases      139
        7.5.5    Other Conditions      141
        7.5.6    Zoonoses      142
7.6      Rabbits      142
        7.6.1    Introduction      142
        7.6.2    Anatomy      142
        7.6.3    Physiology      143
        7.6.4    Behavior      143
        7.6.5    Housing      144
        7.6.6    Nutrition      144
        7.6.7    Reproduction      145
        7.6.8    Diseases      145
        7.6.9    Zoonoses      148
7.7      Mice and Rats      148
        7.7.1    Introduction      148
        7.7.2    Anatomy      149
        7.7.3    Physiology      150
        7.7.4    Husbandry      151
        7.7.5    Common Diseases      152
        7.7.6    Zoonoses      153
7.8      Summary      153
7.9      References      154
7.10     About the Authors      155

■ **CHAPTER 8:** *Records and Case Histories*      *156*

        8.1      Introduction/Overview      156
        8.2      Animal Name and Identification      157
        8.3      Pedigree      157
        8.4      Breeder Records      157

## ■ CHAPTER 10: *Reproductive Biology of Dogs and Cats*    *212*

## ■ CHAPTER 11: *Anatomy and Physiology of Cats and Dogs*    *233*

# ■ CHAPTER 12: *Care, Management, and Training of Dogs and Cats*    *255*

## ■ CHAPTER 13: *Fitting, Grooming, and Showing*     *276*

## ■ CHAPTER 14: *Companion Animal Health*     *299*

## CHAPTER 23: *Career Opportunities Associated with Companion Animals*    *492*

## CHAPTER 24: *Managing Unwanted Companion Animals*    *510*

## ■ CHAPTER 25: *Trends/Future of Companion Animals and Related Functions*    *520*

# PREFACE

The all-important purpose of this resource book is to engage the attention of readers seeking a textbook and useful reference devoted to companion animals. The book provides underpinning principles and time-tested practical information that will prove insightful and useful to those who have fixed their sights and focused their dreams and commitment on a career related to the health and quality of life of *all creatures great and small* with special emphasis on companion animals.

*Companion Animals: Their Biology, Care, Health, and Management* is intended especially as a text for college students enrolled in companion animal biology courses, veterinary technician/technology programs, and companion animal management courses; for those wishing to pursue a professional career in veterinary medicine; persons planning a career in the rapidly expanding petfood, pet care products and services, and related industries; breeders, exhibitors, groomers, owners, and trainers of companion animals; current and prospective owners/managers of pet shops, boarding kennels and catteries; future professionals preparing to work as animal-assisted therapists or representatives of pharmaceutical companies; persons interested in teaching and/or conducting research related to life enhancement of humans and animals through contributions to science by companion/laboratory animals; as well as those preparing for other career opportunities associated with companion animals. The latter includes governmental agencies and private consultants (Chapter 23).

The first two chapters present materials related to the history and value of "Companion Animals" (Chapter 1) in the lives of children and adults, as well as the "Companion Animal Industry" (Chapter 2). Because of the special bonding of dogs and cats with their owners, and their long history as the most widely kept domestic pets, we emphasize these two species throughout the book. Chapters 3 and 4 highlight "Dog/Cat Breeds and Their Characteristics" and "Choosing a Dog or Cat." Chapters 5 and 6 are devoted to "Companion Birds" (Chapter 5) and "Companion Reptiles and Amphibians" (Chapter 6). "Companion Rodents, Ferrets, and Lagomorphs" (Chapter 7) deals with characteristics and the care and management of rodents—chinchillas, gerbils, guinea pigs, hamsters, mice, and rats; lagomorphs (especially rabbits); and ferrets. Chapter 7 also addresses the care, health, housing, management, nutrition, and other pertinent topics related to common laboratory animals. Chapter 8 is devoted to the importance and usefulness of "Records and Case Histories." We then discuss the important topics of "Feeding and Nutrition" (Chapter 9) and "Reproductive Biology of Dogs and Cats" (Chapter 10). The next three chapters include the "Anatomy and Physiology of Cats and Dogs" (Chapter 11); the "Care, Management, and Training of Dogs and Cats" (Chapter 12); and the "Fitting, Grooming, and Showing" of companion animals (Chapter 13).

"Companion Animal Health" is discussed in Chapter 14, and "Parasites and Pests of Companion Animals" are discussed in Chapter 15. Chapter 16, "Common Diagnostic and Therapeutic Procedures and Terms," is of special interest to those preparing for professional careers as veterinary technicians and veterinary technologists. For students and others with special interest in "Pet Sitting, Pet Motels, and Other Boarding Arrangements" (Chapter 17), "Kennel/Cattery Design and Management" (Chapter 18), and the "Business/Financial Aspects of the Companion Animal Enterprise" (Chapter 19), the practical information given in these three chapters will be useful.

Extensive research related to "Companion Animal Behavior and Social Structure" (Chapter 20) has brought to light interesting examples of heretofore secrets of nature. Many of these are of practical significance in the humanizing and socializing role companion animals play in human interactions.

"Therapeutic and Service Uses of Companion Animals" (Chapter 21) are becoming more and more important in enriching the health, lives, and security of people (e.g., as explosive detection, guide, hearing, inspection of imported goods and commodities, medical alert, mobility assistance, personal/premise protection, police, psychiatric, search and rescue dogs). Discussion of "Animals in Biomedical Research" (Chapter 22) and the care, housing, and management of rodents and other laboratory animals (Chapter 7) provide pertinent, useful information on these topics. Numerous examples such as the discovery of the vaccine for poliomyelitis attest to invaluable contributions made by companion, laboratory, and other animals to the health and quality of life of humans worldwide as well as to the health and quality of life of animals. Discussion of animal welfare/animal rights issues and challenges is also included. As mentioned above, Chapter 23 highlights numerous personally rewarding Career Opportunities Associated with Companion Animals.

For a plethora of reasons, many wanted—later unwanted—companion animals are abandoned by their owners and often become a problem, especially in urban settings. Managing Unwanted Companion Animals is discussed in Chapter 24. Finally, pertinent Trends/Future of Companion Animals and Related Functions are topics included in Chapter 25.

Other features of the book include the use of more than 400 photographs, line drawings, and other illustrative materials and more than 100 tables to aid readers in comprehending concepts as well as adding to reading pleasure. Included are full-color photos of breeds of dogs, cats, birds, and other companion animals. To better understand materials presented, an extensive glossary is included.

Authors of a book of this magnitude benefit from the talents and professional expertise of many people. Indeed, scholarly scientists and other dedicated professionals who contributed to research, teaching, and public service over the years—those who generated much of the data and information recorded here—are too numerous to name, impossible to repay. Major contributing authors to the book are named and acknowledged in Chapters 5, 6, 7, and 16. These important chapters were written by invited authors who are world-renowned experts on the biology, care, health, management, and environments of the animals discussed. Other contributors are cited in the credit line of photographs/illustrations they shared, and others are named in the acknowledgments.

Finally, those who have experienced the personal pleasure associated with looking into the affectionate eyes of cuddly cats, devoted dogs, and other admired pets know why they are interested in companion animal biology and why it is increasingly being included in the study of animal sciences. We wrote this resource book for people—most of whom we do not know—but with whom we share a common love: companion animals. It was a genuine pleasure to conceptualize and write the book. It is our sincere wish that readers will find it an enjoyable read and future reference as well as a useful base to help launch and sustain a rewarding and successful career!

*Karen L. Campbell, DVM*
*James E. Corbin, Ph.D.*
*John R. Campbell, Ph.D.*

# ACKNOWLEDGMENTS

Conceptualizing, authoring, and finalizing the manuscript and illustrative materials for a book of this magnitude has been a cooperative team effort.

We are deeply indebted and grateful to Cynthia Coursen Alexander, Cynthia R. Garnham, John E. Harkness, Steele F. Mattingly, Dorcas P. O'Rourke, Jill A. Richardson, Douglas W. Stone, and Rosemary M. Tini who kindly shared their time and professional expertise in authoring/ co-authoring Chapters 5, 6, 7, and 16. And we express appreciation to J. Michael Bale, Stanley E. Curtis, John Cwaygel, Jean Fisher, Frank C. Robson, Kay Stewart, and George Whitney for their important contributions to Chapters 6, 13, 19, 22, and 23.

Numerous manuscript reviewers made helpful suggestions for revision of one or more chapters. We thank all of you! A special thanks to Paul L. Nicoletti, DVM, Professor Emeritus of Pathobiology, University of Florida College of Veterinary Medicine, who kindly reviewed all the manuscript and shared his invaluable teaching/research/service perspectives/ recommendations based on five decades of domestic and global experiences in veterinary medicine.

We are most grateful to those who kindly shared photographs—individuals, corporate, foundation, and governmental leaders. Many are named in figure captions, others requested anonymity. Three photographers were particularly helpful: Sandra Grable, Paul E. Miller, and Jeff Rathmann. We appreciated the level of interest they demonstrated in the book. And we gratefully acknowledge the illustrative materials prepared by Diana Nicoletti—a dedicated, highly professional art teacher!

We are abundantly grateful to Benita Bale for her professional assistance—both clerically and editorially—in preparing the manuscript and to Eunice J. Campbell for the important contributions she made through editorial suggestions and proofreading. We appreciate the patience of our families who saw less of us and helped in numerous ways. A special thanks also to Adam P. Patterson, DVM, who took over extra days of clinic teaching to allow us to concentrate on completing this project.

We were fortunate in having the superb support of a first-class team of Prentice Hall professionals including Kimberly Yehle, Deborah Yarnell, and Janice Stangel who, together with Trish Finley of Carlisle Publishers Services, made invaluable contributions to clarity, consistency, design, and other important aspects of the book. We are grateful for their exemplary services.

Finally, the book's contents, cross-referencing, and readability were made better by the useful suggestions and wise feedback of experienced teachers of much of the subject matter who kindly reviewed the manuscript. For their helpful recommendations, we thank Cynthia A. McCall, Auburn University; Anita M. Oberbauer, University of California–Davis; Stuart Porter, Blue Ridge Community College (VA); John E. Warren, West Virginia University; and numerous other reviewers who chose to remain anonymous.

# 1 COMPANION ANIMALS

*Your dog is your friend, your partner, your defender.*
*You are its life, its love, its leader. It will be yours, faithful and true,*
*to its last heartbeat. As its owner, be worthy of such devotion.*

George Eliot (1819–1880)
*English novelist*

## 1.1 INTRODUCTION/OVERVIEW

For centuries animal pets, particularly dogs and cats, have played a humanizing and socializing role in interactions among humans. Pets have also contributed to the overall health and well-being of people.

**Figure 1.1** The mutual pleasure of being together is readily apparent between this child and her dog. *Courtesy Mike Nelson and Vern Horning, Graphics, American Nutrition Inc.*

The attachments humans develop for companion animals are rooted in two prominent qualities of most dogs and cats: their inherent ability to provide love and tactile reassurance without criticizing or grumbling, and their consistent maintenance of a form of infantile innocent dependence, which in turn provides people an opportunity to feel needed and valued.

The domestic dog (*Canis familiaris*) and domestic cat (*Felis catus*) have become an integral part of human society as evidenced by the large number of animal-owning households worldwide. Nearly two-thirds of American families and over half of all households in English-speaking countries share their homes with one or more companion animals. Relationships between people and companion animals begin at an early age (Figure 1.1), are long-lasting, and are often intense (Figure 1.2).

For many people, living with a companion animal is living with another family member. Animals provide a means of establishing contact, communication, and interaction with people. A dog on a leash, just as a baby in a carriage, invites attention, admiration, and conversation. And wise are the politicians who include a Collie or a Golden Retriever in the family portrait. Folklore is replete with anecdotes of heroic companion animals such as those who have alerted neighbors that their owner needed help, saved families from household fires, rescued drowning children, and alerted parents that their infant had stopped breathing. Media mythology of Lassie—a Collie that made television an early success with children and adults—and Rin-Tin-Tin—a German Shepherd that appeared in more than 40 films— has made the exploits of dogs special conversation topics. Dogs have improved the quality of life for the blind; deaf; and physically, mentally, or emotionally impaired as well as countless others (Chapter 21). Dogs also perform all-important services as police dogs, tracking dogs, herding dogs, search and rescue dogs, guard dogs, and sled dogs (e.g., McGruff, the crime-prevention dog; Sparky, the fire-prevention dog); and as institutional mascots (e.g., the Georgia Bulldog; Texas A&M Revelry; and the Washington Huskies).

The adoption of companion animals in millions of homes has provided impetus for the development of a sophisticated, service-oriented, $12.4 billion petfood industry in the

**Figure 1.2**   The special bonding between this 22-year-old cat and 98-year-old lady has enriched the lives of both. *Courtesy Paul E. Miller (photographer), Sullivan, MO; Elizabeth Bruns; and Mary Bruns Vieten.*

United States (Chapter 2). This has resulted in a petfood market that provides a large variety of commercially prepared, nutritionally complete petfoods that vary in composition and format to provide balanced diets for companion animals (Chapter 9).

Dogs are frequently used in advertisements, television shows, and movies. Examples include using the second most commonly registered breed of dog, the Golden Retriever, in advertising Oreo® products and the greatly admired Border Collie in an aluminum foil commercial. The television comedy show *Frasier* highlighted Eddie, a small Parson Russell Terrier.[1] The *101 Dalmatians* movie significantly increased the sale of spotted dogs.

Snoopy, Charlie Brown's dog in cartoons by Charles Schulz, fits beautifully into the world of children with their fun, loyalty, and warmth. Think of children clutching their favorite stuffed animal as books about animals are read to them. Such a scene serves to impress on us the importance of companion animals in their lives. Many children sleep with a stuffed animal and include companion animals in their drawings.

Examples of famous dogs include Balto,[2] an Eskimo dog who led a series of dog relay teams in 1925 that carried life-saving diphtheria serum 700 miles through an Alaskan blizzard from Nenana to ice-bound Nome (serum was taken the 400 miles from Anchorage to Nenana over the newly completed Alaska railroad); Barry, a Saint Bernard who rescued 40 persons when they became lost in the winter snows of Switzerland's Saint Bernard Pass; Buddy, a German Shepherd who in 1928 became the first seeing-eye dog for the blind; Caesar, a Terrier pet of King Edward VII of Great Britain, who walked ahead of kings and princesses in his master's funeral procession in 1910; Igloo, a special pet of American admiral and polar explorer Richard E. Byrd (1888–1957) who flew with Byrd on flights over the North and South Poles (Figure 1.3); and Laika, the world's first space traveler who was sent aloft in an artificial earth satellite in 1957 by Russian scientists.

---

[1] The Parson Russell Terrier is the official name of the breed formerly (prior to 2003) known in the United States as Jack Russell Terrier. The name change reflects the origin of the breed as fox hunting dogs bred by Parson John Russell (1795–1883) in the United Kingdom.

[2] Balto became famous as the lead dog on the team that mushed into Nome. A statue of Balto was erected a year later in New York's Central Park (www.iditarod.com/background.html).

**Figure 1.3** Igloo with Admiral Richard E. Byrd (*l*) and W. H. Danforth (*r*) who provided Ralston Purina dog food for Byrd's flights over the North and South Poles. *Courtesy Dr. James E. Corbin.*

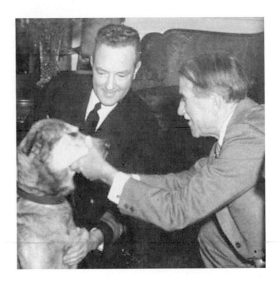

Cats have also excited the interest of children. Examples include the children's classic *Alice's Adventures in Wonderland* in which Lewis Carroll featured the famous Cheshire Cat, whose smile showed a number of sharp teeth; Walt Disney's animated movie of cat adventure in the *Aristocats*; the cartoon characters from *Felix the Cat* and Tom of *Tom and Jerry*; and the fairytale "Puss in Boots" featuring a cat getting away with all sorts of things and ending up eating an ogre (monster) after he talked the ogre into becoming a mouse.

A pleasing scene to parents (also to those passing by who admire cats and enjoy children) is that of the family cat alongside a child waiting for the morning school bus. That special send-off provides youngsters with a memory to which they can refer throughout the day. Knowing their favorite cat will be watching through the window when they return from school is equally gratifying (Figure 1.4). In associating with and caring for pets children learn the personal pleasures of animal companionship as well as responsibility, respect for life, gentleness, and experience in life processes including the transition between life and death (Chapter 9).

**Figure 1.4** The family cat anxiously awaits the school bus bringing its favorite friends home for petting and playtime. *Courtesy Paul E. Miller.*

The cat was a symbol of liberty in Ancient Rome. Roman artists often showed Libertas, the goddess of liberty, in their paintings with a cat lying at her feet.

Opportunities to *make a difference* in the lives of animals and their owners have never been greater than they are for today's veterinarians, veterinary technologists, veterinary technicians, and other members of the professional companion animal support team (Chapter 23). Few people have the opportunity in their lives to genuinely help as many people as the small animal practitioner. According to the 2000 U. S. Census data, there are approximately 46,000 veterinarians practicing small animal medicine in 28,000 veterinary clinics in the United States. They attend to more than 190 million pet visits annually. That number, multiplied by an average of 1.7 persons accompanying each pet to the clinic, is a total of approximately 323 million human contacts annually by the veterinary profession. Other important opportunities in animal healthcare include a career as a veterinary technician, veterinary technologist, and veterinary assistant; pet sitting, pet motels, or other boarding arrangements (Chapter 17); or in kennel/cattery design and management (Chapter 18) as well as many other related professional pursuits (Chapter 23).

## 1.2   COMPANION ANIMALS DEFINED

Companion animals are also called pets and include cage and aviary birds, cats, chinchillas, chipmunks, dogs, ferrets, fish, gerbils, guinea pigs, hamsters, insects, lizards, mice, other pocket pets,[3] rabbits, rats, snakes, turtles, exotic pets, sometimes large animals (calves, goats, horses, sheep, swine), and numerous others (many of the above pets are discussed in Chapters 5, 6, and 7). Companion animals are those species that have a *special* relationship with humans, are partially or totally dependent on them, live in close proximity with people, and go a step further than exotic pets in *bonding* with their owners. That is why they are referred to as *companion* animals.

The term *companion animals* has come to mean those animal species that have a special association with people—those that interact with their owners—those with which there is a mutual or reciprocal action or influence between animal and owner.

The word *companion* is derived from the Latin *com* (with) and *panis* (bread) meaning that as humans we share our bread (food and resources) with them and in return we receive their companionship. Because dogs and cats are the most widely kept domestic pets that fit the above definition of companion animals, they are the two species to which we give the greatest attention throughout this book. This in no way is intended to deemphasize or depreciate the dozens of other companion animals and pets, including the horse, that are the favorite animal to millions of children, youth, and adults. Interestingly, of the 5.1 million horses owned as pets in 2001 (up from 4.0 million in 1996), approximately 80% of the owners under 29 years were female.[4]

## 1.3   DOMESTICATION OF COMPANION ANIMALS

Domestication is defined as the point at which the care, feeding, and breeding of a species come under the control of people. It occurs when humans, as opposed to natural forces, become the primary selection agents in the evolution of a specie. The estimated date of domestication tells us when the effects of humans—selective mating and breeding—first became manifest.

---

[3]Pocket pets is a term used to describe small animals such as hamsters, gerbils, rats, mice, guinea pigs, and rabbits. They are cute, usually inexpensive, and commonly kept as pets. Pocket pets are especially popular with children.

[4]American Veterinary Medical Association. 2002. *U.S. Pet Ownership and Demographic Sourcebook.* Schaumburg, IL: AVMA, www.avma.org.

**Figure 1.5** A small-sized Yorkshire Terrier beside the large-sized Great Dane. As discussed in Chapter 4, adult dog size is frequently a factor in choosing a dog for a specific purpose (e.g., one would not likely select a Great Dane for a lap dog). *Courtesy Dr. James E. Corbin.*

## 1.3.1 Dogs

Dogs exhibit enormous variability from their wild ancestors. Human selection for certain traits resulted in major variations in hair color and length. Another common genetic trend has been a gradual foreshortening of the head and muzzle resulting in certain flat-faced breeds such as the Bulldog, Pekingese, and Pug. Other breeds have been selected for relatively large and frontally directed eyes. Utilitarian considerations such as size, speed, and strength for hunting or working have also played a part in mating and breeding domestic dogs (Chapters 3 and 4). Readily apparent, as well, is their wide range in size from the diminutive 2 to 3 lb Chihuahua and 4 to 6 lb Yorkshire Terrier to the 110 to 125 lb Great Dane (Figure 1.5) and the mighty 180 to 200 lb Mastiff or Irish Wolfhound. Dogs also possess a wide variety of coat types, temperaments, and morphologies (Chapters 3 and 4).

Dogs were domesticated, bred, and developed to work with and for people (Figure 1.6). The concept of animal companionship was slow to evolve in a world in which people hunted and farmed to feed their families and depended on dogs to guard home and hearth, rid premises of pests, and provide power in pulling small carts and sleds. Other chores included bed warming and serving as prestige symbols in palaces and estates around the world. Today most people in Western countries value dogs more as companions and creatures of great beauty than as scavengers or servants.

Being the first animal domesticated, dogs have been living with humans for at least 15,000 years, probably longer. And although their popularity has varied with time and location, they are found in numbers ranging from approximately 5 per 100 people in Germany and Japan to 10 in the United Kingdom and 18 in the United States. Data are unavailable for many countries. Cats have a shorter history of domestication (about 6,000 years), but in countries in which data are available their numbers approximate those of dogs (Table 1.1).

Numerous explanations have been advanced as reasons for domesticating animals. The stimulus for domesticating meat-producing animals (e.g., cattle, goats, sheep) likely included the need for food and clothing. Because dogs and cats are predominately carnivorous and compete for human food resources, there would appear to be less impetus for their domestication.

**Figure 1.6**   The personal pleasures and experiences associated with sled dog races and treks are many, multifarious, and memorable. Here Iditarod racing teams of Alaskan sled dogs are ready to be harnessed and hitched to the sleds that will be pulled over the 1,150-mile endurance-testing journey from Anchorage to Nome, Alaska. Approximately 1,200 dogs compete in the world's most famous dog sled race each March. *Courtesy Dr. James E. Corbin.*

An early motive for domesticating dogs and cats was likely to utilize their hunting and retrieving skills. Both dogs and cats are equipped with an exceptionally keen sense of smell. A dog's sense of smell is more than a thousand times as sensitive as that of a human, thereby enabling it to detect scents through concrete and scents of other animals for considerable distances.

Some biologists believe wolves became domesticated by spending time in prehistoric camps to obtain scraps, and perhaps to seek help in sharpening their hunting or herding skills. They could have lost their wolfish looks and ways, acquiring the social savvy of people. Others believe humans created the canine social sense by selecting and mating the canniest wolves.

Another popular theory of the domestication of dogs goes beyond utilitarian considerations. Indeed, numerous authors have postulated that the earliest domestication of dogs

**Table 1.1**

Number of Dogs and Cats per 100 Humans in Selected Countries (listed in ascending order of dogs)

| Country | Dogs | Cats | Country | Dogs | Cats |
|---|---|---|---|---|---|
| Japan | 3.9 | 2.0[*] | Sweden | 8.9 | 9.5 |
| Germany | 5.5 | 5.8 | UK | 10.0 | 9.6 |
| Switzerland | 6.2 | 12.5[*] | Belgium | 11.5 | 1.0[*] |
| Norway | 6.8 | 9.9 | Canada | 13.0 | 14.0 |
| Italy | 7.8 | 8.4 | Denmark | 13.3 | 10.8 |
| Austria | 8.0 | 14.7[*] | Australia | 15.2 | 13.9 |
| Netherlands | 8.4 | 10.6 | France | 17.0 | 12.6[*] |
| Finland | 8.6 | 9.0 | USA | 17.8 | 21.0 |

*Source: Adapted from J. F. D. Greenhalgh (UK) and J. E. Corbin (USA), Proc. 8th World Conference on Animal Production (1998), Seoul, Korea.*
[*]*Markedly different from a dog:cat ratio of 1:1.*

arose as a natural consequence of the universal human tendency to adopt young wild animals as pets. They pictured young wolf cubs being brought to villages by returning hunters or children. Under such circumstances the new pet may have acquired a personal name and family status.

Present evidence indicates that the human–companion animal bond is of great antiquity and that it was this bond and people's capacity to generalize their social responses to include wild animals of other species that lies at the root of domestication.

Why were wolves and wild cats among the first wild species domesticated? Numerous authors have proposed that these species, especially the wolf, were strongly preadapted in various ways to live as members of human social groups.

Interestingly, neither dogs nor cats differ markedly in behavior from their wild ancestors (Chapter 20). The most obvious behavioral modifications resulting from domestication are changes in response threshold to certain stimuli, increased docility and adaptability, perpetuation of infantile behavior patterns (which enable owners to talk to their pets and treat them as children), and a trend to promiscuity rather than pair-bond matings. In most breeds of domestic dogs, bitches demonstrate estrus twice rather than once annually (Chapter 10).

When a dog jumps up on its owner in greeting, it is behaving in a way similar to the way wolf pups greet adults. Both wolves and dogs roll over and display their underbellies when they are acting submissive and both raise their hackles and tail when they are feeling aggressive (Chapter 20).

Although the earliest dog fossils date to more than 15,000 years ago in Germany, Iraq, and Israel, clues in dogs' DNA reveal that they may have first been domesticated in East Asia, perhaps China. That conclusion is based on results obtained from a worldwide survey of 654 dogs by Peter Savolainen and his professional colleagues of the Royal Institute of Technology in Sweden. Their DNA sequencing of hair samples shows that dogs in East Asia are the most varied genetically. This may indicate that dogs have lived there the longest, mixing and matching their genetic makeup.

Wherever dogs first emerged they apparently followed humans in their ancient treks around the globe. At least one study indicates that dogs trotted across the Bering Land Bridge to Alaska with the first human settlers more than 12,000 years ago. Robert Wayne of the University of California–Los Angeles studied DNA sequences from ancient dog remains unearthed in Latin America and the Alaskan permafrost. Those DNA samples all closely matched European and Asian dogs, suggesting that they, too, descended from a single common progenitor, the Old World Eurasian gray wolf (both the dog and wolf have 39 pairs of chromosomes).

Interestingly, humans and dogs are the most widely dispersed mammals on the earth. Naturally the question arises, "Is this a mere coincidence or were humans and dogs mutually helpful in their respective journeys?" Although human and wolf remains have frequently been found at the same sites, bones alone do not prove whether the canines were scavengers or partners. The use of animal archaeology to study human development and well-being is likely to increase as the scientific tools of DNA sampling and sequencing become more sophisticated and commonplace. One thing is certain, the adage "a dog is man's best friend" has a long-term history and well-founded basis (we believe this is true for women as well!).

## 1.3.2 CATS

Origins of the domestic cat are less easily traced than those of the dog because its morphology differs only slightly from its wild ancestor, the *lybica* group of *Felix sylvestris*, which occurred throughout North Africa and the Near East an estimated 9,000 years ago. The chief modifications have been variation in facial features, color and length of hair coat, and minor variations in body size.

Cats were first domesticated in Egypt approximately 6,000 years ago. The importance of cats to the culture of Egyptians was seen in many ancient tombs and temples. The Cleveland Museum of Art has a fine collection of Egyptian artifacts including the coffin of a cat (cats were often mummified along with their masters).

Domestication of cats is believed to have coincided with the period in which agriculture developed and flourished in the Fertile Crescent known as the Mesopotamia of the Middle East. Houses, barns, and grain stores provided a new environmental niche that was rapidly exploited by mice and other small mammals—favorite prey of many wild cats that evolved eating a carnivorous (meat) diet. As carnivores, cats have sharp claws and teeth to hold their prey. Their canine teeth aid in killing while carnassial teeth are used for tearing their prey apart. From these early developments a mutually beneficial relationship developed in which the cat received an abundant food supply in return for controlling bothersome rodent pests.

It seems likely that these wild African bush cats were encouraged early to stay near human dwellings by giving them scraps of food. And as with wolves, the more docile wild cats were gradually absorbed into human society. This early domestication of cats was probably accelerated when the Egyptians brought kittens into their houses for children to pet and play with. The inner mysticism of the cat—the obscure wisdom and secrecy that seems to dwell in its eyes—attracted the attention of peasants, priests, and nobles alike.

The cat's role soon became diversified. Domestic cats trained to hunt birds were popular as partners of Egyptian aristocrats indulging in this sport. Phoenician trading vessels brought cats into Europe about 900 BC. Cats were in great demand in Europe as killers of flea- and lice-ridden, disease-spreading rodents. During the Middle Ages cats became associated with witches and warlocks and linked with Satan worship and demonology. Superstitious people consider black cats to be familiars of Lucifer. Thousands of cats were killed by religious zealots during the Dark Ages. Killing large numbers of cats likely contributed to the spread of bubonic plague (Black Death) throughout Europe in the 14th century. Interestingly, during one period of time Egyptians restricted the spread of cats to other countries by making it illegal to export them, resulting in an overabundance of cats.

Fortunes of the domestic cat have come and gone, but through it all the cat has never lost its mystique or its role as a controller of wild rodent populations. In the burst of creative thinking that followed the French Revolution and the Enlightenment, artists, novelists, poets, painters, and composers discovered the aesthetic delights of the cat: its grace, agility, beauty, charm, keen mind, and humorous idiosyncrasies.

Unlike dogs, cats in human societies have retained a high level of independence and can with great ease and minimal adjustment move back to the feral (wild) environment from which they were domesticated.

## 1.4  CONTRIBUTIONS OF COMPANION ANIMALS TO HUMANS AND SOCIETY

The overall quality of human life has many mutually interactive dimensions with companion animals. Animals improve human health and well-being by reducing stress, alleviating depression, positively affecting the cardiovascular system, and much more (Section 1.6). Animals teach caring behavior and elevate self-esteem in children and reduce the dependency for medical services for the elderly.

Delta's Pet Partners Program (www.deltasociety.org/) helps train volunteers and their animals to work as professionals in hospitals, nursing homes, schools, and treatment centers. Such programs bring joy and therapy to lonely seniors, abused children, and adults recovering from strokes and burn injuries. Delta's Service Dog Centers deal with advocacy and issues for dogs trained to help people with disabilities. Service dogs assist people who have visual, hearing, mobility, and other impairments (Chapter 21). Only about 15,000 of more than 50 million people with physical impairments are benefiting from service dogs. With more trained volunteers and private monies to support such programs, ways and means of enabling companion animals to change and enhance the lives of more people could become a reality (Figure 1.7).

Groups and societies that sponsor special services/programs of interest include the Delta Society, Pet Partners, Therapy Animals, Paws With a Cause, Pet Food Institute, and the American Pet Products Manufacturers Association (APPMA). We live in a society where it is

**Figure 1.7**  A string of 25 class-ready dogs in a community run at Guide Dogs for the Blind, San Rafael, CA. *Courtesy Malinda Carlson, B.S. in Companion Animal Biology.*

becoming increasingly difficult to establish close friendships with many people. In fulfilling the desire to be accepted, needed, and wanted many individuals turn to the most widely kept companion animals—dogs and cats—to fulfill these basic human needs. Indeed, companion animal ownership has significant mental hygienic value for people of all ages—children, adolescents, mature adults, and seniors.

Research has shown that close relationships between people and companion animals can increase longevity and decrease the incidence of degenerative cardiovascular disease. For example, companion animals provide people with a means of decreasing loneliness; something to care for and keep them busy; to touch and fondle, watch and enjoy; to make them feel safe and provide a stimulus for exercise. These functions contribute to the health and well-being of adults and concurrently enhance their quality of life and extend life expectancy.

A study conducted of clients at a veterinary college found that more than 99% of the clients talked to their animals and over 94% revealed that they talked to their companion animals as a person not as an animal. Interestingly, 28% said they confided in their animals and talked about events of the day. Approximately 81% believe their companion animal is sensitive to their feelings. Many spent as much as 4 hours a day in contact and conversation with their favorite companion animal.

## 1.5    HUMAN–COMPANION ANIMAL BONDING

Throughout recorded history companion animals have played an important role in the lives of individuals and families. But not until the 20th century was the human–companion animal bond discussed and written about extensively.

Companion animals frequently bring forth sentimental responses from people. When one reads stories about favorite pets in newspapers, children's books, or magazines the tone is almost always sentimental. Companion animals bring about a special bonding of friendship that unites people—a compassion and concern of children and adults. Written accounts of animals helping people bring warmth and an emotional uplift to readers.

The human–companion animal bond is similar throughout most of the world. Many cultures have a positive attitude among citizens toward pets in the general sense and companion animals in particular. Even most persons not owning a pet have a remarkably favorable attitude toward pets and the personal pleasure they give friends and people of all ages.

In an age of high-tech research, when it is tempting to reduce emotions to biochemical reactions and to rely more heavily on the modern technology of medicine, it is refreshing to note that a person's health and well-being may be improved through bonding with companion animals. Indeed, living with pets confers emotional and physiological benefits to human health. Overall, quality of life is enriched by experiencing a strong human–companion animal bond.

Characteristics displayed by companion animals as reported by lifelong animal-bonded elderly persons in one study included the following: 91% of the pets greeted their owners when they came home, 84% seemed to understand when their owners talked with them, 78% of the pets communicated to their owners, and 63% of the pets were sensitive to moods of their owners. Participants in the study cited their three principle reasons for owning companion animals as companionship, pleasure, and protection. Three of four owners in the study considered their pets to be family members.

Valuable roles of companion animals include providing companionship for children, adults, and lonely people; a sense of family to elderly, retired people and to younger childless couples; unconditional love and a sense of always feeling accepted for humans of all ages; reinforcement to a person's emotional well-being; a decreased level and occurrence of depression among senior citizens in retirement facilities; a feeling of being needed and wanted; a special alive something to love; loyalty, belonging, and devotion needed by an insecure child or adult; a nonthreatening entity; a sense of unconditional acceptance and compassionate understanding; an important link with nature; natural, uninhibited exchange of honest emotions and responses; a safeguard for the paranoid; a reason to be active by exercising and playing with the animal; and a bridge between an emotionally withdrawn patient and therapist. More than half of Americans take their dog with them on vacations. This is another means of keeping the human–animal bond at a high level.

Veterinarians are encouraged to make available in their waiting room a publication containing precautions and safeguards to take in traveling with a pet (Chapter 19). Such a publication should address the caution of leaving a pet in a closed vehicle when the ambient temperature is high, and the dangers of parasites (Chapter 15) and diseases (Chapter 14) prevalent in certain geographical areas of the United States (e.g., the incidence of Lyme disease in the northeast, coccidiomycosis or "valley fever" in the southwest, and salmon disease for those traveling to the northwest). For additional information related to traveling with pets, visit Web site www.aphis.usda.gov/oa/pubs/petravel.pdf or www.aphis.usda.gov/oa/pubs/petravel.html.

## 1.6   THERAPEUTIC USES OF COMPANION ANIMALS

Healthcare costs for humans in the United States increased approximately 430% during the past two decades (from $245.8 billion in 1980 to $1.3 trillion in 2000). Annual hospital charges exceeded $412 billion in 2000. Utilizing companion animals to reduce the length of stay and/or the intensity of treatment may reduce the escalation and speed of spiraling healthcare costs.

### 1.6.1   ANIMAL–ASSISTED THERAPY (AAT)

Animal-assisted therapy was used by the U.S. Army in convalescent centers during World War II. AAT is based on providing a reliable source of compassion, warmth, and dignity for stressed patients. The search for alternative treatment modalities to heighten the level of individual psychological well-being and save money has led to the increasing acceptance of the AAT program, hospice programs, and an increase in outpatient procedures (a survey of nursing homes and retirement communities revealed that nearly half of them are using AAT visitation programs). Concurrently there has been a timely acknowledgement of a more realistic attitude toward quality-of-life factors. Companion animals improve the quality of life in virtually every household and become especially important to people in households where there is stress. The recognized need for an improved quality of life, rather than simply increased longevity as a therapeutic goal, bodes well in applying the human–companion animal bond in healthcare delivery as a welcomed adjunctive therapy resource.

The proportion of senior citizens in the U.S. population is increasing faster than that of any other age group. The challenges and opportunities of meeting the long-term healthcare needs of the elderly with cost-effective measures are enormous. AAT is perceived both as quality care and as beneficial to the entire patient-family-provider group (the holistic approach).

It has been said that the doctor's preferred prescription for senior citizens is "take two pets and call me tomorrow." Such a philosophy embraces AAT, which is the use of companion animals to provide interaction and relationships with the therapeutic purpose of eliciting beneficial physical, psychosocial, and emotional responses. Other terms sometimes used for AAT are *pet-facilitated therapy, animal-mediated therapy, animal-facilitated psychotherapy,* and *animal-oriented psychotherapy.* Companion animals serve as mediators in therapy as reinforcers, socializing catalysts, co-therapists, pet companions, and in other psychological support roles. The Veterans Administration was an early adopter of AAT.

A significant component of AAT is the utilization of positive, reassuring, nonverbal communication signals between carefully selected, well-trained dogs and cats to benefit socially withdrawn, self-effacing, depressed humans. AAT is not a panacea and should be utilized as an adjunct to—not a substitute for—other forms of therapy. Studies to date indicate that AAT can help decrease excessive reliance on pharmacotherapy. Indeed, the best medicines are often those perceived by patients to make them feel better. Companion animals frequently fit this perception.

Companion animals have been utilized in the stress/grief process. Roles animals may play in the resolution of stress/grief include companionship, nonjudgmental love, safety, security, neutral communication topic, stress reduction, mood distraction, environment stabilization, physical detractor, reality anchor, third-party triangulation, and an alternative demand on mental thought processes.

The value of pet visitation and service dogs has been demonstrated in hospitals, care homes, and institutions for the physically and mentally impaired; the unique effect of animals on the autistic; the relief of loneliness felt by the elderly; the benefits for disturbed and disadvantaged children; and the unique ability to lower blood pressure and increase survival rate among coronary disease patients.

Numerous studies have been made of the influence companion animals and pets have on the emotional well-being of noninstitutionalized persons. In one study, elderly persons living alone were given a parrot, a house plant, or nothing (control group). Extensive follow-up over a 5-month period revealed significant improvement among those given the parrots as measured by self-esteem and emotional well-being.

Although initial use of AAT was primarily in treating patients with psychiatric disorders, recent studies indicate that pets may positively influence survival from other medical illnesses. Most people with depressive reactions to serious illnesses such as myocardial infarctions live outside institutional settings. Their medical care is commonly provided through outpatient contact with doctors, other caregivers, and day use of institutional facilities. Numerous cases have been recorded in which adopting a dog or cat aided in noteworthy improvement of attitude and overall health among depression-prone persons.

Petting a dog or cat produces a dramatic cardiovascular response in the pet; the act of friendship also lowers heart rate and blood pressure in those doing the petting. Simply observing animals can be sufficient to lower blood pressure as shown in one dentist's surgery practice, where watching a tank of tropical fish lowered patients' blood pressure, especially among people with above-average initial blood pressure.

Companion animals, especially well-trained dogs and cats, offer residents of nursing homes a form of nonthreatening, reassuring, nonverbal communication and tactile comfort, which helps break the cycle of loneliness, helplessness, and social withdrawal.

The present population of people over 65 years of age in the United States is projected to double by the year 2030 and to represent more than 20% of the total population. Cost-saving strategies are urgently needed to slow the increase in healthcare costs among that age group. We believe greater use of AAT can assist in achieving that important goal. Further information on AAT is provided in Chapter 21.

# 1.7  TRENDS IN COMPANION ANIMAL POPULATIONS

Long before the first dog biscuit was conceptualized, formulated, manufactured, and marketed, the ownership of companion animals was a wise investment. Indeed, the histories of many cultures are replete with anecdotes and evidence that people have loved, cherished, and benefited from companion animals for many millennia.

More recently, behavioral scientists, including Dr. Boris M. Levinson (author of the signature books, *Pet-Oriented Child Psychotherapy* and *Pets and Human Development*), have investigated why people choose to make pets a permanent part of their lives. Dr. Levinson and others have helped us understand the motivation of pet ownership—why people want pets and why they need and benefit from them.

The fact that so many people in Western countries own companion animals strongly suggests that these animals provide important companionship and social functions. In the United States the total 2002 dog and cat populations were approximately 65 and 77 million, respectively—a total of 142 million. The 10 states having the largest cat and dog populations are given in Figures 1.8 and 1.9, respectively. Expenditures for veterinary services in 2001

**Figure 1.8**   Cat population (millions) for top ten states, 2001. *Source: U.S. Census Bureau.*

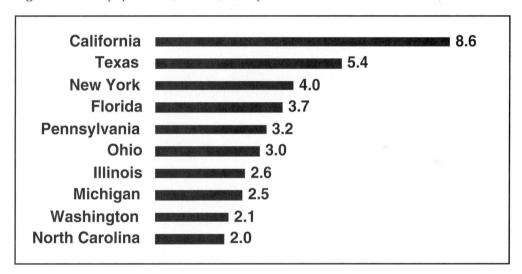

**Figure 1.9**   Dog population (millions) for top ten states, 2001. *Source: U.S. Census Bureau.*

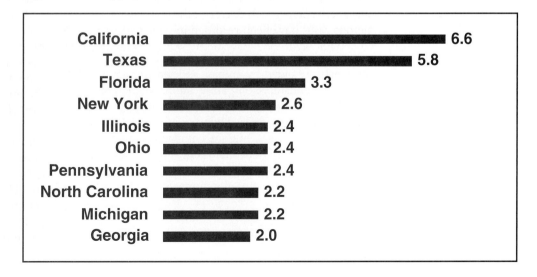

were approximately $11.6 billion for dogs and $6.6 billion for cats (a total of $18.2 billion compared with expenditures for veterinary services of approximately $608 million for horses).[5] Annual monetary expenditures for petfoods are approximately $12.4 billion (Chapter 2).

Reasons for an increasing number of pets and pet owners in the United States include (1) greater personal security, (2) smaller and more single-person households, (3) the increasing desire of many people to experience more engagement with animals and nature, (4) changes in the types of pets owned (more cats and smaller breeds of dogs), (5) an increasing level of disposable income, and (6) greater use of pets and companion animals for therapy purposes among an increasingly larger population of senior citizens.

Approximately three-fourths of all dog and cat owners in the United States are female.[6] One survey found that more than 90% of the cats in apartments were owned by women. The number of dogs increased from approximately 52.5 million in 1991 to approximately 65.0 million in 2002, a 23.8% increase. Approximately two-thirds of farm households own dogs, compared with about a third of those in cities. Yet because of the concentration of people in urbanized areas, more than 90% of all dogs in the United States reside in urban settings, less than 10% on farms (Chapter 24).

## 1.8   SUMMARY

Companion animals have enriched the lives of hundreds of millions of people through the years. To the delight of dog enthusiasts, their favorite pet was the first animal to be domesticated. Indeed, it has been an estimated 15,000 years since canines advanced from predator to beloved pet. Although the dog's early role was mostly utilitarian, strong bonds of affection were quickly forged with humans, as proven by recent fossil finds in Israel and elsewhere. The symbiotic relationship between dogs and cats and humans has grown stronger and more meaningful with advancing time. This occurred because the disposition and responses of dogs and cats are well received and greatly appreciated by humans of all ages.

Cats were domesticated in Egypt about 6,000 years ago. This coincided with the development of agriculture in the Fertile Crescent where the production of food grains flourished. Stored grain provided an abundant source of food and housing for mice, rats, and other rodents, which multiplied rapidly and soon attracted carnivorous cats. Because of their quickness, cats were able to catch the unwelcome pests, which in addition to eating the grain and leaving unwanted excrement, were carriers of fleas and lice that spread disease. Domestication of cats was accelerated when people brought kittens into their houses for children to pet and play with.

Dogs and cats will love and provide companionship whether a person is short or tall, underweight or overweight, attractive or homely, pleasant or difficult. Attachment and loyalty of dogs and cats are given unreservedly, unconditionally, predictably, and nonjudgmentally. Companion animals provide an unbounded friendship that is a source of warmth and companionship. They demonstrate overt innocence, love, trust, and dependency.

Companion animals are an important link with nature and can serve as a bridge between patient and therapist. They can often decrease anxiety in a clinical setting. Human therapists are aided by animal-assisted therapy (AAT) support through a triad of mutual trust and friendship. Companion animals can be selected and bred for affection (particularly cats) for use in AAT. Palliative care is grounded in patient dignity. The healthcare team that utilizes the human–animal bond as an adjunct therapy soon recognizes important benefits. The positive effects of human–companion animal bonds are substantial and noteworthy.

The number of households enjoying the favorable benefits of companion animals is increasing, as is the number of dogs and cats privileged to share family homes. Sharing the family home with companion animals results in many instances from a child's compelling

[5]American Veterinary Medical Association. 2002. *U.S. Pet Ownership and Demographic Sourcebook.* Schaumburg, IL:AVMA, www.avma.org.
[6]Ibid.

nudge. Children's interest in and sensitivity to animals are developed, expanded, and intensified through storytelling and reading. Lassie the dog, Flicka the horse, Miss Piggy of the *Muppets*, Snoopy the dog, and other interest-building stories of memorable animals touch the hearts and minds of children through compelling books, magazines, movies, television, and other media.

Many classrooms contain animal life for children's delight and observation. Cages of gerbils or guinea pigs or a tank of fish create a high level of interest and excitement among children in the home, classroom, and pet store. The "petting zoo" at fairs and zoos is also a popular place for children to visit and touch animals.

The social, societal, and personal benefits derived from companion animals are priceless. Most readers of this chapter have personally experienced the closeness and one-on-one bonding that commonly exists and the affection that is frequently exchanged between people and their favorite animal friends. All of life is a learning experience. Companion animals add immeasurable depth and enrichment to our lives.

The number of animal species and categories included in the companion animal/pet atrium and repository (e.g., amphibians, birds, cats, dogs, ferrets, fish, gerbils, goats, guinea pigs, hamsters, hedgehogs, horses, insects, pigs, rabbits and other lagomorphs, reptiles, rodents, sheep, and numerous others) has expanded in both number and scope in recent years. This has benefited animals, owners, businesses and industries, and other groups who service the whole of the companion animal/pet conglomerate.

Selecting and obtaining a companion animal involves assuming major responsibilities for another living creature—responsibilities that continue for the balance of the animal's life and/or that of its owner (Chapters 4 and 24). Understanding and undertaking this responsibility is to learn about the biology, care, health, and management of companion animals. Much of this useful information is shared in this book's 24 chapters that follow. Acquiring new information is exciting, rewarding, and useful. We hope you enjoy the reading journey!

## 1.9    REFERENCES

1. Anderson, R. S. 1975. *Pet Animals in Society*. New York: Macmillan Publishing.
2. Barloy, J. J. 1978. *Man and Animals: 100 Centuries of Friendship*. New York: Gordon and Cremonese.
3. Bustad, L. K. 1985, December. The Importance of Animals to the Well-Being of People. *Trends:* 54–57.
4. Fogle, B., ed. 1981. *Interrelations Between People and Animals*. Springfield, IL: Charles C. Thomas.
5. Gonzalez, M. L. 2002. *U.S. Pet Ownership and Demographic Sourcebook*. Schaumburg, IL: American Veterinary Medical Association.
6. Institute for Interdisciplinary Research on the Human–Pet Relationship, eds. 1983. The Human–Pet Relationship. *International Symposium Proceedings on the 80th Birthday of Nobel Prize Winner Prof. Dr. Konrad Lorenz.* Vienna, Austria.
7. Hart, L. A., B. Hart, and B. Mader. 1985. Effects of Pets in California Governmentally-Assisted Housing for the Elderly. *Journal of the Delta Society* 1: 65–66.
8. Levingson, B. M. 1972. *Pets and Human Development*. Springfield, IL: Charles C. Thomas.
9. Lynch, J. J. 1977. *The Broken Heart: The Medical Consequences of Loneliness*. New York: Basic Books.
10. Morey, D. F. 1994. The Early Evolution of the Domestic Dog. *American Scientist* 82: 336–347.
11. Ross, S. B., Jr. 1981. Children and Companion Animals. *Ross Timesaver: Feelings and Their Medical Significance* 23(4): 13–16.
12. Schweitzer, A. 1953. *Out of My Life and Thought*. New York: The New American Library.
13. Wayne, R. K. 1993. Molecular Evolution of the Dog Family. *Trends in Genetics* 9: 218–224.

# 2  COMPANION ANIMAL INDUSTRY

*Loneliness is a kind of hunger—hunger for warmth
and affection. And this hunger is more difficult to quench
than the hunger for a piece of bread.*

Mother Teresa (1910–1997)
*Roman Catholic nun of Calcutta*

16

## 2.1  INTRODUCTION/OVERVIEW

Feeling the gentle, sustaining power of companion animal bonding and experiencing the personal pleasures associated with ownership and care of these special animals are integral parts of human society as evidenced by the increasingly large number of households worldwide owning companion animals. As the human–animal bond grows stronger, pets are being treated more and more like family members. And is it going fishing or simply enjoying the companionship of an adoring young friend that is most appealing to the family dog (Figure 2.1)?

The adoption of dogs and cats in tens of millions of homes globally, coupled with a continuing trend toward the increasing movement of people from rural areas to cities where most companion animals are now housed and fed indoors, rather than roaming for prey and food outdoors, has provided impetus for the development of a product-oriented, market-driven, petfood enterprise. This progressive, sophisticated industry provides an expanding variety of commercially prepared, nutritionally complete petfoods, snacks, and treats; and pet care products, pet supplies, and other related items for an increasing global population of dogs and cats (Table 2.1).

**Figure 2.1**  This child feels more secure and contented fishing with his German Shepherd dog at his side. *Courtesy Dr. James E. Corbin.*

**Table 2.1**
Global Dog and Cat Population (millions)

| Companion Animal | 1998 | 2002 | % Increase |
|---|---|---|---|
| Cats | 237 | 264 | 11.4 |
| Dogs | 235 | 263 | 11.9 |
| Totals | 472 | 527 | 11.7 |

*Source: Petfood Forum Proceedings, April 2003, Chicago, IL.*

## 2.2 ANNUAL EXPENDITURES FOR CARE OF COMPANION ANIMALS

Those who say "you cannot buy love" are forgetting about puppies (Figure 2.2). Conversely, many people are surprised at the costs associated with pet ownership (Table 2.2). The purchase price of an animal is only a small part of what the average owner will spend on her/his companion animal.

Two of the largest expenditures in caring for a pet are its food and healthcare costs. As discussed in Chapters 14 and 15, preventive healthcare includes neutering, annual physical examinations, vaccinations, fecal and heartworm tests, heartworm preventive, flea and tick preventive, and regular grooming. Obedience training is highly recommended for dogs. Owners of dogs are strongly encouraged to enclose their yards with secure fencing. Pet sitters, a reliable boarding kennel, or pet motel may cost from $10 to $50 per day for care of the pet when the owner is traveling or on vacation (Chapter 17). The American Pet Products Manufacturers Association estimated that pet owners in the United States spent a total of $31 billion in 2003 to care for, feed, spoil, and pamper the American pet population.

## 2.3 THE PETFOOD INDUSTRY

### 2.3.1 FIRST COMMERCIAL DOG FOOD AND SELECTED PRODUCTS THAT FOLLOWED

James Spratt—an ingenious, enterprising Cincinnati electrician who in 1860, while in London selling lightning rods he had invented—is credited with devising the first baked "dog cake" containing meat, vegetables, wheat flour, and salt (Figure 2.3). He contracted with a London baker to make 2 × 7 inch biscuits—the first "Spratt Patent Meat Fibrine Dog Cake." The idea for these biscuits emanated from the observation of his dogs relishing old biscuits about to be discarded from a British sailing ship. Utilizing the world-class British ships, Spratt was soon exporting his novel dog biscuits globally (Figure 2.4).

Spratt introduced Dog Cakes® into New York City in 1870 as Spratt's English-Produced Biscuit Patent America (Ltd). The business flourished and expanded to a plant in Newark, New Jersey, in 1895. That plant was sold to General Mills in 1959, and the manufacturing facilities were moved to Toledo, Ohio.

**Figure 2.2** Eight puppy littermates anxious to exchange their love and affection for attention, play, and petting. *Courtesy Dr. James E. Corbin.*

Table 2.2
Estimated Annual Costs Associated with Ownership of Selected Pets (dollars)

| Pet | Food | Litter/Bedding | Healthcare | License | Collar/Leash | Crate/Carrier/Cage | Grooming | Toys | Miscellaneous | Total Annual Cost |
|---|---|---|---|---|---|---|---|---|---|---|
| Small Dog | 150 | | 225 | 15 | 25 | 80 | 200 | 50 | 35 | 780 |
| Medium Dog | 250 | | 275 | 15 | 30 | 140 | 300 | 60 | 45 | 1,115 |
| Large Dog | 350 | | 325 | 15 | 35 | 240 | 400 | 70 | 65 | 1,500 |
| Cat | 120 | 175 | 225 | | 10 | 30 | | 50 | 30 | 640 |
| Rabbit | 110 | 400 | 200 | | | 135 | | 25 | 15 | 885 |
| Guinea Pig | 75 | 400 | 50 | | | 80 | | 25 | 15 | 645 |
| Gerbil or Hamster | 50 | 220 | | | | 35 | | 10 | 15 | 330 |
| Small Bird | 50 | | | | | 75 | | 30 | 15 | 170 |
| Fish | 20 | | | | | 150 (aquarium) | | | 15 | 185 |

*Source: The American Society for the Prevention of Cruelty to Animals, www.aspca.org.*

**Figure 2.3** An early advertisement of Spratt's dog cakes. Note the health-promoting assertions of the ad. *Courtesy Dr. James E. Corbin.*

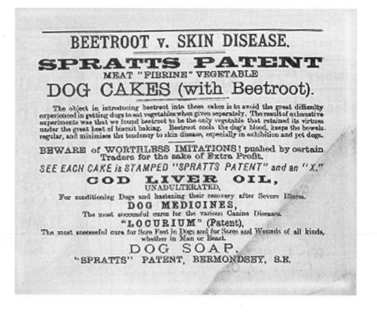

Spratt's first employee in England was Charles Cruft, an accountant/salesperson/promoter who was enlisted to provide the leadership needed to expand the English business. As an advocate of dog showing, Cruft expanded Spratt's into two sections in 1885: one to manufacture Spratt's Dog Cakes, the other to promote their foods through dog shows. Charles Cruft became manager of the show department of Spratt's, and they acquired sufficient hardware to bench 1,000 dogs simultaneously.

Cruft's early influence and successes in superintending highly popular dog shows resulted in naming the world's most prestigious show (Cruft's), which attracts approximately 10,000 entries annually. It still carries Cruft's name. Spratt's participated in sponsoring the first Westminster Kennel Club Dog Show at Madison Square Garden in New York City in

**Figure 2.4** This advertisement infers early global acceptance of Spratt's dog biscuits by noting their fast-spreading use in other nations. *Courtesy Dr. James E. Corbin.*

1877 (www.westminsterkennelclub.org/history.html) and continued doing so until World War II, when the George Foley organization assumed leadership and responsibility for the shows. At the 100th anniversary celebration of the founding of Spratt's, it was still the largest-selling dog food company in Great Britain, and also had manufacturing plants in Germany, the United States, and other countries.

The second dry dog food manufacturer, the F. H. Bennett Biscuit Company, entered the market in 1908. Bennett was a health-food advocate and experimented with specialties. He began making a bone-shaped biscuit for dogs with nutriment-rich ingredients including meat products, cereals, milk, and added minerals. The new "Bennett's Milk-Bone Dog and Puppy Foods" were made in various sizes to meet the physical needs of different sized dogs. National Biscuit Company (Nabisco) purchased Milk-Bone® in 1931. Harry T. Wissler is credited with pioneering petfood sales in the 1930s with 3,000 Nabisco biscuit and cracker salespersons calling on food stores throughout the United States.

Harry Wissler's friend, Dr. Leon Whitney, was a geneticist and nutritionist who conducted research with a colony of Beagles at Yale University. Whitney believed firmly that dogs need scientifically balanced diets. He had already achieved an outstanding reputation based on his research and development of Bal-O-Ration, which was later sold to Tioga Mills in Waverly, New York. It was later acquired by the Quaker Oats Company of Chicago, Illinois. Wissler funded Whitney's research to study the effects of Milk-Bone in Beagles. Whitney's scientific investigations proved Milk-Bone to be nutritionally sufficient in supporting growth, gestation, lactation, and maintenance. Additionally, Whitney observed that Milk-Bone minimized tartar on the teeth of test dogs. This was followed by Milk-Bone being advertised for decades as an effective cleaning agent of dog's teeth. Dr. Leon Whitney graduated from the Auburn University College of Veterinary Medicine (with his son George, contributing author to this book's Chapter 6) in 1943. He authored 52 books; most were about dogs.

The next generation of dry dog foods was pelleted. Spratt's and Milk-Bone led to pelleted dog foods, which included cooked grains, proteins, and supplemental vitamins and minerals. Later, corn flakes, baked biscuits, and sprayed-on fat were added to the pellets.

Canned petfoods appeared in the 1920s. Dry versus canned petfoods differ greatly in cost per unit of nutrients provided. In the United States, canned petfoods contain 70% to 80% moisture compared with 10% to 12% in dry petfoods. Water is expensive when purchased in canned petfoods. Commercial petfoods in the United States vary greatly in the range of guaranteed levels of major ingredient categories, as noted in the following: protein 15% to 30%, fat 5% to 30%, and fiber 1.5% to 8.0%. A "complete" dog food may, on a dry matter basis (Chapter 9), have a price up to 20 times (or more) greater than that of the low range. Price is not always a reliable indicator of product quality. Highly advertised brands can be of less nutritional value than lower-priced, nonnationally advertised petfoods.

The next major development in petfoods occurred when in 1954 Dr. James E. Corbin, then manager of the Ralston Purina Biological Testing Laboratories in St. Louis, formulated and extruded the world's first expanded foods for dogs, cats, fish, guinea pigs, and monkeys. Extruded petfoods are made by cooking the ingredients using high heat, elevated moisture, and pressure and then forcing them through a die (Figure 2.5). The product expands into the desired shape and size as it passes through the die (Figure 2.6). Cooking increases the speed and degree of digestibility of dietary starches and is highly acceptable to dogs and cats. Extruded petfoods are rendered practically sterile during processing and are more nearly free of microorganisms than most foods sold for human consumption. Today virtually all dog, cat, and many other dry-expanded companion animal petfoods, as well as numerous foods for human consumption, are extruded.

After extrusion, fat and/or *digest* (a flavor enhancer produced by enzymatic degradation of animal tissues) is sprayed on the petfood's surface to increase palatability and caloric content. The process of spraying substances on the exterior of food is called *enrobing*. Hot air drying is then used to reduce the food moisture content to < 12%. Dry petfoods are the most

**Figure 2.5**    Depicted below is a 10-ton-per-hour extrusion and dryer system used in the commercial production of dry-expanded petfoods. *Courtesy Wenger Mfg., Inc.*

popular for both dogs (representing over 60% of dog food sales) and cats (Chapter 9). Forms and methods of processing and producing extruded foods and medications for humans and companion animals are depicted in Figure 2.7.

Semimoist (up to 35% moisture), moist, and high-moisture dog foods were introduced in the 1960s. Semimoist dog foods rely on pH control, high sugar content, and additives to minimize fungal and bacterial growth. General Foods's Gainesburgers® high-moisture dog food was introduced in the 1960s. This product resembled hamburger patties and was advertised as "canned dog food without the can." These foods and cubed meat-type products called Choice Cuts were introduced in the 1960s by the Quaker Oats Company.

**Figure 2.6**    Example of an extruded product (*l*) and a pelleted product (*r*). *Courtesy Wenger Mfg., Inc.*

**Figure 2.7**  Food and medication forms and methods of their processing and production. *Upper left*: Tablets are single materials pressed with a uniform composition throughout. They are the most common and least expensive method of producing large quantities of precise dosages (e.g., traditional aspirin). *Lower left*: Pills are produced by rolling and have a composite (ingredient) distribution of two or more layers of medication that may be separated from each other to provide different absorption and delivery rates of each ingredient within the gastrointestinal tract (e.g., coated aspirin). *Upper right*: Pelleting is the relatively inexpensive methodology for producing large quantities of high-density animal foods. Component size varies per unit and may provide individual ingredients or nutritionally complete foods in a product. *Lower right*: Extrusion utilizes cooking, elevated moisture, and high pressure to give an expanded product. This is followed by drying to a common moisture of 10% to 12% and the addition of fat for caloric and flavor enhancement. This well-accepted process is used globally in the manufacture of more than 95% of dry petfoods. *Courtesy of and illustration by Carl James Corbin.*

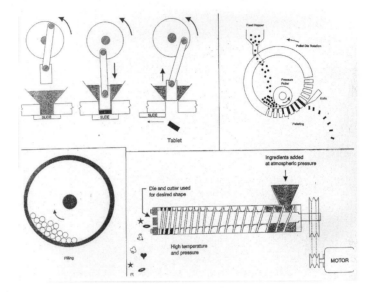

## 2.3.2  SIZE AND SCOPE OF PETFOOD AND PET CARE– RELATED INDUSTRIES[1]

Global sales of petfoods and pet care products exceeded $46.5 billion in 2002 and are growing in both value and volume. North America, Western Europe, and Asia-Pacific had approximately 41%, 33%, and 11%, respectively, of the global petfood market share in 2000.

Petfood sales in North America in 2002 were $19.1 billion. The United States has approximately 330 manufacturers of petfoods producing 3,300 different brands. These represent over 8 million tons of finished products that were valued at more than $12.4 billion in 2002. The largest outlet for petfoods in the United States is grocery stores (the average grocery store carries 271 dog and 239 cat food items) followed by pet specialty stores.

The Asia-Pacific region ranks third globally in petfood sales ($7.8 billion in 2002). Overall, this is the fastest growing market for petfood and pet care products due to the dominance of the Japanese market. Latin America is the fastest growing regional market (sales were $3.9 billion in 2002).

---

[1]Data of this section are from the *Petfood Forum Proceedings*, April 2003, Chicago, IL.

Global sales by product type in 2002 were dog ($19.8 billion) and cat ($12.3 billion) foods, totaling $32.1 billion (68%); pet care products, $11.1 billion (24%); and foods for other pets, $3.9 billion (8%).

A typical pet aisle in grocery stores includes the following: *dog food* (dry-expanded, canned, semimoist, soft-expanded, snacks, and treats); *cat food* (dry, canned, semimoist, snacks and treats); *pet supplies* (cat litter/deodorants, rawhide dog chews); *pet accessories* (collars, leashes, grooming supplies, flea/tick remedies, toys, small cages, utensils); foods and products for other companion animals. Western Europe ranks second with approximately one-third of the global petfood sales ($14 billion). It ranks third in sales of pet care products. More than 450 companies produce petfoods in Europe. They employ more than 50,000 people and manufacture over 5 million metric tons annually.

### 2.3.3   Petfood Treats and Snacks

These products represent the most dynamic and diverse segment of the petfood industry. North America, Europe, and Japan are the strongest markets for pet treats and snacks. Products for dogs dominate the pet treat market (about 89%), although treats are also available for cats (approximately 8%), birds, fish, and horses. Categories of pet treats include baked products, natural treats derived from animal parts, rawhide products, and extruded products. Recently introduced treats include Frosty Paws™, the ice cream for dogs, and Cool Claws™, ice cream for cats. Both of the above products are distributed by Associated Ice Cream.

Global sales of pet treats increased from approximately $2.5 billion in 1998 to $3.2 billion in 2002. Annual sales of the pet treat market exceed $1.1 billion in the United States and are increasing about 3.5% per year. Most pet owners (81%) buy their pets luxury items and other nonessential things (toys, health supplements, and bath supplies).

### 2.3.4   Trends in Petfoods and Related Product Sales

The trend toward proportionately more single-person households benefits pet ownership, as does an aging U.S. population. Persons living in small houses, flats, or tower blocks tend to select smaller dogs and proportionately more cats, whereas those with more household space tend to own larger dogs and a rather stable number of cats. Approximately 70 million of the nation's 108 million households purchased commercial dog or cat food in 2001. The number of households in the United States owning pets has increased approximately 1 million per year for each of the past 10 years.

Because the number of cats in U.S. households is increasing faster than that of dogs, the percent increase in snack and treat products aimed at cats will probably increase faster during the next several years than the percentage of those aimed at dogs. Global sales by product type in 2002 were dog and cat foods, 68%; pet care, 24%; and other petfoods, 8%. For additional information and data related to projected trends and future of the petfood industry, visit the American Pet Products Manufacturers Association Web site at www.appma.org.

### 2.3.5   Providing Standards and Quality Assurance in Petfood and Related Enterprises

The National Academy of Sciences (NAS) was created in 1863 by an act of Congress as a nonprofit national institution of distinguished scholars to serve as official advisers to the government of the United States on matters of science and technology. The National Research Council (NRC) was established in 1916 by NAS to provide technological and scientific information on various nutrition-related topics. The NRC comprises some 700 working committees, panels, and boards on which approximately 9,000 scientists, engineers, and other professionals serve without financial compensation.

The Committee on Animal Nutrition (CAN) was organized in 1928 under the auspices of the NAS/NRC to provide advice on the nutritional management of economically important domestic animals. (Names of its members can be seen on the Web site www.nationalacademies.org.) Reports of the NRC (including an important 2004 publication on dogs and cats) are translated into several languages including Chinese, Japanese, Russian, Spanish, and Turkish.

Historically the Association of American Feed Control Officials (AAFCO), the industry's advisory body responsible for establishing petfood label requirement policies, has relied on publications of the NRC as its recognized authority on companion animal nutrition with regard to concentrations of specific nutrients that constitute a complete and balanced dog or cat food. Additionally, NRC documents are used extensively as the regulatory standards for animal care and use in research and teaching (Chapters 9 and 22). Currently AAFCO does not require feeding trials for diets designed for birds, reptiles, or many species of pocket pets.[2]

## 2.4   PET SERVICES

Important providers of pet services include veterinarians, veterinary technologists, and veterinary technicians; animal behavioral counselors and trainers; animal rescue organizations and humane societies; animal control offices and shelters; breed societies; professional photographers; dog show superintendents and judges; pet groomers; managers of pet shops, kennels, catteries, and boarding facilities; and those involved in the research and development, manufacturing, and marketing of products for companion animals.

Important activities of the companion animal service industry include caring for companion animals (feeding, exercising, grooming, healthcare), diagnosing and treating health problems, providing advice to companion animal owners, training companion animals, and providing counseling to owners dealing with behavioral problems of their pet, or with grief following the loss of a beloved companion animal.

Many of the career opportunities associated with the companion animal industry are discussed more fully in Chapter 23.

## 2.5   PET SUPPLIES

With the growing trend to treat companion animals as members of the family, pet owners are increasingly purchasing products perceived to enhance the quality of life for their pet. This has resulted in a thriving market for the pet supplies and products of the companion animal industry.

The types of products being produced are almost endless—from a simple leash to an elaborate pet car seat; a dog whistle to a microchip that provides permanent identification and tracking of a pet; a dog blanket to a three-story "penthouse" for cats; an exercise wheel to an elaborate maze for pocket pets. An abundance of products is available to satisfy the pet owner's desire to "spoil" a much-loved pet. New products are launched each year at elaborate trade shows such as the American Family Pet Expo® and SuperZoo® (see www.wwpsa.com).

---

[2]*Pocket pets* is a term used to describe small animals such as hamsters, gerbils, rats, mice, guinea pigs, and rabbits. They are cute, usually inexpensive, and commonly kept as pets. Pocket pets are especially popular with children.

## 2.5.1 CORPORATE LEADERS IN THE PET SUPPLIES INDUSTRY

One of the earliest companies to enter the pet supplies business was The Hartz Mountain Corporation. This company began in 1926 when a penniless young German named Max Stern sought new fortunes in the United States by importing singing canaries from Germany to sell in New York City. He soon had a thriving bird import business. In 1932 he decided to expand his business by selling packaged bird foods; this marked the official establishment of the Hartz line of pet products. In 1959 the company expanded into the pet fish food and aquatic pet supply business. In the mid-1960s Hartz Mountain expanded into the dog and cat supply market. In December 2000 the Stern family sold Hartz to a fund managed by J. W. Childs Associates. Today the company markets over 1,500 different products to meet the needs of a vast array of household pets including dogs, cats, parakeets, canaries, parrots, cockatiels, finches, goldfish, tropical fish, reptiles, ferrets, chinchillas, hamsters, and rabbits. Hartz Mountain products are sold primarily through grocery stores and department stores. The Hartz mission is "to enhance the Human-Animal Bond by providing quality products with innovative and technological solutions developed with proven science and love for companion animals." For additional information, see www.hartz.com.

In the late 1930s Dr. Mark L. Morris, Sr., a veterinarian working in Edison, New Jersey, discovered that many dogs with kidney disease showed improved health when fed diets low in protein and salt. One of the first patients to receive this special diet was Buddy, the first Seeing Eye Inc. guide dog in the nation. Buddy and his owner, Morris Frank, traveled around the United States speaking for Seeing Eye Inc. The original diet was home-cooked and Mrs. Morris canned the food for Buddy. The diet soon became popular and the Morris's could not prepare enough of it in their home to meet the demand. Dr. Morris signed a contract with Burton Hill from Topeka, Kansas, to produce the special diet as a product of the Hill Packing Company. The special diet for dogs with kidney disease was patented in 1943 as Hill's k/d® diet. Soon Dr. Morris began developing other foods designed to help manage a number of other diseases of dogs and cats. Dr. Morris and his son, Dr. Mark Morris, Jr., developed over 40 specialized diets for dogs and cats with various medical diseases; these are marketed as Hill's Prescription Diet products for dogs and cats. They also developed the Hill's Science Diet® brand of diets formulated to meet the nutritional needs of healthy dogs and cats. In 1948 Dr. Morris founded the Morris Animal Foundation to sponsor research that would improve the lives of companion animals. Dr. Morris said, "For years animals have been used for medical research into human ills, and now it is time that something was done for the animals themselves."[3] The Morris Animal Foundation has sponsored more than 1,100 studies with funding exceeding $32 million.

PETCO Animal Supplies Inc. is a specialty retail leader of premium petfoods, supplies, and services (www.petco.com). PETCO was founded in 1965 and has grown to operate over 630 stores in 43 states and the District of Columbia. PETCO offers more than 10,000 pet-related products for sale and also provides many pet services including grooming, obedience training, vaccination clinics, and pet photography. The PETCO Foundation was established in 1999 to promote charitable, educational, and other philanthropic activities for the betterment of companion animals. Its programs serve the "Four Rs": Reduce, Rescue, Rehabilitate, and Rejoice. PETCO provides assistance to over 1,200 animal welfare groups.

A fourth giant in the pet supplies industry is PETsMART, Inc. (www.petsmart.com). Founded in 1987, PETsMART has prospered and currently owns more than 550 pet superstores in the United States and Canada. PETsMART stores feature pet "styling salons," pet training classes, and in some locations are associated with full-service veterinary clinics within the same building. PETsMART has acquired several catalog retailers including Sporting Dogs Specialties Inc. and State Line Tack Inc. (an equine supply company). Many PETs-

---

[3]http://www.morrisanimalfoundation.org/foundationinfo/founder.asp.

MART stores include an adoption center for homeless pets from shelters; in 2001 the millionth homeless pet was adopted from a PETsMART in-store adoption center.

### 2.5.2  PET HEALTHCARE COMPANIES

Companies focusing on healthcare products are another rapidly growing segment of the companion animal industry. Merial is currently the world's leading animal health company. Merial produces a wide range of pharmaceutical products and vaccines to help keep pets healthy. Leading products for dogs and cats include fipronil-containing flea preventives (Frontline®) and ivermectin-containing heartworm preventives (Heartguard®). Merial has 17 research and development centers worldwide and operates 16 manufacturing sites. Merial sells products in more than 150 countries; its sales in 2001 were over 1.7 billion euros.

Bayer Corporation is another world leader in animal health products; its sales in 2000 were in excess of $10.1 billion. Popular products produced by the Bayer Corporation include the Advantage®/Advantix® line of flea control products as well as anthelmintics, other parasiticides, antibiotics, and vaccines. The Fort Dodge Company is probably best known for its extensive line of vaccines and as a leader in the development of new vaccines; however, its line of products also includes many pharmaceuticals. Other important companies involved in the pet healthcare industry include 3M Animal Care Products, Abbott Laboratories, Activon Products, Addison Biological Laboratory Inc., Bio-Derm Laboratories, Biopure Corporation, BioZyme Incorporated, Boehringer Ingelheim Vetmedica Inc., Dermapet Inc., DVM Pharmaceuticals Inc., EVSCO Pharmaceuticals, Farnam Companies Inc., Happy Jack Inc., IDEXX Laboratories Inc., Intervet Inc., Novartis Animal Health US Inc., Pfizer Inc., Schering-Plough Animal Health Corporation, Sergeant's Pet Products Inc., Veterinary Solutions, Virbac Corporation, Zinpro Corporation, and many more.

The types of products being produced to enhance the health and quality of life of companion animals is constantly being expanded by active research and development programs of the many companies involved in the companion animal health industry.

### 2.5.3  HERBS AND BOTANICALS

In whole form, herbs and other botanicals provide some nutritive value in the form of fiber, sugar, oil, and certain vitamins and minerals. Their intended use in petfoods is either to provide flavor, or more often, to have an effect on the animal other than by nutritional means. Herbal preparation for animal and human foods is depicted in Figure 2.8.

## 2.6  SUMMARY

The values and emphasis people place on ownership of companion animals vary from person to person and from culture to culture. Yet whether young or old, female or male, people will continue to be drawn by and attracted to the beauty and pleasures derived from companion animals. This bodes well for the product-oriented, market-driven petfood, pet care, pet supplies, and related industries.

As the number of companion animals continues to increase, the range and variety of petfoods continues to expand; and as the knowledge base concerning companion animal nutrition, health, and disease continues to grow, providing safe, adequate, and appropriate nutrients for the nation's 77 million cats, 65 million dogs, 17.3 million companion birds, 16.8 million pocket pets, 8.8 million companion reptiles, 7.0 million saltwater fish, and 185 million freshwater fish is a tremendous challenge and marketing opportunity for product- and service-oriented petfood, pet care, pet supplies, and related business enterprises.

**Figure 2.8**    Preparation of herbal ingredients for animal and human foods. *Courtesy Dr. James E. Corbin.*

## 2.7    REFERENCES

1.  Adams, J. 2002. Managing Product Development for Effectiveness. *Petfood Forum Europe Proceedings,* Amsterdam, Netherlands, 178–187. Mt. Morris, IL: Watt Publishing.

2.  Booth, D. 2000. Nutraceutical Concerns. *Petfood Industry* 42(5): 4–14.

3.  Crossley, A. 2003. Global Sales of Petfoods and Pet Care Products. *Petfood Forum Proceedings,* Chicago. Mt. Morris, IL: Watt Publishing.

4.  Kvamme, J. L., and T. D. Phillips, eds. 2003. *Petfood Technology.* Mt. Morris, IL: Watt Publishing.

5.  *Nutrient Requirements of Dogs and Cats.* 2004. Report of the National Research Council of the National Academies. Washington, DC.

6.  *Petfood Forum Europe Proceedings.* 2003. Amsterdam, Netherlands. Mt. Morris, IL: Watt Publishing.

7.  Swiers, J. 2002. Innovative Dry Processing Technologies. *Petfood Forum Europe Proceedings,* Amsterdam, Netherlands, 188–222. Mt. Morris, IL: Watt Publishing.

# 3 DOG/CAT BREEDS AND THEIR CHARACTERISTICS

*Animals are such agreeable friends—they ask no questions; they pass no criticism.*

George Eliot (1819–1880)
*English novelist*

## 3.1   INTRODUCTION/OVERVIEW

A breed may be defined as a homogenous grouping of animals within a species that was developed by humans through selection and mating to perpetuate particular hereditary qualities. Breeds must be carefully developed over many generations to fix desirable and minimize undesirable breed characteristics. Breeds, then, represent groups of genetically related animals by virtue of having descended from common ancestors. This means they share similar genotypes (genetic makeup) and similar phenotypes (visibly and recognizably similar physical characteristics).

Breeds of dogs were developed by genetically modifying dogs through selection and breeding for particular purposes. British sheep farmers along the England–Scotland border developed Border Collies to help move flocks of sheep from field to field and from the hills. European sheep farmers developed dogs to guard their flocks from wolves.

Additional examples include hunters developing game dogs and waterfowl dogs to hunt birds; terriers to hunt pests; and hounds to chase rabbits, hares, pikas, and gazelles. Nomadic people developed breeds to pull sleds and small carts loaded with food and other goods. Warriors traveled with guard dogs; noblemen utilized fierce dogs to protect their palaces. People living in cold climates developed breeds with coats designed for inclement weather; those living in deserts developed dogs that could tolerate the heat of day and coolness of night; those dwelling near the sea developed dogs that could swim in cold water with no apparent adverse effects.

Although many of these special-purpose breeds no longer travail at their original assignments, they still retain many skills required to perform those tasks and responsibilities. Indeed, puppy purchasers can still select a breed based on traits that enabled their ancestors to fulfill the original functions (Chapter 4). For example, Labrador Retrievers (ranked first in American Kennel Club registrations in the United States) are in high demand by active families as companion animals because they are gentle, playful, active, and relatively easy to train (Table 3.1).

Dogs are special animals in the lives and households of tens of millions of people. Why else would reference be made to them in so many adages and metaphors as noted in the following thirteen uses of words through short phrases to compare the perceived fortunes of humans with those of dogs? (1) *Going to the dogs* is commonly accepted to mean the state of being ruined; (2) *to put on the dog* means to make an ostentatious display of elegance, wealth,

Table 3.1
Twenty Most Frequently Registered Dog Breeds in 2002 with Brief Descriptions

| 2002 Rank | Breed | Average Mature Height (inches) | Description/Notes |
|---|---|---|---|
| 1 | Labrador Retriever | 21–23 | Playful, loving to people, hardworking; respected as guide dog, search and rescue, and narcotics detection. First recognized by AKC in 1917. |
| 2 | Golden Retriever | 20–24 | Easy to train, strong, outgoing and devoted companions, happy, and trusting. First recognized by AKC in 1925. |
| 3 | German Shepherd | 23–25 | Courageous, keen senses, active, dignified, steadfast heart, excellent companion. First recognized by AKC in 1908. |
| 4 | Beagle | 13–16 | Gentle, happy, loving companions; clever, quick, curious, active; enjoys people and other animals. First recognized by AKC in 1885. |
| 5 | Dachshund | 12–14 | Lively, upbeat personality; spunky, curious, friendly. Developed to hunt badgers. First recognized by AKC in 1885. |
| 6 | Yorkshire Terrier | 9 | Enjoys playing and investigating; high energy as puppies, quiet and settled as adults; splendid coats. First recognized by AKC in 1885. |

| 2002 Rank | Breed | Average Mature Height (inches) | Description/Notes |
|---|---|---|---|
| 7 | Boxer | 21–25 | Playful, fun–loving; twinkling black eyes show intelligence and emotions; curious, happy; loves children; patient, strong, defensive; early obedience training is important. First recognized by AKC in 1904. |
| 8 | Poodle | | Smart, loyal, proud, fun; understands owner moods; enjoys challenges and obedience training; intelligent, playful sense of humor, politely reserved around strangers. First recognized by AKC in 1887. |
| | Toy | up to 10 | |
| | Miniature | 10–15 | |
| | Standard | over 15 | |
| 9 | Chihuahua | 6–9 | Small, adorable, apple-shaped head and large eyes; charmer; healthy but fragile; enjoys barking and scampering around; mainly indoor dog, unable to tolerate cold weather. First recognized by AKC in 1904. |
| 10 | Shih Tzu | 9–11 | Silky coats; pleasant, playful; high energy level; gets along well with people and other animals; dislikes hot weather; coat needs grooming daily. First recognized by AKC in 1969. |
| 11 | Miniature Schnauzer | 12–14 | Pleasant, proud, devoted, playful; enjoys being center of household; dignity shows in whiskered face; gets along well with people and other animals. First recognized by AKC in 1926. |
| 12 | Pomeranian | 9–11 | Lively, bold, inquisitive; fearless watchdog despite size; smart and responds well to training; sometimes suspicious of strangers and others animals. First recognized by AKC in 1888. |
| 13 | Rottweiler | 22–27 | Bold guardian; intelligent; steady friend; rather aloof; enjoys exercise and challenges of outdoor sports. First recognized by AKC in 1931. |
| 14 | Pug | 10–11 | Confident, friendly, sensitive; good with children; enjoys exercise and playtime. First recognized by AKC in 1885. |
| 15 | Cocker Spaniel | 15–16 | Big baby eyes; soft coat; upbeat personality; sensitive, playful; loves people, forms lifelong bond with owner; easily adapts to city or country living; enjoys exercise and playtime. First recognized by AKC in 1878. |
| 16 | Shetland Sheepdog | 13–16 | Small, beautiful; happy in city or country; intelligent, loving, sensitive; becomes deeply attached to families; good learner; seeks to please; a pleasure to train; gentle and thoughtful with children. First recognized by AKC in 1911. |
| 17 | Boston Terrier | 15–17 | A native American breed; happy, sunny disposition; content in city or country; barks when someone is at the door; highly intelligent. First recognized by AKC in 1893. |
| 18 | Bulldog | 12–14 | Intelligent, trainable; easygoing, although originally a fighting dog; because of body type light exercise supports good health; enjoys an air-conditioned house in summer. First recognized by AKC in 1886. |
| 19 | Miniature Pinscher | 10–12 | Small but does not know it; quick, active, lively curiosity; feels important; looks out for family, bravely challenging intruders with bold bark. First recognized by AKC in 1925. |
| 20 | Maltese | 9–10 | Becomes bonded with entire family; enjoys visitors and other animals; personality has sparkle; enjoys being treated as the adorable baby of the family; respects rules. First recognized by AKC in 1888. |

*Source: American Kennel Club, Inc., 260 Madison Avenue, New York, NY 10016 (Web sites: www.akc.org/breeds/regstats2001.cfm and www.akc.org/breeds/recbreeds/breeds_a.cfm).*

or culture—often being a phony; (3) *to lead a dog's life* is to be perceived as living an unhappy existence (the accuracy of this adage may be questioned because many pet dogs now enjoy pampered lives); (4) *to die a dog's death* is thought of as a miserable, shameful end; (5) *top dog* means being dominant in the hierarchy; (6) *barking dogs seldom bite* describes people who sound more dangerous than they actually are; (7) *barking up the wrong tree* means looking for something in the wrong place; (8) *every dog has its day* is an expression used when something pleasant happens to a person, especially one who has been experiencing bad luck; (9) *let sleeping dogs lie* means to leave a situation undisturbed; (10) *tail wagging the dog* means, for example, that a less important member of a group is directing the activities of others; (11) *you cannot teach an old dog new tricks* is often used to describe people who refuse to change their ways, or to learn new ways of doing things; (12) *dog in a manger* describes people who keep others from doing/using something that they themselves cannot do/use. This saying comes from Aesop's fable of a dog that crawled into a manger of hay and prevented a horse from eating, even though dogs do not eat hay; and (13) what does *a dog is man's best friend* mean? It means something special in terms of commitment, loyalty, and service!

Now let us discuss breeds and characteristics of dogs—carnivores that are commonly called *man's best friend* (we believe this is true for women as well!).

## 3.2    BREEDS AND CHARACTERISTICS OF DOGS

Although it is impossible to describe the precise milieu surrounding the development of species and breeds of companion animals over the past multimillennia, we are confident the mise-en-scène included persons interested in and intrigued by the domestication of dogs. The archaic dog pipe depicted in Figure 3.1 serves to remind us that our ancient ancestors must have admired and been fascinated by dogs.

For many centuries humans have selected and mated dogs to accentuate or eliminate certain characteristics, traits, and behaviors. It is estimated that there are between 700 and

**Figure 3.1**  This archaic dog pipe is symbolic of the respect and admiration cave dwellers had for dogs. *Courtesy University of Illinois archives.*

800 breeds of dogs in the world, although many of them are either scarcely known outside their own country or are not officially recognized.[1] The American Kennel Club's classification of breeds is discussed in Section 3.3.

The basic structure and characteristics of canine conformation reflect the origin of dogs as land-dwelling predators. In general, dogs have efficient locomotion to chase prey, eyes adapted to detect the motion of prey, keen olfaction to aid in tracking and locating prey, keen hearing to detect prey and other predators, and teeth designed to grip, kill, and consume their prey. Many details regarding the anatomy and physiology of dogs (and cats) are discussed in Chapter 11.

*Hair coat length and texture.*   Selective breeding has resulted in a wide variety of coat types and colors. The "correct" coat color or texture in one breed may be viewed as a flaw in other breeds. Many breeds have a double coat consisting of a short, dense undercoat and a longer outer coat. Some breeds have only a single coat—this may be long as in the Maltese breed or short as found in Italian Greyhounds. Many terrier breeds have a crinkly, harsh, wiry coat; these are also described as being a "broken" coat. When "broken" coats grow long, they lose their harsh texture. To encourage the growth of new coats, terrier owners and groomers remove the old coat by hand-plucking or stripping. Breeds such as the Curly-Coated Retriever and the Irish Water Spaniel have a curly coat with thick, tight curls. Corded coats are found on the Hungarian Puli and Komondor; the cords are formed by the intertwining of both the outer and undercoats. Chinese Shar-Pei dogs have a short, bristly coat.

*Hair coat colors.*   Hair coats show a wide variety of basic colors and variations. Basic colors include black, brown, chocolate, liver, tan, red, yellow or gold, blue, gray, and white. Sable is a specific coat color pattern produced by black-tipped hairs overlaid on a different colored background. Gold-sable is the most common sable color; others include gray-sable and silver-sable. The undercoat of a sable is usually light in color. Merle coloring is due to a dominant gene that produces irregular dark blotches of color against a lighter background. Blue merles are the most common; however, sable merles, liver merles, and red merles can also be found. Ticking is the presence of small areas with dark hairs interspersed with lighter color hairs throughout the coat; this color pattern is common in hound breeds and gundogs. Brindle is a coat with black or dark hairs interspersed with lighter hairs in a subtle striped pattern. A splashed coat is one with irregular markings, usually white, on a more deeply colored background. Tri-colored dogs are black, tan, and white. A parti-colored dog has white and another color in approximately equal proportions. Pinto or piebald refers to a coat with irregular colored patches *superimposed* on a white background. The inheritance of coat color is complex and varies with different breeds, thus it is advisable to contact breeders or other knowledgeable individuals for information on a specific breed.

# 3.3   AMERICAN KENNEL CLUB GROUPINGS OF DOG BREEDS

The American Kennel Club (AKC, www.akc.org) was established in 1884 as a nonprofit organization devoted to the advancement of purebred dogs. In 2003, the AKC recognized 150 breeds classified into seven groups and a miscellaneous class. The latter includes rare breeds having active enthusiasts but whose dogs are not yet established sufficiently to be admitted to the official Stud Book as an AKC-recognized breed. Total new dog registrations by the AKC in 2002 were approximately one million. Labrador Retrievers ranked first with 154,616 registrations. Table 3.1 gives the top 20 breeds listed in descending order of total registrations in 2002.

The seven group categories and miscellaneous breeds are given in Table 3.2.

---

[1]Juliette Cunliffe. 2003. *The Encyclopedia of Dog Breeds.* United Kingdom: Parragon Publishing.

**Table 3.2**
Breeds of Dogs by AKC-Recognized Groups

## Sporting Group

- Brittany
- Pointer
- German Shorthaired Pointer
- German Wirehaired Pointer
- Chesapeake Bay Retriever
- Curly-Coated Retriever
- Flat-Coated Retriever
- Golden Retriever
- Labrador Retriever
- English Setter
- Gordon Setter
- Irish Setter
- American Water Spaniel
- Clumber Spaniel
- Cocker Spaniel
- English Cocker Spaniel
- English Springer Spaniel
- Field Spaniel
- Irish Water Spaniel
- Nova Scotia Duck Tolling Retriever
- Spinone Italiano
- Sussex Spaniel
- Welsh Springer Spaniel
- Vizsla
- Weimaraner
- Wirehaired Pointing Griffon

## Hound Group

- Afghan Hound
- American Foxhound
- Basenji
- Basset Hound
- Beagle
- Black and Tan Coonhound
- Bloodhound
- Borzoi
- Dachshund
- English Foxhound
- Greyhound
- Harrier
- Ibizan Hound
- Irish Wolfhound
- Norwegian Elkhound
- Otterhound
- Petit Basset Griffon Vendéen
- Pharaoh Hound
- Rhodesian Ridgeback
- Saluki
- Scottish Deerhound
- Whippet

## Working Group

- Akita
- Alaskan Malamute
- Anatolian Shepherd
- Bernese Mountain Dog
- Boxer
- Bullmastiff
- Doberman Pinscher
- German Pinscher
- Giant Schnauzer
- Great Dane
- Great Pyrenees
- Greater Swiss Mountain Dog
- Komondor
- Kuvasz
- Mastiff
- Newfoundland
- Portuguese Water Dog
- Rottweiler
- Saint Bernard
- Samoyed
- Siberian Husky
- Standard Schnauzer

## Terrier Group

- Airedale Terrier
- American Staffordshire Terrier
- Australian Terrier
- Bedlington Terrier
- Border Terrier
- Bull Terrier
- Cairn Terrier
- Dandie Dinmont Terrier
- Irish Terrier
- Kerry Blue Terrier
- Lakeland Terrier
- Manchester Terrier (Standard)
- Miniature Bull Terrier
- Miniature Schnauzer
- Norfolk Terrier
- Norwich Terrier
- Parson Russell Terrier
- Scottish Terrier
- Sealyham Terrier
- Skye Terrier
- Smooth Fox Terrier
- Soft Coated Wheaten Terrier
- Staffordshire Bull Terrier
- Welsh Terrier
- West Highland White Terrier
- Wire Fox Terrier

## Toy Group

- Affenpinscher
- Brussels Griffon
- Cavalier King Charles Spaniel
- Chihuahua
- Chinese Crested
- English Toy Spaniel
- Havanese
- Italian Greyhound
- Japanese Chin
- Maltese
- Manchester Terrier
- Miniature Pinscher
- Papillon
- Pekingese
- Pomeranian
- Poodle
- Pug
- Shih Tzu
- Silky Terrier
- Toy Fox Terrier
- Yorkshire Terrier

**Non-Sporting Group**

| | | |
|---|---|---|
| • American Eskimo Dog | • Dalmatian | • Poodle |
| • Bichon Frise | • Finnish Spitz | • Schipperke |
| • Boston Terrier | • French Bulldog | • Shiba Inu |
| • Bulldog | • Keeshond | • Tibetan Spaniel |
| • Chinese Shar-pei | • Lhasa Apso | • Tibetan Terrier |
| • Chow Chow | • Löwchen | |

**Herding Group**

| | | |
|---|---|---|
| • Australian Cattle Dog | • Border Collie | • German Shepherd Dog |
| • Australian Shepherd | • Bouvier des Flandres | • Old English Sheepdog |
| • Bearded Collie | • Briard | • Pembroke Welsh Corgi |
| • Belgian Malinois | • Canaan Dog | • Polish Lowland Sheepdog |
| • Belgian Sheepdog | • Cardigan Welsh Corgi | • Puli |
| • Belgian Tervuren | • Collie | • Shetland Sheepdog |

**Miscellaneous Class**

| | | |
|---|---|---|
| • Beauceron | • Glen of Imaal Terrier | • Plott Hound |
| • Black Russian Terrier | • Neapolitan Mastiff | • Redbone Coonhound |

*Source: American Kennel Club, Inc., 260 Madison Avenue, New York, NY 10016 (Web site: www.akc.org/breeds/recbreeds/group.cfm).*

## 3.3.1   SPORTING GROUP

The Sporting Group comprises dogs used in hunting game birds and waterfowl. Pointers and setter breeds are "forerunners" that locate the game. Spaniels are adept at flushing game. Retrievers work "behind-the-gun" to retrieve downed birds. Weimaraners are multipurpose bird dogs. Naturally active and alert, Sporting dogs make likeable, well-rounded companions. Noted for their innate instincts in water and woods (some work in water, others on land, still others in both), many of these breeds actively continue participating in hunting and other field activities. Prospective owners of Sporting dogs should appreciate that most of these breeds have easy-care coats and thrive on regular, invigorating activity and exercise. The most popular of this group are the Labrador Retriever and Golden Retriever.

## 3.3.2   HOUND GROUP

Most hounds share the common ancestral trait of being used for hunting. Some use astute scenting powers to follow a trail (their keen sense of smell is so powerful, most dogs can detect 1 part of urine in 60 million parts of water). Others demonstrate a phenomenal gift of stamina as they relentlessly run down quarry (prey). The Hound Group probably has the greatest variance in size, ranging from the diminutive Miniature Dachshund to the towering Irish Wolfhound. Most hounds were bred to pursue four-legged game and are divided into those that hunt by sight (sighthounds) and those that follow scent (scenthounds). The sighthounds are long-legged, swift dogs such as the Greyhound, Afghan Hound, Basenji, Borzoi (Russian Wolfhound), Ibizan Hound, Irish Wolfhound, Whippet, Scottish Deerhound, and Saluki.

Scenthounds are given more to perseverance than to rapid pursuit. The "trailers" include breeds such as the low-slung Dachshund and Basset Hound, sad-eyed Bloodhound, perky Beagle, Black and Tan Coonhound, American and English Foxhounds, Harrier, Norwegian Elkhound, Petit Basset Griffon Vendéen, and Otterhound. Because of their sporting nature, most of these breeds (except the Afghan Hound) have coats that are easy to keep groomed. Some hounds possess the distinct ability to produce a unique sound known as baying (barking with deep, prolonged tones). Baying alerts the huntsmen that the dog is following a scent trail.

**Figure 3.2** A typical dog sled team participating in the famed Iditarod race. *Courtesy Champion Petfoods.*

### 3.3.3 WORKING GROUP

The Working Group comprises breeds that have been selected and bred to pull sleds, guard property, and serve as draft workers and in water rescues. They have been invaluable assets to humans for many centuries. The Doberman Pinscher and Great Dane are included in this group, as are the Rottweiler and Bullmastiff, which have been bred for guard/security work and are naturally protective of family and home property. Sledding breeds such as the Siberian Husky and Alaskan Malamute are robust and active. Indeed, much can be expected of a Siberian Husky if it is adequately fed and properly trained. They can consistently travel 20 or more miles a day on a 5,000 kcal diet (Figure 3.2). Several rescue breeds, including the Newfoundland (excellent in water rescue efforts), tend to be respectably strong but usually gregarious.

Working dogs are intelligent, quick to learn, highly capable animals that make devoted companions. Their large size and strength make them marginally suitable as pets for children. For safety of owners and others, as well as for service, these large dogs must be trained properly.

### 3.3.4 TERRIER GROUP

The Terrier Group comprises breeds that traditionally hunt and dig for vermin. They are generally feisty characters (after all, they could not be cautious and cowardly and face rats with enthusiasm) and range in size from fairly small, as in the Norfolk, Cairn, or West Highland White Terrier, to the larger Airedale Terrier. People familiar with this group invariably comment on the distinctive terrier personality. Typically, they have minimal tolerance for other animals, including other dogs. Many project the attitude that they are eager for a spirited confrontation.

Considerable grooming is required on some of the wire-coated terriers such as the West Highland White Terrier, Wire Fox Terrier, and Scottish Terrier. Soft-coated terriers such as the Kerry Blue Terrier, Bedlington Terrier, and Soft Coated Wheaten Terrier require some scissoring and combing to maintain a trim appearance. Most terriers make engaging pets but may, at times, test the owner's determination to match their dog's energy level and lively character.

## 3.3.5 TOY GROUP

The Toy Group comprises breeds that were selected and bred primarily as lap dogs and pampered pets. As the group name implies, Toys are small in stature. Many have been favored by the royalty and nobility over the years. The group includes such breeds as Pekingese, which were kept by empresses and emperors of China; Shih Tzu of Tibet, which were treasured as house pets by members of China's famed Ming Dynasty; Papillon (named by Marie Antoinette) with its butterfly ears; snowy and showy Maltese; Pomeranian (a favorite of Queen Victoria); and several others.

Toy breeds are often excellent companion animals for older and/or infirm people because they require less exercise than many of the larger breeds. Some breeds are a bit fragile to be pets of rough-and-tumble children. Do not permit their small stature to mislead you though; many Toys are tough and tenacious as those who have experienced the barking of an angry Chihuahua can attest. Toy dogs are especially popular with city dwellers and people with limited living space. They make ideal apartment dogs and terrific lap warmers on nippy nights. (Incidentally, small dogs can be found in most groups, not merely in the Toys.) Small dogs have less hair to shed, create less mess to clean up, and require less food. And, training aside, a 10-lb dog is easier to control than one 10 times that size. On the other hand, small dogs are often quick, fast, and difficult to seize if they choose not to be caught.

## 3.3.6 NON-SPORTING GROUP

Non-Sporting dogs are an unusually diverse group. Included are sturdy animals with such different appearances and personalities as the Dalmatian and two of the three Poodles (the smallest Poodle is in the Toy Group). Non-Sporting breeds also include the Bulldog, French Bulldog, and Boston Terrier. The aloof Chow Chow, the profusely coated Keeshond, and the fluffy Bichon Frise are also included. The group also includes several long-coated Tibetan breeds. With the exception of the smooth-coated breeds, this group generally requires dedicated grooming to keep the long, soft hair coat free of tangles and mats.

The Non-Sporting Group includes substantial differences in size, coat, personality, and visage. Some such as the Schipperke and Tibetan Spaniel are uncommon sights in most neighborhoods. Others, including the Poodle and Lhasa Apso, have a large following.

## 3.3.7 HERDING GROUP

Created in 1983, the Herding Group is the newest AKC classification; its members were formerly part of the Working Group. All breeds of this group share the fabulous ability to control the movement of other animals. A remarkable example is the low-set Pembroke Welsh Corgi—only 10 to 12 inches tall at the shoulders—that can drive cows many times its size to pasture by leaping and nipping at their heels.

Although breeds of the Herding Group were selected and bred to drive or herd cattle, sheep, and other livestock, the vast majority are household pets that never come close to a farm/ranch animal. Nevertheless, pure innate instinct prompts many to gently move their owners, especially children of the family, and many will chase objects that move (e.g., cars and bicycles).

Being as small as the petite Shetland Sheepdog or as large as the wooly Old English Sheepdog, Bouvier, and Briard, several of the herding breeds have thick double coats and require regular grooming to remain presentable. These active, intelligent, courageous, and determined dogs make excellent companions and respond beautifully to training exercises. Several of these breeds have excelled in police work, search and rescue (as in locating victims of the September 11, 2001, terrorist attacks on the New York City World Trade Center), tracking, service to physically impaired owners, as well as sentries and couriers during war (Chapter 21).

### 3.3.8 MISCELLANEOUS CLASS

Canine authorities concur that worldwide there are hundreds of distinct breeds of purebred dogs, most of which are not AKC recognized. Breeds officially recognized for AKC registration appear in the Stud Book of the American Kennel Club. The AKC provides a methodical, readily understood pathway for development of a new breed, which may result in a breed receiving full recognition and appearance in the AKC Stud Book.

Briefly stated, the requirement for admission to the Stud Book is clear and categorical proof that a substantial, sustained nationwide interest and activity exists in the breed. This includes an active parent club, with serious and expanding breeding activity over a wide geographic area.

When in the judgment of the AKC Board of Directors such interest and activity exists, a breed is admitted to the Miscellaneous Class (see Table 3.2 for names of the six breeds included in the 2003 listing). Breeds in the Miscellaneous Class may compete and earn titles in the AKC Obedience, Tracking, and Agility events. Owners of miscellaneous breeds are eligible to compete in Junior Showmanship. They may also compete in conformation shows, but are limited to competition in the Miscellaneous Class and are ineligible for championship points (Chapter 13).

Selected popular breeds of dogs are depicted in Color Plates 4 to 6.

## 3.4 UNITED KENNEL CLUB

The United Kennel Club (UKC) was founded in 1898 to emphasize "the total dog." At the time it was founded, the organizers were concerned that the AKC was concentrating on conformation (appearance) and ignoring the working heritage of the various breeds of dogs. The UKC has grown to become the second-largest dog registry in the world. It currently registers 302 breeds of dogs. Events sponsored by the UKC include agility, conformation, obedience, weight pulling, water races, night hunts, and field trials. To encourage participation by owner handlers, the UKC does not allow professional handling in UKC conformation shows. Offspring of UKC-registered dogs are eligible for registration. Requirements for registering other purebred dogs are regulated by any applicable UKC breed clubs. There may be a requirement of performance and/or health testing prior to granting approval for individual registration.

## 3.5 CHARACTERISTICS OF CATS

In general, cats are active, affectionate, clean, easy to care for, protective, and tolerant. Their smooth, flowing movements are graceful, precise, and much admired. As natural carnivores, their finely honed utilization of keen senses of hearing, sight, and smell coupled with their patience and quickness aid them in locating, observing, stalking, pouncing on, and killing prey. Indeed, cats are credited with putting several species of birds and small mammals in jeopardy of extinction during the past three centuries. Whereas dogs expend large amounts of energy running about trying to locate prey in the wild, cats are much more meticulous, single-minded, economical, and effective in conserving their energy for the last stages of the hunt.

*Eyes.* The color of cat eyes is of special interest. When kittens open their eyes at 7 to 10 days, they are blue. The range of eye colors that develops throughout kittenhood varies from orange to shades of amber and green to blue. Breed standards specify which colors are acceptable for registration. In the Abyssinian, eye-color requirements are related to coat color, whereas Siamese and Birmans must have blue eyes (deep blue is preferred) for breed registry.

Odd eye color (one blue, one yellow or orange) is fairly common, especially in white cats. White cats with blue eyes are often deaf, this is true whether one or both eyes are blue. During embryogenesis the dominant white gene, W, inhibits the migration of auditory nerve cells required for hearing.

**Figure 3.3**  Dentition of the cat: incisors (1, 2, 3), premolars (P2, P3, P4), and molars (M1). The maxilla (upper jaw) has the dental formula 3131. The mandible (lower jaw) has the formula 3121. *Courtesy University of Illinois.*

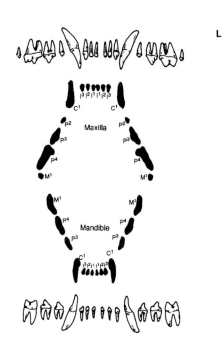

Although a cat's eyes are slightly smaller than those of humans, they can see an estimated six times better in the dark. This results, in part, because cats have larger pupils to let in more light in periods of darkness. Cat's eyes glow (shine) in the dark because a layer of cells behind the retina called the tapetum optimizes the available light and reflects it back into the vitreous humor within the eyeball. The tapetum is pink, gold, blue, or green and reflects a different color as the amount of light changes. Some Siamese cats have a genetic flaw that produces double vision rather than the usual stereoscopic image.

**Teeth.** All domestic cats have 30 teeth (16 upper and 14 lower). Their kittens have only 26 milk (deciduous) teeth. The "dental formula" is 3131/3121 (incisors, canine, premolars, molars; Figure 3.3). The incisors (in the front part of the mouth) are the smallest and are used for cutting. The long, sharp canine teeth are used to seize and kill prey. The fourth upper premolar and the first lower molar teeth are known as the carnassial teeth and are used to shear through food. The carnassial teeth function similarly to scissors in slicing through foods. Cats have very limited ability to grind foods.

**Balance and orientation.** Cats have unusually dependable innate balancing mechanisms enabling them to land on their feet. This trait results from their having an especially flexible spine, which, in concert with a balancing organ in their inner ears, enables them to orient their bodies during a fall or jump from a high place. A cat's tail is also a balancing mechanism and can be used as a rudder to trim, or adjust, the direction of a jump down into a confined space. The tail is useful in maintaining balance when climbing trees or other woody structures. The cat's tail can also serve as a mechanism for balancing when pursuing prey with speed. Cats generally come down a tree backwards because their rear legs do not rotate very well and their claws point in the wrong direction to descend head first. It is interesting that cats can jump up to five or six times their own height.

**Vocal cords.** Unlike most animals, cats have two sets of vocal cords in the throat. One set is above the other and each set produces different sounds. Most scientists believe cats use the lower vocal cords to produce the "meow" and other ordinary sounds cats make. They use the upper vocal cords to purr and growl.

**Foot pads.** Cats have rubberlike pads on their paws that enable them to tread softly and quietly. This is an important asset bestowed by nature to assist them in stalking wary prey.

**Figure 3.4** These cats have developed a habit of following this young boy to individual bowls where they know he will divide the warm milk for their enjoyment. *Courtesy Dr. James E. Corbin.*

*Life expectancy.* The average life expectancy of domestic cats is 10 to 14 years. Ten years in a cat is commonly equated to 50 years in a human. This is a 1:5 ratio so 20 years in a cat would be equivalent to 100 years in a person. The greatest danger resulting in short life expectancy for outdoor cats is that of road traffic. Thousands of cats are killed or maimed annually by traffic in cities as well as on highways and rural roads (Chapter 24). Coyotes, eagles, and owls also pose dangers for outdoor cats. Older cats are prone to a variety of metabolic diseases including renal (kidney) failure, diabetes mellitus, and hyperthyroidism. These diseases are often fatal if not well-controlled (Chapter 14).

*Habits and change.* Cats are creatures of habit (Figure 3.4). They enjoy stability and status quo and dislike change. They are highly sensitive to changes in their environment, which they interpret as leading to danger or disturbance. Perhaps this is one reason most cats do not enjoy traveling by car on the family vacation, whereas many dogs welcome the journey. For most cats, putting them in a car is depriving them of the sights, smells, and sounds of their familiar environment, replacing those accepted experiences with an unacceptable level of noise and the odor of fuel and oil fumes.

*Cat whiskers (vibrissae).* Most cats have approximately 30 long whiskers that grow in four rows from the side of the mouth to above the eyes. These whiskers serve as a device (antennae) to feel the immediate environment (Figure 3.5). For example, they are used to assess the dimensions of spaces in the dark as well as to provide important information when prey are caught and too close to observe. Cat whiskers are also valuable in navigation as indicated by the fact that cats with damaged or lost whiskers (as in a fight) move about with far less assurance, whereas cats with particularly fine, spreading sets of whiskers commonly explore vigorously with supreme confidence. Cats also use their whiskers in communication (Chapter 20).

*Cat hair coat colors and patterns.* The variety of coat colors and patterns provides a colorful assortment of cats for prospective owners to choose from (Figure 3.6). Some breeds have only one color (e.g., Bombay—black, Chartreux—grey blue, Havana Brown—chestnut brown, Korat—silver blue), whereas others may have a wide variety of coat colors and patterns.

    The basic coat colors are white, black, red (orange, ginger), blue (diluted black), cream (diluted red), brown (chocolate), lilac (lavender, a light gray-brown with pink overtones), cinnamon (light brown with red overtones), and fawn (a dilute version of cinnamon). The six basic coat patterns are solid, bicolor, calico, tabby, tortoiseshell, and colorpoint. Bicolors are generally white plus one other color. Calicos are tricolor—white, black, and red (orange) or

**Figure 3.5** Whiskers are one of numerous intriguing features of cats. *Courtesy Dr. James E. Corbin and Marcelo Palmeira.*

the diluted versions white, blue, and cream. These cats are almost always females.[2] Tabby is a very common coat pattern where "ticked" or agouti[3] hairs alternate with solid-colored hairs. This pattern has four varieties (subtypes): (1) mackerel tabby—ticked hairs alternate with

**Figure 3.6** The black and white cat at left (*a*) and the silver tabby at right (*b*) exemplify the variation and beauty in coat colors and patterns of domestic cats. *Courtesy of Mike Nelson and Vern Horning, Graphics, American Nutrition Inc. (a) and Champion Petfoods (b).*

(a)                                                                (b)

---

[2]In cats, the genes for red (orange) or black coat color are carried on the X chromosome. Females have two X chromosomes and thus can have both red and black hairs (the pattern is determined by separate genes that control coat color patterns). Males have only one X chromosome and thus can have only red or black, not both. When a male has both red and black hairs, it has an extra X chromosome (XXY) and is almost always sterile.
[3]Agouti or ticked hairs have bands of light and dark pigmentation. The agouti gene allows full pigmentation when hairs start to grow, inhibits pigmentation during the middle phase of hair growth, and then allows pigment production to resume for a short time prior to stopping pigment production as hairs reach their maximum length. The result is hairs that are darkly pigmented at the tips with a yellow to orange middle band, then a dark band just above the yellow to orange root.

stripes, as on a tiger; (2) classic tabby—ticked hairs alternate with solid hairs in a bold swirling pattern, also known as a blotched tabby; (3) ticked tabby—ticked hairs are found over the entire coat, also known as agouti tabby; and (4) spotted tabby—ticked hairs alternate with spots or rosettes of solid color, as on a leopard or jaguar. The tortoiseshell is a mix of red (orange) and black or the diluted versions of cream and blue. This coat pattern is seen almost exclusively in females for the same reason as discussed previously (see footnote 2). There may also be an underlying tabby pattern, and this combination is sometimes called "torbie." In the colorpoint pattern, the face, paws, and tail are darker in color than the rest of the body; this is due to a temperature-dependent enzyme involved in pigment synthesis with cooler body parts (e.g., the extremities) having a greater production of pigment (Figure 3.7).

*Adages, idioms, legends, and sayings about cats.* The adage that *cats have nine lives* is derived from their remarkable ability in escaping from injury associated with serious falls and their knack of fleeing from a pack of mean-spirited dogs.

The proverbial intense *curiosity of a cat* is truly descriptive of the intelligent cat. A high level of curiosity is a natural tool of a cat's instinct for survival in the wild, where it is important for them to know as much as possible about their environment, including things that might be a threat or provide hunting rewards.

The idiom *curiosity killed the cat* can come true if a cat explores the warm engine of a car and gets fatally injured when the motor is started. Being overcome with curiosity can also claim the life of cats that get into insect sprays or sacks of rodent poisons. Such items should be secured away from cats (Chapter 14).

The idiom to *not let the cat out of the bag* is said to have derived from an old market trick where a cat was substituted for a piglet in a sack following a sale or barter agreement. Thus when a person *lets the cat out of the bag*, they have shared a secret. The legendary *cat of nine tails* was a whip used to deliver punishment in the British navy. The whip had nine leather lashes resembling the tails of cats.

The saying *when the cat's away, the mice will play* suggests that a person is more inclined to participate in mischief when no one is watching. The saying *honest as a cat when the cream is out of reach* refers to a person who is not wholly trustworthy.

*Even a cat may look at a king* means that people have certain rights, even with those having much more power than they do. *Being made a cat's paw* refers to a person who is made a tool of someone else.

**Figure 3.7** The dark color points with light body color typical of Siamese cats (also seen in Himalayans and other breeds) are a result of a temperature-sensitive enzyme that only synthesizes pigment at temperatures below 95°F; this lower temperature is found in the extremities. *Courtesy Dr. James E. Corbin.*

# 3.6   HISTORICAL PERSPECTIVE OF CAT BREEDS

Scientists have found it difficult to map the evolution of cats, partly because in the wild they tend to live in forested areas, which reduces the probability of fossilization. The earliest recorded use of big cat skins for human clothing dates from ca. 6500 BC.

The domestication of cats and dogs is discussed in Chapter 1. It is well established that dogs have been selected and bred for specific purposes (sheepdogs, hunting/gun dogs, racing dogs, lapdogs, and other uses) for hundreds of years, whereas the breeding of cats for specific purposes is little more than a century old.

The domestic cat is one of 38 species of the cat family (*Felidae*). The largest species of cat is the Siberian tiger, which attains an average length of 10 ft 5 inches and a height of 42 inches at the shoulder. The smallest member of the cat family in Africa is the Black-Footed Cat; males weigh a maximum of 4.5 lbs and mature females often weigh under 3.25 lbs.

The Governing Council of the Cat Fancy, the sole registration body of Britain, recognizes about 100 breeds. The Cat Fanciers' Association (CFA, www.cfainc.org), founded in the United States in 1906 and the world's largest registry of pedigreed cats, recognizes 41 breeds (Table 3.3).

## 3.6.1   PEDIGREED CATS

In general a pedigreed cat is one whose ancestry is in a recognized breed that has been recorded and registered through four or more generations. According to the CFA, fewer than 3% of all owned cats worldwide are pedigreed. This does not account for millions of homeless stray and feral cats. Factoring in the "nonowned" cats that run at large, probably fewer than 1% of all domestic cats result from purebred pedigree breeding.

We should differentiate between *breed* and *trait*. A particular breed of cat almost always has certain traits, but not all cats with those traits are members of that breed. For example, the trait of solid blue color is common to four breeds: British Shorthair, Chartreux, Korat, and Russian Blue. However, these breeds differ greatly from one another in body type, boning, facial structure, eye color and shape, and general body conformation.

**Table 3.3**
Cat Breeds Recognized by the CFA

**CFA Breeds**

| | | |
|---|---|---|
| • Abyssinian | • Egyptian Mau | • RagaMuffin |
| • American Bobtail | • European Burmese | • Ragdoll |
| • American Curl | • Exotic | • Russian Blue |
| • American Shorthair | • Havana Brown | • Scottish Fold |
| • American Wirehair | • Japanese Bobtail | • Selkirk Rex |
| • Balinese | • Javanese | • Siamese |
| • Birman | • Korat | • Siberian |
| • Bombay | • LaPerm | • Singapura |
| • British Shorthair | • Maine Coon | • Somali |
| • Burmese | • Manx | • Sphynx |
| • Chartreux | • Norwegian Forest Cat | • Tonkinese |
| • Colorpoint Shorthair | • Ocicat | • Turkish Angora |
| • Cornish Rex | • Oriental | • Turkish Van |
| • Devon Rex | • Persian | |

*Source: The Cat Fanciers' Association, Inc., P.O. Box 1005, Manasquan, NJ 08736–0805 (Web site: www.cfainc.org/breeds.html).*

Blue color is common in the mixed-breed cat population. Genetically it is the dilute form of black. Thus a blue cat is basically (genetic-wise) a black cat with one gene that changes the color to an attractive shade of blue-gray. Black is the most common genetic color of cats. The vast majority of blue cats are from a mixed-breed population. Another common trait often mistaken for a purebred is the so-called Manx trait (partial to complete taillessness). This dominant trait is found in the random-bred population as well as in the purebred Manx population.

With reference to cats, breed per se is more about pedigree than traits. Cats are members of a certain breed because their parents were registered members of that breed. Without breed organization papers, one can only say that a cat looks like a certain breed. Unlike breeds of dogs, specific breeds of cats are a relatively recent result of planned breeding.

Although enormous pleasure can be derived from the companionship of a pedigreed cat, many ardent cat lovers have never owned a pedigreed cat. Many of the most beautiful, exquisitely patterned, good-natured cats—those making attractive, delightful pets—derived their pleasing appearance and personality/temperament from chance matings rather than as the result of a breeder's selection and mating skills.

Nonpedigreed cats can often be shown in a class called "household" or "domestic" cats in which prizes are awarded for good health, good temperament/disposition, showiness, and charm rather than adherence to specific breed standards.

## 3.7 CAT BREEDS MOST FREQUENTLY REGISTERED BY THE CAT FANCIERS' ASSOCIATION

Cats—with their clean and quiet ways, beauty, and charm—have become increasingly popular pets (Figure 3.8). A study sponsored by the Pet Food Institute (Chapter 2) indicated that the 2002 cat population in the United States exceeded 77 million. So it is not surprising that pedigreed cats have an enthusiastic and devoted following of admirers and faithful fans. But, even though purebred cats have never achieved the popularity enjoyed by purebred dogs, the "cat fancy" (group of people involved in breeding and showing pedigreed cats) has an increasing number of fervent followers.

**Figure 3.8** These children enjoy playing with their lovable kittens. *Courtesy Dr. Karen L. Campbell.*

Each year the CFA compiles registration totals for the 41 breeds it recognizes. Those breeds are listed in Table 3.3. The 10 cat breeds most frequently registered are given in Table 3.4 and discussed in the following sections. Selected popular breeds of cats are depicted in Color Plate 3.

## 3.7.1   PERSIAN CATS

With long lovely hair, sturdy conformation, and a pleasing disposition, the Persian cat has been a popular breed in the United States since cat fancy began in 1871. Persians represent approximately one-half of all purebred cats registered in the United States.

The Italian traveler Pietro dellaValle (1586–1652) spent 4 years in Persia (1617–1621) and is credited with having taken the first Persian cats to Europe nearly four centuries ago. Travelers often gave Persians—considered the ultimate luxury cat—to nobles as gifts. So the Persian became recognized as a cat of the aristocracy. In the second half of the 19th century Queen Victoria acquired a pair of blue Persian cats, which was perceived to greatly enhance the royal household.

The popular Persian has short, thick legs, a rounded head, short nose, and large eyes. Its long, flowing, thick coat of variable colors is as soft as silk.

Persians commonly have small litters (often two to three kittens), and an above-average mortality rate in the first 6 months. A coat-color-related genetic flaw sometimes appears in white Persians. This occurs because the dominant white gene W carries with it the liability of deafness, especially in blue-eyed white Persians (see Section 3.5; this occurs in all breeds in which white color is associated with the W gene).

The longhaired Persian breed is officially known in Britain as Longhairs.

## 3.7.2   MAINE COON CATS

The Maine Coon is known for its large size (mature males weigh 12 to 18 lbs, females 10 to 14 lbs); easygoing, good-natured temperament; and rugged muscular appearance. These cats are agile, brave, tough, independent, and resourceful—good workers that excel in controlling rodent populations on farms and ranches.

This native New England breed is well adapted to that climate with its heavy insulating shaggy but silky coat, bushy tail, and tufted ears and toes. The alleged origin of this American-bred cat is an interesting story. According to some cat historians, Marie Antoinette planned, at the start of the French Revolution in 1789, to flee to the United States and sent ahead certain of her possessions including her prized Persian cats. Legend has it that these cats escaped and interbred with American domestic cats giving rise to the new breed, Maine Coon.

These fun-loving, quiet, adaptable cats are good household pets that are available in a variety of coat colors and patterns. Typical litter size is two to three kittens.

**Table 3.4**

Ten Most Frequently Registered Cat Breeds in 2003

| 2003 Rank | Breed | 2003 Rank | Breed |
|-----------|-------|-----------|-------|
| 1 | Persian | 6 | Birman |
| 2 | Maine Coon | 7 | Oriental |
| 3 | Exotic | 8 | American Shorthair |
| 4 | Siamese | 9 | Tonkinese |
| 5 | Abyssinian | 10 | Burmese |

*Source: The Cat Fanciers' Association, Inc., P.O. Box 1005, Manasquan, NJ 08736–0805 (Web site: www.cfa.org/news.html).*

### 3.7.3 EXOTIC CATS

Sometimes referred to as "a Persian in pajamas," the Exotic breed was developed by crossing Persians with shorthaired breeds such as American Shorthair, Burmese, British Shorthair, and Russian Blue. Exotics are available in the same rainbow of colors and patterns as the Persian breed. They have a round head with small rounded ears. Exotics have a medium-large body, are deep-chested, and are low on their legs. Their coat is thick and dense, somewhat longer, softer, and more plush than most other shorthaired breeds. Their average litter size is four kittens.

### 3.7.4 SIAMESE CATS

This former Royal Cat of Siam (now Thailand) lived in palaces and temples for over 200 years. It was brought to the United States in 1895. The Siamese cat is distinguished by its brilliant eyes and colored points (ears, face, tail, and feet), which provide a striking contrast to its usual light-colored body. They are famous for their ability to vocalize. Their loud, raspy yowls are used to gain attention and express their many moods.

Siamese cats have long legs and tail, a slender body, and a long wedge-shaped head with large, pointed ears. They are intelligent and have a rather boisterous temperament. Many are distrustful of strangers, yet they are affectionate and respond well to training. Siamese cats often have large litters (often six or more) and live up to 15 or more years. Their short hair should be brushed twice weekly.

Siamese cats have four classic colors: seal, blue, chocolate, and lilac (frost). The distinctive light body color with darker points is due to a temperature-sensitive enzyme involved in pigment production. Pigment is produced only at temperatures below approximately 95°F. Kittens are born all white due to the uniform temperature inside the mother's uterus. After birth, new hairs growing on the ears, tail, and limbs are pigmented due to the lower skin temperature in these regions (see Figure 3.7).

### 3.7.5 ABYSSINIAN CATS

This shorthaired, older breed of cats probably originated in Egypt, but was bred and raised for thousands of years in Abyssinia (now Ethiopia). Active and agile, playful and inquisitive, "Abys" are prized for their unusual coat, athletic build, and dynamic personality. They are distinguished by a ticked tabby coat pattern—a pattern more commonly seen in wild cats. Though warm brown or reddish brown is the color most often associated with the breed, they are also available in blue, fawn, and red (sometimes called sorrel).

Abyssinians were brought to the United States in 1909. They are medium sized and have fine bones, large ears, and almond-shaped eyes. They are energetic and present a regal bearing. Although they are often shy with strangers, they are intelligent and easy to train. Abys commonly have small litters (they often bear only one or two kittens), but usually have kittens two (and sometimes three) times a year. They usually require grooming only once a week.

### 3.7.6 BIRMAN CATS

Also known as the "Sacred Cat of Burma," the Birman is a large, devoted cat with long, straight, silky hair; pointed pattern; colored mask; blue eyes; and a matching set of four white boots.

Members of this old breed are gentle and quiet natured, placid, with great beauty and dignity. They have numerous coat colors, are playful but not boisterous, and make excellent household pets. Their litter size averages four to five kittens.

### 3.7.7 ORIENTAL CATS

Oriental cats originated from crosses with Siamese cats. They have the body and personality of the Siamese, but their coat colors include myriad solid and tabby colors and patterns (often striped with colored patches). They are active and slender bodied, have a people-oriented temperament, and have above-average-sized litters (seven to nine kittens). The Oriental Longhair is known as the Angora breed in Britain.

### 3.7.8 AMERICAN SHORTHAIR CATS

The American Shorthair was developed from sturdy, native American working cats. Their ancestors earned their daily keep by ridding colonial barns of rodents. They are moderately stocky, even-tempered cats with a short coat. Although the breed is accepted in a wide variety of colors and patterns (e.g., calico and tortoiseshell), the silver classic tabby and black markings set off on a vivid silver background is probably best known.

American Shorthairs are good hunters, jumpers, and climbers with a large head, quiet voice, and powerful, muscular body. They are agile, hardy, affectionate, intelligent, and home-loving. Adult females weight an average of 10 lbs, adult males 14 lbs. They make excellent house pets that require only light grooming. Their average litter size is four kittens.

### 3.7.9 TONKINESE CATS

Tonkinese cats were produced by crossing the Burmese and Siamese breeds. They were established in Canada and the United States in the mid-1960s. This playful, people-oriented breed has a moderate body type and a sleek, soft coat. It features a unique pattern known as "mink"; it is pointed like the Siamese, its body is colored in a shade that harmonizes with the point color, and its eyes are aqua in shade. "Tonks" range in colors between their Burmese and Siamese parent breeds.

Tonkinese cats are well-muscled and have long, slim legs and a long tail. They are active, adventurous yet cautious, and require minimal grooming. They are affectionate pets that need considerable human contact. They are not trustworthy with birds and hamsters. Typical litter size is four kittens.

### 3.7.10 BURMESE CATS

The Burmese is an affectionate, even-tempered, playful cat with a glossy, short, soft, satin-like coat that requires minimal care. Solidly built, they are athletic, playful possessors of many Siamese traits. Though their original color is solid sable brown, other colors include blue, champagne, platinum, and tortoiseshell. Their head and eyes are round. Burmese are sociable cats that seek attention and are lonely when left alone. They are good parents and often have litters of eight to ten or more kittens. Burmese cats enjoy long lives, commonly reaching 20 or more years of age.

## 3.8 SUMMARY

This chapter is devoted to breeds of dogs and cats and their characteristics. Breeds represent groups of animals selected and mated by humans to emphasize certain traits and minimize or eliminate others. Dogs or cats of a particular breed are related genotypically (similar genetic makeup) by having descended from common ancestors. They are phenotypically (visibly) similar in most characteristics.

The American Kennel Club (AKC) classifies dogs into seven groups: Sporting, Hound, Working, Terrier, Toy, Non-Sporting, and Herding. Each of these categories is discussed as well as the AKC Miscellaneous Class. The United Kennel Club (UKC) is a second registry of purebred dogs that places an emphasis on preserving the working heritage of the various dog breeds.

Dogs are highly versatile. They range in size from the tiniest Chihuahua (~2 to 3 lbs) to the largest Irish Wolfhound or Mastiff (~180 to 200 lbs). They come in all coat types from the bare skin of the Chinese Crested to the thick double coat of the Newfoundland and the silken tresses of the Shih Tzu. They represent all colors from black to white and patterns from spotted to brindle. Some have bushy, others smooth tails; some have lop, others tipped and upright ears. They come in a wide range of temperaments from mild and joyful to tough and aloof.

Regardless of the façade, no matter the size, shape, or attitude, dogs have a history as old as civilization and a companionship with humans that has endured through the ages. Indeed dogs of today may not be asked to herd as many sheep, guard as many palaces, or serve as royal hunting companions, but they are nonetheless just as valuable and worth preserving as breeds, companions, and life partners as in yesteryear!

Although many breeds no longer ply their original assignments, they contribute to other important tasks (Chapter 21). German Shepherds herd few sheep but they assist police, security efforts, and the armed forces in many ways. Other breeds use their innate skills in finding lost, ill, or injured people; sniffing out contraband; working in arson investigations; helping physically impaired owners; and visiting patients in hospitals and nursing homes.

As natural pack animals, dogs have an instinct to share their lives with others—especially canines and humans. Companionship is highly important to dogs (Chapter 20).

A defining trait of dogs is their everlasting devotion and fidelity to their owners. The dog's unconditional commitment to its master is well known, admired, and greatly appreciated. Properly fed and cared for, dogs will lay down their lives to protect the life of their master, even against hopeless odds with a full-grown lion or tiger, thereby demonstrating an immeasurable sum of love and fidelity.

Cats are the most populous pets in the United States—an all-important group of companion animals now 77 million strong and growing.

Although cats may be viewed as being aloof, independent, and stand-offish to some visitors, they have earned the respect, admiration, and affections of children and adults alike.

In this chapter we discuss important characteristics of cats as well as the 10 breeds most often registered by the Cat Fanciers' Association (CFA).

Because cats are commonly easier to care for and are less demanding of time and space than dogs, their popularity and numbers will probably continue to increase in the years ahead.

A discussion of choosing a dog or cat follows in Chapter 4.

## 3.9   REFERENCES

1. The American Kennel Club Staff. 1998. *The Complete Dog Book*. 19th ed. New York: Howell Book House.

2. Coile, D. Caroline, and Michele Earle-Bridge. 1998. *Barron's Encyclopedia of Dog Breeds*. Hauppauge, NY: Barron's Educational Series.

3. Cunliffe, Juliette. 2003. *The Encyclopedia of Dog Breeds*. Bath, UK: Parragon Publishing.

4. Helgren, J. Anne. 1997. *Barron's Encyclopedia of Cat Breeds: A Complete Guide to the Domestic Cats of North America*. Hauppauge, NY: Barron's Educational Series.

5. Pollard, Michael. 1999. *The Encyclopedia of the Cat*. Bath, UK: Parragon Publishing.

# 4 CHOOSING A DOG OR CAT

*Money will buy a good dog, but it won't buy the wag of its tail.*

George Bernard Shaw (1856–1950)
*British (Irish-born) author*

## 4.1    INTRODUCTION/OVERVIEW

Buying a dog or cat is similar to buying most things. The more one knows about them before making the purchase, the more confidence one will have in the decision. Moreover, because puppies and kittens are long-term investments, they deserve an extensive process of thoughtful searching and selecting.

Choosing a cat or dog is, in many instances, an opportunity to choose a family member, so the final choice is an important decision. Indeed, more dogs and cats in the United States sleep on their owner's bed than sleep in dog houses or on cat beds—part of the reward given special companion animals that bond with and enhance the lives of people.

Virtually all kittens and puppies are affectionate and adorable (Figure 4.1). Often one has an entire litter to choose from, and each one is cuter than a Cabbage Patch Kid®. So, how does one decide?

The all-important thing to remember is to not make a spur-of-the-moment decision. Although most kittens and puppies are virtually irresistible, fast forward and be ever-mindful that they will be fully grown in only a few months. With the increasing number of once-wanted, later-unwanted pets, it is especially important that "love at first sight" be tempered with abundant thought and deliberation (Chapter 24). Moreover, sage advice says one should not buy a dog or cat until emotionally, financially, mentally, and physically prepared for that long-term experience and expense.

Once the personal decision and commitment is made to select a companion animal, consideration should be given to choosing one that fits the owner's needs and adapts to the complexities of modern living. Does the dog or cat exhibit good health and temperament as well as promise of ease in training? A beautiful, well-behaved, housebroken, quiet, loyal dog or cat does not simply emerge out of nowhere; your new pet will need training.

In this chapter we discuss numerous factors that should be considered in the wise choice of a cat or dog. Readers are encouraged to also peruse Chapter 12, Care, Management, and Training of Dogs and Cats.

Before proceeding, let's review the reasons dog and cat owners say they acquired pets in the first place. This will be helpful as readers review points to ponder in choosing a dog or cat. According to polls conducted by the American Pet Association (www.apapets.com), America's dog and cat owners acquire pets for the following reasons (dog %/cat %): someone to play with (90%/93%), companionship (83%/84%), help children learn about pets and responsibility (82%/78%), someone to communicate with (57%/62%), security (79%/51%).

**Figure 4.1**    Cute kittens (*a*) and adorable puppies (*b*) soon to be available for a new home. *Courtesy Paul E. Miller (a) and Champion Petfoods (b).*

(a)                                                        (b)

## 4.2 RESPONSIBLE PET OWNERSHIP

Providing adequate food, water, shelter, and healthcare are important; however, responsible pet ownership requires much more than just fulfilling the pet's basic needs. Prospective pet owners should be familiar with local and state laws concerning pets. In most locales dogs and cats are required to be registered (e.g., obtaining a city or county registration tag or license) and to be vaccinated against rabies (Section 4.2.1). Many cities have leash laws requiring dogs to be kept under the owner's immediate control—on a leash or in a secure confined area (kennel run or fenced yard). In some areas, "nuisance" laws also apply to cats; complaints from neighbors that a cat is allowed to roam may result in impoundment of the cat and fines. Other nuisance laws may result in fines to owners of dogs allowed to bark incessantly. Covenants of homeowner associations may limit the number and size of pets allowed in a subdivision or housing complex. Some communities have banned certain breeds of dogs (e.g., pit bulls). Investigate all applicable laws prior to obtaining a new pet.

### 4.2.1 Vaccinate and License Your Dog or Cat

Vaccination for rabies and other diseases is low-cost insurance. A licensed dog or cat is much easier to return to the owner if it becomes lost. A collar tagged with the owner's name and address aids greatly in reuniting lost pets with the owner. Additionally, owners have the option to microchip and/or tattoo their dogs and cats. These provide permanent identification of the pet and are essentially painless, relatively inexpensive, and difficult to remove.

## 4.3 CHOOSING A DOG OR CAT

Prior to selecting a cat or dog, prospective owners should think through carefully whether they are fully prepared for and totally committed to companion animal ownership. Is the living situation favorable? Are there any restrictions regarding pets in the place of residence? Is the prospective owner prepared to make the time commitment necessary to enjoy and properly care for the pet (consider regularity of work hours, travel schedule, time at home)? Is the lifestyle (activities level, hobbies) flexible and compatible with responsible companion animal ownership (Chapters 12 and 24)?

Assuming the answers to those and other related questions are in the affirmative, the prospective owner is prepared to proceed in considering factors to contemplate in choosing a dog (Subsections 4.3.1 through 4.3.11). Many discussion points mentioned in connection with choosing a dog also apply when selecting a cat or other companion animal. Special factors to ponder in choosing a cat are discussed in Section 4.4 and in Subsections 4.4.1 through 4.4.3.

### 4.3.1 Male or Female

Generally, either sex will make a good companion animal. Dogs are faithful friends; cats are mystical, quick, and smart. Males tend to be somewhat larger and heavier than females of the same breed. In some breeds males may also be more aggressive.

Which makes the best pet dog—a male or a female? A common misconception is that females make better pets than males. Why? Perhaps people believe the maternal instinct of females gives them an advantage. But there is no scientific proof that female dogs make better pets than males. Actually the idea that one sex makes a better companion than the other

**Figure 4.2**   This young girl is pleased with her choice of a dog and is fully committed to providing the care and attention her dog seeks and deserves. *Courtesy Dr. James E. Corbin.*

is an old adage. That is somewhat like saying women make better cooks. If that is true, why are most chefs males? What matters most is whether the person—male or female—received the training and knowledge needed to become an outstanding chef. The same is true with dogs. When a puppy—male or female—is brought into a place of residence, it must be cared for, loved, and trained. One can expect to realize the positive results of those efforts regardless of sex if the dog is cuddled and learns to cuddle with its owner. The more one cares and shows affection for the puppy, the more it is going to know the kind of love and affection the owner seeks in return (Figure 4.2).

Another reason people seem to prefer females is that they may be less likely to display "mounting behavior" or "mark" areas of the residence as males often do. On the other hand, some people prefer males because they need not be concerned about estrous cycles, and the suitors that accompany estrus, or the possibility of unwanted litters. Of course, spaying and neutering pets equalizes those concerns. Furthermore, neutering and spaying aid greatly in preventing dogs from developing cancers of reproductive organs (Chapter 10). This is one reason many breeders require that the dogs they sell as pets be spayed or neutered.

The most important factor to remember in choosing a puppy as a pet is not the sex or color, but rather, its personality. After all, that is what the future owner must live with. This is not to imply that color is unimportant, but it should not be the main factor on which a choice is made in selecting a new puppy. Notwithstanding the above discussion, gender is often considered to be important by some prospective pet owners. Consider, for example, the woman who called a breeder inquiring about the availability of a female puppy. Her reason for wanting a female was because her favorite hobby was developing and cultivating specialty roses. It can be appreciated that female dogs squat when they urinate, whereas males may spray roses and other plants. This woman believed a male dog would not be compatible with her lifestyle and hobby. However, males neutered before 5 months of age almost never exhibit urine marking and rarely lift their legs to urinate. Neutered males rarely exhibit mounting behavior, and urine marking largely ceases once testosterone levels recede following neutering.

Another myth is that females are more docile and attentive than males; that females do not participate in fighting over dominance. Truth be told, females do demonstrate social dominance and most of them compete to maintain and/or alter that order. As a result, females may even be more independent, stubborn, and territorial than their male counterparts. Moreover, females are more intent on exercising their dominance by participating in alpha behaviors (Chapter 20) such as mounting. Numerous dog owners report that most of the fights they break up are between females.

Conversely, males are commonly more affectionate, exuberant, attentive, and demanding of attention than females. Indeed, males are very attracted to people. They also tend to be more steadfast, reliable, and less moody. Males are frequently more outgoing, more accepting of other pets, and bond quicker with children. Most males are easily motivated by praise and food and are eager to please. This facilitates training; however, males are more easily distracted during training. Regardless of age, most male dogs are fun-loving throughout their lives. Females tend to be more reserved and dignified with advancing age. Perhaps the human equivalent is the twinkle-eyed grandfather still playing catch at age 70 while the grandmother quietly observes from the porch.

Male dogs are usually larger than females but only by an average of 1 to 2 inches and 3 to 5 lbs in medium-sized breeds. Most breeds have a range of sizes, hence the difference between sizes of males and females is often minimal.

Unless spayed, females will have periods of estrus (heat) approximately every 6 months (Chapter 10). During estrus, a bitch will have a hemorrhagic (bloody) vaginal discharge that may stain carpets, floors, and furniture. Walking females outdoors when they are in estrus can become hazardous if male dogs are in the vicinity. During estrus, females leave a scent for wandering intact males to follow directly to the owner's front porch or yard. For these and many other reasons (Chapters 10 and 24), female dogs not being used for breeding should be spayed. The cost of spaying a female dog is higher than neutering a male dog (the surgery is slightly more difficult and requires more time in females).

In sum, it may not matter so much which one chooses—male or female—as both can be excellent pets. The more one cares for her/his dog, the better companion it will be. This is especially true when the pet dog is spayed or neutered, thereby eliminating much of the concern related to which sex to choose.

## 4.3.2   BREED

Did you ever walk into a bakery, breathe the delightful taste-tantalizing aroma of freshly baked pastries, and stand there struggling over whether you want a rich chocolate cupcake, a mouthwatering *mille-feuille*, an irresistible éclair, or simply a glazed donut? Prospective pet owners may experience a similarly difficult decision when faced with the overwhelming array of dog and cat breeds available (Chapter 3).

Because the dog or cat will likely be a family member for 10 or more years, selecting the "best-fit" breed is important. The commitment to feed, groom, house, attend to health measures, and nurture this nongenetic family member should be based on the most appropriate breed for the family circumstances and purpose of ownership.

Each breed of dog has evolved based on selecting and mating for a certain purpose. For example, one would not likely select a Greyhound for a lapdog to cuddle and love; a Toy Poodle might be more desirable. Conversely, a Toy breed would not likely be one's first choice if the dog will be expected to accompany its owner on a jog each morning (Figure 4.3) or go on extended hikes in wooded, hilly terrain.

The prospective dog owner may love the lithe look of the Greyhound, the sturdiness of the Bernese Mountain Dog, the elegance of the Standard Poodle, the cuddly puffiness of the Bichon Frise, the bounciness of the Shetland Sheepdog, the confident composure of the Collie, or admire other breeds for other reasons. How much time and interest is available for grooming? One can brush a Beagle in only a few minutes or spend hours on an Old English Sheepdog. Prospective owners should talk with friends who own the breed(s) in which they

**Figure 4.3** A happy owner enjoying a health-friendly swim with her devoted dog. *Courtesy Champion Petfoods.*

have the most interest. Somewhere, there is a breed of dog that is just right. Doing the research necessary to find the right one will be worth it!

### 4.3.3   Temperament

This trait is defined differently by different people and for different reasons, but most agree that"good"temperament or disposition is closely associated with a dog being well-suited for the owner's preferences and purposes. A desirable disposition in a dog for a quiet, inactive senior living in an apartment has a different meaning than that of a good temperament in a dog to be used in police or corrections work.

The socially attracted dog exhibits a strong interest in people, enjoys being petted and receiving attention, follows humans easily, and wants to be with people. Poodles have been selected for these traits and make good companion animals. This explains why they have ranked among the top 10 breeds in registrations for more than three decades in the United States.

Terriers are inclined to be scrappy, tough, and independent, although Airedale Terriers bond well with people and are often protective. Hounds, on the other hand, follow eyes and noses and are commonly oblivious to the presence of people.

Dogs with good temperament get along well with children and seniors alike. Moreover they train easily. Socially attracted dogs are more cuddly and friendly. This interest in people is useful in training.

Finally, does the temperament of the breed(s) being considered promise to be tolerant yet protective, dominant or submissive, and what will be its reaction to other animals, especially other dogs and cats? In short, how well do they accept and accommodate people and other pets? Many cats and dogs get along well with each other (Figure 4.4).

**Figure 4.4** This cat and dog pair display a high level of comfort as they enjoy the companionship of each other. *Courtesy Noel and Mary Vieten (pet owners), and Paul E. Miller (photographer).*

## 4.3.4 EASE IN TRAINING

Do not assume that breeds with the reputation of being smart instinctively know how to behave. Regardless of their breed, all will need training. In fact, the so-called smart dogs are not always the most easily trained. Intelligent dogs often have their own agenda. For example, Hounds and Terriers are intelligent but may be difficult to train because of their independent nature. Different breeds of dogs learn different tasks at different rates that usually correlate to the original purpose their breed was developed for (e.g., it is easier to train a Bloodhound than a Miniature Poodle to do tracking work).

Popular breeds that traditionally function and train well with humans include Golden Retrievers, Labrador Retrievers, Border Collies, German Shepherds, and Shetland Sheepdogs (Chapter 20).

## 4.3.5 PUPPY APTITUDE TEST (PAT)

Numerous PATs have been developed and applied in evaluating prospective pets. Traits evaluated and considered include social attraction, following, restraint, social dominance, elevation dominance, retrieving, touch sensitivity, sound sensitivity, sight sensitivity, and physical structure/conformation. Performing a PAT is most effective when administered and interpreted by a professional. Several examples may be seen on the Internet (e.g., www.golden-retriever.com/puppy_aptitude_test.htm). However, these tests do not reliably predict adult behavior. Environmental factors, experiences, and training have a tremendous impact on the temperament of adults (Chapter 20).

## 4.3.6 HEALTH ASPECTS

In purchasing a pet, select one that appears healthy, clear-eyed, active, and energetic. A large number of hereditary defects can adversely affect the health and life expectancy of dogs and cats (see Section 4.3.7 and Tables 4.1 and 4.2).

**Table 4.1**
Databases of Genetic Diseases Maintained by the Orthopedic Foundation for Animals

| Genetic Disease OFA Databases | |
|---|---|
| • Hip dysplasia | • Von Willebrand's disease |
| • Elbow dysplasia | • Progressive retinal atrophy |
| • Patellar luxation | • Copper toxicosis |
| • Autoimmune thyroiditis | • Cystinuria |
| • Congenital heart disease | • Phosphofructokinase deficiency |
| • Legg-Calve-Perthes disease | • Congenital stationary night blindness |
| • Sebaceous adenitis | • Pyruvate kinase definciency |
| • Congenital deafness | • Renal dysplasia |
| • Craniomandibular osteopathy | |

*Source: www.offa.org/history.html.*

**Table 4.2**
DNA Tests, Breeds, and Laboratories

| Disease | Breeds | Testing Laboratories |
|---|---|---|
| Canine Cyclic Neutropenia | Collie | HealthGene Corp |
| Canine Globoid Cell Leukodystrophy | West Highland White Terrier | HealthGene Corp |
| Canine Leukocyte Adhesion Deficiency | Irish Setter, Irish Red and White Setter | HealthGene Corp, OptiGen |
| Canine Myotonia Congenita | Miniature Schnauzer | HealthGene Corp |
| Cone Degeneration/Day Blindness | German Shorthaired Pointers | OptiGen |
| Congenital Stationary Night Blindness | Briard | HealthGene Corp, OptiGen |
| Copper Toxicosis | Bedlington Terrier | HealthGene Corp, VetGen |
| Cystinuria | Newfoundland, Labrador Retriever | VetGen, HealthGene Corp, PennGen, OptiGen |
| Fanconi Syndrome | Basenji, Norwegian Elkhound | PennGen |
| Fucosidosis | English Springer Spaniel | PennGen |
| Glycogenosis Type IV | Norwegian Forest Cat | PennGen |
| Hemophilia B (canine) | Labrador Retriever, Bull Terrier, Lhasa Apso, Airedale Terrier | HealthGene Corp |
| Mannosidosis | Domestic shorthair cats, Persian | PennGen |
| Methylmalonic Aciduria | Border Collie, Giant Schnauzer, Shar-Pei | PennGen |
| Mucopolysaccharidosis | Domestic shorthair cats (Type VII), Miniature Pinschers (Type I), mixed-breed dogs (Type VII), Schipperke (Type IIIB), Siamese cat (Type VI) | PennGen |
| Muscular Dystrophy (canine) | Golden Retriever | HealthGene Corp |
| Narcolepsy (canine) | Labrador Retriever, Dachshund, Doberman Pinscher | HealthGene Corp, OptiGen |
| Phosphofructokinase Deficiency | American Cocker Spaniel, English Springer Spaniel, Chihuahua, Dachshund, mixed-breed dogs, domestic shorthaired cats | PennGen, OptiGen, VetGen, HealthGene Corp |

| Disease | Breeds | Testing Laboratories |
|---|---|---|
| **Progressive Retinal Atrophy** | American Eskimo, Australian Cattle Dog, Chesapeake Bay Retriever, English Cocker Spaniel, Labrador Retriever, Miniature Poodle, Nova Scotia Duck Tolling Retriever, Portuguese Water Dog, Toy Poodle, Irish Setter, Cardigan Welsh Corgi, Bullmastiff, Mastiff, Miniature Schnauzer, Samoyed, Siberian Husky, Sloughi | OptiGen, VetGen, HealthGene Corp |
| **Pyruvate Kinase Deficiency** | Basenji, Beagle, West Highland White Terrier, Eskimo dog, Somali cat, Abyssinian cat | HealthGene Corp, PennGen, OptiGen, VetGen |
| **Renal Dysplasia** | Lhasa Apso, Shih Tzu, Soft Coated Wheaten Terrier | VetGen |
| **Severe Combined Immune-Deficiency** | Basset Hound, Welsh Corgi | PennGen |
| **Von Willebrand's Disease** | Bernese Mountain Dog, Doberman Pinscher, Drentsche Patrijshond, German Pinscher, Kerry Blue Terrier, Kooikerhondje, Manchester Terrier, Papillon, Pembroke Welsh Corgi, Poodle, Scottish Terrier, Shetland Sheepdog | VetGen |

*Sources: HealthGene Corporation, www.healthgene.com; OFA, www.offa.org/dnainfo.html; OptiGen, www.optigen.com; PennGen, www.vet.upenn.edu/research/centers/penngen/services/alldiseases_breed.html; VetGen, www.vetgen.com.*

Large breeds are often more prone to joint problems including hip and elbow dysplasia (Chapter 14), whereas smaller breeds of dogs and cats are prone to "loose kneecaps" (luxating patella[1]). Dalmatians may be more susceptible to kidney problems. Large breeds tend to have shorter lives than small ones.

Seek advice from veterinary professionals. Most veterinarians, veterinary technologists, and veterinary technicians are knowledgeable about the reputation and disposition of popular breeds of dogs and cats. Additionally, veterinary professionals will advise you on vaccinations recommended as well as other companion animal health-related matters.

The purchase contract for a dog or cat should include a health guarantee that gives the buyer 3 or more days to have the dog or cat examined by a veterinarian. Further, if a serious problem is found, the contract should have a provision that entitles the buyer to receive a full refund upon returning the animal.

## 4.3.7 GENETIC SCREENING

Many diseases in dogs are genetic in origin. Over 500 genetic diseases have been identified in purebred dogs and more than 100 in mixed-breed dogs.[2] Unfortunately, the identification and diagnosis of genetic diseases is very complex. Many diseases (e.g., hypothyroidism, sebaceous adenitis, progressive retinal atrophy, some forms of heart disease) do not show up until late in life when a dog is past prime breeding age. Other diseases are greatly influenced by environmental factors (e.g., hip dysplasia). Dogs may appear normal yet carry and transmit genes capable of causing disease (e.g., canine dermatomyositis). This is because some genes are recessive (meaning identical genes must be inherited from both parents for the disease to be apparent), other genes have variable expressivity (meaning signs of the disease

---

[1] The patella is commonly called the kneecap. In many breeds of dogs, the grooves and ligaments that hold the kneecap in its proper location are malformed, resulting in an unstable joint and slippage of the kneecap to the inside of the leg.

[2] Gary F. Mason, "Eliminating Genetic Diseases in Dogs: A Buyer's Perspective," www.workingdogs.com/eliminating_gen.htm.

can vary from unapparent to severe), and in some cases, multiple genes are involved in causing the disease. The elimination of genetic diseases can be accomplished only through selective breeding—but first it is necessary to know the inheritance pattern of a given disease and to have tests available to detect the presence of that disease.

Several registries and databases have been established to help dog breeders identify and reduce the incidence of genetic diseases in purebred dogs. One of the first of these was the Orthopedic Foundation for Animals (OFA). The OFA was established in 1966 to provide radiographic evaluation, database maintenance, and breeding advice to purebred dog breeders with the goal of reducing the incidence of hip dysplasia. The OFA expanded as additional hereditary diseases were recognized in dogs. In 2003, the OFA was managing 17 separate databases of canine genetic diseases (Table 4.1). Some of these diseases continue to rely on the detection of clinical or laboratory signs of disease (phenotypic expressions) whereas DNA tests have been developed for positive identification of a few hereditary disorders. DNA tests are the gold standard because these tests identify the presence of the disease-causing genes in the genetic makeup of the animal. Several laboratories are now providing DNA tests for a variety of canine and feline diseases. These include VetGen (www.vetgen.com), OptiGen (www.optigen.com), PennGen (www.vet.upenn.edu/research/centers/penngen/services/alldiseases_breed.html), and HealthGene (www.healthgene.com) (Table 4.2).

CERF is the Canine Eye Registration Foundation. CERF was founded by purebred dog breeders who sought to reduce the incidence of hereditary eye diseases by working with American Veterinary Medical Association (AVMA) board certified veterinary ophthalmologists and maintaining a national registry. Dogs are examined and, if free of hereditary eye diseases, a CERF certificate is issued to the owner. Information from all examination forms is recorded in the national Veterinary Medical Database and is maintained by CERF and also by the OFA.

More than 150 hereditary disorders have been reported in cats and more are being discovered each year. DNA testing is available for a few of these disorders. For more information, see the Web site of the University of Pennsylvania's section on genetic diseases (see PennGen Web site provided above).

Researching the hereditary diseases affecting a breed of dog or cat being considered for purchase is worth the effort. Requesting genetic testing, when available, before purchase can prevent the heartbreak of later discovering a beloved pet has a life-threatening genetic disease. In 1994, the Association of Veterinarians for Animal Rights published "Canine Consumer Report, a Guide to Hereditary and Congenital Diseases in Dogs." This guide is one source for information on congenital diseases affecting various breeds of dogs (http://members.aol.com/PugsUK/webpage/listpurb.htm).

## 4.3.8 PARENTS AND SIBLINGS

Seeing both parents provides prospective owners the opportunity to observe firsthand what to expect size- and conformation-wise when the puppy or kitten matures. Observing the entire litter and how they interact with each other can also be indicative of future behavior of the one selected. However, each animal is an individual and there is no guarantee that offspring will resemble their parents at maturity.

Marketing studies indicate that few people select automobiles because of paint color, but many people purchasing a cat or dog are influenced by coat color and markings. Therefore, if coat color and/or markings are of special interest, observing the parents and siblings should be helpful. DNA testing is available (VetGen and HealthGene Corporation) to provide definitive proof of the coat color genotype for owners wanting to breed for specific colors.

## 4.3.9 WHERE TO PURCHASE A DOG OR CAT

**Breeders.** Many experienced dog and cat owners prefer to buy from an established reputable breeder. One or both parents can commonly be seen there. Also, the prospective owner can expect animals to be healthy, free of parasites, and vaccinated. Many breeders will agree to sell on a trial basis. Breeders are knowledgeable about popular breeds of dogs and cats and can provide the genetic history of the animal(s) being considered. Reputable breeders have selected and bred for specific purposes and away from genetic problems. Most breeders are very concerned about their reputation and the welfare of the pets they raise. The majority of breeders are conscientious about investing time socializing puppies and kittens (Chapter 20) during the critical period between 21 days and the placement of the pet in its new home (generally between 8 and 12 weeks of age). Thus, puppies and kittens purchased from breeders have a high likelihood of being well-adjusted and sociable as adults.

Because reliable, responsible breeders incur sizeable costs in breeding their dogs and cats—costs associated with good facilities and equipment, veterinary expenses, health certification (including testing of breeding stock for genetic diseases), registration papers, advertising, and promotion—prospective owners can expect to pay more for a puppy or kitten from them than if they choose a dog or cat from the local animal shelter or city pound.

The majority of serious breeders are members of the respective national breed organizations and adhere to a rigorous code of ethics, such as those published by Breeders.net (www.breeders.net/code_of_ethics.html) and the United Kennel Club (http://ukcdogs.com/rg/ethics.shtml).

*Newspaper ads.* Although pets purchased through newspaper ads may cost less (some may even be free) than buying from a reputable breeder, they may have genetic or other problems. Also, these pets may not come with a contract or guarantee, and often, no health or other types of records are available. One can, however, usually see one or both parents and siblings.

*Pet stores.* These are popular places for children to observe pets and provide ready access in buying one. They are also convenient in purchasing pet supplies and publications. The disadvantages of purchasing from a pet store include not seeing a parent and having little, if any, genetic history or socialization background. People purchasing puppies and kittens from pet stores may pay premium prices for genetically inferior animals. The majority of pet stores obtain their puppies from commercial breeding facilities often termed "puppy mills." All too often, dogs in these facilities are viewed solely for their economic value and are overcrowded, bred at every opportunity, provided with minimal veterinary care, and inadequately socialized. Although the puppies may often have AKC registration papers, these papers are no guarantee of animal quality. Many of these same concerns apply to purchasing animals over the Internet. There are many stories of fake and misleading Internet advertisements (including ones where photographs of champion dogs were scanned from advertisements and posted on Web sites as being the sire or dam of animals offered for sale).

*Animal shelter/humane society.* Obtaining a cat or dog from an animal shelter or the local pound is often the least expensive route to companion animal ownership. Those who give a home to an unwanted pet are to be commended. For many pets, it is a life saved—a welcomed second opportunity to be wanted and loved.

Disadvantages of this means of companion animal acquisition include not seeing or knowing about the pet's parents or their heritage and not having information about the health, training, and previous treatment of the animal. As noted in Chapter 20, dogs and cats not socialized at an early age (prior to 12 to 16 weeks) tend to be fearful of humans and new environments throughout their life. Dogs from shelters also have an increased risk of developing separation anxiety and other behavioral problems (Chapter 20).

*Breed rescue organizations.* There are numerous breed rescue organizations devoted to finding new homes for adult purebred dogs and cats. Rescue organizations can provide prospective owners with excellent information regarding the breed being considered for adoption. These organizations generally place abandoned dogs in foster homes while waiting for adoptive homes to be identified. There are numerous Internet sites to search for breed-specific rescue groups, (e.g., www.akc.org/breeds/rescue.cfm, www.infodog.com/ads/rescue/rescue.htm, and www.siameserescue.org).

## 4.3.10 THE SALE/PURCHASE CONTRACT

At a minimum, the contract should completely identify the puppy, preferably including tattoo or microchip information. If the decision is to purchase a purebred, the written agreement should contain all AKC requirements as well as the guarantee and stipulations of the sale. The latter should include provisions for return and money back within 3 or more days if a veterinarian finds serious health problems.

If the puppy is a female being purchased from a breeder, it is not uncommon for the price to be reduced somewhat in exchange for one or more pups from a future litter. In such cases, the contract may specify that the buyer agrees to mate the female to the breeder's choice of stud dog, raise the litter, and then permit the breeder her/his top pick. Any such agreements should be included in the contract to protect both parties.

Finally, the contract should include any overall pet health guarantee—what it covers, its date of expiration, and what the compensation or other consideration includes.

***Spay-neuter contracts.*** Some breeders sell puppies on a spay-neuter contract, which requires that the dog be spayed or neutered by a certain age. Spay-neuter contracts may be used in conjunction with, or independent of, Limited Registration designation. Breeders and owners should be aware that the AKC does not and cannot enter into arbitration when an understanding such as a spay-neuter contract between buyers, sellers, or co-owners goes awry. The AKC will abide by the decision of the court if the case is litigated. The only way breeders can be assured that the offspring of their puppies will not be AKC-registrable is by specifying Limited Registration on the registration application (Section 4.3.11).

The AKC encourages breeders and owners to discuss in full, and reach agreement on, all conditions of the sale of puppies before the purchase is finalized. Ideally, a new owner's relationship with the puppy's breeder should be as long and happy as the relationship with the puppy itself.

## 4.3.11 FULL AND LIMITED AKC REGISTRATION

Most breeders sell pet puppies on a Limited Registration basis. The AKC has always recognized the role of responsible breeders in preserving the integrity of its registry and the quality of purebred dogs. In June 1989, delegates of the AKC voted to give breeders a valuable tool to protect their breeding programs—the option of selling puppies under Full or Limited Registration. Dogs with Full Registration privileges can compete in all AKC events and their offspring can be registered with the AKC. Dogs with Limited Registration privileges are allowed to compete in all AKC Companion and Performance events but not in Conformation events (dog shows). Additionally, no puppies produced by dogs with Limited Registration are eligible for AKC registration. The choice to register a dog with Full or Limited privileges is solely at the discretion of the dog's breeder. The breeder indicates the Full or Limited designation by completing the designated section of the dog's individual registration application. The breeder's designation is entered into the AKC registration system along with the dog's name, sex, color, and number and becomes part of the dog's permanent registration record. If the dog is entered in an AKC event, or if it appears as a sire or dam on a litter application, it is checked for eligibility against its Full or Limited Registration status.

***Benefits for breeders.*** Limited Registration honors the prerogative of knowledgeable, responsible breeders to decide which dogs in a litter may later be bred to produce AKC-registered dogs and which ones may not. By indicating Limited Registration on a dog's application and explaining the conditions to the new owner(s), the breeder can be confident that the owner(s) cannot use the dog to produce and sell AKC-registrable puppies that may not meet the standards of the breeder's program. If the breeder later decides that the dog has developed in such a way that its registration status should be changed, the breeder has the authority to change the status from Limited to Full Registration. Only the breeder has the authority to make that change, and in the case of multiple breeders, all breeders must concur in the decision to make the change. The owner cannot make any changes to the dog's registration status.

By using Limited Registration, breeders can provide AKC papers to all owners of their puppies while still controlling which ones will be bred. This is an excellent tool to demonstrate breeders' concern for the future of their bloodlines and their commitment to preserving excellence within the breed.

***Benefits for owners.*** Owners should know that the Limited Registration status has no bearing on a dog's potential to be an excellent representative of its breed, a wonderful pet, or a standout participant in AKC events. In fact, having a Limited dog can be liberating because owners are not required to wrestle with the question of whether to breed or not; that decision is made by the breeder when the Limited option is indicated on the registration application. Dogs with Limited Registration can compete in every AKC event except Conformation (because the latter is an evaluation of breeding stock). These include the Companion Events—Obedience, Rally, Agility, and Tracking—which are open to all breeds, and the Performance Events—Field Trials, Hunting, Lure Coursing, Herding, Earthdog, and Coonhound—which are open to specific breeds only. Training for, and participating in, these events promotes a special bond between dogs and owners. Together dogs and owners learn new skills, spend quality time together, and may achieve considerable success in their chosen sport. When it comes to being a great competitor or simply a lifelong friend, a dog with Limited Registration has no limits.

### 4.3.12 ADOPTING AN OLDER PET

The unwanted pet population is a serious concern, especially in large cities (Chapter 24). An estimated 20 million pets end up in animal shelters annually. A significant number are purebreds.

Adopting a previously owned companion animal, if health and other records/considerations are acceptable, can be a good decision for several reasons. First, one can feel good about the decision to provide love and attention to an animal that otherwise might have been euthanized (Chapter 24). Given proper care and attention, previously owned dogs and cats are often some of the best companions. They will usually provide loyalty, love, and companionship in their own individual, distinctive way. However, some animals adopted from animal shelters suffer from separation anxiety and socialization problems (Chapter 20). These animals may require behavioral modification therapy and a great deal of patience on the part of the new owner(s) before adjusting to a new home.

## 4.4 OTHER FACTORS IN CHOOSING A CAT

Cats are the most populous pet in the United States. Approximately one-third of all households own one or more of the 77 million cats in the nation.

People choose cats as companion animals for numerous reasons. They have a natural beauty and grace that is both pleasing and exciting, and although they may appear aloof, cats usually make affectionate, lifelong companions. Moreover they do not need to be taken for walks. Most of them are clean, inexpensive to feed, and normally quiet (Figure 4.5).

Although cats can be challenging at times, occasionally being guided by their own set of rules, they are likeable, devoted companions. Their personalities vary as much as those of their owners. Yet, choosing a cat having a personality that is compatible with the lifestyle and personality of its owner will help ensure a happy relationship. A careful study of the breed characteristics prospective owners prefer will aid greatly in the selection process. Several questions should be asked in deciding which cat will be chosen to join the residence of the individual owner or family. Selected points to ponder are discussed in the following subsections.

**Figure 4.5** A charming, well-adjusted, and well-behaved cat. *Courtesy Champion Petfoods.*

## 4.4.1 TYPE OF CAT

Does the prospective owner prefer a cuddly, lively kitten or would she/he rather skip that rambunctious stage and purchase an adult cat? Is a particular breed of cat preferred? That answer will affect the choice of a purebred or simply a domestic/household cat. Is a shorthaired or longhaired cat preferred? Although most longhaired cats require more grooming than shorthaired ones, some longhaired breeds require minimal maintenance.

## 4.4.2 HEALTH ASPECTS

Selecting a healthy, playful cat—one free of fleas and sores caused by biting or scratching at these parasites—is important. If the decision is made to select a kitten, watch it play to be certain it walks and runs with vigor and without interference. Also, kittens should not be separated from their mother before they are 7 or 8 weeks of age. As soon as possible after joining its new owner, the kitten (cat) should be taken to a veterinarian for an examination and to set up a vaccination schedule (Chapter 14).

Once a decision is made regarding which member of the feline family will move to one's household, the new environment should be made as pleasant as possible and thereby minimize "new home" anxiety. Place a clean litter box, shallow bowl of water, accessible cat food, and toys in the room with the new companion animal (Chapter 12).

## 4.4.3 SHOPPING FOR A CAT

A reliable source of information related to locating a satisfactory kitten or adult cat is available from responsible breeders at cat shows. Dates and places of shows in the area can usually be found in publications such as *Cat Fancy* or *Cats*. These shows provide a good opportunity to learn about breeds of cats as well as obtain pertinent information and advice from breeders/exhibitors.

Prospective buyers of cats can usually obtain names of ethical, responsible breeders from the Web sites of professional cat organizations (e.g., www.simplypets.com/pet-directory/Pets/Cats/Breeds/Breeder_Directories/). Some prospective owners prefer to shop at local pet stores, animal shelters, and humane societies as well as with private individuals identified via local newspapers, public bulletin boards, neighbors, and friends. Whatever the source, the prospective pet should be examined by a veterinarian.

In purchasing a kitten, numerous people have observed advantages in buying two. Kittens are charming and playful, providing company for each other, and entertainment for the proud owner. This association benefits the cats by making them happier and healthier.

## 4.5 SUMMARY

To better appreciate the joys, personal pleasures, and benefits that accrue from ownership of dogs and cats, and because the newly acquired companion animal may share one's home for its lifetime, one should do the research needed and be deliberate in choosing a dog or cat.

In this chapter, we discussed many factors to consider in selecting a companion animal. These included plans for a dog or cat. Does the owner want a watch animal, a jogging partner, a show animal, or simply a companion? We noted the importance of temperament/disposition in the selection process as well as health, expected ease in training, and alternative options for acquisition. Decisions related to breed characteristics, adult animal size, coat type, conformation, need for exercise, behavior, attitude, activity level, and numerous other considerations were also discussed.

The authors commend those who befriend their favorite pets that in turn share continuous, long-lasting affection and companionship with their owners—a mutual, life-enriching experience for children, adults, and companion animals!

## 4.6 REFERENCES

1. Dunbar, Ian. 2001. *Before You Get Your Puppy*. Berkeley, CA: James & Kenneth.

2. Hart, Benjamin L., and Lynette A. Hart. 1988. *The Perfect Puppy: How to Choose Your Dog by Its Behavior*. New York: W. H. Freeman.

3. Robinson, Roy. 1990. *Genetics for Dog Breeders.* 2nd ed. Oxford, UK: Butterworth Heinemann.

4. Scott, John P., and John L. Fuller. 1998. *Genetics and the Social Behavior of the Dog: The Classic Study*. Chicago: University of Chicago Press.

# 5 COMPANION BIRDS

Jill A. Richardson, Rosemary M. Tini,
and Cynthia R. Garnham[1]

[1]Please turn to the end of the chapter for complete author affiliation information.

*We marvel at the sheer physical appearance of animals—their eyes, coats, fins, feathers, and their movement. We feel enriched by the obvious pleasure our pets take in our companionship, and we are touched by their sensitivity to our moods. . . . In caring for our fellow creatures, we express some of our finest qualities, and it is normal to develop bonds of love with animals that are as powerful as those we experience with people.*

Leo K. Bustad, DVM, PhD (1920–1998)
*Former Dean, College of Veterinary Medicine*
*Washington State University*

Knowing that pet birds rank third among companion animals kept by children, youth, and adults in North America, we wanted to identify a prominent avian authority to lead the authorship of this important chapter. Fortunately we found that person in Dr. Jill Richardson. Her passion for and knowledge of companion birds coupled with the experience and counsel of her professional colleagues, Rosemary Tini and Cynthia Garnham, have added immeasurably to the book. We gratefully acknowledge the contributions of these professionals in authoring this chapter.

*Campbell-Corbin-Campbell*

# 5.1  INTRODUCTION/OVERVIEW

Birds are increasingly popular companion animals (Figure 5.1). According to the National Pet Owner Survey there are an estimated 17.5 million pet birds in the United States. Color photographs of major pet birds are shown in Color Plate 7 as well as inside the front cover. Although pet birds make wonderful pets, owning a bird requires a great deal of responsibility.

**Figure 5.1** Birds are popular pets for children. This boy is admiring his parrotlet. *Courtesy Questhavenpets.com.*

Pet birds have very specific nutritional, housing, and social requirements. Additionally, a lifetime commitment to a bird may be longer than prospective owners expect; many species live 50 to 100 years.

## 5.2 COMMON TYPES OF COMPANION BIRDS

The order Psittaciformes—the most common type of companion birds—includes more than 300 species. Psittaciformes includes parrots, macaws, and cockatoos. Finches and canaries are classified as passerines, which belong to the largest order of birds (Passeriforms). There are about 5,400 species of passerines, which include more than half of all living birds. Passerines consist chiefly of songbirds of perching habits.

### 5.2.1 AFRICAN GREY PARROTS

African Greys are medium-sized birds with an average length of 13 inches. They are grey in color with white areas around the eyes and have a short red tail (see Figure 5.2 and Color Plate 7D). They are very popular as pets and are considered to have a high level of intelligence and good talking capability. The two most common types are the Timneh Grey and Congo Grey. The average life span for an African Grey is 50 to 65 years.

### 5.2.2 AMAZON PARROTS

Amazons are medium-sized birds with thick bodies and short wings (see Figure 5.3 and Color Plate 7A). The average length of an Amazon is between 10 and 18 inches. Amazons are native to the rainforests of Central and South America. The most commonly recognized species include the Yellow-naped, White-fronted, Tucuman, Lilac-crowned, Red-lored, Blue-

**Figure 5.2** Congo African Grey parrot. *Courtesy Sailfin Pet Shop, Inc., Champaign, IL.*

fronted, Double-yellow headed, Panama yellow-front, Yellow-crowned, Orange-winged, and Mealy Amazon parrots. The average life span of an Amazon is 20 to 30 years but some live more than 50 years.

### 5.2.3 BUDGIES

Also known as budgerigars and parakeets, budgies are native to Australia. Budgies are relatively small, with an average length of 7 inches (Figure 5.4). These birds are very popular as pets, often being recommended for first-time bird owners. The budgie usually has green plumage on the chest, green and black wings, and a long tapered tail. Heads of budgies are yellow with black lines. However, there are several variations of color including yellow, blue, violet, and albino.

**Figure 5.3** Amazon parrot. *Courtesy Sailfin Pet Shop, Inc., Champaign, IL.*

**Figure 5.4** Budgies are small and relatively hardy birds. They are excellent pets for first-time bird owners. *Courtesy Sailfin Pet Shop, Inc., Champaign, IL.*

## 5.2.4 CANARIES

Canaries are a type of finch that originated from the Canary, Madeira, and Azores Islands in the Atlantic. Canaries are available in many colors and patterns (see Figure 5.5 and Color Plate 7B). They are small birds weighing about 0.5 to 0.9 oz (15 to 25 g). Their life span is 6 to 16 years.

**Figure 5.5** Yellow canary. *Courtesy Sailfin Pet Shop, Inc., Champaign, IL.*

**Figure 5.6**  White pearl faced cockatiel.
*Courtesy Paul E. Miller, Sullivan, MO.*

## 5.2.5  COCKATIELS

Cockatiels are closely related to cockatoos and are native to Australia. Cockatiels are small birds about 12 inches long (see Figure 5.6 and Color Plate 7E). The most commonly recognized cockatiels have gray bodies with white patches on the wings and yellow/white faces with a bright orange patch on the cheeks. Variations include pied (white patches over body), pearl (grey feathers with white mottling), lutino (white body with yellow/white/orange face), and albino (an all white bird). The life span of a cockatiel varies from 18 to 30 years.

## 5.2.6  COCKATOOS

Cockatoos are also native to Australia and are often thought to be the most lovable of all companion birds. There are several varieties of cockatoos; the most common pets are the Moluccan, Umbrella, Bare-Eyed, Citron, Sulfur Crested, and Lesser Sulfur Crested. Sizes range from the medium Goffin (approximately 12 inches) to the larger Moluccan (approximately 26 inches). Cockatoos are primarily white birds with peach, salmon, or yellow shading. They are most often recognized by their impressive crests (see Figure 5.7 and Color Plate 7C). The life span of a cockatoo can reach 75 or more years.

## 5.2.7  CONURES

Conures range in size from small (approximately 9 inches in length) to medium (approximately 12 inches). They originate from Central and South America and have a variety of colors. Conures are the largest family of parrots (see Color Plate 7K).

## 5.2.8  ECLECTUS PARROTS

Eclectus parrots are from the South Pacific Islands. The male Eclectus is green with a yellowish beak; females are a red color with a black beak. Their average length is about 14 inches. The average life span of an Eclectus parrot is 30 years.

**Figure 5.7** Citron cockatoo named Merlin, owned by Lori Costello. *Courtesy Sailfin Pet Shop, Inc., Champaign, IL.*

## 5.2.9    LOVEBIRDS

Lovebirds include nine species of small parrots of the genus *Agapornis* (see Color Plate 7I). Eight of these species are native to the African continent and the ninth from Madagascar (island off southeast Africa). The three most common species are Peachfaced Lovebird, Masked Lovebird, and Fischer's Lovebird. They are small birds ranging in size from 4 to 7 inches. Their life span is usually 5 to 15 years.

## 5.2.10    MACAWS

Macaws are large birds with an average length of 33 inches. They are native to South America. Common types include the blue and gold (see Figure 5.8 and Color Plate 7H), green wing, scarlet, military, and hyacinth. Mini macaws are about half the size of other macaws. The life span of a macaw can reach 75 to 100 years.

## 5.2.11    QUAKER (MONK) PARAKEETS

Quakers are natives of South America. They are a small parrot, approximately 10 to 11 inches long. They usually have a green body with grey feathers on the neck and forehead (see Color Plate 7J). Variations include blue and yellow (rare). The average life span of a Quaker is 25 to 30 years.

## 5.2.12    PARROTLETS (DWARF PARROTS)

Parrotlets are native to South America. In the wild, they roam in forested areas, nesting in hollow trees. They are small parrots, averaging 4.3 to 5.5 inches long (see Figures 5.1 and 5.9 and Color Plate 7G). They were introduced into the United States as pets in the 1980s. These are smart birds that can easily be taught many tricks and can also learn to talk. Parrotlets

**Figure 5.8** Young blue and gold macaw. This 5½-week-old bird is being hand reared. *Courtesy Sailfin Pet Shop, Inc., Champaign, IL.*

have been nicknamed "pocket parrots" because they fit into people's shirt pockets and seem to enjoy riding around that way. They are curious and playful and should have multiple toys in their cages to keep them entertained. Their life span in captivity is between 20 and 30 years.

## 5.3   BIRD ANATOMY AND PHYSIOLOGY

### 5.3.1   FEATHERS

Feathers are made of keratin and are unique to birds. Feathers provide a waterproof covering for birds that facilitate flight. They are arranged in feather tracts or pterylae; unfeathered areas are called apteria. The part of the hollow central stalk of the feather below skin level is

**Figure 5.9** Parrotlets, also known as "pocket parrots," are active, easily trained birds. *Courtesy Questhavenpets.com.*

called the calamus or quill; the part above the skin is called the rachis. Barbs and barbules grow out from the rachis and interlock to enable waterproofing. Developing feathers, also called blood feathers, contain a nutrient artery that regresses as feathers mature. As feathers become worn, they are replaced regularly through molting (once or twice a year) depending on the species.

Contour feathers are those that cover the body and give the bird its shape and color. They include flight feathers (remiges) and tail feathers (retrices). Down feathers are smaller, lack the barbules, and provide most of the insulation. Powder feathers are scattered throughout the plumage and grow continuously. The barbs of these feathers break down into a powder and are thought to help keep the plumage clean. Uropygial glands are found above the base of the tail in most psittacine. These glands secrete a sebaceous (oily) fluid that spreads over feathers during preening.

Molting occurs when the old worn feathers are replaced with new ones. Most species of birds molt for a period of 2 to 3 weeks a year; an exception is the Amazon parrot, which seems to molt year-round. During molting, new feathers replace the old ones starting at the head and neck and finish with the tail and wing feathers. Throughout molting, the metabolic rate of birds increases about thirtyfold.

## 5.3.2 BEAKS

Beaks consist of bones of the upper maxilla and lower mandible and their keratinized covering called the rhamphotheca. The beak continues to grow throughout the life of birds. This is necessary to replace the wearing that occurs at the tips. The upper jaw is the rhinotheca; the lower jaw is the gnathotheca. Birds do not have teeth. The hard and sharp edges of the beak help crush items before swallowing. The nostrils of most birds are found at the base of the top mandible.

## 5.3.3 SKELETAL SYSTEM

Birds have very lightweight bones, which help enable flight. The bones in bird legs and wings are hollow, thus providing space for tiny air sacs; these bones are referred to as pneumatic and assist with respiration.

## 5.3.4 RESPIRATORY SYSTEM

Birds have a unique respiratory system. The lungs are paired; they are not lobulated like those in mammals. Birds do not have a diaphragm. The lungs are firmly attached and have very little ability to expand. Most birds have eight air sacs, which facilitate air movement through the respiratory tract. Many of the air sacs extend into pneumatic bones.

## 5.3.5 DIGESTIVE SYSTEM

Birds have a two-part stomach. The first part is called the proventriculus (glandular stomach). The proventriculus secretes chemicals that aid in digestion. The second part is called the gizzard, which is the muscular stomach that helps grind food. Some birds ingest grit that remains in the gizzard and assists in grinding food. The terminal end of the digestive tract, reproductive tract, and urinary tract is the cloaca.

### 5.3.6   CIRCULATORY SYSTEM

Birds are warm-blooded animals with a normal body temperature of 106°F (41°C). Birds, like humans, have a four-chambered heart. The difference is that the aorta arches to the right in birds, to the left in humans. Also, the heart beats much faster in birds than in humans.

### 5.3.7   REPRODUCTIVE SYSTEM

Male birds have two internal testes located above the kidneys. The testes produce both spermatozoa and testosterone. Psittacine birds do not have a phallus. The process of mating requires cloacal contact between male and female birds for sperm transfer.

In females of most bird species, only the left ovary and oviduct are functional. Spermatozoa fertilize the egg in the funnel portion of the left oviduct. Females lay eggs even if there has been no mating to fertilize them. The egg travels through the oviduct and remains in the uterus (also called the shell gland) for 20 to 26 hours. The egg passes out the final part of the female reproductive tract through the cloaca.

### 5.3.8   NUTRITION

Nutritional deficiencies are a leading cause of disease among birds. Therefore proper nutrition is a key factor for ensuring overall good health. Although specific nutritional requirements for companion birds are not well defined, birds require dietary protein, fat, carbohydrate, vitamins, and minerals. Because birds have a high rate of metabolism, they require more food per unit of body weight than larger animals. (In general, there is an inverse relationship of relative metabolic rate and body size.)

In nature, birds originating from tropical areas eat a wide range of fruits and seeds, whereas those native to arid regions, such as cockatiels and budgerigars, eat mostly seeds. A well-balanced diet for most companion birds includes fresh vegetables and fruit, grains, nuts, pellets, and fortified seeds (see Color Plate 7L).

A diet based solely on unfortified seeds lacks the proper nutrition for optimum health in birds and may cause problems such as mineral deficiency and obesity. Instead, diets containing fortified seeds are recommended because they contain additional calcium and other minerals, vitamins, and digestive enzymes.

It has become popular to hand raise neonatal birds to increase their bonding with people, thereby making them better pets. Newly hatched birds should be fed four to eight times a day. By 3 to 4 weeks of age, this can be reduced to three feedings per day (Figure 5.10). A number of commercial hand-rearing mixes are available. The diets should contain approximately 18% to 20% protein and 5% fat. Fresh gruel must be prepared for each feeding and warmed to approximately 104.9°F before feeding. Chicks are fed until their crops are full (do not overfill). Newly hatched chicks will consume 25% to 28% of their body weight for the first 4 days, decreasing to 15% by 12 days of age. The diet should be 93% water for the first 3 days (consistency of a thick soup), then gradually decreasing to 75% water (consistency of yogurt). Weaning should be done gradually after the bird has started to eat on its own (generally around 12 weeks of age).

## 5.4   COMPANION BIRD HUSBANDRY

### 5.4.1   CAGES

Most bird owners prefer a cage for their feathered friend. They are readily available at most pet shops in a variety of colors, styles, and sizes. Selecting the correct size and type of cage

**Figure 5.10** This 5½-week-old blue and gold macaw receives three feedings daily. *Courtesy Sailfin Pet Shop, Inc., Champaign, IL.*

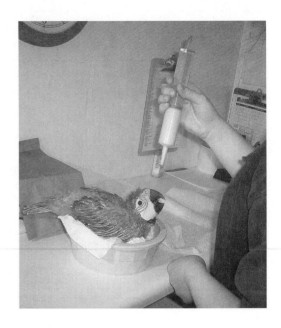

is very important. A general rule is that adult birds should be able to fully stretch their wings and exercise them vigorously while on their perch without harming themselves on fixtures or the cage itself. The head of the bird should not touch the top of the cage nor should the tail touch the bottom or sides. An example of the minimum requirements for cockatiels is 39.5 in H × 23.5 in W × 23.5 in D (100 cm × 60 cm × 60 cm) for a pair, or 19.5 in H × 17.5 in W × 17.5 in D (50 cm × 45 cm × 45 cm) for a single bird.

Canaries and other nonclimbing birds do fine with vertical bars. Cockatiels, lovebirds, parakeets, and other climbing birds utilize horizontal bars for climbing (Figure 5.11). The spacing of bars must be narrow enough to ensure the bird cannot put its head through the bars (to prevent accidental hanging) but wide enough apart to allow the bird to utilize the side of the cage for climbing. This also permits the owner to see the bird clearly and facilitates easy cleaning. The bars should also be an appropriate thickness. Canaries, parakeets, and finches do well with a thinner bar, whereas stronger birds, such as cockatiels and parrots, require a thicker bar to prevent the bird from destroying its new home.

The construction of the cage bottom is equally important. Plastic cage bottoms are hygienic and recommended because they do not rust, which not only is unsightly but also can

**Figure 5.11** Amazon parrot using the horizontal bars of its cage for climbing. *Courtesy Sailfin Pet Shop, Inc., Champaign, IL.*

provide a place for disease organisms to multiply. The bottom should also have high sides to help prevent excess feed, husks, and droppings from falling to the floor. Most cages marketed today have a slide-out bottom or clamps on the side of the cage for ease in cleaning.

Perches should have a comfortable grip for the bird. A perch that is too thin will result in discomfort for the bird and require frequent nail clippings. Different widths of perches should be placed in the cage to exercise the feet and relax the muscles. The bird's toes should not wrap completely around the perch. Perches should be placed in front of the seed and water hoppers. Because most birds enjoy being as high as they can, a perch placed high in the cage is a welcomed addition. Again, ensure that the head of the bird does not touch the top of the cage. Avoid putting perches directly under one another as droppings will accumulate on the lower perch.

*Water.* Most cages purchased from pet shops have food and water hoppers. Additional hoppers can be added for special treats. Fresh water must be provided at all times. Birds can dehydrate very easily, especially under stress.

Water hoppers are great for smaller birds such as canaries and finches; water bottles are preferred for larger birds such as lovebirds, cockatiels, parakeets, and parrots. Although more difficult, many people have also trained smaller birds such as canaries and finches to use a water bottle. A glass water bottle similar to the type used for small animals is preferred for larger birds because they are more hygienic and dishwasher safe.

The water bottle must be filled daily and checked to ensure that it is working properly. Some birds learn they can take a shower with the water bottle; others seem to enjoy sticking seeds in the water dispenser thereby blocking the flow of water. When this happens, moving the water bottle to a new location or adding a second water bottle on the other side of the cage may be necessary to correct the behavior.

Special attention must be given to water hoppers and bottles. They should be cleaned daily even if they are not visibly dirty. Bacteria grow readily in the water, resulting in a sick bird.

*Toys.* One of the most important parts of any cage arrangement is entertainment for the bird. After all, the bird spends most of its time there. Toys keep the bird busy and mentally stimulated. Birds left alone with nothing to do become bored. This can result in unwanted behavior such as screaming or feather plucking.

Many toys are available commercially for birds including ladders, bells, swings, mirrors, hard plastic toys, rawhide ropes, and chains (Figure 5.12). Always examine the toy thoroughly before buying it. If the toy is made from chains make sure the links are large enough that toes and beaks cannot become stuck but not so large the bird can place its head through.

**Figure 5.12** A variety of toys provide entertainment for this Congo African Grey parrot. *Courtesy Sailfin Pet Shop, Inc., Champaign, IL.*

Rope toys have pros and cons. Rope toys are available in many shapes and colors that are attractive to birds. However, the owner should keep a close eye on rope toys; frayed ones should be removed from the cage.

*Placement.* The cage should be placed as high as possible while still allowing the owner to view the bird. By nature birds enjoy being as high as possible. Height gives them a sense of security and allows them to see what is occurring in their environment. Never place a cage in front of a window with direct sunlight. The bird will be unable to escape from the heat, which may result in heatstroke. Drafty areas such as open doors and windows should be avoided as well. If the owner wishes to air out a room, the bird should be placed in another room until the window or door is closed again.

*Cleaning.* Keeping the cage clean is one of the most important parts of bird husbandry. Cleaning the cage bottom every 2 to 3 days decreases the level of bacterial growth. Placing newspaper under the cage before the bottom is removed makes cleaning easier. The cage bottom should be scrubbed with soapy water and diluted chorine bleach each time it is emptied.

Maintaining a clean cage is healthy not only for the bird, but for the owner as well. Particles of dried bird dropping can easily be dispersed from the bird flapping its wings. Once airborne these potential disease-carrying particles can easily be inhaled by humans or other pets. Every 2 weeks, the entire cage should be disinfected using soapy water and diluted chorine bleach (diluted as 5% bleach and 95% water). Following cleaning, it is recommended to allow the cage to dry and air out thoroughly. Some owners allow free flight during cage cleaning, whereas others use a bird carrier to hold the bird temporarily. Always keep cleaning products out of the reach of pets and avoid cleaning in the same room because birds can be sensitive to chemical odors.

## 5.4.2 AVIARIES

An aviary is an option for bird lovers who live in an area with mild winters and have the time, money, and space an aviary requires. Birds kept in an aviary have the space needed to fly freely and enjoy the world around them. Aviaries make a wonderful addition to backyards or gardens. Many aviculturists have different species of birds in one aviary with success. Care must be exercised when choosing the species. For example, parakeets tend to peck at the toes of other species in acts of aggression. For these reasons, first-time aviculturists should begin with a single species.

When setting up an aviary, keep in mind the number of birds you want to keep and the space available. An aviary incorporated into the natural surrounding is more pleasing and eye catching. Aviaries can be made in many shapes; however, straight lines are easiest to work with. In colder areas, the ideal aviary is divided into three parts: a completely enclosed night shelter, an open flight, and a partially covered flight. An enclosed night shelter provides protection from inclement weather, as well as a safe place for bird privacy and peaceful rest. If temperatures drop below freezing, the birds must be moved indoors. This requires the owner to have indoor cages that can accommodate the number of birds kept in the aviary. In warmer areas, the aviary can be completely open as long as there is a shaded area to which the birds can retreat.

The use of ornamental plants enhances the appearance of the aviary. Whether to perch or chew on, all birds enjoy plants. Caution must be used when selecting plants. Table 5.1 lists safe and unsafe plants.

## 5.4.3 GROOMING

The claws of a bird can be maintained by using different-sized perches and providing toys that require the bird to grab hold. However, from time to time, the claws may still need trim-

**Table 5.1**
Safe and Unsafe Plants for Birds

| Safe Plants | Unsafe Plants |
| --- | --- |
| *Acacia* spp | Oleander |
| Bougainvillea | Azalea |
| *Ficus* spp | Rhododendron |
| Guava (*Psidium guava*) | Yew, *Taxus* species |
| Asparagus | Foxglove |
| Mulberry | Castor Bean |
| Strawberry tree | Kalanchoe |

*Source: Harrison, G. J., and L. R. Harrison. 1986. Clinical Avian Medicine and Surgery. Philadelphia: W. B. Saunders.*

ming. A good rule of thumb is that claws are too long when they begin pinching the owner while perching on a finger or arm. An owner can also tell visually if the claw is too long by observing the bird on a perch. If the toes are being held up by the claw or the claw has begun to curl under, it is time to trim.

A bird's beak can be maintained by providing cuttlebone, mineral blocks, or lava rock. Chewing on these items helps keep the bird's beak in good condition and helps wear down the new growth. Birds with overgrown beaks should be taken to an avian veterinarian or professional bird groomer for trimming.

Whether to clip the flight feathers crosses the mind of every bird owner. Clipping flight feathers is painless and helps keep the bird from flying away during training, although birds with clipped wings can still take flight for short distances. A veterinarian, bird breeder, or trained pet shop employee can easily show owners how to clip the feathers. When birds molt, new flight feathers replace old ones.

Most birds love water and enjoy taking baths. For smaller birds such as canaries, finches, parakeets, and lovebirds, a birdbath can be placed inside the cage for bath time fun. For larger birds, a heavy ceramic dog bowl can be used as a birdbath. Alternatively, birds can be misted with lukewarm water. During bath time always make certain there are no drafts and the birds are kept warm while drying.

# 5.5   COMMON DISEASES OF BIRDS

Similar to mammals, companion birds can have bacterial infections, cancer, infestations with parasites, and hormonal disorders. Unfortunately, birds may not show signs of disease until they are quite sick. Knowing which diseases are common in a particular species of birds often helps identify signs of illness.

## 5.5.1   ASCARIDS

Ascarids are a type of parasitic worm commonly referred to as roundworm (Chapter 15). Ascarids can affect any body system including the digestive and respiratory tract, heart and blood vessels, brain, eyes, and connective tissue. They are a common parasite among companion birds, especially cockatiels, budgies, and imported macaws. They are more common among birds kept outdoors with access to the ground. Birds with ascarid infestations may develop diarrhea, anorexia, and weight loss. Some birds vomit or have a decreased amount of feces. In a severe infestation, worms can cause a partial or complete obstruction of the intestine, which can result in death.

## 5.5.2   ASPERGILLOSIS

Aspergillosis is a respiratory disease of birds caused by the fungus *Aspergillus*. This type of fungus is found virtually everywhere in the environment and grows readily in warm and moist environments. Poor ventilation, poor sanitation, dusty conditions, and close confinement can increase the probability that the spores will be inhaled. Aspergillosis appears to be more common in parrots and mynahs than in other pet birds.

Birds with acute aspergillosis have difficulty breathing, decreased or loss of appetite, frequent drinking and urination, cyanosis (blue color gums caused by inadequate oxygenation), and even sudden death. The fungus usually affects the trachea, syrinx, and air sacs. Diagnosis is made through endoscopic examination of the airways (Chapter 16), fungal cultures, or postmortem tissue examination.

Chronic aspergillosis is more common and much more deadly due to its nature. The bird may not become symptomatic until the disease has progressed too far for a cure. White nodules appear and erode through the tissue and large numbers of spores enter the bloodstream. The spores may travel throughout the body infecting multiple organs including kidneys, skin, muscle, gastrointestinal tract, liver, eyes, and brain.

## 5.5.3   BEAK DISORDERS

Many diseases and conditions can affect beak health. In budgies and finches an abnormal growth can be caused by liver disease, a vitamin D3 deficiency, soft food, malnutrition, or a previous trauma. In cockatoos, a shiny surface to the beak (as opposed to a powdery appearance) can be a sign of Psittacine beak and feather disease (PBFD). In cockatiels, a soft, pliable beak (rubber beak) can be caused by a calcium or vitamin D deficiency. Avian pox, mites, or a bacterial infection can cause crusty lesions on the beak.

## 5.5.4   CHLAMYDIOSIS

Chlamydiosis, also known as psittacosis or parrot fever, is a serious disease of birds not only because it can be difficult to diagnose and treat, but also because it is zoonotic (transmissible to humans). The disease is caused by the bacteria *Chlamydia psittaci*. Pet birds, especially budgies, cockatiels, and lovebirds, are frequently infected. Some birds may not exhibit obvious clinical signs of illness but can still transmit the disease to humans and other birds. Stress can cause an infected asymptomatic bird to become sick. Young birds are most susceptible to *Chlamydia* and may develop inappetence, conjunctivitis (inflammation of eyelids), breathing difficulty, sinusitis, tremors, emaciation, dehydration, and death.

## 5.5.5   EGG BINDING

Egg binding is a term for the obstruction of the uterus or oviduct by an egg. It can occur even if a mate is not present. Egg binding can be life threatening, especially in smaller species. Yolk peritonitis may result if an egg breaks inside the female. One of the most common causes of egg binding is a diet deficient in calcium. Other causes include obesity, lack of exercise, excessive egg laying, and infection.

## 5.5.6   GOUT

Gout is a common disease among humans, birds, and reptiles. It is more common in budgies, waterfowl, and poultry. Gout occurs when the blood level of uric acid, a product of protein breakdown, exceeds the ability of the kidneys to remove it. Enlarged, stiff, and painful joints with the bird shifting from one foot to the other and having a shuffling gait may be symptomatic of gout. The bird may be unable to perch and as a result stays on the bottom of the cage. If the wings are affected, the bird may be unable to fly. There may also be a decrease in appetite, lethargy, weight loss, and abnormal droppings.

### 5.5.7   MITES

The most common mite affecting pet birds is *Cnemidocoptes (Knemidocoptes) pilae.* This sarcoptiform mite (Chapter 15) causes scaly leg and scaly face in parrots and budgerigars (parakeets). It burrows into the skin causing a honeycomb-like proliferation. In severe cases, the beak becomes deformed and interferes with eating. *Cnemidocoptes* spp also affects the feet of canaries producing a syndrome called tasselfoot. A related species, *Knemidocoptes laevis,* affects the feathers of birds. The feather mites burrow into the feather quill below the surface of the skin resulting in scratching and feather picking. Mites can be found on scrapings of the lesions (Figure 5.13) and treated with ivermectin. *Ornithonyssus* spp (northern fowl mite and tropical fowl mite) and *Dermanyssus gallinae* (red mite) primarily affect poultry and wild birds.

### 5.5.8   NEWCASTLE DISEASE

Often referred to as avian distemper, Newcastle disease is one of the most serious of all avian diseases. First identified in 1926 in England, the disease was found in the United States in 1944. Now the disease has spread to include worldwide avian populations, affecting birds of all ages. It is more common among birds imported from Southeast Asia and Central America. The virus is highly contagious and can spread in droppings and through nasal discharge by direct contact, on airborne particles, or on contaminated items including bottoms of shoes, food, or infected dishes and cages. The virus can also penetrate eggshells and infect embryos. It can survive outside a host for several weeks in a warm and humid environment and indefinitely in frozen material. Newcastle disease virus affects the respiratory, nervous, and digestive systems. Infected birds may sneeze, cough, have a nasal discharge, gasp for breath, develop a greenish watery diarrhea, have muscle tremors, drooping wings, twisting of the neck, swelling of the tissue around the eyes and neck, or die suddenly. There are no effective treatments for this disease.

### 5.5.9   PACHECO'S PARROT DISEASE

Also known as parrot herpes virus, Pacheco's parrot disease causes acute viral hepatitis. It is often seen in shipments of imported birds. The close proximity during shipment and stress

**Figure 5.13**  *Cnemidocoptes pilae,* the sarcoptiform mite that causes scaly leg and/or scaly face in budgies and parrots. *Courtesy University of Illinois.*

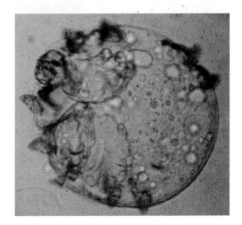

associated with shipping contribute to transmission of the virus. The incubation period for Pacheco's disease is 3 to 14 days. The most common symptom is sudden death with diagnosis confirmed at necropsy. Other symptoms may include diarrhea with a rapid progression to death within 48 hours. Vomiting, yellow-green stools, and acute central nervous symptom signs such as tremors, ataxia (incoordination), or seizures may also be seen.

## 5.5.10  PSITTACINE PROVENTRICULAR DILATATION SYNDROME

Psittacine proventricular dilatation syndrome (PPDS) is a disease of the digestive tract of birds affecting the nerves that supply the muscles of the gizzard, or muscular stomach. As a result, the food is not ground up before passing through the digestive tract. Birds affected with this disease are unable to absorb nutrients due to the destruction of gastric motility and secondary improper digestion. Symptoms of PPDS include depression, vomiting, and passing whole seeds in the feces. Some birds show progressive central nervous system signs including ataxia, head tremors, and paralysis. Seizures can be the first sign before any gastric upsets become evident. Almost any type of bird is at risk, but a higher incidence of PPDS is observed in African Grey parrots and blue and gold macaws.

## 5.5.11  PSITTACINE BEAK AND FEATHER DISEASE

Psittacine beak and feather disease (PBFD) is a contagious and fatal viral disease that affects the beak, feathers, and immune system of birds belonging to the Psittacidae family. PBFD has been diagnosed in over 40 species of psittacines including South American parrots. PBFD is most often seen in cockatoos, Eclectus parrots, lovebirds, budgies, and African Grey parrots. PBFD is caused by a virus that affects the immune system and also causes beak and feather development problems (Figure 5.14).

The acute form of the disease most commonly occurs in very young birds. Affected birds are most often depressed and regurgitate. They may develop enteritis, causing diarrhea or pneumonia, and may die without displaying any lesions of the feathers or beak. Affected young birds losing their feathers may have lesions on feathers including circular bands around the feathers. The bands constrict the base of the feathers. These feathers are often loose, break easily, may bleed, and are extremely painful.

**Figure 5.14**  Feather loss due to psittacine beak and feather disease in a young cockatiel. *Courtesy University of Illinois.*

Clinical signs of the chronic form of PBFD include loss of feather dust, abnormal feather development, abnormal beak growth or deformities, beak lesions, and secondary infections. The disease is progressive and very few birds recover. Most birds die within 6 months to 2 years after developing PBFD.

### 5.5.12 SALMONELLA

Salmonella is a bacterial infection that can affect people, birds, reptiles, amphibians, and mammals. Birds may become infected through ingestion of contaminated food or through salmonella penetrating the egg shell. Clinical signs in pet birds include watery diarrhea, breathing difficulty, depression, inappetence, and sudden death. Some birds may remain carriers for life.

Most human cases of salmonella are acquired by eating contaminated food items rather than directly from pet birds. The incubation period is 6 to 72 hours in people. Humans carrying salmonella can infect their pet birds.

## 5.6 COMMON TOXICOLOGICAL HAZARDS

Birds are curious in nature and certain dangerous objects may be attractive to them. Because most pet birds have clipped wings, remain caged, or have limited activity outside their cages, toxicoses are not common. However, birds with free household access or free-ranging birds are at risk of becoming exposed to toxicants.

### 5.6.1 ZINC

Sources of zinc include hardware such as wire, screws, bolts, and nuts and U.S. pennies. Pennies minted after 1982 contain 96% to 98% zinc and about 2.5% copper coating; one penny contains approximately 2,440 mg of elemental zinc. The process of galvanization involves the coating of wire or other material with a zinc-based compound to prevent rust. Owners are often unaware of galvanization on the wire used for making cages. Food and water dishes may also be galvanized and sufficient zinc may leak into the water or food to create toxicity (Table 5.2). Although the exact toxicological mechanisms of zinc in birds or other animals are unknown, zinc toxicosis can adversely affect the kidneys, liver, and red blood cells. Clinical signs of zinc poisoning in birds may include increased urination, increased thirst, diarrhea, weight loss, weakness, anemia, seizures, and death.

### 5.6.2 LEAD

Sources of lead may include paint, toys, drapery weights, linoleum, batteries, plumbing materials, galvanized wire, solder, stained glass, fishing sinkers, lead shot, foil from champagne

**Table 5.2**
Possible Sources of Zinc

---

U.S. pennies (those minted after 1982 contain 96% to 98% zinc)
Galvanized wire cages
Galvanized toys or chains
Galvanized food or water bowls
Zinc hardware—nuts, bolts, screws, metal buttons

---

**Table 5.3**
Possible Sources of Lead

Lead paint chips
Some types of artist paints or tubes containing artist paint
Lead weights (e.g., curtain weights)
Lead fishing sinkers
Lead-containing window blinds
Some types of wine or champagne bottle foil
Plumbing material
Soldering products
Linoleum tiles

bottles, and improperly glazed bowls (Table 5.3). Lead is the most commonly reported avian toxicosis, with acute toxicities being more common in captive birds and chronic toxicities in free-ranging birds.

Lead affects multiple tissues, especially the gastrointestinal tract, kidneys, and nervous system. Clinical signs seen in psittacine birds are often vague and may include lethargy, weakness, anorexia, regurgitation, increased urination, ataxia, circling, and convulsions. In some species such as Amazons, bloody urine may also be noted.

## 5.6.3 NICOTINE PRODUCTS

Tobacco products contain varying amounts of nicotine, with cigarettes and cigars containing 13 to 30 mg and 15 to 40 mg, respectively. Butts contain about 25% of the total nicotine content. Nicotine is also found as a natural form of insecticide. Signs develop quickly following ingestion in most avian species, usually within 15 to 45 minutes and include excitation, panting, salivation, and vomiting. Muscle weakness, twitching, depression, increased heart rate, breathing difficulty, collapse, coma, and cardiac arrest may follow. Death from nicotine toxicosis occurs secondary to respiratory paralysis. A less serious but common response to cigarette smoke deposition on the feathers is feather destructive behavior.

## 5.6.4 INHALANTS

The avian respiratory tract is extremely sensitive to inhalants. Any strong odor or smoke could be potentially toxic (Table 5.4). Cookware or cooking utensils coated with polytetrafluoroethylene (PTFE) can emit toxic fumes when overheated (> 280°F). Clinical signs may

**Table 5.4**
Examples of Noxious Inhalants

Some nonstick surfaces (e.g., pots, pans, cookware, irons, ironing boards), self-cleaning
 ovens, and drip pans for ranges
Gasoline or other volatile gas fumes
Any source of smoke
Automobile exhaust
Carbon monoxide
New carpet
Aerosol sprays
Cleaning products (e.g., ammonia or bleach)
Paint fumes
Fumigants

include acute death, breathing difficulty, ataxia, depression, and restless behavior. Hemorrhage (bleeding) and fluid accumulation in pulmonary tissues leads to respiratory failure and death.

## 5.6.5  Avocado

The toxic principle in avocado (*Persea americana*) is persin. Leaves, fruit, bark, and seeds of the avocado have been reported to be toxic to birds and various other species. Several varieties of avocado are available and not all varieties appear to be equally toxic. In birds, clinical effects observed with avocado toxicosis include respiratory distress, generalized congestion, and death. The onset of clinical signs usually occurs about 12 hours after ingestion with death occurring within 1 to 2 days of the time of exposure. Small birds such as canaries and budgies are considered to be more susceptible; however, clinical signs have been observed in other species as well.

## 5.6.6  Plants

Table 5.5 is a partial list of plants that have been shown to cause toxicity in small animals. The severity of signs or toxicity of these plants in birds has not been studied thoroughly. Plants containing calcium oxalate crystals can cause mechanical irritation of the oral cavity and tongue of birds when plant material is ingested. Clinical signs usually include regurgitation, oral pain, dysphagia (difficulty swallowing), and anorexia. The signs are rarely severe and birds usually respond to supportive care.

**Table 5.5**
Examples of Toxic Plants

| **Cardiotoxic Plants** |
|---|

Lily of the Valley–*Convallaria majalis*
Oleander–*Nerium oleander*
*Rhododendron* species
Japanese, American, English, and Western Yew (*Taxus* species)
Foxglove–*Digitalis purpurea*
*Kalanchoe* species
*Kalmia* species

| **Plants that Could Cause Kidney Failure** |
|---|

Rhubarb (*Rheum* species)–leaves only

| **Plants that Could Cause Liver Failure** |
|---|

Cycad, Sago, Zamia Palm (*Cycad* species)
Amanita mushrooms

| **Plants that Can Cause Multisystem Effects** |
|---|

Autumn Crocus (*Colchicum* species)
Castor Bean (*Ricinus* species)

| **Plants Containing Calcium Oxalate Crystals** |
|---|

Peace lilies (*Spathiphyllum* spp)
Calla lilies (*Zantedeschia* aethiopica)
Philodendron (*Philodendron* spp)
Dumbcane (*Dieffenbachia* spp)
Mother-in-Law plant (*Monstera* spp)
Pothos (*Epipremnum* spp)

## 5.7 SUMMARY

Companion birds rank third in the number of companion animals in North America. Data and information presented in this chapter highlight the most common types and species of pet birds as well as proper nutrition, husbandry practices, common diseases, and toxicological hazards of these pleasure-giving pets.

For students and others desiring additional information about companion birds, the chapter concludes with 18 reference publications and 6 Web sites. Please continue to enjoy the beauty and companionship of birds!

## 5.8 REFERENCES

1. Bauck, L., and J. LaBonde. 1997. Toxic Diseases. In *Avian Medicine and Surgery,* edited by R. B. Altman, S. L. Clubb, G. M. Dorrestein, and K. Quesenberry, 604–613. Philadelphia: W. B. Saunders.

2. Beynon, Peter H., Neil A. Forbes, and Martin D. C. Lawton, eds. 1996. *Manual of Psittacine Birds.* Ames: Iowa State University Press.

3. Boussarie, D. 2002. Anomalies and disorders in beak and feathers. *World Small Animal Veterinary Association, Congress Proceedings,* October 3–6. Granada.

4. Dumonceaux, G., and G. J. Harrison. 1994. Toxins. In *Avian Medicine: Principles and Application,* edited by B. W. Ritchie, G. J. Harrison, and L. R. Harrison, 1047–1049. Lake Worth, FL: Wingers Publishing.

5. Gallerstein, G., and H. Acker. 1994. *The Complete Bird Owner's Handbook.* New York: Simon & Schuster Macmillan.

6. Harcourt-Brown, N. H. 2000. Psittacine Birds. In *Avian Medicine,* edited by Tully, T. N., Jr., M. P. C. Lawton, and G. M. Dorrestein, 112–143. Oxford, England: Butterworth Heinemann.

7. LaBonde, J. 1995. Toxicity in Pet Avian Patients. *Seminars in Avian and Exotic Pet Medicine* 4(1): 23–31.

8. Lightfoot, T. L. 2002. Common Avian Medicine Presentations II. *Western Veterinary Conference Proceeding Notes.*

9. Macwhirter, P. 2002. Basic Anatomy, Physiology, and Nutrition. In *Avian Medicine*, edited by Tully, T. N., Jr., M. P. C. Lawton, and G. M. Dorrestein, 1–25. Oxford, England: Butterworth Heinemann.

10. Oglesbee, B. 1997. Mycotic Diseases. In *Avian Medicine and Surgery,* edited by R. Altman, S. Clubb, G. Dorrestein, and K. Quesenberry, 611–612. Philadelphia: W. B. Saunders.

11. Olsen, G. H., and S. E. Orosz. 2000. *Manual of Avian Medicine.* St. Louis, MO: Elsevier.

12. Roudybush, T. E. 1999. Psittacine Nutrition. *Husbandry and Nutrition, The Veterinary Clinics of North America, Exotic Animal Practice* 2(1): 111–126.

13. Romagnano, A. 1999. Examination and Preventive Medicine Protocol in Psittacine. *Physical Examination and Preventive Medicine, The Veterinary Clinics of North America, Exotic Animal Practice* 2(2): 333–356.

14. Wells, R. E., R. F. Slocombe, and A. L. Trapp. 1982. Acute Toxicosis of Budgerigars (Melopsittacus undulatus) Caused by Pyrolysis Products from Heated Polytetrafluoroethylene: Clinical Study. *American Journal of Veterinary Research* 43: 1238–1242.

15. Ritchie, B. W. 2003. Diagnosis, Management and Prevention of Common Gastrointestinal Diseases. *Western Veterinary Conference Proceedings Notes.*

16. Ritchie B. W. 2003. Management of Common Avian Infectious Diseases. *Western Veterinary Conference Proceedings Notes.*

17.  Richardson, J. A., L. A. Murphy, S. A. Khan, and C. Means. 2001. Managing Pet Bird Toxicoses. *Exotic DVM* 3(1): 23–27.

18.  Vriends, M. M. 1999. *The Cockatiel Handbook*. Hauppauge, NY: Barron's.

# 5.9   WEB SITES

1.  www.hartz.com
2.  www.veterinarypartner.com
3.  www.birdsnways.com
4.  www.wingedwisdom.com
5.  www.parrotclubs.com
6.  www.petfinder.org

# 5.10   ABOUT THE AUTHORS

Jill A. Richardson, DVM, is Associate Director of Consumer Relations and Technical Services for The Hartz Mountain Corporation. She co-authored this chapter with Rosemary M. Tini and Cynthia R. Garnham. Dr. Jill A. Richardson is a graduate of Tuskegee University College of Veterinary Medicine. After graduation she practiced as a small animal practitioner in Knoxville, Tennessee, and in Charleston, West Virginia. She became a Consulting Veterinarian in Clinical Toxicology with the American Society for the Prevention of Cruelty to Animals (ASPCA)-sponsored Animal Poison Control Center (APCC) in 1996. In 2000 she became the APCC's Coordinator for Professional and Public Relations and led multiple nationwide pet safety campaigns. In 2002 she left the ASPCA to accept her current position with The Hartz Mountain Corporation. Dr. Richardson is a 2002 graduate of the University of Illinois Executive Veterinary Program, a Contributing Editor for the Veterinary Information Network (VIN), a member of the Cat Fanciers' Association (CFA) Health Committee, and a member of the Cat Writer's Association. She also is a columnist for "Hartz Best Friends," a monthly online pet column. Previously she was an adjunct instructor in the University of Illinois College of Veterinary Medicine. Her goal is to increase awareness of pet safety and health through education of both the public and veterinary health professionals.

Rosemary M. Tini earned a B. S. degree in Business Administration from New York University. She joined The Hartz Mountain Corporation in 1998 as a Consumer Relations Analyst where she found that her love for animals and love for business could be combined and enhanced. She received her diploma as a Certified Veterinary Assistant in 2003. Her goal is to continue working with the Hartz Mountain family in their efforts to educate pet owners in matters related to the health and quality of life of animals in general and pet animals in particular.

Cynthia R. Garnham is a Certified Veterinary Assistant and a Consumer Relations Specialist with The Hartz Mountain Corporation. She has more than 16 years of experience in the pet industry including working in veterinary clinics, with pet products manufacturers, and at the retail level. Her goal is to help increase pet owner awareness of ways to enhance pet animal health, especially that of exotic pets, through education.

# 6 COMPANION REPTILES AND AMPHIBIANS

*Dorcas P. O'Rourke[1]*

[1]Please turn to the end of the chapter for complete author affiliation information.

*Our task must be to free ourselves ...*
*by widening our circle of compassion to embrace all*
*living creatures and the whole of nature and its beauty.*

Albert Einstein (1879–1955)
*American (German born) physicist*

In many ancient cultures, reptiles and amphibians were feared and degraded. Although this abhorrence continues in certain contemporary societies, an ever-increasing number of people have come to recognize that these fascinating creatures are vital co-inhabitants with humans and many ecosystems. For the authorship of this chapter on reptiles and amphibians, we were fortunate in being guided to Dr. Dorcas P. O'Rourke. With respect and appreciation, we gratefully acknowledge the invaluable contributions of this eminent scholar, teacher, and herpetologist.

*Campbell-Corbin-Campbell*

# 6.1    PET FROGS

## 6.1.1    INTRODUCTION

Frogs are members of the class Amphibia, which also includes salamanders. There are almost 4,000 species of amphibians worldwide in a variety of habitats. Frogs and toads are members of the order Salientia. There are 20 families, 234 genus groups, and 3,494 species of frogs and toads.

In recent years, wild amphibian populations have suffered significant reductions in numbers. Conservationists continue to struggle to understand and prevent this decline. Responsible pet owners have an obligation to do their part to protect these valuable animals. Therefore, when choosing a pet frog, every effort should be made to purchase animals from healthy, captive-bred colonies.

## 6.1.2    COMMON PET FROGS

Several species of frogs are commonly kept as pets. These include fire-bellied toads, poison dart frogs, South American horned frogs, African clawed frogs, tree frogs, bullfrogs, leopard frogs, and toads.

*Fire-bellied toads.* Fire-bellied toads (*Bombina orientalis*) are hardy toads, approximately 2 inches long, and considered to be good choices for a new hobbyist. Their native habitat is southern Asia. These toads have bright green and black colored backs and orange and black undersides. The bright colors warn predators that the toads are poisonous. These toads have a life span of 10 to 15 years.

*Poison dart frogs.* Poison dart frogs (*Dendrobates pumilio*) received their name because some native South Americans rub tips of darts in the poisonous mucous of the frog's skin and then use the darts to kill game. These small, brightly colored rainforest frogs (see Color Plate 9Q) have glands in the skin that produce strong toxins. Adult frogs are approximately 0.75 inch long. Their natural habitat is the tropical rainforests of South and Central America.

*South American horned frogs.* South American horned frogs are large (up to 5.5 inches long), fat frogs with a wide mouth (see Color Plate 9N). They have been nicknamed "PacMan"

**Figure 6.1** The African clawed frog (*Xenopus*) is an aquatic species commonly kept as a pet. *Courtesy Dr. Dorcas P. O'Rourke.*

frogs. Females weigh up to one pound; males are smaller. Horned frogs live approximately 6 years. The name comes from folds of skin located over their eyes. Coloration is varied; the more colorful varieties are bright green with red markings, others are duller green with black markings. They are passive hunters and spend most of their time sitting or half-buried in leafy, muddy vegetation. They are native to tropical rain forests of Argentina, Uruguay, Paraguay, and Brazil.

*African clawed frogs.* African clawed frogs (*Xenopus laevis*) are 4 to 5 inches long and live up to 15 years. They are strictly aquatic and good pets for beginning frog hobbyists (Figure 6.1). Their forelegs are too weak to ambulate on dry land.

*Tree frogs.* Tree frogs are found in North, South, and Central America and also in Central Asia and Europe. There are over 637 species of tree frogs (see Color Plate 9X). The red-eyed tree frog (*Agalychnis callidryas*) has red eyes with slitlike pupils. Its body is green with yellow markings on its sides and red toes. The American green tree frog (*Hyla cinerea*) is found in the Mississippi Valley and southeastern United States. It has a yellow-green body with a yellow stripe running from the lower jaw and along its side.

*Bullfrogs.* Bullfrogs are found as native species in most countries. Two of the most popular varieties kept as pets are the American bullfrog (*Rana catesbeiana*) and the African bullfrog (*Pyxicephalus adspersus*). These are large frogs. The American bullfrog (Figure 6.2) may exceed 6 inches in length; its color varies from a dull green to brown, dark grey, or black with a white to yellow ventral surface. It is native to the central and eastern United States and has also been introduced to many areas in the western United States and southern portions of Canada. African bullfrogs may exceed 9 inches in length. They are olive-green with darker skin ridges. They have loud voices (deep bellows) and are aggressive. These frogs may live 35 or more years.

*Leopard frogs.* Leopard frogs are slender greenish to brown frogs with dark spots edged with a slightly lighter color. Northern leopard frogs (*Rana pipiens*) are native to the northern half of the United States and southern parts of Canada. Southern leopard frogs (*Rana sphenocephala*) are most abundant in the southern parts of the United States, but can be found as far north as Iowa and parts of New York. These frogs may grow to 3.5 inches in body length. Southern leopard frogs are green or brown on top with a few round black spots scattered randomly about the back and on the sides, but none on the snout. The belly is plain white. There are two light complete dorsolateral ridges down the back. There are spots on the forelimbs and tiger stripes or bars on the hind legs. The groin and thighs have a greenish (very rarely yellowish) tinge.

**Figure 6.2** Young American bullfrogs (*Rana catesbeiana*) are popular pets. These excellent jumpers are used in many frog jumping contests. *Courtesy Maria Lang (photographer) and Dr. B. Taylor Bennett.*

*Toads.* Found in most parts of the world, true toads are members of the family Bufonidae, which includes 339 species. Toads have short, thick bodies and short legs; they cannot jump as far as frogs. The American toad (*Bufo terrestris*) is found in the eastern parts of the United States. The marine toad (*Bufo marinus*) is native to parts of Texas and the northern parts of South America. It is poisonous to dogs, cats, and many other animals.

## 6.1.3   BIOLOGY AND BEHAVIOR

Frogs have smooth, glandular skin. In most species (toads are an exception) the skin is moist. Because the skin does not have a stratum corneum,[2] it is highly permeable to water. Two kinds of skin glands are present in frogs—mucous and granular. Mucous glands secrete a slimy coating that protects the skin and helps retain moisture. Granular glands can be very specialized and secrete a variety of substances from antimicrobial and analgesic (pain relieving) compounds to pheromones and antipredator toxins. Toads in particular have clusters of these glands located behind the eyes. These parotid glands secrete toxins that prevent ingestion by predators. Consequently, when handling toads, care must be taken not to rub these secretions into eyes or onto other mucous membranes. Marine toads can actually eject this toxin quite a distance, so extreme care must be taken when handling them. Poison dart frogs secrete a curare-like substance, which was applied to spear tips by South American natives and used to paralyze hunted animals. Interestingly, captive-bred poison dart frogs often lose the ability to secrete this toxin.

Many frogs have enlargements in the lymphatic system that collect lymph and drain it into the circulatory system. In some species, these lymph sacs are located below the skin of the back adjacent to the hind legs. Therapeutic drugs can be injected into these lymph sacs

---

[2]The stratum corneum is the outermost layer of the epidermis in most species of animals; it is fully keratinized and consists of "horny cells" composed of keratin (an insoluble protein) within an intercellular lipid matrix that protects the skin from desiccation and from invasion by microorganisms.

and will drain directly into the venous circulation, thereby enabling a veterinarian to give essentially an intravenous injection to these frogs.

Frogs are designed to jump and have several structural modifications to permit this. Various fused bones and heavy muscling on the legs enable these animals to jump 2 to 10 times their body length.

Many frogs have a bladder, which also serves as a storage site for water. When frightened during handling, they will empty bladder contents onto the handler.

Frogs undergo metamorphosis. Most species begin life as tadpoles, which closely resemble fish. Tadpoles breathe through gills, have a two-chambered heart, no limbs, a tail designed for swimming, and a lateral line system (series of pores in the skin that allow tadpoles to detect changes in water currents and assist in locating food). Many tadpoles are herbivorous and some, like the African clawed frog, are filter feeders. During metamorphosis, the heart changes to a three-chambered organ, gills are replaced by lungs, limbs grow, and the tail is resorbed. Adult frogs may be aquatic (e.g., African clawed frog), semiaquatic (e.g., bullfrogs and leopard frogs), or terrestrial (e.g., tree frogs, poison dart frogs, and toads).

## 6.1.4   HUSBANDRY (CARE AND MANAGEMENT)

*Cages.*   The biology and behavior of the species being housed must be considered when selecting or designing a cage for frogs. Many frogs can be successfully housed in glass aquaria. Aquatic species can be kept in water of appropriate depth (water should be deep enough to allow the frog to swim freely when submerged). Frogs, and in particular tadpoles, are extremely sensitive to toxic compounds in water, because these are directly absorbed through their permeable skin. Chlorine is very toxic to frogs and water must be dechlorinated. Many municipal water supplies use chloramines to treat drinking water; chloramines are more toxic and harder to remove than chlorine. Information about chlorination and amounts of toxic trace metals (e.g., lead, copper) in water should be obtained before exposing frogs to municipal water. Alternatively, bottled spring water can be used to fill aquaria.

Aquatic frogs can be kept in static (fill/dump/refill), flow through (constant trickling of fresh water into tank with constant slow draining), or recirculating (filtered) systems. Stocking density, ease of cleaning, and cost of systems will dictate the most appropriate system to use.

Semiaquatic species can be housed in aquaria with slightly slanted floors. These allow animals to emerge from the water onto a solid surface. More elaborate designs can include a large, deep "pond" area and a solid substrate (sphagnum moss or heat-treated soil) that slopes down to the "shoreline."

Terrestrial frogs must be housed on a solid surface; however, this substrate should be kept moist. Sphagnum moss or heat-treated soil or leaves work well. Clean, saturated foam or sponge material is also satisfactory. A water dish or small "pond" should also be placed in the cage; this water source should be easily accessible and large enough for the frog to crawl into and submerge.

In all cases, water quality is critical. Buildup of ammonia from excreta or other toxins in the water and environment will be absorbed through the frog's skin and can quickly become toxic. Cleanliness is essential.

Aquaria should have tight-fitting lids to prevent escape. Many frogs can jump great distances, and some species such as tree frogs can climb up the sides. Lids should be made of nonabrasive substances (such as coated wire mesh or smooth, solid, impervious material with holes for air exchange) to prevent skin damage if the frog becomes frightened and jumps.

Cages should contain structures that allow the frogs to engage in natural behaviors. Pieces of heat-treated bark or PVC (polyvinyl chloride) pipe allow the frogs to hide from view and decrease stress. Climbing species such as tree frogs should have branches and artificial vegetation to facilitate normal climbing and hiding activities. Even aquatic species such as African clawed frogs prefer having a dark hiding place to retreat from view. For these animals, submerged PVC pipe, terra cotta pots, or floating artificial "lily pads" work well.

Because frogs are secretive by nature and tend to live in ground cover or moist environments, drafts and high airflow should be avoided. Increased air turnover will rapidly desiccate (dry out) the environment and the animal.

*Cleaning.*  When cleaning tanks, remember that frogs are extremely sensitive to chemicals. Consequently, many common cleaning agents are toxic to them. Frequent cleaning and flushing with clean, dechlorinated water will help maintain an appropriate environment. Soil and other substrates and structures should be replaced when contaminated. For sanitizing the tank, a dilute chlorine bleach solution can be used; however, the tank must be thoroughly rinsed and air dried prior to replacing the water and substrate. Soapy agents should never be used around amphibians and phenolic compounds such as Lysol are highly toxic.

*Temperature.*  Amphibians are ectotherms (poikilothermic) and therefore rely on heat from the environment to maintain their optimal body temperature. Frogs tend to prefer fairly warm temperatures; however, many species are adapted for cooler environments. Scientific literature should be consulted to ascertain the appropriate temperature for the species being housed. Most frogs do well in a relative humidity of approximately 80%. Allowing the environment to become too dry will result in desiccation and death of the frog.

*Light.*  Natural light should be provided in a normal day/night cycle. Many frogs, however, are nocturnal and will suffer if lighting is too intense or too prolonged. Care must be taken to avoid overheating of glass aquaria—lethal temperatures can be reached within minutes if a closed aquarium is placed in direct sunlight.

*Feeding.*  Although many tadpoles are herbivorous, adults are carnivorous. Most frogs visually orient on their prey. Consequently, they must be fed live food. Crickets and mealworms are among food items commonly offered to medium- to large-sized frogs. If these are purchased commercially, care should be taken to ensure the insects have been maintained on a high-quality diet or they will not be nutritious for the frogs. Insects should be dusted prior to feeding with a vitamin/mineral mixture or can be fed highly nutritious food with vitamins/minerals immediately before feeding them to the frogs (gut loading). Alternatively, they can be purchased and maintained in an optimal environment in the owner's home. Several excellent references on maintaining mealworms and cricket colonies are available. Very small species such as poison dart frogs can be fed commercially purchased wingless fruit flies. African clawed frogs are relatively easy to maintain because they will eat commercially available reptile floating food sticks or floating fish pellets.

*Handling.*  Frogs require careful handling to avoid injuring the animal and to avoid disrupting the protective mucous skin covering. Large frogs have powerful hind limbs, which must be securely restrained to prevent jumping or pushing away. They should also be held around the shoulders. Small frogs can be cupped carefully in the hands. Hands should be thoroughly washed after handling any frog, to ensure that irritating or toxic secretions are removed.

*Breeding.*  Amphibian reproduction involves quite complex and fascinating behaviors. Before attempting to breed any species of frog, scientific and husbandry references should be consulted to ensure provision of appropriate habitat, food, and environmental conditions. Additionally, information on housing and feeding tadpoles should be obtained and closely followed to successfully raise these fascinating, delicate creatures.

## 6.1.5   COMMON DISEASES

*Redleg.*  One of the most common diseases identified in frogs is redleg. It is precipitated by stress and subsequent immunosuppression. *Aeromonas hydrophila* is the bacterium most often associated with the disease. *Aeromonas* is normally found in aquatic environments, but does not cause the disease unless the frog is immunosuppressed. Poor water quality, unclean environments, crowding, and inadequate nutrition stress frogs and make them susceptible to infection. The organism enters the body and colonizes internal organs leading to septicemia. This septicemia results in hemorrhages and ulcerations, most visible on the legs and abdomen (hence the term *redleg*). Antibiotics may be helpful in treating the disease, but correcting the underlying cause is also necessary. This disease is best prevented by appropriate husbandry and nutrition.

*Mycobacteriosis.* Mycobacteriosis can also result from immunosuppression of frogs. The responsible bacteria, *Mycobacterium xenopi, M. fortuitum,* and *M. marinum,* are commonly found in soil and water. Once they colonize a susceptible animal, they can cause emaciation, pneumonia, or nodules in the skin and internal organs. Treatment with antibiotics is difficult, and because this disease can be transmitted to humans, affected frogs are usually euthanized. The disease can be prevented by providing appropriate husbandry and nutrition and preventing stress in frogs.

*Parasites.* Frogs can also be affected by parasites. One of the more common parasite infections identified in African clawed frogs is caused by the nematode *Pseudocapillaroides xenopi.* These worms burrow under the skin and cause a roughened, pitted appearance. Patches of skin may also slough. *Pseudocapillaroides* infestations can be treated with drugs such as ivermectin.

*Fungi.* *Saprolegnia* and other fungi may cause secondary infections in frogs suffering from other diseases or stress. These organisms are common in soil and water and usually cause no problem unless an animal is immunosuppressed or has an existing skin lesion. *Saprolegnia* forms characteristic cottony tufts at the site of a preexisting wound. It can be treated with salt water baths or benzalkonium chloride. This disease can be prevented by keeping frogs healthy and avoiding stress.

*Lucke tumor herpesvirus.* Lucke tumor herpesvirus (LTHV) is a viral disease unique to leopard frogs. The virus replicates in the frog during cool winter temperatures and causes kidney tumors that grow rapidly during warm summer months. Frogs may not show signs of depression, muscle loss, and abdominal distension until the disease is advanced. There is no known treatment. Purchasing captive-reared, healthy leopard frogs will diminish the likelihood of acquiring infected animals.

*Chlamydia psittaci.* *Chlamydia psittaci,* the causative agent of psittacosis in birds, can also infect African clawed frogs. The organism, when swallowed, colonizes the internal organs of clawed frogs and causes depression, abdominal distension, and redness and blanching of the skin. Tetracycline may be effective against the organism but most affected frogs are euthanized because this disease is zoonotic. Frogs should always be purchased from reputable suppliers and should be nonstressed and disease-free.

## 6.1.6 ZOONOSES

The most common zoonotic agent associated with frogs is atypical mycobacteriosis, also known as "aquarists' nodules." Aquatic and semiaquatic frogs and their tank water can contain *Mycobacterium xenopi, M. fortuitum,* and *M. marinum.* These organisms typically enter preexisting wounds on the hands or arms of humans and cause self-limiting skin nodules and ulcerations. Unfortunately, immunosuppressed individuals are more susceptible to these organisms and can develop severe systemic disease. To prevent this disease, gloves should be worn when handling frogs. This is especially true if individuals have sores on their hands or are immunosuppressed. Hands should always be thoroughly washed immediately after handling frogs.

*Salmonellosis.* *Salmonellosis* associated with handling amphibians has been reported; however, this disease is much more frequently associated with reptiles. Wearing gloves and thorough hand washing offer protection against salmonellosis.

*Chlamydia psittaci.* *Chlamydia psittaci* has been identified in frogs and is a known zoonotic agent (most commonly contracted from birds). Although reports of frog-to-human transmission have not been currently identified, appropriate precautions should be taken especially when handling sick frogs with suspicious signs.

## 6.2    PET LIZARDS

### 6.2.1    INTRODUCTION

Lizards are members of the class Reptilia. Other common reptiles kept as pets include snakes (the closest relatives of lizards), turtles, and crocodilians. Lizards and snakes are referred to as squamates and there are over 5,000 species occurring worldwide.

### 6.2.2    COMMON PET LIZARDS

Lizards commonly kept as pets include iguanas, dragons, anoles, skinks, geckos, monitors, and tegus.

*Iguanas.* Iguanas are members of the family Iguanidae, which includes over 650 different species. The most frequently kept pet iguana is the common green iguana (*Iguana iguana*; see Color Plate 8L). It is native to Central and South America where it reaches lengths of up to 6.5 ft. The desert iguana (*Dipsosaurus dorsalis*) is native to the southwestern United States. It is light brown in color and may have black lines and white spots on its body (Figure 6.3). Another desert-dwelling member of the family Iguanidae is the chuckwalla (*Sauromalus ater*). Chuckwallas have a flattened body shape (see Figure 6.12 and Color Plate 8G) that enables them to hide in the cracks and crevices of boulders. They can inflate their lungs to further wedge their body between the rocks making it almost impossible for a predator to pull them out. Chuckwallas are the second-largest lizard found in the United States and may be as long as 18 inches. Their native habitats are boulder-covered areas in the deserts of southern California, Nevada, Utah, western Arizona, and south to Sonora, Mexico.

**Figure 6.3**   Desert iguanas are native to the southwestern United States and adapt well to captivity. *Courtesy Michael J. Adkesson, DVM.*

**Figure 6.4** The bearded dragon (*Pogona vitticeps*) is a medium-sized lizard known for its docile temperament. Native to the hot desert of Australia, this omnivorous species is easy to keep and breed in captivity. *Courtesy Jeff Rathmann (photographer) and James Brumley, Exotic A.R.C., St. Louis, MO.*

*Dragons.* Dragons are mid-sized stocky lizards. The most commonly kept pet dragon is the Australian bearded dragon (*Pogona vitticeps*), a native of Central Australia. Basic colors vary from brown, grey, reddish-brown to bright orange. They can change their body color to help in regulating body temperature. Dragon lizards (Figure 6.4) have dark beards and when frightened "puff out" their jaws and open their mouths, giving them the appearance of a dragon. Bearded dragons are easily tamed as pets (see Color Plate 8B).

*Anoles.* Anoles are also members of the family Iguanidae. The green anole (*Anolis carolinensis*) is found in the southern United States. It lives in trees, shrubs, and around houses and is commonly sold in pet shops. Green anoles are 6 to 8 inches long as adults (see Figure 6.13) and can change color to various shades of grey and brown depending on the environment. The Cuban brown anole (*Anolis sagrei sagrei*) is found in southern Florida. It is brown in color with light brown to grey markings. Knight anoles (*Anolis equestris*) are natives of Cuba. Adults are 18 inches long and are green with white markings. The eyes of Knight anoles are surrounded by a circle of blue.

*Skinks.* Skinks are members of the family Scincidae, which includes over 1,275 species. Skinks usually live among leaves and underbrush of forest floors. The five-lined skink (*Eumeces fasciatus*) is found in the eastern United States. They are reddish-brown and have white or yellow stripes; the tail is blue in adults. The Australian blue-tongued skink (*Tiliqua scincoides*) is native to Australia but is easily bred and maintained in captivity and commonly sold as a pet (see Color Plate 8C). Its name comes from its broad, blue tongue. The Solomons giant skink (*Corucia zebrata*) has a prehensile tail and lives in trees. It is a native of San Cristobal Island (in the Solomon Island chain). These skinks are easy to breed in captivity and are commonly found in pet stores (Figure 6.5).

*Geckos.* Geckos are members of the family Gekkonidae, which includes approximately 800 species. Geckos are found in tropical and semitropical habitats. The feet of geckos have adhesive pads known as lamellae that enable these animals to climb almost any surface including ceilings! Most species are nocturnal and have loud voices. The tokay gecko (*Gekko gecko*) is native to the rainforests of Southern Asia, India, and New Guinea. Their name comes from their call, which sounds like "Tō-kay! Tō-kay!" These lizards are greyish-blue with small orange spots and have large yellow eyes (see Color Plate 9W). Adults reach a length of approximately 12 inches. Leopard geckos (*Eublepharis macularius*) are found in dry regions such as northern India and Iraq. It is grey-brown in color with dark patches and spots and small wartlike projections on its epidermis (skin). Leopard geckos (Figure 6.6 and Figure 6.11) are nocturnal, ground-dwelling lizards that have

**Figure 6.5** The prehensile-tailed skink is a diurnal lizard that is an interesting pet. *Courtesy Jeff Rathmann (photographer) and James Brumely, Exotic A.R.C., St. Louis, MO.*

claws rather than adhesive pads and thus cannot climb well. They are usually yellow to white with black spots or stripes. Adults are 8 to 10 inches long and may live 20 or more years. House or chitchat geckos (*Hemidactylus frenatus*) are natives of Pacific islands and Southeast Asia. These geckos are sandy-brown in color with darker patches and spots. The African fat-tailed gecko (*Hemitheconyx caudicinctus*) is dark to light brown; some have a white stripe down the back, whereas others do not. Fat is stored in their large, thick tail. These lizards may live 15 or more years. Day geckos (belonging to the genus *Phelsuma*) include over 60 species that are active during daytime hours (see Color Plate 8J). Species popular as pets include the giant day geckos, gold dust day geckos, frog eyed geckos (Figure 6.7), and lined day geckos.

*Monitors.* Monitors are large lizards with powerful jaws, tails, and well-developed claws that have caused some local communities to propose bans on their ownership as pets. The most common monitor in pet stores is the African Savannah monitor (*Varanus exanthematicus*). Savannah monitors (see Color Plate 9T) may reach 5 ft in length as adults and live 10 to 12 years. Those raised in captivity tend to be docile and easy to handle. Other monitors

**Figure 6.6** The leopard gecko is a nocturnal, ground-dwelling lizard that may live over 20 years in captivity. *Courtesy Jeff Rathmann (photographer) and James Brumley, Exotic A.R.C., St. Louis, MO.*

**Figure 6.7**   The frog eyed gecko (*Teratoscincus scincus*) is a desert species that loves to dig. Their small size makes them fairly easy to keep, but they need hot temperatures and have delicate skin, so handle with care. *Courtesy Jeff Rathmann (photographer) and James Brumley, Exotic A.R.C., St. Louis, MO.*

occasionally kept as pets include African Nile monitors (*Varanus niloticus*; Figure 6.8), Asian water monitors (*V. salvator*), Dumeril's monitors (*V. dumerilii*), black rough-necked monitors (*V. rudicollis*), mangrove monitors (*V. indicus*), and numerous others (see Color Plate 9U).

***Tegus.***   Tegus are fairly large lizards (Figure 6.9, see also Color Plate 9V) that can be aggressive and are not recommended for beginning hobbyists. The Argentine red tegu (*Tupinambis rufescens*) is the most popular tegu kept in captivity. Tegus are terrestrial lizards that are most active during the day (diurnal) and are good swimmers. Adults can reach 4 ft in length, weigh 20 lbs, and live approximately 15 years.

## 6.2.3   BIOLOGY AND BEHAVIOR

Lizards have smooth, dry skin covered by scales. The skin is shed in cycles with large patches (Figure 6.10) coming off during each shed (several species will ingest this shed skin). Most lizards have eyelids; geckos are an exception. The heart has three chambers consisting of two atria and a single ventricle. A large midventral abdominal vein lies just inside the abdominal wall. The tongue of lizards may be fleshy (iguanas) or forked like a snake tongue (monitors). All lizards have teeth. Only two species of lizards—Gila monsters and beaded lizards—are

**Figure 6.8**   A baby African Nile monitor can grow to over 7 ft in length. *Courtesy Gonzalo Barquero, Brazil, and Dr. James E. Corbin.*

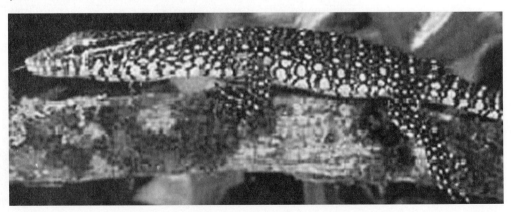

**Figure 6.9** Tegu lizards are dull in color as hatchlings and darken as they mature. *Courtesy Michael J. Adkesson, DVM.*

venomous. Although many lizards can be housed in groups, males of some species are quite territorial and will display aggressive behaviors if housed together (e.g., anoles). Several species of lizards have a predator avoidance mechanism called autotomy. If these animals are restrained by the tail, the tail will break off and the lizard can escape. Although the tail will eventually regrow, it will typically appear off-color from the rest of the animal.

## 6.2.4 HUSBANDRY

*Cages.* Virtually all lizards must be kept in enclosures with tight-fitting lids. Glass aquaria work well for smaller animals. Large individuals may be kept in wire cages. Appropriate substrates include newspaper, indoor/outdoor carpet, or hardwood shavings. Softwood shavings such as pine contain compounds that can be toxic to reptiles and should not be used. Desert species can be kept on sand. Lizards tend to be secretive by nature, so animals should be provided with a retreat. Hide boxes are commercially available. PVC pipe segments, large pieces of heat-treated bark (Figure 6.11 ), and broken pieces of terra cotta pots also work well. Species such as iguanas, chuckwallas (Figure 6.12 and Color Plate 8G), anoles, and some geckos climb and should be provided with branches, artificial vegetation, and other elevated perches.

*Water.* Fresh water should always be available. Most lizards will drink from bowls and several species will soak in their water dishes. Certain lizards lap dew in the wild (e.g., chameleons); these animals require frequent misting of the vegetation in their cages in addition to having a water bowl. Water bowls should be cleaned daily as many lizards defecate in their bowls.

*Temperature.* In general, lizards prefer warm climates; therefore cages should be kept at appropriate temperatures. Several references are available detailing temperature preferences of various species. Lizards are ectotherms (poikilothermic) and therefore

**Figure 6.10** Lizards such as this day gecko normally shed their skin in patches. *Courtesy Dorcas P. O'Rourke.*

**Figure 6.11** Reptiles such as this leopard gecko should be provided with a retreat like this piece of bark. *Courtesy Dr. Dorcas P. O'Rourke.*

must utilize external sources of heat to regulate body temperature. Lizards engage in behavioral thermoregulation. Cages should be provided with a focal heat source to allow animals to move from warmer to cooler areas as needed. A low-wattage incandescent bulb focused on a rock or other basking site will provide a thermal range. Care must be taken to ensure that animals do not come into contact with the heat source or fatal burns can result. "Hot rocks" can be hazardous if they malfunction; an intense hot spot in the rock will result in thermal burns.

*Humidity.* Most lizards tolerate 30% to 70% relative humidity. Certain species require a moister environment (up to 100% for tropical species) whereas desert species do well in a low relative humidity (e.g., 30% to 40%). Appropriate references should be consulted to determine the specific needs for the species being kept. Airflow should be relatively low, particularly with young animals prone to desiccation.

**Figure 6.12** Chuckwalla lizards are excellent climbers—this is why the name "chuckwalla" is given to many challenging rock climbing routes and walls. Chuckwalla climbing a tree limb (*a*), and chuckwalla climbing a wall (*b*). *Courtesy Michael J. Adkesson, DVM.*

(a)

(b)

*Light.* Some lizards, particularly young iguanas, require exposure to ultraviolet light of an appropriate wavelength to properly metabolize vitamin D. Full spectrum[3] fluorescent bulbs should be placed approximately 18 to 24 inches above the cage. Light from these bulbs must not pass through glass or plastic, as these absorb the wavelengths of light needed by the lizard. Bulbs should be changed approximately every 6 months as the light emitted from older bulbs does not contain the proper wavelengths required for vitamin D metabolism.

*Cleaning.* Cages should be cleaned frequently enough to prevent buildup of fecal material and bacterial pathogens. Diluted chlorine bleach solutions can be used to disinfect cages; however, care must be taken to thoroughly rinse and dry the cages before replacing the animals. Lizards and other reptiles are extremely sensitive to phenolic compounds and these should never be used around lizards.

*Feeding.* Lizards fill a variety of niches in the wild; consequently, diets are quite variable. Iguanas do well on a diet of dark leafy greens, other vegetables, occasional fruit, and commercially available high-quality iguana diet. Anoles, skinks, and some gecko species eat crickets, mealworms, and other insects. Other geckos, such as day geckos, require fruit or fruit baby food in addition to insects. Crickets should be dusted with vitamin/mineral powder or "gut loaded" (fed highly nutritious food) just prior to feeding them to the lizard. Insects are low in calcium levels and mineral imbalances will occur if the lizard is not provided with other foods and/or supplements. Monitors are voracious feeders. Large specimens will eat whole killed rodents and may be trained to accept commercially prepared diets. When determining an appropriate diet for the species being kept, reputable scientific references should be consulted to ensure that an optimum diet is chosen.

*Handling.* When handling lizards, the animal's entire body must be supported (Figure 6.13). Small species can be held in one hand, but larger animals require two hands for adequate support. One hand is used to hold the lizard around the shoulders behind the head and the other hand supports the hind legs and tail. Lizards should never be restrained by the tail as it can break off. Most lizards will bite to defend themselves (they typically bite and hold on) and bites from large lizards can be quite painful. They will also scratch with sharp claws and some species will slap with their tails. The key to restraint is to firmly but gently restrain the animal while supporting its entire body. Most lizards will readily adjust to this type of gentle handling.

*Breeding.* Some lizard species are sexually dimorphic[4] with males generally larger and more colorful than females. Male anoles have a large dewlap, which is displayed during courtship. Male iguanas have a prominent row of femoral pores on the inside of the hind legs (females

---

[3]Optimal metabolism of vitamin D requires light wavelength of 290 to 315 nm (UVB).
[4]Sexual dimorphism refers to male and female differences in physical characteristics such as size and coloration.

**Figure 6.13** When restraining lizards like this green anole, the entire body must be supported; do not restrain by the tail. *Courtesy Dr. Dorcas P. O'Rourke.*

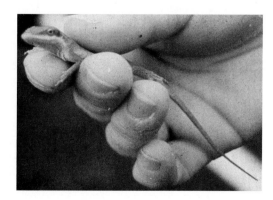

have small to no pores). Male lizards also have paired sexual organs, called hemipenes, which are used during copulation.

Lizards engage in a wide variety of reproductive strategies. Most lay eggs but several species are live bearers. Some, such as the whiptail lizard, are parthenogenetic, with females producing genetically identical offspring. Iguanas may lay up to 40 eggs in a single clutch, whereas anoles lay 1 or 2 eggs at a time throughout the entire breeding season. Eggs may be placed in a single large nest or hidden in small groups in various locations. These eggs may be collected and artificially incubated. Specific information on incubation substrates and temperatures is available in scientific literature.

## 6.2.5   COMMON DISEASES

*Infectious stomatitis.*   Commonly called "mouth rot," infectious stomatitis is a bacterial infection that occurs in lizards secondary to stress and immunosuppression. Animals kept in substandard environments (low temperatures, crowded unclean cages) or fed inappropriate diets can become severely immunosuppressed and susceptible to infection by bacteria such as *Aeromonas* and *Pseudomonas*. Lesions begin as reddened areas in the oral cavity. These can progress to ulcerations with caseous deposits. Affected animals experience severe pain and will not eat. The disease can progress invading the bone and causing infection and degeneration. Treatment consists of aggressive antibiotic therapy combined with good supportive care and nutrition. Recovery is usually prolonged. Mouth rot can be prevented by ensuring adequate husbandry, sanitation, and nutrition.

*Amebiasis.*   Amebiasis is a protozoal disease affecting lizards and other reptiles. *Entamoeba invadens* is the responsible organism. Affected animals may regurgitate, stop eating, lose weight, become dehydrated, and have bloody diarrhea. Treatment consists of administration of appropriate antiprotozoal drugs and supportive care. All cages and equipment must be thoroughly disinfected or replaced. Good hygiene will help prevent spread of the disease.

*Metabolic bone disease.*   Metabolic bone disease (MBD) is commonly seen in young iguanas and chameleons fed a diet low in calcium (e.g., primarily insects) and not provided with appropriate wavelengths of light to synthesize vitamin $D_3$. As calcium-deficient animals attempt to maintain appropriate serum levels of calcium, their bodies remove calcium from the bones. The bones weaken, and the jaw becomes swollen and rubbery (rubber jaw). The earliest symptom of MBD is a lack of truncal lifting—normal iguanas lift their trunk and proximal tail off the ground when walking. Early in MBD, iguanas lift the front half of the body and drag their pelvis and tail. Eventually the lizard will be unable to move. Limbs may break from normal activity and vertebral fractures and abnormal curvatures of the spine may develop. Treatment with calcium and vitamin D supplementation and fracture fixation is not always rewarding. This disease is best prevented by ensuring that young iguanas are fed a high-quality diet with appropriate levels of calcium and that they have appropriate exposure to full spectrum light.

*Thermal burns.*   Any lizard that is allowed access to a radiant heat source may develop thermal burns. Burns are more likely if the ambient cage temperature is below the animal's thermal preference. Cold lizards will sit against a heat source in an attempt to warm themselves and will not move even when the skin begins to blister. Burns that become infected with *Pseudomonas* are particularly difficult to manage. All light sources used to provide radiant heat should be covered with a screen or placed outside the enclosure to help prevent thermal burns. "Hot rocks" are generally not recommended because these may malfunction, resulting in burns.

## 6.2.6   ZOONOSES

Salmonellosis in humans has been associated with handling reptiles including lizards. *Salmonella* can be carried in the gastrointestinal tract of lizards. When stressed, lizards may shed the organism in their feces. Humans can become infected when handling lizards or their contaminated cages and equipment. The disease can cause vomiting, diarrhea, and fever in humans. Very young, old, or immunosuppressed individuals are at greater risk of developing salmonellosis. Elimination of the carrier state from reptiles by antibiotic treatment has not been successful. People should wear gloves or thoroughly wash hands after handling lizards. Persons at increased risk should avoid handling reptiles.

## 6.3  PET SNAKES

### 6.3.1  INTRODUCTION

Snakes, like lizards, turtles, and crocodilians, are members of the class Reptilia. Snakes and lizards are the most closely related groups of reptiles and are collectively referred to as squamates. There are many colorful and interesting species of snakes. Among the ones more commonly kept as pets are boas, pythons, garter snakes, king snakes (including milk snakes), rat snakes (including corn snakes), and hognose snakes. The three basic groups of venomous snakes are vipers (e.g., rattlesnakes, copperheads, gaboon vipers), elapids (e.g., cobras, coral snakes, mambas), and some colubrids (e.g., boomslangs, brown tree snakes, mangrove snakes). Vipers have large, erectile fangs located near the front of the upper jaw. Elapid fangs are in the same location but are smaller and fixed. Venomous colubrids are referred to as rear-fanged snakes, with these enlarged teeth located farther back in the upper jaw. In general, viper venom causes tremendous tissue damage, whereas elapid venom attacks the central nervous system. Many species of both vipers and elapids can deliver fatal bites. Rear-fanged snakes tend to be more mildly venomous; however, some of these species can also inflict lethal bites. People are fascinated by venomous snakes, and some individuals keep them as pets. Venomous snakes require special permits, caging, and handling practices and should be kept only by trained professionals. Even nonlethal bites can be extremely painful and can cause tremendous tissue damage and debilitation.

### 6.3.2  COMMON PET SNAKES

*Boas.*  Boas are generally natives of Central and South America, although a couple of species are found in North America. The boa constrictor (*Boa constrictor*) is one of the most popular pet snakes as it tames quickly and can be bred in captivity. Colors vary from light brown to yellow or orange with dark bars and markings (Figure 6.14 and Color Plates 8D and 8F). Boa constrictors can grow to 18 ft in length. They are excellent swimmers and in nature are found in trees and on the ground. The two subspecies most often found in pet stores are the Mexican red-tailed boa and the Peruvian red-tailed boa.

*Pythons.*  Pythons are large snakes. Ball pythons (*Python regius*) are considered to be a good snake for beginning hobbyists. The name is derived from the snake's habit of rolling itself into a ball and tucking its head inside its coils when it feels threatened. Ball pythons are 3 to 5 ft long and most are docile pets. Burmese pythons (Figure 6.15) can reach over 20 ft in length and weigh more than 200 lbs; these are powerful snakes and are not recommended as pets for novice owners. Reticulated pythons are also large and have poor temperaments making them potentially dangerous to keep as pets.

*Garter snakes.*  Garter snakes (see Color Plate 8A) are widespread throughout North America. The common garter snake (*Thamnophis sirtalis*) and others can be tamed and generally adapt well to captivity. These snakes have a dark body; some are solid colored and others are striped or have a checkerboard appearance.

**Figure 6.14** Boa constrictors are beautiful snakes that grow fast and need lots of room. *Courtesy Michael J. Adkesson, DVM.*

***King snakes and milk snakes.*** Beautiful and relatively docile, king snakes and milk snakes can be found throughout the United States as well as the southern parts of Canada and parts of South America. Adults may reach 6 to 7 ft in length and live to 20 years. They are constrictors, suffocating their prey before eating. Many have color bands of red, black, and yellow (Figure 6.16, see also Color Plate 9M). They are members of the genus *Lampropeltis*.

***Rat snakes and corn snakes.*** Generally docile, rat snakes and corn snakes can be excellent pets. These snakes have also been called chicken snakes. Adults are 3 to 5 ft long and live 10 to 15 years. Several species are brightly colored (Figure 6.17) although some are primarily black. Rat and corn snakes are members of the genus *Elaphe*.

***Hognose snakes.*** Hognose snakes are medium-sized, with adults reaching approximately 46 inches in length. When threatened these snakes inflate their bodies and fatten their heads; they will strike at intruders but rarely bite. Their name was given because of their characteristic pointed, upturned nose (Figure 6.18). The three varieties found in the

**Figure 6.15** Burmese pythons are light brown with dark blotches. These powerful snakes are usually gentle but are not recommended as pets for beginning hobbyists. *Courtesy Michael J. Adkesson, DVM.*

**Figure 6.16** King snakes are generally docile and easy to handle. *Courtesy Jeff Rathmann (photographer) and James Brumley, Exotic A.R.C., St. Louis, MO.*

United States are the eastern hognose snake (*Heterodon platyrhinos*), southern hognose snake (*Heterodon simus*), and western hognose snake (*Heterodon nasicus*).

### 6.3.3 BIOLOGY AND BEHAVIOR

All snakes share a common, elongated body and no limbs. Size is remarkably variable, ranging from very small species such as the red-bellied snake (usually less than 12 inches long) to large pythons such as the reticulated python (individuals can reach well over 20 ft long). Snakes have smooth, scale-covered skin. They lack eyelids, but a modified scale, called the spectacle, covers and protects the eyes. Snakes shed their skin (ecdysis) in cycles, and the entire outer layer is shed as a single piece (including the spectacles). A patchy shed is often indicative of environmental problems or underlying disease. Approximately 2 weeks prior to a shed, a lymphlike fluid is produced between the old and new layers of skin. This gives the snake an opaque hue most visible over the eyes.

Snakes have a forked tongue, which withdraws into a sheath at the front of the mouth. The tongue is used to pick up scent molecules and deposit them into the vomeronasal organ, which is an olfactory organ located in the roof of the mouth. Normal, alert snakes flick their tongues frequently while exploring their environment.

Most snake teeth are quite small; however, boas and pythons have long, inward-curving teeth that are used for holding prey. These teeth can inflict painful bites.

The opening of the trachea in snakes is located immediately behind the tongue sheath in the front of the mouth. This adaptation allows the animal to breathe freely while holding and swallowing large prey.

**Figure 6.17** Corn snakes are good climbers and will vibrate their tails and strike if cornered. *Courtesy Michael J. Adkesson, DVM.*

**Figure 6.18** Hognose snakes received their name because of their pointed, upturned nose. *Courtesy Jeff Rathmann (photographer) and James Brumley, Exotic A.R.C., St. Louis, MO.*

Most snake species have an elongated right lung (the left lung is rudimentary) that may end in an air sac. Other organs, like the liver and kidneys, are also elongated. The heart has three chambers, and there is a ventral tail vein that is easy to access in some species.

Snakes tend to be solitary by nature. Compatible animals can be housed together but must be separated when feeding. Several species, such as king snakes, will eat other snakes (including members of their own species); therefore, these animals should be housed alone.

## 6.3.4 HUSBANDRY

*Cages.* Snakes must be housed in cages with tightly fitted lids as virtually all species are escape artists (Figure 6.19). Lids should be free of sharp edges and protrusions to prevent damage to the snake's nose and body. Glass aquaria are most commonly used for snake housing; however, fiberglass and Plexiglas cages also work well. As a general rule, the diagonal of the enclosure should be approximately the same length as the snake. A variety of substrates can be used for the flooring. Hardwood shavings afford burrowing species the opportunity to

**Figure 6.19** Lids on snake cages should be fitted securely to prevent escape. *Courtesy Dr. Dorcas P. O'Rourke.*

**Figure 6.20** Snakes can be housed on hardwood shavings, which allows them to burrow and engage in natural behaviors. This is a grey band king snake on aspen shavings. *Courtesy Dr. Dorcas P. O'Rourke.*

engage in this behavior (Figure 6.20). Indoor/outdoor carpet provides traction, which facilitates locomotion. Newspaper and brown paper, although not as attractive, make adequate substrates. Specialized substrates such as sand may be preferred by desert species. Aromatic softwood shavings such as pine should be avoided.

Snakes are secretive by nature and require a hide box or other type of retreat. Commercially available plastic hide boxes, pieces of bark, and large terra cotta pot fragments are examples of retreats. Species that spend time climbing such as rat snakes should be provided with branches, dowels, or elevated resting surfaces.

*Water.* Water should be provided in bowls large and sturdy enough for the snake to crawl into and soak (Figure 6.21). Soaking is especially important prior to shedding. Water bowls should be cleaned frequently to prevent buildup of harmful bacteria such as *Pseudomonas.*

*Temperature.* Like other reptiles, snakes are ectotherms (poikilothermic) but have a preferred temperature zone in which physiological processes such as digestion, reproduction, and optimal immune function occur. Most snakes prefer warm temperatures and do well if kept about 80°F. Snakes will not eat if the ambient temperature is too cool. Focusing a low-wattage incandescent bulb on part of the cage will create a thermal gradient and allow the animal to engage in behavioral thermoregulation. Care must be taken to prevent the snake from coming into direct contact with the heat source or serious burns can result. For this reason, if "hot rocks" are used, they should be buried under a nonflammable substrate. Heat tapes or lamps provide a more reliable and safer source of heat.

**Figure 6.21** A rat snake soaking in a water bowl. *Courtesy Dr. Dorcas P. O'Rourke.*

*Humidity.*  An average relative humidity of about 50% is acceptable for most snakes. Low relative humidity can result in desiccation and difficult shedding, whereas too much moisture can result in blister disease. Higher levels of humidity can be helpful during shedding.

*Cleaning.*  Cage cleaning frequency varies with species and amount of waste generated. Snakes will explore and frequently defecate when placed into a clean cage thereby marking their territory. Cages should be cleaned frequently enough to prevent accumulation of fecal material and pathogens while allowing for territorial marking. A diluted chlorine bleach solution can be used for cleaning. Cages should be thoroughly rinsed and air dried before replacing animals. Phenolic compounds are toxic to snakes and should never be used for cleaning cages or animal rooms.

*Feeding.*  All snakes are carnivorous. Diets vary widely and can range from insects and earthworms to frogs, fish, rodents, and rabbits. Many species are nonselective whereas other species are quite fastidious and require one specific prey item. Reference texts should be consulted to determine the appropriate prey for the species being kept.

Prey items should be fed nutritionally complete diets to ensure they are providing snakes with appropriate nutrients. Insects can be dusted with calcium powder or "gut loaded" (fed high vitamin diets immediately prior to feeding to the snake) to ensure that the snake is getting the required levels of vitamins and minerals.

Most snakes will readily eat prekilled prey items. Frozen prey should be thawed and allowed to warm prior to feeding as cold food will putrefy and not be digested. Certain individual animals will eat only live food. If feeding live rodents, the snake should be monitored until the rodent is killed and consumed. Snakes that do not feed will often not defend themselves, and rodents can subsequently gnaw through the snakes' flesh. Snakes are particularly susceptible to injury during the shedding process when it is normal for them to be anorexic (not eating). Rodent bites can be difficult to manage and, in many cases, fatal to the snake.

Snakes swallow their prey whole (Figure 6.22). Unique body structures (flexible lower jaw ligaments, joints, and attachments) allow them to open their mouths extremely wide and ingest prey items often larger than their heads.

**Figure 6.22**   Snakes swallow their prey whole. This pit viper is being hand fed. *Courtesy Gonzalo Barquero, Brazil, and Dr. James E. Corbin.*

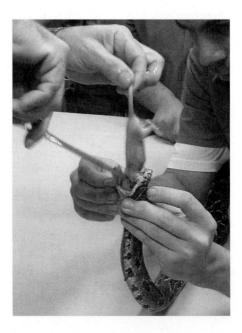

Following feeding, snakes should be disturbed as little as possible or they may regurgitate as a defensive mechanism. Depending on the snake species, food items can require several days to digest. Many snakes do well if fed approximately once every 2 weeks; young and highly active species may require more frequent feeding.

*Handling.*   When snakes are handled, the body should be supported as much as possible. Most snakes can be approached calmly and picked up gently. Even nervous animals will generally calm down if carefully supported in this manner. If an animal must be restrained behind the head, it should be held with a firm but gentle grasp behind the jaws (Figure 6.23). Excess or harsh restraint will cause the animal to become agitated and more prone to strike. The snake's demeanor should be assessed before picking it up; aggressive or frightened animals will often assume a defensive posture. Snake hooks are quite useful for transferring animals from one enclosure to another. Snake tongs can cause damage if used improperly and should only be used in extreme cases by trained individuals.

*Breeding.*   Reproductive strategies vary with species. Most species require a period of physiological rest (dormancy or hibernation) with decreases in temperature and day length and a period of fasting prior to breeding. Water is provided throughout the period of dormancy. The dormancy ends when the temperature is gradually increased and feeding resumes. All male snakes have paired reproductive organs, called hemipenes. Boas and pythons have spurs, which are vestiges of hind limbs. In general, spurs are larger in males than in females. The most accurate method of sexing snakes is a technique known as probing. A blunt instrument is gently inserted into the genital opening near the tail (Figure 6.24). Females can be probed to a depth of approximately 2 to 6 ventral scales and males to a depth of 7 to 15 ventral scales. Many snakes lay eggs, although several species bear live young (e.g., boas, garter snakes, water snakes, ribbon snakes). Eggs are generally deposited in the cage and abandoned. Some species will coil around the eggs and others (some pythons) will actually shiver thereby increasing body temperature and incubating the eggs. For most snake species, the eggs can be removed after deposition and artificially incubated in a substrate such as vermiculite or sphagnum moss. Unlike bird eggs, reptile eggs should not be rotated during incubation and care should be taken to maintain the same orientation when recovering and placing them into a new container.

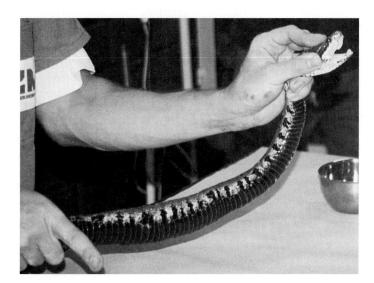

**Figure 6.23** Snakes should be restrained by firmly grasping the head behind the jaw and supporting the rest of the snake's body. *Courtesy Gonzalo Barquero, Brazil, and Dr. James E. Corbin.*

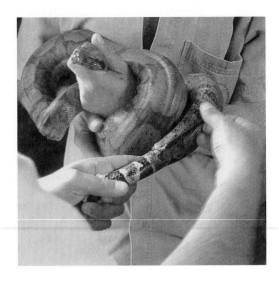

**Figure 6.24** Sexing of a male boa constrictor; the probe inserted into the genital pore extends for a distance of 10 ventral scales. *Courtesy Gonzalo Barquero, Brazil, and Dr. James E. Corbin.*

## 6.3.5 COMMON DISEASES

*Infectious stomatitis.* Infectious stomatitis, or mouth rot, is a common disease of snakes that typically occurs secondary to suboptimal environmental conditions, poor nutrition, stress, and immunosuppression. *Aeromonas* and *Pseudomonas* are the bacteria commonly involved in this disease. Lesions begin as subtle to more obvious reddened areas in the oral cavity. As the disease progresses, caseous deposits and ulcerations can develop. Eventually, if untreated, the organisms can invade bones of the mouth. Snakes with infectious stomatitis are in much pain and will not eat. Treatment consists of removing the caseous deposits, aggressive topical and systemic antibiotic therapy, and excellent supportive care. Animals often must be tube fed. Recovery can be quite prolonged. The disease is prevented by providing appropriate temperature, clean environment, low stress, and adequate nutrition.

*Amebiasis and cryptosporidiosis.* Two protozoal diseases commonly affecting snakes are amebiasis and cryptosporidiosis. Amebiasis is caused by *Entamoeba invadens,* and cryptosporidiosis is caused by *Cryptosporidium.* Both diseases can cause loss of appetite and vomiting. Amebiasis is usually associated with bloody diarrhea, whereas cryptosporidiosis can cause a firm swelling in the snake's midbody region due to thickening of the gastric (stomach) wall. Amebiasis is treatable with metronidazole; there is no effective treatment for cryptosporidiosis.

*Inclusion body disease.* Inclusion body disease is a recently identified viral disease of snakes in the boa and python family. The responsible agent most closely resembles a retrovirus. Affected animals develop chronic regurgitation, disorientation, and paralysis. The disease is very contagious among boas and pythons, and no effective treatment is known.

*Parasites.* Snakes are susceptible to a variety of internal nematode parasites including the hookworm *Kalicephalus.* Infestations of this parasite can cause anorexia, intestinal ulceration, and anemia in snakes. Fenbendazole has been effective in treating *Kalicephalus* spp in snakes. The snake lungworm (*Rhabdias*) causes inflammation of the lower airways and is often complicated by a secondary bacterial pneumonia.

Ticks and mites are external parasites that are commonly found on snakes. Ticks can be manually removed. The most common mite affecting snakes is *Ophionyssus natricis.* The adult mites are black and feed at night. Mites can cause anemia and also interfere with shedding, especially of the spectacles. Ivermectin, topically or injected, has been recommended for mite treatment. The environment must also be treated to achieve eradication; a dilute solution of bleach will kill the mites. Treatment should be repeated a minimum of two times at monthly intervals to prevent reinfestation.

***Dysecdysis and retained spectacles.*** Abnormal shedding is most commonly a result of too low humidity in the snake's enclosure. Other causes include mites, ticks, and systemic diseases. Soaking the snake in a shallow water bath for several hours is usually effective; care must be taken to prevent drowning. Placing the snake in a pillowcase with a wet towel is another alternative. Retained spectacles will usually be shed with the next molt if mites are present and eliminated and proper humidity is maintained.

***Toxicosis.*** Pesticides have been associated with toxicosis in snakes. Organophosphate (OP) impregnated strips (e.g., dichlorvos strips) have sometimes been placed in or near snake enclosures to kill mites. However, this is dangerous to some species of snakes that are highly susceptible to OP poisoning. Signs of OP toxicity include tremors, spasms, paralysis, and death. Overdosing with ivermectin (used to kill mites and other parasites) and metronidazole (used to treat intestinal infections) can result in ataxia (incoordination), tremors, and paralysis. References should be consulted prior to using these compounds to be certain they are safe for that particular species.

***Health concerns.*** Thermal burns can be a significant problem in pet snakes. If the ambient temperature is too low, and a reachable heat source is offered, snakes will often maintain direct contact with the heat source until severe burns occur. These burns are easily infected with *Pseudomonas* and are extremely difficult to manage. Snakes should not be allowed to contact heat sources.

Snakes can develop gout from starvation, kidney disease, or inappropriate use of nephrotoxic antibiotics (e.g., aminoglycosides). Uric acid will form crystalline deposits in internal organs. Treatment is a combination of drug therapy, nutritional and environmental changes, and good supportive care. This disease can be very severe, and the prognosis is not good.

Obesity is a common problem in snakes due to overfeeding and a lack of exercise. Overweight snakes are predisposed to constipation, dystocia, atherosclerosis, and liver failure.

### 6.3.6 ZOONOSES

Snakes routinely carry *Salmonella* in their digestive tract. When stressed, the organism is shed in the feces. Humans generally become infected when handling animals or their contaminated equipment and cages. Signs of salmonellosis in humans can include vomiting, diarrhea, and fever. Immunosuppressed, old, and very young individuals are at greater risk of developing this disease. *Salmonella* cannot be effectively eliminated from snakes; therefore, appropriate precautions (gloves, thorough hand washing) should be taken when handling or caring for snakes.

## 6.4 PET TURTLES AND TORTOISES[5]

### 6.4.1 INTRODUCTION

Like crocodilians, snakes, and lizards, turtles are members of the class Reptilia. Collectively, turtles are known as chelonians. There are over 240 species of chelonians worldwide broadly

[5]Appreciation is expressed to John Cwaygel, president of Sailfin Pet Shop, Inc., Champaign, IL, for his contributions to this section.

divided into two groups: those which withdraw their heads into their shells in a vertical S shape (cryptodirans; hidden-neck turtles) and those who fold their heads in a sideways fashion onto their shoulders (pleurodirans; side-neck turtles). Many species are kept as pets. The most familiar of these are the aquatic sliders, painted, and map turtles. Less commonly, aquatic mud, musk, softshell, snapping, and various side-neck turtles are kept. Box turtles and tortoises are terrestrial species that are also enjoyed by the hobbyist.

## 6.4.2 COMMON PET TURTLES

Turtles are some of the easiest pets to keep, although different species will have different requirements for feeding and husbandry.

*Mud and musk turtles.* Members of the family Kinosternidae, mud and musk turtles are aquatic turtles with flattened shells, retractable heads, and webbed feet.

*Map turtles.* Map turtles are found in streams and rivers throughout the United States. The name was given because of the intricate maplike designs, or hieroglyphics, created by patterns on their upper shell or carapace (Figure 6.25 and Color Plate 9P). In the south, some species are also known as sawbacks. The scientific name for their genus, *Graptemys*, literally means "map turtle" in Greek.

*Painted turtles.* Painted turtles (*Chrysemys picta*) have an olive or black shell and olive, yellow, or red markings along the seams of their scutes (see Figure 6.25, see also Color Plate 9P). They also have red bars or crescents on the outermost scutes. Red and yellow stripes are present on the neck, legs, and tail. Adults are 4 to 9 inches long. They become tame in captivity and are good pets.

*Pond sliders.* Pond sliders (*Chrysemys scripta*) are one of the most common species sold by pet stores in the United States. There are several subspecies including the red-eared slider (*Chrysemys scripta elegans*). Adults range from 5 to 12 inches in length with the females being longer than males (Figure 6.26, see also Color Plate 9R).

*Softshell turtles.* Softshell turtles (*Trionychidae* spp) have a leathery carapace. They have paddle-like webbed feet and a snorkel-like nose. These turtles are active swimmers and like to bury in sandy gravel. They can become tame but have sharp jaws that inflict painful bites and should be handled with caution.

*Asian box turtles.* Asian box turtles (*Cuora*) include several varieties commonly sold as pets. These turtles all require high levels of humidity. Examples include the Malayan box turtle, yellow-margined box turtle, flowerback box turtle, three-striped box turtle, and keel box turtle.

**Figure 6.25** Map and painted turtles are aquatic turtles that use "haul out" areas to bask in the sun. *Courtesy Michael J. Adkesson, DVM.*

**Figure 6.26** A male red-eared slider. Males have long claws that are used to stroke a female during courtship. *Courtesy Jeff Rathmann (photographer) and James Brumley, Exotic A.R.C., St. Louis, MO.*

*Native box turtles.* In the United States, native box turtles are members of the genus *Terrapene*. They include the common or Eastern box turtle (*Terrapene carolina carolina*), the Florida box turtle (*Terrapene carolina bauri*), and the Gulf Coast box turtle (*Terrapene carolina major*). Box turtles (see Figure 6.29, see also Color Plate 8E) have a domed shell with a hinged plastron and can totally withdraw their heads, legs, and tail within their shell for protection from predators.

*Tortoises.* A wide variety of terrestrial tortoises are kept as pets. Examples include the Chaco tortoise, desert tortoise, leopard tortoise (see Color Plate 9O), Mediterranean tortoise (Hermann's tortoise), pancake tortoise (Figure 6.27), red footed tortoise (redleg or Savanna tortoise; see Color Plate 9S), African spurred tortoise (Sulcata tortoise), and Russian tortoise (Horsfield's tortoise; Figure 6.28). These are predominantly herbivorous grazers that may live 50 or more years.

## 6.4.3 BIOLOGY AND BEHAVIOR

All turtles have bony shells, which are fused to their vertebrae and ribs. In most species, the shell is covered by scales called scutes. The upper shell is called the carapace and the lower shell is known as the plastron. "Hinges" are mobile sutures (seams) in the plastron and/or carapace. Turtle jaws are covered with horny plates of skin and referred to as beaks. Turtles can withdraw into their shells for protection when threatened. Species such as box turtles have hinges on their bottom shell which allow it to close and protect their heads and forelegs. Turtles lack teeth but have a horny beak that can tear off chunks of food. All turtles

**Figure 6.27** The pancake tortoise has a flat carapace that allows it to fit into narrow crevices in rocks. *Contributor requested anonymity.*

**Figure 6.28** The Russian tortoise (*Testudo horsfieldii*) is a small, easy-to-keep tortoise that is great for an indoor vivarium. *Courtesy Jeff Rathmann (photographer) and James Brumley, Exotic A.R.C., St. Louis, MO.*

have a three-chambered heart and a urinary bladder. Although most turtles are not social by nature, individuals of many species can be housed together. Care should be taken, however, because some males can be quite territorial and aggressive to other individuals.

## 6.4.4   HUSBANDRY

*Cages.* Turtles vary greatly in size, and cages should be large enough to accommodate the specific animal's needs. Glass aquaria work well for smaller turtles. Larger individuals can be kept in wading pools, livestock troughs, and pens (terrestrial species). If turtles are kept outdoors in pens or enclosures, fencing should extend deep enough below the surface of the ground to prevent digging and escape. If kept indoors, terrestrial species can be housed on shredded hardwood, chipped bark, or other substrate which allows the animal to dig. With the exception of softshell turtles (which bury themselves in the sandy substrate of the tank bottom), aquatic turtles do not require a substrate in their aquaria.

Turtles require a retreat or hiding place to decrease stress. Terrestrial species can be provided with deep substrate for burrow digging. Alternatively, hide boxes or large pieces of heat-treated bark can be used as retreats. Aquatic species benefit from live or artificial vegetation, rocks, or other underwater objects that can provide a visual barrier. It is critical to ensure that all structures are sturdy and will not trap the animal.

*Water.* Water in aquaria and tanks should be deep enough to allow the turtle to swim freely and right itself if turned over. Terrestrial species should have water provided in large, shallow, heavy bowls. The edge of the bowl must be low enough to allow the animal to crawl into the dish and soak. Many turtles will immerse their entire head (and sometimes body) in the water bowl when drinking. Water must be changed frequently to prevent accumulation of fecal material, uneaten food, and bacterial and fungal pathogens.

Some aquatic turtle species (e.g., sliders, painted turtles, map turtles) require a "haul out" area. These animals need to emerge from the water to bask and dry (see Figure 6.25). A platform with a gentle slope into the water will provide a means to easily crawl out. Placement of a low-wattage radiant heat source above the basking site will encourage the animal to engage in this critical behavior. Likewise, a basking site should be provided for terrestrial species. In all cases, the animal must not be able to contact the heat source or serious burns can result. Provision of a thermal gradient is important for turtles to engage in behavioral thermoregulation.

*Heat.* Turtles, like all reptiles, are ectotherms (poikilothermic) and engage in behavioral thermoregulation to maintain their body at the preferred body temperature. They will seek warm surfaces to gain heat by conduction and will bask in sunlight (or heat lamps) to gain radiational heat. If too warm, they will seek a shady area. If body temperature increases above the preferred body temperature, the turtle may become stuporous (estivation). Low environmental temperatures may result in hibernation.

*Humidity.*  Normal relative humidity levels of 30% to 70% are adequate for many turtle species. Some turtles prefer increased humidity, and desert species require much drier environments.

*Light.*  Many species of turtles require exposure to appropriate wavelengths of ultraviolet light to ensure adequate vitamin D metabolism. Full spectrum lights containing the necessary wavelengths of ultraviolet light (290 to 315 nm) should be placed within 18 inches of the animals. Because glass and most Plexiglas will absorb the critical wavelengths of ultraviolet light, the fixture should be positioned such that the light directly contacts the animals. Ultraviolet lights should be changed every 6 months to ensure the needed wavelengths are effectively produced.

*Cleaning.*  Turtle tanks and cages should be cleaned frequently enough to prevent buildup of waste materials and uneaten food. Even if filtration systems are used in aquaria, frequent water changes are necessary to prevent filter overload. Tanks and cages can be broken down and sanitized with a diluted chlorine bleach solution; however, care must be taken to thoroughly rinse and dry tanks and cages before replacing animals. Phenolic compounds are toxic to all reptiles and should never be used around turtles.

*Feeding.*  Different turtle species require distinctly different diets. Aquatic turtles such as sliders, painted, and map turtles will eat prekilled minnows, earthworms, floating fish pellets, or high-quality commercial diets. If commercial diets are fed, careful attention must be paid to nutritional contents to ensure that appropriate levels of protein, calcium, vitamin A, and other essential nutrients are provided. These species will also eat aquatic vegetation, especially when young. Aquatic turtles will only eat in water; moving them to a separate, easily cleaned pan of water for feeding is helpful in keeping the primary habitat clean and also allows owners and caretakers to more easily monitor food intake. Box turtles do well on a diet of canned or moistened high-quality dog food, fresh vegetables, and fruits and berries. Strawberries, tomatoes, and earthworms are particularly relished. Box turtles prefer yellow, orange, and red-colored foods. It is important that they be fed a variety of foods (Figure 6.29) to ensure that all nutritional needs are met. Vitamin A is essential for good health; excellent sources include kale, collard greens, carrots, squash, sweet potatoes, and cantaloupe. Misting the enclosure to simulate rainfall can stimulate activity and feeding.

Tortoise diets can vary considerably depending on species. Many herbivorous tortoises will eat fresh or frozen (must be thawed) mixed vegetables, dark leafy greens, and fruit. Most tortoises are healthiest when fed diets high in fiber and relatively low in protein. For all reptile species, appropriate texts must be consulted to determine the appropriate diet for the species being kept. Most turtles can be fed two to three times a week; young animals should be fed daily.

*Handling.*  Many turtle species can be handled by firmly grasping the sides of the shell. Turtles have claws and can scratch, and some will bite if they feel threatened. Species such as softshell, snapping, mud, musk, and side-neck turtles have very long necks and can reach around and bite the handler if restrained by the sides of the shell. These species should always be held by the back portion of the shell (between the hind legs) only. Turtles have no teeth; however, the horny beak can deliver painful and (in the case of snapping turtles) extremely serious bites.

**Figure 6.29** A variety of foods are available for consumption by these box turtles. *Courtesy Michael J. Adkesson, DVM.*

*Breeding.* Male turtles can be distinguished from females in some species. Adult male sliders have long claws on their forelegs (see Figure 6.26), which are used during courtship. Male box turtles and some male tortoises have a depression in their bottom shell, which facilitates mating. All turtles lay eggs. Even aquatic species emerge from the water to dig nests and lay their eggs on land. Hatchlings have an egg tooth, much like a chicken's, which is used to break out of the shell. Newly hatched turtles appear as miniatures of their parents.

## 6.4.5 COMMON DISEASES

*Metabolic bone disease.* Metabolic bone disease is commonly seen in young turtles fed diets low in calcium and/or deprived of exposure to ultraviolet light of appropriate wavelength. Without adequate dietary vitamin D and calcium, the turtle's body begins to resorb calcium from the bones, most notably from the shell. Affected animals typically have pliable, "soft" shells and are vulnerable to fractures and other injuries. Treatment consists of calcium supplementation, exposure to full spectrum light, and dietary correction. If caught early, the disease can be corrected; however, advanced cases have a poorer chance of recovery. Metabolic bone disease is best prevented by feeding appropriate diets and ensuring adequate exposure to full spectrum light (290–315 nm).

*Vitamin A deficiency.* Another problem seen predominantly in young turtles fed diets containing mostly insects or poor quality fish is vitamin A deficiency. Insufficient levels of vitamin A result in damage to cells lining the respiratory tract and the eyelids. Affected animals have swollen eyes and a discharge from the nose and eyes. Bacteria can invade the damaged tissues and result in serious infection and pneumonia. Treatment includes vitamin A injections, antibiotics if secondary bacterial infections are present, and correction of the diet to ensure adequate levels of vitamin A. Topical antibiotics may be needed to treat eye infections. Overdosing with vitamin A is also harmful and can result in dry scaly skin, blisters, and liver damage. The safest way to ensure adequate levels of vitamin A is to feed diets containing beta-carotene rich vegetables, vegetation, or pellets containing these ingredients. Diets containing fresh whole fish (e.g., minnows) are also suitable for feeding aquatic turtles.

*Respiratory tract infections.* Clinical signs of upper respiratory tract disease (URTD), common in turtles, include nasal discharge, bubbling from the nares, poor appetite, and dehydration. *Mycoplasma* and a variety of bacteria can be isolated from turtles with URTD. Infectious organisms are spread by direct and indirect contact with nasal exudates. Treatment involves flushing the nasal cavity and systemic antibiotics. Pneumonia is a common problem in newly acquired turtles. Crowded conditions in holding facilities, compromised water quality, chilling, poor nutrition, and parasitism weaken the immune system, predisposing them to pneumonia. Clinical signs may include anorexia, lethargy, wheezing, sneezing, mouth gaping, and nasal exudates. Attention to husbandry and good nutrition as well as antibiotic therapy are necessary for recovery.

*Shell fractures.* Shell fractures can result from improper handling, dog bites, or other trauma. When shell damage occurs, it is critical to assess the level of internal injuries and infection in underlying tissues. The shell wound should be thoroughly flushed and cared for as an open wound (bandages, antibiotics) until it is certain no infection is present. Repair can then be accomplished by stabilizing with surgical wires or bonding with epoxy and fiberglass.

*Middle ear infections.* Middle ear infections are common in box turtles. The eardrum (tympanic membrane) appears swollen and the turtle becomes anorexic. A variety of bacterial organisms may be involved; these are thought to ascend to the middle ear through the Eustachian tube (Chapter 11). Treatment involves perforating (incising) the tympanic membrane, flushing the middle ear cavity with an antiseptic, and using both topical and systemic antibiotics.

### 6.4.6 ZOONOSES

Turtles, like all reptiles, can transmit *Salmonella* to humans. The organism is carried in the turtle's intestinal tract and is shed in the feces during periods of stress. Transmission occurs through handling infected animals, their cages, or other contaminated objects. *Salmonella* can cause vomiting, diarrhea, and fever in humans. Immunosuppressed, old, and young individuals are at greatest risk. Wearing gloves and thorough hand washing are the best means of preventing salmonellosis. Very young children (less than 5 years old), pregnant women, and other persons at risk (e.g., those with AIDS or receiving immunosuppressive drugs) should avoid handling turtles and other reptiles.

## 6.5 SUMMARY

Interest in keeping reptiles and amphibians as pets is growing. Reptiles and amphibians vary considerably in their environmental, behavioral, nutritional, and other requirements. These species, even those that are captive-bred and -raised, are still essentially wild animals that do best when provided with a habitat similar to the one in which the species evolved. Failure to meet these needs is the major cause of disease in pet reptiles and amphibians. Basic information on the biology, behavior, husbandry (care and management), common diseases, and zoonotic concerns associated with major classifications of reptiles and frogs is discussed in this chapter. Many species were illustrated in this chapter, and pictures of others can be viewed in Color Plates 2, 8, 9, and 11. Owners of reptiles and amphibians should become informed about these pets, including the species' place of origin and natural history. Understanding an animal's natural history is helpful in determining the best way to house and feed it. Pet owners seeking more information can consult the listed references, a member of the Association of Reptilian and Amphibian Veterinarians (http://www.arav.org), or a local herpetology society.

## 6.6 REFERENCES

1. Beynon, Peter H., Martin P. C. Lawton, and John E. Cooper, eds. 1992. *Manual of Reptiles*. Ames: Iowa State Press.
2. Boyer, Thomas H. 1993. *A Practitioner's Guide to Reptilian Husbandry and Care*. Lakewood, CO: American Animal Hospital Association.

3. Boyer, Thomas H. 1998. *Essentials of Reptiles, A Guide for Practitioners*. Lakewood, CO: American Animal Hospital Association Press.

4. Mader, D. R., ed. 1996. *Reptile Medicine and Surgery*. Philadelphia: W. B. Saunders.

5. McArthur, Stuart. 1996. *Veterinary Management of Tortoises and Turtles*. Oxford, England: Blackwell Science.

6. Meredith, Anna, and Sharon Redrobe, eds. 2002. *BSAVA Manual of Exotic Pets*. 4th ed. Quedgeley, Gloucestershire, England: British Small Animal Veterinary Association.

7. O'Rourke, D. P., and J. Schumacher. 2002. Biology and Diseases of Amphibians. In *Laboratory Animal Medicine*, 2nd ed., edited by J. G. Fox, L. C. Anderson, F. M. Loew, and F. W. Quimby, 793–823, 827–861. San Diego, CA: Academic Press.

8. Rossi, John V., and Roxanne Rossi. 1995. *Snakes of the United States and Canada: Keeping Them Healthy in Captivity*. Malabar, FL: Krieger Publishing.

9. Wright, Kevin M., and Brent R. Whitaker, eds. 2000. *Amphibian Medicine and Captive Husbandry*. Malabar, FL: Krieger Publishing.

# 6.7   ABOUT THE AUTHOR

Dr. Dorcas P. O'Rourke is Attending Veterinarian for the University of Tennessee, Knoxville's animal care and use program. She is Director of the Office of Laboratory Animal Care and Associate Professor in the Department of Comparative Medicine at UT's College of Veterinary Medicine. Dr. O'Rourke received her B.S. in Zoology and M.S. in Neuroanatomy from Louisiana State University. She also was awarded her DVM and residency certification in Laboratory Animal Medicine from that same institution. Dr. O'Rourke is board certified by the American College of Laboratory Animal Medicine. She is a member of the Association for Assessment and Accreditation of Laboratory Animal Care International (AAALAC) Council on Accreditation and of the Board of Trustees of the Scientists Center for Animal Welfare. Dr. O'Rourke has authored chapters on anesthesia and analgesia of nontraditional species, amphibian and reptile biology and medicine, and rodent diseases in laboratory animal medicine and exotic pet texts. She has been co-editor of publications on reptile and amphibian biology and care and research in zoos. Dr. O'Rourke has given numerous national presentations on a variety of topics including field research, amphibian and reptile care, and environmental enrichment for reptiles, amphibians, and fish.

# 7 COMPANION RODENTS, FERRETS, AND LAGOMORPHS

*Douglas W. Stone, John E. Harkness, and Steele F. Mattingly[1]*

[1]Please turn to the end of the chapter for complete author affiliation information.

*Pets upgrade our quality of life, bring us closer to nature, provide companionship and emphasize the fact that animals are readily accepted as desirable participants in society.*

Boris M. Levinson (1907–1984)
*American child psychologist*

Numerous species of rodents and lagomorphs make significant contributions to the health and well-being of people—some through their invaluable role in the advancement of science and medicine; others through their priceless relationship as companion animals in enriching the lives of children and adults worldwide. In this chapter, three eminent, scholarly professionals discuss eight species that serve both of the above important fundamental functions. Color photographs illustrating many of these species can be seen in Color Plates 1, 2, 10, and 11.

*Campbell-Corbin-Campbell*

# 7.1    CHINCHILLAS

## 7.1.1   INTRODUCTION

The chinchilla is a hystricomorph rodent related to the guinea pig. There are two species of chinchilla, *Chinchilla laniger* and *C. brevicaudata* (Table 7.1). The *C. brevicaudata* differs in having smaller ears and a shorter tail. *C. brevicaudata* have 20 tail vertebrae, whereas the *C. laniger* have 30. *C. laniger* is the most common. Adult males are called bucks, the females are does, and the young are kits. The chinchilla is a native of the Andes mountains in South America living at elevations of 10,000 to 15,000 ft. In their natural habitat, chinchillas dwell in rocky burrows and crevices in groups of up to 100 animals. Family groups of 2 to 5 share the same burrow. Their diet consists of available vegetation. Few chinchillas remain in their natural habitat. Their valuable furs made them a target of trappers who hunted them to near extinction by 1917. The first chinchillas in the United States were 11 animals imported by Mathias F. Chapman from South America in 1923. All chinchilla furs sold in the United States today originate from animals raised commercially. The *C. brevicaudata* has been listed as an endangered species by the United States Endangered Species Act since 1976.

## 7.1.2   BIOLOGY/HUSBANDRY

Chinchillas are clean and nearly odorless. They are friendly and rarely bite. These characteristics make them excellent pets (Figure 7.1). Their fur is soft and dense; their natural coat color is a smoky blue grey. Color variations are numerous, ranging from white, silver, beige, and black (see Color Plates 10A and 10G). They have a broad head (Figure 7.2) with a large external pinna (the largely cartilaginous projecting portion of the external ear). Chinchillas are often used in biomedical research of the human ear because their anatomical ear structure and hearing range are more similar to those of humans than any animal except primates. They are crepuscular (active in twilight) and nocturnal, but are also active during the day. Each foot has four digits with no fur on the palmar and plantar surfaces (soles). Their

**Table 7.1**
Biodata of Chinchillas

| | |
|---|---|
| **Order** | Rodentia |
| **Family** | Chinchillidae |
| **Genus** | *Chinchilla* |
| **Species** | *laniger* and *brevicaudata* |
| **Birth weight** | 1.0–1.8 oz (30–50 g) |
| **Weight, adult male** | 14–17 oz (400–500 g) |
| **Weight, adult female** | 4–21 oz (400–600 g) |
| **Body temperature** | 98.6°F–100.4°F |
| **Heart rate** | 100–150/min |
| **Respiratory rate** | 40–80/min |
| **Buck, age of puberty** | 8 months |
| **Doe, age of puberty** | 8.5 months |
| **Mammary glands** | 3 pairs |
| **Breeding season** | November–May |
| **Estrous cycle** | 41 days (30–50 range) |
| **Gestation** | 111 days (105–118 range) |
| **Litter size** | 1–5 kits (2 average) |
| **Weaning age** | 6–8 weeks |
| **Life span** | over 10 years |
| **Dental formula (upper/lower)** | 1/1 incisors, 0/0 canines, 1/1 premolars, 3/3 molars |

**Figure 7.1** Chinchillas are excellent pets for children because they have a soft coat, are friendly, and rarely bite. *Courtesy Questhavenpets.com.*

incisor teeth are yellow colored. As with other rodents, their teeth grow continuously throughout life with the incisors growing approximately 2 to 3 inches (5.0 to 7.5 cm) annually. Females have two pair of thoracic and one pair of inguinal mammary glands. The female is dominant and larger than the male. When threatened, chinchillas will chatter, sit erect, and urinate at the threat with pelvic thrusts.

Chinchillas are active animals and large cages are preferred. They can be group housed, but aggression may occur. They enjoy climbing and, if possible, cages should be tall enough to permit shelving to allow them to climb. Large pieces (4 to 5 inches in diameter) of polyvinyl chloride (PVC) pipe sections can be used for hiding. Either solid-bottom or wire-bottom caging can be used. When wire-bottom caging is used, the mesh should be at least 0.5 × 0.5 inches (15 × 15 mm) to prevent injury. Females with litters should be housed on solid-bottom caging. As a typical rodent, they will chew on most items. They enjoy taking dust baths. A pan large enough for the chinchilla to roll in, filled to a depth of 1 inch (2 to 3 cm) with silver sand (9 parts) and Fuller's earth (1 part), placed in the cage for a few minutes each day provides welcomed entertainment and improves coat condition. The dust should be discarded periodically to help prevent bacterial or fungal infections.

**Figure 7.2** A white chinchilla. Note the broad head and large ears that are characteristic of this species. *Courtesy Questhavenpets.com.*

The optimum temperature for chinchillas is 68°F. They can tolerate cold temperatures (as low as 35°F) better than warm ones. Temperatures above 80°F approach their upper comfort/tolerance level. Light cycles of 12 hours on and 12 off are recommended.

Chinchillas are "hind gut" fermenters, meaning that most fermentations occur in the posterior portion of the alimentary canal. Their measured intestinal tract is up to 11.5 ft (3.5 m) long. Their colon is highly sacculated (has a series of saclike expansions). Chinchillas normally practice coprophagy (eat their feces). They tend to be sensitive to dietary changes. Therefore, dietary changes should be made gradually over several days. A high-fiber diet is an important requirement of these herbivorous animals. A commercial chinchilla diet such as Mazuri® Chinchilla Diet containing 16% to 20% protein and 15% to 35% fiber is recommended. The diet can be supplemented with grass hay to add fiber. Water is best provided using glass bottles with sipper tubes. Chinchillas cannot vomit.

Restraint can be accomplished by grasping the base of their tail. For short distances, the animal can be carried by the base of the tail. Other means of carrying chinchillas include placing one hand under the abdomen while holding the base of the tail with the other. For more restricted restraint, gripping one hand over the back of the neck with the other hand supporting and holding the rear legs is recommended (Figure 7.3). When restraining, it is important not to grip the fur as it will pull out in patches (fur slip) requiring 6 to 8 weeks to regrow.

*Fractures.* Fractures of the tibia are common. These occur most often from inappropriate handling or from housing on suspended cage flooring.

## 7.1.3  BREEDING

Male chinchillas are sexually mature at 8 months of age, females at 8.5 months. However, because they are seasonal breeders, the breeding age varies depending on the time of year they are born. The breeding season is November to May in the Northern Hemisphere. Chinchillas are polyestrous with cycles averaging 41 days (range of 30 to 50). Their gestation is 111 days (range of 105 to 118). Sexing chinchillas is accomplished by measuring the anal genital distance (males have twice the length of females). No scrotum is present in males. Instead, the testicles remain in the inguinal canal, which requires surgical closure if they are castrated. The vagina of females is kept closed by a closure membrane except during estrus and parturition. The urogenital papilla of the female is distinct and can be mistaken for a penis. Females are larger and more aggressive than males and may fight when not receptive to the male. This is one reason many commercial chinchilla ranchers use a unique breeding system. This special system has several individually housed female cages side by side with a connecting tunnel system running behind and into each cage. The female is fitted with a

**Figure 7.3**  This 1-day-old chinchilla kit can be easily carried by holding it in cupped hands. *Courtesy Questhavenpets.com.*

metal or plastic collar around her neck that prevents her from entering the tunnel; the male has no collar. This permits the male to freely enter the cage of several single-housed females. If a female is aggressive, the male is free to escape.

Chinchillas do not make a nest for parturition. The average litter size is two (range one to five) kits with two litters possible annually. The kits are born precocious, fully furred, with eyes open, and weigh approximately 43 g. Kits can eat solid food within a few weeks and be weaned at 6 to 8 weeks of age.

## 7.1.4  DISEASES

*Gastrointestinal disorders.* Gastrointestinal disorders are the most common disease observed in chinchillas. This results in part from their sensitive digestive system. Constipation is a common problem and can be difficult to diagnose in animals housed in cages with bedding. Laxatives can be used for treatment. Prevention includes increasing the percentage of fiber in the diet. Diarrhea is also commonly seen, often being associated with feeding inappropriate diets, excess treats, or making sudden dietary changes.

*Bloat.* Bloat is seen in all ages of chinchillas and results in part from their inability to vomit. Symptoms include a swollen abdomen, lateral recumbency, and dyspnea (difficult breathing). Treatment requires emergency veterinary decompression. Gastric trichobezoars (hairballs) are also seen.

*Malocclusion.* As with all rodents, chinchillas have open-rooted, continuously growing teeth. Malocclusion of their incisors is a commonly encountered problem. Malocclusion of molars growing across the mouth, thus obstructing the tongue, may also occur. Clinical signs include anorexia (not eating) and excessive salivation (slobbers). Examination of the molar teeth requires viewing the entire mouth cavity using an otoscope or other similar device. Holding the mouth open can be accomplished using gauze strips over the upper and lower incisors while a second person examines the mouth. When malocclusion occurs, the animal will need its teeth trimmed at monthly intervals throughout the animal's life. Because this is hereditary, animals with malocclusion should not be bred.

*Integumentary diseases.* *Dermatophytosis.* Dermatophytosis (ringworm) is a commonly observed disease of chinchillas. *Trichophyton mentagrophytes* is the most common causative agent, but *Microsporum canis* and *M. gypseum* can also cause dermatophytosis. Symptoms include scaly patches of alopecia (baldness) typically at the nose, behind the ears, and on the front feet.

*Bite wounds.* Group-housed animals will fight at times, particularly when females are not receptive to the male and the male is unable to escape. Common bacteria associated with bite wounds include *Streptococcus* spp and *Staphylococcus* spp.

*Fur rings.* Male chinchillas are prone to paraphimosis (prolapsed penis), which occurs when a ring of fur forms around the penis. This fur ring can prevent the penis from retracting into the prepuce and result in urinary obstruction and death.

*Bacterial diseases.* Chinchillas are susceptible to a large number of bacterial diseases. Bacterial agents causing pneumonia include *Pasteurella* spp, *Bordetella* spp, and *Streptococcus* spp. Gastrointestinal bacterial pathogens include *Clostridium* spp, *Escherichia coli*, *Proteus* spp, *Pseudomonas* spp, *Salmonella* spp, *Listeria monocytogenes*, and *Yersinia pseudotuberculosis*. *Listeria* can cause numerous disease syndromes including gastrointestinal disorders, abortion, encephalitis, and septicemia.

Chinchillas are highly susceptible to *Listeria monocytogenes*. Both neurological and gastrointestinal forms of infection may occur. Neurological signs include ataxia (uncoordinated movements), circling, and convulsions. Antibiotic treatment is usually ineffective once clinical signs are observed.

*Conjunctivitis.* Dust baths often contribute to conjunctivitis (inflation of eyelids). Topical treatment is usually effective. Dust baths should be discontinued until the malady is resolved.

*Parasitic diseases.* Chinchillas are susceptible to a number of parasitic diseases including *Cryptosporidium* spp and *Giardia* spp, which are often found in fresh fecal smears and can be associated with diarrhea. Debate continues as to whether *Giardia* are part of the normal flora or a pathogen.

*Neoplasia.* There are few reports of neoplasms (cancer) in chinchillas despite their longevity.

### 7.1.5 ZOONOSES

Chinchillas are subject to numerous diseases of public health significance. These pathogens include *Lymphocytic choriomeningitis*, *Listeria monocytogenes*, *Trichophyton mentagrophytes*, *Microsporum canis*, *M. gypseum*, *Baylisascaris procyonis*, and rabies.

## 7.2 GERBILS

### 7.2.1 INTRODUCTION

Although there are about 90 species of gerbils, this discussion will focus on the Mongolian gerbil, *Meriones unguiculatus*, the most common pet gerbil and first described in the literature in 1867 (see Color Plate 10K). It is also referred to as the jird, clawed jird, sand rat, or desert rat. It is native to the desert regions of northeastern China and eastern Mongolia. The most common coat color is agouti (light brown with individual hairs having a grey base and black tip); however, black, grey, white, dove, cinnamon, piebald, and hairless varieties are also seen. Gerbils were first brought to the United States in 1954 for medical research. A breeding colony was established at Tumblebrook Farms in Massachusetts. They are used as animal models for brucellosis, tuberculosis, stroke, leprosy, rabies, and rickettsial disease research. It is illegal to possess gerbils as pets in the state of California. Biodata of neonatal and older gerbils are presented in Tables 7.2 and 7.3.

### 7.2.2 PHYSIOLOGY, BIOLOGY, AND BEHAVIOR

Gerbils are active, friendly animals (Figure 7.4 and Color Plate 10K). Although gerbils are friendly to people, adult gerbils do not tolerate new cage mates and will fight to death. They

Table 7.2
Biodata of Neonatal Gerbils

| Age | Appearance |
| --- | --- |
| 1 day | Eyes closed, ears closed, hairless |
| 5–7 days | Hair begins to appear, ears open |
| 7–10 days | Nipples appear on female pups, not present on males |
| 10–16 days | Incisors erupt |
| 14 days | Fully haired |
| 16–20 days | Eyes open |
| 21–24 days | Weaning age |

**Table 7.3**
Biodata of Gerbils

| | |
|---|---|
| Order | Rodentia |
| Superfamily | Muroidea |
| Family | Cricetidae |
| Genus | *Meriones* |
| Species | *unguiculatus* |
| Birth weight | 0.08–0.1 oz (2.5–3.0 g) |
| Weight, adult male | 2.3–4.8 oz (65–135 g) |
| Weight, adult female | 2–3 oz (55–85 g) |
| Body temperature | 98.6°F–101.3°F |
| Heart rate | 360/min |
| Respiratory rate | 90/min |
| Food consumption, adult | 0.2–0.3 oz (5–8 g) daily |
| Water consumption, adult | 0.1–0.2 oz (4–7 ml) daily |
| Puberty, male | 6–7 weeks |
| Puberty, female | 6–7 weeks |
| Breeding age, male | 11–12 weeks |
| Breeding age, female | 10–12 weeks |
| Mating scheme | 1:1 to 1 male:3 females permanent pairing |
| Mammary glands | 4 |
| Estrous cycle | 4–6 days |
| Gestation | 24–26 days |
| Litter size | 3–7 pups |
| Weaning age | 21–24 days |
| Urine production | drops/day |
| Preferred light cycle | 12–14 hours/day |
| Relative humidity | 30% |
| Temperature | 60.8°F–80.6°F |
| Life span | Over 4 years |

are crepuscular and diurnal, staying awake during the day, but also active at night. In their natural habitat, gerbils live in a dry, harsh environment with temperature extremes ranging from −58°F (−50°C) in the winter to 77°F (25°C) in the summer, where rainfall is often brief and infrequent. In this environment, the gerbil must adapt to wide temperature variations and conserve water efficiently. Gerbils normally burrow and spend a considerable amount of time attempting to burrow in captivity. Audible foot stomping is commonly seen when they are alarmed or excited.

**Figure 7.4** This Mongolian gerbil has the typical agouti coat color; hairs are light brown with a grey base and black tip. *Courtesy Dr. Douglas W. Stone.*

Anatomically, gerbils are unique among rodents in that they have an incomplete Circle of Willis on the ventral surface of the brain. The Circle of Willis is a collection of arteries that come together at the base of the brain in most animals. This circle of arteries ensures a constant collateral supply of blood to the brain, even if one artery is obstructed (e.g., from a stroke). The lack of this unique anatomical feature enables medical researchers to use the gerbil for stroke research in which other animals would be less valuable research subjects. Another anatomical feature of gerbils is the ventral marking gland. This gland is on the ventral abdomen near the umbilicus. It is present in both males and females but is larger in males. The gland is controlled by sex hormones and is used by both males and females to mark territory. Gerbils rub this ventral marking gland over objects to mark their territory.

Gerbil hematology is also unique. Adult gerbils normally have polychromasia and basophilic stippling of red blood cell (RBC) reticulocytes (immature RBCs). Some gerbils have marked blood serum lipemia[2] even when fed normal diets. The adrenal glands of gerbils are relatively larger than those of most rodents. The large size of the adrenal glands may be related to their ability to efficiently conserve salts and water required for survival in their natural habitat.

## 7.2.3 BREEDING

Gerbils are monogamous. Monogamous breeding pairs should be permanently established before they are 10 to 12 weeks of age. Females are sexually mature at approximately 10 to 12 weeks; males reach sexual maturity at approximately 12 weeks of age. The presence of mature animals will retard sexual development of the same-sex offspring. The male and female should remain together at all times and never be separated. Both the male and female build nests and care for the young. The gestation period is approximately 25 days with four to five young being born in each litter.

Even when not establishing breeding pairs, gerbils should be placed with their permanent cage mates by 10 to 12 weeks of age. New gerbils should not be placed in cages with other gerbils after 12 weeks of age or fighting will ensue. If a cage mate is removed at any time for prolonged periods, fighting may result when it is reintroduced to the cage. If either the male or female adult of a breeding pair dies, it may not be possible to reintroduce the surviving animal to a new mate because severe fighting usually occurs.

## 7.2.4 HUSBANDRY

Gerbils are friendly, active animals and, in comparison to most rodents, require minimal care. They drink little water and excrete little urine. Gerbils drink only about 0.1 to 0.2 oz (4 to 7 ml) of water daily and can survive as long as 45 days without water. Although gerbils can survive for some time without water, free access to water is recommended. When necessary, they can obtain most of their daily water requirement from foods they ingest; however, this is not recommended as mortality is increased, especially among nursing females. Gerbils tolerate a wide range of temperatures. Although room temperature (72°F) is optimal, gerbils easily tolerate ambient temperatures ranging from 60.8°F to 80.6°F. They are more temperature tolerant than many mammals and can survive temperatures of 95°F continuously for some time. A relative humidity of 30% is recommended.

Commercial rodent chow should be fed *ad libitum*. In their natural habitat, gerbils eat seeds, grain, and roots. Breeding is improved when some seeds are fed. If given a choice, gerbils will consume sunflower seeds exclusively to their own detriment. Sunflower seeds, being low in calcium and high in fat, are nutritionally incomplete and should be fed only in limited amounts as a treat. Suspended food hoppers are the preferred method of providing

---

[2]Lipemic serum has high concentrations of circulating triglycerides and cholesterol.

food to adults. If suspended food hoppers are used, make certain young gerbils can reach the food. Unlike most rodents, gerbils do not normally practice coprophagy.

Gerbils are best housed in solid-bottom caging with bedding. Cages must be constructed of durable materials because gerbils will spend a considerable amount of time attempting to burrow. Plastic piping or opaque tubes provide environmental enrichment.

Handling gerbils can be done by grasping the base of the tail. However, care must be exercised not to hold the tip of the tail because the skin at the tip is easily pulled off resulting in exposure of the tailbone. For more confined restraint, gerbils can be gripped by firmly holding the skin over the back of the neck.

### 7.2.5   COMMON DISEASES

One reason gerbils are good household pets is that they have few health problems. They are resistant to many illnesses. One of the more common health challenges in gerbils is overgrown incisors. As with all rodents, the incisors grow throughout their life. If the incisors are maloccluded, they will need to be trimmed about every 2 weeks.

Gerbils are prone to epileptic seizures. The incidence of seizures is genetically related (black gerbils are the most commonly affected). Seizures can be induced by perceived threats or if the gerbil is placed in an unfamiliar environment. The intensity of the seizure can vary from hypnotic-like seizures to intense grand mal seizures. Recovery is spontaneous (usually within 2 minutes). Medication or preventive treatment is not usually needed. Spontaneous seizures are very common in gerbils. The incidence of seizures increases with novel stress. The stress can be as minor as handling or weighing them.

Nasal dermatitis (sore nose) is a condition that can be observed in gerbils. Diagnosis is easily made by noting reddened hairless nares. Treatment for this condition can be as simple as housing the gerbil on sand.

The tail of the gerbil is long and covered with thin skin (see Figure 7.4). Although a gerbil can be restrained by grasping the base of the tail, if restrained or picked up by the tip of the tail, the skin will usually pull off exposing the bone (degloving). Amputation of the tail is necessary if this occurs.

Aged gerbils are prone to tumors but at a lower incidence than many other rodents. One of the more common tumors seen in aged gerbils is ovarian cysts, which can occur at a rate of 20%.

Gerbils exposed to a relative humidity of over 50% may develop rough hair coats. They do best at humidity levels of approximately 30%, which is considerably lower than that recommended for most species of rodents.

Bacterial, viral, and parasitic infections in gerbils are uncommon. The most significant bacterial disease of gerbils is Tyzzer's disease caused by *Clostridium piliforme*. Signs of Tyzzer's disease include depression, ruffled hair coat, and watery diarrhea. Infected animals typically die within 5 days. Treatment is usually ineffective.

### 7.2.6   ZOONOSES

Gerbils present very few zoonotic disease threats to people. However, allergies to gerbils may occur among humans.

## 7.3   GUINEA PIGS

### 7.3.1   INTRODUCTION

The domestic guinea pig (*Cavia porcellus*) originated from South America. Guinea pigs have been bred for food consumption for more than 3,000 years. The source of domestic guinea pigs is unknown. It is thought they originated from wild ancestors of the *Cavia aperea* that are still widespread throughout South America, ranging in the north from Peru and Brazil to the south in Argentina. In the wilds of South America, cavies occupy a wide range of habitats including mountains, swamps, and open grasslands, where guinea pigs are year-round breeders and live in groups of 5 to 10 individuals. Domestic guinea pigs were brought from South America to Europe in the 1500s by Spaniards. During this time, they were bred as pets and further domesticated into what are now referred to as *fancy* types. Guinea pigs are rodents, but they are more closely related to chinchillas and porcupines than to rats or mice. They are often referred to as hystricomorphs, a name originating from the Latin word *hystrix* meaning "porcupine." Guinea pigs are crepuscular, nonburrowing, and strictly herbivorous. They are common household pets because of their friendly disposition, low odor, and quiet behavior. Guinea pigs are not known, however, for ease of care because they are less fastidious and more messy than many other rodents. They are used in biomedical research for studies involving immunology, genetics, infectious diseases, nutrition, and other important disciplines. The term *guinea pig* is often used to describe people who volunteer for experiments, dating to 1913 when British author George Bernard Shaw used the term to describe humans participating in experiments.

## 7.3.2  BIOLOGY AND BEHAVIOR

Guinea pigs are short, stocky animals (Figure 7.5 and Color Plate 10I) with no tail, short legs, unfurred ears, four digits on the front feet and three on the rear. Females have a single pair of inguinal mammal glands (Table 7.4). The teeth of guinea pigs grow throughout the animal's life. This can lead to malocclusion of the lower premolar teeth preventing the animal from eating. Guinea pigs rarely bite, which is one reason they make excellent pets for children. The three main breeds of guinea pigs are short haired (American and English), long haired (Peruvian), and rosette haired (Abyssinian). The American Cavy Breeders Association currently identifies at least 13 breeds or varieties.

Guinea pigs are fussy eaters and may stop eating if their diet is changed. Care must be taken when introducing new foods as the animal may starve to death rather than eat an unfamiliar food. The guinea pig is a monogastric herbivore; its digestive system functions much the same as that of the horse.

Guinea pigs form male-dominated hierarchies when housed in groups. Adults rarely fight; however, if fighting occurs it is usually with newly caged males. Other guinea pig behaviors include a tendency to freeze and then scatter in response to noise or when startled.

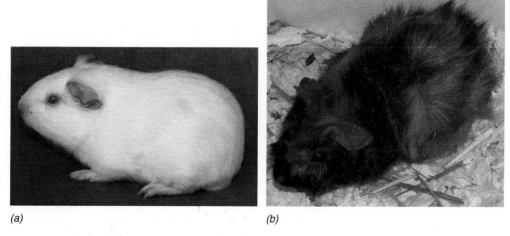

*(a)*  *(b)*

**Figure 7.5**  Guinea pigs are available in a variety of coat colors. Shown here are a white guinea pig (*a*) and a brown and black one (*b*). *Photograph (a) by Jim Roy. Courtesy Hilltop Lab Animals, Inc.; and (b) by Jeff Rathmann. Courtesy PetMarket Place pet store, Webster Groves, MO.*

**Table 7.4**
Biodata of Guinea Pigs

| | |
|---|---|
| **Order** | Rodentia |
| **Suborder** | Hystricomorpha |
| **Family** | Caviidae |
| **Genus** | *Cavia* |
| **Species** | *porcellus* |
| **Birth weight** | 80–95 g |
| **Weight, adult male** | 0.9–1.5 kg |
| **Weight, adult female** | 500–800 g |
| **Body temperature** | 99.0°F–103.1°F |
| **Heart rate** | 150–400/min |
| **Respiratory rate** | 42–150/min |
| **Food consumption, adult** | 0.2 oz (6 g)/3.5 oz (100 g) body weight daily |
| **Water consumption, adult** | 0.3–0.5 oz (10–14 ml)/3.5 oz (100 g) body weight daily |
| **Boar puberty** | 8–10 weeks |
| **Sow puberty** | 4–5 weeks |
| **Boar breeding age** | 3–4 months |
| **Sow breeding age** | 2–3 months |
| **Mammary glands** | 1 inguinal pair |
| **Milk composition** | 3.9% fat, 8.1% protein, 3.0% lactose |
| **Gestation** | 68 day average (range 59–72) |
| **Estrous cycle** | 15–17 days |
| **Litter size** | 2–5 |
| **Weaning age** | 14–21 days |
| **Breeding duration** | 18–48 months |
| **Life span** | 4–8 years |
| **Dental formula (upper/lower)** | incisors 1/1, canines 0/0, premolars 1/1, molars 3/3 |

This can result in animals being injured by jumping out of cages or falling off objects. Another interesting response is the Preyer or pinna reflex. This response is seen when a guinea pig hears a noise and responds by turning its head and directing its ears to the sound. Failure to respond can be a sign of hearing impairment.

### 7.3.3 HUSBANDRY

Pet guinea pigs are best housed in solid-bottom cages with bedding. Cages should be made of polycarbonate plastic or other durable materials that are easily cleaned and maintained. Adults need at least 101 square inches of floor space and 7 inches of cage height. Cages should be sanitized at least once every 2 weeks. Guinea pigs are messy. They scatter feed as well as defecate into their feed and water if these are not dispensed appropriately. Food is best provided from an elevated feeder. Water bowls should not be used because the animal will contaminate the water with food, bedding, and feces. Water is best provided using water bottles with sipper tubes or an automatic water system. Water bottles and sipper tubes should be cleaned weekly, or more frequently if needed, because guinea pigs constantly play with them and introduce food and debris into the water, thereby contaminating it and sometimes clogging the sipper tube.

### 7.3.4 BREEDING

Male guinea pigs are called boars and females are sows. They can be bred using either monogamous or polygamous systems. With polygamous systems, one male with up to 10 females is recommended. The female can remain with the male or be removed during late gestation. Cage size and type are usually critical factors in determining whether to keep the male with the female as parturition nears. Males become reproductively active in about 30 days; however, breeding should be postponed until the male is 3 to 4 months of age. Females can be reproductively active as early as 30 days, too; however, they should not be bred until 2 to 3 months of age. They should be bred before 7 months of age or their pubic symphysis may be fused and not separate during birth resulting in dystocia. The gestation of guinea pigs averages 68 days (Table 7.4).

## 7.3.5  COMMON DISEASES

*Bacterial.* Pneumonia caused by *Bordetella bronchiseptica* is a common disease in guinea pigs. This bacterium is part of the normal flora of rabbits and can be found in some dogs and cats. The disease can be transmitted to guinea pigs from these animals. Symptoms are more common in young animals and can include sneezing, nasal discharge, conjunctivitis, and death.

*Antibiotic-induced enterotoxemia.* Many antibiotics including penicillin and related compounds, erythromycin, cefazolin, clindamycin, gentamicin, and others can cause a fatal enterotoxemia in guinea pigs. These antibiotics alter the intestinal flora in the guinea pigs, resulting in an overgrowth of gram-negative bacteria and toxin production by others including *Clostridium* species. Diarrhea may or may not be present. Symptoms of enterotoxemia include anorexia, rapid weight loss, dehydration, acute death, and a history of recently received antibiotic treatment. Safer antibiotics that have not been associated with this condition in guinea pigs include enrofloxacin, trimethoprim, and sulfonamides.

*Viral.* Viral diseases in guinea pigs are uncommon.

*Parasitic.* *Trixacarus caviae* is a sarcoptic mite and is one of the most common ectoparasites of guinea pigs. It can cause severe pruritus (itching) with self-mutilation. Guinea pigs are susceptible to two biting lice: *Gliricola porcelli* (Figure 7.6) and *Gyropus ovalis*. A very common protozoal parasite, *Cryptosporidium wrairi*, is found in the intestine. It can cause weight

**Figure 7.6** Shown here is the slender guinea pig louse (*Gliricola porcelli*). Lice are species specific and are not transmissible to humans. *Courtesy Dr. Karen L. Campbell.*

loss and diarrhea in young guinea pigs. Nonpathogenic protozoa including *Balantidium caviae* and *Eimeria caviae* are part of the normal intestinal flora of guinea pigs and seen microscopically in fecal examinations.

## 7.3.6    MISCELLANEOUS DISEASES

*Vitamin C deficiency (scurvy).* Guinea pigs are one of only a few animals that require dietary vitamin C. For this reason, guinea pigs should not be fed diets prepared for other rodents such as rats, mice, and chinchillas. A daily source of vitamin C should be provided to prevent scurvy. Because vitamin C deteriorates with time, ensuring that commercially prepared guinea pig food is fresh is important. Supplementing the diet with fresh fruits and vegetables containing vitamin C is also recommended. Foods that contain vitamin C include green and red peppers, oranges, cabbage, and parsley. Symptoms of vitamin C deficiency are rough hair coat, anorexia, swollen and painful joints, reluctance to move, and spontaneous hemorrhages (bleeding) in joints. Spontaneous bleeding in guinea pigs is a hallmark of vitamin C deficiency. These symptoms can develop in as little as 2 weeks on a diet deficient in vitamin C.

*Malocclusion.* As noted previously, guinea pigs are rodents whose teeth grow throughout their life. At times, the mandibular premolars and first molar teeth grow medially (toward the middle) to the point they cover the tongue. Examination can be difficult because of the small mouth and may require sedation. Examination requires viewing the entire mouth and all teeth including the molars. Malocclusion is usually observed in older animals. Symptoms include weight loss and excessive salivation; hence the common name for this disease is *slobbers*.

*Pododermatitis.* Pododermatitis is a subacute or chronic inflammation of the feet. Symptoms are obvious with swelling of the plantar (bottom) portions of feet. It is most frequently seen in animals housed on rough surfaces such as wire. Staphylococcal infections are often associated with this condition. Treatment should include changing the flooring or bedding of the cage.

*Alopecia (baldness).* Hair loss can be seen in group-housed animals from dominant animals barbering (removing hair from) their submissive cage mates and in frequently bred sows.

*Kidney disease.* Chronic interstitial nephritis is common in aged animals. It is a frequent cause of death but usually presents few other symptoms.

*Dystocia.* Dystocia (difficult parturition) is observed most frequently in females bred for the first time after 7 months of age.

## 7.3.7    ZOONOSES

Fortunately, guinea pigs are infrequently associated with zoonotic diseases. Guinea pigs can become infected with *Salmonella*, which can be transmitted to humans. Symptoms of salmonellosis in guinea pigs include lethargy and anorexia for a few days followed by death. It is not uncommon for the entire colony to die. Lymphocytic choriomeningitis is a serious viral disease that can infect both guinea pigs and humans, although it is rarely found in guinea pigs. Streptococcal pneumonia from guinea pigs can affect immunocompromised people.

*Streptococcus pneumoniae.* Part of the normal flora of the upper respiratory tract of many people, *Streptococcus pneumoniae* can cause disease in guinea pigs. Infected guinea pigs can infect immunocompromised humans.

*Allergies.* People can develop allergies to guinea pig dander and proteins. Bathing the guinea pig may help reduce the allergen load in the home environment. Additionally, bedding materials can cause human allergies.

# 7.4    HAMSTERS

## 7.4.1    INTRODUCTION

The most common pet hamster is the *Mesocricetus auratus,* also known as the golden hamster and Syrian hamster (see Color Plate 10F). The four species in the Mesocricetus family are *M. auratus* (Syrian hamster, golden hamster), *M. newtoni* (Romanian hamster, Newton's hamster), *M. brandti* (Brandt's hamster), and *M. raddei* (Georgian hamster, Radde's hamster). Other hamster species include *Cricetus griseus* (Chinese hamster) and *Phodopus sungorus* (Djungarian hamster). However, *M. auratus* is by far the most common pet and the discussion here will focus on that specie. Selected biodata of neonatal and older hamsters are presented in Tables 7.5 and 7.6, respectively.

The golden hamster originated from the Middle East in the area of Syria. In their natural environment, hamsters live in dry rocky areas, often within tunnels deep underground. The tunnels provided a cooler temperature and higher humidity in an otherwise dry and at times hot environment. The golden hamster may be near extinction in the natural environment of Syria where it was last captured in 1980.

The earliest recorded mention of the golden hamster was in *The Natural History of Aleppo* by Alexander Russell in 1797. In 1839 George Robert Waterhouse, curator of the London Zoological Society, introduced the golden hamster as a new species at the Society meeting with a single specimen.

The source of golden hamsters in captivity today dates back to 1930. During that period, Dr. Saul Alder, a parasitologist at the Hebrew University in Jerusalem, Israel, was conducting research on Leishmaniasis, a protozoal disease transmitted by certain species of sand flies (Chapters 14 and 15). Dr. Alder had been using the Chinese hamster, which was susceptible to the protozoa. However, he had problems successfully breeding Chinese hamsters in captivity and depended on shipments from China to conduct his research. Dr. Alder reportedly learned about the golden hamster after reading articles published by Waterhouse. Hoping to find an alternative animal model, he persuaded Professor Israel Ahroni, a zoologist from the Hebrew University, to visit Syria in search of the golden hamster. Dr. Ahroni led a zoological expedition in Syria in 1930 capturing one female with 11 neonates from a

Table 7.5
Biodata of Neonatal Hamsters

| Age | Appearance |
| --- | --- |
| 1 day | Eyes closed, no hair, ears closed, incisor teeth present |
| 4–6 days | Skin pigment appears |
| 5–7 days | Hair appears, eyes open |
| 9–10 days | Short fur present, young start wandering from their mother, begin eating solid food |
| 14–16 days | Eyes open, covered with fur |
| 21 days | Weaning age |

**Table 7.6**
Biodata of Hamsters

| | |
|---|---|
| Order | Rodentia |
| Family | Cricetidae |
| Genus | *Mesocricetus* |
| Species | *auratus* |
| Birth weight | 0.07–0.1 oz (2–3 g) |
| Weight, adult male | 3.0–4.5 oz (85–130 g) |
| Weight, adult female | 3.3–5.3 oz (95–150 g) |
| Body temperature | 98.6°F–100.4°F |
| Heart rate | 280–500/min |
| Respiratory rate | 35–135/min |
| Food consumption, adult | 0.35–0.5 oz (10–15 g) daily |
| Water consumption, adult | 1 oz (30 ml) daily |
| Puberty, female | 36–84 days |
| Puberty, male | 30–40 days |
| Breeding age, female | 56–90 days recommended |
| Breeding age, male | 60–90 days recommended |
| Estrous cycle | 4 days (range 3–5) |
| Mammary glands | 14 |
| Gestation | 16 days |
| Litter size | 6 (range 5–11) |
| Weaning age | 21 days |
| Life span | 2.0–2.5 years (average) |
| Urine production | 0.2 oz (7 ml)/day |
| Preferred light cycle | 14 hours light/10 hours dark |
| Temperature | 65°F–79°F |

tunnel reported to be 8 ft below ground. Soon after her capture, the mother cannibalized one of the litter. Dr. Ahroni, reportedly in anger, killed the mother, choosing to risk hand raising the remaining 10 neonates. He successfully hand raised 9 of the 10 remaining littermates, returning to the Hebrew University with them in July 1930.

Upon returning, Dr. Ahroni presented the nine hamsters to Ben-Menachen, director of the animal facilities at Hebrew University. Shortly thereafter, five escaped and were never found. Of the remaining four hamsters, one female was killed by a male leaving only three survivors. Despite the difficulties, Ben-Menachen successfully raised from these three animals the first laboratory-bred golden hamsters. After this original colony of three animals was successfully bred, Dr. Alder transported some of the hamsters to England in 1931 where he gave them to E. Hindle of the Zoological Society of London. In 1938 the first golden hamsters were exported to the United States.

## 7.4.2   ANATOMY AND SELECTED CHARACTERISTICS

The most common coat color of golden hamsters is reddish golden brown, but variations of coat color are common (Figure 7.7). They are nocturnal but may awake for short periods of time during the day. Adult golden hamsters weigh 3 to 5.3 oz (85 to 110 g) with adult females being larger than adult males. Hamsters have abundant loose skin, a blunt nose, four digits on the front feet, five on the rear, and a short tail. Hamsters have well-developed cheek pouches located on either side of their head. In their natural habitat, these cheek pouches are used to transport food back to their tunnel nests for storing. Cheek pouches are noticeable when full and can be easily everted. Female hamsters can place their entire litter within their cheek pouches. Check pouches are "immunologically privileged sites" and have been

**Figure 7.7** The golden hamster is an attractive pet. *Courtesy Dr. Douglas W. Stone.*

used by researchers to study immune tolerance. Paired flank glands, used for territorial marking, may be seen as dark patches on either flank. They are larger and more evident in male hamsters.

### 7.4.3  BIOLOGY AND BEHAVIOR

In their natural habitat, adult hamsters live singly in a burrow. They are territorial and will fight other hamsters encountered. Females are typically larger, more aggressive, and dominant to males. Female hamsters will fight other females and males, whereas males will fight but not as often. Pregnant and lactating females are particularly aggressive. For this reason, hamsters should be separated after they are 6 to 10 weeks of age. Although hamsters are aggressive to other hamsters, they will respond favorably to frequent gentle handling. When handling one, it is best to give warning prior to picking it up because a startled hamster is more likely to bite. This is especially true when picking up a sleeping hamster or one that is nursing. It is advisable to permit a nursing female to move away from her litter before attempting to pick her up. Picking up a hamster can be done by cupping the hands under the animal and lifting it from the cage, by using a tin can, or by gripping the skin over the neck. For intensive restraint, gripping the loose skin over the neck is the preferred method. When doing intensive restraint, grip as much loose skin as necessary to prevent the hamster from turning around and biting, while still ensuring the animal can breathe normally. Frequent gentle handling will help reduce the risk of being bitten.

### 7.4.4  HUSBANDRY

Hamsters are notorious escape artists. They will chew plastic, wood, and soft metals such as aluminum. Cages must be constructed of materials that cannot be chewed (e.g., polycarbonate or stainless steel). Solid-bottom caging with bedding is preferred over suspended wire flooring. Suggested bedding materials include commercially available corn cob, hardwood, or paper products. Cedar or pine wood bedding should not be used. Pregnant animals should be provided nesting material such as paper products. Should a hamster escape, it will not return to its cage. One reported successful method of recapturing an escaped hamster is to place a tall container such as a bucket partially filled with bedding on the floor. Next, place a wooden ramp from the floor to the edge of the bucket. A hamster's natural tendency is to climb the ramp and attempt to walk the rim of the bucket. Once on the bucket's edge, they will often fall into the bucket where they can be captured.

Hamsters are classified as granivorous, meaning they eat seeds and grain. They routinely hoard food, typically placing the food in one corner of a cage and using the opposite corner for urination and defecation. Food should be placed on the floor of the cage. A commercially prepared rodent diet containing 17% to 23% crude protein, approximately 4.5% crude fat, and 6% to 8% crude fiber is recommended. Hamsters are coprophagous and will consume their own feces several times daily. Because hamsters have a blunt nose, wire feeders must have adequate spacing between the bars to allow the hamster to reach the food. Hamsters housed in cages with wire bar feeders that restrict feed access often have alopecia at their nares (nose). They are susceptible to vitamin E deficiency, which causes symptoms of muscle weakness. A freshly milled (within the past 6 months) commercially prepared rodent diet usually provides adequate quantities of vitamin E.

Hamsters tolerate cold well. Indeed, they are one of the most cold tolerant rodents. Although not a true hibernator, at temperatures below 48°F hamsters can enter a state of pseudohibernation. This state can be stimulated to some degree by lack of food or light, solitude, and temperatures below 48°F. While in this state of pseudohibernation, their heart rate may decrease to 4 to 15 beats per minute with respiratory rates of one every 2 minutes. If the temperature remains low, they may continue in this state for up to 1 week. Care should be exercised to not mistake a dead hamster for one simply hibernating.

## 7.4.5　BREEDING

Sexing hamsters is accomplished by measuring the anal-genital distance. Males have the longest measurement. Because hamsters are solitary in their natural environment, breeding can be difficult. Females come into estrus once every 4 days in the evening. Estrus lasts about 24 hours. Prior to coming into estrus, females emit a strong musky odor with a white vaginal discharge that can cause the urine to appear cloudy. When breeding, it is best to place the female in the male's cage. Males and females should not be housed together for extended periods or serious fighting may occur. They should be closely observed and separated should fighting occur. If the female is receptive when placed in the presence of a male, she will usually display lordosis (an arching of the back). The optimal time for females to accept males is 2 hours after the onset of darkness. When lordosis is displayed, copulation usually occurs in about 30 minutes. If mating has not occurred within 4 hours, they should be separated. Following successful breeding, a copulatory plug can be seen as a white gelatinous material in the female's vagina. After breeding, the female should be moved to her own cage. The gestation period is 16 days (ranges 15 to 18 days). Pregnancy can be determined about the 10th day of gestation by observing the enlarged abdomen. The cage should be cleaned shortly before the mother gives birth so the newborn pups need not be disturbed by cage cleaning for several days following birth. Most births occur during the evening. Once parturition begins, a baby hamster is born approximately every 10 minutes. Pups are born naked, blind, and with teeth. Cannibalism is more common in hamsters than any other laboratory rodent. It is most common among first-time mothers but can occur with more experienced mothers. For this reason, a nursing female should not be disturbed nor should the babies be handled for several days or even weeks following birth. Light cycles are important when breeding hamsters. A light cycle of 14 hours on, 10 hours off is recommended for optimal breeding. Hamster breeding is affected by season, with a lower conception rate during the winter.

## 7.4.6　COMMON DISEASES

*Viral.* Hamsters are prone to lymphoma (cancer involving lymphocytes) caused by hamster polyoma virus. Epizootic outbreaks approaching 80% in some colonies have been reported. Symptoms include emaciation, weakness, lethargy, diarrhea, and palpable abdominal tumors. This viral agent is very stable in the environment and easily transmitted to other hamsters.

*Bacterial.* Hamster Enteritis Complex is a term used to describe several diarrhea-causing diseases in hamsters. Hamster Enteritis Complex includes the diseases proliferative ileitis, Tyzzer's disease, and clostridiosis. Proliferative ileitis (Wet Tail, Hamster Enteritis, and others) is one of the more common diseases observed. The cause of this disease is believed to be the bacterium *Lawsonia intracellularis*. Symptoms are most often observed in hamsters 3 to 10 weeks of age and include diarrhea (fecal matting of the perineum is commonly seen), dehydration, and stunted growth. In its fatal form, death occurs 24 to 36 hours after onset. Treatment consists of supportive care. Control measures are unknown. Tyzzer's disease is caused by the spore-forming bacteria *Clostridium piliforme*. Symptoms include a hunched posture, rough hair coat, and a watery, foamy diarrhea. Sudden death is often the only sign observed. Administration of several antibiotics to hamsters can induce a fatal intestinal bacterial overgrowth by *Clostridium perfringens, C. difficile,* and possibly others. Antibiotics associated with causing clostridiosis enteritis (antibiotic toxicity) include lincomycin, erythromycin, penicillin, ampicillin, gentamicin, vancomycin, and cephalosporins.

## 7.4.7 PARASITES

Hamsters are not normally prone to any significant parasites. However, those more than 18 months of age may have dermatitis and alopecia over the back resulting from demodectic mites (Chapter 15).

## 7.4.8 KIDNEY FAILURE

Kidney failure is a common age-related cause of death in hamsters. Causes of kidney failure can include amyloidosis, polycystic disease, and chronic nephropathy. In amyloidosis, a very common disease in hamsters over 1 year of age, deposits of white proteinaceous material develop in organs throughout the body. Amyloidosis is more common in females than males. Renal insufficiency often occurs as the deposits increase with age. The cause of amyloidosis is unknown. Chronic nephropathy is another age-related cause of kidney failure in hamsters, which is followed by death. Polycystic kidney disease is inherited and can affect up to 75% of a colony resulting in death secondary to kidney failure. Symptoms of kidney failure include generalized edema.

## 7.4.9 ATRIAL THROMBOSIS

Atrial thrombosis is another common cause of death in aged hamsters. It results when a blood clot develops in the heart. Females are more commonly affected than males. Atrial thrombosis symptoms include difficulty in breathing followed by sudden death.

## 7.4.10 NEOPLASIA/TUMORS

Tumors are relatively rare in hamsters; however, they are susceptible to both benign and malignant tumors. Tumors seen in hamsters include lymphomas, leukemias, and melanomas. Except for melanomas, skin tumors are uncommon.

## 7.4.11 ZOONOSES

One of the more serious zoonotic diseases is the viral disease lymphocytic choriomeningitis (LCM). Although this disease is rare in humans, severe forms can be fatal. Infected hamsters spread the virus via urine. It is also transmitted transplacentally from mothers to offspring. Salmonellosis has also been associated with exposure to hamsters on rare occasions. Zoonotic diseases transmitted by hamster bites include *Actinobacter* spp, *Leptospira* spp, and LCM. Allergies to hamsters appear to be less common than rat or mouse allergies.

## 7.5    FERRETS

### 7.5.1    INTRODUCTION

The domestic ferret, *Mustela putorius furo*, is a carnivore whose highly adaptable mustelid ancestry dates back at least 40 million years and whose domestication from a wild Eurasian or North African antecedent precedes 2400 BC. Living relatives—all wild animals—include North American black footed ferrets, skunks, weasels, mink, and the European polecat. Domestic ferrets kept as pets or hunters of small animals have been in North America for about 300 years. Currently, they constitute a pet and research animal population of over 7 million. Ferrets have been commercially bred for pets and research for more than 50 years. Ferrets are small, usually gentle, clean, and easily housebroken animals that are popular pets and subjects for studying infectious diseases, physiology, endocrinology, and embryology.

Despite many concerns that ferrets may develop feral populations in North America, none has. Nevertheless, despite the long domestication, an exaggerated fear of ferrets killing wildlife, biting people, or spreading disease has condemned these animals to the "wild animal" category and to frequently changing but usually discriminatory regulations affecting where ferrets may or may not be kept as pets. It is illegal to keep ferrets as pets in California, Hawaii, and many other areas. Therefore, no one should obtain, possess, or (as a veterinarian) treat or vaccinate a ferret without first checking regulations applicable to ferrets. Fisheries and wildlife units of the various states have the necessary information for prospective owners.

### 7.5.2    BIOLOGY

Ferrets, in accordance with their lives as obligate, carnivorous predators, are adapted for rapid, convoluted movements in narrow spaces. The body is long without protruding head or hips, and with a thick neck and short legs. Nonneutered male ferrets weigh 2 to 4 lb (1 to 2 kg), about twice the 1 to 2 lb (0.5 to 1.0 kg) weight of intact females. Neutered males weigh 1 to 2 lb (0.5 to 1.0 kg) and have slimmer bodies than do intact males. Ferrets gain subcutaneous fat in the fall and winter. Annual weight fluctuations, at least in younger intact animals, may vary by 20% to 40%. Selected biodata of ferrets are presented in Table 7.7.

The common "wild-type," fitch or sable coloration, consists of black hair on feet, tail, face "mask," and guard hair elsewhere over a cream colored undercoat. Other color variants, of which there are more than 20, include white (albino), cinnamon (sandy), grey, chocolate, panda, silver, and Siamese colors or patterns (Figure 7.8 and Color Plate 10E). Colors may vary as ferrets age, and older ferrets may have oily fur discolored to yellow from accumulated sebaceous secretions. Ferrets molt (whether obvious or not) each spring and fall when body weight changes. Hair regrowth occurs over months, depending on photoperiod, and hair is typically blue-black as it emerges from follicles.

Male ferrets are called hobs, females are jills, and young are kits. Male ferrets have a conspicuous preputial opening on the ventrum just caudal to the umbilicus, whereas the female urogenital opening is ventral to the anus and elongated in nonestrus and round, donut-shaped, and swollen in estrus, which may last weeks. Males have a greater anogenital distance than do females, and males have a J-shaped os penis. Testes descend into the scrotum at 5 to 6 weeks of age. Each foot has five digits with nonretractable claws. These require frequent trimming (ferrets are not declawed because declawing reduces traction on smooth floors). Deciduous teeth appear in 20 to 28 days, and the long-rooted (extraction is difficult) permanent teeth erupt at 50 to 74 days. Ferrets have the simple, distensible stomach and

**Table 7.7**
Biodata of Ferrets

| | |
|---|---|
| **Order** | Carnivora |
| **Family** | Mustelidae |
| **Genus** | *Mustela* |
| **Species** | *putorius furo* |
| **Birth weight** | 0.2–0.4 oz (6–12 g) |
| **Weight, intact hobs** | 2–4 lb (1.0–2.0 kg) |
| **Weight, jills** | 1–2 lb (0.5–1.0 kg) |
| **Rectal temperature** | 100°F–104°F (37.8°C–40°C) |
| **Heart rate** | 180–400/min |
| **Respiratory rate** | 33–36/min |
| **Food (dry) consumption, adult** | 1.5 oz/lb (43 g/kg) body weight daily |
| **Water consumption, adult** | 2.5–3.4 oz (75–100 ml) daily |
| **Sexual maturity, hobs (males)** | over 8 months |
| **Sexual maturity, jills (females)** | 4–12 months |
| **Estrous cycle** | induced ovulators |
| **Gestation period** | 40–44 days |
| **Litter size** | 8–10 kits (average) |
| **Weaning age** | 6–8 weeks |
| **Breeding life** | 2–5 years |
| **Life span** | 5–10 years |

short intestinal tract (75 inches, 190 to 200 cm) typical of carnivores. There is no cecum, and the divisions of the small intestine are not obvious grossly (macroscopically; not visible with the naked eye). Gastrointestinal transit time is 3 to 4 hours. Ferrets can vomit and rarely experience antibiotic-induced dysbiosis. They have a relatively long trachea and large lung capacity. Ferrets have well-developed, paired anal glands from which a scented, yellow liquid is expressed in response to a threat. This liquid, however, contributes little to the well-known musky odor of male ferrets. Sebaceous glands in the skin, under androgen control, contribute significantly to that odor, which is reduced following castration. Ferrets have few sweat glands, which contributes to their inability to tolerate ambient temperatures above 80°F.

Male ferrets reach puberty around 8 to 9 months and females at 4 to 12 months of age. The breeding season in the Northern Hemisphere in a normal photoperiod (increasing then

**Figure 7.8**   An albino ferret. *Courtesy Jeff Rathmann (photographer) and PetMarket Place pet store, Webster Groves, MO.*

decreasing reflecting daily available light) is March through August, but alteration of the photoperiod can produce receptivity throughout the year. Sperm are produced from December to July during which time the testes enlarge. Females are seasonally polyestrous (may have two litters a year) and ovulate 5 to 13 ova 30 to 40 hours after copulation. The gestation period averages 41 to 42 days (ranges 40 to 44 days), but if pregnancy does not occur, a pseudopregnancy of 40 to 43 days follows. Approximately one-half of pseudopregnant jills remain in estrus and develop an estrogen-caused bone marrow suppression and thrombocytopenia that may be fatal, even if treated. Litters consist of 1 to 18 kits (usually 8 to 10). The young weigh 0.2 to 0.4 oz (6 to 12 g) at birth. Eyes and ears open at 30 to 35 days, and young are weaned at 6 to 8 weeks of age. Fertility lasts 2 to 5 years, depending on number of litters. With light cycle and nutritional manipulation, a ferret can have 3 to 4 litters a year. The first litter is born in a ferret's second year of life, unless light cycles are artificially altered. The vulvar edema diminishes within 2 weeks of breeding, but pseudopregnant jills retain the swelling and may show the melena and subcutaneous hemorrhage characteristic of a clotting disorder. Sebaceous secretions and resulting odor increase in the breeding season.

### 7.5.3   HUSBANDRY

Ferrets respond to gentle handling and may be picked up and held as one would a small dog or cat. Young ferrets or ferrets unaccustomed to handling may be more active and may have to be held more securely. Ferrets restrained for examination are held so that one hand grasps the neck skin (scruffing), or otherwise restrains head and forelimbs, and the other hand restrains the rear legs (Figure 7.9). Digital pressure on the temporal mandibular joint or on the philtrum will cause a ferret to release a bite. Also, ferrets held by the scruff of the neck will usually yawn, allowing examination of the oral cavity.

Ferrets may be housed in a cage part- or full-time, or allowed to roam the house. They will use a litter pan or covered floor area for urination and defecation. Risks associated with allowing a ferret access to the house and furniture include their tendency to chew and perhaps ingest cloth, foam, rubber, plastic, and significant components of furniture. Ferrets also climb into furniture crevices and may become trapped in recliner chairs. Wire-walled cages with a 0.5 × 1.0 inch mesh, usually with a solid bottom, an enclosure for sleeping, and at least 2 ft square and 1.5 ft high are suitable for housing one or two ferrets. Wire cages allow for

**Figure 7.9**    Ferrets can be restrained by gripping the back of the neck with one hand while the other hand holds the rear legs. *Courtesy Dr. Douglas W. Stone.*

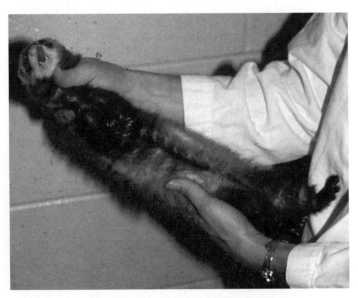

better ventilation than do those with glass or other solid sides, but because ferrets lick and chew cages, they must have no sharp points or edges. Wooden cages may be chewed, absorb urine, and are not cleaned easily. Dust from pan litter such as clay may irritate their skin and respiratory tract. Chemicals released by cedar and pine shaving litter or bedding may also irritate the respiratory tract. Bedding and litter should be changed at least daily as ferrets, like cats, may use the litter pan for resting or sleeping. Sexually mature, intact, same-sex ferrets should be housed separately to prevent fighting.

The comfort range for adult ferrets is between 39°F and 64°F (4°C and 18°C) and 40% and 65% relative humidity. Young ferrets should be housed at temperatures 59°F (15°C) and above. Shade, fresh water, and good ventilation are needed if ferrets must be housed in a warm area, as ferrets may overheat at 80°F (27°C) or higher. Breeding ferrets respond best to 16 hours of light per day.

Ferrets are strict carnivores with a characteristic simple stomach, short intestinal tract, and relatively rapid transit time. Ferrets will eat several times a day and can be fed *ad libitum* unless obesity becomes a problem. They do not digest carbohydrates or fiber well. Sweet carbohydrate foods should be avoided as they may lead to periodontal disease and exacerbate insulinoma effects (see Section 7.5.5). Ferrets will eat some low-fiber fruits and vegetables, but may develop gastrointestinal blockage if they overeat large quantities of vegetables. Bananas have been reported to cause gastric bloat. Withholding food for periods as short as 24 hours or changing food during pregnancy may lead to a fatal ketoacidosis in jills.[3]

Ferret feed should contain about 30% to 40% meat protein and 15% to 30% fat, which is available in quality commercial ferret feed or in quality cat food mixed appropriately with animal protein such as liver. Ferrets may prefer poultry meat over beef and fish and must have ample fresh water to facilitate consumption. Pregnant ferrets require at least 35% protein and 25% fat in their diet. A high meat diet results in an acidic urine, which prevents urolith formation. Pelleted food for 3- to 4-week-old ferrets should be softened with milk or water. Food should be cooked to reduce bacterial contamination. Ferrets may play in their drinking water or spill contents from bowls.

Breeding cages should be in a quiet area and contain a nest box with bedding and at least 6-inch high sides. Two weeks after full vulvar enlargement, males are placed into the jill's cage and left for up to 2 days. Coitus is active with the male dragging the jill while biting her neck. This normal activity, combined with intromission, elicits ovulation. Pregnancy can be confirmed by manual palpation at 2 weeks. Gestation lasting beyond 42 days may result in dystocia, in which case Cesarean section may be needed. Parturition lasts 2 to 3 hours. The jill stimulates neonatal urination and defecation by licking the perineum. Orphaned young can be raised using orphaned kitten or puppy milk replacers.

## 7.5.4    DISEASES

Ferrets may develop a great variety of diseases of the gastrointestinal tract, respiratory system, endocrine organs, and skin. Many diseases manifest as a common set of signs including anorexia, weakness, nasal and ocular discharges (secondary bacterial infection), abnormal breathing, hair loss, skin disorders, and blood in the feces. Because signs may be nonspecific, at least on first encounter, specific diagnosis usually requires physical examination and examination of blood components. Disease prevention is effected by providing the ferret with proper food, a noninjurious living environment, and routine veterinary care including vaccinations for canine distemper and rabies, diagnostic tests for parasitism, and examination for abnormal signs. Such preventive care is essential because so many diseases common in ferrets become rapidly fatal including canine distemper, intestinal disorders, pneumonia, hyperestrogenism, and neoplasia.

---

[3]Ketoacidosis is a metabolic disease caused by catabolism (breaking down) of body fat stores.

*Viral diseases. Canine distemper.* The virus that causes canine distemper is an inevitably fatal but preventable disease in ferrets caused by a paramyxovirus related to the human measles virus. Affected ferrets exhibit anorexia, fever, sensitivity to light, skin rash, and a watery nasal discharge, unless secondary bacterial infection results in thick nasal and ocular discharges. This so-called catarrhal phase may progress to a neurological phase with ataxia (incoordination), tremors, paralysis, and death. Infectious virus is shed in discharges from the skin and in feces. Prevention is accomplished by vaccinating with a ferret-approved modified live virus canine distemper vaccine given every 2 to 3 weeks between 6 and 14 weeks of age and annually thereafter.

*Influenza.* Influenza, essentially the same disease and same agent that occur in humans, is caused by an orthomyxovirus that can be passed from ferrets to humans and from humans to ferrets. The virus usually affects the nasal epithelium but can cause pneumonia. Signs include anorexia, fever, sneezing, nasal and ocular discharges, and usually spontaneous recovery in 4 days or less. Antibiotic use can prevent complicating secondary bacterial infections.

*Rabies.* A rhabdovirus causes rabies, signs of which include anxiety, lethargy, and posterior paralysis. Ferrets that bite or scratch a person should be quarantined for at least 10 days and public health officials notified. A ferret-approved killed virus rabies vaccine should be given between 3 months and 1 year and repeated annually thereafter.

*Other viral diseases.* Aleutian disease is caused by a parvovirus, and known affected ferrets, which should be removed from the colony, may show no signs or have chronic weight loss, weakness, blood in stool, and central nervous system signs. Rotavirus infection can cause a rapidly fatal diarrhea in nursing kits.

*Bacterial infections.* **Clostridium perfringens** *type A.* *Clostridium perfringens* type A is a common component of ferret's intestinal flora, but may become toxigenic following dietary changes, including overeating. Affected weanling ferrets exhibit bloating (gas), difficult breathing, watery diarrhea, cyanosis,[4] and death, or death with no other obvious signs. Prevention is achieved by providing nutritious feed without rapid changes in quality and quantity.

*Helicobacter mustelae.* Helicobacter mustelae is a common inhabitant of the ferret's gastrointestinal tract, and infected ferrets usually show no or mild signs of infection. Clinically affected ferrets, however, exhibit vomiting, gastric and duodenal ulceration, weight loss, and melena (blood in feces). Treatment includes administration of bismuth subsalicylate and antibiotics.

*Proliferative bowel disease.* *Lawsonia intracellularis* causes a pronounced thickening of the bowel wall, which can be detected on abdominal palpation. Affected ferrets may have chronic, green diarrhea with mucus and blood and rectal prolapse.

Other bacteria infections reported in ferrets include tuberculosis, salmonellosis, pneumonia, bite wound abscesses, and mastitis.

*Parasitic diseases. Otic acariasis.* Otic acariasis is caused by *Otodectes cynotis*. These ear mites are also found in dogs and cats (Chapter 15). Infestation is often asymptomatic but heavier infections elicit head shaking, self-mutilation (from pruritus), otic discharge and scabbing, and secondary bacterial otitis media and interna. Mites are detected by direct or microscopic examination of detritus (exudate). Treatment is with an otic topical acaricide or by injected ivermectin.

---

[4]Cyanosis refers to blue mucous membranes caused by inadequate oxygenation of arterial blood.

*Sarcoptes scabiei.* A pedal (foot) or generalized alopecia and crusted skin with pruritus (scratching) may be caused by the mite sarcoptes scabiei.

Other ectoparasites encountered on ferrets are fleas and *Demodex* mites.

*Intestinal parasites.* Internal parasites that occasionally cause diarrhea in young ferrets are coccidia and *Cryptosporidium*. Signs of coccidiosis may include diarrhea and rectal prolapse in the young.

*Heartworm disease.* *Dirofilaria immitis*, a microfilaria transmitted by mosquitoes from canines to ferrets (aberrant hosts) can cause weakness, difficult breathing, cyanosis, anorexia, and death. Preventive measures include keeping ferrets indoors, routine heartworm preventive medication, and annual heartworm antigen testing.

*Fungal diseases.* Ferrets are susceptible to systemic mycoses such as blastomycosis and histoplasmosis and to superficial dermatophytoses (ringworm) such as *Trichophyton mentagrophytes* infection.

## 7.5.5 OTHER CONDITIONS

Pregnancy toxemia occurs primarily in primiparous jills carrying eight or more later-term fetuses. An additional precipitating factor is fasting during late pregnancy. Large litters coupled with consumption of diminished nutrients may result in inadequate perfusion and nutrition of the uterine-placental unit with resulting fetal changes and death. Maternal signs include lethargy, loose hair, and melena (bleeding into the gastrointestinal tract). Cesarean section and intense supportive therapy (calories, fluids, and warmth) are required treatments. Surviving young, if any, may have to be fostered or hand raised. Pregnant jills must have clean water and quality food available at all times.

Hyperestrogenism occurs commonly in jills experiencing an estrous period of 1 month or longer. If the estrus is not terminated by pregnancy, luteinizing hormone use, or ovariohysterectomy, the resulting estrogen stimulation and induced bone marrow hypoplasia result in reduction of cellular blood components including thrombocytes (platelets). Clinical signs of hyperestrogenism include vulvar enlargement, bilateral symmetrical alopecia, weakness, anorexia, weight loss, anemia, and multisite secondary bacterial infections. Unless reversed by ovulation, blood transfusion, and various increasingly desperate palliative measures, the low numbers of platelets results in hemorrhage, coagulopathy, anemia, pallor, melena, and death.

*Neoplasia.* Insulinomas or pancreatic islet cell tumors are the most common neoplasm in ferrets between the ages of 4 and 5 years but may also occur in younger ferrets. Affected ferrets have vomiting, weight loss, ataxia, weakness, drooling, and pawing at the mouth. These tumors produce large amounts of insulin, which causes profound hypoglycemia (low blood sugar). Supportive care and surgical removal of the neoplasm are recommended treatments.

Adrenal adenomas and adrenal cell carcinomas constitute the second most common tumor type and usually occur in ferrets from 2 to 6 years old. Signs, indicative more of androgenic than corticotropic influences,[5] include weight loss, symmetric alopecia, vulvar swelling, and occasional pruritus. Hair loss may come and go and apparently depends on other, perhaps seasonal, influences. Adrenal hyperplasia and neoplasia occur in up to 90% of older ferrets and are detected by ultrasound and abdominal palpation. Treatment involves unilateral or, as treatment or prophylaxis, bilateral adrenalectomy or androgen-blocking drugs.

---

[5]Adrenal tumors in cats, dogs, horses, and humans generally produce excessive amounts of cortisol, whereas those in ferrets produce excessive amounts of sex hormones (androgens or estrogens).

Lymphoma (cancer of lymphocytes) appears in various forms, some perhaps caused by viruses or even bacteria, and occurs in both young and old ferrets. Signs are often nonspecific disabilities. Tumors of the skin, which are relatively common, include nodular mast cell tumors, rapidly growing and ulcerated basal cell tumors, and several others.

Other notable diseases seen in ferrets include heat stress, cardiomyopathy, splenomegaly, eosinophilic gastroenteritis, megaesophagus, and gastric hairballs.

### 7.5.6   ZOONOSES

Pathogens of ferrets with a potential to affect humans are influenza virus, rabies virus, *Clostridium perfringens* type A, *Sarcoptes scabiei*, and *Trichophyton mentagrophytes*.

## 7.6    RABBITS

### 7.6.1   INTRODUCTION

The domestic or European rabbit, *Oryctolagus cuniculus,* evolved in recent times in Europe and North Africa. It was domesticated from a wild population (which still exists) around 1,400 years ago. Fossil rabbits are thought by some to date back 30 million years. Feral populations of this rabbit still exist, most notably in Australia. All other lagomorphs, including cottontail rabbits and hares, are wild animals and do not breed with European rabbits.

Rabbits are social animals, females (does) more than males (bucks). They burrow into soft earth, are herbivorous and crepuscular, and through domestication represent various weight groups, hair coat types, hair colors, and breeds (Figure 7.10 and Color Plates 10J and 10L). Small breeds (e.g., Netherlands dwarf, Polish, and Dutch belted) weigh 2 to 4 lb (1 to 2 kg). Medium-sized breeds (e.g., New Zealand whites, California rex, and chinchilla) weigh 4 to 14 lb (2 to 6 kg), whereas giant breeds (e.g., Flemish giant and checkered giant) weigh 14 lb or more. Rabbits may have one, two, or three hair colors or be albino. Hair coat variants include the dense-furred rex and the long-haired angora. Small breeds are popular as pets (Figure 7.11). New Zealand whites are used commonly for pelt, meat, production of biologicals, and medical research.

### 7.6.2   ANATOMY

Rabbit anatomy has apparently changed little over millions of years. Therefore, it must confer considerable environmental adaptability. Rabbits have lightweight bones (nearly half as dense as cat bones of the same size), a functional third eyelid, a wide field of vision, and good

**Figure 7.10**  Domestic rabbits are available in a variety of sizes and coat colors. *Courtesy Dr. James E. Corbin.*

**Figure 7.11**  Small breeds of rabbits are especially popular as pets. *Courtesy Questhavenpets.com.*

light sensitivity. Rabbit teeth are all open rooted (continuously growing) and are constantly worn down by contact with opposing teeth and by abrasion on foods. The thoracic cavity and heart are relatively small, and the intestine is capacious and long (10 times the body length of rabbits). The stomach is simple and glandular, whereas the cecum is capacious. Lymphoid masses occur at the distal tip of the cecum (the lymphoid appendix) and at the ileocecal junction (sacculus rotundus). Inguinal canals remain open for life, and the cervix has two internal and external ora. Female rabbits have two bands of mammary tissue (four to five glands per band) extending from ventral neck to groin.

### 7.6.3  PHYSIOLOGY

Rabbits are strict herbivores and chew food to a powder size before swallowing, except for ingested cecotrophs, which are swallowed directly from the anus. Rabbits are induced ovulators. They ovulate about 9 to 13 hours following copulation. Rabbits housed under natural photoperiods experience an autumnal infertility (reluctance to breed and reduced ovulatory response). Receptivity is retained year-round under constant, artificial light cycles of 14 hours light and 10 hours dark. Rabbits usually live 5 to 6 years but may live 10 to 12 years when properly fed as pets. A doe may produce 7 to 25 litters with 7 to 8 young per litter during peak production. Rabbits may breed within hours of kindling and lactate and be pregnant simultaneously. Such production requires healthy animals and a proper diet. Rabbit urine is cloudy (calcium carbonate crystals), alkaline, and ranges in color from light yellow to orange-red. Selected biodata of rabbits are presented in Table 7.8.

### 7.6.4  BEHAVIOR

Rabbits make gentle pets but can inflict severe scratches with their rear feet. They may bite if a protruding finger resembles a familiar food item such as a carrot stick. Male rabbits may

**Table 7.8**
Biodata of Rabbits

| | |
|---|---|
| Order | Lagomorpha |
| Family | Leporidae |
| Genus | *Oryctolagus* |
| Species | *cuniculus* |
| Birth weight | 40–90 g |
| Body weight | 1000–6000 g |
| Rectal temperature | 101.3°F–104°F (38.5°C–40°C) |
| Heart rate | 130–325/min |
| Respiratory rate | 30–60/min |
| Food consumption, adult | 1.7 oz/lb (50 g/kg) body weight daily |
| Water consumption, adult | 1.7–3.0 oz/lb (50–100 ml/kg) body weight daily |
| Sexual maturity, bucks | 6–10 months |
| Sexual maturity, does | 4–9 months |
| Estrous cycle | induced ovulators |
| Gestation period | 29–34 days |
| Litter size | 4–10 kits |
| Weaning age | 4–6 weeks |
| Breeding life | 1–4 years |
| Life span | 5–12 years |

castrate one another, but sibling females maturing together show less agonistic behavior than do other rabbits. For mating, bucks are taken to a doe's cage and mating usually occurs within minutes (about 85% of the time), after which the buck is returned to his cage. Pseudopregnancies may occur. Rabbits may thump with one rear foot if fearful or aroused and may spray urine on persons standing outside the cage.

## 7.6.5   HOUSING

Pet rabbits can be house trained (to a litter pan or latrine area) and have the run of the house, or they may be caged individually in wire cages ranging from 1.5 sq ft (small breeds) to 5.0 sq ft (large breeds). Breeder rabbits should have approximately 6.0 sq ft of floor space and a nest box for delivery. Rabbits require a feeder and waterer (preferably sipper-tube type); they also will interact with many types of objects or toys.

Rabbits are most comfortable at ambient temperatures of 40°F to 85°F (15°C to 29°C), but can tolerate colder temperatures if acclimatized and if provided an enclosed shelter. Rabbits do not, however, tolerate heat well, and heat stress is common, especially if rabbits are in direct sunlight or are deprived of water. Bucks in high ambient temperatures become infertile. Rabbit housing must be kept clean and well ventilated to reduce ammonia gas.

## 7.6.6   NUTRITION

Rabbits digest some fiber, but growth is optimal at around 15% to 17% crude fiber. Diarrhea and gastric atony (and hairballs) are reduced significantly when dietary fiber levels are increased up to 22%. Fiber is essential to stimulate intestinal motility, but the higher the dietary fiber, the lower the energy and protein components, which may reduce capacity for intensive breeding and production. Rabbits digest about 70% of plant-origin protein, and ingest directly from the anus vitamin- and protein-rich cecotrophs. Cecotroph ingestion occurs in early morning and is not retarded when rabbits are housed on wire floors. Commercial rabbit pellets are usually alfalfa based and contain (per label) about 15% to 17% crude fiber and 15% to 19% crude protein, although feed-content labels may be misleading with actual levels being below or above percentages listed. Alfalfa-based feed, according to some specialists, may contain excessive calcium, which presumably contributes to increased blood plasma calcium (already high in most rabbits) and consequent urolithiasis. Food quantity consumed daily by a medium-sized, nonpregnant, adult rabbit is about 3 to 6 oz (pelleted rabbit feed), but lactating rabbits (young nurse 4 to 6 weeks) should be fed *ad libitum*. Food quantity and quality changes must be effected gradually or intestinal substrates and microbial populations may change and cause fatal enteropathies. Administering certain antibiotics (e.g., all penicillins, lincomycin) may lead to dysbiosis (disruption of intestinal flora) and precipitate a clostridial (or coliform) enterotoxemia. Rabbits may be fed small quantities of "treat" foods. Many pet rabbit owners compose, at some risk of malnutrition, diets of vegetables and grains. Rabbits seem to prefer oat and strawberry flavors, among others.

## 7.6.7   REPRODUCTION

Male rabbits have a rounded, donut-shaped, protruding urinary opening. The penis can be everted by applying digital pressure anterior to the prepuce. In adult bucks, hairless scrotal pouches are evident. Does have an elongated vulvar opening near the anus and may have a pronounced fold of throat skin, the dewlap. Bucks have a bulky head. Both sexes have paired, inguinal pouches lateral to the anogenital midline.

Bucks are sexually mature or suitable for breeding at 4 to 5 months of age and are mature at 6 to 10 months. Does are receptive for cycles of 7 to 10 days then are nonreceptive (flatten) for a few days while a new wave of ovarian follicles matures. During receptivity, the vulva is often swollen and reddened. Mating occurs when does are housed for minutes to half an hour in a buck's cage. The doe then enters a 29- to 34-day gestation. A nest box should be provided at day 25 post breeding. The box must be dry and contain materials such as shavings or straw onto which the doe builds a nest lined with hair plucked from her forequarters. Fetuses can be palpated *in utero* between 9 and 14 days of gestation. Young are delivered at night, hairless and blind, and group together within the warm, humid nest. The doe nurses the young once a day for less than 3 minutes total time. The young begin consuming maternal feces at 2 weeks of age and are weaned at 4 to 6 weeks (lactation peaks at about 3 weeks). Rabbit milk composition resembles that of bitch milk (high protein, fat, and minerals), but kitten milk replacers are usually used to hand raise orphaned bunnies. Weaned rabbits, before drinking, may need to be directed to a sipper tube and can be group housed until nearing sexual maturity (3 to 4 months). Fryer (meat) rabbits are commonly slaughtered at 8 weeks of age. Intensively bred rabbits may produce 40 or more young (alive at 8 weeks) each year, yielding a quantity of meat exceeding that produced by most other mammals. Rabbits rank fifth in the world's livestock meat production.

## 7.6.8   DISEASES

*Viral diseases.*   There are no common viral diseases in domestic rabbits, which is fortunate because the rare or uncommon pox virus (myxomatosis) and calicivirus infections (hemorrhagic fever) are highly fatal. Other still uncommon but less lethal viral diseases are cutaneous and oral papillomatosis, and pox-virus-caused fibromas and myxofibromas spread by insects from various wild cottontail rabbits. No commercially available vaccines are used in rabbits.

*Bacterial diseases.*   Two common bacterial diseases in rabbits, pasteurellosis and clostridial enterotoxemia, can cause major health problems. Several others occur sporadically.

Pasteurellosis is caused by the gram-negative rod *Pasteurella multocida*, and clinical signs are as diverse as the sites in rabbits where infection and resulting tissue damage occur. The primary colonization site is apparently the nares (nose) and oropharynx (division of pharynx lying between the soft palate and the upper edge of the epiglottis), but the organisms can extend readily to (1) the nasal passage, causing rhinitis (inflammation of the mucous membrane of the nose; also called snuffles); (2) the inner ear, causing irritation of the semicircular canals on one side resulting in torticollis (also called wry neck) reflected in muscle spasms, which in turn result in abnormal carriage of the head; (3) the eyes, causing conjunctivitis (an inflammation of the delicate membrane that lines the eyelids and covers part of the sclera, Chapter 11); (4) the lungs, causing pneumonia (inflammation of the lungs); (5) uterus or testes; and/or (6) any other organ or tissue. Pus associated with pasteurellosis is usually yellow-white. Treatment of pasteurellosis is by prolonged administration of antimicrobials (e.g., quinolones) safe to use in rabbits and by drainage and excision of abscesses. Prevention includes use of *Pasteurella*-free breeding stock, good ventilation (ensuring ammonia removal) and sanitation, and culling of clinically abnormal rabbits.

Other bacteria causing disease outside the gastrointestinal tract include *Treponema cuniculi* (venereal spirochetosis), *Staphylococcus aureus* (abscesses), *Pseudomonas aeruginosa* (pyoderma), and *Fusobacterium necrophorum* (anaerobic abscesses).

*Clostridium spiroforme* is a C-shaped anaerobic organism and is a normal resident of the rabbit's intestinal tract (cecum). The organism under circumstances of optimal pH, glucose and specific cation concentration, lowered host resistance, intestinal hypomotility, or diminished maternal immunity produces a potent enterotoxin that can kill rabbits with few preceding clinical signs (weakness, watery diarrhea) or cause prolonged diarrhea in older rabbits with moderate to high mortality. Entertoxemia usually affects weanling rabbits (4 to 8 weeks) but may cause clinical disease in older rabbits. Clostridial enterotoxemia, as with most other enteropathies in rabbits, is prevented rather than treated because few treatments, beyond fluid administration and other supportive care, are efficacious. Prevention involves selection of quality stock, use of higher fiber (18% to 22%) feed, avoidance of antibiotic administration to rabbits, slow changes of food quality or quantity especially at the beginning of lactation or cold weather, and in certain circumstances, use of food with copper or probiotic supplementation.

Clostridial enterotoxemia may occur in association with intestinal coccidiosis, rotavirus infection, and colibacillosis, all potentially separate diseases with similar clinical signs, but treatment remains the same except that intestinal coccidiosis may be prevented through administration of a coccidiostat such as sulfaquinoxaline in the water. Another enteric condition is Tyzzer's disease, *Clostridium piliforme,* which causes small focal lesions in the liver as well as the usual enteropathy signs of intestinal hyperemia (inflammation), hemorrhage, and edema.

***Parasitic diseases.*** Protozoal diseases in rabbits include hepatic and intestinal coccidiosis, encephalitozoonosis, and the rare conditions toxoplasmosis and sarcosporidiosis. The severity of clinical signs of coccidiosis in rabbits varies with host immunity (infection by an *Eimeria* species confers immunity to that species), virulence of organism, and number of sporulated oocysts ingested. In hepatic coccidiosis (*Eimeria stediae*), there may be no signs, or weanlings may have an anterior abdomen distended by an enlarged liver in which biliary tracts are grossly distended by coccidia-induced biliary hyperplasia and exudation. Intestinal coccidiosis, caused by many *Eimeria* species, causes epithelial and lamina propria damage in any of several sections of the intestinal tract. Sudden death or acute or chronic diarrhea are common signs of coccidiosis, which has become, through good sanitation, an uncommon disease in American rabbitries.

*Encephalitozoon cuniculi* is spread by the urine of affected adults and passes through the intestine to the brain and kidneys where focal lesions are produced by sporozoites. The only clinical signs reported are central nervous system signs (circling, head tilt, convulsions) seen in young, heavily infected rabbits. Toxoplasmosis and sarcosporidiosis are reported rarely in rabbits and may produce neural or muscular signs.

Intestinal nematodes and cestodes may occur in domestic rabbits but are rare except for the rabbit pinworm, *Passalurus ambiguus,* a bacterial feeder, which in very heavy populations can cause a rabbit to appear in "poor condition," perhaps due to weight loss and a rough hair coat.

Ectoparasites may be common in poorly maintained rabbitries. Fleas (several types can be treated with the insecticide imidacloprid, Advantage®),[6] ticks, lice, flies, and mites may infest rabbits. Ear mites, *Psoroptes cuniculi,* are a common problem. *Psoroptes cuniculi* is a sucking mite with a predilection for the ear canal and inner surface of the pinna. The mites, often present in populations of many thousand, ingest serum, blood cells, and epithelial debris and create and live among brown, crusty material (serum, tissue, and mite debris) adhering to tender, reddened skin. Removal of these scabs (and the mites) without prior softening with oil or an oil-based acaricide causes rabbits considerable pain. Two or three subcutaneous injections of ivermectin given at 2-week intervals will eliminate mites and allow the rabbit to remove the crusts. *Psoroptes* mites are seen easily with the naked eye or by hand lens.

*Cheyletiella parasitovorax* is a nonburrowing fur mite (Figure 7.12) that causes superficial, crusty, oily, nonpruritic dermatitis and hair loss over the dorsum and is easily treated and cured by using ivermectin. It can be visualized as a small white speck moving among the hairs—hence its nickname "walking dandruff."

Fungal disease is confined primarily to ringworm caused by *Trichophyton mentagrophytes*, which causes a characteristic hair loss and flaky dermatitis on the face, forelimbs, and feet of young or debilitated rabbits. Diagnosis and treatment is as for ringworm in other small animal species (Chapter 14).

Neoplastic disease is, with three exceptions, rare in rabbits. The most common neoplastic condition, seen often in older does of certain breeds, is the uterine adenocarcinoma, which is suspected in cases of infertility and vulvar discharge. If detected early by clinical signs or ultrasound, a hysterectomy can remove the tumors (they are often multiple) before they spread by extension or metastasis. Another relatively common tumor is the nephroblastoma, which involves one or both kidneys but is detected only on necropsy. Lymphosarcoma is the third most common rabbit neoplasm. Less common neoplastic conditions include osteosarcoma (bone cancer), skin tumors, and several other types including the virus-caused fibromas, myxofibromas, and papillomas.

*Miscellaneous conditions.* Ulcerative pododermatitis or "sore hocks" occurs on the plantar surface (sole) of one or more feet in confined rabbits with thin or continuously wet foot pads or in rabbits that thump excessively. Lesions may occur on any or all four feet and are treated by housing the rabbit on soft, dry bedding.

Pregnancy toxemia in does in late gestation results in death of does or young and is caused by reduced blood and nutrient supply to the utero-placental unit. Pregnant does should be fed quality rabbit feed and should not be obese. There is no specific, reliable effective treatment.

Incisor malocclusion (and consequent overgrowth) occurs in most cases because in affected rabbits the maxilla is too short for the mandible; this causes upper incisors to fall behind lower incisors and results in their not being continuously worn down. Treatment is by frequent, careful tooth trimming or by extraction of all six incisor teeth. Malocclusion may result also if teeth are broken or if tooth roots are infected and as a result teeth do not occlude properly.

Splay leg is a clinical sign caused by several abnormal conditions including broken lumbar spine, fractured limb bones, and malformed bones and joints. Time of onset is related to a traumatic event or to genetic influences. Posterior paralysis usually is permanent if the spinal cord is damaged; however, rabbits with cord contusion but not tearing may with time regain use of legs and anal and urinary functions.

Gastric hairballs (trichobezoars) and atony of the gastric musculature are seen commonly in long-haired rabbits or in rabbits fed a low-fiber diet. Signs of gastric immotility or blockage include inappetance, depression, weight loss, absence of feces, and, if lactating, starvation and death of young. Many rabbits, however, may have a gastric hairball and presumably a functional stomach, so no clinical signs occur. Treatment of rabbits showing clinical signs involves hydrating the rabbit and judicious use of drugs to stimulate gastric muscle

---

[6]Imidacloprid (Advantage®) is safe for use in rabbits. However fipronil (Frontline®, Topspot®) is toxic and may kill rabbits.

**Figure 7.12** *Cheyletiella* mite at 400× magnification. This common fur mite can infest rabbits, dogs, cats, and humans. *Courtesy University of Illinois.*

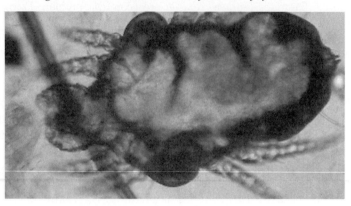

contraction. Surgery may be necessary in some cases to remove the hair mass. Hairball formation is best prevented by feeding a high-fiber diet.

### 7.6.9 ZOONOSES

Humans often develop allergies to rabbit hair. Zoonotic agents (most rare) found in rabbits include *Cheyletiella parasitovorax*, *Salmonella* spp, *Listeria monocytogenes*, and occasionally *Pasteurella multocida*.

## 7.7 MICE AND RATS

### 7.7.1 INTRODUCTION

Mice and rats are members of the class Mammalia, subclass Myomorpha, order Rodentia. Common species of rats are in the family Muridae. Color Plates 10C and 10D depict the Brat rat (Long Evans) and Brown Norway rat, respectively. There are over 1,100 species of mice. The house mouse, *Mus musculus,* is in the family Muridae. Interestingly, deer mice and harvest mice belong to the family Cricetidae—the family that also includes pack rats, voles, hamsters, lemmings, muskrats, and gerbils. Pocket mice and kangeroo rats belong to the family Sciromorpha.

The laboratory rat, *Rattus norvegicus,* is the rat most commonly selected as a pet. The organized production of rats is thought to have started in Europe where they were used in a sporting game in which people wagered on identifying the terrier dog that could kill the most rats confined in a pit. People who bred these rats selected individuals for hair color, length of hair, curly or straight hair, and other preferred traits. They called their pets "Fancy," which is an old English word meaning a hobby. The mouse, *Mus musculus,* was also produced as Fancy Mice (Figure 7.13 and Color Plates 10B and 10H). Albino Fancy Rats and

**Figure 7.13** Pet mice are available in a variety of colors. *Photograph by Jim Roy. Courtesy Hilltop Lab Animals, Inc.*

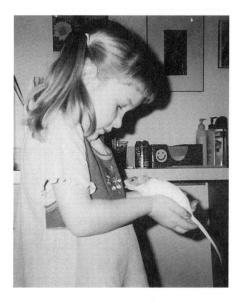

**Figure 7.14** Kelsey Elbel enjoys interacting with "Girlfriend," a rat belonging to Lee's Summit Animal Hospital. Kelsey is the daughter of Kathy Elbel, RVT, who has worked at the hospital for more than 20 years. *Courtesy Robert E. "Bud" Hertzog, DVM, Lee's Summit, MO.*

Mice contributed to the blood lines of today's laboratory rats and mice. Most of today's laboratory rats and mice can be domesticated and make good pets.

The idea of selecting a mouse or rat as a companion pet might disgust some people. They know that rodents are destructive, and people routinely kill rats and mice that invade their home. However, a growing number of people have found that mice and rats really do make good pets (Figure 7.14). Valid reasons for people to consider rats and mice as pets are because apartment owners may not allow renters to keep dogs and cats in the unit, homeowners may not have sufficient yard space for dogs, and small animals are less expensive to maintain. Children commonly have the responsibility of feeding the animals, cleaning the cages, and making certain the pets are comfortable and happy. A disadvantage of keeping mice and rats as pets is that they have a short life span (only about 2 to 3 years), which will introduce children to the unpleasant experiences associated with death of a pet. However, this lesson has the advantage of helping prepare them for the inevitable loss of family members.

The relationship that develops between the pet mouse or rat and its owner is indeed similar to that of other species; people love their pets! After enjoying their first pet mouse or rat, some people might consider becoming members of groups such as the Rat & Mouse Club of America (www.rmca.org). Members of these clubs have been breeding and exhibiting mice and rats for more than 100 years.

After deciding to purchase rodents for pets, one should learn about feeding, housing, and disease control to keep the animals healthy and happy. An excellent reference is *The Biology and Medicine of Rabbits and Rodents* (Harkness and Wagner, 1995).

Pet stores usually offer for sale a number of animal species they obtained from an equally large number of providers. An animal in this environment is constantly being exposed to sick or diseased animals. Animals that have been recently exposed to a disease agent may not show signs of illness for a few days. It is best to obtain mice and rats from rodent producers that sell disease-free animals to the research community.

## 7.7.2   ANATOMY

The anatomy of mammals has many similarities but a comparative anatomist may cite a multitude of differences among species. The rat's incisor teeth grow continuously and must be checked frequently to detect abnormal growth that could result in malocclusion, which may reduce or even prevent food intake. A special gland, the harderian gland, is located behind the rat's eyeball. When the rat is stressed or sick, the gland produces a red porphyrin secretion that resembles blood. If the red secretion is seen around the rat's eyes and nose, quickly determine if it has water and food. Lack of water for a few hours can be very stressful. Rats have no gallbladder but do produce bile.

The mouse has a large surface area relative to body weight. This anatomical difference changes the way mice respond to cold room temperatures and water loss. Avoiding wide variations in room temperature and providing an unlimited water supply will add to mouse comfort and health.

### 7.7.3 PHYSIOLOGY

Pet owners should become familiar with many physiological facts to properly raise and care for rats and mice. Useful biological and physiological facts about rats and mice are given in Table 7.9.

The gestation periods of rats and mice are short, usually 21 to 23 days. They begin breeding at about 60 days of age. Litter size averages 6 to 12 pups. The mother usually has 12 teats. Unless a house full of rodents is wanted, the owner must learn to manage the breeding habits of these pets. Many people purchase only one sex, which solves the prolific procreation potential of mice and rats.

At birth, the pups are hairless, have a beautiful pink color, and are called "pinkie" mice (Figure 7.15). A small white spot can be seen posterior to the left ribs when the stomach is full of milk. Most pups will be fully haired in about 7 to 10 days. The eyelids are closed at birth and open when the pups are about 14 to 17 days old. The mother usually weans the pups at about 3 weeks of age.

Newborn pups should be handled gently. When rats are weaned, handling is a welcomed experience with certain exceptions. Cotton rats do not enjoy being handled, so it is important to select Sprague Dawley or Long Evans rats for pets (Figure 7.16 and Color Plates 10C and 11L). With additional experience, one will be able to properly care for the aggressive stocks and strains. Docile stocks of rats will bite only when they are hurt by improper handling; a rat should be picked up by putting a hand over its shoulders and gently closing fingers around its chest. A rat should never be picked up by the tip of the tail. The tail is not strong enough

**Table 7.9**
Selected Biological and Physiological Information Pertaining to Mice and Rats

| Value | Mouse | Rat | Miscellaneous tips |
|---|---|---|---|
| **Gestation period** | 20–30 days, usually about 21 days | 20–23 days | Nursing increases the gestation period |
| **Weaning age** | 21–28 days | 21–25 days | |
| **Birth weight** | 0.02–0.07 oz (0.75–2.0 g) | 0.17–0.2 oz (5–6 g) | |
| **Food consumption, adults** | 0.5 oz/3.5 oz (15 g/100 g) body weight daily | 0.35 oz/3.5 oz (10g/100 g) body weight daily | Overeating shortens life span |
| **Water consumption, adults** | 0.5 oz/3.5 oz (15 g/100 g) body weight daily | 0.3–0.4 oz/3.5 oz (10–12 ml/100 g) body weight daily | Mice and rats will soon die without water |
| **Breeding age, male** | 50 days | 65–110 days | |
| **Breeding age, female** | 50–60 days | 65–110 days | |
| **Estrous cycle** | 4–5 days | 4–5 days | |
| **Litter size** | 4–12 pups | 6–12 pups | |
| **Life span** | 1.5–3.0+ years | 2.5–3.5+ years | Freedom from disease increases life span |
| **Breeding duration** | 200–400 days | 350–450 days | Varies with stock and strain |

**Figure 7.15** Newborn "pinkie" mice. *Courtesy Dr. Steele F. Mattingly.*

to support the weight of the rat. Occasionally, improper handling will result in the skin being pulled off the tail. Most stocks and strains of mice do not enjoy being cuddled in the hand. For children or adults who want to pet their animals, the mouse is not a good choice.

## 7.7.4 HUSBANDRY

Producers of food animals have long recognized that success depends on sound husbandry practices and the same is true for rodents. They must be provided proper food, shelter, and protection from diseases.

As with food animals, the daily food intake is called a ration. The ration must be balanced and contain the proper amounts (not too much or too little) of carbohydrates, fats, minerals, protein, and vitamins needed for growth and good health of all pets. Rodents not provided proper nutrition develop deficiencies and may soon become unhealthy. Pet stores usually stock rodent diets that have been developed for research rodents. This is a good choice if the diet is fresh—ideally less than 6 months since it was milled and certainly less than a year. Additionally, it must have been properly stored, because feed exposed to high temperatures loses much of its vitamin potency. During storage, feed must be protected from stray or wild rodents looking for food, because they may be diseased and contaminate the feed with deadly pathogens.

Water must be available at all times for rats and mice. A good choice is a bottle with a sipper tube from which the mouse or rat drink their needed supply of water (Figure 7.17). These tubes must be checked to determine whether they are patent (open) so water can

**Figure 7.16** Sprague Dawley rats. *Photograph by Jim Roy. Courtesy Hilltop Lab Animals, Inc.*

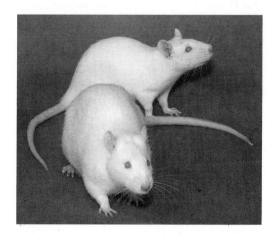

**Figure 7.17** Sipper bottles provide a convenient, sanitary source of water for mice. *Courtesy Dr. James E. Corbin.*

come out the opening, and be positioned so the animal can readily access it. Rats and mice housed without water will die in a few days.

Shelter or housing for rodents is analogous to human residences. Homes come in many shapes and sizes but all have some common qualities that keep the inhabitants happy. Rats and mice must be maintained in an enclosure made of wire, stainless steel, special plastic, or glass. Wood would be destroyed by the gnawing of rats or mice and is difficult to keep clean. Pet stores frequently offer a good selection of cages. Cage size depends on the number of animals to be housed and whether they are expected to reproduce. A single adult mouse requires a minimum cage space of 15 sq inches, but a female with a litter should have at least 65 sq inches. Adult rats should be provided at least 40 sq inches of floor space. It is advisable to select the largest cage the pet store stocks for pet mice and rats. They will enjoy the additional space. Some owners provide two levels in the cage. Adding furniture such as flowerpots, empty coconut shells, PVC pipes, and connectors keep mice and rats busy and add to the pet's quality of life. Paper houses have been put in cages. Interestingly, some pets will tear up the house in short order, whereas others will use it as a bedroom. The cage floor should be covered with shredded paper, wood shavings, or processed corn cobs as bedding. These will absorb moisture and not be considered food by the mouse or rat. To keep odors at a minimum, bedding is usually changed 2 to 3 times a week.

Room temperature should be maintained between 65°F and 85°F and humidity in the range of 30% to 70%. Low humidity (less than 30%) will remove moisture from the skin of the tail causing it to contract and restrict its flow of blood. If not corrected within a few hours, the rat's tail will drop off.

The rat needs a light cycle of 12 hours of light and 12 hours of dark for most successful breeding.

## 7.7.5 COMMON DISEASES

Parasites, bacteria, and viruses can create major health problems for the rat. Purchasing only disease-free rats and preventing them from being exposed to disease-causing agents will help minimize major causes of disease.

People who care for research rodents practice preventive medicine by wearing sterile gowns and gloves when they handle the animals (see Figure 22.7). This reduces or prevents the spread of diseases from humans to rats or mice. When owners are ill, they should not handle their pets. Some human respiratory and enteric diseases will infect rats and mice. Washing hands before and after handling pet rodents is a good practice.

Respiratory diseases are the most common health problems encountered in mice and rats.[7] At least 14 specific agents have been classified as respiratory pathogens of mice and

[7]National Research Council. 1991. *Infectious Diseases of Mice and Rats.* Washington, DC: National Academy Press.

rats. Clinical signs frequently observed include snuffling, chattering, rapid breathing, weight loss, hunched posture, ruffled coat, inactivity, and head tilt. Bacteria and viruses are the main causes of these conditions.

The digestive system is attacked by at least 21 infectious agents and a number of parasites. Unexpected deaths, watery diarrhea, yellow stools, oily coats, stunting, and even respiratory changes may be associated with diseases of the digestive system.

When pets show some of these signs, the owner should call a veterinarian. Because the cost of diagnosis and/or treatment will frequently be more than the cost of new animals, many veterinarians engaged in private practice have limited experience treating sick mice and rats. Some refer pet owners to veterinarians who work with research rodents. The veterinarian should write out detailed treatment instructions when sick mice or rats are to be treated by the owner.

External and internal (hair coat and intestinal) parasites can be a problem. The neck and anterior trunk are favorite locations for mites to make their home on mice and rats. Mites feeding on animals will cause ulcerative lesions on the neck and anterior trunk. Cestodes, or tapeworms, and their intermediate forms can be found in poorly managed rodent colonies. If the mice or rats come from unknown sources, a veterinarian should check for these parasites because *Hymenolepis nana* is pathogenic for humans and can lead to enteric disease.

Tumors, especially mammary tumors, are commonly observed in aged animals. The mammary tissue may extend to the top of the back. Obtain professional advice if a mass appears on the abdominal wall or near the animal's back. Many tumors are easy to remove surgically, and unless they are removed, the unsightly tumor will become a burden for the animal to carry.

When an animal stops eating, it is frequently due to the lack of water or a dental problem. Malocclusion can make it difficult for an animal to eat.

## 7.7.6 ZOONOSES

Zoonoses refer to those diseases that may be transmitted from animals to humans or from people to animals. The Black Plague was carried throughout Europe by rodents, and a frequently held view is that all rats are carriers of deadly diseases. Pet rodents are not dangerous unless they have been exposed previously to disease agents. Therefore, people who purchase a healthy, disease-free rat need not worry about their pet giving them a zoonotic disease unless they or their human friends first give the pet rat a zoonotic disease.

Rats, as with many other species of pets, can be a source of allergic diseases. These diseases are the result of individuals being exposed to proteins present in rat urine or saliva. The immune system of some humans recognizes these proteins as foreign substances that could be dangerous and responds by developing antibodies. In a few individuals, these antibodies cause an allergic reaction that occasionally is sufficiently severe to cause death. We should remember that although allergic conditions can be caused by pets, there are many other sources of allergens (e.g., house dust, insects, mites, pollens, and mold spores).

## 7.8    SUMMARY

The important, informative, and interesting information about chinchillas, gerbils, guinea pigs, ferrets, hamsters, rabbits, mice, and rats in this chapter has given students and other persons interested in the expanded family of pets a broader perspective of the biology, care, health, and management of companion animals.

## 7.9   REFERENCES

1. Aiello, Susan E. 1997. *The Merck Veterinary Manual.* 8th ed. Whitehouse Station, NJ: Merck & Co.

2. Canadian Council on Animal Care. 1980–1984. *Guide to the Care and Use of Experimental Animals* (2 volumes). Ottawa, ON: CCAC.

3. Cheeke, P. R., ed. 1987. *Rabbit Feeding and Nutrition.* San Diego, CA: Academic Press.

4. Field, Karl J., and Amber L. Sibold. 1998. The Laboratory Hamster and Gerbil. In *A Volume in the Laboratory Animal Pocket Reference Series,* edited by Amber L. Sibold and Mark A. Suckow. Boca Raton, FL: CRC Press.

5. Fox, J. G. 1988. *Biology and Diseases of the Ferret.* Philadelphia, PA: Lea and Febiger.

6. Fox, James G., Lynn C. Anderson, Franklin M. Lowe, and Fred W. Quimby, eds. 2002. *Laboratory Animal Medicine.* 2nd ed.: 286–291. San Diego, CA: Academic Press.

7. Hanes, Marti. 1999. Diseases of Gerbils (*Meriones unguiculatus*). In *Pathology of Laboratory Animals.* Washington, DC: Armed Forces Institute of Pathology. www.afip.org/vetpath/POLA/99/1999-POLA-Meriones.htm

8. Harcourt-Brown, F. 2002. *Textbook of Rabbit Medicine.* Boston, MA: Butterworth Heinemann.

9. Harkness, John E., and Joseph W. Wagner. 1995. *The Biology and Medicine of Rabbits and Rodents.* 4th ed. Baltimore: Williams and Wilkins.

10. Hayes, Pilar M. 1999. Diseases of Chinchillas. In *Kirk's Current Veterinary Therapy XIII: Small Animal Practice,* edited by John Bonagura, 1152–1157. Philadelphia: W. B. Saunders.

11. Hillyer, E. V., and K. E. Quesenberry, eds. 1997. *Ferrets, Rabbits, and Rodents: Clinical Medicine and Surgery.* Philadelphia: W. B. Saunders.

12. Hoefer, Heidi L. 1994. Chinchillas. In *Veterinary Clinics of North America: Small Animal Practice* 24: 103–111. Philadelphia: W. B. Saunders.

13. Hrapkiewicz, Karen, Leticia Medina, and Donald D. Holmes. 1998. *Clinical Laboratory Animal Medicine.* 2nd ed. Ames: Iowa State Press.

14. Jenkins, J. R., and S. A. Brown. 1993. *A Practitioner's Guide to Rabbits and Ferrets.* Lakewood, CO: American Animal Hospital Association.

15. Krauss, H., et al. 2003. *Zoonoses. Infectious Diseases Transmissible from Animals to Humans.* 3rd ed. Washington, DC: ASM Press.

16. Manning, P. J., D. H. Ringler, and C. E. Newcomer. 1994. *The Biology of the Laboratory Rabbit.* 2nd ed. San Diego, CA: Academic Press.

17. Marini, R. P., G. Otto, S. Erdman, L. Palley, and J. G. Fox. 2002. Biology and Diseases of Ferrets. In *Laboratory Animal Medicine,* 2nd ed., edited by J. G. Fox, L. C. Anderson, F. M. Loew, and F. W. Quimby, 483–517. San Diego, CA: Academic Press.

18. National Research Council. 1991. *Infectious Diseases of Mice and Rats.* Washington, DC: National Academy Press.

19. Nowak, Ronald M. 1997. *Walker's Mammals of the World.* Baltimore: Johns Hopkins University Press.

20. Oberman, L. 1994. *Diseases of Domestic Rabbits.* 2nd ed. Boston, MA: Blackwell Scientific Publications.

21. Phipatanakuil, Wanda. 2002. Rodent Allergens. *Current Allergy and Asthma Reports* (2): 412–416.

22. Purcell, K. 1999. *Essentials of Ferrets: A Guide for Practitioners.* Lakewood, CO: American Animal Hospital Association.

23. Richardson, V. C. G. 1994. *Rabbits: Health, Husbandry and Diseases.* Malden, MA: Blackwell Science, Ltd.

24.  Suckow, M. A., and F. A. Douglas. 1997. *The Laboratory Rabbit.* Boca Raton, FL: CRC Press.

25.  Universities Federation for Animal Welfare Staff. 1999. *The UFAW Handbook on the Care and Management of Laboratory Animals,* 7th ed. (1), edited by Treavor Poole. Ames: Iowa State Press/Blackwell Publishing.

26.  Van Hoosier, G., and Charles McPherson, eds. 1987. *Laboratory Hamsters.* San Diego, CA: Academic Press.

27.  Weir, B. J. 1970. Breeding Techniques: Chinchillas. In *Reproduction and Breeding Techniques for Laboratory Animals,* edited by E. S. E. Hafez, 209–223. Philadelphia: Lea & Febiger.

# 7.10   ABOUT THE AUTHORS

Dr. Douglas W. Stone, BS, DVM, MS, DACLAM, authored the sections on chinchillas, gerbils, guinea pigs, and hamsters. Dr. Stone is Director of the Office of Animal Research Services at the University of Connecticut. He has served as a member of the Association for Assessment and Accreditation International (AAALAC) Council on Accreditation, including two terms as president. Dr. Stone has received numerous honors for his scholarly research and professional service including The Ohio State University College of Veterinary Medicine Distinguished Service Award and the Central Ohio Branch of the AALAS Distinguished Service Award. He is a member of Phi Zeta National Honor Society of Veterinary Medicine.

John E. Harkness, DVM, MS, MEd, DACLAM, authored the sections on ferrets and rabbits. Dr. Harkness is University Laboratory Veterinarian and Professor, College of Veterinary Medicine, Mississippi State University. He earned his DVM degree from Michigan State University and his MS and residency in laboratory animal medicine and an MEd in Higher Adult Education from the University of Missouri–Columbia. His professional career has been in the employment of Kansas State University (USAID project in Nigeria), the University of Missouri, The Pennsylvania State University, and Mississippi State University. He is co-author of the textbook titled *The Biology and Medicine of Rabbits and Rodents.* Dr. Harkness authored the book titled *Pet Rodents—A Guide for Practitioners* (originally published as *A Practitioner's Guide to Domestic Rodents*), co-authored the book titled *Exotic Formulary,* and contributed chapters to numerous books and dozens of scientific papers in professional journals and popular articles in trade magazines and elsewhere. His leadership at the national/international level is noteworthy by his having been elected to serve as president of three major professional organizations: American Society of Laboratory Animal Practitioners, American College of Laboratory Animal Medicine, and Council on Accreditation of the Association for Assessment and Accreditation of Laboratory Animal Care International.

Dr. Steele F. Mattingly, BS, DVM, DACLAM, authored the section on mice and rats. Following his earning the DVM degree from Auburn University in 1955, he worked for the Dow Chemical Company (1955–1962) caring for animals used to test human and veterinary biologics. From 1962–1965, he managed the Laboratory Supply Company in Indianapolis that produced mice and rats. In 1965, Dr. Mattingly was recruited to establish the Department of Laboratory Medicine in the University of Cincinnati College of Medicine. Following an illustrious 26-year tenure, he took early retirement in 1991 from the University of Cincinnati to serve as a consultant for industrial and academic institutions. Dr. Mattingly is known internationally for the professionally enriching continuing education courses he organized at the University of Cincinnati. He served as President of the American Society of Laboratory Animal Practitioners and of the American Association for Laboratory Animal Science (AALAS). He has served on numerous Boards of Directors including his 1999–2003 service on the Board of Trustees of the United Theological Seminary. Berea College Alumni Association named him the 1997 recipient of its Distinguished Alumnus Award. Dr. Mattingly was the 1999 recipient of the Griffin Award, the highest award given by AALAS.

# 8 RECORDS AND CASE HISTORIES

*The study of the causes of disease must be preceded by the study of diseases caused.*

Sir William Osler (1849–1919)
*Canadian physician*

## 8.1 INTRODUCTION/OVERVIEW

Clear and accurate record keeping is extremely important for those involved in boarding, breeding, grooming, providing medical care, marketing, showing, training, or other businesses and hobbies with companion animals. In an emergency, having records showing an animal's vaccination status and past medical problems is important to ensure that proper care is given for the current problem. As noted in Chapters 14 and 16, an accurate history provides important clues required to diagnose many medical problems. Inaccurate record keeping can result in suspension of registration privileges for breeders. Veterinarians found in violation of acceptable standards[1] for medical records will be subjected to disciplinary

---

[1] In the United States, each state has its own Veterinary Practice Act, which includes the acceptable standards for medical records and other regulations that veterinarians and veterinary technicians must follow to maintain their licensure.

measures and possible loss of their professional license. Similarly, incomplete or missing business records may result in severe penalties from the Internal Revenue Service.

## 8.2   ANIMAL NAME AND IDENTIFICATION

Owners of dogs and cats should keep the following information for every animal they own: (1) breed; (2) name (called name and registered name, if applicable); (3) registration number (if applicable); (4) permanent identification (if applicable—identification tag, tattoo, or microchip number); (5) name and registration number of sire (if applicable); (6) name and registration number of dam (if applicable); (7) date of birth; (8) sex; (9) color and markings; (10) name, address, and telephone number of the person from whom the animal was obtained; (11) breeder; (12) any transactions involving this animal [breeding record (Section 8.4), performance record, health record (Section 8.5), final disposition of the animal (dead, sold, given away—if sold or given away record the name, address, and telephone number of that person)].

## 8.3   PEDIGREE

A pedigree is a record of an individual's ancestry (genealogy). Pedigrees frequently include the registration number of the individual and its ancestors. Other information that may be included on pedigrees includes hair coat color(s), special recognitions awarded to an individual (e.g., championship, various performance titles, and others), and results of tests providing genetic-related information (DNA registry number, Orthopedic Foundation for Animal's hip dysplasia registry classification, Canine Eye Registration Foundation registry classification, and others). Evaluation of ancestors for four or more generations is helpful in formulating a breeding plan to improve a breed. The sire's family is listed at the top of a pedigree and the dam's family is listed at the bottom of a pedigree (Figure 8.1).

## 8.4   BREEDER RECORDS

Breed registries require breeders of dogs and cats to keep complete and accurate records. Those breeding dogs and cats as a business must also keep a record of all financial transactions made by their kennel or cattery (Chapter 19).

At the time of a breeding, records should include (1) complete name and registration numbers of both the male and female, (2) dates the female was bred, (3) name(s) of person(s) handling the mating, (4) whether the breeding was natural or used artificial insemination, (5) names and addresses of the owners of both animals, and (6) a copy of the breeding contract (if applicable). Litter records should include (1) breed, (2) date of birth, (3) dam's name and registration number (if applicable), (4) sire's name and registration number (if applicable), (5) dam owner's name and address, (6) sire owner's name and address, (7) litter registration number (if applicable), and (8) number of males and females born in the litter. Individual records for puppies or kittens should include (1) sex, (2) identification (number, tattoo, microchip, or name), (3) colors and markings, (4) birth weight, (5) disposition (sold, dead, or given away) with date and name/address/telephone number of new owner (if applicable), and (6) date and type of papers provided to new owner (purchase contract, registration certificate or application, whether limited or full registration).

## 8.5   HEALTH RECORDS FOR OWNERS

Owners should maintain basic information as part of the health history of their companion animals (Figure 8.2). These health records should include (1) a complete record of all vaccinations (and any adverse reactions to vaccines), (2) a record of the animal's weight, (3) any

**Figure 8.1**    This five-generation pedigree includes breed and performance titles awarded to each animal (see Chapter 13 for information on the meaning of these titles). *Courtesy Jane Rothert.*

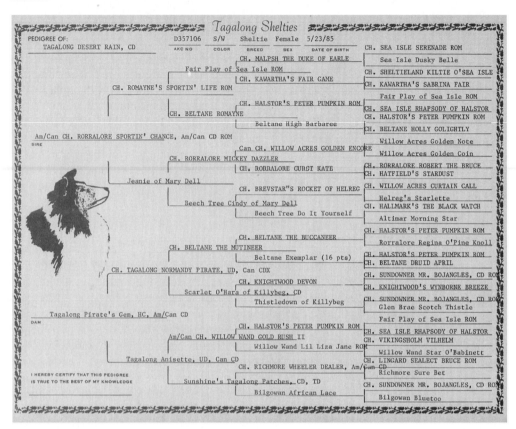

medical problems or surgeries and the animal's response to all treatments (including any adverse reactions to medications), (4) a complete dietary history (this is helpful should the animal develop a food allergy later in life), (5) copies of test results (e.g., fecal examinations, radiograph reports, heartworm and other blood tests, biopsies or other diagnostic tests), and (6) legal documents (e.g., county registration and insurance policies).

## 8.6    MEDICAL RECORDS

Medical records are legal documents that must be kept by veterinarians for a minimum of 6 years (may vary by state depending on the applicable Veterinary Practice Act regulations). Veterinary medical records should provide information documenting adequate medical care that is in accordance with professional standards. Medical records should include (1) name, address, and telephone number of the animal's owner; (2) name and identity of the animal; (3) birth date, sex, and breed of the animal; (4) dates the animal has been under the veterinarian's care; (5) history of the animal; (6) lesion distribution forms recording the location of any skin lesions or tumors (Figure 8.3); (7) list of problems and diagnoses made; (8) medications (including doses and frequency) given; (9) test results and progress reports (if applicable); (10) consent forms (if applicable, e.g., consent for treatment, endorsement of estimated bill, consent for euthanasia); (11) log of any surgeries (including anesthesia records); and (12) log of communications with the client (e.g., release forms, instruction forms, summary of discussions and telephone calls). Medical records must be maintained by the attending veterinarian. Owners may request copies of the medical records (veterinarians may charge a fee for copying records).

**Figure 8.2**   Example of a pet health record. *Courtesy University of Illinois.*

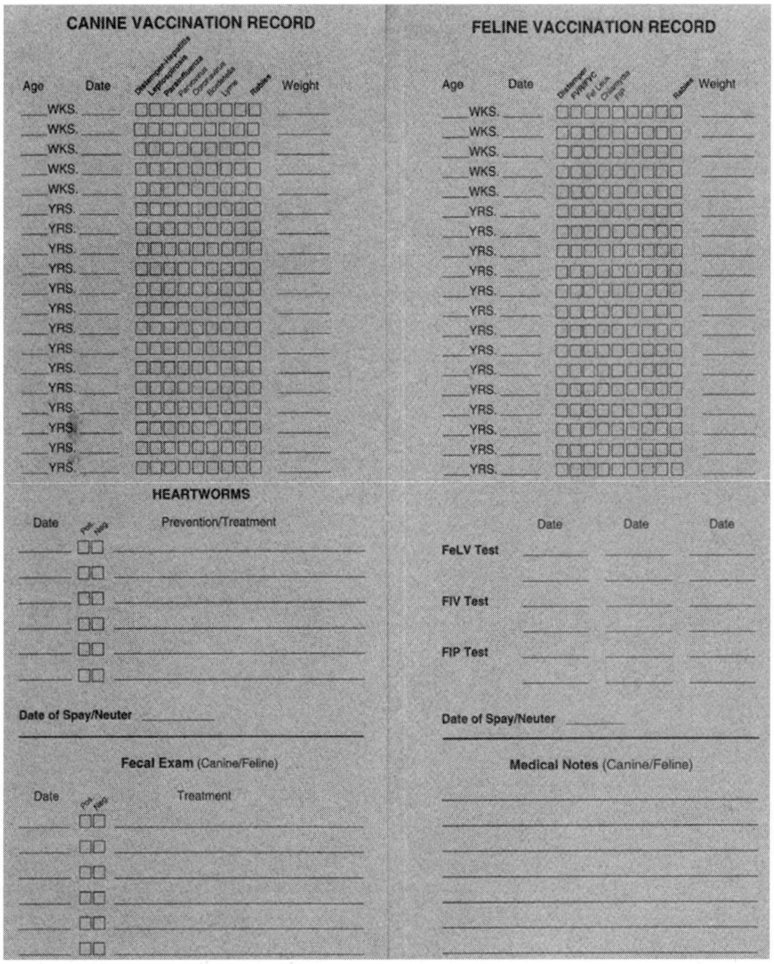

## 8.6.1   PROBLEM-ORIENTED MEDICAL RECORD

Problem-oriented medical record (POMR) is a widely used format for the organization of information in medical records. The method was developed by Lawrence Weed, MD, in the 1960s and soon became popular as a way to organize information in a useful format that results in concise, complete, and accurate records. In POMRs, each "complaint" is identified as a problem (e.g., due for vaccinations, possible parasitism, vomiting, diarrhea, lameness, hair loss, and others). Information is typically collected and recorded on history and physical examination forms (Figures 8.4 and 8.5). When the problems have been identified, notes concerning each problem are recorded using the SOAP format. "S" stands for subjective information (e.g., the history of the problem). "O" stands for objective information (e.g., findings of physical examination or diagnostic tests). "A" stands for assessment of the problem (e.g., a diagnosis or a list of possible diagnoses—this is called a list of differentials or differential diagnoses and is often abbreviated as DDx). "P" stands for the plan (e.g., plans for further diagnostic tests—abbreviated as $P_{DX}$; and/or treatment plan—abbreviated as $P_{TX}$; and plan for client education—abbreviated as $P_{CE}$). Use of the POMR and SOAP structure is helpful in documenting the veterinarian's line of reasoning. The POMR for a hospitalized patient includes daily progress notes organized using the SOAP format. The patient's record will also include copies of all test results and of discharge summaries.

**Figure 8.3** Example of a topography form used to record the location (distribution) of lesions on an animal. *Courtesy University of Illinois.*

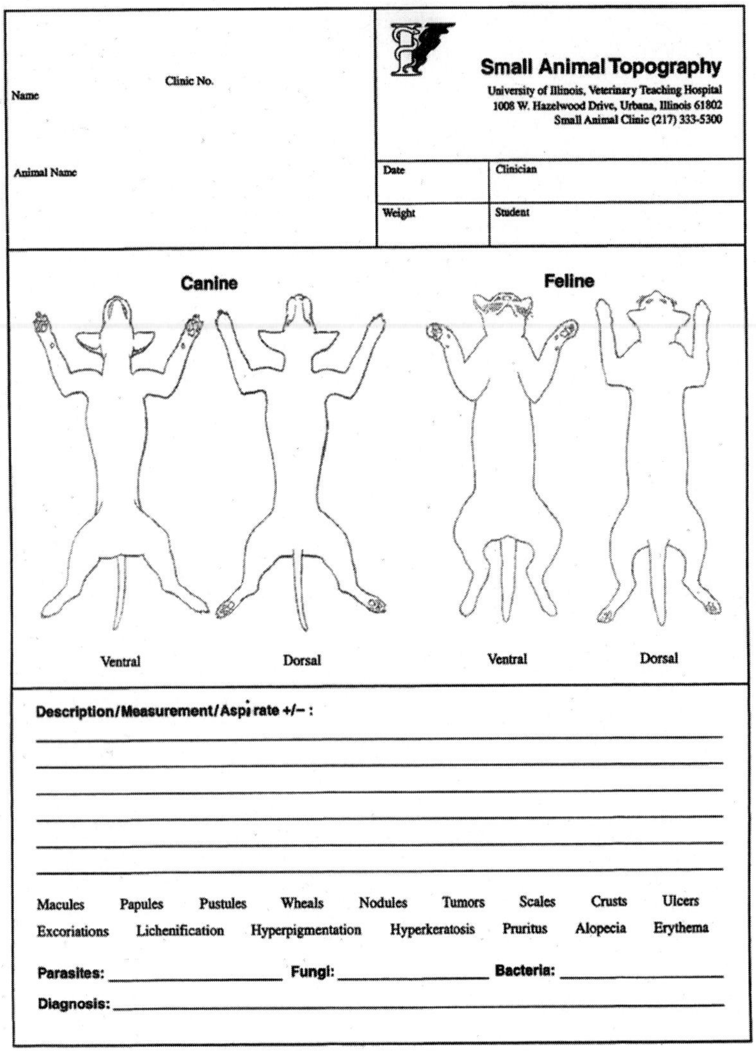

Name
Clinic No.
Animal Name

**Small Animal Topography**
University of Illinois, Veterinary Teaching Hospital
1008 W. Hazelwood Drive, Urbana, Illinois 61802
Small Animal Clinic (217) 333-5300

| | |
|---|---|
| Date | Clinician |
| Weight | Student |

**Canine** **Feline**

Ventral Dorsal Ventral Dorsal

**Description/Measurement/Aspirate +/– :**

Macules Papules Pustules Wheals Nodules Tumors Scales Crusts Ulcers
Excoriations Lichenification Hyperpigmentation Hyperkeratosis Pruritus Alopecia Erythema

**Parasites:** ____________ **Fungi:** ____________ **Bacteria:** ____________
**Diagnosis:** ____________

## 8.7 CASE HISTORIES (EXAMPLES)

### CASE 1: "CAPTAIN."

Chief Complaint: Pruritus (pulling out hair) and scabs on back

S History: The owners have noticed clumps of white hair lying throughout the house for the past 2 weeks and observed scabs on Captain's back (Figure 8.6). Captain is a 6-year-old castrated male domestic shorthair cat. All vaccinations are current, and there have been no previous health problems observed. He is kept indoors. He has been fed Friskies® cat food for the past 5 years (no changes in diet). The family has one other pet that was obtained 1 month ago. It is a 3-month-old intact female black Labrador Retriever puppy. The puppy is taken for walks and allowed to socialize with other dogs at a local dog park; she has been scratching intermittently but has no sores on her skin. Neither animal receives any form of parasite preventive.

**Figure 8.4** Example of a history form used in obtaining information for an animal's medical record. *Courtesy University of Illinois.*

SMALL ANIMAL HISTORY

University of Illinois
Veterinary Medicine
Teaching Hospital

| | | Admission Date | Discharge Date | Clinician |
|---|---|---|---|---|
| | | | | |

**HISTORY AND OWNER'S DESCRIPTION OF PROBLEM:**

What is your animal's problem (why are you visiting the clinic)? _____

_____

How long has this problem been present? _____

How frequently does this problem occur? _____

What previous treatment for this problem has your animal received? _____

_____

What was the response? _____

List all medications your animal is taking: _____

Any adverse reactions or allergies to medications or anesthetics? _____

How long have you owned your animal? _____

Is there any previous illness/injury? _____

List any other animals in your household _____

Has you animal eaten today?   No__   Yes__            (continue history on reverse side)

Normal diet: _____

| | | | |
|---|---|---|---|
| Skin disease: | No__ Yes__ Unknown__ | | |
| Vomiting: | No__ Yes__ Unknown__ | | |
| Difficulty Swallowing: | No__ Yes__ Unknown__ | | |
| Diarrhea: | No__ Yes__ Unknown__ | | |
| Spayed/Castrated: | No__ Yes__ Unknown__ | | |
| Coughing: | No__ Yes__ Unknown__ | | |
| Sneezing: | No__ Yes__ Unknown__ | | |
| Tires Easily: | No__ Yes__ Unknown__ | | |
| Weak/Lame: | No__ Yes__ Unknown__ | | |
| Abnormal Behavior: | No__ Yes__ Unknown__ | | |
| Convulsions: | No__ Yes__ Unknown__ | | |
| Head Shaking: | No__ Yes__ Unknown__ | | |
| Ear Scratching: | No__ Yes__ Unknown__ | | |
| Abnormal Eyes/Vision: | No__ Yes__ Unknown__ | | |
| Weight Loss: | No__ Yes__ Unknown__ | | |

| Vaccination | Date |
|---|---|
| D-H-L Para | |
| Parvo | |
| Rabies | |
| FVRCP | |
| FELV | |
| Other | |
| **Heartworm Check:** | |
| Date of last check | |
| List heartworm preventative(s) given: | |
| | |
| | |

Attitude:     Alert____     Depressed____

| | | | | |
|---|---|---|---|---|
| Appetite: | Increased____ | Decreased____ | Normal____ | Unknown____ |
| Water Consumption: | Increased____ | Decreased____ | Normal____ | Unknown____ |
| Urination: | Increased____ | Decreased____ | Normal____ | Unknown____ |
| Defecation: | Increased____ | Decreased____ | Normal____ | Unknown____ |

Date of last estrus (heat cycle): _____

Signed:_____        _____        _____
           Student                       Clinician                       Date

(medrec)sah.ks/05/19/99

O   Physical examination findings: Captain has multiple scabs and hair loss that is dorsally distributed. Many hairs appear to have been broken off. Black "specs" can be found throughout his hair coat. These "specs" have a reddish tinge when placed on a moistened cotton ball. Acetate tape skin cytology (Chapter 16) shows the presence of *Staphylococcus* spp bacteria in the scabs. Skin scrapings are negative for mites. A fecal examination is negative (no parasites or ova). There is a moderate amount of dental calculus on the teeth. Rectal temperature 101°F, heart rate 166 beats/minute, respiratory rate 28 breaths/minute.

A   Assessment: Captain's clinical signs of itching, hair loss, and scabs can be attributed to fleas and a secondary bacterial skin infection. The black specs were flea feces (the reddish tinge when moistened is due to digested blood). Other differential diagnoses for the itching and hair loss could include allergies (environmental or dietary), ringworm (dermatophyte infection), or stress from introduction of the new puppy to the household. Captain is at risk for periodontal disease.

**Figure 8.5**   Example of a physical examination form used to record abnormalities detected on the examination of an animal. *Courtesy University of Illinois.*

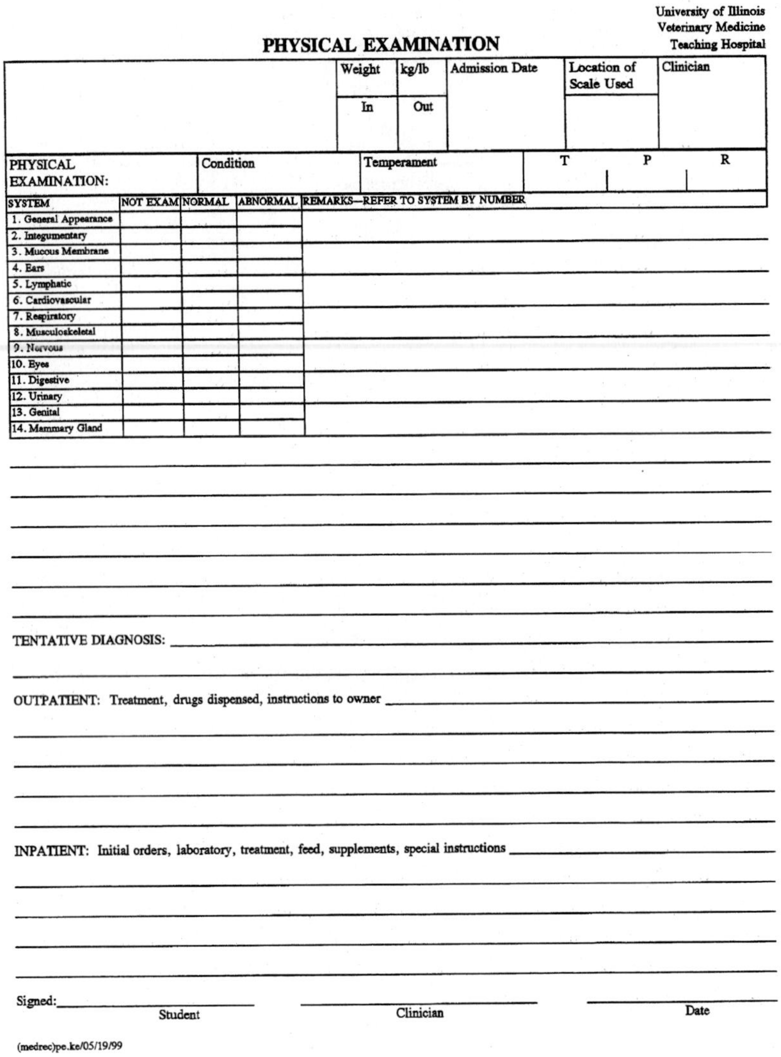

PHYSICAL EXAMINATION

University of Illinois
Veterinary Medicine
Teaching Hospital

|  | Weight | kg/lb | Admission Date | Location of Scale Used | Clinician |
|---|---|---|---|---|---|
|  | In | Out |  |  |  |

| PHYSICAL EXAMINATION: |  | Condition |  | Temperament |  | T | P | R |
|---|---|---|---|---|---|---|---|---|

| SYSTEM | NOT EXAM | NORMAL | ABNORMAL | REMARKS—REFER TO SYSTEM BY NUMBER |
|---|---|---|---|---|
| 1. General Appearance |  |  |  |  |
| 2. Integumentary |  |  |  |  |
| 3. Mucous Membrane |  |  |  |  |
| 4. Ears |  |  |  |  |
| 5. Lymphatic |  |  |  |  |
| 6. Cardiovascular |  |  |  |  |
| 7. Respiratory |  |  |  |  |
| 8. Musculoskeletal |  |  |  |  |
| 9. Nervous |  |  |  |  |
| 10. Eyes |  |  |  |  |
| 11. Digestive |  |  |  |  |
| 12. Urinary |  |  |  |  |
| 13. Genital |  |  |  |  |
| 14. Mammary Gland |  |  |  |  |

TENTATIVE DIAGNOSIS: _____

OUTPATIENT:   Treatment, drugs dispensed, instructions to owner _____

INPATIENT:   Initial orders, laboratory, treatment, feed, supplements, special instructions _____

Signed:_____       _____       _____
            Student                        Clinician                           Date

(medrec)pe.ke/05/19/99

P    P$_{TX}$: Recommend bathing Captain using an antiseptic shampoo and then starting monthly applications of a flea control product (e.g., Frontline™, Merial, Inc.). Dispense a systemic antibiotic (e.g., Clavamox™, Pfizer Animal Health, Inc.) to be given orally for the next 21 days.

P$_{CE}$: Schedule a recheck appointment for 3 weeks; if Captain's itching and skin lesions have not resolved, further assessment will include a culture for ringworm and allergy testing. The house should be thoroughly vacuumed and the vacuum cleaner bag destroyed to help remove fleas from the house. Captain's bed should be washed. An appointment should also be scheduled for Captain's teeth to be cleaned; regular brushing of his teeth would also help prevent future dental disease. The puppy should be examined and started on monthly flea and heartworm preventives.

*Discussion.*   The historical clue important in diagnosing fleas as the probable cause of itching and hair loss was the recent introduction of a new pet in the household. Also important was information related to walking the puppy in parks allowing it to socialize

**Figure 8.6** Case 1. A 6-year-old castrated male domestic shorthair cat with hair loss and scabs on the back. *Courtesy Dr. Karen L. Campbell.*

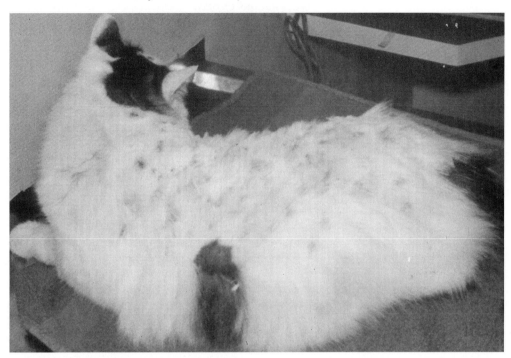

with other dogs. It was concluded that the puppy is likely to get fleas during this socialization and bring them home to infest the cat. The puppy's black hair coat made it difficult to see fleas. The clues for the diagnosis of fleas from the physical examination were the dorsal distribution of the lesions and the presence of flea dirt. The identification of black specs as flea dirt was made by moistening them and finding the reddish tinge characteristic of digested blood. The hair loss was attributed to the cat scratching and biting at fleas and also a possible allergic reaction to flea bites. Differential diagnoses of the scabs included self-trauma from biting and scratching and a secondary bacterial skin infection. Cytological examination of the scabs demonstrated the presence of bacteria in the lesions. The owner was provided with information regarding the treatment required for the current problems (scratching, itching, loss of hair, scabs) and also with recommendations for treating an unrelated problem—dental calculus. The owner was advised of the importance of flea preventive programs for the cat and puppy and also of the need to start the puppy on heartworm preventive. The documentation of this information in the POMR–SOAP format makes it easy for other persons knowledgeable about veterinary medicine to understand the thought processes used in managing this case.

## CASE 2: "BARNEY."

Chief Complaint: Recent onset of severe vomiting and diarrhea

S   History: Barney is a 9-week-old intact male Springer Spaniel puppy. He was obtained from a family friend 2 weeks previously and appeared healthy until last night when he began vomiting. He vomited four times last night and this morning has a watery diarrhea. He had no interest in playing last night and no interest in eating or drinking today. His diet is Purina Puppy Chow™ supplemented with milk; this has been his only diet (started at 3 to 4 weeks of age). He has received no vaccinations or other healthcare (no "dewormers").

O   Physical examination findings: Rectal temperature 102.8°F, heart rate 168 beats/ minute, respiratory rate 32 breaths/minute. Depressed attitude, pale mucous membranes, capillary refill time 3 seconds, estimated as 5% dehydrated, painful on abdominal palpation.

A   Assessment: The most probable differential diagnoses for the fever, abdominal pain, depression, vomiting, and diarrhea include internal parasites (hookworms, roundworms, coccidia, and/or protozoa), a gastrointestinal foreign body (puppies enjoy chewing on objects!), lactose intolerance, and gastroenteritis from toxins/ poisons or infectious agents (parvovirus, coronavirus, canine distemper virus, *Campylobacter, Salmonella,* and others). The pale mucous membranes and slow capillary refill time could result from poor profusion due to dehydration and/or blood loss from parasitic or other gastrointestinal infections. The dehydration is secondary to the vomiting and diarrhea.

P   $P_{DX}$: Initial diagnostic tests should include a fecal examination, a complete blood cell count, a biochemical profile, and a urinalysis. If a diagnosis is not readily apparent, additional diagnostic tests would include a fecal ELISA test for parvovirus (Chapter 16), fecal electron microscopy, fecal cultures, and abdominal radiographs.

$P_{TX}$: After samples are obtained for the initial diagnostic tests, the puppy will be started on intravenous fluid treatment and given an antiemetic (to prevent further vomiting). Additional treatments will depend on the results of diagnostic tests.

$P_{CE}$: Laboratory tests will be needed to determine the cause of Barney's vomiting and diarrhea. He is a very sick puppy and will need aggressive supportive care including intravenous fluids. Once a diagnosis is made, a better estimate can be given on his prognosis. After he recovers, he should be started on a preventive healthcare program involving vaccinations and parasite preventives.

*Follow-up.* The initial test results showed a mild anemia and a severe decrease in the number of white blood cells plus numerous hookworm and roundworm ova. The low white blood cell count was highly suggestive of parvovirus infection; this was confirmed by a positive fecal ELISA test for parvovirus. Concurrent parasitism was contributing to the gastrointestinal disease. Supportive care of intravenous fluids was continued and antibiotic treatment was used to prevent secondary bacterial infections. Barney's condition improved, and after 3 days of treatment, he was discharged from the veterinary hospital, placed on oral antibiotics, and given an anthelmintic to kill the hookworms and roundworms. The owners were instructed to bring him back to the clinic in 3 weeks for a second anthelmintic treatment and vaccinations. Instructions at the follow-up visit would include recommendations for flea and heartworm preventives.

*Discussion.* Important clues in the dog's health history included its age (puppies are highly susceptible to viral infections) and the lack of prior vaccinations. The physical examination findings indicated that aggressive treatment would be required (resulting from dehydration, pale mucous membranes, and slow capillary refill time). Initial test results revealed more than one problem contributed to the puppy's gastrointestinal disease. Using the POMR–SOAP approach to this case facilitated a rapid and accurate diagnosis of the underlying problems and helped ensure a successful outcome.

## 8.8 SUMMARY

Maintaining accurate, complete records related to the biology, care, health, and management of companion animals is fundamental to successful veterinary practices as well as for other related business enterprises.

In this chapter, we discussed the key components of problem-oriented medical records (POMRs). When health problems are identified, notes pertinent to each problem are recorded using the SOAP format, where "S" stands for subjective information, "O" objective information, "A" assessment of the problem, and "P" the plan (e.g., further diagnostic tests planned as well as plans for treatment and client education). Using POMRs and SOAP formats, two histories were discussed.

## 8.9 REFERENCES

1. Chauvis, Sharee A., Judith L. Hutton, and Joanna M. Bassert. 2002. Medical Records. In McCurnin, Dennis M., and Bassert, Joanna M., *Clinical Textbook for Veterinary Technicians,* 5th ed., edited by Dennis M. McCurnin and Joanna M. Bassert, 812–838. Philadelphia: W. B. Saunders.

2. Hannah, Harold W. 1988. Practice and the Law. In *Veterinary Practice Management,* edited by Dennis M. McCurnin, 285–307. Philadelphia: J. P. Lippincott.

3. Poffenbarger, Ellen M. 1991. The Health History. In *Small Animal Physical Diagnosis and Clinical Procedures,* edited by Barbara H. McGuire and Howard B. Seim III, 6–15. Philadelphia: W. B. Saunders.

4. Poffenbarger, Ellen M. 1991. The Physical Examination as a Diagnostic Tool. In McGuire, Barbara H., and Seim, Howard B. III, 1991, *Small Animal Physical Diagnosis and Clinical Procedures,* edited by Barbara H. McGuire and Howard B. Seim III, 16–21. Philadelphia: W. B. Saunders.

# 9 FEEDING AND NUTRITION

*Currently more is known about the nutrition of dogs and cats than is known about the nutrition of humans. America's dogs receiving good commercial dog foods receive a better balanced diet than is consumed by America's children.*

James E. Corbin, petfood nutritionist
*from* Pet Foods Around the World Then and Now
*Proceedings "Pet Food Forum," May 1993, Chicago, IL*

## 9.1   INTRODUCTION/OVERVIEW

Proper and adequate nutrition is the keystone of a normal life for companion animals. It is impossible to raise healthy animals, even under perfect environmental conditions, if their diet is deficient in quality or quantity. Indeed, malnutrition can have detrimental effects on a companion animal's immune system and, consequently, impede its ability to heal, maintain homeostasis, and even survive.

As the longevity of human and pet populations continues to increase, attention of pet owners is increasingly shifting to wellness, fueling further development of health-sustaining petfoods. Concurrently, new formulations of petfoods sparked an increased interest in supplements, antioxidants, and other nutrients/compounds that affect immunity and other aspects of animal health (Chapter 14). Nutrients including amino acids; omega–3 fatty acids; vitamins A, E, and $B_6$; and the minerals copper, iron, selenium, and zinc have been shown to have immune-enhancing effects.

Research indicates that approximately 20% of adult dogs experience some form of osteoarthritis. Knowledge of the effects of mineral balance as well as vitamins D and K on bone and joint health has resulted in these micronutrients receiving increased attention in formulating petfoods. Calcium, phosphorus, sodium, and magnesium are associated with renal and urinary tract health. Several minerals, including calcium, sodium, and magnesium and the amino acids taurine and carnitine, are related to heart health in pets. Magnesium deficiency contributes to hypokalemia, a well-recognized problem in older cats.

More than 50 nutrients are believed to be essential to the health and quality of life of companion animals. We discuss most of these and related topics in this chapter.

## 9.2   FROM SEARCHING FOR FOOD IN THE WILD TO SCIENTIFIC FEEDING AND NUTRITION

In their natural habitat, carnivorous cats and feral dogs sought out small mammals (e.g., mice, rats, and rabbits) for food. As consumed, small mammals have an average body composition of approximately 69% water, 16% protein, 12% fat, and 3% minerals and vitamins. Most vitamins were obtained from the liver and other internal organs of prey. Based on this heritage, it is understandable why cats are not fond of sweet substances but rather prefer diets associated with meat-based components. Some persons describe cats as finicky, selective eaters when, in fact, their choice of food is simply in keeping with the sources of nutrition and sustenance they depended on in the environment from which they evolved.

Over the past three centuries, sporting dogs have been bred to hunt game and others to perform tasks such as pointing and retrieving. While sporting dogs were bred for these dedicated tasks, other breeders produced dogs with very different physical attributes such as large prominent eyes, long flowing coats, short faces, or wrinkled skin (Chapter 3). For the most part, similar diets are used to feed these dozens of types of dogs. One question for us to consider is, "Can all of these renderings of genetic manipulations be properly fed by one basic dog food?" The answer resides in animal nutritionists and veterinarians working hand-in-hand as they apply research findings in formulating scientifically based petfoods to supply the nutritional needs of the various dogs and cats. The resulting diets frequently include modified formulations for growth, pregnancy, lactation, and high levels of activity.

Trainers of dogs (Chapter 12) know the importance of positive reinforcement of the relationship between pet and owner (and members of the owner's family). For example, many dogs have been taught to sit and beg to obtain a morsel of meat from the table at dinner time. More recently, dogs and cats have been fed specially prepared treats to pamper pets and enhance the positive reinforcement factor. These treats include Beef Jerky Strips™, Pepperoni™, T-bone Steaks™, Pig-in-a-Blanket™, Marbled Meat Pieces™, Bacon Strips™, and dozens of other treats for dogs (Figure 9.1). For examples of the variety of treats available, search the Internet for *treats for pets*.

**Figure 9.1** Examples of treats for dogs. From left to right: Marbled Meat Pieces, T-bone Steaks, and Pig-in-a-Blanket. *Courtesy Dr. James E. Corbin.*

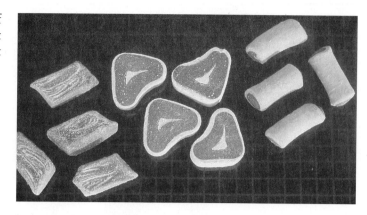

# 9.3   NUTRIENT INTAKE AND TERMINOLOGY

An important means of comparing dietary nutrient intake is to determine the amount of nutrient per unit of dry matter (DM). This permits comparison of nutrients in foods of differing moisture content, such as dry and canned petfoods. For example, a dry diet containing 20% protein and 9% water ( = 91% DM) and as fed contains 20/91 × 100 = 22% protein on a DM basis; similarly, a canned diet containing 5% protein and 77% water ( = 23% DM) and as fed contains 5/23 × 100 = 22% protein also on a DM basis.

## 9.3.1   FOOD PREFERENCES: PALATABILITY AND ACCEPTABILITY

*Palatability* of a diet is how well animals like a particular food (demonstrated by taste preference). *Acceptability* is an indication of whether an animal will eat a sufficient amount of food to meet its nutritional requirements. If a diet is acceptable and contains the proper amounts and balance of nutrients, it can nourish an animal for maximum health and longevity even though the animal might show a preference for other food(s) in taste tests. Palatability receives considerable marketing attention because owners are happy when pets are enthusiastic about eating food. However, highly palatable diets may promote overeating and subsequent obesity (Figure 9.2).

Food acceptability is measured with the one-pan test. Access to a single food is provided and the amount of food consumed is measured. Comparative food palatability is determined with a two-pan test where excess amounts of two foods are offered concurrently to several animals and the amount of each food consumed by each animal is measured daily.

Many variants affect food preference. Although food flavor is the most important factor affecting whether a dog or cat accepts a food, the texture, moisture, temperature, and form in which the food is offered also affect selection. Most dogs prefer a canned meat or semisoft diet to a dry food. Texture is especially important in cat foods. Adult cats have 30 teeth, whereas adult dogs have 42. Perhaps because cats are less adept at chewing and grinding, most cats prefer relatively soft foods—those with higher water content. Most dogs also prefer foods with high moisture content. Some owners believe neophobia (fear of new objects or rejection of new foods) is a factor in companion animal food preference. Actually, neophobia is uncommon in dogs. Cats tend to be individualistic, preferring either a canned or dry food, and generally prefer the type they were fed as kittens. Some cats develop a preference for a certain shape of dry food (e.g., nuggets, stars, rings, triangles), whereas other cats avoid sharp points on their food. Thus, familiarity can affect food preferences, especially in cats. Other factors influencing food preference include taste (bitter, salty, sour, and sweet), aroma, temperature, and visual (this may also be affected by owner/feeder attitude or bias). Dogs and cats prefer warm foods (approximately body temperature) over refrigerated or hot foods. Cats prefer foods with an acid pH (3.5 to 5.0) over neutral or alkaline foods. Most dogs and cats prefer diets with high protein and fat content over those with high fiber or carbohydrate content.

**Figure 9.2** These four English Setters were participants in a study to determine the long-term effect of full-feeding on body weight. All dogs were the same age, size, and body weight when the study began. The dogs were housed, exercised, and cared for under the same conditions. After being fed a highly palatable food *ad libitum* (free choice) for 24 months, their body weights varied from 43 to 105 lbs (left to right). This demonstrates that dogs will consume a palatable food in excess of their nutritional needs, and to minimize obesity many dogs should not be fed free choice. *Courtesy Dr. James E. Corbin.*

Given the choice of selecting from a variety of foods, dogs and cats are inept at selecting a balanced diet. Instead, palatability is the primary factor influencing their selection of foods. The only nutrients dogs and cats seek out based on nutritional needs are water, sodium, and sufficient food to meet their energy requirement. To ensure that animals receive the correct amount of each nutrient requires that each nutrient be properly balanced with respect to each other and the caloric density of the diet. When the diet is properly balanced, an animal consuming enough food to meet its energy (caloric) requirements will also obtain the correct amount of all other required nutrients. Dogs essentially "taste" with the olfactory buds of their nose. That explains why the olfactory intensity of most palatable products designed for dogs is important. This is also true for cats but to a lesser extent. Moreover, because many pet owners also smell the foods fed to their pets, the odor perceived by humans is important in evaluating the appeal and sales of petfoods.

Raw materials including cereals, fibers, animal by-products, meat meals, fish, amino acids, animal fats, vegetable oils, fish oils, and vitamin mixes contribute to the odor of petfoods. Furthermore, the smell of each new raw material changes with different processes used in manufacturing, storage, and preservation. Additionally, the odor of most ingredients, such as poultry fats, will vary among suppliers depending on the freshness, age of poultry, dietary fat consumed by the poultry, and degree of fat oxidation.

## 9.4   NUTRIENTS

Companion animal nutrition deals with foods and the nutrients these foods provide. Plant and animal tissues are composed of water, carbohydrates, proteins, lipids (including fat and related substances), vitamins, and minerals. Plants usually contain large amounts of carbohydrates; animals only traces. The cell walls of plants are composed primarily of fibrous carbohydrates; cell membranes of animals consist mostly of proteins and lipids. Plants store most of their reserve food as starch; animals as fat. Animals depend on plants for energy;

plants derive energy from the sun, and through photosynthesis, manufacture nutrients not otherwise available in nature to animals.

## 9.4.1 WATER

Animals have three sources of water: (1) that which they drink; (2) that ingested as a component of food and other drinks; and (3) *metabolic water,* which is derived from the digestive breakdown of carbohydrates, fats, and proteins. Metabolic water is the primary source of water for animals during hibernation.

Water has many functions in companion animals. It transports nutrients throughout the body. It is used in most biochemical reactions and is essential in performing many body functions. Water helps regulate body temperature. It facilitates the elimination of body wastes. Water is the principal constituent of the synovial fluid that lubricates joints. Animals will die more rapidly from lack of water than from lack of any other dietary substance.

## 9.4.2 CARBOHYDRATES

Carbohydrates are the major energy storage and structural constituent of plants (60% to 90% of DM). They include monosaccharides, disaccharides, oligosaccharides, and polysaccharides. Plant polysaccharides, starch (plant cell contents), and fiber (plant cell walls) are the principal carbohydrate constituents in manufactured petfoods, commonly representing up to one-third of the ingredients in canned and up to nearly two-thirds in dry petfoods.

Carbohydrates contribute the major source of energy utilized for many body functions and are essential for the metabolism of other nutrients. They are plentiful in cereals and tubers, representing approximately 70% to 80% of the energy in cereal grains.

Monosaccharides are simple sugars that can be absorbed directly from the gastrointestinal (GI) tract. These include glucose, fructose, and galactose (Figure 9.3). Glucose is the principal carbohydrate used for energy within the body; it is the primary end-product of starch digestion. Fructose is found in honey, fruits, and some vegetables. Galactose is derived from the digestion of lactose (milk sugar).

Disaccharides are composed of two monosaccharide units linked together. Sucrose (table sugar) is composed of one molecule of glucose linked with one of fructose. Milk sugar (lactose) is composed of a molecule of glucose linked to a molecule of galactose. The digestion of these disaccharides requires the enzymes sucrase and lactase. Young animals have high levels of lactase but may lose this enzyme with age (this is the primary reason feeding milk to older dogs or cats often causes diarrhea). Young animals have low levels of sucrase and thus should not be fed formulas containing table sugar during the first few weeks of life.

Oligosaccharides are short chains of sugar molecules. Fructooligosaccharides and inulin are composed of short chains of fructose molecules and are found in many vegetables and in chicory root. Galactooligosaccharides are composed of short chains of galactose and are

**Figure 9.3** Chemical structure of three simple sugars. *Courtesy Dr. Karen L. Campbell.*

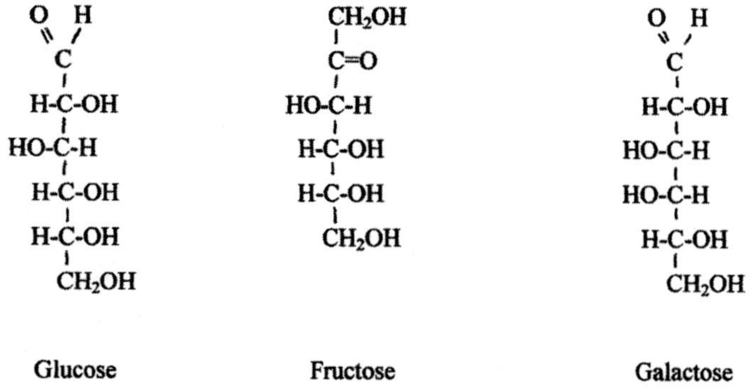

Glucose   Fructose   Galactose

found in soybeans. These are only partially digested by mammals, and the undigested portion promotes the growth of beneficial colonic bacteria such as the genera *Bifidobacteria* and *Lactobacillus* while decreasing numbers of harmful bacteria (e.g., *Campylobacter, Clostridia,* and coliforms). These oligosaccharides are marketed commercially as probiotic agents to promote the growth of friendly bacteria in the intestines (Section 9.14.5). Other health claims include improving triglyceride levels in humans and also improving the control of blood sugar in diabetics.

Polysaccharides consist of long, complex chains of monosaccharide units linked together. The two primary types of polysaccharides are starch and fiber. Starch is composed of soluble "alpha" monosaccharide units that may be readily digested by dogs and cats. Fiber is composed of insoluble "beta" monosaccharide units, which are resistant to mammalian digestive enzymes. The GI systems of herbivores contain large numbers of microorganisms that secrete cellulases; however, microbial digestion of starches and fiber is limited in dogs and cats.

Starch digestibility of cereal grains is enhanced by heat through toasting, cooking, baking, or other processes that cause gelatinization or dextrinization. In 1954, Dr. J. E. Corbin, then Director of Petfood Research with the Ralston-Purina Company of St. Louis, Missouri, developed the extrusion processing of cereal grains (Figure 9.4). This innovative technique greatly increases the surface area of cereal grains through exposure to heat, pressure, and penetration of water, which swells starch molecules and facilitates enzymatic degradation and thereby enhances the speed and completeness of starch digestion. Because of its versatility and practical significance, the petfood industry began producing extruded petfoods in various forms. It is estimated that extruded petfood currently comprises approximately 95% of the dry-type commercial companion animal food sold globally.

## 9.4.3  FATS

Dietary fat provides a concentrated source of energy, essential fatty acids, and fat-soluble vitamins; enhances palatability; and adds a desirable texture to dog and cat foods. Dietary fats are part of the group of compounds referred to as lipids. These compounds are soluble in organic solvents such as ether and insoluble in water. Simple lipids are composed of triglycerides and waxes. Compound lipids have fatty acids linked to a nonlipid molecule such as a

**Figure 9.4**  Example of a commercial extruder used in processing petfoods. The extrusion process greatly increases the surface area of cereal grains through exposure to heat, pressure, and penetration of water. This innovative technique swells starch molecules and facilitates enzymatic degradation thereby enhancing the speed and completeness of starch digestion. *Courtesy Wenger Manufacturing, Inc.*

protein (e.g., a lipoprotein). Lipids with a high percentage of short-chain or unsaturated fatty acids are liquid at room temperature and referred to as oils. Lipids with a high percentage of saturated fatty acids and longer chained fatty acids are solids at room temperature and called fats.

Fatty acids are classified by their size (number of carbon atoms) and number of double bonds. Fatty acids with no double bonds are saturated, whereas the presence of one double bond results in a monounsaturated fatty acid (e.g., oleic acid). Polyunsaturated fatty acids (PUFAs) have more than one double bond. Dogs, cats, and other mammals are unable to synthesize PUFAs with double bonds located at the 3- and 6-carbon position from the carboxyl end; these fatty acids are referred to as omega-3 and omega-6 families, respectively. Fatty acids of the omega-3 and omega-6 families are essential components of plasma cell membranes throughout the body. These fatty acids are important to the maintenance of cell membrane integrity, fluidity, and permeability. Additionally, omega-3 and -6 fatty acids have roles in inflammation and in immune regulation. Two members of the omega-6 fatty acids (linoleic acid and arachidonic acid) plus one member of the omega-3 fatty acid family (alpha-linolenic acid) are particularly important for normal health and are referred to as essential fatty acids (EFAs) (Figure 9.5). Inadequate dietary intake of these EFAs will lead to a fatty acid deficiency characterized by poor growth, a dry scaling skin, hair loss, weight loss, infertility, and unthriftiness. The most important EFA in the diet of dogs is linoleic acid, which is abundant in vegetable oils. In addition to linoleic acid, cats also require a dietary source of arachidonic acid, which is found in fats of animal origin.

Most commercial dry dog foods for adult maintenance contain 5% to 10% fat. EFAs should constitute at least 1% of the diet dry matter or 2% of caloric intake. The level may increase to 15% to 20% of DM in diets formulated for gestation, lactation, or performance. Poultry is the most common source of fat used in dog and cat foods, although beef tallow and pork lard are also used. Corn, soybean, and safflower oils are the most commonly used vegetable fats in petfoods. Fish oils and flaxseed are rich in omega-3 fatty acids. Omega-3 fatty acids are beneficial in decreasing inflammation in some types of disease (e.g., allergies and kidney disease) and may improve survival in some cancer patients.

Although dogs are more efficient than cats in digesting fats, cats can digest and utilize relatively high levels of dietary fat. However, because 1 g of fat provides 2.25 times more calories than 1 g of carbohydrate, and nearly twice as many calories as 1 g of protein, excess dietary fat may promote obesity. This is especially true in neutered cats. Overweight cats are more likely to develop diabetes mellitus, experience lameness, have skin disorders, and have a shorter life expectancy (Chapter 14).

**Figure 9.5** Chemical structure of essential fatty acids (EFAs). Note that linoleic acid (LA) contains 18 carbon molecules and 2 double bonds. The first double bond is at the sixth carbon; therefore, LA is an omega-6 fatty acid. Alpha-linolenic acid (ALA) also contains 18 carbon molecules, however it has 3 double bonds with the first being at the third carbon; thus, ALA is an omega-3 fatty acid. Arachidonic acid has 20 carbon molecules and 4 double bonds. The first double bond is at the sixth carbon molecule; therefore, arachidonic acid is an omega-6 fatty acid. Dogs can synthesize arachidonic acid from LA through two desaturation and elongation enzymatic reactions. However, cats lack the desaturation enzymes required for the conversion of LA to arachidonic acid; thus, arachidonic acid is an EFA required in diets for cats.

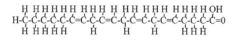

Linoleic Acid

Alpha-Linolenic Acid

Arachidonic Acid

## 9.4.4   Protein

Protein constitutes nearly one-half of the dry matter of an animal's body. Approximately one-third of the protein is in muscle, one-fifth in bone and cartilage, and one-tenth in the skin. The balance is found in other tissues and body fluids.

Proteins are composed of amino acids attached to each other by peptide bonds. There are 22 amino acids important as structural components of body tissues. Ten of these amino acids cannot be synthesized by companion animals and thus are dietary essentials; 12 of the amino acids can be synthesized provided an animal has an adequate intake of nitrogen-containing compounds (Table 9.1). The essential amino acids can be remembered using the mnemonic PVT MAT HILL.

Phenylalanine

Valine

Threonine

Methionine

Arginine

Tryptophan

Histidine

Isoleucine

Leucine

Lysine

Taurine is a unique amino acid that is not incorporated into proteins but is found as a free amino acid in tissues. Taurine has important roles in feline reproduction, bile acid conjugation, retinal function (vision), and normal function of the myocardium (heart). Cats cannot synthesize taurine and require a continual dietary source to replace taurine excreted in the bile. Signs of taurine deficiency in cats include central retinal degeneration (resulting in blindness) and dilated cardiomyopathy (heart failure). Taurine is present only in animal tissues; high concentrations are found in red meat, poultry, fish, and even higher concentrations in shellfish.

**Table 9.1**
Essential and Nonessential Amino Acids for Dogs and Cats

| Essential Amino Acids | Nonessential Amino Acids |
| --- | --- |
| Arginine | Alanine |
| Histidine | Asparagine |
| Isoleucine | Asparatate |
| Leucine | Cysteine |
| Lysine | Glutamate |
| Methionine | Glutamine |
| Phenylalanine | Glycine |
| Taurine[*] | Hydroxylysine |
| Threonine | Hydroxyproline |
| Tryptophan | Proline |
| Valine | Serine |
| | Tyrosine |

[*]*Taurine is essential in the diet of cats but not dogs fed adequate quantities of sulfur-containing amino acids. Taurine is one of the most abundant free amino acids in mammals, being particularly high in brain, heart, and skeletal muscle.*
Source: *National Research Council of the Council of the National Academies. 2004.* Nutrient Requirements of Dogs and Cats. *Washington, DC: The National Academies Press.*

In addition to their structural functions in body tissues, proteins participate in numerous body activities. The structural and functional roles proteins play in dogs and cats include growth, tissue and cellular repair, enzymes, hormones, antibodies, transport proteins, and as a source of energy. Symptoms of advanced protein deficiency, particularly of the essential amino acids, include decreased food intake, growth retardation and/or weight loss, lowered levels of blood proteins, muscular wasting, emaciation, and even death. Protein requirements are in large part dictated by the physiologic state. Growth, pregnancy, lactation, and geriatric age are examples of physiologic states requiring protein intake above that needed for maintenance.

Protein is the most expensive major component of companion animal diets. Use of raw materials of animal origin not suitable for human consumption constitutes an increasing proportion of the protein used in formulating petfoods. These commercially processed animal meats are by-products (functional products) of the meat packing, poultry processing, and fish canning industries. Animal-based proteins are important sources of high-quality protein (based on amino acid profile and bioavailability), energy, and minerals.

Plant proteins have been a reliable source of nutrients in the manufacture of petfoods for decades. Soybean meal is the most common plant protein used in dog diets. Most other cereal proteins are low in the amino acids lysine, methionine, leucine, and tryptophan.

Protein quality, as measured by the dietary amino acid profile and bioavailability/digestible amino acids per unit of ingredient relative to the animal's requirement, is especially important in the nutrition and quality of life of dogs and cats. Animal proteins are preferred in companion animal diets, especially those of cats. Approximately 25% to 40% of the dietary DM in premium dog foods is of animal origin.

Fish meal contains 60% to 72% crude protein (CP) and has a high biological value[1] (BV) in animal diets. It is rich in essential amino acids (especially lysine and sulfur-containing amino acids), and its presence in a complete diet compensates for the deficiencies in amino acids present in vegetable protein. Fish meal is commonly incorporated in cat diets at levels between 25% and 33% and in some cases represents as much as 60% of the dietary protein.

## 9.4.5 VITAMINS

Vitamins are essential organic elements in companion animal diets (Table 9.2). They are involved in more than 30 metabolic reactions in cellular metabolism. Vitamins are divided into two categories: fat soluble (A, D, E, and K) and water soluble (B-complex and C).

*Vitamin A.* Vitamin A metabolites, especially retinoic acid, are important in the maturation of skin and other tissues. In nature, vitamin A is found in precursors including β-carotene and other carotenoids. Most mammals convert β-carotene into two molecules of vitamin A. Cats lack the dioxygenase enzyme involved in this conversion and thus must have a dietary source of preformed vitamin A. Forms of preformed vitamin A include retinyl palmitate, retinyl acetate, and other retinols found in animal organs (especially liver and kidneys) and fish liver oils. Vitamin A is a component of visual pigments of the eye and is required for normal vision. Vitamin A is also required for normal fetal development. In its absence, fetuses develop congenital defects (a common one in kittens is cleft palate).

Excesses of vitamin A are as harmful as a deficiency. In countries where a common source of liver fed to cats is from cattle grazing on pasture[2] before slaughter (e.g., Australia and Argentina), a continued diet of this liver may cause skeletal changes such as exostosis (bony growths projecting outward from the surface of a bone) of the spine. Exostosis is one of the manifestations of vitamin A toxicity.

---

[1]Biological value of a protein is the percentage of absorbed protein retained by the body and used for metabolic purposes (largely growth and maintenance).
[2]Green grass is rich in carotenes, which are converted to vitamin A and stored in the livers of animals grazing on green pasture.

**Table 9.2**
Roles of Vitamins in the Health of Dogs and Cats

| Vitamin | Functions | Signs of Deficiency | Signs of Toxicity |
|---|---|---|---|
| Vitamin A | Important in cellular metabolism. Regeneration of visual purple in eyes. Required for normal maturation of the skin and hair follicles. Essential for normal epithelial tissue lining the digestive, respiratory, and reproductive tracts. Required for proper functioning of the immune system. | Scaly skin, poor hair coat, reproductive failure, retinal degeneration and night blindness, increased susceptibility to infections. | Anorexia, weight loss, bone decalcification, liver damage. |
| Vitamin D | Required for normal calcium absorption and metabolism. Essential for normal bone development. | Rickets in young, osteomalacia in adults, chest deformity, and poor eruption of teeth. | Increased blood calcium levels, soft tissue calcification, diarrhea, kidney failure, death. |
| Vitamin E | Antioxidant. Protects cells from oxidative damage. May have a role in normal immune function. | Muscular dystrophy, pansteatitis, steatitis, reproductive failure, intestinal lipofuscinosis, impaired immunity. | Anorexia. |
| Vitamin K | Required for formation of clotting factors and normal blood clotting. | Hemorrhage, increased bleeding times (poor blood clotting). | None reported. |
| Vitamin C | Antioxidant. Formation and maintenance of matrix of bone, cartilage, and connective tissue. | Rickets, impaired wound healing, bleeding, anemia, increased susceptibility to infections. | None reported. Note that dogs and cats do not require a dietary source of vitamin C (synthesize adequate levels). |
| Thiamine ($B_1$) | Component of two coenzymes, essential in carbohydrate metabolism and energy transfer. Promotes normal health and digestion and normal nerve function. | Anorexia, weight loss, vomiting, dehydration, ventral flexion of neck, paralysis, incoordination. | Nontoxic. |
| Riboflavin ($B_2$) | Forms parts of two coenzymes with roles in energy transfer and protein metabolism. Component of xanthine oxidase required for epithelial cell maturation. | Retarded growth, dry scaly skin, erythema, posterior muscle weakness, anemia, ocular lesions (pannus), glossitis, reduced fertility, testicular hypoplasia, fatty liver. | Nontoxic. |
| Niacin | Component of two coenzymes with roles in energy transfer. Required for metabolism. | Blacktongue (pellegra), dermatitis, diarrhea, dementia, anorexia, anemia, emaciation and death. | Cutaneous flushing, itching. |
| Pyridoxine ($B_6$) | Part of enzyme involved in protein metabolism. Essential for normal metabolism of tryptophan. | Dermatitis, seizures, anemia, high serum iron, anorexia, weight loss, impaired growth. | Nontoxic. |
| Pantothenic acid | Constituent of coenzyme A required for normal metabolism of carbohydrates, fats, and proteins. | Anorexia, stunted growth, hypoglycemia, uremia, gastroenteritis, seizures, fatty liver, coma, death. | Nontoxic. |

| Vitamin | Functions | Signs of Deficiency | Signs of Toxicity |
|---------|-----------|---------------------|-------------------|
| **Folic acid** | Required for normal red blood cell development and DNA synthesis. | Anemia, leukopenia, stunted growth, glossitis. | Nontoxic. |
| **Biotin** | Required for metabolism of fats and amino acids, essential for skin and hair health, functions in enzyme systems. | Scaly dermatitis, alopecia, anorexia, weakness, diarrhea, progressive spasticity, and posterior paralysis. | Nontoxic. |
| **Cobalamin (B$_{12}$)** | Required for synthesis of nucleic acids. Involved in purine synthesis and carbohydrate and fat metabolism. | Anemia, impaired growth, posterior incoordination. | Nontoxic. |
| **Choline*** | Component of phospholipids. Essential role in cell membranes, nerve impulse transmission, and fat metabolism. | Fatty liver, renal tubular degeneration, neurological dysfunction, impaired blood coagulation. | Diarrhea. |

*Choline is often listed with vitamins; however, it is not needed for metabolism and is not a vitamin. It is required as a structural component of fat and nerve tissue.*

*Vitamin D.* This fat-soluble vitamin becomes a dietary essential only when companion animals are not exposed to sunlight. Dogs and cats have a predilection for lying in the sun; one benefit of sunbathing is the synthesis of vitamin D from 7-dehydrocholesterol within the skin. Vitamin D is essential in preventing rickets by regulating calcium and phosphorus metabolism in animals. Regulating the balance of calcium and phosphorus within the body is important not only for healthy bones but also for muscle strength and normal neuromuscular transmission. Excesses of vitamin D can result in calcification of connective tissues, lungs, kidneys, and stomach. Acute toxicity can result in hypercalcemia and death; this is most commonly seen following accidental ingestion of cholecalciferol-containing rodenticides.

*Vitamin E.* A deficiency of vitamin E was first reported in 1953 as a result of feeding kittens a canned fish diet containing primarily red tuna. The condition of steatitis (inflammation of adipose tissue) or "yellow fat disease," in which a yellow-brown or orange pigment is deposited in stored fat, results from peroxidation of body fat. Vitamin E acts as an antioxidant and is important in maintaining the stability of cell membranes. An animal's requirement for vitamin E varies with the amount of fat and selenium in the diet. Vitamin E and selenium function synergistically in helping prevent lipid peroxidation. Vitamin E scavenges free radicals formed during oxidation of fats and selenium reduces peroxide formation. Vitamin E concentrated in an oil base is used as an antioxidant. In addition to steatitis, other signs associated with vitamin E deficiency include infertility (fetal resorption), muscular dystrophy, retinal atrophy (causing blindness), brown gut disease (intestinal lipofuscinosis), and impaired immunity. Adding 4 g of alpha-tocopherol daily to feline diets prevents signs of deficiency. Wheat germ oil, most cereal grains, egg yolk, and beef liver are excellent sources of vitamin E.

*Vitamin K.* Vitamin K is composed of a group of compounds known as quinines, which are essential for blood coagulation. Deficiencies of vitamin K are very rare in dogs and cats as intestinal bacteria synthesize it. Supplementation may be necessary when dogs and cats are on prolonged oral antibiotic therapy that disrupts the normal intestinal flora (bacterial organisms that synthesize vitamin K in the absence of antibiotics).

***Water-soluble vitamins.*** Most water-soluble vitamins play roles in metabolic processes. Members of this group include the B vitamins and vitamin C.

- *Thiamine (vitamin $B_1$)* is a sulfur-containing compound and has an important role in carbohydrate metabolism. In humans, thiamin deficiency is known as beriberi and is characterized by anorexia, vomiting, weight loss, and an abnormal posture with ventral flexion of the neck (Chastek paralysis). Several raw freshwater fish and some plants contain thiaminases, which can destroy thiamine.

- *Riboflavin (vitamin $B_2$)* is important as a component of coenzymes involved in oxidative metabolism. Deficiency of riboflavin may result in dry scaly skin, erythema (red skin), muscular weakness, glossitis (tongue inflammation), and reduced fertility.

- *Pantothenic acid* is a component of coenzyme A and is involved in metabolism of carbohydrates, amino acids, and fats. Deficiencies may result in depression, retarded growth, gastric ulcers, fatty liver, and hair loss.

- *Niacin* is important in oxidation and reduction reactions and the metabolism of nutrients. The cat is unique among mammals as it cannot convert tryptophan into niacin. Niacin deficiency is characterized by inflammation and ulceration of the mouth and thick blood-stained saliva. This syndrome is called blacktongue in dogs and pellegra in humans.

- *Pyridoxine (vitamin $B_6$)* is essential for amino acid metabolism. Deficiencies result in anemia and weight loss. Pyridoxine is essential for the conversion of oxalate to glycine and pyridoxine deficiency in cats has been associated with kidney damage from deposits of calcium oxalate crystals in renal tubules.

- *Biotin* is essential in the metabolism of fats and amino acids. A deficiency of biotin is rare because it is synthesized by intestinal bacteria. Raw egg white contains avidin, a protein that inhibits biotin absorption from the GI tract. Thus animals fed diets containing excessive quantities of raw egg whites may develop a scaly dermatitis, alopecia (hair loss), anorexia (poor appetite), and a progressive spasticity and posterior paralysis as a result of biotin deficiency. Cooking eggs at 196°F for 5 minutes coagulates egg albumin and inactivates avidin.

- *Folic acid* is found in nature conjugated with the amino acid glutamate. It has important roles in coenzymes required for the synthesis of thymidine in DNA and is also required for red blood cell development. Deficiency is rare as it is synthesized by intestinal bacteria; however, sulfonamide antibiotics inhibit its action and can result in bone marrow suppression, anemia, and glossitis.

- *Cobalamin (vitamin $B_{12}$)* is unique as the only vitamin that contains a trace element (cobalt). It is involved in the synthesis of myelin (surrounds nerves) and also has roles in fat and carbohydrate metabolism. It is also important in red blood cell maturation; a deficiency of cobalamin results in pernicious anemia and neurological dysfunction.

- *Ascorbic acid (vitamin C)* is required for the synthesis of collagen, an important structural protein. Dogs and cats synthesize vitamin C from glucose and do not require a dietary source. Some reports indicate that supplemental vitamin C may be beneficial for dogs that are highly active (e.g., sled dogs) or under stress.

- *Choline* is often included in lists of vitamins although it is not needed for metabolism and is technically not a vitamin but rather a structural component of phospholipids in cell membranes and a precursor of the neurotransmitter acetylcholine. Deficiencies result in hypoalbuminemia, neurological dysfunction, and a fatty liver.

## 9.4.6   MINERALS

Minerals are important inorganic elements, many of which are needed in companion animal diets. They constitute about 3% to 5% of the animal body, and because they cannot be synthesized by animals, they must be obtained from food. Macrominerals are those required in the diet in levels expressed as parts per hundred (%). The macrominerals include calcium, phosphorus, magnesium, sodium, chloride, potassium, and sulfur (Table 9.3). Trace or microminerals are those required in the diet in levels expressed as parts per million (ppm). Trace minerals include iron, zinc, copper, manganese, iodine, chromium, and selenium.

**Table 9.3**
Roles of Minerals in the Health of Dogs and Cats

| Mineral | Functions | Signs of Deficiency | Signs of Toxicity |
|---|---|---|---|
| Calcium | Bone and tooth formation, blood clotting, enzyme activation, muscle contraction, nerve impulse transmission. | Rickets in young and osteomalacia in adults, lameness, stiffness, constipation, anorexia, loss of teeth; tetany with acute deficiency. | Impaired skeletal development; secondary deficiencies of other minerals especially zinc, phosphorus, and copper (interference with absorption); bloat. |
| Phosphorus | Bone and tooth formation, component of enzyme systems, involved in energy transfer (component of high energy bonds), part of RNA and DNA. | Rough hair coat, pica, anorexia, slow growth, rickets in young, osteomalacia in adults. | Impaired skeletal development, secondary deficiency of calcium, kidney damage. |
| Sodium | Muscle contraction; maintenance of body fluid volumes; component of bile, muscle, and nerve function. | Salt hunger, pica, weight loss, fatigue, impaired milk secretion, polyuria, circulatory failure. | Thirst, pruritus (itching), constipation, anorexia, seizures, hypertension (these are unlikely if water is freely available). |
| Potassium | Required for muscle and nerve function, maintenance of electrolyte balance. | Anorexia, weakness, lethargy, decreased muscle tone. | Hyperkalemia (unlikely unless animal is in renal failure), cardiotoxicity, and death. |
| Chloride | Required for acid–base balance, body fluid volume, nerve and muscle function, energy metabolism. | Anorexia, craving for salt, circulatory failure. | Unlikely if normal kidney function. |
| Magnesium | Enzyme activator, constituent of skeletal tissue, required for muscle and nerve function, roles in energy metabolism and protein synthesis. | Calcification of soft tissues, retarded growth, spreading of toes, hyperirritability, seizures, excess salivation. | Acute excesses may cause diarrhea. Chronic excesses may contribute to urolithiasis, cystitis, and FLUTD in cats. |
| Iron | Component of hemoglobin and myoglobin, component of enzymes involved in energy metabolism. | Anemia, fatigue, weakness, diarrhea, anorexia. | Excess is unlikely due to regulation of absorption. |
| Zinc | Essential component of many enzyme systems including those involved in protein and carbohydrate metabolism, required for maturation of skin cells and healthy hair coat, required for normal immune function. | Impaired growth, scaly skin with parakeratosis, depigmentation of hair, infertility, testicular hypoplasia, impaired wound healing, increased susceptibility to infections. | Excesses may interfere with absorption of calcium and/or copper. Acute toxicity may result in hemolytic anemia. |
| Copper | Roles in erythropoiesis, coenzymes, hair pigmentation, reproduction, collagen and elastin synthesis, iron utilization. | Pica, stunted growth, diarrhea, depigmentation of hair, anemia, impaired bone growth. | Inherited disorder of metabolism in some breeds leads to liver damage. |

*continued*

**Table 9.3**
Roles of Minerals in the Health of Dogs and Cats (Continued)

| Mineral | Functions | Signs of Deficiency | Signs of Toxicity |
|---|---|---|---|
| Manganese | Involved in carbohydrate and lipid metabolism, formation of cartilage. | Infertility, enlarged stiff joints, short brittle bones. | Infertility, partial albinism (rare). |
| Iodine | Required for thyroid hormone synthesis (involved in regulation of metabolism). | Hypothyroidism, goiter, alopecia, infertility, lethargy, myxedema. | Excesses can also result in decreased thyroid function and signs similar to deficiency. |
| Selenium | Component of glutathione peroxidase, involved in prevention of peroxide formation, protects cell membranes from oxidative injury, functions closely with vitamin E. | "White muscle disease," dysfunction of skeletal and cardiac muscles. Deficiencies are unlikely in dogs and cats. | Anorexia, vomiting, weakness, incoordination. This is rare in dogs and cats. |
| Sulfur | Required for synthesis of chondroitin-sulfate in cartilage and insulin; component of glutathione. | Deficiencies not observed in dogs and cats due to sulfur content in the amino acids methionine and cysteine. Research indicates that inorganic sulfur may be useful in bird diets. | Toxicity not observed in dogs and cats. |

*Calcium and phosphorus.* The major minerals involved in the structural formation of bones (Figure 9.6) and teeth are calcium and phosphorus. Calcium is also involved in blood clotting, muscle contraction, nerve conduction, and many metabolic reactions. Phosphorus is involved in many enzyme systems and is essential for the storage and transfer of energy in the body. The ratio of calcium to phosphorus in the diet is important as excesses of either interfere with the absorption of the other. The optimum ratio is generally considered to be between 1.2:1 and 1.4:1 with ratios between 1:1 and 2:1 being acceptable. Dietary levels between 0.5% and 0.9% of readily available forms of calcium and phosphorus[3] are acceptable for adult maintenance; levels for growth and lactation should be approximately double this amount.

*Magnesium.* Magnesium plays a key role in many enzyme reactions. Heart and skeletal muscle contraction is dependent on the ratio between calcium and magnesium in the blood. A deficiency in magnesium results in muscle weakness, abnormal skeletal development (e.g., weak carpal joints), and in severe cases convulsions. Chronic intake of high amounts of magnesium may contribute to urolithiasis and feline lower urinary tract disease (FLUTD) in cats (Chapter 14).

*Sodium.* Sodium is found primarily in extracellular fluids and is important for the maintenance of normal blood volume and pressure. Symptoms of deficiency include pica (abnormal appetite), weight loss, fatigue, increased urination (from inability to concentrate urine), and impaired milk secretion. Excess sodium results in excessive thirst and may contribute to hypertension.

*Potassium.* The highest concentration of potassium is found inside cells. Potassium is required for nerve transmission, fluid balance, and muscle metabolism. Deficiencies result in anorexia, weakness, lethargy, and in severe cases, paralysis. Excesses do not cause problems

---

[3]Many plants contain phytins, which are poorly absorbed sources of phosphorus for animals.

**Figure 9.6** Bones of the pelvis and rear limbs from a dog fed adequate levels of calcium (left) and from a dog fed a calcium-deficient diet (right). Note especially the deformed femur bone. *Courtesy Dr. James E. Corbin.*

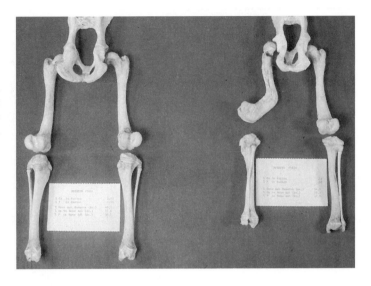

unless the animal is in kidney (renal) failure; hyperkalemia can cause cardiac arrhythmias (irregular heartbeat) and death.

*Chloride.* Chloride is important in balancing ionic charges and in metabolism; it is also critical for normal nerve transmission. It is combined with sodium in common salt (sodium chloride). Adequate amounts are present in most diets. Diets containing high levels of phosphoric acid (as in cat foods formulated to lower urine pH and minimize crystal formation in urine) result in an increased dietary requirement for chloride.

*Iron.* Iron is a trace element required for the formation of hemoglobin and myoglobin. It is essential for oxygen transport within the body and is also an essential component of many enzymes involved in cellular respiration. Deficiencies result in anemia, weakness, and fatigue. Iron toxicity results in anorexia and weight loss. The absorption of iron is influenced by many factors. Iron contained in foods of animal origin has the highest availability, whereas soy and other vegetable proteins may decrease iron absorption. Ferrous iron is absorbed more readily than ferric iron or iron oxide. Iron oxide has a low solubility and its primary use in petfoods is as a coloring agent to imitate the color of fresh meat.

*Copper.* Closely linked with iron metabolism, copper is also involved in the formation of melanin (brown-black pigment of skin and hair) and is a component of several enzyme systems. A deficiency of copper results in slow growth, pica, anemia, and hair depigmentation. Certain breeds of dogs (e.g., Bedlington Terriers) have an inability to mobilize hepatic copper efficiently. In these dogs, copper accumulates in the liver causing hepatitis, cirrhosis, liver failure, and a shortened life span. A similar syndrome of hepatic copper accumulation and liver failure has been reported in some West Highland White Terriers and Doberman Pinschers.

*Manganese.* Manganese is involved in many metal-enzyme systems and thus has important roles in metabolism and other body functions. Symptoms of deficiency are rare in dogs and cats but can include infertility, stiffness, enlarged joints, and brittle bones. Toxicity has been reported to cause partial albinism and infertility in Siamese cats.

*Zinc.* Zinc has roles in enzyme function and in protein synthesis. Zinc deficiency results in poor growth (Figure 9.7), anorexia, scaly dermatitis (Figure 9.8), hair depigmentation, testicular atrophy, impaired wound healing, and immunodeficiency with secondary infections. The availability of zinc in the diet is influenced by the presence of other minerals (e.g., calcium) and phytates (found in plants), which decrease zinc absorption. Certain breeds of dogs (e.g.,

**Figure 9.7** Pointer puppy fed a zinc-deficient diet showing poor body condition and stunted growth. *Courtesy Dr. James E. Corbin.*

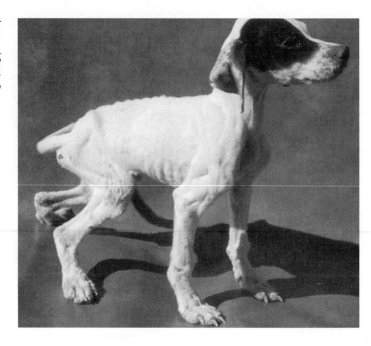

Alaskan Malamute and Siberian Husky) have increased requirements for zinc (Chapter 14). Zinc toxicity is usually associated with the ingestion of metals containing zinc (e.g., the nuts and bolts holding cages together), pennies (those minted after 1982 contain 96% to 98% zinc), or tubes of zinc oxide (e.g., sunscreens and diaper rash ointments). Acute zinc toxicity may result in hemolytic anemia and death.

*Iodine.* Iodine is an essential dietary nutrient for all animals. Although iodine is present in most body cells, 70% to 80% of the total body iodine is found in the thyroid gland. The kidney is the second major site of iodine concentration. The primary function of iodine is in the synthesis of thyroid hormones, especially thyroxine. Thyroid hormones play a critical role in the regulation of many metabolic functions including thermoregulation, reproduction,

**Figure 9.8** Effect of supplemental zinc on the health and integrity of skin and hair surrounding the eye. (*a*) When fed a nonpurified diet containing 35 ppm of zinc, the dog developed scaling around its eyes after only 3 weeks. (*b*) After supplementing the diet with 100 ppm of zinc, the same dog demonstrates rapid response with normalization of the skin. *Courtesy Dr. James E. Corbin.*

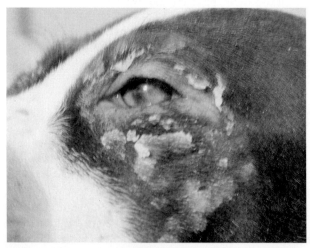

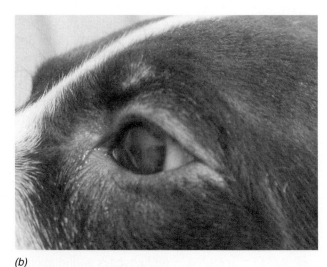

(a)                                                    (b)

growth and development, intermediary metabolism, hematopoiesis (formation and development of blood cells) and circulation, and neuromuscular functioning. Approximately 80% of thyroxine, the iodine-containing hormone, entering circulation is metabolized in the liver and other tissues and recycled.

When the amount of iodine available to the thyroid gland is insufficient, the thyroid gland cannot synthesize sufficient thyroxine and the blood concentration declines causing the release of thyrotropin (Chapter 11). This hormone of the anterior pituitary gland in turn increases activity of the thyroid gland causing it to enlarge. This condition is called goiter (compensatory hypertrophy, or an enlargement involving the formation of more tissue in an effort to secrete more thyroxine). Iodized salts (calcium iodate and potassium iodide) are commonly used in petfoods as a preventive measure in ensuring an adequate dietary intake of iodine.

*Selenium.* Selenium is an obligatory component of the enzyme glutathione peroxidase, which protects cells from lipid peroxidation. It is closely interrelated with vitamin E in its function of protecting the body from oxidative damage. It is also essential for normal muscle function. A deficiency in selenium results in white muscle disease (skeletal and cardiac myopathy). Large doses are highly toxic causing vomiting, weakness, ataxia (incoordination), pulmonary edema, and death.

*Cobalt.* Cobalt is a component of vitamin $B_{12}$. This is thought to be its only biological function in dogs, cats, and other pets. When adequate amounts of vitamin $B_{12}$ are present in the diet, there is no dietary requirement for cobalt.

*Other trace elements.* These include chromium (role in carbohydrate metabolism, linked with insulin function), fluoride (involved in tooth and bone development and reproduction), nickel (involved in membrane function and the metabolism of RNA), molybdenum (involved in several enzyme systems including the metabolism of uric acid), silicon (involved in skeletal and connective tissue development), vanadium (involved in growth, reproduction, and fat metabolism), and arsenic (may have roles in hemoglobin production and growth). The daily requirements for these trace minerals are very low. Adequate amounts are found in most diets.

# 9.5    NUTRIENT REQUIREMENTS OF DOGS AND CATS

Many research studies have been conducted in which dogs and cats were fed purified diets and water to determine specific requirements for each nutrient. This is accomplished by eliminating individual nutrients in a basic diet and adding them back in a purified chemical form at graded levels. By feeding a control diet that contains 0.0% of the nutrient in question and test groups that are fed varying known amounts of the nutrient, research nutritionists can pinpoint the precise level required of the specific nutrient being studied under controlled conditions. This technique is used to evaluate the bioavailability of nutrients from commercial petfoods as well as study the effects of deficiencies or excesses in the diets of dogs and cats.

Optimal health and performance require an adequate intake of nutrients. But how is "adequate intake" determined? Requirements for energy and nutrients vary significantly at different stages of an animal's life and also with differing environmental and activity levels. Currently, there are two primary sources of published standards with information on nutrient requirements for dogs and cats. The National Research Council (NRC) of the National Academies compiles lists of minimum nutrient requirements for the various species of animals. In addition, the Association of American Feed Control Officials (AAFCO) has developed standards of practical nutrient profiles for dog and cat foods based on data related to commonly used ingredients. The AAFCO Nutrient Profiles provide recommendations for minimum and some maximum levels of nutrients in commercial petfoods intended for growth and reproduction and those intended for adult maintenance.

The Committee on Animal Nutrition of the NRC revised the 1985 and 1986 publications on *Nutrient Requirements of Dogs* and *Nutrient Requirements of Cats*. These were published as a single report in 2004. It provides updated scientific estimates of requirements for essential nutrients and contains discussions of nutrient metabolism, toxicity, deficiency, and nutritionally related diseases of dogs and cats. Information on the impacts of physiological status, temperature, breed, age, and environment on nutrient requirements are also included.

Most formulated petfoods reflect current state-of-the-art companion animal nutrition. For example, the 2004 NRC publication of dog/cat nutrient requirements provides companion animal nutritionists with guidelines to scientifically design petfoods that fulfill the dietary requirements of dogs and cats and promote normal growth and good health, as well as to minimize the risk of nutritional deficiencies. Table 9.1 listed essential and nonessential amino acids for dogs. The minimum quantity of an individual nutrient that must be supplied daily for proper body metabolism is referred to as the *minimum daily requirement* (MDR). It is important to remember that these are minimums and allowances need to be made for individual variations in metabolism, activity levels, life stages, disease states, nutrient bioavailability, and other factors, which may alter the levels needed by a specific individual animal. Allowances for such factors are considered in formulating *recommended daily allowances* (RDA). The RDA is almost always greater than the MDR. Some nutrients have significant toxicity if supplied in excess (e.g., vitamin A). The maximum tolerable levels must also be considered for these nutrients. The ratios of certain nutrients in the diet are also important (e.g., Ca:P).

## 9.5.1   WATER REQUIREMENTS

Normal adult dogs and cats require approximately 1 oz of water per pound of body weight daily for maintenance of good health. Much of this water is consumed by drinking; however, some is acquired from moisture in food and a small amount of water is generated during the metabolism of food for energy (metabolic water). Other methods used to estimate normal water consumption are: (1) water consumption should be approximately 2.5 times the amount of dry matter consumed, and (2) the ml/day of water should be approximately the same as the daily energy intake in kcal/day. Water requirements are increased when body water losses increase due to increased panting (e.g., thermoregulation, stress), increased ambient temperature, fever (pyrexia), increased physical exertion (e.g., running, working), pregnancy and lactation, vomiting and diarrhea, or increased urine excretion (e.g., kidney disease, Cushing's disease, drugs, and other causes). Increases in salt or electrolyte levels in the diet will also increase water consumption.

## 9.5.2   ENERGY REQUIREMENTS

After water, food for energy is the second most critical nutritional need of animals. Animals derive energy from the metabolism of carbohydrates, fats, and proteins. Energy is measured by determining the amount of heat generated by the complete combustion of a food. This represents the gross energy (GE) of food. The amount of heat required to raise the temperature of 1 kg (liter) of water from 14.5°C to 15.5°C is defined as 1 kilocalorie (kcal, C). One kcal is equivalent to 1,000 calories. The word *calorie* written with a lowercase "c" is the amount of heat required to raise the temperature of 1 ml of water; this is such a small quantity of energy that when nutritionists speak of "Calories" they are referring to kilocalories written as a capital "C." Animals cannot use all of the GE in a food; some is lost in feces, urine, and gases produced during digestion. Additional energy is lost in digestion, absorption, and utilization of food. Only the remaining net energy (NE) is available for maintenance and production (work, growth, pregnancy, and lactation) (Figure 9.9).

Calculations of energy requirements include the maintenance energy requirements (MER) plus additional energy needed for varying levels of physical activity, growth, reproduction, lactation, stress, and coping with adverse environmental conditions (Table 9.4). Basal energy expenditure is the amount of energy used by an awake animal in a thermoneutral environment 12 to 18 hours following food consumption. The MER includes basal energy expenditure plus

**Figure 9.9**  Partition and utilization of food energy.

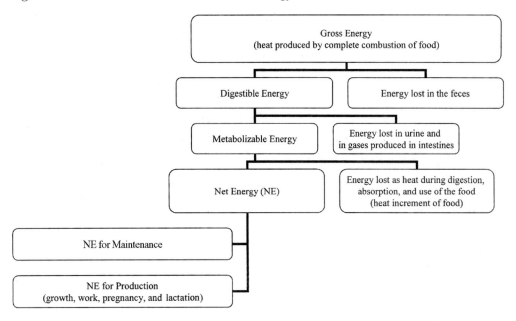

the energy expended in obtaining and using food in amounts necessary to maintain body weight. Formulating diets for dogs is a great challenge because of the differences in size, level of activity, and body weight. Indeed, from the tiny Chihuahua to the large St. Bernard the energy and other nutritional requirements of dogs vary significantly. Energy expenditure is proportional to the body surface area of an animal. Smaller animals have a greater body surface area per unit of body weight (e.g., a dog weighing 2.5 kg has 300% more body surface area per kilogram than a dog weighing 50 kg). Body surface area is directly related to the body weight in kilograms to the 0.75 power ($Wt_{kg}^{0.75}$); this is called the metabolic body size. For dogs, the MER (metabolizable kcal/day) = $2[70\,Wt_{kg}^{0.75}]$. For cats, the MER (metabolizable kcal/day) = $1.4[70\,Wt_{kg}^{0.75}]$. The determination of "final" daily energy requirements can be made by applying adjustment factors to the MER based on the extra needs of the animal.

**Table 9.4**
Determining Daily Energy Requirements for Dogs and Cats

| Level of Activity, Function, Condition | Daily Energy Required |
|---|---|
| Inactivity | $0.8 \times MER^*$ |
| Work—1 hour of light work (e.g., hunting) | $1.1 \times MER$ |
| Work—full day of heavy work (e.g., sled dog) | $2–4 \times MER$ |
| Gestation—last 3 weeks | $1.1–1.3 \times MER$ |
| Peak lactation | $[1 + 0.25 \times (\text{\# offspring})] \times MER$ |
| Growth—birth to 3 months | $2 \times MER$ |
| Growth—3–6 months (3–9 months in giant breeds of dogs) | $1.6 \times MER$ |
| Growth—6–12 months (9–24 months in giant breeds of dogs) | $1.2 \times MER$ |
| Cold (wind chill of 32°F) | $1.75 \times MER$ |
| Heat (tropical climate) | Up to $2.5 \times MER$ |
| Trauma | $1.35–1.5 \times MER$ |
| Serious infections | $1.5–1.7 \times MER$ |
| Post surgery, stress | $1.25 \times MER$ |

*Maintenance energy requirement.*

To determine the amount of food an animal must consume to meet its energy requirements, one must know both the daily energy requirements and the energy density of the food. If the caloric density of a food is unknown, it can be calculated by using the formula:

$$\text{caloric density} = \text{energy provided by each nutrient} \times \text{the amount of that nutrient in the food}$$

The net energy derived from each nutrient varies among species due to differences in the efficiency of digestion and absorption. The net energy (kcal/g) from dietary protein, fat, and soluble carbohydrates are calculated as 3.5, 8.7, and 3.5 kcal, respectively, for dogs, and 3.9, 7.7, and 3.0 kcal for cats. Thus, calculation of the energy density of a cat food containing 30% protein, 12% fat, and 48% soluble carbohydrates (CHO) would be 3.9 kcal/g protein × 30 g protein/100 g diet + 7.7 kcal/g fat × 12 g fat/100 g diet + 3.0 kcal/g CHO × 48 g CHO/100 g diet = [117 + 92.4 + 144] kcal/100 g diet = 353.4 kcal/100 g diet. The average dry diet weighs between 85 and 100 g/cup (using a standard 8-oz measuring cup). For accurate calculations, three individual measuring cups of the diet should be weighed in grams and averaged. The final estimation of the amount of food needed to meet the animal's daily energy requirements is made by dividing estimated energy requirement by the energy density of the food. For example, a 7-lb inactive adult cat with a total daily energy requirement of 210 kcal/day consuming the above diet (353.4 kcal/100 g) needs 59 g of the diet daily. If the diet weighs 85 g/cup, the cat will need 0.69 cup of the diet per day. Feeding should start with this estimate and then be adjusted as needed to maintain optimal body weight. The concentrations of all nutrients contained in the diet should be balanced so an animal consuming enough of the diet to meet its energy needs automatically receives the amounts it requires for other nutrients.

Most dogs need about 2.5% of their body weight in dietary dry matter for body maintenance. A 40-lb dog, for example, will consume about 1 lb of dry food daily. The same dog, while working hard, will consume between 5.0% and 7.5% of its body weight (2 to 3 lbs of dry food daily).

### 9.5.3   PROTEIN REQUIREMENTS

Protein requirements are based on nitrogen balance studies and growth rates of animals fed different protein levels. Nitrogen balance is equal to nitrogen intake minus nitrogen excreted in urine and feces. When the nitrogen balance is zero, there is no net gain or loss of total body protein. A positive nitrogen balance occurs when new tissue is being synthesized in the body, such as during growth or gestation or in recovery from surgery or animals in training building muscles. Negative nitrogen balance occurs when protein excretion exceeds intake. This may occur (1) when energy intake is insufficient and body tissues are being catabolized, (2) with severe illness or injury, or (3) with increased protein loss in the urine from kidney disease or in the feces from GI disease. Factors affecting protein requirements include (1) protein quality (higher quality means less is needed), (2) amino acid composition (less is required when the protein contains a balance of essential amino acids), (3) protein digestibility (increasing digestibility decreases quantity of protein needed), and (4) energy density of the diet (high energy density requires a high percentage of protein in the diet). AAFCO recommendations for minimum levels of proteins in the diets for dogs are 18% of ME for adult maintenance and 22% of ME for growth and reproduction. AAFCO recommendations for minimum levels of proteins in the diets for cats are 23% of ME for adult maintenance and 26% of ME for growth and reproduction.

### 9.5.4   UNIQUE NUTRIENT REQUIREMENTS OF CATS

Cats require the same 10 essential amino acids needed by other mammals. Additionally, cats require taurine. Cats fed a diet deficient in arginine develop severe hyperammonemia, which may result in death. A deficiency in the amino acid taurine results in numerous clinical symptoms, including fetal abnormalities, delayed growth and development, formation of abnormal blood platelets, central retinal degeneration (CRD), lethargy, and dilated cardiomyopathy (DC).

Although taurine-deprived adult cats may maintain normal weight and food intake, come into estrus, and conceive, they commonly resorb or abort fetuses or produce low–birth-weight or stillborn kittens.

Clinical cases of CRD have resulted from feeding cereal-based dog foods to cats for prolonged periods. Cats with CRD are commonly treated with 250 to 500 mg of taurine twice daily. Taurine occurs only in animal and fish proteins. The proteins of cheese and eggs are marginal sources of taurine. Fish meal is usually a dependable source.

## 9.6 DIGESTION AND ABSORPTION OF NUTRIENTS

Most foods as consumed are too complex to be absorbed into the blood and lymph without digestive changes. Glucose, soluble salts, water, and a few other nutrients are exceptions.

Digestion involves a number of mechanical, chemical, and microbial activities to degrade foods into simple molecular compounds that can then be absorbed across the intestinal mucosa. The dog is classified as a monogastric omnivore (Figure 9.10), the cat as a carnivore.

Digestion in the mouth is mainly mechanical; chewing (mastication) breaks down large pieces of food into sizes that can be swallowed. Dogs and cats have four pairs of salivary glands: the sublingual glands (located under the tongue), the mandibulary (or submaxillary) glands located on each side of the lower jaw, the parotid glands (located in front of each ear), and the zygomatic glands (located in the upper jaw below the eyes). Some saliva is always present; however, secretion increases with the sight and smell of food. (Readers probably remember that Dr. Pavlov received a Nobel prize for describing conditioned secretion of saliva in dogs accustomed to receiving food in conjunction with ringing of a bell.) Saliva is 99% water and 1% mucus, inorganic salts, and enzymes. Mucus lubricates the mouth and esophagus and aids swallowing. The saliva of dogs and cats, unlike humans, does not contain the enzyme amylase and thus there is no digestion of starches in the mouth. Dentition reflects the type of diet normally consumed by the species. Cats have 30 permanent teeth with 12 incisors, 4 canines, 10 premolars, and 4 molars; their teeth are useful for cutting and tearing prey. In addition to 12 incisors and 4 canines, adult dogs have 16 premolars and 10 molars for a total of 42 permanent teeth (see Figure 11.6). The additional premolars and molars in the dog provide effective grinding and crushing actions needed for the consumption of fibrous foods in an omnivorous diet.

**Figure 9.10** The digestive organs of a dog. *Illustration by Diana Nicoletti.*

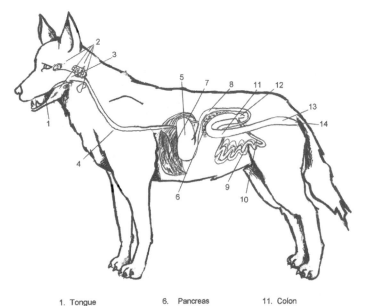

| | | |
|---|---|---|
| 1. Tongue | 6. Pancreas | 11. Colon |
| 2. Salivary glands | 7. Liver | 12. Cecum |
| 3. Pharynx | 8. Duodenum | 13. Rectum |
| 4. Esophagus | 9. Jejunum | 14. Anus |
| 5. Stomach | 10. Ileum | |

The esophagus of both dogs and cats is lined with mucus-secreting cells to facilitate passage of food from mouth to stomach. The stomach has several functions. It stores food and controls the rate of passage of food into the intestines. Gastric muscles mix food with digestive secretions and promote the mechanical breakdown of food particles. Gastric juices are produced by the gastric mucosal cells and contain hydrochloric acid (HCl), pepsinogen, mucus, and intrinsic factor. HCl promotes optimum activity of gastric digestive enzymes. Pepsinogen is an inactive precursor of pepsin. HCl cleaves small peptides from pepsinogen to form pepsin. Once pepsin is activated, it cleaves food particles into intermediate-size polypeptides. Mucus produced by the gastric mucosal cells is important in protecting the gastric mucosa from "autodigestion" by pepsin and HCl. *Intrinsic factor* is a mucoprotein produced by the cells of the gastric mucosa, which is required for the binding and absorption of vitamin $B_{12}$.

Food leaving the stomach consists of a thick, semifluid mass called chyme. Once in the duodenum (first segment of the small intestines) chyme is mixed with pancreatic secretions and bile from the liver. Pancreatic secretions include large volumes of bicarbonate and a variety of digestive enzymes. Bicarbonate changes the pH of chyme from acidic to the alkaline pH required for optimal activity of the pancreatic enzymes. Pancreatic enzymes include amylases for digesting carbohydrates, proteases (e.g., trypsin, chymotrypsin, carboxypeptidases, and elastase) for digesting proteins, and lipases for digesting fats. Cells lining the intestine are called luminal cells or brush border cells (name derived from the brushlike appearance of the microvilli on their surface). These produce brush border enzymes (e.g., sucrase, maltase, isomaltase, lactase) important in the final stages of digestion of dietary carbohydrates.

Absorption is the process whereby food nutrients, following digestion, are transferred from the lumen of the gastrointestinal tract to the blood or lymph system. Most absorption in dogs and cats occurs in the small intestine. Except for water absorption from the colon, very little absorption occurs from the large intestine of carnivores. The brush border cells lining the small intestine are richly supplied with minute, finger-shaped projections called *villi,* which absorb food nutrients. Within the villi are tiny capillary blood vessels and lymph ducts, which collect and absorb nutrients. The brush border of the small intestines is lipophilic and relatively impermeable to monosaccharides. Absorption of monosaccharides is therefore dependent on an active transport mechanism where energy is used to move sugars across the membrane. Amino acids are absorbed by a similar active transport process. Absorbed sugars and amino acids are then transported to the liver through the portal blood system.

Fat absorption is a multistep process. Pancreatic lipase and colipase are responsible for breaking fats into free fatty acids, monoglycerides, and diglycerides. Bile salts secreted by the liver emulsify fats, which creates an interface whereby the hydrophilic lipase and hydrophobic lipids can interact and also form micelles (small aggregates of lipids emulsified with bile salts). Fatty acids and monoglycerides can diffuse directly from micelles into the lipid-rich layer of intestinal cell membranes. Fatty acids are transported to the endoplasmic reticulum of the intestinal mucosal cells where they are resynthesized into triglycerides and then incorporated into chylomicrons. Chylomicrons are secreted from the inner cell membrane into the lymphatic system for delivery to the systemic circulation.

Fat-soluble vitamins are absorbed with dietary lipids. Water-soluble vitamins are absorbed primarily by simple diffusion (some exceptions are vitamin $B_{12}$, which must first be bound to intrinsic factor, and folic acid, which is absorbed by active transport in the upper small intestine). The absorption of many minerals is dependent on the body's need for the mineral and also the form present in the diet. For example, calcium absorption is regulated by parathyroid hormone concentrations and vitamin D and influenced by the form of calcium in the diet as well as the concentrations of phosphorus, phytates, oxalates, and other substances.

The large intestine is relatively short in dogs and cats. Its principle nutrition-related role in these species is the absorption of water and salts. The large intestines contain high concentrations of bacteria, the majority of which are anaerobes. The predominate species of bacteria in the large intestines of dogs and cats include streptococci, lactobacilli, *Bacteroides,* and clostridia species. The transit time of undigested food in the large intestines of dogs and cats is approximately 12 hours but varies with fiber content of the diet.

# 9.7 PRACTICAL FEEDING MANAGEMENT FOR DOGS AND CATS

Proper feeding requires a basic understanding of nutrient requirements and development of appropriate feeding management programs for the various phases of the life cycle as well as consideration for special-needs situations including nutritional management of various diseases. In this section, we discuss composition and types of food, methods of feeding, and the special needs associated with different life stages and performance levels. Nutritional management of diseases is discussed in Section 9.12.

## 9.7.1 TYPES OF FOOD: SOURCES AND QUALITY

Commercially available petfoods can be divided into popular, premium, and private label or generic foods. Popular brands include foods that are developed by large corporations devoting substantial time and resources into the development and national marketing of their foods to ensure high brand-name recognition. These foods are generally of moderate to high quality and have been formulated to be highly palatable (often the primary focus of marketing is "taste appeal" of the diet). Many popular diets have undergone rigorous diet testing and contain label claims conforming to AAFCO nutrient profiles for growth and adult maintenance.

Premium brands are often sold through specialty pet stores, veterinarians, and feed stores. These diets often have a wide range of formulas available for the various life stages (growth—with different formulas for large and small breeds of dogs—gestation and lactation, performance, senior, and "lite" or calorie restricted). Premium brands may also have different nonprescription formula diets targeted toward the control of nutritionally responsive diseases (e.g., hairball formulas for cats, tartar control formulas for dogs and cats, novel protein-based diets for food allergic animals, and others). These are typically "fixed-formula diets" with consistent use of high-quality ingredients. Although these diets are more expensive than the "popular" diets, the digestibility is usually higher resulting in lower stool volume and less food being fed. Because less food is needed, the per-serving cost may be similar to that of popular commercial brands.

A primary consideration of manufacturers of generic and store-brand petfoods is the production of a low-cost diet. This may result in the use of lowest-cost ingredients, some of which may be of poor quality. The least-cost goal is often met by purchasing from a range of sources resulting in subtle but sometimes significant differences in the contents of the diet (e.g., one month the animal protein source might be beef by-products and the next month it might be pork by-products). Lower quality ingredients may result in lower digestibility and larger stool volumes. In some cases, the nutrient content is inadequate to meet the needs of the pet resulting in deficiency symptoms such as hair loss and skin disease (a syndrome termed "generic dog food dermatosis").

Some pet owners prefer homemade diets. Homemade diets should be evaluated by a nutritionist to ensure that the diet is complete and balanced. Once a recipe for a balanced diet has been selected, the exact ingredients specified in the recipe should be used. Substituting different ingredients may unbalance the diet. Remember, most humans do not eat a diet as well-balanced as a dog or cat consuming a commercial diet. Foods enjoyed by humans may not be the best ones for pets.

## 9.7.2 TYPES OF PETFOODS: PHYSICAL FORMS

There are two broad categories of commercially produced petfoods: dry and canned. Dry petfood can be further subdivided into dry-expanded (10% to 12% moisture), semimoist (25% to 35% moisture), or soft-expanded (27% to 32% moisture). The latter are also called soft dry petfoods.[4] The quality of a food is not related to its form. High-quality and low-quality products are available in both categories.

Dry petfoods may be produced as kibbles, pellets, baked products (e.g., biscuits), meals, or expanded particles (extruded). Kibbled food is made by baking on a sheet and then breaking it into small pieces called kibbles. Baked products are made by shaping the food into the desired shape (e.g., biscuits) prior to baking. Meals are prepared by mixing together a number of dried, flaked, or granular ingredients; meal petfoods are more accurately described as "textured" food because they are often a blend of heat-processed grain (flaked or extruded) and a pelleted protein supplement that also contains vitamins, minerals, trace minerals, and amino acids. Extruded dog foods are made by cooking the ingredients using high heat and pressure and forcing them through a die. The product expands into the desired shape and size as it passes through the die. Cooking increases the digestibility and utilization of dietary starches. After extrusion, fat and/or digest[5] is sprayed on the food's surface to increase palatability and caloric content. The process of spraying substances on the exterior of the food is called "enrobing." Hot air drying is used to reduce food moisture content to < 12%. Dry petfoods are the most popular for both dogs and cats (Figure 9.11). Advantages of dry foods include (1) less expensive (lower shipping and handling costs per unit fed), (2) may be fed free choice, and (3) their abrasive effect reduces the accumulation of tartar on teeth. Disadvantages of dry foods are (1) lower palatability for most dogs and cats, (2) the required drying limits the use of fresh animal tissues in the diet, (3) the requirement for stability in packaging and storage limits the types and quantity of fat that can be included resulting in lower levels of PUFAs and a low nutrient density, and (4) dry foods are generally higher in fiber and lower in digestibility when compared with canned foods.

Semimoist dog foods have a moisture content between 25% and 35%. These diets contain high amounts of corn syrup and polyhydric alcohols (e.g., propylene glycol) as inhibitors of microbial growth and to give the food its desired texture by binding water. Some products also contain acids (e.g., phosphoric, hydrochloric, or malic acid), which further inhibit microbial growth. These diets do not require refrigeration and can be fed *ad libitum*. The high content of sugars and soft texture of the diets make these diets highly palatable. A variety of ingredients, including fresh animal tissues, can be used in manufacturing semimoist diets. Semimoist diets appeal to many owners because of the physical similarity to ground beef. Disadvantages of semimoist foods include higher cost and lack of abrasive action on teeth. In addition, the high sugar content makes these diets unsuitable for use in feeding diabetic cats and dogs.

Canned foods are usually the most palatable and thus are particularly useful in feeding finicky eaters. Approximately 70% to 75% of the content of canned diets is water. The high moisture content helps ensure adequate water intake in cats and thereby may decrease the incidence of FLUTD (see also Chapter 14). An advantage of canned diets is that a variety of wet and dry ingredients can be used. The mixture may be homogenized and put into the can as a "loaf" or may be formed into chunks combined with gravy. The chunks resemble chunks of meat (the manufacturer's objective) but are often composed of textured vegetable proteins such as extruded soy flour mixed with red or brown dye. Canned foods typically have the highest concentrations of protein and fat. A disadvantage is that these foods are usually the most expensive on a per-meal basis resulting from high costs in manufacturing, transportation, and handling. Some types of canned foods are not complete balanced diets—these "gourmet" diets are intended to be used as supplements (e.g., to "top-dress" dry foods) and should not be used as the sole food fed to a cat or dog.

---

[4]Board on Agriculture and Natural Resources of the National Research Council. 2004. *Nutrient Requirements of Dogs and Cats*. Washington, DC: The National Academies Press.

[5]Digest is a flavor enhancer produced by the enzymatic degradation of animal tissues. Proteolytic enzymes are used to partially digest ground poultry viscera, fish, liver, and beef lungs. Degradation is stopped by the addition of an acid (usually phosphoric acid). The resulting liquid solution is called "digest." The digest is sprayed on the exterior of dry foods at a level of 3% to 5%. Digest may also be dried and dusted onto the surface of a food following the application of fat.

**Figure 9.11** One example of the various shapes available in commercial dog and cat foods. *Courtesy Dr. James E. Corbin.*

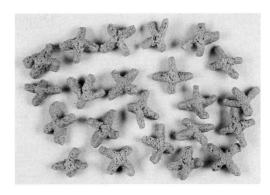

### 9.7.3 SNACKS AND TREATS

Snacks and treats represent one of the fastest-growing segments of the petfood industry. Owners purchase treats as a way of showing their love for pets. Treats are useful as training aids to reinforce desired behaviors. Another purpose for treats is to promote dental health. In addition, owners use treats to provide variety to the pet's diet. Many treats are made to resemble human foods such as sausages, hamburgers, bacon, or cheese. Other treats are made to resemble bones or other items dogs enjoy chewing (see Figure 9.1). Most treats are intended to be used as dietary supplements and as such should not constitute more than 10% of the pet's daily caloric intake.

### 9.7.4 METHODS OF FEEDING

Three methods of feeding dogs and cats are free-choice (*ad libitum*) feeding, time-restricted meal feeding, and portion-controlled meal feeding. Each feeding method has advantages and disadvantages that the owner should consider.

Free-choice feeding requires the least amount of work. All that is required is making certain a reasonably fresh food is always available to the pet. Other advantages of free-choice feeding are that kenneled dogs are quieter when food is always available, it discourages co-prophagy (eating stools), and there is less competition for food among dogs or cats housed in groups. Disadvantages are that changes in food intake may not be apparent until a dog or cat has lost or gained considerable weight. Overeating and obesity are common problems for pets fed free choice (see Figure 9.2). To decrease weight gain, it may be necessary to use a high-fiber, low-calorie food for free-choice feeding; however, such foods may not meet the nutritional needs of growing or working animals. It is preferable to use one of the meal-feeding systems for feeding kittens and puppies until they reach at least 90% of their adult weight.

Time-restricted meal feeding involves giving the cat or dog free access to food for 15 to 20 minutes one or two times daily. This method relies somewhat on the pet's ability to regulate its daily energy intake and has the advantage that fresh or canned foods can be incorporated into the diet. Pets that are finicky eaters (nibblers) or easily distracted may not consume enough food in the limited time allotted. Other pets may quickly learn that they must "beat the clock" and eat voraciously throughout the allotted time resulting in digestive upsets and a higher risk of bloating.

Food-restricted meal feeding is the optimal method of feeding most pets. The amount of food needed by the animal is measured and fed in one or several divided portions. This allows the owner to immediately observe any changes in the pet's appetite and food intake. A variety of food types can be fed easily and the pet's weight can be readily controlled. The primary disadvantage of food-restricted meal feeding is the owner must determine the amount of food to feed. Guidelines for feeding are given on most petfoods; however, additional factors discussed in Section 9.5 should be considered.

## 9.7.5   PREGNANT/LACTATING ANIMALS

Ensuring the best possible start in life for newborn animals begins with the correct nutrition for the pregnant and lactating bitch or queen.

*The bitch.*  Prior to breeding, the bitch should be examined and, if necessary, treated for internal and external parasites and vaccinated for all diseases that are problems in the geographical area to ensure she has optimal immunity to transmit to puppies. The bitch should be at optimal body weight prior to being bred. The nutritional needs of the pregnant bitch are similar to those for maintenance during the first several weeks of gestation. Many pregnant bitches have a decrease in appetite during weeks 3 to 5 of pregnancy (perhaps experiencing a syndrome similar to morning sickness in women). Approximately 70% of fetal weight gain occurs in the final 3 to 4 weeks of gestation; it is recommended that caloric intake be increased by 15% to 25% during that period. The greatest nutritional needs occur during lactation. During peak lactation (weeks 3 to 6 following parturition), the bitch needs approximately 100 kcal ME/lb of puppies daily in addition to the energy required for maintenance. During lactation, it is advisable to feed a good-quality growth/lactation diet free choice, anticipating that the bitch will consume approximately 25% more food over her normal maintenance amount for each puppy nursed.

*The queen.*  Before breeding, the queen should be given booster vaccinations and, if needed, treated to eliminate internal and external parasites. She should be at optimal weight and fully mature (wait until at least the second estrous cycle before breeding a young cat). Unlike the bitch, the queen gains weight linearly throughout gestation and thus her food intake should be gradually increased starting immediately following breeding, reaching 25% above maintenance by parturition. The best feeding method for pregnant queens is free-choice feeding of a growth/lactation cat food. The lactating queen should consume two to three times more food than that needed for maintenance.

## 9.7.6   EARLY DEVELOPMENT AND FEEDING OF PUPPIES

A challenge to newborn puppies is that they are born with limited protection against disease resulting from the low level of passive immunity acquired from their mother when antibodies crossed the placental membranes. Fortunately, the bitch concentrates disease-fighting antibodies in the colostrum (first milk following whelping). It is important that newborn puppies receive colostrum because it provides a nutritious source of food and also provides antibodies at a critical time. The ability to absorb the high molecular weight antibodies decreases steadily throughout the first 24 hours and after 48 hours the puppy digests the antibodies much like other proteins, thus destroying their immunologic properties.

Puppies and kittens are among the fastest-growing mammals. They double their birth weight in about a week, triple it in 2 weeks, and quadruple it in 3 weeks. As the time required to double birth weight decreases, there is an increase in the percentage of protein and minerals in the mother's milk (Table 9.5). This is needed to support rapid muscle and bone development.

Puppies at 3 weeks are beginning to consume adult dog food in addition to the mineral- and protein-rich mother's milk. Gradually increasing their consumption of the bitch's food, puppies can obtain their entire nutritional needs from solid food by 6 weeks of age. The bitch should be fed in a shallow pan thereby enabling puppies to have easy access to solid food prior to weaning. Also, never use a deep pan for water as the puppies may tumble in and drown.

Weaning is easy. Simply remove the bitch and provide a commercially prepared solid puppy chow and fresh water *ad libitum* (free choice in accordance with the puppy's desire). (It is desirable to withhold food from the bitch on the day of weaning to decrease her milk production and lessen her discomfort from milk engorgement when the puppies are removed.)

**Table 9.5**
Percentage of Protein and Minerals in Mother's Milk and Time Required for Newborn to Double Birth Weight

|  | Milk Fat | Protein | Lactose | Minerals | Total Solids | Time for Newborn to Double Birth Weight (days) |
|---|---|---|---|---|---|---|
| **Dog** | 8.3 | 7.1 | 3.7 | 1.3 | 20.4 | 8 |
| **Cat** | 3.3 | 9.1 | 4.9 | 0.6 | 17.9 | 9 |
| **Composition Compared with Milks of the Cow and Woman** | | | | | | |
| **Cow** | 4.0 | 3.3 | 5.0 | 0.7 | 13.0 | 47 |
| **Woman** | 3.7 | 1.6 | 7.0 | 0.2 | 12.5 | 180 |

The goal of a feeding program for puppies is to promote the optimal growth that minimizes developmental orthopedic bone diseases (Chapter 14) and obesity while maximizing the growing dog's overall health. The goal should be to have puppies grow at the average rate for their breed. Feeding for maximum growth rates is associated with increased risks of skeletal deformities and developmental bone diseases. As a general guideline, small- and medium-sized dogs (adult weights up to 50 lbs) should reach 50% of their adult weight around 4 months of age. Large- and giant-sized breeds should reach 50% of their adult weight around 5 months of age. Puppies should be assessed every 2 weeks and dietary adjustments made as needed to maintain optimum body condition and growth. Foods for puppies of large- and giant-sized dog breeds should contain 0.7% to 1.2% calcium (dry matter basis), 0.6% to 1.1% phosphorus, 8% to 12% crude fat, 20% to 32% crude protein, and approximately 3.5 kcal metabolizable energy per gram of dry matter. Foods for puppies of small- to medium-sized breeds should contain 0.7% to 1.7% calcium, 0.6% to 1.3% phosphorus, 10% to 25% crude fat, 22% to 32% crude protein, and approximately 4.0 kcal metabolizable energy per gram of dry matter (Figure 9.12). Many petfood manufacturers produce different foods formulated specifically for puppies of small/medium-sized breeds and large/giant-sized breeds. Owners should be cautioned to limit treats fed to puppies unless the treats are formulated specifically with the recommended nutrient content for puppies of various breed sizes.

**Figure 9.12**   Rickets in a Shetland Sheepdog puppy that was fed a calcium/phosphorus-restricted diet following a misdiagnosis of kidney disease. This illustrates the importance of adequate levels of calcium and phosphorus in the diet of growing dogs. *Courtesy University of Illinois.*

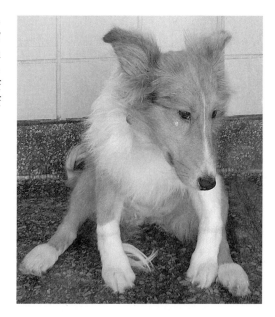

## 9.7.7  KITTEN CARE AND FEEDING

Like newborn puppies, kittens require adequate amounts of colostrum shortly after birth to provide passive immunity against infectious diseases and also to contribute to postnatal circulatory volume. Similar to puppies, the newborn kitten is unable to regulate its body temperature or to eliminate without stimulation from its dam. Kittens should be weighed daily for the first 2 weeks of life and then every 3 days until they are 1 month old. A steady weight gain (approximately 100 g/week) and normal stools are the best indicators of adequate nutrition and good health. Kittens should be encouraged to begin eating a kitten growth formula at 3 weeks of age. Mix solid food with a kitten milk replacer to a mushlike consistency and place on the kitten's lips. In licking the gruel off their lips the kitten will find they like the taste and should soon begin eating the gruel. After a few days, the amount of milk can be reduced until the kittens are eating the dry food. The queen will usually begin weaning the kittens when they are 6 to 10 weeks old. If the queen has not weaned the kittens by 10 weeks, withhold food from the queen to decrease her milk production, then remove her from the kittens and gradually return to normal maintenance feeding.

## 9.7.8  ORPHAN FEEDING AND CARE

Commercially prepared formulas are preferred for feeding orphaned puppies (e.g., Esbilac®, Borden) or kittens (e.g., KMR®, Borden). These formulas have been developed to meet the unique nutritional needs of the respective neonates. Two alternative emergency formulas for puppies are (1) 1 cup whole milk, 1 teaspoon vegetable oil, 1 drop infant multiple vitamins; and (2) evaporated whole milk diluted 3:1 with water. An emergency formula for kittens is 4 oz of whole milk + 1 egg yolk + 1 drop infant vitamins blended together. Puppies and kittens overfed milk can develop diarrhea. Orphans should be fed at least four times daily; the amounts to be fed are listed in Table 9.6.

Methods of feeding orphaned puppies and kittens include using nipple bottles, droppers, spoons, and tube feeding. Commercial nipple bottles are made specifically for feeding orphan puppies or kittens (Figure 9.13); toy baby bottles or premature infant bottles may also be used. The hole in the nipple should be such that when the bottle is inverted milk oozes out slowly. Tube feeding is the fastest way to feed orphans. A number 8 or 10 French infant feeding tube is measured and marked at a point equal to three-fourths of the distance from the orphan's nose to its last rib (Figure 9.14). When the tube is inserted this distance, it will reach the caudal esophagus. If any resistance is felt in inserting the tube, it is probably in the trachea and should be withdrawn. The amount of formula to be fed is drawn into a syringe and slowly (over 2 minutes) administered through the tube (Figure 9.15); stop if any resistance is felt as the stomach is probably full. Following feeding, a moistened cotton ball should be rubbed around the genital area to simulate defecation and urination (Figure 9.16). The orphaned puppy or kitten should be kept in a warm environment and handled only during feeding for the first 2 weeks when most of its time should be spent sleeping. Do not feed a

**Table 9.6**
Feeding Orphaned Puppies and Kittens

| Age (weeks) | Amount of Formula to Be Fed (ml/100 g body weight/day)[*] |
|---|---|
| 1 | 13 |
| 2 | 17 |
| 3[**] | 20 |
| 4[**] | 22 |

[*]*Divide and feed four times daily.*
[**]*Offer additional growth-formula solid food or gruel free choice.*

**Figure 9.13**   Bottle feeding an orphaned kitten. *Courtesy Paul E. Miller.*

milk formula if the neonate's body temperature is below 95°F[6]; instead give 1 ml/30 g body weight of a nutrient-electrolyte solution (e.g., equal parts of 5% glucose and Ringer's lactate solutions) every 15 to 30 minutes subcutaneously and/or orally.

## 9.7.9   NUTRITION AND ENDURANCE IN DOGS

Healthy, well-nourished dogs provided with adequate water have enormous stamina. Research has shown they can work continuously for 30 or more hours. Interestingly, dogs can use their body stores of fat, which, unlike humans in similar situations, they can catabolize without becoming ketotic. Herding, guide, and police dogs are valued for their intelligence, service, strength, and endurance. Following the September 11, 2001, terrorist attacks on the New York City World Trade Center, more than 350 canine search specialists worked 12-hour shifts to locate victims. Workers were impressed with the endurance displayed by the dogs as well as their desire to work without rest.

## 9.7.10   EXERCISE PHYSIOLOGY AND NUTRITION IN DOGS

Since the dawn of athletic competition, athletes and trainers have sought ways of optimizing performance through nutrition. Writings of Plato (428–348 BC) suggest that diets of ancient Olympic athletes were higher in protein than those of the general populace. Research conducted by A. J. Reynolds of Cornell University,[7] designed to mimic the type of work performed by sled dogs on the Iditarod Trail or Yukon Quest, showed that dogs fed 40% of calories as protein maintained a larger plasma volume of red blood cell mass during strenuous training than dogs fed only 16% of calories as protein. Additionally, dogs fed 16% of calories as protein were

---

[6]Digestive enzymes do not function at temperatures below 94°F; feeding milk to hypothermic neonates results in bloating and diarrhea.

[7]A. J. Reynolds, DVM, PhD, *Exercise Physiology and Nutrition,* 1997 *Petfood Forum,* Chicago, IL, April 14–16, 1997, p. 203–209.

**Figure 9.14** Measuring for tube feeding a puppy. *Courtesy Dr. James E. Corbin.*

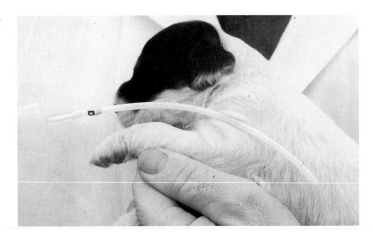

more prone to injury. Based on those studies, scientists concluded that providing optimal dietary protein enhances blood volume and helps prevent injury during intense training of dogs.

In additional studies with 16 Alaskan Huskies, 8 dogs were fed a high-fat diet (15% carbohydrate, 60% fat, 25% protein) and 8 were fed a high-carbohydrate diet (60% carbohydrate, 15% fat, 25% protein) during a 6-month experimental trial of intensive exercise, the researchers concluded that diet plays an important role in endurance and specifically that feeding high levels of dietary fat increases $VO_2$ max[8] and the maximal rate of fat oxidation by stimulating mitochondrial growth. Reynolds concluded further that by feeding a high-fat diet while training, dogs are training their muscles to metabolize fat more efficiently thereby sparing the use of carbohydrates, especially the polysaccharide glycogen. Although a glycogen-sparing strategy may be sounder for endurance than glycogen supercompensation in canine endurance athletes, glycogen depletion is still associated with a deterioration in performance.

---

[8]$VO_2$ max is the maximum volume (V) of oxygen ($O_2$) in milliliters that can be used by mammals per kilogram of body weight 1 minute while breathing air at sea level. Because $O_2$ consumption is related linearly to energy expenditure, measuring $O_2$ consumption is indirectly measuring an animals's maximal capacity to perform work aerobically.

**Figure 9.15** Tube feeding a puppy. *Courtesy Dr. James E. Corbin.*

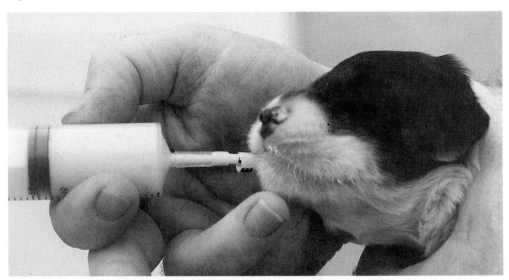

**Figure 9.16** Stimulating urination following feeding of a neonatal puppy. *Courtesy Dr. James E. Corbin.*

When the researchers examined the effects of postexercise carbohydrate supplementation on muscle glycogen repletion in an Iditarod-winning sled dog team, they found that dogs supplemented with carbohydrates immediately following exercise showed significantly greater muscle glycogen repletion than nonsupplemented dogs during the first 4 hours of recovery. Carbohydrate supplementation was also associated with an increase in plasma glucose concentration 100 minutes postexercise, which may have promoted glycogen storage.

Finally the scientists concluded that oversupplementation of any given nutrient will displace the intake of other nutrients and may lead to deficiencies. These nutrient deficiencies can be manifested in problems such as anemia, inappetence, poor coat quality, and lethargy. Hence, the feeding of diets that are complete and balanced for all nutrients is recommended.

## 9.7.11    STRESS/PERFORMANCE DIETS

Working dogs have higher energy requirements than nonworking adult dogs (Figure 9.17). The exact requirements vary with the type of intensity of the work and environmental conditions. Examples of stressful work situations include police and guard duty, guiding the blind, and performance in show circuits. Physically strenuous activities include those previously mentioned plus racing, herding, hunting, and search and rescue work. In general, stress and performance diets should contain a minimum of 1,900 kcal ME/lb of dry matter

**Figure    9.17** Working dogs have increased protein and energy needs. In preparation of sled dog competition, these dogs were fed energy-dense diets with increased protein levels. Energy density was achieved in large part by formulating the diets with an increased fat content. *Courtesy Champion Petfoods.*

(4.2 kcal/g). Working dogs may derive 70% to 90% of the energy used for work from fat metabolism. Energy density and diet digestibility are the two most important nutritional factors affecting performance in working dogs. The easiest method of supplying an energy-dense diet is to increase the fat content. Performance diets should contain a minimum of 23% fat on a dry matter basis. Performance and working dogs should have continual access to clean fresh water. A small meal should be provided 0.5 to 2 hours prior to working and small amounts of food should be offered frequently during endurance events. The largest meal of the day should be fed after the day's training or work is completed.

### 9.7.12 FEEDING GERIATRIC DOGS AND CATS

The objectives of feeding geriatric dogs and cats are to extend and improve their quality of life. The average life span of domestic dogs is approximately 12 years with reports of small breeds living as long as 29 years. The average life span of domestic cats is approximately 14 years with reports of cats living as long as 36 years. Longevity can be increased with good care and nutrition throughout life. The dog and cat at 1 year of age are at the same physiologic stage of life as a 15-year-old person. The 2-year-old cat or dog is physiologically similar to a 24-year-old person and aging for each successive year is approximately equivalent to 4 years in humans for most breeds of dogs and cats (7 years for giant breeds of dogs). For nutritional purposes, most dogs and cats are considered geriatric at 7 years of age and giant breeds of dogs at 5 years. There are many body changes with aging that have nutritional implications.

*Factors that may decrease energy requirements in geriatric animals.* An animal's basal metabolic rate decreases with aging; this, combined with decreased activity, results in decreases in energy requirements of 20% to 40% for geriatric animals. To avoid weight gain, the elderly pet should be fed a diet moderately restricted in caloric density (e.g., moderately high in fiber with lowered levels of protein and fat).

*Factors that may decrease food intake in geriatric pets.* Olfactory and taste perceptions are decreased in older animals, which may result in decreased interest in eating. Decreased eyesight, dental calculus, periodontal disease, loss of teeth, oral ulcers, and decreased saliva production may all contribute to decreased food intake and weight loss. Additionally, older animals frequently have decreases in their digestion and absorption of nutrients. Geriatric animals experiencing weight loss should be fed highly palatable diets warmed to near body temperature to maximize their odor and appeal to the pet.

*Disease considerations in feeding geriatric pets.* Renal (kidney) failure and heart disease are two of the leading causes of death in older animals. Many studies have shown favorable responses to feeding older pets diets with moderate restrictions in protein, phosphorus, and sodium. These diets lessen the workload of the kidneys and heart and help prevent or slow the progression of disease in these organs. Section 9.12 includes additional information on the role of dietary therapy in the management of animals with impaired heart or kidney function.

## 9.8 LABELING DOG AND CAT FOODS

Because petfoods are marketed widely in supermarkets and elsewhere, and are frequently shelved in close proximity with human foods in the home, it is conceivable that people could consume petfoods. Therefore, it is especially important that accurate labeling ensures proper use of any petfoods containing medicated additives such as antibiotics, those used in controlling internal parasites, or those fed to control reproduction.

Proper labeling is also important in providing information that prospective buyers can use in comparing petfoods as well as in ensuring a sound basis for fair competition among manufacturers. Numerous private and public groups/organizations/administrations/ agencies involved with regulations and public policy assist in the whole of proper petfood labeling.

- *Food and Drug Administration (FDA).* The Federal Food, Drug, and Cosmetic Act (FFDCA) defines "food" as an article used for food or drink for man or other animals, and a "drug" as an article intended for use in the diagnosis, cure, mitigation, treatment, or prevention of disease. Additional information is available at the FDA Web site http://vm.cfsan.fda.gov/~redbook.red-toca.

- *Hazard Analysis of Critical Control Points (HACCP).* The United States Department of Agriculture's (USDA) Food Safety and Inspection Service (FSIS) is the agency within the USDA responsible for ensuring the safety, wholesomeness for humans, and accurate labeling of meat, poultry, and egg products. HACCP is designed to ensure product and public safety. It is aimed at consumer safety in all aspects of manufacturing, storage, and distribution. These control points typically include, for example, thermal treatments or acidity to prevent microbial contamination, and physical mechanisms such as filters and metal detectors to avoid foreign body contamination. Additional information is available at the FSIS and USDA Web site http://www.fsis.usda.gov/OA/codex/system.htm.

- *Generally Recognized as Safe (GRAS).* This is terminology based on the history of use prior to 1958 as well as research based on data/information gleaned using tools of the scientific method. Including GRAS in considerations of whether to license a food/feed product is aimed more at animal feeds/foods than those intended for human consumption. The term *GRAS* is commonly used by the FDA and others when referring to food and feed additives.

- *Association of American Feed Control Officials (AAFCO).* This group was created by the petfood industry to prepare definitions/terms related to packaging and labeling petfoods as well as to develop guidelines/standards/underpinnings to ensure overall quality of petfoods. For additional information, see http://www.aafco.org.

## 9.8.1 NUTRITIONAL ADEQUACY ON LABEL

A dog or cat food, other than one clearly identified as a snack or treat, must bear a statement of nutritional adequacy (e.g., "complete and balanced" with a range of animals to which the petfood can be fed safely). If the nutritional adequacy of the product is based on its meeting the AAFCO dog or cat food nutrient profile(s), the following statement must appear on the label: "(name of product) is formulated to meet nutritional levels established by the AAFCO dog (or cat) food nutrient profile for (stage or stages of the pet's life such as gestation, lactation, growth, maintenance, or the words 'all life stages')."

## 9.8.2 LABEL INFORMATION

The label on the petfood should indicate the specie (e.g., dog or cat) for which it is intended as well as the following information: name and address of the producer/manufacturer; name of the product; treatment of ingredients (e.g., processed, pressed, heated, milled, pelleted); name and content of protein, starch, fiber; net weight (or net volume); directions for use of product; list of raw ingredients arranged in decreasing order of preponderance by weight (such a listing could be misleading regarding nutrients contributed by having a high-water/low-bioavailability-nutrient content versus a high-bioavailability-nutrient/low-water content); guaranteed analytical caloric content; percent protein, fat, fiber, starch and other carbohydrates (e.g., sugar), moisture, minerals, vitamins; descriptive terms (e.g., light, lean, low- or reduced-calorie or fat. An affidavit for calorie count must be completed for all labels that contain a calorie content statement); best-to-use-before date; lot number; and/or date of production.

Several points should be considered when looking at a petfood label. First, manufacturers are required to list only the minimum percentages for crude protein and fat and the maximum percentages for moisture and fiber. These numbers are minimums and maximums only and may not reflect the actual content of the diet. For example, a diet that lists a minimum crude protein of 22% may have an actual crude protein content of 30%. In addition, crude percents do not equate to the actual digestible quantities of nutrients in the diet. Thus, the guaranteed analysis panel on a petfood provides only rough estimates of the nutrient content of a food. Likewise, ingredient lists do not provide information on the quality or digestibility of the ingredient. As previously mentioned, the order rankings of ingredients in different diets may not be comparable due to differences in moisture content of the individual items within the diets. For these reasons, the ingredient list is of limited help in comparing two petfoods.

### 9.8.3 FEEDING DIRECTIONS

Feeding directions must be listed on dog or cat food labeled as "complete and balanced." Directions must include the amount to feed (weight/unit of product) per weight of the dog or cat.

### 9.8.4 USE OF THE TERM *NATURAL*

Use of the term *natural* in petfoods has increased in recent years. This has resulted in increased sales of formulated "natural" petfoods. In keeping with AAFCO regulations and policy documents, the term *natural* may be used on petfood labels providing (1) that all ingredients and components of ingredients other than vitamins and minerals have not been chemically synthesized; and (2) that chemically synthesized vitamins or minerals present in a petfood using the term *natural* have a disclaimer juxtaposed in the same size, style, and color print as the term *natural*, such as "Natural with Added Vitamins and Minerals."

There is no specific petfood law per se in Europe, only an animal feedstuffs law, which includes pets and other animals bred for economic purposes. Of all European legislation, two important directives relate to additives (e.g., antibiotics, coccidiostats, and growth promoters) and the marketing of finished products (here the concern is that no substance may be used in processing feed materials that may leave residues that could present a health risk to pets and other animals).

### 9.8.5 QUALITY ASSURANCE OF PETFOODS

Petfood quality includes carefully selected raw ingredients, sound manufacturing procedures, high nutritional value, product palatability, the use of mostly "natural" ingredients, and an acceptable shelf life (many petfoods are cooked to improve their digestibility and concurrently kill bacteria that could cause disease and/or shorten shelf life).

The start-to-finish "quality systems process" in the petfood (and other consumer goods) industry begins with identifying the needs and expectations of consumers and ends with fulfilling them. Quality assurance is founded on getting things done right the first time—*and all other times*. The challenge is doing things right every step of the process: identifying customers' needs and expectations (consumer research), developing products that match or exceed those needs and expectations (product development), persuading consumers to buy the petfoods (marketing), manufacturing the exact product that was developed (operations and production), persuading petfood shops and retailers to offer the product (sales), and delivering the petfoods to points of sale (distribution).

## 9.9   NUTRITIONAL COUNSELING AND MARKETING PETFOODS

Veterinary client education should include nutritional counseling in addition to information regarding vaccinations, internal and external parasite examinations, preanesthesia laboratory testing, physical examinations, dental hygiene examinations, and other related health-care interests. The health and quality of life of companion animals reflects in large part the diet they receive.

Periodontal (gum) disease is the most common disease in companion animals. It is caused by dental plaque and is exacerbated by accumulation of dental calculus (tartar). Several petfoods and treats designed to aid in minimizing the accumulation of dental calculus are now available.

The Veterinary Oral Health Council (VOHC) was established by the American Veterinary Dental College in 1997. This organization awards its *Seal of Acceptance* to company products that have been determined to meet the requirements as stated in the VOHC protocols. These requirements are specific and statistical data must demonstrate efficacy. The VOHC awards its Seal in two claim categories: *Helps Control Plaque* and *Helps Control Tartar*. For additional information see the Web site http://www.VOHC.org/.

The petfood industry is an effective leader in implementing evolving technologies and marketing approaches to advance market presence and capture additional market share. Specialty pet shops and many veterinary practices are also involved in selling premium brands of health-promoting petfoods (Chapter 19). The growth in pet store and veterinary markets has aided in expanding sales of petfoods, pet treats, and accessories. Marketing concepts form the basis on which a product is advertised and sold. There are nine basic concepts used in marketing petfoods: (1) specific-purpose foods (e.g., targeted life stages, weight control, and performance diets), (2) all-purpose diets (e.g., one food for the life of the pet), (3) low-price (e.g., to appeal to price-conscious consumers), (4) people-food appeal (e.g., stews and meat-chunks-and-gravy type diets for owners who believe their pet will prefer the same types of foods that humans like), (5) flavor varieties (e.g., for owners who believe their pet will prefer a food containing a variety of flavors), (6) presence of a food item (e.g., lamb or other meats in dog foods and fish in cat foods with the premise that these are the foods preferred by the pet), (7) absence of a food item (e.g., "no corn" or "no wheat" or "no soy" with the premise that these items are not optimal ingredients for pets to consume despite the fact that dogs and cats can safely utilize these food sources), (8) nutrient content (e.g., "high protein" with the premise that performance dogs benefit from high-protein diets despite the fact that energy is their greatest need and that is suitably supplied by fat), and (9) product name (e.g., use of names that are amusing, easy to remember, or sound authoritative—examples include Strongheart™ and Happy Cat™).

When evaluating petfoods and marketing claims consider the following: (1) Is the concept of the product scientifically valid? (2) Does the concept of the product relate to the purpose of the product? (3) Does the product adhere to the concept or just promote it as a sales gimmick?

## 9.10   ANALYSIS OF COMMERCIAL PETFOODS

Health-sustaining commercial dog and cat foods contain all the required nutrients: proteins and essential amino acids, fats and essential fatty acids, carbohydrates, fibers, minerals and vitamins, plus nonnutrient additives in sufficient quantities to meet the nutritional needs of dogs and cats without any nutritional deficiencies or significant excesses occurring during all phases of the pet's life. Water is the only addition needed. Desirable petfoods must be palatable, free of sharp points, soft enough to be easily consumed, and include sufficient fiber to ensure a desirable stool firmness.

All dog and cat foods marketed in the United States must have guaranteed analysis listing on the package/container. The guaranteed analysis panel includes the minimum percentages of crude protein and crude fat and maximum percentages of crude fiber and moisture. As previously mentioned, the guarantees do not indicate ingredient quality or nutrient availability to the dog or cat.

Proper labeling of dog and cat foods is important in providing information that prospective customers can use in comparing petfoods as well as in ensuring a reliable basis for fair competition among manufacturers. Labeling petfoods was discussed in Section 9.8.

# 9.11 COMMON PROBLEMS AND ERRORS IN FEEDING DOGS AND CATS

Feeding instructions commonly found on the label of packaged commercial dog and cat foods are often ignored by pet owners. Common errors in feeding dogs and cats include:

- *Feeding cat foods to dogs and dog foods to cats.* Although dogs relish cat foods, they do not need a diet containing 30%, or more, protein nor do they need the extra fat and flavor enhancers found in cat foods. The primary adverse effect of allowing dogs to eat cat foods are economic (cat foods are generally more expensive than dog foods) plus the increased risk of obesity (due to the high palatability and high energy density of cat foods).

  Cats should not consume dog foods as they *do not* contain all of the nutrients needed by cats. Moreover, dog foods are not formulated/adjusted to produce an acidic urine needed by some cats to help prevent the development of FLUTD (Chapter 14).

  Additionally, cats cannot (1) convert cystine into taurine (an amino acid required in cat diets, but not in dog diets); (2) utilize β-carotene from green plants as a source of vitamin A (as can dogs), which means cats require a supplemental source of vitamin A; (3) efficiently convert tryptophan (an amino acid) into niacin (a vitamin); (4) desaturate linoleic acid (therefore, cats require dietary supplementation with arachidonic acid); or (5) utilize protein as efficiently as dogs and other animals.

  Although cats enjoy the taste of many dog foods, based on the above facts, they should not consume dog food as their sole diet over a long period of time. When cats live outdoors and can catch mice and other small animals that provide varying amounts of the nutrients discussed previously, they can consume more dog food safely (healthwise) than when confined to indoor living.

  Cats are deficient in the enzyme, cysteine sulfinic acid decarboxylase, needed in the biosynthesis of taurine. Taurine is normally derived from high-protein diets through biosynthesis from the sulfur-containing amino acids. It can be produced by commercial synthesis and added when formulating cat foods. Taurine is associated with bile acids and, if not present in ample quantities, results in retinal and cardiac lesions. This may lead to blindness and enlarged, weak heart muscles and subsequent heart failure. Because taurine assays of foods do not accurately reflect the availability or the amount of taurine that will be used by the cat, it is best to determine the blood concentration of taurine. Testing blood levels of taurine is one means of testing commercial cat foods.

- *Feeding sugar to dogs and cats.* Many dogs and cats can tolerate moderate amounts of sugar in their diets, whereas others may be hyperactive when they consume foods high in sugar. Some say sugar tends to be addictive. This is often reflected in difficulty getting cats and dogs to accept a significant change to a diet lower in sugar.

- *Failing to ensure that cats drink plenty of water.* Cats dislike drinking stale or unclean water, so it is important to give them fresh water frequently. Interestingly, cats tend to consume only about two-thirds as much water per unit of food consumed as dogs. This contributes to cats having a higher risk of mineral precipitation in the urinary tract. Many mineral precipitates (crystals) are composed of magnesium-ammonium-phosphate (struvite). Struvite is associated with about 95% of the urinary-tract blockage and FLUTD in cats (Chapter 14). The water-intake problem is one reason canned cat foods (containing about 78% water) can help minimize struvite formation in cats, because water dilutes the concentration of minerals in their urine.

■ *Not monitoring urine pH levels in cats.* To reduce the probability of forming struvite crystals, many veterinarians recommend feeding cats diets that produce an acidic urine. This may be accomplished by including pH-lowering ingredients such as ammonium chloride, phosphoric acid, and methionine in commercial cat foods. Decisions to add urinary acidifiers to diets should be made only after consultation with a veterinarian and evaluation of urine sediment from the individual cat because crystalluria can also be caused by calcium oxalate, which is most soluble in alkaline urine.

# 9.12   NUTRITION AND DISEASE: THERAPEUTIC PETFOODS

Therapeutic diets may be recommended by veterinarians in dealing with obesity, diarrhea, liver damage, heart disease, urinary obstruction, stress, and other companion animal health disorders. Most therapeutic foods for dogs and cats are recommended and sold by veterinarians (Sections 9.9 and 19.9).

*Liver disease.* Liver disease is common in dogs and cats. Numerous injuries and diseases can cause liver damage (Chapter 14). A variety of toxins, such as chemicals, drugs, and microbial toxins, can damage the liver. The normal, healthy liver removes ammonia and converts it to urea; it is subsequently excreted in urine. Ammonia tolerance testing can be performed to evaluate this function. Diets formulated for feeding animals with liver disease should contain increased levels of the branched chain amino acids important for normal neurological function. Proteins in the diet should be of high quality to aid in regenerating liver tissue. Highly digestible carbohydrates should be included in the diet to supply energy needs and minimize protein catabolism. High levels of arginine help support liver protein and energy metabolism. Carnitine helps in fat transport and metabolism within the liver. Dogs should be fed diets low in copper and high in zinc to help avoid copper accumulation in the liver. Moderate restrictions of sodium help minimize fluid retention in both dogs and cats with liver disease.

*Chronic renal disease (CRD).* CRD is common in older dogs and cats. Diets with low levels of protein and phosphorus help reduce waste products that require excretion through the kidneys. High omega–3 fatty acids help improve blood flow to the kidneys. Restricted levels of sodium help decrease blood pressure thereby slowing the progression of kidney disease. Increased levels of dietary fiber help increase fecal excretion of nitrogen thereby decreasing the workload on the kidneys.

*Diabetes mellitus.* Semimoist diets should not be fed to diabetic cats and dogs due to their high sugar content. Diets high in fiber are generally recommended to help minimize blood glucose fluctuations and help decrease insulin requirements. Nutraceuticals, including the trace elements chromium and vanadium, may also aid in improving control of blood glucose levels in diabetic animals. Diets for diabetic animals should be kept constant (assuming the animal's weight is at the optimum level). The diet should contain a high-quality protein and be moderately restricted in fat (< 20% of calories). A minimum of 40% of the calories should be supplied by complex carbohydrates.

*Cardiac disease.* The goal of dietary management of heart disease is to reduce the workload on the heart while meeting the nutritional needs of the animal. To accomplish this (1) ensure that sodium intake is low, (2) provide sufficient protein to prevent hypoproteinemia, (3) provide potassium supplementation to replace losses associated with the administration of diuretics, (4) provide sufficient caloric density to prevent weight loss or gain, (5) make moderate reduction in protein to reduce the level of catabolic waste products from protein digestion requiring renal excretion, (6) provide additional B vitamins to replace those lost in diuresis (increased urine output), and (7) provide additional levels of taurine to support cardiac muscle function in cats.

*Gastrointestinal disease.* Dogs and cats with gastrointestinal diseases are best supported by feeding highly digestible foods plus a mixture of soluble and insoluble fibers. Small amounts should be fed frequently (three to six meals daily).

***Struvite stone and crystal dissolution.*** Struvite crystals and stones are composed of ammonium, magnesium, and phosphorus. Diets containing very low levels of protein, magnesium, and phosphorus and those promoting an acidic urine have been shown to aid in the dissolution of struvite crystals and stones.

*Feline lower urinary tract disease (FLUTD).* See Chapter 14 for more information. Cats with struvite crystalluria should be fed diets that promote an acidic pH. In addition, it is helpful to feed diets with minimal magnesium and phosphorus levels. Cats with oxalate crystalluria should be fed diets that promote an alkaline pH. Potassium citrate is often added to these diets as it helps produce a more alkaline urine. Restricted levels of animal protein and low sodium levels also reduce oxalate formation in urine.

*Food allergens/food hypersensitivity.* It is estimated that one of every seven dogs suffers from allergic disease. Flea allergy dermatitis and atopy (allergic dermatitis associated with environmental allergens) are the two most common types of canine allergic disease. Food hypersensitivity ranks third, contributing to pruritus in nearly two-thirds of dogs with non-seasonal allergic skin disease.

Food allergy (food hypersensitivity) is an immunologically mediated adverse reaction to an ingested food or food additive. Virtually all food allergens are proteins. Most dietary proteins are potentially allergenic because they are recognized as foreign by the immune system. A common requisite for a substance to be an allergen is the ability to stimulate an immune response. The ability of an allergenic protein to induce an IgE-antibody mediated hypersensitivity apparently depends on the size and structure of the protein (Chapter 14). Most protein food allergens have molecular weights of 18,000 to 36,000 Daltons.[9] The major allergens in soybean protein are 26,000 to 91,000 Daltons.

Recent research indicates that it is possible to utilize enzymatic proteolysis to produce proteins having hypoallergenic characteristics. The key is to reduce the molecular weight of the modified protein below the smallest known allergen in the parent protein. By enzymatically reducing the molecular weight of soy protein, it is less likely to elicit the immune-mediated hypersensitivity. Methods used in diagnosing and managing food allergies are discussed in Chapter 14.

*Obesity.* Obesity is a national human health issue; it is also a problem among companion animals (Figure 9.18). A recent study involving more than 23,000 dogs and 10,000 cats seen at 60 private primary-care veterinary clinics in 30 states found that approximately 20% of dogs and 23% of cats 0 to 5 years old, and 40% to 45% of dogs and 40% to 50% of cats 5 to 12 years old were overweight[10] (see also Chapter 14, Section 14.4).

Numerous factors influence the development of obesity in companion animals. They include (1) breed or strain differences in the propensity to deposit fat or lean tissue at the same level of energy intake (examples of breeds having a high prevalence of obesity are Cairn Terrier, Cocker Spaniel, Dachshund, Basset Hound, Beagle, Parson Russell Terrier, and Corgi; examples of breeds with a low prevalence of obesity are Greyhound, Lurcher, Weimaraner, and

---

[9]Dalton is a unit of mass; 1 Dalton is 1/16 of the mass of the oxygen atom.
[10]Armstrong, P. J., and E. M. Lund. 1997. *Obesity: Research Update, Proc. of Petfood Forum.* Chicago, IL, April 14–16, 1997, pp. 112–114.

**Figure 9.18** An obese dog. *Courtesy Dr. James E. Corbin.*

Whippet); (2) a higher frequency of pet obesity when the dog or cat owner is overweight; (3) *ad libitum* feeding of dry foods and/or liberal access to highly palatable and/or high-fat foods; (4) a form of diet that can influence the amount of food consumed (e.g., canned or semimoist diets are preferred by many companion animals); (5) reduced metabolic rate associated with genetic factors or acquired from clinical disease (e.g., low blood thyroid hormone function, hypothyroidism) or secondary to neutering (spaying/castration results in an estimated 15% to 20% reduction in metabolic rate); (6) variation in biofeedback mechanisms that regulate satiety and appetite; (7) hypertrophy or hyperplasia of fat cells (adipocytes); (8) psychological factors causing overeating (polyphagia); (9) reduced ability or desire to exercise relative to the level of food/energy ingested; and (10) the age-related reduction in lean body mass (muscle tissue) where the bulk of energy is utilized (more true of aging dogs than of aging cats).

Interestingly, it has been observed in laboratory animals that the dietary protein:energy ratio affects the composition of body weight gain. Because differences are seen in the prevalence of obesity among breeds of dogs, additional research is needed to determine if changing the dietary protein:energy ratio could be used to help control obesity in breeds with a prevalence for above-average obesity.

In addition to specific disorders worsened by obesity, it can also complicate routine veterinary care including physical examination, imaging (radiography and ultrasonography) of body cavities, blood circulation, cystocentesis (collection of sterile urine using a needle and syringe by aspiration of the urinary bladder through the body wall), collection of cerebrospinal fluid (CSF) necessary for diagnosis of certain neurological problems, and collection of joint fluid. Important as well, obesity may also render routine anesthetic and surgical procedures difficult or hazardous. Obesity complicates calculation of drug dosages and fluid administration rates because these are based in large part on lean body mass or weight. Diabetes mellitus and skin problems are more common among obese dogs and cats.

A nutrient profile optimized for successful weight loss in dogs and cats promotes safe, efficient, and effective long-term reduction in body weight, resulting in maximal body fat loss and minimal reduction in lean body mass. In general, weight-reducing diets should provide a minimum of 60% of the recommended caloric level of dogs and cats at their optimal weight. (Too rapid weight loss increases the likelihood of weight being regained.)

## 9.13    ROLE OF DIETARY FUNCTIONAL FIBER IN PETFOODS

Dietary functional fibers (DFF) are plant substances not digested by mammalian intestinal enzymes. These include carbohydrates (e.g., cellulose, hemicellulose, and pectin) and non-carbohydrate (lignin) plant and cell wall components. Algal polysaccharides (e.g., seaweeds and many cellular fresh-water plants) and plant gums are also classified as dietary fiber. Also included are plant intracellular polysaccharides such as gums and mucilages. Other carbohydrate components that share indigestibility with DFF and, therefore, behave as DFF include oligosaccharides (e.g., fructooligosaccharides [FOS] and lactosucrose) and resistant starches (e.g., crystal structure, physically enclosed, or retrograded amylose). In addition to lignin, other noncarbohydrate materials (e.g., cutin, suberin, waxes, protein fractions, and inorganic constituents) that are indigestible components of plant cellular tissues serve as DFF.

Various sources of DFF can affect the health and quality of life of companion animals through physical or chemical interactions with the intestinal tract. For example, DFF can influence output and consistency of dog and cat feces through their bulk and water-holding capacity. DFF can also affect cat and dog colonic microbial cell production as well as intestinal transit time. The latter can be an important factor in managing conditions such as diabetes mellitus and hyperlipidemia. Through fermentation, DFF can provide an alternate route for nitrogen excretion. Fermentation of various types of fiber results in production of different amounts and proportions of short-chained fatty acids that benefit the colon environment and support the health and integrity of colon epithelium.

Dietary fiber source influences stool consistency of dogs and cats. For example, dietary inclusion of a blend of highly fermentable fibers (e.g., pectin, carbo bean gum, locust bean gum, guar gum, talha, and citrus pectin) causes liquid, unformed stools; whereas inclusion of a poorly fermentable fiber source (e.g., cellulose, soybean and peanut hulls) results in hard, dry stools. Dietary inclusion of a moderately fermentable fiber source (e.g., beet pulp, rice bran, and gum arabic) results in moist, well-formed stools. Thus, selecting DFF sources or blends of sources that are moderately fermented can be expected to provide bulk and maintain optimal stool consistency in dogs and cats.

High-fiber diets are being used increasingly in commercial petfoods (10% to 30% DM) to decrease energy density and to aid in treating certain diseases and disorders of dogs and cats including diabetes mellitus, hyperlipidemia, and colonic disorders as well as obesity.

The practice of blending several sources of fiber in formulating petfoods is increasing. Reliable sources of dietary fiber for companion animals include beet pulps, peanut hulls, pomaces (e.g., apple and citrus), soybean hulls, wheat bran, and wheat middlings. Adding canned pumpkin and gums with high viscosity is useful in blending other sources of dietary fiber while formulating health-promoting petfoods.

High levels of DFF (5% to 25%) can be included in dog and cat foods to reduce dietary energy density. This practice can be useful in diets designed to promote weight loss in obese animals as well as to help maintain ideal body weight when dogs and cats are fed *ad libitum*. An overweight client with an overweight companion animal will be unlikely to embrace statements about obesity in their pet, but may respond favorably to the veterinarian who wisely comments about the need for a higher dietary fiber level to help extend the life span of a favorite companion.

## 9.14    MYCOTOXIN CONCERNS IN PETFOODS

Several mycotoxins are found in grains and other products used in the manufacture of petfoods.

*Aflatoxins.* Aflatoxins occur in a variety of food sources. Those most frequently contaminated are corn, cottonseed, peanuts, and certain tree nuts. The primary target organ of aflatoxins is the liver, and liver disease has been found in dogs ingesting aflatoxin-contaminated dog food. Aflatoxins are carcinogenic as well as potent immunosuppressive compounds. Aflatoxins can be conveyed through the food chain via meat, milk, and eggs. Thus, these animal products, as well as grains used in petfoods, must be considered potential sources of aflatoxin. A sound mycotoxin testing program is an important preventive measure.

*Deoxynivalenol.* Deoxynivalenol (DON) is a frequently occurring mycotoxin in grains. Certain climatic conditions during the growing season of wheat, barley, and oats foster contamination with DON as a result of infection of the grain by the toxigenic fungus *Fusarium graminearum*. The resulting head blight/scab is a serious disease of the crops. DON is immunosuppressive and may cause unexplained vomiting by pets.

*Zearalenone.* Zearalenone can co-occur with DON in commodities because it is produced by the same species of *Fusarium*. This mycotoxin exerts its effect primarily on the reproductive system of pets. Zearalenone has estrogen-like activity and can cause infertility (effects are most severe in swine).

# 9.15 XENOBIOTIC CONCERNS IN PETFOODS

Xenobiotics are chemical compounds that are foreign to the biological systems of animals. These include naturally occurring compounds, drugs, environmental agents, carcinogens, and insecticides. Dogs and cats differ in their ability to metabolize many xenobiotoics. For example, cats can tolerate levels of cocoa and chocolate that would cause poisoning in dogs. Conversely, the liver of cats is deficient in enzymes required to detoxify certain food additives such as benzoic acid and propylene glycol while the liver of dogs has these enzymes.

# 9.16 PETFOOD ADDITIVES

## 9.16.1 ANTIOXIDANTS

Several antioxidants including BHA, BHT, TBHQ, propyl gallate, and ethoxyquin (Section 9.17.7) are used in petfood ingredients and in complete diets to prevent oxidative losses of fat-soluble vitamins and pigments and to prevent loss of metabolizable energy value in the fat component of petfoods. Antioxidants are used also to prevent or delay the onset of flavor deterioration, off-flavor formation (rancidity), nutrient loss, textural changes, color changes, and formation of by-products believed to be detrimental to animal health.

Most petfood additives improve product appearance and/or stability. Emulsifiers and surface-active agents prevent the separation of water and fat. Antimicrobial agents reduce spoilage. Added flavors improve product acceptance by dogs and cats. Certain additives are nutrients that some companion animals absorb and metabolize. Examples include sorbitol and propylene glycol, which retard spoilage. However, these additives can cause diarrhea in dogs and cats. Additionally, cats lack the enzyme required to detoxify propylene glycol, and large amounts of this sweet, viscous liquid can cause damage to their red blood cells, liver, and kidneys (Section 9.17.8).

## 9.16.2 WHY COLORING MATERIALS ARE ADDED TO PETFOODS

Because dogs are color blind and cats are nearly so, dyes and coloring materials are added to commercial companion animal foods for the benefit of pet owners. Most people believe that colors make the petfood appear more "meatlike." Colors added to petfoods are harmless to dogs and cats.

## 9.17   NUTRACEUTICALS/HERBS/BOTANICALS

The American Veterinary Medical Association (AVMA) has recognized the increasing trend of alternative and complementary medicines in the profession. It has offered guidelines for these components of plants associated with provisions as to the mode of their use. Many have medicinal value. Included are guidelines for veterinary homeopathy, veterinary botanical medicine, and nutraceutical medicine. Nutraceutical medicine is defined by the AVMA as "the use of micronutrients, macronutrients, and other nutritional supplements as therapeutic agents."[11]

The North American Veterinary Nutraceutical Council (NAVNC) was formed in 1995 by interested individuals in industry, practice, and academe. Although nonregulatory (i.e., un-affiliated with a government regulatory agency), the council's primary mission is to promote and further enhance the quality, safety, and long-term effectiveness of nutraceutical use in veterinary care. The council has defined a veterinary nutraceutical as "A substance which is produced in a purified or extracted form and administered orally to patients to provide agents required for normal body structure and function and administered with the intent of improving the health and well being of animals."[12] The Nutraceutical Regulatory Advisory Panel (NRAP) was formed in AAFCO to define the scope of issues related to regulation of these compounds.

The Dietary Supplement Health and Education Act of 1994 (DSHEA) defines a *dietary supplement* as

- intended to supplement the diet
- contains at least one: vitamin, mineral; herb or other botanical; amino acid, dietary substance for use by man to supplement the diet by increasing total dietary intake; concentrate, metabolite, constituent, extract, or combination of the above.

For more information see http://vm.cfsan.fda.gov/~dms/dietsupp.html.

With an increasing interest in "natural" products, nutraceuticals and herbal medicines have grown in popularity in the United States. Herbal supplements represented the highest growth category in the pharmaceutical market during the past decade. Consumers spend more than $15 billion annually on natural supplements. Dietary supplements are being sold in health food stores, supermarkets, pharmacies, and through mail order catalogs. Becoming increasingly aware of alternative medical options, Americans are beginning to explore these alternatives for pets. It is important that members of the veterinary profession learn how to better evaluate these remedies to be able to advise clients in product selection and utilization.

Following are some of the nutraceuticals currently being promoted for use in dogs and cats.

### 9.17.1   CREATINE MONOHYDRATE

Creatine monohydrate is an amino acid with important roles in the regulation of skeletal muscle energy metabolism (increases production of ATP) and in buffering lactic acid buildup in muscles during exercise. Supplements containing creatine monohydrate are marketed for use in performance pets (to promote muscle strength and endurance), injured pets (to aid in

---

[11]*JAVMA*, vol. 209, No. 6, September 15, 1996. Please see http://www.vet-task-force.com/guidelines.htm or www.aafco.org/NRAPR.htm.

[12]Guidelines for Alternative and Complementary Veterinary Medicine. Approved by AVMA House of Delegates, 1996, http://ag.arizona.edu/ANS/courses/am/am.html (accessed March 16, 2004).

rebuilding and strengthening muscles), and geriatric patients (to help maintain muscle strength and endurance).

### 9.17.2 METHYLSULFONYLMETHANE

Methylsulfonylmethane (MSM) supplements contain the sulfur-rich amino acids methionine and cysteine, which are important in muscle strength and the health of cartilage and joints. MSM supplements are promoted for use in animals with joint disease (e.g., osteoarthritis, OA). It is reported to relieve pain and stiffness. It is often combined with vitamin C and glucosamine HCl.

### 9.17.3 GLUCOSAMINE

Glucosamine is an amino sugar present in articular cartilage, synovial fluid, and connective tissue. It has also been shown to inhibit free radical production in joints. It is marketed for use in performance dogs (to minimize joint injuries) and dogs with OA.

### 9.17.4 OMEGA-3 FATTY ACIDS

Omega-3 fatty acids compete with arachidonic acid for metabolism to prostaglandins, thromboxanes, and leukotrienes. The result of this metabolic competition is a modulation (decrease) in the inflammation associated with allergies and tissue injuries. High levels of omega-3 fatty acids also improve survival of animals with cancer and kidney disease.

### 9.17.5 PROBIOTICS

Probiotics[13] are live (viable) microorganisms found in the gastrointestinal (GI) tract of healthy animals (e.g., *Lactobacillus, Bifidobacterium,* and *Aspergillus*). They are intended to influence gut microflora by preferentially populating it with nonpathogenic organisms to the exclusion of *Salmonella, Escherichia coli,* and other potential pathogens. These health-friendly microorganisms also synthesize vitamins, enzymes, and volatile fatty acids, which may have a beneficial effect on GI health and function as well as aid in nutrient absorption—all to exert health benefits in the host animal.

### 9.17.6 PREBIOTICS

Numerous scientific studies in dogs and cats have shown beneficial effects from administration of "prebiotic" fibers such as inulin and short-chain fructooligosaccharides. The primary benefit of prebiotics in companion animals (and humans) is an increase in gastrointestinal tract health by increasing the number of beneficial bacteria (*Lactobacillus* and *Bifidobacterium* spp), decreasing the number of pathogenic bacteria (especially *Clostridium* spp), and increasing the production of short-chain fatty acids (SCFA) providing fuel for colonocytes. Other potential benefits include an increase in calcium absorption, a decrease in the production of potentially carcinogenic compounds such as amines and ammonia, and stimulation of the gut's immune response. Improvement in glucose homeostasis and modulation of blood lipid concentrations have also been reported as beneficial responses to prebiotics.

---

[13]Board on Agricultural and Natural Resources of the National Research Council. 2004. *Nutrient Requirements of Dogs and Cats.* Washington, DC: The National Academies Press.

### 9.17.7 ETHOXYQUIN

A synthetic antioxidant preservative used in animal feeds over the past four decades, ethoxyquin functions to retard degradation of fatty acids and fat-soluble vitamins and formation of peroxide compounds. It is especially useful in preserving the quality and acceptability of ingredients containing large quantities of polyunsaturated fats such as fish meal and vitamin premixes.

Early research suggested that petfoods containing ethoxyquin might induce cancer and other maladies in dogs. According to the 2004 Report of the National Research Council of the National Academies,[14] further research data from controlled studies have not corroborated these purported effects. Regardless, to ensure safe use of the preservative in foods intended for all life stages including gestation-lactation, the FDA asked manufacturers in 1997 to voluntarily reduce the maximum levels of ethoxyquin in dog foods from 150 to 75 ppm.

### 9.17.8 PROPYLENE GLYCOL

Propylene glycol is a humectant (promotes moisture retention) and an antimicrobial preservative codified as a GRAS (generally recognized as safe; Section 9.8)[15] general-purpose food additive for use in animal feed. Its main purpose in petfoods is to allow them to retain larger percentages of water than would be possible in a dry dog or cat food without resulting in microbial growth and spoilage. It is especially useful in the formulation of semimoist foods. (Propylene glycol is frequently used therapeutically as a source of energy in ruminants suffering from ketosis.)

In semimoist petfoods, propylene glycol is typically used at levels between 6% and 12% of the formulation. Because research indicates that propylene glycol may be associated with oxidative damage in cats, which included a decrease in red blood cell half-life, increased susceptibility to damage to other oxidants and disease processes, and changes in the liver and other organs, propylene glycol was subsequently prohibited from use in the manufacture of cat foods. An informal exception to this prohibition is made for its use as a carrier in fat-soluble vitamin premixes—providing the resultant level of propylene glycol in the final food is well below the known no-adverse-effect level.

### 9.17.9 YUCCA SCHIDIGERA

Yucca Schidigera extracts have no known nutritive value; however, studies have shown that these extracts help mitigate offensive fecal odor of dogs and cats.

## 9.18 GENETICALLY MODIFIED ORGANISMS (GMOs)

There is substantial public debate about whether GMOs are safe or whether they have long-term adverse effects on the environment and/or on the health and longevity of consumers. There is no scientific evidence that they are unsafe, yet a lively debate continues.

We believe all foodstuffs, plants, and animals have been genetically modified by environmental forces and natural selection. Apart from safety, important questions related to marketing products containing GMO ingredients include "what information must be printed on the label, and under which requirements?" The test for proving the presence of

[14]Board on Agricultural and Natural Resources of the National Research Council. 2004. *Nutrient Requirements of Dogs and Cats.* Washington, DC: The National Academies Press.
[15]Ibid.

GMOs has been twofold. First, it must be tested to see if the protein contains GMOs. This must be followed by a DNA test.

If scientific investigations find no proof that GMOs pose a safety or health hazard, discussion may focus on free consumer choice to be guaranteed by clearly understood labeling. Already, in certain markets, foods containing GMOs must be appropriately labeled. Other markets are virtually certain to follow with specific requirements related to labeling pertaining to GMOs.

According to the World Health Organization/Food and Agriculture Organization (WHO/FAO), a hazard in the context of food safety is defined as "a biological, chemical, or physical agent, in or condition of food with the potential to cause an adverse health effect."[16]

In his book titled *Facing Starvation: Norman Borlaug and the Fight Against Hunger*, Lennard Bickel quotes Dr. Norman Borlaug (recipient of the 1970 Nobel Peace Prize for his monumental applied research efforts that gave a hungry world the "Green Revolution" that has saved tens of millions of people from starvation) as saying, "We deal with opposing forces: the scientific power of food production and the biological power of human reproduction."[17]

Recognizing that the world population quadrupled from 1.5 billion to 6.0 billion during the 20th century, we appreciate the fact that biotechnology holds enormous promise and potential benefits for producers and consumers, and their animals, as well as for the environment.

## 9.19 SUMMARY

The nutritional qualities of companion animal diets are fundamental to the maintenance of good health, quality of life, and longevity. Indeed, the feeding and nutrition of dogs and cats are enormously important aspects of the biology, care, health, and management of these companion animals. Feeding and nutrition also underpin the ever-expanding petfood industry, whose major focuses are on safety, health/nutrition, quality, and palatability (Chapter 2).

## 9.20 REFERENCES

1. Burger, I. H. 1993. *The Waltham Book of Companion Animal Nutrition*. Oxford: Pergamon Press.

2. Case, Linda P., Daniel P. Carey, and Diane A. Hirakawa. 1995. *Canine and Feline Nutrition*. St. Louis: Mosby.

3. Hand, Michael S., Craig D. Thatcher, Rebecca C. Remillard, and Phillip Roudebush. 2000. *Small Animal Clinical Nutrition*. 4th ed. Topeka, KS: Mark Morris Institute.

4. Lewis, Lon D., Mark L. Morris, and Michael S. Hand. 1987. *Small Animal Clinical Nutrition*. 3rd ed. Topeka, KS: Mark Morris Institute.

5. National Research Council of the Council of the National Academies. 2004. *Nutrient Requirements of Dogs and Cats*. Washington DC: The National Academies Press.

---

[16]*Agriculture 21*. March–April 2004. Department of Agriculture. Food and agriculture organization of the United Nations, www.fao.org/ag. "FAQ" page: http://www.fao.org/ag/againfo/subjects/documents/lps/dairy/faq/faq.htm.
[17]Bickel, Lennard. 1974. *Facing Starvation: Norman Borlaug and the Fight Against Hunger*. New York: Readers Digest Press, distributed by E. P. Dutton & Co., p. 351.

# 10 REPRODUCTIVE BIOLOGY OF DOGS AND CATS

*No matter how much the cats fight
there seems to always be plenty of kittens.*

Abraham Lincoln (1809–1865)
*16th President of the United States of America*

## 10.1   INTRODUCTION/OVERVIEW

An understanding of the reproductive biology of dogs and cats is fundamental to achieving the goal of producing healthy litters of puppies and kittens. Animals selected for breeding should be excellent representatives of their breed, free from hereditary defects, and have excellent temperaments. Soundness and temperament should never be compromised. Only females that have had uncomplicated births and have good maternal behavior should be used for repeated breeding. Due to the growing problems of overpopulation (Chapter 24), the decision to breed a dog or cat should not be made without a specific goal in mind and the knowledge that good homes can be found for each of the offspring.

## 10.2   MYTHS AND CONSIDERATIONS REGARDING BREEDING OF DOGS AND CATS

People breed dogs and cats for many reasons. The decision to bring new life into the world should be made following serious consideration and only by those willing to accept the responsibility of ensuring that each offspring will be of sound health and have a good home. Being a responsible breeder requires deliberate thought and effort. There are many myths and "faulty" reasons for breeding dogs and cats.

■ *Breeding dogs/cats is a good way to make money.* Few responsible breeders make money from sales of puppies/kittens. Responsible breeding requires health testing and medical care as well as careful consideration of the pedigrees and temperaments of the parents and whether the offspring would improve the breed. Medical tests recommended before breeding include extensive health testing (radiographs of hips, blood profile, tests for infectious diseases, tests for genetic diseases, ultrasound examinations of the heart, eye and hearing examinations, and other special tests depending on the breed). Medical costs after breeding include pregnancy tests, perhaps emergency Cesarean surgery, vaccinations, deworming, and other veterinary care for the offspring. Other costs include stud fees, registration fees, travel costs, expenses associated with showing, advertising costs, and the time and money invested in feeding and caring for the litter.

■ *My children should see the miracle of birth.* Most animals give birth in the middle of the night when the children are asleep. The female prefers to have quiet and privacy during birth. If disturbed by too many observers, she may become stressed and have difficulty giving birth or may abandon the newborns. Renting a video is a safer and more reliable way of showing children the miracle of birth.

■ *My dog/cat is really great, and I want to have another dog/cat just like her.* Each animal is a unique individual; there is no guarantee that puppies will resemble their parents in looks, personality, or temperament. A significant number of dogs in shelters are purebreds; just because an animal is registered does not mean that it should be bred.

- *Raising a litter is so much fun.* It is exciting to watch the development of puppies and kittens. However, the same satisfaction can be found by fostering a litter from a humane society or by adopting a kitten or puppy that would otherwise be destroyed. Raising a litter takes much more time than most people anticipate. Feeding and attending to a nursing litter, monitoring weight gains of the newborns, and ensuring that each is healthy requires at least 30 to 60 minutes a day. By the time the litter is 3 weeks old, each puppy or kitten should be handled and played with individually to provide the early socialization needed to become good family pets that enjoy interacting with people (Chapter 20).

- *My pet will miss out on an important part of life if it is not given the opportunity to reproduce.* Dogs and cats do not share the same priorities in life as humans. In the wild, only the most dominant individuals reproduce. Pets do not feel deprived if they do not produce offspring; they are much more interested in spending time with their human companions. Animals do not share the same social norms as human society—animals will breed with littermates, parents, and offspring. In addition, pregnancy and whelping pose dangers to the health of females with the risk of certain metabolic disorders (e.g., hypocalcemia resulting in eclampsia, Section 10.3.9), infections (uterine and mammary), and dystocia (Section 10.3.9).

- *We can find good homes for the puppies/kittens.* There are simply not enough homes for all the dogs and cats currently being produced. Are you willing to follow the puppies or kittens throughout their lives and take them back if their new home does not work out? Are you willing to screen the potential homes to be certain they have the time to love, train, and care for the animal? Will the new home provide a safe environment? Will they value the animal and be willing and financially able to provide it with veterinary care as needed? What will you do if one of the animals shows up with a heart defect at 8 weeks? Conscientious breeders must be willing to accept these responsibilities.

## 10.3   CANINE REPRODUCTION

The breeding and raising of dogs has always been a popular hobby and business. Many newcomers to dog breeding lack experience and understanding of the unique physiology associated with canine reproduction. In this section, we will review the basics of canine reproduction.

### 10.3.1   REPRODUCTIVE ANATOMY OF THE BITCH

The major organs of the reproductive tract of the bitch include the ovaries, oviducts, uterus, vagina, vulva, and the mammary glands (Figure 10.1). The ovaries are small, bean-shaped organs located in the dorsal abdomen near the kidneys. The ovaries produce ova (eggs) and reproductive hormones (estrogen and progesterone). The oviducts are small, thin tubes that function to transport the ova from the ovaries to the uterus; fertilization usually occurs near the end of the oviducts (near the uterus). The uterus is a muscular, hollow, Y-shaped organ consisting of two horns, a short body, neck, and cervix. The cervix is a fibrous/muscular structure that serves as the channel between the uterus and vagina. The vagina is lined with stratified squamous epithelial cells that vary in thickness and degree of cornification (maturity) during different stages of the estrous cycle. The vulva is the external portion of the reproductive tract and includes the clitoris and the external folds of the labia. The mammary glands consist of four to six pairs of glands located in parallel rows along the ventral abdomen. There is little development of these glands prior to puberty (thus bitches spayed before puberty are extremely unlikely to develop mammary cancer later in life). The production of estrogen during proestrus and estrus stimulates the development of the duct system and secretory cells within the mammary glands.

**Figure 10.1** Anatomy of the female urogenital system (carnivores). *Illustration by Diana Nicoletti.*

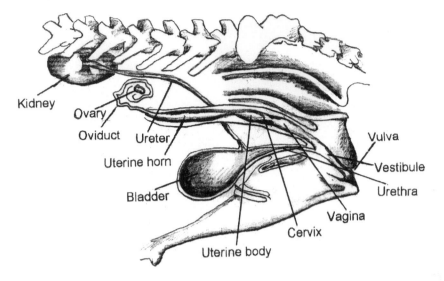

## 10.3.2 PHYSIOLOGY OF REPRODUCTION IN BITCHES

Reproductive physiology of bitches differs from that of most other mammals in having estrous cycles with long periods of ovarian inactivity between cycles. The age of onset of puberty varies with breed. In general, bitches have their first estrous cycle a few months after the time they achieve adult height and body weight. Small breeds typically go through puberty between 6 and 10 months of age, whereas giant breeds may not do so until 18 to 24 months of age or later. It is recommended to wait until the second or third estrous cycle for the first breeding. The ideal age for breeding is between 2 and 6 years.

*The canine estrous cycle.* The estrous cycle of dogs has several stages: anestrus, proestrus, estrus, and diestrus or metestrus. It is commonly believed that dogs have two estrous cycles per year, although the interval between cycles is highly variable and there is no definite seasonal effect—bitches cycle, breed, and whelp litters in all months of the year. The normal interval between estrous cycles is between 5 and 10 months with an average of 7 months. There are breed differences: German Shepherd Dogs often cycle every 4 to 4.5 months, whereas Basenjis cycle only once yearly (usually in the fall). Bitches will continue to cycle throughout their life although fertility declines and problems with parturition and birth defects increase in dogs over 8 years of age.

*Anestrus.* Anestrus is a time of reproductive rest (quiescence). It is the interval between the end of the luteal phase (diestrus or whelping) and the beginning of the next proestrus. The length of anestrus is variable, averaging 4.5 months. Factors that initiate the next estrous cycle are thought to include a complex interaction among environment, general health, genetics, ovarian and uterine health, age, and other unidentified factors.

*Proestrus.* Proestrus is a period of ovarian follicular activity that precedes estrus. The anterior pituitary gland secretes follicle-stimulating hormone (FSH) and luteinizing hormone (LH) resulting in the development of ovarian follicles, which in turn secrete estrogen. Rising estrogen levels stimulate growth of glandular tissue in the uterus and vaginal mucosa. The vaginal mucosa becomes highly vascular and edematous (swollen). Fragile capillaries within the uterus and vaginal mucosa leak red blood cells resulting in a bloody vaginal discharge. The appearance of this discharge is considered to mark the first day of proestrus. Behavioral changes include restlessness, frequent urination (urine marking), and a tendency to roam if not confined. As proestrus proceeds, the bitch will become increasingly playful with male

**Figure 10.2** Schematic drawing of changes in hormone concentrations during the canine estrous cycle. Day 0 = the day of ovulation. *Illustration by Diana Nicoletti.*

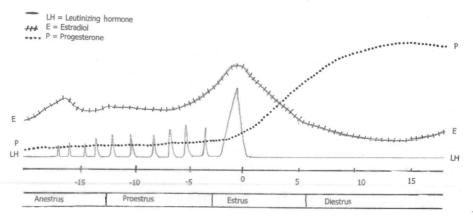

dogs but will not allow mounting. The vulva slowly enlarges throughout proestrus. The vaginal epithelial cells proliferate and the nuclei of the more superficial cells become pyknotic (shrink and become dark) and eventually disintegrate resulting in anuclear cells (squames) lining the surface of the vagina. The duration of proestrus is usually between 6 and 11 days (average 9 days).

*Estrus.* The period of sexual receptivity is called estrus. The first day the bitch will allow a male to mount and breed (standing heat) is considered the first day of estrus; the end of estrus is when she no longer will allow a male to breed her. Estrogen levels peak 1 to 2 days prior to the start of estrus and decline throughout estrus (Figure 10.2). Progesterone levels begin to increase as estrogen levels decline. The falling estrogen and rising progesterone concentrations result in a surge of FSH and LH at the beginning of standing heat. Ovulation occurs about 48 hours after the LH surge. Bitches normally have multiple ovulations during this period. Postovulatory follicles luteinize and develop into corpora lutea, which secrete increasing amounts of progesterone over the next 1 to 3 weeks. The primary oocytes require 24 to 72 hours to mature into secondary oocytes capable of being fertilized. Individual cells are fertile for 12 to 24 hours following maturation. During estrus, the bitch may actively seek males, may crouch and elevate the rump toward the male, and "flag" her tail (hold it to one side). Potent pheromones (odors that serve as sex attractants) are produced, which can attract males over long distances. The vulva becomes soft and flaccid during estrus and the vaginal discharge changes from blood-tinged to straw-colored in most bitches. The duration of standing heat is usually 5 to 9 days, although it ranges from as short as 1 to 2 days to as long as 18 to 20 days.

*Diestrus (metestrus).* Diestrus is the 2-month period following estrus during which progesterone is being produced by the corpora lutea. Dogs differ from most mammals in that the corpora lutea remain functional and secrete progesterone for approximately the same period of time in both pregnant and nonpregnant bitches. In pregnant bitches, diestrus ends abruptly at the time of whelping, approximately 56 to 58 days after the end of estrus. In nonpregnant bitches, the corpora lutea persist between 60 and 80 days before regressing. Diestrus begins when a previously receptive bitch refuses to accept the mounting of a male. The vulva returns to its normal size. There is some development of glandular tissue in the mammary glands from the influence of progesterone. Some nonpregnant bitches will show behavior of "pseudopregnancy" with increased appetite and occasionally exhibit nesting and mothering behavior. Following diestrus, the bitch returns to anestrus and the cycle begins again.

## 10.3.3 CANINE MALE REPRODUCTIVE ANATOMY

The reproductive system of the male dog consists of paired testes (testicles), the epididymis and spermatic cord, the scrotum, prostate gland, penis, and prepuce. The testes of the male dog should descend into the scrotum within a few days following birth. The testes should be oval in shape with a smooth, regular outline and a firm texture. The left testes is usually caudal to the right; however, both should be similar in size. Microscopically, the testes are composed of tightly coiled seminiferous tubules lined with germinal cells, which produce spermatozoa, and Sertoli cells. The latter secrete estrogen. Between the seminiferous tubules are the interstitial cells (cells of Leydig), which produce testosterone. The seminiferous tubules empty into collecting ductules that merge to form the epididymis. Sperm cells spend 10 to 14 days maturing as they travel through the epididymis. The epididymis empties into the ductus deferens (vas deferens) located within the spermatic cord. The spermatic cord is comprised of the ductus deferens, the spermatic artery, the pampiniform plexus of veins, lymphatics, nerves, and cremaster muscle. The prostate gland is the only accessory sex gland of male dogs. It is located near the rim of the pelvis and surrounds the neck of the bladder and terminal portion of the ductus deferens. A medium septum divides the prostate gland into two symmetrical lobes. The prostate gland produces a watery secretion that is ejaculated after the sperm-containing portion of semen. The ductus deferens empties into the urethra.

The penis of the dog has several unique characteristics. Within the penis is a small bone called the os penis,[1] which gives the penis support during intromission. The base of the penis contains a swelling called the bulbus glandis. During coitus, pressure from the vaginal muscles of the bitch stimulate swelling of the bulbus glandis into a spherical shape, which "ties" the bitch and male dog together until deposition of seminal fluid is complete. Once the tie has been made, the male dog generally turns and stands tail to tail with the bitch until ejaculation is complete and the swelling of the bulbus glandis has subsided. The tie may last from a few minutes to more than an hour, the average time being 15 to 20 minutes. The tie ensures that semen is deposited far into the vaginal tract and is prevented from draining out for sufficient time to enable sperm to travel into the uterus and oviducts. The scrotum of a normal dog is sparsely covered with hair and has skin that is freely movable over the testes. The prepuce completely encloses the nonerect penis. The penis should be freely moveable within the prepuce.

## 10.3.4 REPRODUCTIVE ENDOCRINOLOGY OF MALE DOGS

Male reproductive physiology is controlled by LH and FSH secreted by the anterior pituitary gland. The release of LH and FSH are regulated by gonadotrophin-releasing hormone (GnRH) produced by the hypothalamus. Testosterone, estradiol, and dihydrotestosterone act in a negative feedback loop to control secretion of GnRH, FSH, and LH. FSH acts on the germinal cells in the seminiferous tubules to produce spermatozoa. The Sertoli cells are stimulated by FSH to produce estrogens and other products important in providing the environment required for the seminiferous germ cells to produce spermatozoa. Sertoli cells also produce inhibin, a hormone that provides negative feedback to regulate anterior pituitary secretion of FSH. LH binds to receptors on the Leydig cells and stimulates the production of testosterone and other steroid hormones. Testosterone is essential for development of secondary sex characteristics, normal sexual male behavior, function of the accessory sex glands, production of spermatozoa, and maintenance of ductal cells (Table 10.1).

## 10.3.5 BREEDING MANAGEMENT OF DOGS

Once the decision to breed has been made, the health of the dogs should be thoroughly evaluated. Records should be kept of the length of the female's estrous cycles (e.g., number of days the bitch shows signs of proestrus and estrus, and the intervals between estrous cycles). The bitch and prospective stud should be screened for any hereditary disorders common to the breed (e.g., hip dysplasia, hypothyroidism, blood coagulation disorders, heart

---

[1]Many species of mammals share this characteristic; the penis of primates lacks an os penis.

**Table 10.1**
Summary of Reproductive Hormones: Sources and Actions

| Hormone | Secreted By | Action in Males | Action in Females |
|---|---|---|---|
| GnRH | Hypothalamus | On anterior pituitary to secrete LH and FSH | On anterior pituitary to secrete LH and FSH |
| FSH | Anterior pituitary | On Sertoli cells to stimulate and support spermatogenesis | On ovarian follicles to stimulate development and secretion of estrogen |
| LH | Anterior pituitary | On Leydig cells to secrete testosterone | On ovarian follicles to cause ovulation and formation of corpus luteum |
| Testosterone | Leydig cells | On Sertoli cells to stimulate and support spermatogenesis, negative feedback on GnRH, FSH, LH | – |
| Progesterone | Corpus luteum | – | Growth of endometrial and mammary glands |
| Estrogen | Male—Sertoli cells Female—ovarian follicles | Negative feedback on FSH | Cornification of vaginal cells, behavioral estrus |

defects, cataracts, retinal degeneration, deafness, and other breed-specific disorders). Immunization boosters should be given and the bitch should be checked for both internal and external parasites and treated if any are found. Both dogs should be tested for canine brucellosis. Breeding records should be kept showing the dates of breeding and the outcomes. The male dog should have a semen evaluation; for optimal fertility, > 70% of the spermatozoa should show vigorous forward motility and > 70% should be morphologically normal. Prior to breeding, the bitch should have a complete physical examination; the examination of her reproductive tract should include a digital examination of the vulva and vagina to check for the presence of congenital defects such as a vestibulovaginal stricture.

The bitch is usually taken to the male for breeding. The timing of breeding is important for optimal fertility. Ovulation usually occurs around 48 hours following the LH surge. The LH surge is usually on the second day of estrus, thus ovulation typically is on day 4 of estrus. The oocytes require 1 to 3 days to mature and are fertile for 1 to 2 days. Thus, the fertile period is generally days 5 through 9 of estrus (standing heat). Sperm require approximately 7 hours of capacitation (maturation) within the uterus and remain viable for 4 to 6 days. Methods of timing when to breed a bitch include (1) days from start of proestrus, (2) behavior, (3) vaginal discharge/cytology, and (4) measurement of LH or progesterone levels. Each of these methods will be discussed.

*Days from start of proestrus.* Traditionally, many breeders mate or inseminate bitches on days 10 and 14 following the onset of proestrus (alternatives include days 10 and 12; days 9 and 11; days 10 and 13; or days 10, 12, and 14). Although this method is successful in many dogs, the high variability in length of proestrus (may be as short as 1 day—perhaps due to owners failing to observe the earliest signs—or as long as 21 days) makes this method one of the most common reasons for low conception rates with breedings based on days from the onset of proestrus.

*Behavior.* For most dogs, this method is reliable and results in high conception rates. The bitch is brought to the male for "teasing" every other day starting on day 5 of proestrus (day

**Figure 10.3**   Vaginal cytology during the canine estrous cycle. (*a*) Proestrus is characterized by the presence of red blood cells and numerous parabasal and intermediate cells (nucleated epithelial cells with vacuolated cytoplasm). (*b*) Estrus is characterized by the presence of angular epithelial cells and anuclear squames. (*c*) Diestrus is characterized by the reappearance of parabasal and intermediate cells and neutrophils (white blood cells). *Courtesy Dr. Karen L. Campbell.*

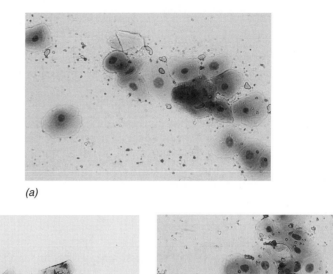

(a)

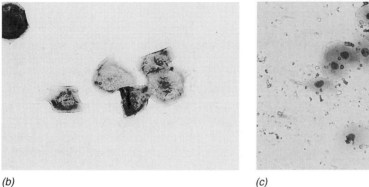

(b)                                                    (c)

1 for bitches with histories of short cycles or previous infertility). When the bitch is willing to stand for mounting, breeding should begin and continue every 2 to 4 days throughout the acceptance period. A few bitches may refuse to allow breeding; for these, artificial insemination may be used with timing based on one of the other methods.

*Vaginal discharge/cytology.*   Some breeders consider the optimal time to begin breeding is when the vaginal discharge changes from bloody to straw-colored; however, this sign is unreliable for many bitches—some never show a bloody discharge and others bleed throughout estrus. However, vaginal exfoliative cytology is strongly influenced by plasma estrogen levels and provides an excellent method of monitoring the bitch's cycle.

The most common method of collecting samples for vaginal exfoliative cytology is the cotton swab technique. A sterile 5- to 7-inch cotton swab is inserted along the dorsal wall of the vagina until it reaches the pelvic canal; it is then rotated a complete revolution in each direction and withdrawn. The swab is rolled the length of a glass slide and air dried prior to staining. The slides are stained and evaluated using a standard light microscope. The smears from a bitch in early proestrus will have numerous parabasal and intermediate cells (nucleated epithelial cells with vacuolated cytoplasm), and many red blood cells and a few neutrophils (Figure 10.3). As proestrus progresses, an increasing number of the epithelial cells will become angular in shape with pyknotic nuclei. As the bitch enters estrus, the neutrophils disappear, red blood cells decrease in number, and > 80% of the epithelial cells are angular with either pyknotic nuclei (superficial cells) or no visible nucleus (anuclear squames). When the bitch enters diestrus, the percentage of angular superficial and anuclear epithelial cells abruptly falls to approximately 20% and neutrophils commonly reappear. "Metestrus" cells may be seen; these are foamy vaginal epithelial cells containing one to two neutrophils within their cytoplasm. When breeding is based on vaginal exfoliative cytology, the bitch should be bred every 2 to 4 days throughout the period when > 80% of the epithelial cells are superficial cells and anuclear squames. Whelping can be predicted to occur on day 57 following the onset of diestrus (the day the epithelial cells abruptly change from anuclear squames and superficial cells to intermediate and parabasal cells).

*LH and progesterone testing.* Progesterone concentrations are low during anestrus and early proestrus and reach 3 nmol/L by late proestrus. The level begins to rise at the time of the preovulatory LH surge and is 6 to 12 nmol/L at the time of ovulation and 18 to 30 nmol/L within another 2 days—coinciding with the onset of the fertile period. If progesterone levels are being monitored to predict timing of optimal breeding, plasma samples should be checked every 1 to 2 days starting in late proestrus. Breeding should be done when the progesterone level reaches 18 nmol/L and repeated in 2 to 3 days. LH levels are more accurate than progesterone levels in predicting the time of ovulation as the surge in LH occurs approximately 48 hours prior to ovulation. Simple "in office" kits have been developed to measure LH levels; serum samples are tested daily starting in late proestrus. Breeding should be planned for days 4 and 6 following the surge in LH.

## 10.3.6 ARTIFICIAL INSEMINATION

Artificial insemination (AI) is required if a bitch refuses to allow the selected stud dog to mate with her. Other reasons for AI include physical problems preventing the male or female to breed naturally (e.g., nonhereditary orthopedic problems, vaginal strictures, major size differences), avoiding venereal diseases (these are uncommon in dogs), and avoiding the stress associated with shipping the bitch to the stud. In 1981, the American Kennel Club approved registration of litters conceived using frozen semen. The primary disadvantage of using AI is lower conception rates, especially when chilled or frozen semen is used. When AI is used, timing is critical and the use of LH or progesterone assays to predict the date of ovulation is highly recommended. AI is then performed on days 2 and 4 postovulation.

*Collection procedure.* Semen is usually collected in an artificial vagina (Figure 10.4); the dog's prepuce is slipped posteriorly to exteriorize the penis past the bulbus glandis, downward pressure and massaging of the bulbus glandis will stimulate ejaculation, the male dog

**Figure 10.4** Artificial vagina and insemination pipette for use in dogs. *Courtesy University of Illinois.*

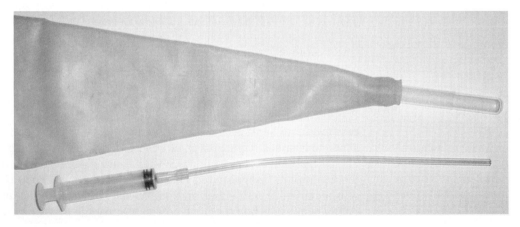

will step over the collector's hand (as if dismounting a bitch), the penis should then be directed between the dog's rear legs during collecting of semen. Collection should be continued for approximately 4 minutes. The sperm-rich fraction is whitish and cloudy and is followed by clear sperm-free prostatic fluid, which does not need to be collected. The semen should be checked for normal morphology and progressive motility (both should be > 80%). It may then be inseminated into the bitch or the semen may be chilled or frozen for shipment.

*Insemination procedures.* (1) *Digital placement of the insemination catheter*: The semen is drawn into a sterile syringe with a sterile insemination pipette attached (see Figure 10.4). Nonspermicidal lubricant is used and the inseminator's index finger is used to guide the pipette over the rim of the pelvis and into the anterior portion of the vagina. The semen is gently expelled into the vaginal vault; stroking of the dorsal wall of the vaginal vault with the index finger will stimulate vaginal contractions that may aide movement of the sperm into the uterus. When contractions have ceased, the finger is withdrawn and the bitch's rear legs are elevated for 10 minutes. (2) *Transcervical insemination*: A videoendoscope is used to guide placement of an insemination pipette through the cervix allowing the semen to be deposited directly into the uterus. This helps ensure that sperm reach the eggs. (3) *Surgical deposition*: This requires general anesthesia and exteriorization of the uterus. The semen is injected into the uterine body and will flow into both uterine horns. The uterus is then repositioned in the abdominal cavity and the incision is closed. The primary advantages of this technique are that the veterinarian can assess the health of the uterus and ovaries at the time of insemination and that it provides maximum opportunity for the sperm to reach the oviducts. The use of either transcervical or surgical intrauterine deposition of semen should be used when inseminating a bitch with thawed frozen semen as thawed sperm have lower motility and may be unable to travel through the cervix.

## 10.3.7   PREGNANCY

Fertilization occurs within the oviducts. The ovum matures to a 16-cell morulae stage prior to entering the uterus 8 to 12 days following fertilization. Implantation occurs 17 to 21 days after fertilization. The fetuses will space themselves equally throughout both horns of the uterus (Figure 10.5). The placentation is endotheliochorial with four layers of cells separating fetal and maternal blood.

*Pregnancy diagnosis.* The diagnosis of pregnancy in dogs can be made via palpation, ultrasound examinations, radiographs, or using a relatively new blood test.

*Palpation.* Palpation of the abdomen is performed by gently examining the abdominal organs using two hands (in larger dogs) or a thumb and fingers (small dogs). By days 20 to 30, individual walnut-size swellings of the uterus, consisting of the fetuses within an amniotic sac, can be felt (palpated) within the bitch's abdomen. Between 30 and 50 days of gestation, the uterus becomes diffusely enlarged making it difficult to distinguish individual fetuses. After 50 days, the fetuses are large enough to again be distinguished from intestines and other internal organs.

*Ultrasound examinations.* The simplest devices are Doppler units that amplify the sounds of the fetal heartbeats and the placental blood flow. These are relatively inexpensive and can be used to diagnose pregnancy after 25 days of gestation. Two-dimensional and three-dimensional ultrasound units can detect embryos as early as day 10 of gestation.

*Radiographs.* Enlargement of the uterus can be seen by 20 days of gestation; however, individual fetuses cannot be counted until their skeletons become calcified around day 45 of gestation (Figure 10.6). Radiographs can be useful in predicting litter size. Early exposure to x-rays can be harmful to developing fetuses, thus other methods of pregnancy diagnosis are generally preferable.

**Figure 10.5** Fetal development in dogs. (*a*) Nonpregnant reproductive tract of a bitch. (*b*) Canine uterus at 21 days of gestation; note the 1.0 to 1.5-inch diameter swellings associated with the embryos. (*c*) Canine uterus at 40 days of gestation; the uterus is now uniformly distended making it difficult to distinguish individual fetuses. (*d*) Canine embryo at 21 days of gestation. (*e*) Canine embryo at 31 days of gestation. (*f*) Canine embryo at 40 days of gestation. *Courtesy Dr. James E. Corbin.*

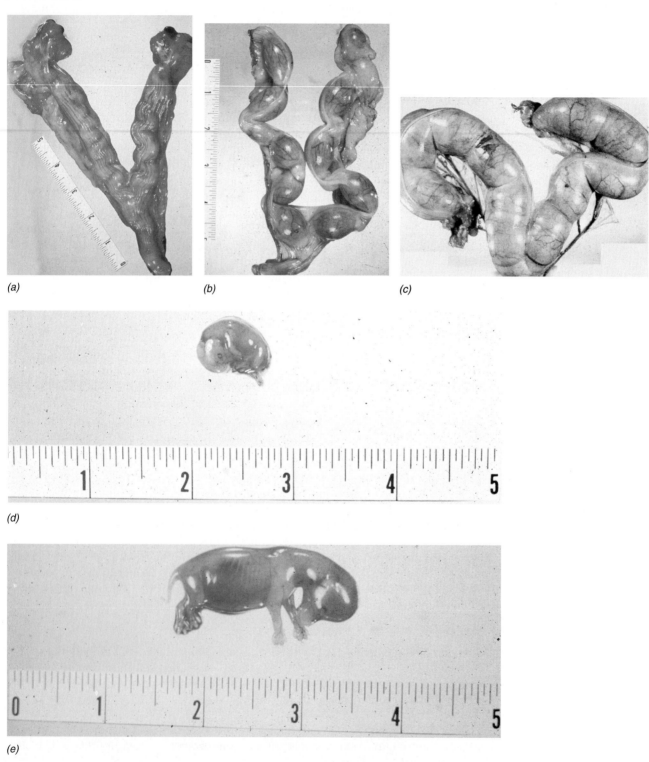

(a)

(b)

(c)

(d)

(e)

**Figure 10.5**  Continued

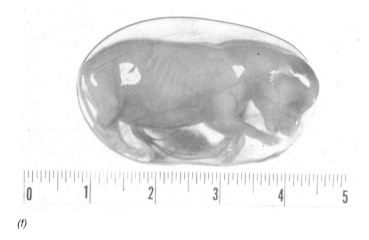

*(f)*

*Hormonal pregnancy tests.*  Progesterone levels remain elevated for over 6 weeks following estrus in both pregnant and nonpregnant bitches; thus, human pregnancy tests are not useful for pregnancy diagnosis in dogs. However, serum relaxin concentrations are significantly increased in pregnant bitches by 24 to 30 days after ovulation. A canine pregnancy diagnosis kit (ReproCHEK RELAXIN®) is available commercially.

*Gestation length.*  The traditional length of gestation in dogs is 63 days from the date of first breeding (Figure 10.7). However, the actual length of time from first breeding to parturition ranges from 58 to 68 days. Reasons for this variation include variations in timing of ovulation, variations in timing of breeding, variations in estrus length, and variations in litter size (single pup litters tend to be born 1 to 2 days later than large litters). The most accurate

**Figure 10.6**  Radiograph of a pregnant bitch showing skeletons of the puppies. *Courtesy Dr. Stephen Kneller.*

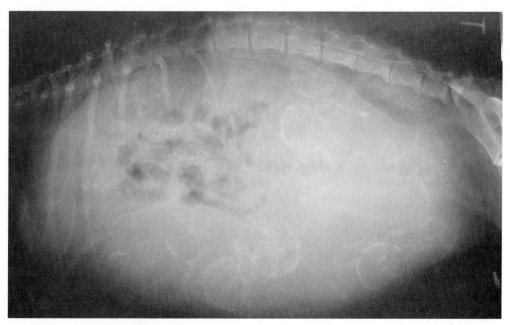

**Figure 10.7** Pregnant bitch at day 63 of gestation. *Courtesy Dr. James E. Corbin.*

predictor of whelping date is timing from the first day of diestrus. The first day of diestrus is the day vaginal cytology changes abruptly from predominately cornified to predominately noncornified cells. Whelping will occur on the 57th day ($\pm$1) of diestrus.

An excellent predictor of an imminent delivery is a drop in the rectal temperature of the bitch. Owners should monitor the bitch's rectal temperature twice daily during the final week of gestation. At one week before delivery the bitch's rectal temperature will be between 100°F and 100.5°F; it will drop abruptly to between 98°F and 99°F 8 to 12 hours prior to the onset of labor.

## 10.3.8 PARTURITION

Although most deliveries proceed smoothly, it is helpful to be prepared and to monitor the progress of the bitch and status of the puppies.

*Location.* A clean, disinfected whelping box should be prepared for the bitch a week before delivery. A large appliance box, the bottom half of a large fiberglass crate, a child's plastic wading pool, or a wooden whelping box can be used. A thick layer of newspaper can be used to line the box. Having a heating pad under one end of the whelping box to provide a warm spot for the puppies is recommended (Figure 10.8).

**Figure 10.8** A heating pad placed in the whelping box is helpful in preventing hypothermia in newborn puppies. *Courtesy Dr. James E. Corbin.*

*Supplies.* Useful supplies to have available during parturition include (1) disinfectant soap such as chlorhexidine or povidone iodine; (2) washcloths or towels; (3) disposable gloves; (4) water-based lubricant (e.g., KY jelly); (5) dental floss, thread, or surgical suture for tying umbilical cords; (6) sterile scissors and clamp (if needed to cut the umbilical cord); (7) tincture of iodine or chlorhexidine solution for disinfecting umbilical cords; (8) cotton balls; (9) large stack of newspapers; (10) large garbage bag; (11) clock; (12) record book; (13) postage or food scale for weighing the puppies; and (14) liquid glucose to give weak puppies. The whelping box and supplies should be kept in a quiet room and located close to a telephone in case a call for professional assistance is necessary.

*Prelabor.* Many bitches show "nesting" behavior a day or two before going into labor, with scratching in the whelping box, pacing, and whining. Often, refusal to eat occurs in conjunction with the rectal temperature drop marking the onset of prelabor. When the rectal temperature drop is noted, any long hair should be trimmed from the bitch's hindquarters. The hindquarters and ventral abdomen should be washed with disinfectant soap. The prelabor stage may last 2 to 36 hours (average is about 6). The bitch may pant, shiver, vomit, or appear restless. The vulva will enlarge and soften and there may be a thick, clear, sticky discharge. A little exercise may be beneficial but stay with the bitch in case she delivers a puppy in the yard.

*Delivery.* Delivery is preceded by "hard labor." The first contractions may appear as mild tensing of the abdominal muscles, these will be followed by harder contractions and straining. A puppy should be delivered within one hour of hard labor. The chorioallantoic membrane (water bag) usually breaks when a puppy enters the birth canal. Puppies may be born either head first or hind feet first (breech). Breech presentation is normal and not associated with dystocia (difficult birth) in dogs. The puppy is usually still within its amniotic sac at birth; the bitch will instinctively lick off the sac and eat it. If the bitch does not remove the amniotic sac from the puppy, the attendant should tear it off to free the puppy's nose then wipe the puppy's head, nostrils, and mouth and rub it to stimulate breathing. The placenta is usually delivered within 15 minutes following the puppy's birth. If left alone, the bitch will usually eat the placenta and sever the umbilical cord. If the bitch does not do this, the attendant should crush and tear the umbilical cord about an inch from the puppy's abdomen. If blood oozes from a torn umbilical cord, clamp and tie it off using dental floss, thread, or surgical suture. Iodine or chlorhexidine should be dabbed on the end of the cord to help prevent infection.

A greenish-black fluid is discharged from the vulva during parturition and for up to 3 weeks following whelping. This fluid is from the breakdown of placental attachment sites in the uterus. Intervals between deliveries of puppies may vary from a few minutes to an hour; if the bitch is in pain or actively straining for more than an hour, a veterinarian should be consulted.

*Postpartum care.* The bitch can be allowed to eat a small meal as soon as delivery is complete. The bitch should be observed carefully, and her rectal temperature checked twice daily; if the temperature rises above 102.5°F, or if she becomes listless or other abnormalities are noted, a veterinarian should be consulted.

Puppies often seek out a nipple and start nursing within a short time following birth. If a puppy has not nursed within an hour, it should be held to a nipple and encouraged to suckle. A weak puppy can be given a drop or two of liquid glucose (dextrose) on its tongue; this often gives the puppy enough energy to start nursing. In newborn puppies, the reflex to urinate and defecate is stimulated by rubbing and moisture.

The first bowel movement is called meconium and should be passed within the first hour following birth. If the bitch is not licking the puppies to stimulate urination and defecation, the attendant should use a cotton ball moistened with warm water to rub around the anus and prepuce or vulva (see Figure 9.16). A healthy puppy will be either sleeping or eating for the first 2 weeks of life (Figure 10.9). During sleep, the puppy will twitch and jerk its legs; this

**Figure 10.9** Healthy puppies are either nursing or sleeping during their first 2 weeks of life. *Courtesy Dr. James E. Corbin*

is called "activated sleep" and helps strengthen muscles (Table 10.2). The weight of puppies should be recorded daily to be certain they are gaining (for additional information on feeding of neonates see Section 9.7). A sick puppy is often chilled, is dehydrated, and has low blood sugar. A chilled puppy should be slowly warmed and given 1 to 2 cc of a 5% glucose solution every 30 minutes. If a puppy is showing signs of illness, a veterinarian should be consulted.

## 10.3.9   COMMON REPRODUCTIVE PROBLEMS OF DOGS

***Failure to observe estrus.*** Explanations for failing to observe estrus in a bitch include (1) prepuberty—although the majority of bitches reach puberty between 6 and 24 months of age, some breeds such as Greyhounds may not reach puberty until 4 years of age; (2) breed differences—Basenjis and other African breeds have long inter-estrous periods commonly lasting one year; (3) stress—racing Greyhounds may not cycle until strenuous physical training is stopped; (4) "silent heat"—vaginal discharges may be scant in normal dogs and may not be observed if the bitch licks herself frequently and/or has a long hair coat; (5) ovarian cysts or neoplasia—these may secrete progesterone or other hormones that inhibit estrus; (6) endocrine disease—hypothyroidism, hyperadrenocorticism, and other endocrine diseases or hormonal treatments being given to the bitch can interfere with the normal reproductive cycle; and (7) genetic—a rare phenotypic bitch may have chromosomal abnormalities or hermaphroditism resulting in a lack of ovarian activity.

**Table 10.2**
Signs to Look for in Newborn Puppies and Kittens

| Signs of Good Health | Signs of Illness |
| --- | --- |
| Gains weight | Does not gain weight |
| Sleek, smooth skin and coat | Rough, dull coat, wrinkled skin |
| Quiet | Cries or mews frequently |
| Good muscle tone | Limp muscle tone |
| Activated sleep | Little movement |
| Pink gums | Reddish, blue, or purple gums |
| Yawns frequently | Listless |
| Plump, firm body | Pot-bellied or tucked up abdomen |
| Seeks dam or littermates | Scattered around nest |
| Nurses strongly | Does not nurse |
| Pasty or firm stools | Diarrhea |

*Abnormal estrous cycles.*  A variety of abnormalities in the estrous cycle may affect fertility. These include (1) split heats—proestral or early signs of estrus are seen; however, the bitch does not ovulate and has a second proestrus in 2 to 8 weeks, which is usually associated with ovulation and normal fertility; (2) ovulatory failure—this is most common when the bitch has a short inter-estrous period (less than 4 months), which may not allow time for normal endometrial repair between cycles; and (3) prolonged proestrus or estrus—proestrus or estrus are considered to be prolonged if either lasts more than 21 days. Causes include ovarian cysts and tumors. These can often be identified via an ultrasound examination of the ovaries.

*Failure to conceive.*  Possible causes for failure to conceive include (1) incorrect timing of mating/insemination—due to the variability in time of ovulation, arbitrary breeding on days 10 and 14, as is commonly practiced, may not be optimal for conception to occur; (2) stress—shipment of a bitch just prior to breeding may result in failure to conceive; (3) infertile male—a semen analysis should be done to evaluate the viability of the spermatozoa; or (4) uterine infections/endometritis—these will result in a failure to conceive or death of the embryos.

*Dystocia.*  Difficulties in giving birth (dystocia) may result from (1) uterine weakness (may result from poor condition, nervousness impairing contractions, fatigue in trying to deliver a large pup or a large number of puppies); (2) pelvis being too narrow (common in brachycephalic breeds such as Pugs, Bulldogs, Boston Terriers; these breeds also have puppies with large heads and wide shoulders) or abnormally shaped due to previous fractures;[2] (3) fetus being too large (single pup litter) or abnormally positioned (e.g., head back, a leg back, transverse presentation); or (4) obstructions in the vaginal canal (e.g., strictures or tumors). Medical treatment with oxytocin (and occasionally calcium) will strengthen uterine contractions and may result in successful delivery in dogs with uterine weakness. Cesarean section is usually required to save the lives of the puppies and/or bitch when dystocia is due to problems with the pelvis, fetus size/position, or vaginal strictures/tumors.

*Pyometra.*  An infection of the uterus, pyometra is most common in older (8 to 10 years) bitches during late diestrus (4 to 8 weeks after estrus), although it can also develop in younger bitches. Progesterone stimulation of the endometrial glands of the uterus results in an environment favorable for growth of bacteria. Clinical signs of pyometra include depression, fever, decreased appetite, increased thirst, frequent urination, and sometimes a purulent vaginal discharge (discharge is absent if the cervix is closed). Diagnosis is based on history of a recent heat period; finding an enlarged pus-filled uterus on palpation, ultrasound, or radiographs; and an increased white blood cell count. Surgical removal of the diseased uterus as part of a complete ovariohysterectomy (spay) is the treatment of choice. Pyometra is often fatal if treatment (surgery) is delayed. If the bitch is a valuable breeding animal, medical treatment using prostaglandins and systemic antibiotics may be successful if the cervix is open. Prostaglandins cause powerful uterine contractions to help expel pus and bacteria from the uterus. Prostaglandins are contraindicated if the cervix is closed. Uterine contractions against a closed cervix may result in rupture of the uterus. The success rate in treating open-cervix pyometra is 75% to 90%. Unfortunately, recurrence is common (50% to 80%), so the bitch should be bred at her next estrous cycle or ovariohysterectomized.

*Eclampsia.*  Eclampsia (puerperal tetany) is a life-threatening condition caused by low blood calcium concentrations. It is most common in small breeds of dogs during the first 3 weeks of lactation; it is occasionally seen in bitches during late pregnancy. Clinical signs include nervousness, panting, whining, drooling, a stiff gait, muscle twitching—and if not treated—seizures, and death. Treatment includes intravenous administration of calcium (this must be

[2]Bitches that have had their pelvis fractured (e.g., following a vehicular accident) may have a narrowed pelvic canal or partial obstructions from callus formation.

administered carefully as too much calcium will cause irregular heartbeats and death) and decreasing milk production by hand-feeding the puppies.

*Cryptorchidism.* Cryptorchidism is the failure of one or both testes to descend into the scrotum. Cryptorchidism is an inherited trait affecting up to 10% of male dogs. Breeds most commonly affected include the Toy Poodle, Pomeranian, Yorkshire Terrier, Miniature Dachshund, Cairn Terrier, Chihuahua, Maltese, Boxer, Pekingese, English Bulldog, Old English Sheepdog, Miniature Poodle, Miniature Schnauzer, Shetland Sheepdog, Siberian Husky, Standard Poodle, and German Shepherd Dog. In addition to being a hereditary defect, cryptorchidism predisposes dogs to the development of testicular cancer; therefore, castration of affected dogs is strongly recommended.

*Prostatic diseases.* Prostatic diseases are common in older intact male dogs. Benign enlargement of the prostate gland results in difficult urination and defecation. Infections of the prostate gland result in fever, blood or pus in the urine, abdominal pain, a stilted gait, and dripping of blood or pus from the prepuce. Prostatic cancer can result in clinical signs similar to those of prostatic enlargement and infection. Neutering (castration) is the treatment of choice for prostatic enlargement. Neutering is combined with antibiotics for the treatment of prostatic infections. Prostatic cancer is treated with combinations of surgery and chemotherapy or radiation therapy.

## 10.4   FELINE REPRODUCTION

Domestic cats reach sexual maturity between 6 and 9 months of age. The optimum breeding age is between 1.5 and 7.0 years of age. Cats are classified as seasonally polyestrous-induced ovulators. Queens in the Northern Hemisphere begin to cycle in January or February and, if not bred, will cycle every 4 to 30 days with the intervals increasing throughout the summer. Cycling ceases in September or October. Ovulation is induced by copulation.

### 10.4.1   REPRODUCTIVE ANATOMY OF THE QUEEN

The major organs of the reproductive tract of the queen are very similar to those of the bitch (see Figure 10.1). Queens almost always have four pairs of mammary glands.

### 10.4.2   PHYSIOLOGY OF REPRODUCTION IN QUEENS

Short-haired queens usually reach puberty at a body weight of 5 to 6 lb and at a time when daylight is increasing; this is often between 4 and 9 months of age. Long-haired and Manx queens tend to be older (11 to 21 months) before their first estrous cycle.

*The feline estrous cycle.* The stages of the estrous cycle of the queen include anestrus, proestrus, estrus, postestrus, and diestrus. As noted above, photoperiods influence this cycle.

*Anestrus.* Anestrus is a seasonal absence of cycling activity that occurs during short daylight hours (October, November, and December in Northern Hemisphere).

*Proestrus.* Proestrus in queens is associated with rising concentrations of blood estrogen; there may be no outward signs or the queen may show an increase in head and neck rubbing behavior. Proestrus lasts 1 to 2 days in cats.

*Estrus.* The period of sexual receptivity is estrus. Queens in estrus frequently vocalize to call males. In the presence of a male, the queen will crouch with forequarters pressed to the ground, tail held off to one side, and vulva elevated; this posture is referred to as "lordosis." In the absence of a male, the queen will show rolling, head rubbing, treading with the hind legs, lordosis, and tail deviation. Duration of estrus behavior when not bred ranges from 2 to 19 days. Some reports indicate the duration of estrus is shorter in queens that are bred; however, other studies have not confirmed this difference.

*Postestrus.* The period between estrous cycles in queens that have not ovulated is postestrus. The duration of postestrus averages 8 to 10 days. During this phase, blood estrogen levels are low and there is no sexual behavior.

*Diestrus.* Diestrus occurs in queens that have been induced to ovulate. The postovulatory follicles luteinize and secrete progesterone for approximately 40 days in a nonpregnant queen and 60 days in a pregnant one. At the end of diestrus (or pregnancy) the corpora lutea undergo luteolysis, serum progesterone levels decrease, and estrus follows in 7 to 10 days in a photoperiod with 14 hours of daylight.

*Induction of ovulation.* Copulation is followed by release of GnRH from the hypothalamus of the queen. Gonadotropin induces a surge in serum LH resulting in ovulation. The quantity of LH released is correlated with the number of copulations; maximum LH release requires 8 to 12 copulations. Queens in estrus will mate multiple times; reports of 20 to 36 copulations in 36 hours are common. Glass rod stimulation of the vagina of an estrous queen may be used to induce ovulation; however, this does not reliably shorten the length of estrus.

## 10.4.3 FELINE MALE REPRODUCTIVE ANATOMY AND PHYSIOLOGY

The tom reaches sexual maturity around 9 months of age. The feline penis is ventral to the scrotum and is directed backwards (Figure 10.10). The penis has a band of 120 to 150 spines, 0.1 to 0.7 mm in length; these are directed caudally away from the tip of the glans penis. The testes are usually in the scrotum at birth. Cryptorchidism (failure of the testes to descend into the scrotum) is very rare in cats. Long-haired cats occasionally have a problem with hairs accumulating around the base of the glans penis, which interfere with intromission; this can be corrected by retracting the sheath and sliding the ring of hair over the glans. As in other mammals, reproductive function is mediated by LH and FSH. LH stimulates testosterone secretion by the interstitial cells of the testes. FSH stimulates secretion of estrogens by the Sertoli cells of the testes and also stimulates spermatogenesis.

**Figure 10.10** Reproductive anatomy of the male cat (tomcat). *Illustration by Diana Nicoletti.*

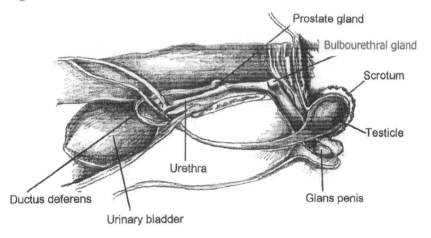

## 10.4.4    BREEDING MANAGEMENT OF CATS

The estrous queen should be taken to the male. In a colony, one male can be placed with a group of 15 to 20 queens. If multiple males are available, a queen often mates with several and produces a litter sired by more than one male. Mounting and neck biting occur within a few minutes; the tom will straddle the queen and quickly complete intromission, ejaculation, and dismounting. The queen emits a loud cry, leaps away from the male and rolls, stretches, and licks her vulva. This "after-reaction" lasts 1 to 17 minutes. The queen may refuse to allow a male to mount for up to 5 hours after a successful intromission. Multiple breedings over 2 to 3 days help ensure optimal conception.

## 10.4.5    ARTIFICIAL INSEMINATION

Toms can be trained to ejaculate into an artificial vagina. Electroejaculation of anesthetized males can also be used to obtain semen for fertility evaluations and artificial insemination. Semen may be mixed with extenders and chilled or frozen for shipment to inseminate queens in other locations. A lubricated catheter is inserted as deep as possible into the queen's vaginal vault and the semen deposited. Following insemination, the queen's hindquarters should be elevated for at least 10 minutes. The queen should be injected with LH or human gonadotrophin to induce ovulation. The insemination and injection of LH should be repeated in 24 and 48 hours.

## 10.4.6    PREGNANCY

Fertilized ova remain in the oviducts for 5 to 6 days. Implantation in the uterus occurs 11 to 14 days following breeding. Gestation length averages 66 days from the first breeding (range 64 to 69 days). Individual fetuses can first be felt using abdominal palpation from days 17 to 25 of gestation. Ultrasound confirmation of pregnancy can be made as early as day 14; individual beating hearts can be seen around day 22. Radiographs can be used to identify fetuses after day 45. There are no reliable hormonal markers of pregnancy in the queen.

## 10.4.7    PARTURITION

Queens rarely have difficulties giving birth (queening). Owners should provide the queen with a box or cage containing nesting materials (blankets or towels). The queen commonly seeks a secluded area to give birth. Nesting behavior is usually seen 12 to 48 hours prior to parturition. The queen's body temperature often drops below 99°F approximately 12 hours prior to labor, although this is not always observed and is not considered as reliable a sign of imminent parturition as was noted by the temperature drop of a bitch. Signs of prelabor include restlessness, pacing, panting, nesting behavior, and antagonism toward strangers and other cats. When contractions begin, the queen usually assumes a semisquatting position. The allantochorion membrane (water bag) usually breaks when the kitten enters the vagina (birth canal). Head first and hind feet first presentations are both normal; time from appearance of the head or feet at the vulva and complete delivery is usually 3 to 5 minutes. The queen will vigorously lick remaining membranes off the kitten and will then chew through the umbilical cord and eat the placenta. Litter size ranges from 1 to 10 kittens, the average being 4. The entire queening process may range from 4 to 42 hours; with 14 hours being average.

## 10.4.8    CARE OF THE POSTPARTURIENT QUEEN

The queen may refuse to leave her kittens for the first day following their birth. Provide a litter pan, food, and water next to the queen's box. A normal queen has a very scant postpartum discharge. A veterinarian should be consulted if a purulent vaginal discharge is observed or if the queen is febrile, depressed, anorexic (poor appetite), or has any signs of mastitis (heat, pain, swelling, or discoloration surrounding a nipple). Peak milk production occurs in 20 to 30 days. The queen may consume two to four times her normal food intake until the kittens begin eating solid food.

### 10.4.9  CARE OF THE NEWBORN KITTENS

An average kitten weighs 100 g at birth. The newborn kitten will crawl slowly toward the warmth of the queen and will move its head side to side until it finds a nipple. The kitten should nurse within the first few hours after birth for maximum absorption of colostrum (Chapter 9). The newborn kitten's eyes and ears are closed for the first 1 to 2 weeks. It is difficult to sex newborn kittens because the vulva and preputial openings both appear as narrow slits. The anogenital distance is the best way to distinguish the sex of newborns; it is 6 to 9 mm in females and 11 to 16 mm in males. Kittens should be weighed daily; they should gain 7 to 10 g/day for the first few weeks. Signs of good health and illness are similar to those of newborn puppies (see Table 10.2). Care and feeding of orphaned kittens was presented in Section 9.7.

## 10.5  PREVENTION OF ESTROUS CYCLES

The preferred method of preventing estrous cycles is to perform an ovariohysterectomy (surgical removal of both ovaries and the uterus). When this surgery is performed prior to puberty, the risk of the bitch or queen developing mammary cancer later in life is reduced to near zero. Animals are anesthetized for this surgery and recover rapidly from the procedure. Many people fear their pet will be prone to obesity or other health problems related to ovariohysterectomy, but the health benefits of the surgery greatly exceed the risks. (*Benefits*: prevents mammary cancer; prevents uterine cancer; prevents uterine infections; prevents unwanted pregnancies; prevents the mess, odor, unwelcome attention of visiting males, and behavioral problems associated with estrous cycles; prevents estrous cycles from interfering with showing or work of the female; neutered females have a lowered risk of developing diabetes mellitus; and many others. *Risks*: slight risk of complications associated with anesthesia or surgery; may have slight tendency to gain weight due to lack of increased caloric demands of estrus and pregnancy/lactation.)

Although traditionally ovariohysterectomies were performed at approximately 6 months of age, earlier neutering programs now perform these surgeries on puppies and kittens as young as 8 weeks of age. Early neutering programs have proven to be safe and ensure that animals being adopted from shelters do not perpetuate the problem of pet overpopulation (Chapter 24).

Androgens (e.g., mibolerone) and progesterones (e.g., megestrol acetate) are hormones that can be administered to female dogs and cats to prevent estrous cycles. These hormones are effective only during the time they are being given and have serious potential side effects (including liver disease, kidney disease, diabetes mellitus, and increased risk of pyometra). Thus, they are not recommended for routine use in preventing or suppressing estrous cycles in dogs or cats.

## 10.6  STERILIZATION OF MALES

Castration, also known as orchiectomy, is neutering male dogs and cats via surgical removal of the testicles. Bilateral orchiectomy is the method of choice of sterilizing male dogs and cats. In addition to sterilization, castration prevents testicular and epididymal disorders (torsion, infection, cancer), decreases the risk of perineal hernias, and decreases objectionable behaviors including roaming, mounting, urine marking, urine spraying (cats), urine odor (cats), and aggression.

Traditionally, castration was postponed until dogs and cats were near the age of puberty (6 to 9 months). However, early neutering programs have increased in popularity and now many shelters and humane societies neuter puppies as young as 8 weeks. An alternative to surgery for the neutering of young puppies is the use of a testicular sclerosing agent. Neuter-sol® is a zinc-containing compound that results in atrophy of the testicles. It is 99% effective in producing sterility and has had few reported side effects.

## 10.7    SUMMARY

It is readily evident that companion animals must be born before they can share companionship with and provide services to children and adults. This makes having an understanding of the fundamental concepts and applications of reproduction in dogs and cats one of the most important aspects of companion animal biology, care, health, and management.

In this chapter we discussed the anatomy and reproductive physiology of female and male dogs and cats. We learned to differentiate among the several stages of the canine and feline estrous cycles as well as the hormonal regulation of reproduction in male and female dogs and cats.

Artificial insemination is sometimes an important adjunct to successful pregnancy in the bitch and queen. Procedures related to the collection of semen and its handling and deposition in female dogs and cats were discussed.

We learned about the most common methods used to diagnose pregnancy in the bitch and queen as well as the precautions associated with parturition and postpartum care. Common reproductive problems of dogs were discussed as well.

The prevention of estrous cycles in the bitch and queen can be accomplished surgically or hormonally. These methods, along with surgical and hormonal sterilization of canine and feline males, were discussed.

## 10.8    REFERENCES

1. Feldman, Edward C., and Richard W. Nelson. 2004. *Canine and Feline Endocrinology and Reproduction*. 3rd ed. Philadelphia: W. B. Saunders.

2. Hoskins, Johnny D. 2001. *Veterinary Pediatrics*. 3rd ed. Philadelphia: W. B. Saunders.

3. Johnston, Shirley D., Margaret V. Root Kustritz, and Patricia N. S. Olson. 2001. *Canine and Feline Theriogenology*. Philadelphia: W. B. Saunders.

4. Simpson, Gillian. 1998. *Manual of Small Animal Reproduction and Neonatology*. Cheltenham, UK: Shurdington.

# 11 ANATOMY AND PHYSIOLOGY OF CATS AND DOGS

*It is important to study bones, blood, muscles, nerves, and individual organs;
it is equally as worthwhile to investigate the physiological functioning of
body parts in conjunction with systems of the whole.*

Elizabeth Blackwell (1821–1910)
*American physician*

## 11.1　INTRODUCTION/OVERVIEW

Anatomy is the branch of science concerned with the structure and form of the tissues and organs that comprise the body. Physiology is the branch of science that studies functions of various body parts with emphasis given to the chemical processes involved in creating, developing, and maintaining life. A systems approach is commonly used in studying anatomy and physiology. The major body systems are the integumentary, skeletal, muscular, circulatory, lymphatic, digestive, nervous, urinary, reproductive, and endocrine systems plus the organs of special senses (sight, smell, taste, hearing). All systems must work together for an animal to enjoy an active and healthy life. The anatomy and physiology of the reproductive systems of dogs and cats were presented in Chapter 10. Other important physiological functions are included in Chapter 9 (Feeding and Nutrition) and Chapter 14 (Companion Animal Health).

Having knowledge of the anatomical and physiological characteristics of dogs and cats is fundamental in describing and communicating about these most widely kept companion animals. This knowledge is also fundamental to understanding and applying the precepts of management, veterinary diagnostic testing, and therapy.

Concurring with Dr. Blackwell's succinctly expressed, medically related precepts shared above, in this chapter we first highlight the external body parts of dogs and cats. This is followed by an emphasis on the basic physiological systems that enable these special pets to function as companion animals and perform many other services enjoyed and appreciated by humans.

## 11.2　EXTERNAL BODY PARTS

It is important to become familiar with the terms used to refer to the various parts of the bodies of dogs and cats (Figures 11.1 and 11.2). These terms are used by dog and cat breeders and judges to describe the conformation of an animal. Animal healthcare professionals use these terms to refer to the areas of the body affected by disease processes. Each breed of dogs and cats has a written breed standard that describes the "ideal" animal of that particular breed (Chapters 3 and 4). Breed standards include references to the ideal proportions, angles, and other features of the various external anatomical parts. Adhering to the breed standard as the "ideal" or "blueprint" to strive for in breeding animals helps ensure that the distinctive characteristics of each breed are maintained and improved over time. Table 11.1 includes a brief definition of terms used to describe the external anatomy of animals.

## 11.3　THE INTEGUMENTARY SYSTEM

The skin is the largest and most visible organ of the body. Skin and hair coat represent approximately 24% of the weight of newborn puppies and kittens and 12% of their adult weights. Important functions of skin include (1) an enclosing barrier preventing the loss of water, electrolytes, and cells; (2) protection from the environment (physical barrier, chemical barrier, microbiological barrier); (3) allowing motion (flexibility); (4) temperature regulation

**Figure 11.1** External anatomical terms used for cats. *Illustration by Diana Nicoletti.*

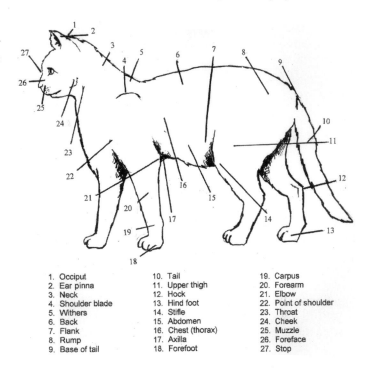

| 1. Occiput | 10. Tail | 19. Carpus |
|---|---|---|
| 2. Ear pinna | 11. Upper thigh | 20. Forearm |
| 3. Neck | 12. Hock | 21. Elbow |
| 4. Shoulder blade | 13. Hind foot | 22. Point of shoulder |
| 5. Withers | 14. Stifle | 23. Throat |
| 6. Back | 15. Abdomen | 24. Cheek |
| 7. Flank | 16. Chest (thorax) | 25. Muzzle |
| 8. Rump | 17. Axilla | 26. Foreface |
| 9. Base of tail | 18. Forefoot | 27. Stop |

(hair coat, cutaneous circulation, sweat); (5) storage (water, electrolytes, vitamins, fats, carbohydrates, proteins); (6) pigmentation (protection from solar damage, camouflage, social roles); (7) immunosurveillance (protection from infections and tumors); (8) production of vitamin D; (9) sensory perception (touch, pressure, pain, itch, heat, cold); and (10) excretory secretions (sebum, sweat). In addition, the skin and hair coat serve as a mirror to the overall health of an animal. Many diseases affecting dogs and cats are reflected by changes in the skin and hair coat prompting the diagnosis and treatment of these disorders (Chapter 14).

**Figure 11.2** External anatomical terms used for dogs. *Illustration by Diana Nicoletti.*

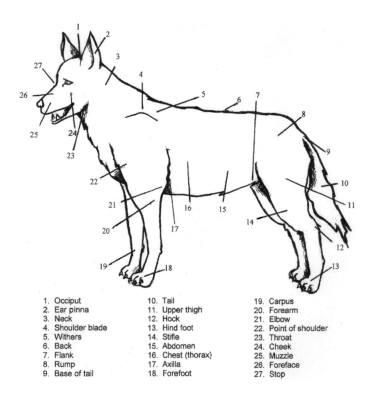

| 1. Occiput | 10. Tail | 19. Carpus |
|---|---|---|
| 2. Ear pinna | 11. Upper thigh | 20. Forearm |
| 3. Neck | 12. Hock | 21. Elbow |
| 4. Shoulder blade | 13. Hind foot | 22. Point of shoulder |
| 5. Withers | 14. Stifle | 23. Throat |
| 6. Back | 15. Abdomen | 24. Cheek |
| 7. Flank | 16. Chest (thorax) | 25. Muzzle |
| 8. Rump | 17. Axilla | 26. Foreface |
| 9. Base of tail | 18. Forefoot | 27. Stop |

**Table 11.1**
Terms Used to Describe the External Anatomy of Animals

| Term | Description |
|---|---|
| Abdomen | The part of the body between the chest and the hindquarters |
| Axilla | The "armpit" region; area of the chest under the shoulder and elbow |
| Bite | In anatomy, this term is used to describe the apposition of teeth; some standards call for a "level bite" in which the teeth meet evenly at the edges, others call for a scissors bite |
| Carpus | The equivalent of the human wrist; the joints between the radius/ulna and metacarpal bones |
| Cheek | The fleshy portion of the side of the face |
| Chest | The trunk part of the body including between the elbows and along the ribs; also called the thorax |
| Coat | The dog's hair; may be used in referring to the quantity, quality, or color of the coat |
| Conformation | The structure and appearance of the dog or cat; the quality as compared to the breed standard |
| Croup | The part of the topline from the loin to the tail; extends from the last lumbar vertebra to the first coccygeal vertebra |
| Dewclaw | The extra toe (vestigial) on the inside of the leg; is absent or removed in many breeds of dogs |
| Elbow | The joint between the forearm and the upper arm; the point of the elbow is the area over the top of the ulna's anconeal process (also called the olecranon) |
| Foreface | The head from the nose to the stop; includes the muzzle and the underjaw |
| Flank | The ventral part of the abdomen in front of the hind legs |
| Forearm | The area between the shoulder and carpus; over the radius and ulna |
| Front | In descriptions of breed standards this is referring to the forelegs and shoulders plus the chest |
| Hock | The joint between the stifle and metatarsus; lower hind leg |
| Loin | The portion of the back between the ribs and the croup; also referred to as the lumbar region |
| Muzzle | The upper part of the foreface |
| Occiput | The back or top of the head |
| Overshot | Top teeth protrude beyond the lower teeth (brachygnathus, having a short lower jaw) |
| Pads | The soles of the feet |
| Pastern | The lower portion of the legs; metacarpal and metatarsal bones |
| Pinna | The external part of the ear |
| Rump | The upper hindquarters; gluteal region |
| Scissors bite | The top teeth overlap the bottom teeth and touch on the inside edges (this is the preferred conformation in many breeds) |
| Stifle | The knee; the joint between the femur (upper leg) and tibia (lower leg), also refers to the leg between the knee joint and the hock |
| Stop | The break or depression located between the top of the skull and nose as viewed from the side; many breed standards call for a definite stop rather than a gradual transition between the skull and nose |
| Thigh | The part of the hind limb over the femur; upper bone of hind leg |
| Undershot | The teeth on the lower jaw protrude beyond the top teeth (prognathous, protruding jaw) |
| Upper arm | The area from the shoulder blade to forearm |
| Whiskers | Long hairs on the muzzle (vibrissae) |
| Withers | The area of the back between the shoulder blades; this is the highest part of the back of most animals and the standard location used in measuring height |

The skin consists of two layers—the epidermis and the dermis—and epidermal appendages (Figure 11.3). The upper layer is the epidermis. It is composed of epithelial cells containing keratin (a filamentous protein) and lipids, which together form the "brick and mortar" structure of the skin surface. The surface layer of epidermal cells is constantly being shed and replaced by new cells; the epidermal turnover time averages 22 days in dogs. The lower layer is known as the dermis. It is composed of a network of connective tissue fibers, blood and lymph vessels, nerves, specialized muscles in-

**Figure 11.3** Cross section of the skin of dogs and cats. *Courtesy DVM Pharmaceuticals, Miami, Florida.*

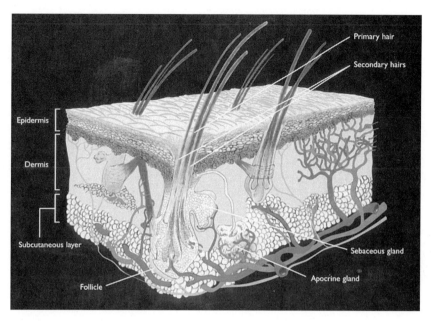

volved in piloerection (raising hairs to stand erect), and epidermal appendages (hairs, sweat glands, sebaceous glands). Cats and dogs have compound hair follicles with primary hairs sharing a single pore with 5 to 20 secondary hairs. Hair follicles are positioned at a 20° to 60° angle to the skin surface; this angle directs hairs caudally and ventrally to shed water. Dogs produce 60 to 180 g hair/kg and cats 30 to 40 g hair/kg of body weight annually. Hair growth is influenced by the photoperiod, ambient temperature, nutrition, hormones, genetics, and overall health of the animal. Poodles have a very long hair-growth cycle and are classified as a nonshedding breed. Most breeds of cats and dogs regrow a normal-appearing hair coat within 3 months after being shaved. Long-coated breeds, such as Afghan Hounds, may require 18 months for shaved areas to blend in with the rest of the hair coat.

## 11.4   THE SKELETAL SYSTEM

Functions of the skeleton include protecting internal organs, supporting the body, and providing attachment sites for muscles used in movement. The cat's skeleton (Figure 11.4) contains approximately 245 bones, whereas a dog's skeleton (Figure 11.5) includes up to 319 bones.

The skull of the cat has large eye sockets to accommodate the large eyes required for a visual hunter. Its jaw is short and strong to capture and hold prey. There are three major types of skulls in dogs: doliocephalic, mesaticephalic, and brachycephalic. The doliocephalic shape has a narrow skull base and an elongated muzzle; the narrow head is accompanied by a long neck, a forward position of the center of gravity, and a body build that is agile and fast. Examples of doliocephalic breeds include Collies, Greyhounds, Whippets, German Shepherds, and numerous others (see examples in Color Plates 4 to 6). Mesaticephalic skulls have a medium ratio of skull-base width to muzzle length and are characteristic of the spaniel and retriever breeds. Brachycephalic skulls have a broad skull base and short muzzle and are seen in breeds such as the Boxer, English Bulldog, Boston Terrier, Pug, and Pekinese.

Dogs have 28 teeth as puppies, 42 as adults (Figure 11.6). Cats have 26 teeth as kittens, 30 as adults (see Figure 3.3). Both species have 12 incisors (6 upper, 6 lower) plus 4 canine teeth (2 upper, 2 lower). Adult cats have only 10 premolars (6 upper, 4 lower) and 4 molars

**Figure 11.4**   Major bones of the cat skeleton. *Illustration by Diana Nicoletti.*

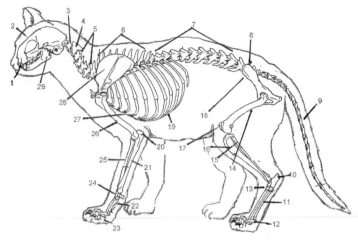

| | | |
|---|---|---|
| 1. Canine tooth | 11. Metatarsal bones | 21. Ulna |
| 2. Skull | 12. Phalanges | 22. Metacarpal bones |
| 3. Atlas | 13. Tarsal bones | 23. Phalanges |
| 4. Axis | 14. Femur | 24. Carpal bones |
| 5. Cervical vertibrae | 15. Fibula | 25. Radius |
| 6. Thoracic vertibrae | 16. Tibia | 26. Humerus |
| 7. Lumbar vertibrae | 17. Patella | 27. Sternum |
| 8. Sacrum | 18. Pelvis | 28. Scapula |
| 9. Coccygeal vertibrae | 19. Ribs | 29. Mandible |
| 10. Calcaneus | 20. Olecranon | |

(2 upper, 2 lower), whereas adult dogs have 16 premolars (8 upper, 8 lower) and 10 molars (4 upper, 6 lower). The larger number of teeth in dogs enables the crushing and grinding of a variety of foods. Cats are true carnivores and have a limited ability to grind food. The fourth upper premolar and the first lower molar teeth function together like the blades of scissors to shear through foods and tear flesh; these teeth are also called the carnassial teeth.

**Figure 11.5**   Major bones of the dog skeleton. *Illustration by Diana Nicoletti.*

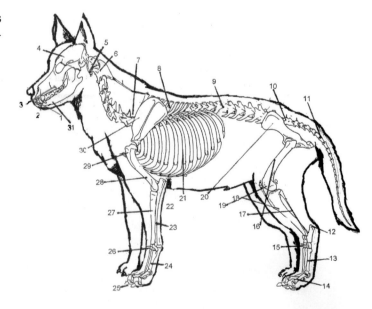

| | | |
|---|---|---|
| 1. Incisors | 12. Calcaneus | 23. Ulna |
| 2. Canine tooth | 13. Metatarsal bones | 24. Metacarpal bones |
| 3. Fourth premolar | 14. Phalanges | 25. Phlanges |
| 4. Skull | 15. Tarsal bones | 26. Carpal bones |
| 5. Atlas | 16. Femur | 27. Radius |
| 6. Axis | 17. Fibula | 28. Humerus |
| 7. Cervical vertibrae | 18. Tibia | 29. Sternum |
| 8. Thoracic vetibrae | 19. Patella | 30. Scapula |
| 9. Lumbar vertibrae | 20. Pelvis | 31. Mandible |
| 10. Sacrum | 21. Ribs | |
| 11. Coccygeal vertibrae | 22. Olecranon | |

**Figure 11.6**   Dentition of the dog: incisors (1,2,3), premolars (P1, P2, P3, P4), and molars (M1, M2, M3, M4). The maxilla (upper jaw) has the dental formula 3143. The mandible (lower jaw) has the formula 3142. *Courtesy University of Illinois.*

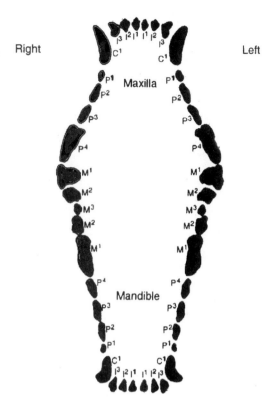

All mammals have 7 cervical vertebrae; the length of individual bones varies with the breed (short in brachycephalic breeds, longer in doliocephalic breeds). Cats and dogs have 13 thoracic vertebrae, 7 lumbar vertebrae, 3 sacral vertebrae (these fuse to form the sacrum), and 0 to 23 coccygeal vertebrae (varies with length of the tail). The vertebrae are important in protecting the spinal cord and also serve as sites for the attachment of muscles used in locomotion. The first cervical vertebra is the atlas. Its articulation with the base of the skull (occiput) provides up-and-down mobility giving the atlantooccipital joint the nickname of the "yes joint."

Toes (phalanges) are comprised of three bones (phalanxes). Dogs have four functional toes on each foot and one that is vestigial (the dewclaw). Cats generally have five toes on the front feet and four on the back. Extra toes are fairly common in cats; this condition is known as polydactyly and is inherited as a dominant trait. The third phalanx (distal phalanx, P3) produces claws that are important for traction (dogs and cats), digging (dogs and cats), climbing (cats), hunting (cats), and defense (cats). Cats have an elastic ligament between the distal end of P3 and the second phalanx (middle phalanx, P2). At rest, this elastic ligament pulls the claw upward so it is sheathed within the skin covering the toe (Figure 11.7). Extending the claws involves contraction of tendons that lock P2 and P3 in a rigid extended position; this stretches the elastic ligament and unsheathes the claws.

The forequarters consist of the shoulder blade (scapula), upper arm (humerus), forearm (radius and ulna), carpus (wrist), metacarpal bones (pastern), and feet. The forequarters are attached to the rest of the body by muscles. The angulation of the shoulder blade affects stride and forward movement. Shoulder angulation is measured by the angle between the long axis of the scapula and a line drawn parallel to the ground; 45° to 55° is considered optimal, whereas slopes greater than 55° are steep and associated with short strides and a choppy gait.

The hindquarters consist of the pelvis (sacrum, ilium, ischium, pubis, acetabular bones), femur (upper thigh bone), tibia and fibula (shank), tarsus (hock), metatarsals (pastern), and feet. The articulation between the head of the femur and the acetabular bones is a ball-and-socket

**Figure 11.7** Sheathed and extended positions of the cat's claw. *Illustration by Diana Nicoletti.*

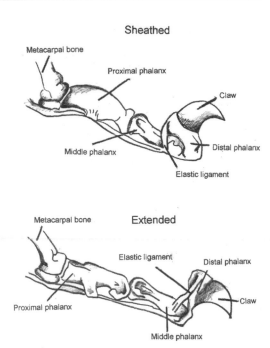

joint known as the acetabulum. This joint is abnormal in dogs and cats with hip dysplasia (Chapters 9 and 14). The joint between the lower femur and tibia and fibula is known as the stifle. The patella (kneecap) is embedded in tendons and ligaments connecting the femur with the tibia. The horizontal slope of the pelvis affects the rear stride; a slope of approximately 30° from horizontal is considered optimum.

## 11.5   THE MUSCULAR SYSTEM

Muscles contribute to the outward appearance of animals and are essential for movement, posture, breathing, circulation, digestion, and many other functions. The functional cell unit is known as a muscle fiber. There are three primary types of muscle fibers: smooth, cardiac, and skeletal. Muscles are classified as being voluntary or involuntary. Voluntary muscles are those that can be contracted or relaxed at will, whereas involuntary muscles are regulated by the nervous and endocrine systems.

Smooth muscles are found in the walls of hollow organs (e.g., digestive, reproductive, urinary, respiratory, and circulatory systems) and also in association with numerous glands. Other names used to describe smooth muscles are unstriated, plain, involuntary, white, or visceral muscles. Smooth muscles are controlled by the autonomic nervous system and also influenced by hormones.

Cardiac muscle fibers form the heart. These fibers are striated (striped) and arranged in networks to provide the strong coordinated contractions required for efficient pumping of blood. Cardiac muscle is controlled by the autonomic nervous system and is also influenced by hormones.

There are hundreds of skeletal muscles, each having its own function. The ends of many skeletal muscles are attached to bones through fibrous, cordlike connective tissue called tendons. Skeletal muscle fibers are arranged in bundles and layers that overlap and are organized within connective tissue envelopes. Other names used to describe skeletal muscles are striated (striped), voluntary, or somatic muscle. Skeletal muscles comprise one-third to one-half of the body weight of dogs and cats. The major muscles of the dog are shown in Figure 11.8. Skeletal muscles are classified by their functions. Extensor muscles cause bones to straighten; flexor muscles cause them to bend. Abductor muscles cause limbs to move away from the body, whereas adductor muscles move limbs toward the center plane of the body. Table 11.2 lists the major muscle groups of dogs and cats with their functions.

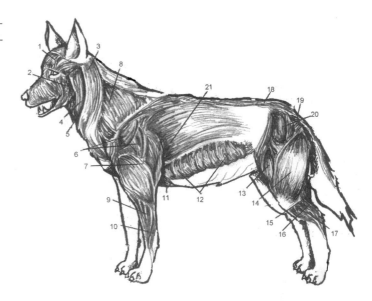

**Figure 11.8** Major skeletal muscles of the dog. *Illustration by Diana Nicoletti.*

1. Temporalis
2. Masseter
3. Cleidocervicalis
4. Sternocephalicus
5. Sternohyoideus
6. Deltoideus
7. Triceps
8. Trapezius
9. Extensor carpi radialis
10. Flexor carpi ulnaris
11. Deep pectoral (axilla)
12. External abdominal oblique
13. Quadriceps femoris
14. Biceps femoris
15. Semimembranosus
16. Semitendinosus
17. Gastrocnemius
18. Sartorius
19. Coccygeus
20. Gluteus
21. Latissimus dorsi

**Table 11.2**
Major Muscle Groups of Dogs and Cats

| Muscle Group | Function |
| --- | --- |
| **Biceps brachii (front leg)** | Flexors of the joint between the humerus and radius (bends the forearm) |
| **Biceps femoris** | Extends the hip, stifle, and tarsal joints |
| **Cleidocervicalis and sternocephalicus** | Involved in pulling the front leg forward and also in bending (flexing) the neck |
| **External abdominal oblique and other muscles of the ventral abdomen** | Compression and support of the abdominal viscera (abdominal press for urination, defecation, parturition) |
| **Gluteal muscles** | Flex and abduct the hind leg; located between the ilium, sacrum, and femur |
| **Latissimus dorsi** | Draws the trunk forward, helps support the front legs |
| **Masseter and temporalis** | Raise the lower jaw to close the mouth |
| **Quadriceps** | Extensors of the stifle joint |
| **Semimembranosus and semitendinosus** | Flexors of the stifle joint at rest; extends the hip, stifle, and tarsus when weight-bearing |
| **Trapezius** | Raises the front leg and pulls it forward |
| **Triceps** | Extensors of the joint between the humerus and radius (straightens the upper front leg) |

# 11.6   THE CIRCULATORY SYSTEM

The circulatory system consists of the heart, arteries, veins, capillaries, and lymphatic vessels. Other important contributors to the circulatory system include bone marrow (produces blood cells), spleen (serves as a reservoir for storage of blood cells and also removes old and damaged blood cells), and lymph nodes (produce lymphocytes and remove foreign material from lymph).

Arteries are thick-walled blood vessels that carry blood from the heart to various body tissues. Arteries branch into progressively smaller vessels known as arterioles; eventually the branching results in vessels known as capillaries that are only one cell thick. Blood traveling through capillaries can exchange gases, nutrients, and waste products with tissue cells. From capillaries, blood flows into venules, which join together to form veins that return blood to the heart for oxygenation. Veins are relatively thin walled; some have valves to maintain the one-way flow of blood back to the heart. Arteries and veins supplying body tissues comprise the systemic circulation. Systemic circulation is complex. The portal system transports venous blood from the stomach, small intestines, and spleen to the liver where it is filtered through a second capillary system before being returned to the heart. The pulmonary circulation includes the arteries and veins of the lungs. The pulmonary circulation differs from the systemic circulation in two major ways: the blood circulates at a much lower pressure and the arteriole blood is unoxygenated whereas the venous blood is oxygenated. The right ventricle pumps at a low pressure because there is not as much resistance to blood flow within the lungs compared with other body tissues.

The heart is located in the ventral thorax (chest) between the third and seventh ribs. The heart is divided into four chambers (Figure 11.9). Blood enters the heart through thin-walled chambers known as atria. The right atrium receives venous blood returning from the systemic circulation through the vena cava and directs it into the right ventricle. Blood is pumped into the pulmonary arteries from the right ventricle. The blood goes into the lungs where carbon dioxide carried by red blood cells is exchanged for oxygen. Blood returns to the heart through the pulmonary veins and enters the left atrium. From the left ventricle, blood is pumped into the systemic circulation through the aorta. Valves are important in maintaining the one-way direction of blood flow. Atrioventricular (AV) valves separate the atria from the ventricles; these are known as the mitral or bicuspid (left AV) and the tricuspid (right AV) valves, respectively. The aortic and pulmonary valves are also known as semilunar valves. Torn or "leaky" valves result in a turbulent flow of blood. This results in a heart murmur that can be heard by auscultation (listening to heart sounds). Strictures or narrowing (stenosis) of either the aorta or pulmonary artery will also cause turbulence in blood flow and a heart murmur. The ventricular pumping phase of the heart cycle is known as systole. The filling phase of the ventricles is known as diastole.

The pacemaker of the heart is the sinoatrial node located in the right atrium. When the sinoatrial node depolarizes (fires), the right atrium contracts followed by contraction of the

**Figure 11.9** Diagram of four chambers of the heart with valves and major vessels. *Illustration by Diana Nicoletti.*

1. Pulmonary arteries
2. Pulmonary veins
3. Semilunar valve (aortic valve)
4. Left atrium
5. Left atrioventricular valve (mitral, bicuspid)
6. Left ventricle
7. Ventricular septum
8. Apex
9. Right ventricle
10. Right atrioventicular valve
11. Caudal vena cava
12. Semilunar valve (pulmonary valve)
13. Right atrium
14. Cranial vena cava
15. Aorta

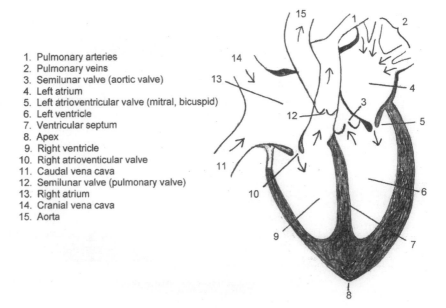

left atrium. The ventricles then contract closing the AV valves (closing of these valves creates the first sound of the heart beat, lub) and forcing blood into the arteries. Following the contraction of the ventricles, the semilunar valves close producing the second heart sound (dub). As the ventricles relax, the AV valves reopen to permit blood in the atria to reenter the ventricles. The left ventricle is the largest and thickest chamber of the heart as it must contract strongly to pump blood throughout the systemic circulation.

Lymphatic vessels function to return fluids that have accumulated in body tissues back to the circulatory system. Lymphatic vessels originate in tissue spaces and join to form larger vessels (ducts) that pass through lymph nodes and eventually empty into large veins near the vena cava. Lymph nodes filter foreign substances from the lymphatic fluid and also produce lymphocytes. Lymphocytes are white blood cells involved in antibody production and other immune functions.

Blood is composed of cells (red blood cells = erythrocytes, white blood cells = leukocytes, and platelets = thrombocytes) and a liquid (plasma). Erythrocytes are the densest component and will settle to the bottom of a tube of blood. When blood is centrifuged, the column of "packed" erythrocytes at the bottom of the tube is measured (Figure 11.10) and compared with the total volume of the tube and expressed as packed cell volume (PCV). The normal PCV is 35% to 52% for dogs and 30% to 45% for cats. Leukocytes and thrombocytes form a "white" layer at the top of the erythrocyte column; this is referred to as the "buffy coat." This is normally a very thin layer, and the numbers of leukocytes and thrombocytes must be measured by specialized cell-counting microscope chambers or cell-sorting machines.

Erythrocytes are produced by bone marrow through a process known as hematopoiesis. Erythrocytes contain hemoglobin, a special iron-containing compound that functions to bind and carry oxygen and carbon dioxide to and from tissues. Oxygenated hemoglobin is red and deoxygenated hemoglobin is blue (thus blood in systemic arteries is a bright red and blood in systemic veins appears blue). Erythrocytes have genetically determined proteins on their surface that differ in individuals with different blood types. Dogs have 11 different blood group systems; however, the most important one is the "A" system. About 60% of dogs are A-positive, 40% are A-negative. Blood transfusions are safest when between individuals of the same blood type; however, A-negative dogs can be used as universal donors in emergency situations. Cats have an A, AB, or B blood group system. Very severe reactions (often fatal) can occur if a group B cat receives a transfusion containing group A blood. Thus, it is important to cross match blood prior to transfusions (Chapter 16).

There are many types of leukocytes. The bone marrow produces neutrophils, basophils, and eosinophils. These cells are also referred to as granulocytes as they have intracytoplasmic granules containing proteolytic enzymes and other substances to kill microorganisms and/or create inflammation. Eosinophils have roles in the killing of parasites and are also involved in many allergic reactions. Monocytes are nongranulated cells that function in phagocytosis, killing and removing microorganisms and diseased cells (when monocytes leave the bloodstream and enter tissues they are called macrophages). Lymphocytes are produced in the thymus and lymph nodes and are important components of the immune system.

**Figure 11.10** Hematocrit tube showing the major components of blood. *Courtesy University of Illinois.*

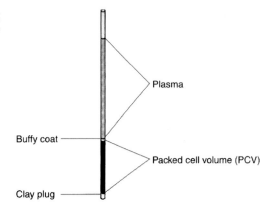

Thrombocytes (platelets) are important in clot formation. They "plug" holes in blood vessels to prevent blood loss and also produce proteins important in blood clotting.

Blood plasma contains approximately 90% water and 10% solids. The solids include electrolytes (sodium, potassium, chloride), calcium, phosphorus, bicarbonate, antibodies, blood clotting factors, hormones, vitamins, enzymes, proteins, lipids, carbohydrates and other nutrients, as well as soluble waste products such as ammonia, lactate, and urea. Levels of the various electrolytes and glucose are regulated by the endocrine system.

## 11.7    THE LYMPHATIC SYSTEM

The lymphatic system includes the tonsils, thymus, spleen, lymph nodes, lymphatic vessels, and lymphocytes. Functions of the lymphatic system include immunosurveillance and the movement of fluids from interstitial spaces to the circulatory system.

### 11.7.1    TONSILS

The tonsils are located in the pharynx near the base of the tongue. They are normally recessed inside the tonsillar crypts (folds of mucosa that cover the tonsils). Tonsils contain lymphocytes and are a site of production of antibodies against microorganisms entering the body through the nose or mouth. Inflammation of the tonsils is called tonsillitis.

### 11.7.2    THYMUS

The thymus is located in the cranial part of the chest (thoracic cavity) within the mediastinal septum (connective tissue separating the left lung lobes from the right lobes). This organ is composed of lobules of epithelial cells containing dense infiltrates of lymphocytes. The thymus is largest in young animals during the time when the immune system is developing and responding to antigens (foreign organisms and substances entering the body). The thymus is the source of many of the lymphocytes found in circulation; these are called T-lymphocytes. T-lymphocytes are essential for antibody production and also for the cell-mediated immunity involved in the destruction of microorganisms, foreign particles, and tumor cells.

### 11.7.3    SPLEEN

The spleen is located in the craniodorsal abdomen near the stomach. The spleen has two main components termed white and red pulp. White pulp is composed of nodules that produce lymphocytes and filter antigens from the blood. Red pulp stores, filters, and removes erythrocytes from circulation.

### 11.7.4    LYMPH NODES

Peripheral lymph nodes are located throughout the body; many are named according to the area of the body in which they are located (e.g., prescapular[1] lymph nodes, mandibular[2]

---

[1]Prescapular lymph nodes are located in front of the scapula (shoulder blade).
[2]Mandibular lymph nodes are located under the mandible (lower jawbone).

lymph nodes, axillary[3] lymph nodes, popliteal[4] lymph nodes, superficial inguinal[5] lymph nodes, and others). Other lymph nodes are located within body cavities (e.g., hilar[6] lymph nodes, mediastinal[7] lymph nodes, mesenteric[8] lymph nodes, sublumbar[9] lymph nodes, and others). Lymph nodes contain germinal nodules where lymphocytes are produced and are located within a network of lymphatic vessels. The lymph nodes filter lymphatic fluid to remove foreign substances and pathogenic organisms.

### 11.7.5   LYMPHATIC VESSELS

Interstitial fluid surrounds all the cells of the body. It originates from capillaries and excess fluid is collected by lymphatic vessels. Lymphatic fluid flows in response to gravity and movement of surrounding muscles and tissues. Lymphatic ducts empty into large veins in the chest and abdomen.

### 11.7.6   LYMPHOCYTES

Sites of lymphocyte production include the bone marrow, thymus, spleen, lymph nodes, and other lymphoid tissues. T-lymphocytes differentiate in the thymus and are important in recognizing foreign materials, pathogenic organisms, and tumor cells. Some classes of T-lymphocytes destroy infected and abnormal cells, and other classes of T-lymphocytes assist B-lymphocytes in antibody production. B-lymphocytes are involved in antibody production. Antibodies, also called immunoglobulins, are glycoproteins involved in protecting an animal from disease. Antibodies bind to foreign substances (or organisms) and initiate reactions to destroy that substance (or organism). Sometimes antibodies do more harm than good; examples of this are allergic diseases and autoimmune or immune-mediated diseases.

## 11.8   THE RESPIRATORY SYSTEM

The respiratory system includes a series of tubes and sacs that function together with the cardiovascular system to provide the body with the exchanges of carbon dioxide and oxygen required for cellular metabolism and life. Air enters the passageways during inspiration and exits during expiration. The nostrils of dogs and cats are called nares. Air entering the nares passes through folds of highly vascular mucous membranes covering the nasal turbinates. During passage through the nasal turbinates, the air is warmed, humidified, and filtered. The epithelial cells covering the nasal turbinates are ciliated and produce mucus. Foreign particles are caught in the mucus and swept caudally by the cilia and swallowed when they reach the pharynx.

When air reaches the pharynx, the epiglottis opens to direct air into the larynx and then the trachea. Vocal folds are located within the larynx; contraction of muscles associated with vocal folds causes them to vibrate during expiration creating audible sounds. The trachea is composed of fibrous connective tissue, smooth muscle, and a series of C-shaped cartilaginous rings. The C shape keeps the airways from collapsing during breathing and also facilitates expansion of the esophagus, which lies over the gap in the C, during eating. The trachea

---

[3] Axillary is equivalent to the human armpit, the area where the forelimb joins the chest.
[4] Popliteal lymph nodes are located between muscles of the caudal thigh, behind the stifle joint.
[5] Inguinal is equivalent to the human groin, the area where hind limb is adjacent to the abdominal wall.
[6] Hilar lymph nodes are located near the bifurcation of the trachea at the base of the heart.
[7] Mediastinal lymph nodes are located in connective tissue between the right and left lungs.
[8] Mesenteric lymph nodes are located in the connective tissue membranes between segments of the intestines.
[9] Sublumbar lymph nodes are located on the dorsal surface of the abdominal cavity, beneath the lumbar muscles.

**Figure 11.11** Dorsal (top) view of the lungs of a dog. *Illustration by Diana Nicoletti.*

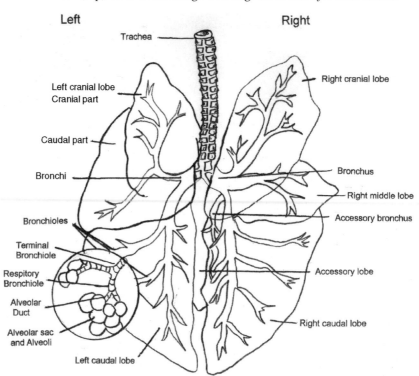

is lined with a ciliated epithelium that moves mucus and debris toward the larynx where it can be expelled during coughing. After passing the level of the heart, the trachea divides into two bronchi and then into a series of bronchioles. These continue to branch to form progressively smaller airways that terminate in an alveolar duct leading into an alveolar sac lined with alveoli (Figure 11.11). Each alveolus has a single layer of simple squamous epithelium surrounded by a dense network of capillaries. The alveolar and capillary walls form the respiratory membrane through which gas exchange occurs. The right lung divides into cranial, middle, and caudal lobes; the left lung consists of a partially divided cranial lobe and a caudal lobe; and an accessory lobe is located dorsocaudal to the heart. Cats have cranial, middle, and caudal lobes bilaterally plus the accessory lobe located dorsocaudal to the heart.

Air follows the path of least resistance. Inspiration involves contraction of intercostal muscles expanding the rib cage and movement of the diaphragm caudally; these movements stretch the lung tissue and air rushes in to fill the expanded cavities. Normal expiration is a passive relaxation of muscles. Lung tissue is elastic, and as the respiratory muscles relax, tension is released, the lungs then collapse forcing air out. Normal respiration is involuntary and is controlled by nerves originating in the medulla of the brain (respiratory center). The respiratory center is regulated by the carbon dioxide content of blood, body temperature, and other factors. Sneezing and coughing are also involuntary respiratory actions in dogs and cats. Barking, meowing, howling, growling, and other forms of vocalization represent voluntary movements of respiratory muscles (primarily involving muscles of the larynx). Panting is a form of open-mouth breathing used primarily to dissipate heat. Rapid movement of air over the moist mucous membranes of the tongue and mouth dissipates heat through evaporation.

Oxygen is essential for cellular metabolism. Cells use oxygen and produce carbon dioxide. In the lungs, oxygen diffuses from air into the blood where it combines with hemoglobin in erythrocytes to form oxyhemoglobin. When blood reaches the capillaries of the systemic circulation, oxygen diffuses from the blood into body tissues. Carbon dioxide is transported bound to hemoglobin (forming carbaminohemoglobin) and in plasma as a component of bicarbonate and carbonic acid. In the lungs, carbon dioxide diffuses into the air in the alveoli and is then expired into the atmosphere.

# 11.9   THE NERVOUS SYSTEM

The nervous system is a complex network that regulates most activities of the body, coordinates movements, and relays sensations. The two major divisions of the nervous system are the central nervous system (CNS) and the peripheral nervous system (PNS). Nerve cells are called neurons. One end of a neuron has small branches called dendrites that receive impulses from other nerves or from specialized receptors such as the sense organs. Impulses pass through a long fiber called an axon, which may pass the impulse to the dendrite of another neuron or to an effector organ such as a muscle cell. The junction between an axon and dendrites is called a synapse. Nerves that receive stimuli are called sensory or afferent neurons. These neurons carry impulses to the CNS. Nerve cells that carry impulses away from the CNS to muscles or glands are called motor or efferent neurons.

The CNS includes the brain and spinal cord. The brain controls most activities of the body, both voluntary and involuntary. The major divisions of the brain are the cerebrum, cerebellum, diencephalon, and brainstem. The composition of the CNS includes both gray and white matter. Gray matter is formed by the aggregation of the cell bodies of neurons within the CNS. White matter is composed of myelinated nerve cell processes (axons).

The cerebrum is the largest part of the brain. The surface of the cerebrum consists of numerous folds; the ridges are called gyri and the depressions sulci. The cerebrum is the decision-making center of the brain and is responsible for controlling voluntary muscles, making decisions, reasoning, and memory. It is divided into two cerebral hemispheres (right and left).

The cerebellum is located between the cerebrum and the brain stem. It is responsible for coordinating muscles involved in eating, vocalizing, body posture, walking, and running. Damage to the cerebellum results in abnormal gaits with exaggerated motions and incoordination.

The diencephalon is the interbrain and provides connections between the cerebral hemispheres and the brainstem. The diencephalon includes the thalamus and the hypothalamus. The thalamus serves as a relay station directing sensory information to the cerebrum. The hypothalamus regulates many body functions including heart rate, blood pressure, and body temperature.

The brainstem is composed of the mesencephalon (midbrain), pons, and medulla oblongata. The mesencephalon connects the diencephalon to the brainstem. The pons and medulla oblongata control breathing, swallowing, and vomiting.

The spinal cord consists of nerve fibers that transmit impulses between peripheral nerves and between peripheral nerves and the brain. Afferent (sensory) nerves enter the spinal cord through the dorsal roots. Efferent (motor) nerve fibers originate in ventral roots of the spinal nerves.

The spinal cord and brain are encased in specialized triple-layered membranes called meninges. The outer layer of the meninges is known as the dura mater. The space between the dura mater and the vertebrae is called the epidural space. Regional anesthesia involves the injection of anesthetic agents into the epidural space. The middle meningeal layer is called the arachnoid mater and the inner meningeal layer is the pia mater. The brain and spinal cord are surrounded by cerebrospinal fluid. This fluid provides cushioning for these delicate structures.

The PNS includes all of the nerves outside the brain and spinal cord. It is designed to receive sensory input, transmit it to the CNS, and transmit the appropriate response directives to organs and tissues of the body.

The cranial nerves (CN) include 12 pairs of nerves originating from various parts of the brain and brainstem: CN I = olfactory nerve (smell), CN II = optic nerve (vision), CN III = oculomotor nerve (eyelid movement), CN IV = trochlear nerve (eyelid movement), CN V = trigeminal nerve (sensory and motor for chewing), CN VI = abducens nerve (eye movement), CN VII = facial nerve (sensory and motor for facial muscles), CN VIII = vestibulocochlear nerve (sensory for equilibrium and hearing), CN IX = glossopharyngeal nerve (motor for swallowing and saliva secretion), CN X = vagus nerve (regulates heart rate; involved in vocalization, swallowing, gastrointestinal motility), CN XI = accessory nerve (motor for throat, neck, shoulder), and CN XII = hypoglossal nerve (motor for tongue movement).

There are numerous peripheral nerves throughout the body. Somatic nerves are those supplying muscles. The brachial plexus supplies the sensory and motor nerves for the forelimb. An important nerve branch of the brachial plexus is the radial nerve. This nerve provides motor input to the extensor muscles of the forelimb. Damage to the radial nerve results in paralysis and dragging of the front leg. The lumbosacral plexus is a bundle of spinal nerves innervating the hind limbs. The sciatic nerve is one of the most important nerves arising from the lumbosacral plexus. Damage to the sciatic nerve results in weakness or paralysis of the corresponding hind leg.

Autonomic nerves are those involved in the regulation of glands, blood vessels, the heart, and smooth muscles of the gastrointestinal system. These nerves control vital functions including breathing, heart rate, and digestion. Autonomic nerves operate in a check-and-balance system provided by sympathetic and parasympathetic nerves. The sympathetic nerves provide for fight-or-flight responses. These nerves increase heart and respiratory rates and also increase blood flow to vital organs and muscles while decreasing blood flow to the skin and digestive tract. The parasympathetic branch of the autonomic nervous system functions in opposition to the sympathetic branch. It slows heart and respiratory rates and promotes activity of the digestive tract.

Reflexes are protective mechanisms and involuntary reactions to stimuli. Examples include the menace reflex where poking a finger toward an eye prompts a reflex to close the eyelid; pinching a toe results in a reflex that pulls the leg up and away from the pincher. This withdrawal reflex occurs at the level of the spinal cord and does not require transmittal of impulses to the brain.

## 11.10 THE URINARY SYSTEM

The urinary system includes the kidneys, ureters, bladder, and urethra. The locations of these organs were shown in Figures 10.1 and 10.10.

The kidneys are paired and located ventral to the lumbar vertebrae. They are attached to the abdominal wall by blood vessels and nerves and are covered by the peritoneum [connective tissue lining of the peritoneal (abdominal) cavity]. The right kidney is located behind the last rib in the right dorsolateral abdomen. It is firmly embedded in connective tissue and can be difficult to palpate (feel) in dogs. The left kidney is more loosely attached and located caudal to (behind) the right kidney in the left dorsolateral abdomen. The kidneys are encased in a renal capsule composed of tough, fibrous connective tissue. The outer part of the kidney is the renal cortex. It is comprised of billions of tiny nephrons, which are functional units of kidneys. The inner part of the kidney is called the renal medulla. It is composed of numerous tubules that join together to form collecting ducts that lead to the renal pelvis. The renal pelvis funnels urine from the collecting ducts into the ureter.

Nephrons function in filtering waste products from the blood and are also essential in regulating the composition and volume of plasma in the body. Nephrons are composed of a glomerulus (a bundle of capillaries) surrounded by a spheroidal structure called Bowman's capsule. Blood entering the kidneys goes through the glomeruli and pressure forces water and dissolved solutes into Bowman's capsule; this fluid is called the glomerular filtrate. From Bowman's capsule, it passes into a series of convoluted tubules leading to the collecting ducts that feed into the renal pelvis. The convoluted tubules are surrounded by peritubular capillaries and venules. As the glomerular filtrate passes through the convoluted tubules, nutrients are reabsorbed into peritubular capillaries and returned to the bloodstream. The proximal part of the convoluted tubules reabsorbs glucose, amino acids, proteins, and electrolytes. The middle portion of the convoluted tubules has a U-shaped region called the loop of Henle; water and sodium are reabsorbed in this loop. By the time the fluid (now called urine) passes into the collecting ducts, the nutrients needed by the body have been reabsorbed leaving only waste products and excess water. One of the most important waste products in the fluid is urea, a by-product of protein metabolism. Urea causes serious problems for the animal if not removed by the kidney.

Each kidney has a single ureter that carries urine to the urinary bladder. The urinary bladder is an elastic sac located in the caudal abdomen. The walls of the urinary bladder are composed of smooth muscle that relaxes as the bladder fills with urine. At the neck of the bladder is a sphincter that prevents urine leakage into the urethra. During micturition (urination), muscles in the walls of the bladder contract, the sphincter relaxes, and urine flows through the urethra. The urethra is an elastic tube that leads from the bladder to and through the penis of males and to the vagina of females. In male dogs, the urethra passes through a bony structure called the os penis.

## 11.11 ORGANS OF SPECIAL SENSES

The special senses of sight, smell, taste, touch, and hearing influence how dogs and cats learn, and their interactions with other animals and the environment. Humans rely most heavily on vision and secondarily on hearing. Cats have well-developed visual acuity for hunting and also possess very sensitive senses of smell and hearing. Dogs tend to focus more on smells and sounds than on sight.

### 11.11.1 OCULAR ANATOMY AND PHYSIOLOGY

The eyeballs (globes) are located within a bony socket in the skull known as the orbit. Adipose (fatty) tissue and loose connective tissue surround the globe and provide cushioning. The eyelids (palpebrae) provide protection to the anterior exposed surface of the globe. The junctions of the upper and lower eyelids are referred to as the medial canthus and the lateral canthus. The epithelial lining of the eyelids is called the palpebral conjunctiva and is continuous with the ocular conjunctiva, which covers the exposed white portion (sclera) of the globe. Palpebral muscles open and close the eyelids. Dogs and cats have a third eyelid, called the nictitating membrane or haw, located near the medial canthus. The nictitating membrane rises and covers the globe during blinking and when the globe is retracted (sunken). Glands in the nictitating membrane contribute to tear production, and movement of this membrane helps spread a film of tears over the cornea. Additional tears are produced by a lacrimal gland located in the dorsal lateral palpebrae. Tears have a watery component that helps carry foreign particles away from the globe and an oily component that helps protect the globe from drying. A nasolacrimal duct located near the medial canthus drains excess tears into the nasal cavity. If the nasolacrimal duct becomes plugged, excess tears will overflow from the eyes staining the hairs near the medial canthus a reddish brown color.

The globe of the eye has a clear anterior portion called the cornea and a white posterior portion called the sclera (Figure 11.12). The iris is a muscle located within the eye; it forms a central hole called the pupil. The iris is the colored part of an animal's eye and

**Figure 11.12** Anatomy of the eye. *Illustration by Diana Nicoletti.*

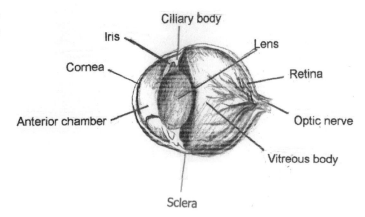

functions to regulate the amount of light entering the eye. When the iris is constricted, only a small amount of light can pass through. The space between the cornea and the iris is called the anterior chamber; it is filled with a fluid called aqueous humor. The lens is located directly behind the iris. The shape of the lens can be changed by contractions of muscles in the ciliary body. Changing the shape of the lens focuses (directs) light waves to different areas in the back of the eye. The posterior chamber is the cavity between the lens and retina. It is filled with a gelatinous substance called the vitreous body; this substance keeps the globe from collapsing. The wall of the globe consists of the retina (innermost layer), a pigmented cell layer, the choroid (vascular layer), and the sclera (outer fibrous layer). The retina is composed of specialized neural tissue with sensory receptors (rods and cones).

Dogs and cats have two types of pigmented tissue in the back of the eye (fundus). The first is the tapetum lucidum, which has brightly colored green and blue iridescent pigments that reflect unabsorbed light back to the rods and cones to increase vision in low light settings. Reflections of light from the tapetum lucidum can be seen as "eye shine" when a flashlight is directed toward dogs and cats at night. The region of the choroid near the ciliary bodies is darkly pigmented and referred to as the tapetum nigrum. Sensory cells of the retina generate a nerve impulse when exposed to light. Different wavelengths of light interact with different receptors. Rods transmit black and white images. Rods are very sensitive to even low levels of light and are the predominate type of visual receptors in both dogs and cats. Cones respond to color-generating wavelengths of light; dogs and cats can see yellow, green, and blue but have very few cones that respond to the longer wavelengths of orange or red light. Nerve impulses generated by the rods and cones travel through the optic nerve to the brain for interpretation of vision.

Cats have large eyes in relation to their body size. Their eyes are set forward on the head to provide a wide field of vision. The forward positioning of the eyes plus the curvature of their large cornea gives cats both binocular and lateral vision. Binocular vision is required for depth perception useful in catching prey. Lateral vision contributes a wide total range of vision to aid in locating prey and avoiding danger. The total range of vision of cats is approximately 280°. In comparison, humans have a range of vision averaging only 100°. Brachycephalic dogs have forward placed eyes with less curvature than those of cats; their field of vision is approximately 200°. Doliocephalic breeds have more laterally placed eyes and a 270° field of vision; these breeds have limited binocular vision, have trouble focusing on objects close to them, and have poor depth perception.

## 11.11.2   Anatomy and Physiology of the Ear

Dogs and cats can detect sounds inaudible to humans. Most humans can hear sounds up to 20,000 cycles per second (cps). Dogs can detect sounds with wavelengths of 40,000 cps and cats may hear those as high as 80,000 cps. The ears of cats and dogs are mobile allowing them to pivot to locate sounds over distances four times farther than humans can detect.

The external ear is composed of the pinna (ear flap), vertical canal, horizontal canal, and tympanic membrane (eardrum) (Figure 11.13). The middle ear cavity contains the openings of the Eustachian tube, three small bones (auditory ossicles: incus, malleus, and stapes), and the oval window. The middle ear cavity is located inside the tympanic bulla; it is filled with air. The Eustachian tubes provide a passageway between the middle ear cavity and the pharynx (throat). Air and fluids can pass through the Eustachian tubes to keep the pressure in the middle ear the same as that of the atmosphere and to prevent fluids from accumulating in the middle ear.

The inner ear is composed of the cochlea, vestibule, and semicircular canals. The cochlea is located adjacent to the oval window. Sound waves traveling through the ear canal vibrate the tympanic membrane. The malleus is attached to the tympanic membrane. As it moves, vibrations are transferred to the incus and stapes. The stapes moves through the oval window and strikes the cochlea generating fluid waves that stimulate sensory receptors and

**Figure 11.13** Anatomy of the ear. *Illustration by Diana Nicoletti.*

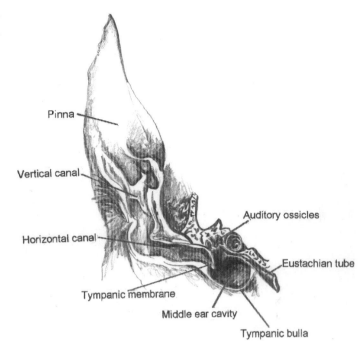

Pinna

Vertical canal

Horizontal canal

Tympanic membrane

Auditory ossicles

Eustachian tube

Middle ear cavity

Tympanic bulla

send nerve impulses to the brain. The nerve impulses arising from the cochlea are interpreted as sounds. The semicircular canals are lined with specialized sensory receptors involved in maintaining postural orientation and balance.

### 11.11.3 OLFACTION

Dogs rely heavily on scents in their relationships with one another. Scent discrimination is also important in cats. Dogs have over 220 million scent receptors in their nose, cats have approximately 200 million, humans only 5 million. Dogs and cats use sniffing to maximize detection of odors. Sniffing is accomplished through a series of short, rapid inhalations and exhalations. During sniffing, air is forced into a nasal pocket instead of flowing into the lungs. Scent receptors in the nasal pocket interact with inhaled molecules generating nerve impulses that are interpreted as smells.

Dogs and cats have an additional olfactory organ called the vomeronasal organ (also called Jacobson's organ). This organ consists of a pair of fluid-filled sacs located above the roof of the mouth. It is connected to the mouth and nasal cavities by the nasopalatine canal located behind the upper front teeth. Olfactory receptor cells in the vomeronasal organ send impulses to the region of the hypothalamus associated with sexual, feeding, and social behaviors. This organ is thought to be important in the detection of pheromones (body scents).

An example of the discriminatory acuity of dog's sense of smell is the ability to follow scent trails. Tracking dogs can detect scents at $10^{-12}$ molar concentrations. This is one million times greater than a human's ability to detect scents. Dogs can reliably discriminate against identical twins and can detect the presence of an odor following its mixture with other compounds.

### 11.11.4 TASTE

The organs of taste are specialized papillae (taste buds) located on the front and sides of the tongue, lips, and mouth. Taste buds of cats respond to foods that are salty, bitter, or acidic but have limited ability to respond to sugars. This may be a result of the low-carbohydrate diet normally consumed by cats. In contrast, the most numerous taste receptors in dogs are those that respond to sugars. The second most abundant are those that respond to acids (abundant in meats and meat products). Dogs seem to enjoy eating sweet foods but also readily consume meats.

## 11.11.5 TOUCH

Sensory receptors for touch respond to pain, pressure, warmth, cold, and vibration. Touch is the best-developed sense in newborn puppies and kittens. Neonates actively seek warmth and tactile contact. Dogs and cats use touch in social interactions (Chapter 20). Dogs frequently lick their owners and seek to be petted. Cats rub their heads on their owner's legs or body. The muzzles of cats and dogs have whiskers (vibrissae) that serve as important touch receptors. Large numbers of touch receptors are also located in the paws of cats; this may be one reason cats use their paws in playing and interacting with other animals.

## 11.12 THE ENDOCRINE SYSTEM

The endocrine system is comprised of glands and other tissues that produce hormones. Hormones are chemicals that stimulate reactions in specific body organs and tissues. Endocrine glands do not have ducts; they release hormones directly into the bloodstream. Hormones are carried in the blood until they reach receptors specific for that hormone. These receptors are located on a target organ or tissue. The interaction of the hormone with its receptor causes the target tissue to perform a specific function. Hormones are not stored in the body; they are produced in response to specific stimuli to produce a specific action. Hormones are involved in all functions of the body (Table 11.3). Concentrations of the various hormones in the blood can be measured by a variety of laboratory tests.

Hormone secretion is regulated by negative feedback systems. These are delicate systems of checks and balances to provide precise control of hormone concentrations and thus actions of the target tissues. For example, low levels of blood calcium stimulate parathyroid hormone secretion. Increased levels of parathyroid hormone increase calcium absorption from the intestines and calcium release from bones. Increasing levels of calcium then inhibit further secretion of parathyroid hormone. Negative feedback is provided by the increased blood calcium concentration.

## 11.12.1 HYPOTHALAMUS AND PITUITARY GLAND

The hypothalamus of the brain regulates and coordinates many endocrine organs by regulating the release of hormones from the pituitary gland. The hypothalamus also regulates body temperature, appetite, and thirst. The pituitary gland has two parts: the anterior pituitary (adenohypophysis) and posterior pituitary (neurohypophysis). Six hormones are secreted by the anterior pituitary and two by the posterior pituitary. These hormones have important roles in regulating growth, metabolism, reproduction, lactation, and water balance. The most common tumor of the pituitary gland is of adrenocorticotropin-secreting cells; this results in Cushing's disease (Chapter 14).

## 11.12.2 THYROID GLAND

The thyroid gland is located on the ventrolateral aspect of the trachea. It produces thyroxine, which regulates metabolism, and calcitonin, which helps regulate calcium and phosphorus concentrations. Two of the most common endocrine diseases in dogs and cats involve the thyroid gland. Dogs commonly develop hypothyroidism from atrophy or destruction of the thyroid gland (Chapter 14). Older cats commonly develop hyperthyroidism from nodular hyperplasia or neoplasia of the thyroid glands (Chapter 14).

**Table 11.3**
Hormones Secreted by the Endocrine Glands and Their Major Functions

| Endocrine Gland | Hormone Secreted | Major Physiological Function |
| --- | --- | --- |
| Hypothalamus | Gonadotropin-releasing hormone (GnRH) | Stimulates release of LH and FSH |
| | Corticotropin-releasing hormone (CRH) | Stimulates release of ACTH |
| | Thyrotropin-releasing hormone (TRH) | Stimulates release of TSH |
| | Growth-hormone-releasing hormone (GHRH) | Stimulates release of growth hormone |
| | Growth-hormone-inhibiting hormone (somatostatin) | Inhibits release of growth hormone |
| | Prolactin-releasing hormone (PRH) | Stimulates release of prolactin |
| | Prolactin-inhibiting hormone (PIH) | Inhibits release of prolactin |
| Anterior pituitary | Growth hormone (GH or somatotropin) | Promotes growth of tissues and bone matrix of the body |
| | Andrenocorticotropin (ACTH) | Stimulates secretion of steroids (especially glucocorticoids) from the adrenal cortex |
| | Thyrotropin or thyroid-stimulating hormone (TSH) | Stimulates thyroid gland to secrete thyroxine |
| | Prolactin (Prl) | Initiates lactation and induces maternal behavior |
| | Gonadotropic hormones | |
| | Follicle-stimulating hormone (FSH) | Stimulates follicle development in the female and sperm production in the male |
| | Luteinizing hormone (LH) | Causes maturation of follicles, ovulation, and maintenance of the corpus luteum in the female; causes testosterone production by the interstitial cells of the testes in the male |
| Posterior Pituitary | Oxytocin | Causes ejection of milk, expulsion of eggs in hens, and uterine contractions |
| | Vasopressin (antidiuretic) | Causes constriction of the peripheral blood vessels and water resorption in the kidney tubules |
| Thyroid | Thyroxine, triiodothyronine | Increase metabolic rate |
| | Calcitonin | Lowers the concentration of calcium in the blood and promotes incorporation of calcium into bone |
| Parathyroid | Parathyroid hormone | Maintains or increases the level of blood calcium and phosphorus |
| **Adrenal Glands** | | |
| Cortex (**shell**) | Glucocorticoids | Mobilize energy, increase blood glucose level, have an antistress action |
| | Mineralocorticoids | Maintain salt and water balance in the body |
| Medulla (**core**) | Epinephrine (adrenalin) | Stimulates heart muscles and the rate and strength of their contraction |
| | Norepinephrine | Stimulates smooth muscles and glands and maintains blood pressure |
| **Ovaries** | | |
| Follicles | Estrogens | Cause growth of reproductive tract and mammary duct system |
| Corpus luteum | Progesterone | Prepares reproductive tract for pregnancy, maintains pregnancy, and causes development of mammary lobule-alveolar system |
| Testes | Androgens (testosterone) | Cause maturation of sperm; promote development of male accessory sex glands and secondary sex characteristics |
| Pancreas (islets of Langerhans) | Insulin | Lowers blood glucose |
| | Glucagon | Raises blood glucose |
| Pineal | Melatonin | Controls seasonal breeding and hair growth |

### 11.12.3    PARATHYROID GLANDS

The parathyroid glands are located on the surface of the thyroid glands. Parathyroid hormone (parathormone) is responsible for regulating blood calcium and phosphorus concentrations. Tumors of the parathyroid hormone result in hypercalcemia.

### 11.12.4    ADRENAL GLANDS

The adrenal glands are located near the kidneys. The adrenal cortex produces cortisol and sex hormones. The adrenal medulla produces epinephrine and norepinephrine. Tumors of the adrenal cortex are another cause of hyperadrenocorticism (Cushing's disease) (Chapter 14). Tumors of the adrenal medulla result in hypertension (high blood pressure).

### 11.12.5    PANCREAS

The pancreas is located along the duodenum. It is unique in having both exocrine and endocrine functions. Its exocrine function is to produce bicarbonate and digestive enzymes (Chapter 9). Insulin and glucagon are hormones produced by the pancreas to regulate blood glucose concentrations. Insufficient insulin production or excessive production of glucagon (as by a glucagon-secreting tumor) results in diabetes mellitus (Chapter 14). Pancreatic tumors secreting insulin (insulinomas) can cause life-threatening hypoglycemia.

### 11.12.6    OVARIES AND TESTES

The ovarian follicles produce estrogens. Following ovulation, the corpora lutea produce progesterone. These hormones prepare the reproductive tract for pregnancy and are also responsible for female sexual behavior and mammary gland development. Testes produce testosterone and estrogens. These hormones are involved in sperm production and maturation as well as the secondary sex characteristics of males. (The anatomy and physiology of these and other reproductive organs are discussed in Chapter 10.)

## 11.13    SUMMARY

In this chapter, we discussed the anatomical structures and physiological functions of the major life-sustaining systems of dogs and cats. For sound pet health, quality of life, and longevity, each system must function in harmony with all other systems of the body. This is possible only if a specific cat or dog is blessed with an absence of genetic flaws and disease, receives a balanced diet, and has life-enhancing environmental conditions.

Equipped with the basic knowledge of anatomy and physiology needed to communicate with those working in the many disciplines and rewarding careers related to companion animals, let us turn to Chapter 12 and learn more about the care, management, and training of dogs and cats.

## 11.14    REFERENCES

1. Anderson, Wesley D. 1994. *Atlas of Canine Anatomy*. Philadelphia: Lea & Febiger.
2. Cunningham, James G. (ed). 2002. *Textbook of Veterinary Physiology*. 3rd ed. Philadelphia: W. B. Sauders.
3. Guyton, Arthur C. 1984. *Physiology of the Human Body*. Philadelphia: W. B. Saunders.
4. Hudson, Lola C. 1993. *Atlas of Feline Anatomy for Veterinarians*. Philadelphia: W. B. Saunders.
5. Kainer, Robert A. 2003. *Dog Anatomy: A Coloring Atlas*. Jackson, WY: Teton NewMedia.
6. Miller, Malcolm E. 1993. *Anatomy of the Dog*. Philadelphia: W. B. Saunders.

# 12 CARE, MANAGEMENT, AND TRAINING OF DOGS AND CATS

*When I play with my cat, who knows whether she is
not amusing herself more with me than I with her.*

Michel Eyquem de Montaigne (1533–1592)
*French philosopher and essayist*

## 12.1   INTRODUCTION/OVERVIEW

Choosing to become an owner of a cat and/or dog should be a decision made concomitantly with accepting the responsibility to provide care, management, training, and upkeep (to include appropriate medical care) essential to owning a healthy, trained, well-behaved pet. This entails making a major personal commitment of time, attention, and financial resources.

Except for primates, dogs and cats are among the most intelligent terrestrial mammals. Companion animal owners admire this level of innate intelligence that helps pets adapt to the idiosyncrasies of life in partnership with humans. Notwithstanding their notable intelligence, dogs and cats benefit from training. Unacceptable behavior is a major reason dogs and cats are surrendered to shelters. The vast majority of dogs and cats taken to animal shelters have not had the benefit of participating in basic training classes.

Companion animals depend on their owners for food, water, shelter, healthcare, affection, and training needed to enjoy lifelong companionship with their caretakers. These and other related topics are discussed in this chapter devoted to the overall care, management, and training of dogs and cats.

## 12.2   RESPONSIBLE DOG AND CAT OWNERS/NEIGHBORS

Companion animals may be kept to fulfill an emotional need for companionship, security, protection, social status, or simply to derive personal pleasure. Maintaining a pet for any of these purposes is ethically acceptable provided the pet receives a proper diet; has unlimited access to clean, fresh water; and does not become psychologically deranged to the point of becoming potentially dangerous or a public nuisance.

Owners should be knowledgeable about the proper care, raising, socializing, and training of their companion animals and also be aware of the social and public responsibilities of animal ownership. Indeed, owners incur an obligation to be compassionate stewards of their companion animals, never abusing, neglecting, or exploiting them.

In most cities, restraining dogs is a law. Still, many owners enjoy letting their dogs run loose, apparently thinking more about the pleasures their dogs are having than the risks that may be incurred. These risks include having their dog attack and/or bite another animal or person, dart into the street and be hit by a vehicle or cause an accident, knock people over from being overly friendly, leave unwelcome excrement on neighbor yards, chase and/or injure livestock, and many others. Additionally, the owners may be fined for failure to observe leash laws.

Because of the high risks associated with dog ownership, many insurance policies refuse to insure owners of certain breeds of dogs. Other insurance companies charge high premiums to insure dog owners. When a loose dog injures a person while violating a leash law, this violation presumptively provides a basis for half of the important elements of a negligence lawsuit. Moreover, in some jurisdictions the dog owner may be found guilty of a misdemeanor. So it is prudent to always leash a dog when walking it outdoors in cities. Additionally, when walking dogs in urban areas—for sanitation, human health, and aesthetic and good-neighbor reasons—owners should pick up their dog's feces. Using a plastic bag or a "pooper scooper" is recommended.

## 12.3 ATTENDING THE YOUNG PUPPY

Neonatal puppies are born toothless, blind, and deaf. Because the newborn puppy cannot generate sufficient heat to maintain normal body temperature, it depends on the warmth produced by its dam and littermates plus supplemental heat from heating pads/lamps to keep warm. During the first 2 weeks, the temperature of the nest box should be maintained at 85°F to 90°F.

Puppies are born with few antibodies to protect them from disease in early life. Antibodies are high molecular weight proteins too large to freely cross the placental barrier during fetal growth and development. This leaves the newborn puppies susceptible to common infectious diseases. Fortunately, colostrum (milk produced the first 2 days following whelping) contains high levels of antibodies that are absorbed across the immature intestinal mucosa of neonates during the first 24 to 36 hours of life. This safety feature of nature provides puppies with maternal immunity against many common infectious diseases including distemper (Chapter 14). Because this passive immunity will be lost early in life, it is important to establish a vaccination program for puppies with the family veterinarian.

Puppies sleep, eat, and stay near their mother and littermates the first 2 to 3 weeks. By about 2 weeks of age, puppies are hearing, seeing, and their deciduous incisors ("baby" teeth) are developing (baby teeth are commonly lost in about 16 weeks). Most puppies are able to stand and do some supplemental eating and drinking by 3 weeks of age. This period of 2 to 3 weeks represents the greatest change that will occur in such a short time during the entire life of a normal dog. The 3-week-old puppy is noticeably different than it was a week earlier or will be a week later. This remarkable change is almost as dramatic as the change of a tadpole into a frog. Puppies at 3 weeks are beginning to play, may growl, and even engage in playful fighting with littermates. They are also beginning to consume solid foods in addition to their mother's milk.

### 12.3.1 DEWCLAW REMOVAL

In breeds not requiring dewclaws (in accordance with each breed's standards), 3 days of age is an excellent time to remove the dewclaws. This will avoid the bloody, painful mess that can result if a dog catches a dewclaw on a bush when running through fields. This vestigial toe serves no useful purpose. In most cases, the dewclaw can be easily removed by clamping the base with a sterilized hemostat and twisting the claw off; any bleeding can be stopped using a silver nitrate stick or pressure bandage.

## 12.3.2    PUPPY SOCIALIZATION (3 TO 16 WEEKS)

Socialization is the process by which animals develop social relationships (bonds) with members of their own and/or other species (Chapter 20). Socialization is the most important behavioral adjustment/adaptation period in a puppy's life. A puppy's exposure to other animals and people and also to a variety of environments and noises during this period will determine in large part how well the puppy will adapt to lifelong living with people. If the puppy is not properly handled and exposed to various conditions during the socialization period, it is extremely difficult to rectify this important developmental need later.

By 3 weeks of age, puppies begin responding to sound and the sight of other animals and people. They explore the immediate area but seldom go more than 10 to 15 ft away from their nest. In the wild, foxes and wolves use a den or similar confined area so their offspring cannot stray. Dog breeders use crates or nest boxes for the same purpose.

## 12.3.3    WEANED PUPPY TO A NEW HOME

Puppies are commonly weaned by 6 or 7 weeks of age (Figure 12.1). Soon thereafter, they are usually sold and moved to a new home. The newly purchased puppy suddenly taken from its mother and littermates to a totally new environment needs a large dose of tender loving care early to prevent the stress of being lonely.

Being unfamiliar with the environment of its new home, the new puppy will be especially lonesome at night. Its natural reaction is to whine or bark throughout the night. The tendency of the owner is to yell "quiet" as a response to the lonesome puppy's request for attention (an owner act that may encourage incessant barking later as an attention getter). This can usually be overcome through the use of a small confinement cage at night, a piece of old bedding or a stuffed toy from the puppy's earlier nest box, and a source of noise such as a ticking clock or radio. A commercial radio station having a variety of noises is usually more effective as a moderator than an FM station broadcasting "peaceful" music.

When the puppy arrives at its new home, the owner should have a bowl of food (ideally the same food the breeder was feeding), clean water, and a few chew toys readily available. Additionally, a safe and comfortable place to sleep and play should be readily accessible. Most breeds of dogs are social animals; they enjoy human attention and welcome being part of a family. Owners of new dogs are encouraged to become a member of a local dog club. Much can be learned about the care, management, training, and related topics through exchanging experiences and by asking questions of club members.

The home should be "dog-proofed" before a puppy is brought home. Breakables and "chewables" should be placed out of the puppy's reach. Make sure electrical cords are inaccessible. Block access to poisonous plants and household chemicals. It is preferable to have

**Figure 12.1**    Collie puppies that have been weaned and are ready to be placed in new homes. *Courtesy Mike Nelson and Vern Horning, Graphics, American Nutrition Inc.*

a secure fence around the yard; if this is not possible, be sure the puppy is never allowed outdoors without careful supervision. Family members should decide who will be responsible for feeding, watering, exercising, cleaning up, and grooming the puppy. Establishing a regular schedule will help the puppy adjust to its new home.

### 12.3.4   GROWTH–MATURING PERIOD

By about 12 weeks, the puppy becomes intensely inquisitive, consumes increasing quantities of food, and develops efficient motions in walking, running, and other activities. The deciduous teeth are lost and replaced with permanent teeth starting around 14 to 16 weeks. Compared with earlier periods, growth is more gradual and the change into maturity is recognized as the age when sexual maturity is reached.

## 12.4   SEXUAL MATURITY OF DOGS

The first signs of sexual maturity in females usually occur at 9 to 11 months with vulvar swelling as the bitch enters proestrus (Chapter 10). A hemorrhagic (bloody) discharge follows, and the female may display considerable effort and ingenuity to meet male dogs. The decision should have been made previously whether to have her ovariohysterectomized (spayed) or kept confined to prevent an unplanned mating.

Making the decision whether to have a female dog spayed deserves serious consideration for several reasons. Although spayed females cannot be entered in American Kennel Club conformation dog shows, one of the most serious problems facing companion animals is that of overpopulation (Chapter 24). Neutering is highly recommended to prevent contributions to pet overpopulation and because of the substantial health benefits associated with neutering (Chapter 10). Neutered dogs can be shown in AKC performance events, trials, and shows.

## 12.5   FEEDING DOGS AND CATS

Less than half a century ago, dogs and cats relied mostly on scraps from the family's table with an occasional saucer of milk. Today, thanks to a progressive service-oriented petfood industry (Chapter 2) that benefits from a wealth of nutritional expertise (Chapter 9), pets enjoy the most convenient and balanced diets ever available to dogs, cats, and other companion animals. Indeed, virtually all commercial petfoods are balanced for essential amino acids, fatty acids, minerals, vitamins, fiber, and other dietary ingredients needed to ensure good health. Many diets are specifically designed for animals of different breeds, ages, reproductive status (e.g., pregnant or lactating), work levels (inactive or working), and caloric needs (e.g., "lite" or "reducing" diets). In addition, a variety of prescription diets are available to meet the specialized dietary needs of animals with certain health disorders (e.g., heart disease, kidney disease, diabetes mellitus, pancreatitis, food allergies, intestinal disorders, cancer, and many others). See Chapter 9 for information on the nutrition and feeding of dogs and cats.

## 12.6   OTHER ASPECTS OF CARE AND MANAGEMENT OF DOGS AND CATS

### 12.6.1   HOUSING

When provided with a secure sleeping space, food, water, and areas to exercise and eliminate, pets can develop into contented, well-adapted companions. Pets should have access to

**Figure 12.2** Cat relaxing in its padded cat bed. *Courtesy Paul E. Miller.*

dry, draft-free housing that enables them to get away from people and rest in peaceful, secure comfort (Figure 12.2). Those living as indoor pets almost always have satisfactory housing (many dogs and cats sleep on their owner's bed). Outdoor housing can be relatively simple and yet satisfactory: a draft-free dry shelter such as a wooden barrel with a raised floor, front cover, and dry bedding. Iron barrels are not recommended because they tend to invite moisture condensation resulting in damp or wet litter/bedding. Most commercial dog houses are satisfactory.

## 12.6.2 Exercise and Attention

Most puppies spend considerable time eating and sleeping and welcome as much attention as possible. Constructive play can include, for example, short sessions of ball retrieval (with balls made for dogs, not children, because ingested pieces of some children's balls can cause digestive problems). Short periods of training should incorporate "sit" and "come" commands linked with the pup's name (Figure 12.3). Exercise on a leash prior to bedtime is healthful for

**Figure 12.3** Dog given a "sit-stay" command. *Courtesy Christopher J. Byrne, Dogs Unlimited L. L. C.* (www.K9one.com).

both owner and pet. Caution: do not permit a child to sneak up and scare/startle your puppy; instinctive protective responses by the puppy may result in biting.

On returning from a walk in rain or snow, and before entering the residence, wipe the puppy's feet to help keep floors clean. Little time is required to teach a dog to participate in this effort; soon it will expect to have its feet cleaned and can be taught to lift one foot at a time to have it wiped.

Walking a puppy on a leash is an excellent way of exposing it to new environments and situations. These experiences help develop confidence and adaptation. Do not encourage contact with other animals as they may be carriers of and transmit diseases.

### 12.6.3   EQUIPMENT AND RELATED ITEMS

Each puppy/dog needs certain items of equipment beginning with food and water containers. Shallow stainless steel, nontippable bowls are the least destructible and are easily cleaned. A nonskid, waterproof, washable mat as a base for food containers saves cleaning time and work. It is also safer and appreciated by the dog.

Leashes are available in lengths varying from 16 inches to 18 feet. Both rolled and flat collars and chains are available. The collar should fit loose enough that a thumb can be placed between the collar and dog's neck. Fit should be checked frequently for proper looseness as the pet grows.

Training collars (often referred to as restraint collars or choke chains) should not be used except when a person is with the dog. Such a collar should be used on an as-needed basis only. Many professional trainers prefer using a Gentle Leader® (Figure 12.4). The Gentle Leader head strap provides immediate humane control and eliminates pulling, lunging, and tug-of-war without choking the dog. Affectionate training is the preferred way to work with a slow learner or even the somewhat stubborn dog. This approach is usually more effective than punishment. Moreover, it is the more likely way to ensure a good relationship between dog and owner and to keep the dog happy throughout its life. Additional aspects of training are discussed in Section 12.7.

Crates/pet carriers are useful in providing dogs and cats with a secure "safe" area during transportation (Section 12.8.7) and when guests are present in the home. Additionally, crates are useful as aids during house training (Section 12.7.7) and protect the owner's home and belongings from destructive behaviors when the owner is not present to supervise the pet. Crates should be large enough for the pet to stand, turn around, and lie down comfortably.

**Figure 12.4** Dog wearing a Gentle Leader head-piece. The Gentle Leader has two soft nylon loops: a nose loop and a neck loop. The nose loop encircles the dog's muzzle to demonstrate to the dog that the handler is its leader (a pack leader demonstrates its authority over subordinates by gently but firmly grasping the lower-ranking member's muzzle in its mouth). The neck strap applies pressure to the back of the dog's neck and calms the dog much like it re-laxed when its mother picked it up by the scruff of the neck. *Courtesy Sandra Grable, CVT.*

## 12.6.4  GROOMING SUPPLIES AND TOOLS FOR DOGS AND CATS

These should include a brush, flea comb, nail trimmer, toothbrush, and finger brush or gauze squares. It is well to keep a bottle of shampoo handy in case your dog or cat has fleas or rolls in the dirt. (Caution: make certain the label specifies that the shampoo is safe for the species. Some human shampoos are toxic to dogs and cats; and shampoos safe for use in dogs can be toxic to cats.) The amount of grooming a dog or cat requires depends largely on the breed. Length and texture of coat are important as well. Grooming is especially important to cats and is discussed in more detail in Chapter 13.

## 12.6.5  HEALTH CARE

The health of companion animals should be monitored continuously. Using the services of veterinarians and members of their professional team (veterinary technologists, veterinary technicians, trainers, groomers, and others) helps ensure the likelihood of good pet health. Routine vaccinations, regular checkups, and other health-related care and management procedures are discussed in Chapters 14, 15, and 16. Examples of important topics discussed in the above chapters include (1) preventing canine herpesvirus and canine parvovirus infections that can be fatal in young puppies; (2) internal and external parasite prevention and treatment (Chapter 15); and (3) how history, physical examinations, and laboratory tests can be used to identify and treat health disorders in dogs and cats (Chapter 16).

## 12.6.6  DENTAL CARE OF DOGS AND CATS

In its original wild state, an animal's teeth were important tools that enabled it to survive. With the assistance of their teeth, wild animals obtained food by attacking prey, dragging it down, killing it, and then biting through hair, hide, and muscle to get the preferred organ meats. Females that were the best hunters and those that maintained the best nutritional state bore more and healthier offspring. Males that were healthy and strong—those having good teeth for fighting—dominated the pack and sired the most offspring.

Most companion animals have teeth that may be grouped into four functions: (1) Incisors (front teeth) serve to nip something or to pick up an object. These teeth are small, not well-anchored in bone, and not very strong. (2) Canines (fang teeth) puncture, grasp, and hold. They are sharp, pointed, and anchored deeply in the upper and lower jaws. (3) Premolars serve to carry items such as bones. They oppose each other in a "pinking shear" pattern. Multiple points of contact ensure a secure grip. The fourth upper premolar and the first lower molar teeth are also called carnassial teeth and are the "shearing" teeth that tear through meats. (4) Molars are for crushing. These teeth have a flat occlusal surface to grind the foodstuff into a digestible consistency (cats have only four molars and lack the ability to grind food). See diagrams of teeth in Figures 3.3 (feline) and 11.6 (canine).

Other than ectoparasites, periodontal disease is the most common health problem in contemporary dogs and cats. Periodontal disease refers to a collection of plaque-induced inflammatory conditions of the periodontium or surrounding tissues. Gingivitis, an early form of periodontal disease, refers to an inflammation of the gingiva. Its clinical signs include red and/or swollen and bleeding gums. It is reversible. However, untreated gingivitis may develop into periodontitis with a loss of supporting tissues surrounding the teeth and can result in gingival pockets, recession of gums, bad breath odor, and tooth loss. (It can also lead to bacterial infections spreading to other parts of the body.)

The pellicle is a film of protein that forms on the tooth's surface after eating. It binds to enamel of teeth. Plaque is further development of the pellicle as bacteria, food debris, and

**Figure 12.5** Canine home dental care kit. This toothbrush is designed for use in dogs. The finger cot has a ribbed surface to "finger brush" the teeth of dogs objecting to the bristle brush. Only toothpaste formulated for use in pets should be used—human toothpaste contains too much fluoride and dogs object to its foaming. *Courtesy Dr. Karen L. Campbell.*

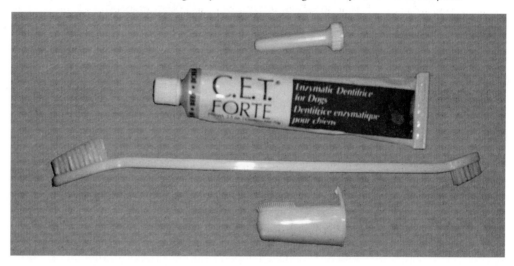

salivary components build up on the pellicle film. When calcium and phosphorus from saliva combine with plaque, it can cause the plaque to mineralize into calculus (commonly called tartar) and be deposited on teeth. This mineralized coral-like substance appears as hard white or light yellow material (may appear light to dark brown if it absorbs stains from food) and continues to accumulate until it is removed. Plaque forms on teeth within a few minutes of eating. People can help minimize the buildup of plaque by brushing regularly (Figure 12.5). Animals do not brush their teeth and less than one-half of owners do it for them. In the wild, dogs and cats had to rip and tear food to obtain their nutrients; this vigorous action helped clean their teeth. Even when owners fail to brush the teeth of pets, they can supplement manufactured diets with biscuits, rope toys, rawhide, and artificial bones to aid in keeping teeth clean.

Periodontal disease is common among cats and dogs. A good home dental care program should include daily tooth brushing, a balanced diet, and treats/toys that help improve oral hygiene (Chapter 13). Most pets also should have their teeth professionally cleaned on a regular basis (similar to people who despite brushing their teeth at least twice daily require regular cleaning by a dental professional). A good time to have the teeth cleaned is in conjunction with the pet's annual wellness check (preventive health examination). Proper dental care ranks among the most important yet frequently ignored preventive health measures among cats and dogs.

## 12.6.7 IDENTIFICATION

All cats and dogs should have a collar for identification. The pet's name, home address, and owner's telephone number should be displayed. Moreover, it is advisable to have an extra ID tag as pets sometimes lose them. Some owners now use microchips implanted by their veterinarian. This provides a permanent, tamper-proof form of ID. A rice-grain-sized chip bearing a special code, which can be cross-referenced with the owner's identity, is painlessly inserted just beneath the skin of the pet's neck. Another long-used form of pet ID is tattooing. Such a positive ID is useful for accurate record keeping, assisting in the return of lost and stolen pets, and in making specific identification of pets in shows, sporting events, and sales. Advantages of microchips and tattoos include their permanency and safety (versus collars and tags that can be lost or may become entangled and choke the pet). Some organizations, including the Orthopedic Foundation for Animals (OFA), require a positive method of identification (microchip, tattoo, or DNA profile) before they will issue a certificate giving results of tests for hereditary diseases (e.g., OFA hip ranking—see Chapter 14 for information on OFA's hip dysplasia and other registries).

## 12.7 DOG DISCIPLINE AND TRAINING

Many pets are abandoned and/or destroyed because of behavioral problems. Professional trainers and animal behaviorists can help owners learn useful techniques for correcting behavioral problems and thereby save the animal. Owners of well-trained companion animals derive great pleasure and take personal pride in the good behavior and performance of their pets.

Effective training of a dog is rooted in its respect for discipline and its proper application. Puppies respond quickly to discipline when it is immediate and the puppy associates the correction with the error. Thus, "in-the-act" discipline is effective; punishment for an act committed minutes before is not. Except for the intrinsic value the owner derives in venting his/her feelings, it does no good to plunge a dog's nose into a pile of "misplaced" excreta and then punish it. Indeed, hitting a puppy or rubbing its nose in feces causes the puppy to be afraid of the owner. Animals simply do not understand punishment after the fact.

Dogs developed from pack animals with stringent social structures (Chapter 20). As such, dogs are happiest when they know their place in the social structure of their households. Someone must be the pack leader (alpha position)—family life is safest and happiest when the owner establishes himself/herself as the alpha individual.

### 12.7.1 REWARDS AND PUNISHMENT

Recognizing the difference between reward and punishment is important. When a puppy fails at first to come when called and then suddenly decides to come, it should not be punished. Because the puppy does not understand that the discipline was for early refusal, it may believe the punishment was for coming. Similarly, praising a dog that is not doing something positive lessens the value of praise.

When a puppy or adult dog does something to please its owner, a reward should be given. The reward may be in the form of praise, which pleases the dog and encourages the same good deed to be repeated, or the reward may be a treat that dogs enjoy. Conversely, if a dog misbehaves, punishment should be administered. However, punishment is in order only when the dog is caught in the act. Otherwise, the dog will likely fail to associate the two and wonder why it is being punished. Punishment at the wrong time confuses dogs. A verbal reprimand is usually sufficient. Simply saying "bad dog" and not petting it for a short while often suffices.

Puppies should be taught at an early age. To seize a dog's attention, clap your hands and shout "no" if it is doing something undesirable. At that point, hold its collar and turn the puppy so it will see you. Then shake your finger in its face and sternly say "no." If a more serious reprimand is necessary, gather the skin on both sides of the neck in your hands and shake the puppy gently from side to side; that will resemble punishment the bitch sometimes gives a young puppy. It is usually effective. Caution: never strike a puppy with a newspaper. It may notice many people carrying newspapers throughout its life, and if newspapers are associated with punishment, the dog can become upset when around newspapers.

The name given a dog can be important. If a simple, easy-to-understand, single- or two-syllable name that does not sound like common commands is selected, the puppy will soon learn its name. It is advisable to use that name often and reinforce the puppy's response behavior with a treat. Names like "Joe" are too similar to "no" to be good choices.

## 12.7.2  TEACHING OLD DOGS NEW TRICKS

The old adage "you cannot teach old dogs new tricks" is simply untrue. For example, a dog named Autumn was labeled "stupid" by her family for more than 10 years. Then she was adopted by an especially caring, patient person who took a great deal of interest in Autumn and resolved to demonstrate that there is above-average potential in seemingly below-average dogs (just as is true in humans). Patiently and persistently she taught this senior dog to excel in her obedience classes in competition with puppies and much younger adult dogs. Indeed, by age 14 Autumn won awards and recognitions for ranking first in obedience classes. Older dogs are frequently calmer, quieter, and more steady and predictable than a puppy. They are also better able to focus on what they are being taught. Many previously owned dogs were neglected and in a new "caring" environment such older dogs respond quickly by learning "new tricks."

## 12.7.3  LEASH TRAINING

For dog safety, owner patience/pride, and neighbor comfort and good relations, dogs should be leash trained early. In most areas, it is well for the dog to walk on the owner's left side, the position required in AKC competition.

When dogs walk near traffic, among people or other dogs, and in public places, they should always be on a secure leash. Establishing a good relationship between trainer and dog is important. Indeed, for the most successful training possible, both the trainer and dog must have complete confidence in each other.

## 12.7.4  CLICKER TRAINING

Clicker training is based on the same principles of animal behavior as whistle training for marine mammals—the clicker (or whistle) is used to shape behavior. A trainer watches for the desired behavior then immediately "marks" the behavior with a click followed by giving a reward (praise and/or treat). Dogs (and other animals) quickly learn that the clicker sound means a reward is coming. When they learn the association between their behavior and the clicker, they become enthusiastic partners in their own training. Clicker training is a positive, motivational method that promotes a good working relationship between the pet and its owner. With deaf dogs, a light flash is substituted for the clicker noise. Clicker training has been widely promoted by Karen Pryor, author of *Don't Shoot the Dog: The New Art of Teaching and Training* (1999, Bantam). Additional information on this method of training can be found at http://www.clickertraining.com/.

## 12.7.5  PROFESSIONAL OBEDIENCE TRAINING FOR DOGS

Most kennel clubs and many veterinary practices offer obedience classes for dogs (Figure 12.6). Several class levels are available.

**Figure 12.6**  Outdoor training of dogs. Willow Creek Pet Center, Sandy, UT. *Courtesy Dr. Rick Campbell.*

**Figure 12.7** Dog watching its handler for a release from the "down" position. *Courtesy Christopher J. Byrne, Dogs Unlimited L. L. C.* (www.K9one.com).

*Puppy class.* This class offers the basics of obedience to dogs 12 to 20 weeks of age. The class teaches sit, down, stand, come, how to walk safely and obediently on a secure leash, how to take food gently from a person's hand, learning not to jump on people, and the basics for discipline that work (e.g., looking dogs directly in their eyes when giving a reprimand).

*Beginning obedience.* This class teaches well-accepted foundation words for dogs 5 months or older. Included are learning to sit, down, wait, come back ("recall"), sit-stay, down-stay (Figure 12.7), heel, come, retrieve/fetch, learning to never jump on people, and how to gently and safely take food from a person's hand. It is an "all on-lead class" for basic control and improved understanding of commands. At the end of these classes, the dog should be trained well enough to pass the AKC Canine Good Citizen (CGC) test and earn the CGC certificate. Included in the CGC test are such things as calmly allowing a stranger to approach, walking naturally on a loose lead, walking through a crowd, sitting for examination, calmly meeting a strange dog, and staying calm when a door slams or a jogger runs by the dog.

*Intermediate novice.* This class includes thoroughly teaching heeling with automatic sits,[1] pace changes, and the beginnings of off-leash training and progressive understanding of stand-stays, sit-stays, down-stays, and recalls with a finish.[2] Students successfully completing this class are expected to develop an excellent understanding of their dog as well as an effective method of communicating and establishing a strong bond between handler and dog.

---

[1]An automatic sit is required in AKC Obedience competitions. When the handler stops, the dog must sit at the "ready" position—adjacent to the handler's left foot—without receiving a "sit" command from the owner. It is called an automatic sit because the dog sits without being told to do so.
[2]A recall with a finish involves the handler leaving the dog in a sitting position and walking to the opposite side of the ring. When the handler calls the dog, it quickly goes to the handler and sits directly in front of him/her. Next the judge tells the handler to "finish," the handler gives a command for the dog to move into the "ready" position—sitting adjacent to the handler's left foot.

**Figure 12.8** Dog sitting in the "ready position" for off-lead heeling exercises. *Courtesy Kennelwood Pet Suites, Kennelwood Village, St. Louis, MO* (www.kennelwood.com).

*Competitive novice training.* Dogs in this class are trained to compete at AKC, United Kennel Club (UKC), and other registry dog obedience trials. Exercises in novice classes include heeling with automatic sits (on- and off-leash), staying in position by the owner during pace changes, standing for examination, 1-minute sit-stays, 3-minute down-stays and coming when called, and sitting in front of the owner and then upon command returning to sit at the "ready" position (Figure 12.8). Dogs receiving a satisfactory score on these exercises at three AKC Obedience Trials earn the title of Companion Dog (CD).

*Open obedience training.* Dogs that have earned their AKC CD title are eligible for competition in Open Obedience classes. Open work adds retrieving a dumbbell, jumping hurdles, and broad jumping. All heeling is done off-leash. Dogs receiving a satisfactory score on these exercises at three AKC Obedience Trials earn the title of Companion Dog Excellent (CDX).

*Utility obedience training.* The third and most advanced level of obedience training is known as Utility. Scent discrimination and responding to hand signals are new exercises at this level. Dogs receiving a satisfactory score on these exercises at three AKC Obedience Trials earn the title of Utility Dog (UD).

*Agility training.* Dog agility is a sport comprised of a course of many different obstacles (Figure 12.9). It was initially patterned after equestrian events and combines handler control, agility, and confidence. Agility began as an exhibition sport in Great Britain and was introduced to the United States in the late 1970s. It is now the most rapidly growing dog sport; it is fun for spectators as well as the competing dogs and handlers. Prerequisites for being accepted in this class include (1) dogs knowing sit and down on first command, (2) 1-minute sit-stay and 3-minute down-stay, (3) recall with distractions, and (4) off-leash control.

Agility training is of special interest to owners of unregistered purebred dogs who wish to participate in this increasingly popular sport. Interestingly, the UKC recognizes American

**Figure 12.9**  Shetland Sheepdog crossing the "dog walk" obstacle in an agility course. *Courtesy Tom Schaefges Photography.*

Mixed Breed Obedience Registration (AMBOR, www.amborusa.org) as the leading mixed-breed registry in the United States. The AKC offers opportunities for unregistered purebred dogs to compete in AKC events. One such event is the sport of agility (visit United States Dog Agility Association, Inc., USDAA, www.usdaa.com). Agility is a popular sport for those who adopt dogs from shelters or who purchase purebred dogs without registration papers and who would like to participate in performance dog events. Many owners of registered dogs also enjoy agility training (Figure 12.10) and showing.

Commonly used obstacles include an A-frame, see-saw (Figure 12.11), pipe tunnel, tire or hoop jump (see Figure 13.15 and Color Plate 11K), weave poles (see Figure 13.16), pause table, dog walk (see Figure 12.9), and various jumping hurdles (see Figures 13.11 and 13.17). Dogs are timed during their runs through the courses.

***Others.*** Dogs can be trained for many other performance dog events including tracking, field trials (retrievers, spaniels, pointing breeds, Basset hounds, Dachshunds, Beagles), herding (breeds in the Herding Group, Rottweilers, Samoyeds), hunting (pointing breeds, retrievers, spaniels), lure coursing (designed for sighthounds), Earthdog (events designed specifically for small terriers and Dachshunds), and Coonhound events (night hunts, field

**Figure 12.10**  Agility jump training of a German Shepherd Dog. *Courtesy Christopher J. Byrne* (www.K9one.com).

**Figure 12.11**  Dog crossing a see-saw obstacle on an agility course. *Courtesy Tom Schaefges Photography.*

trials, and bench shows limited to Coonhounds). Dog Olympics, Flyball, and Frisbee® competitions are other events enjoyed by owners of mixed breeds as well as purebred dogs (Figure 12.12). Many performance events for dogs are described in Chapter 13.

## 12.7.6   TRAINING DOGS FOR SPECIAL PURPOSES

Dogs can be trained individually for many special purposes. Training dogs to serve as *guide dogs* for the sightless is well-accepted, long-standing, and enormously important. *Therapy dogs* are trained for service in nursing homes and hospitals. *Mobility assistance dogs* help the physically impaired (Figure 12.13) and may carry a pack, pull a wheelchair, operate switches (e.g., light or elevator switches), or recover dropped items. *Search and rescue dogs* perform rescue work (as was done in conjunction with the September 11, 2001, terrorist attacks on the New York City World Trade Center). *Protection dogs* are used to patrol property or provide

**Figure 12.12**  German Shepherd Dog showing enthusiasm in catching a Frisbee. *Courtesy Paul E. Miller.*

**Figure 12.13** Mobility assistance dog opening a door for its owner. *Courtesy Paws with a Cause.*

personal protection for their handlers. *Hearing dogs* can alert deaf persons to selected sounds such as alarm devices (e.g., clocks, fire alarms, door bells, telephones, infant distress calls). Many types of specialized service work performed by dogs are described in Chapter 21.

### 12.7.7    HOUSE TRAINING CATS AND DOGS

Cats are easier to house train than dogs. Most cats learn to use an indoor litter box within days. House training cats is discussed in Section 20.6 (Elimination Behavior). One alternative and training technique to use but not discussed in Section 20.6 is the cat door. Teaching a cat to use a cat door involves showing it that each side of the flap can be made to swing to and fro with a paw. The cat is encouraged to go through the hole when the flap is open (in return for a tidbit and lavish praise). It is important to show the cat that it can go through the opening in both directions. Such training can be turned into a game by feeding a scrap of paper on the end of a string through to the cat's side and jerking the paper through. Cats are intelligent, curious pets and soon learn how to access and use the cat door. Interestingly, English mathematician and physicist Sir Isaac Newton (1642–1727), discoverer of the principles of gravity, also invented the cat door.

House training puppies usually requires a longer time than cats but is made easier by practicing the following three tips: (1) because puppies commonly eliminate within 20 minutes after awakening, eating, drinking, or playing, take them to their elimination site after each of the four activities noted above; (2) return your puppy to the same spot each time as the familiar smell is a reminder of what he/she is supposed to do there; and (3) when outdoors, play with your puppy after it has eliminated, not before. This encourages the puppy to "do his/her thing" so playing can begin. Dogs do not like to eliminate in their sleeping areas, thus a useful tool during house training is the confinement of the puppy to a crate when it is not being observed by the owner. Further discussion of house training dogs can be found in Section 20.6 (Elimination Behavior). Many dogs can be trained to eliminate on command.[3]

Carnivores in general are selective and fairly predictable regarding where they leave their excreta. This "naturally tidy" behavior contributes to the popularity of cats and dogs as house pets. Certain other pets such as birds and rodents are less dependable in this regard

---

[3]Smith, M. L. 2003. *You Can Teach Your Dog to Eliminate on Command.* Port Washington, WI: Seaworthy Publications.

and present the problem of undesirable eliminative behavior. This is commonly solved by keeping them in cages. Some herbivores such as goats and horses are regarded as companion animals by many owners but because of their eliminative behavior need to be housed outside the human residence.

# 12.8 ATTENDING TO KITTENS/CATS

The average newborn kitten is 4 to 6 inches long and weighs less than a quarter of a pound. Its eyes remain closed for 7 to 10 days. All kittens' eyes are grey-blue when they do open and begin changing color at about 12 weeks of age. Being unable to regulate their body temperature for 2 to 3 weeks, neonatal kittens move toward warmth—their mother. They are essentially immobile the first 2 weeks, begin crawling at around 3 weeks, and start walking at 3 to 4 weeks of age. It is important for kittens to begin socializing with people by 2 weeks and extend this socialization period for 7 or more weeks.

## 12.8.1 WEANED KITTEN TO A NEW HOME

Kittens are frequently weaned on solid food and taken to a new home by 10 to 12 weeks of age. They are often litter-box trained by that age, although with the move may temporarily forget their litter training and need to be reminded.

Food, clean water (in a shallow container), and a comfortable dry bed should always be available. Separate feeding site from sleeping location and avoid loud noises.

Kittens tend to be somewhat fragile and should be handled with care. They should not be squeezed as their bones are delicate and easily damaged. It is best to pick them up by the scruff of the neck while sliding the other hand under their midsection.

By 12 weeks, kittens should be eating a good-quality commercial kitten growth-promoting diet that provides the nutrients required to grow and develop into healthy cats. Because the kitten's stomach is small, they should be fed small amounts of food (three to four times daily). By 10 months, most kittens are eating adult-size meals twice daily and can be introduced to a wider range of food. See Chapter 9 for additional information on the nutrition and feeding of cats and kittens.

Because young kittens are prone to ear mite infections, this and other health-related functions must be watched carefully and veterinary attention given on an as-needed basis. A vaccination and deworming schedule should be set up early (Chapter 14) and followed closely.

## 12.8.2 INDOOR VERSUS OUTDOOR CATS

Cats are safest indoors. Those roaming freely outdoors frequently fall victim to fighting, predators (e.g., dogs, coyotes, hawks, owls, eagles), and infectious diseases. They can bring home fleas, lice, ticks, and mites and may be injured, poisoned, or killed by cars. Moreover, outdoor cats have a higher probability of being stolen or getting lost. They can decimate a local population of songbirds. Benefits of being outdoors include increased opportunity for exercise and a lower likelihood of obesity. Indoor cats are more likely to become bored and lazy.

Those electing to have an outdoor cat should consider the following steps to help ensure its safety and well-being: (1) Identification via a breakaway collar with ID that bears your name, address, and telephone number. Your veterinarian can implant an ID microchip beneath your cat's skin (in case the collar is lost or stolen). (2) Vaccinations are important for all cats—indoor and outdoor. Cats should be vaccinated and receive parasite preventives. (3) Contraception—have your cat neutered to prevent unwanted pregnancies and undesirable traits such as urine spraying (intact males) and loud vocalization (intact females during estrus). (4) Remove toxins from your garage and lawn. These include antifreeze, insecticides, rodent poisons, herbicides, and other poisons. (5) Protect birds and other animals by placing a bell on your cat's collar to alert them of your cat's presence.

### 12.8.3    YOUR RESIDENCE AND YOUR CAT

Cats are curious creatures with abilities to jump and explore things in high places. It is advisable to protect/secure breakables from kittens and young cats. The same applies to sharp kitchenware, detergents/cleansers, open washing machines/dryers, electric cords and wires that the kitten/cat can chew, fireplaces without screens, fish bowls/aquariums, bird cages, and other cat temptations.

Just as children enjoy quality toys and amusements, the same applies for kittens and cats. It is entertaining and fun to watch kittens/cats exploit their natural instincts to stalk, jump, chase, and pounce. As a safety precaution, small children should not be left alone with a cat.

### 12.8.4    PLANTS AND YOUR CAT

Green plants add a colorful touch to homes but can be dangerous for cats (e.g., lilies). Eating an excessive amount of any plant material is not good for pets and may cause a gastrointestinal upset. If you keep plants that are toxic to cats (Chapter 14), keep them out of the cat's reach. Should your cat eat part of a poisonous plant, take the cat and a portion of the plant to your veterinarian immediately! To learn more about which plants are safe/unsafe for cats, read the "Plants and Your Cat" article (www.cfainc.org/articles/plants.html). This article discusses the symptoms of poisoning and what action to take should they occur.

### 12.8.5    TRAINING CATS

Cats are intelligent, have phenomenal memories of people and places, and are trainable. Unlike dogs that possess an inherent desire to please, cats are more independent and pragmatic, often taking a particular course of action only if they perceive an advantage to themselves. This means that training for cats must be largely reward based.

Although difficult to train, cats are smart. Unlike dogs, cats are not pack animals so they have no need to impress superiors. That may explain why they have no compulsion to obey training instructors.

Training of cats begins in early life when the mother teaches her kittens how to stalk prey, fight, and protect themselves. Kittens practice their mother's teachings by lying motionless in the grass then pretending to spring on prey or an enemy. They attempt to frighten each other by arching their back, puffing themselves up, and walking stiff-legged. Kittens strengthen their bodies by wrestling with littermates and their mother (Figure 12.14).

Cats frequently use furniture to exercise their claws. They enjoy hooking their claws in a cloth-covered chair or sofa and then stretching and/or scratching. Owners can help train cats to leave furniture alone by providing easy access to a scratching post. These are available in most pet stores or a piece of wood can be wrapped with soft carpet.

### 12.8.6    ON NEUTERING CATS

Once a cat is neutered (commonly after 8 to 10 months of age), the personality differences between males and females are minimized. If, however, you do not plan to have your cat neutered, certain distinctive sex-linked behaviors are worthy of consideration. Male cats (toms) may spray or mark their territories with more frequency. Outdoor toms tend to roam more and

**Figure 12.14**   Kittens playing while their mother rests. *Courtesy Paul E. Miller.*

get into fights with other cats. Females (queens) vocalize when they are in estrus (heat), sometimes at all hours of the night. In either gender, these behaviors are modified by neutering.

## 12.8.7   OTHER CAT MANAGEMENT TIDBITS

Many of the care/management aspects of dogs discussed above also apply to cats.

*Exercise.*   It is important that cats—juniors and seniors—exercise regularly to promote good health and decrease the likelihood the pet will become lazy and lethargic. Schedule play sessions for your cat to provide planned exercise and reduce boredom.

*Claw trimming.*   Unless your cat has been declawed, it should have its claws trimmed regularly. By applying light pressure behind the toes you can expose the claws for easier trimming. Snip off the lighter colored tip of the claw—not the darker more sensitive quick.

*Fresh air and sunlight.*   Cats enjoy lying in a sunbeam, and when security and outside temperatures permit, a window can be opened to provide access to fresh air. Placing a bird feeder outside the window provides entertainment for cats. Too much sunlight can be hazardous, especially for cats with white ears—sunlight damage to nonpigmented skin can result in skin cancer (squamous cell carcinoma).

*Toys and treats.*   Cats are fond of toys with fur or feathers, toys that move, toys that contain catnip, and toys they can scratch. They also welcome treats made of meat or meat-flavored substitutes.

*Cat carriers.*   Although cats may dislike being transported in a carrying cage, it is the safest and best means of controlling them. Loose animals in a car can distract the driver, damage the vehicle's interior, and permit an unapproved exit when a door is opened. Caution: never leave a cat (or dog) in a parked car during hot weather; temperatures can soar to over 130°F within minutes, resulting in heatstroke and death (Chapter 14).

   For family travel and ease of transporting to/from the veterinarian, it is advisable to use a cat carrier. For extended drives or trips by air/sea, a carrier approved by the International Air Transport Association (IATA) is desirable.

## 12.9   HUMANE HANDLING, CARE, AND TREATMENT OF DOGS AND CATS

Legally, Code of Federal Regulations (CFR, www.gpoaccess.gov/cfr/index.html), Title 9 (Animals and Animal Products), Chapter 1 (Animal and Plant Health Inspection Service, Department of Agriculture), applies to animals on *display*, those undergoing interstate shipment, or critters used in research and provides veterinary healthcare professionals a baseline for applying their professional experiences and opinions with confidence. The following are selected excerpts from 9CFR3 (Part 3—Standards), Subpart A (Specifications for the Humane Handling, Care, Treatment, and Transportation of Dogs and Cats):

- Dogs and cats must be provided adequate shelter from all elements at all times to protect their health and well-being. Housing must be designed . . . kept in good repair, made of materials that are readily cleaned and sanitized, and protect the animal from injury.
- Cleaning—hard surfaces must be spot-cleaned daily and sanitized, and floors must be spot-cleaned . . . to ensure . . . freedom to avoid contact with excreta.
- Ambient temperatures in the facility, or during transportation, must not drop below 50°F for dogs and cats not acclimated to lower temperatures, nor rise above 85°F, for more than four consecutive hours, except as approved by a veterinarian.
- Dogs and cats must be fed at least once daily; food must be uncontaminated, wholesome, palatable, and of sufficient quantity and nutritive value to maintain the normal condition and weight of the animal.
- An effective program to control insects and external parasites must be established and maintained as to promote the health and well-being of the animals and reduce contamination by pests (includes other birds or mammals).

The original Animal Welfare Act (AWA) was enacted to protect draft animals and was soon applied to sweatshops abusing children. The implementation provisions of the AWA are updated frequently in CFR Title 9 and should be in every veterinary practice library as an operational resource. The following guidelines are excerpts from 9CRF2.131 (Handling of Animals):

- The animal should not show trauma, stress, overheating, excessive cooling, behavioral stress, physical harm, or unnecessary discomfort.
- Deprivation of food or water shall not be used to train, work, or otherwise handle animals . . . animals will receive full dietary and nutritional requirements each day.
- An animal shall not be subjected to any combination of temperature, humidity, and length of time that is detrimental to the animal's health or well-being.

## 12.10   ADDITIONAL CONSIDERATIONS

Become familiar with your pet's normal patterns of eating, drinking, sleeping, eliminating, and exercising. Any deviations in these may indicate illness and should be reported to your veterinarian. You should become familiar with the common health problems of the pet's breed and obtain information to help you recognize the symptoms of these disorders. For example, giant breeds of dogs are prone to bloat, and owners of these breeds must be alert for symptoms such as retching, abdominal distension, or discomfort following eating. Cocker Spaniels are predisposed to ear infections, and owners should examine the ears regularly (including a "sniff" test for odors associated with infections). Older cats are prone to thyroid hyperactivity, and owners should be alert to weight loss despite an increase in food consumption.

Be prepared to take care of your pet in a disaster such as fire, flood, hurricane, tornado, or earthquake. Although human safety should come first, have an evacuation plan that includes your pets and emergency supplies of food and water plus a first aid kit. Make advance arrangements for a friend or family member to take care of your pets if you should have a sudden illness or accident. A list of general care instructions should be available. Also make provisions in your will for someone to take care of your pets.

Most importantly—enjoy your pet! Dogs and cats love to play. Set aside time each day for play sessions. Talk to your pet—they will enjoy your praise and companionship. Include your dog in family activities—it will enjoy going on walks and playing with you on beaches.

## 12.11 SUMMARY

Most persons pursuing a career in veterinary medicine, companion animal biology, and related fields have had personal experiences with dogs and cats and can relate with many aspects of the care, management, and training topics/discussions of this chapter.

Training a well-mannered and well-behaved, loyal, affectionate dog or cat means loving the results. It means having fulfilled a long-held dream. It means taking pride in your pet as well as in yourself.

The health, vigor, and companionship displayed by dogs, cats, and other pets reflect in large part the efforts given by pet owners in applying the best possible accepted practices in the care and management of an important member of *all creatures great and small.* A great deal of personal satisfaction is derived in knowing one has achieved that objective.

Now let us move to Chapter 13 in which we discuss other topics having special interest to most companion animal owners and admirers, namely, fitting, grooming, and showing.

## 12.12 REFERENCES

1. American Kennel Club Staff. 1991. *American Kennel Club Dog Care and Training*. New York: Macmillan Publishing.
2. Pryor, Karen. 1999. *Don't Shoot the Dog! The New Art of Teaching and Training*. New York: Bantam.
3. Monks of New Skete. 2002. *How to Be Your Dog's Best Friend: The Classic Training Manual for Dog Owners*. New York: Little Brown.
4. Brian, S., and Wilson Kilcommons. 1991. *Good Owners, Great Dogs*. New York: Warner Book.
5. Smith, M. L. 2003. *You Can Teach Your Dog to Eliminate on Command*. Port Washington, WI: Seaworthy Publications.

# 13 FITTING, GROOMING, AND SHOWING[1]

*Dogs live to please their owners, cats live to please themselves. That is why there are no obedience classes at cat shows.*

Phillip A. Maggitti (b. 1943)
*American author*

[1]Please turn to the end of the chapter for complete author affiliation information.

## 13.1   INTRODUCTION/OVERVIEW

Whether preparing dogs to participate in the annual 1,100-mile Iditarod Trail Sled Dog Race from Anchorage to Nome, Alaska; or competing in the famed annual Westminster Kennel Club Dog Show in New York City's Madison Square Garden, the second oldest continuous sporting event in the United States (the Kentucky Derby is the oldest); or having one's prized cat be accepted in the world-renowned annual Empire Cat Show held on Mother's Day on Staten Island in New York; or owning dogs or cats simply to enjoy their presence and companionship, certain fundamental prerequisites in the care of dogs and cats require regular attention. All dogs and cats need grooming. Additionally, they will be healthiest and in best condition for work when well-nourished. Indeed, providing a balanced diet—one that meets or exceeds all of a pet's daily nutritional requirements—is essential for proper fitting, and if show ring winning is a goal, many other steps are required to prepare the animal for competition.

In this chapter, we discuss numerous necessities related to fitting, grooming, and showing dogs and cats. Other pertinent topics were discussed in Chapter 12.

## 13.2   FITTING DOGS AND CATS

"Fitting" is properly preparing a dog or cat for a specific assignment, opportunity, or competition; this is also referred to as conditioning. To perform with distinction requires being ready for action. Fitness includes an overall state of good health nutritionally, physically, and mentally. For a dog to compete successfully in an endurance race, in the show ring, and in service assignments, it must be in top condition. This includes physical fitness and a high level of motivation and endurance. Although cats do not compete in physically demanding activities, they too require ample exercise to maintain muscle tone and condition.

Biking the Tour de France would be a really bad idea without physically training and developing stamina. The same is true for dogs in competition. Conformation dogs must have proper muscling and development while not being overweight. The same holds true for performance and working dogs; some say even more so. The conditioning of dogs for Performance events is gaining more interest as the need for top fitness is recognized as being essential for top performance.

Performance dogs are asked to cover distance, jump, twist, change direction, crawl, climb, and full-out run during their activities (Figure 13.1). It is important for both dog and handler to perform warm-up exercises before competition. Warm-up exercises should include stretches. For dogs, start with a 5- to 10-minute muscle massage over the dog's entire body beginning with the neck, back, and working down each leg. Stretch the dog's front legs

**Figure 13.1** Catching a Frisbee involves jumping and twisting. It is important to perform warm-up exercises first to decrease the likelihood of injury to muscles and ligaments. *Courtesy Paul E. Miller.*

over its head while the dog is standing on its back legs. Next, stand the dog on its front legs and stretch the back legs in a "wheelbarrow" position. Then, run with the dog over a span of about 100 yards going back and forth several times. Sometimes handlers will run or go on brisk walks prior to competition. Running is useful in building endurance. Many people throw balls or Frisbees® for the dogs. However, current thinking among well-known trainers is that chasing balls or Frisbees is not the best technique to use in warming up dogs prior to a Performance event. "High drive" dogs especially are at risk for an injury when jumping or twisting to catch a ball or Frisbee and can easily tear tendons and ligaments or develop sprains or minute fractures. Other means of conditioning can entail the use of biking beside a dog on lead or the use of a canine-compatible treadmill. Swimming can be a great way to utilize conditioning and provide therapy concurrently. Special vests can be placed on dogs that might not be thrilled about getting wet and thereby can be held by the handler from straps attached to the vest like a handbag.

Keeping dogs in top condition can also entail a visit to some increasingly popular sports medicine services being offered by both veterinarians and holistic approach specialists. Canine sports medicine diagnosticians can locate a problem source and provide therapy or surgery if necessary. Holistic approaches may include the use of acupuncture, readjustment, and massage therapy sessions. Canine fitness will become increasingly important as the numbers of competitors in Performance sports increase, and our understanding of canine exercise physiology and breed-specific traits are better understood. Keeping our companions healthy enables them to live longer and stronger, which should be our ultimate goal as they provide owners with unconditional love, devotion, protection, improved health, therapy, and enjoyment.

In addition to exercise, addressing the nutritional needs for dogs "at work" or involved in rigorous activity needs to be carefully examined (Chapter 9). A complete and balanced diet may need to be supplemented due to an increase in caloric needs. Likewise, there is usually a need for increased water consumption, especially during the summer to ensure proper hydration. Many Performance exhibitors carry ice chests filled with ice, water, and meat or cheese treats to competitions for their dogs as well as themselves. "Power bars" for dogs are popular with exhibitors and provide a quick source of nutritional energy. They are sold in most pet supply specialty stores.

## 13.3    GROOMING CATS AND DOGS

Daily grooming provides a bond of loving companionship between pets and caretakers. Pets groomed and cared for on a regular basis will readily accept this attention and are usually patient and tolerant of even the somewhat unpleasant activities of trimming claws and brushing teeth.

### 13.3.1    CAT GROOMING

Notwithstanding the inherent cleanliness and tidiness of cats, they benefit from routine grooming. Additionally, removing dead hair from the coat decreases the likelihood of a hairball forming in the cat's stomach or intestines, and it provides an excellent opportunity for regular checks on the condition of the coat, ears, eyes, nose, and paws. It also helps minimize deposits of hair and other debris on carpet and furniture.

Cats are fastidious by nature. Indeed, most cats are meticulous in their cleanliness and hygiene. Being groomed is the neonatal kitten's first experience, commonly occurring before its first feed. This helps explain why cleanliness becomes such an important fundamental throughout a cat's life. As each kitten is born, its mother licks it clean. This act also stimulates breathing.

From birth to about 3 weeks of age, kittens are groomed daily by their mother. This is her way of awakening them to nurse. By about 3 weeks, kittens begin grooming themselves

**Figure 13.2** Mother cat grooming her kitten. *Courtesy Paul E. Miller.*

and each other. Mutual grooming is a method of bonding between mother and kittens (Figure 13.2). Littermates that stay together often continue mutual grooming throughout their adult life. And two adult cats living amicably together often groom each other (Figure 13.3). They may also groom a friendly dog.

Cat grooming typically begins with scratching the neck, head, and ears with a hind leg. Then the cat methodically cleans its flanks, back, tail, and underparts. Licking the paws fore and aft follow. Next, with a saliva-dampened forepaw, the head and ears are cleaned. The teeth and/or claws may be used to clear mud, seeds, or other debris from the coat and between their toes. Throughout the sequence, the cat may return to a body part needing additional attention. Occasionally, a difficult spot such as something sticky may receive repeated cleaning.

**Figure 13.3** Two adult cats grooming each other. *Courtesy Paul E. Miller.*

Grooming removes loose hair and external parasites (e.g., fleas) from the coat. Loose hair is frequently ingested and normally ejected in the form of a hairball (Chapter 9). But self-grooming is not a substitute for owner/caretaker grooming, particularly for long-haired cat breeds.

In warm environments, grooming is an important means by which cats cool themselves via the evaporation of saliva deposited on their coat. Another function of grooming is to enable cats to alleviate stress (e.g., cats frequently groom themselves following an awkward fall or after confrontations with unwelcomed cats).

Lack of attention to grooming by cats is a common response to illness. Spreading cooled, melted butter on a cat's coat—something few healthy cats can resist—is a useful test of illness (if the cat does not remove the butter deposits, it is probably sick). Older cats sometimes decrease their frequency of grooming. This makes regular inspections of older cats for mites and other external parasites especially important (Chapter 15).

Grooming is important in the bonding between kittens and owners particularly in early life when they miss the gentle attention of their mother. Brushing kittens gently with one's hand or a soft brush while the owner talks to them helps familiarize kittens to the pleasures of human companionship and in addition helps form trusting, long-term relationships. Grooming also provides an excellent opportunity to evaluate the condition of the coat, skin, ears, and eyes.

Young kittens are especially susceptible to ear mite infections, often acquiring mites from their mother. The first symptom is dark brown wax in the ear canal. Ear mites often multiply rapidly causing irritation, pain, and sometimes behavioral change (Chapter 14).

Although the cleanliness of cats is one of their most attractive traits as pets, owners should not assume that cats always look after themselves. Grooming is particularly important in spring and early summer when most cats (except breeds such as Rex and Sphinx) shed their coats, often in tufts. Regular grooming helps prevent unwanted deposits of hair on carpets and furniture. The primary tools needed are a brush with long, soft natural bristles (a baby brush can be used) and a fine-toothed metal comb.

***Grooming technique.*** Stand the cat on a newspaper to catch any dead fleas, debris, stray hair, or other extraneous matter. Hold the cat gently with one arm to control the front paws and prevent rolling. Comb the coat lightly and then brush gently with a soft brush, working down the body toward the tail and down the hind legs.

Any matted hair or tangles should be teased out with the fingers or a wide-toothed comb. Slightly dampening the hair may help facilitate the process. If tangles are unusually contrary, it may be necessary to use scissors (cutting out matted hair is not recommended for cats being prepared for showing). When grooming is complete, the newspaper should be folded inward and disposed of appropriately.

## 13.3.2   Dog Grooming

The amount of grooming needed for dogs depends largely on the breed. Everyone agrees that a flashy dog attracts attention. Proper grooming is essential for both show dogs and pampered pets. Tools need for grooming include brushes, combs, claw trimmers, scissors, and cotton balls or squares (Figure 13.4). Grooming some breeds can be very time consuming to sport the appropriate coiffeur, yet other breeds could be termed "wash and wear" for their ease of care. Breeds requiring frequent trimming, combing, or bathing include the English Setter, Old English Sheepdog, Wheaton Terrier, Shih Tzu, Poodle (Figure 13.5), and many others. Short-coated and hairless breeds, such as the Basenji, Chihuahua, Chinese Crested, and numerous others, can be maintained more easily. In all breeds, attention should be given to the eyes, ears, feet, claws, skin and hair coat, and teeth.

***Eyes.*** Eyes should be free from discharge, redness, swelling, cloudiness, or itching. Eyes are one of the most expressive features of a dog and should "sparkle" as a reflection of good

**Figure 13.4** Tools needed for grooming include brushes (left column), combs (middle column), and claw trimmers (right column). *Courtesy Lawrence Motsinger.*

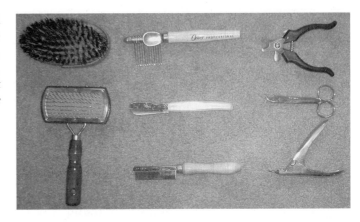

health. A dull or hazy appearance to the eyes is an indication of poor health and the advice of a veterinarian should be sought (Chapter 14). Dust, pollens, or weed seeds will irritate the eyes causing tears to overflow onto the face. Flushing the eyes with an artificial tears solution may help, otherwise seek advice from a veterinarian. Poodles and other toy breeds are prone to excessive tearing due to plugging of the nasolacrimal duct (Chapter 11); this results in a reddish-brown stain on the face. Pugs and Bulldogs may have irritation to their eyes from hairs growing on facial folds rubbing against the cornea. A veterinarian should be consulted for recommendations regarding trimming the hairs or removal of facial folds. Another cause of excessive tearing is "ingrown" eyelashes rubbing against the cornea; these misplaced eyelashes (distichia) can be removed by a veterinarian using cryoepilation or electroepilation.

*Ears.* Ears should be free from discharge or odor, redness, swelling, or itching. Normal ears require minimum care beyond occasional removal of wax from the outer part of the ear canal and inner surface of the ear pinnae (Chapter 11). A quilted cotton square or cotton ball dampened with rubbing alcohol or a pet ear cleaning solution can be used to remove this wax. Consult with a veterinarian if there is any pain, odor, or discharge, or if the pet is scratching or rubbing its ears (Chapter 14).

*Feet.* Feet should be free from discharge or odor, redness, swelling, or itching. Some breeds should have hair between the toes trimmed routinely; other breeds are required to have hair between their toes to be shown in Conformation. Consult with a mentor knowledgeable about the breed you are working with prior to trimming the feet. Check between toes for

**Figure 13.5** Poodles are among the breeds requiring frequent grooming. *Courtesy Kennelwood Pet Suites, Kennelwood Village, St. Louis, MO (www.kennelwood.com).*

**Figure 13.6** Anatomy of the canine claw. Note that the third phalanx (P3) extends past the claw fold; it can be damaged if the claw is trimmed too close to the foot. Cutting into the corium (commonly called the "quick") will result in bleeding. *Courtesy University of Illinois.*

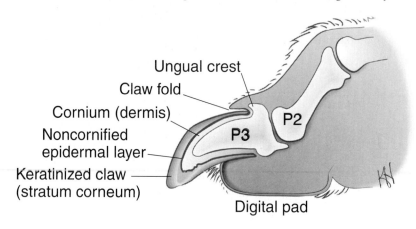

foreign matter, redness, or irritation. Consult with a veterinarian if there is any cracking, swelling, or bleeding of the feet.

*Claws.* Trimming of claws (Figure 13.6) should be done regularly so as not to allow a claw to penetrate a paw pad. If trimming is neglected, long claws will force the dog's weight onto the heel of the foot; this results in splaying of toes and eventually the foot "breaks down" resulting in lameness and a foot that is ruined for competitive showing. Most dogs do not like having their claws trimmed. It is best to start trimming the claws when the dog is a puppy. This helps it learn early to tolerate the procedure without struggling. Ideally, the claws should be trimmed every 7 to 10 days with the claws kept short enough that the tip does not touch the ground when the dog is standing squarely. Do not forget to also trim any dewclaws (non-weight-bearing claws on the sides of feet). Trimming can be performed at home by using a cutting type of scissors, guillotine cutters, or a grinding tool (battery or electric; e.g., Dremel™ grinder). The "quick" or blood line grows out with the claw, thus a very long claw must be trimmed carefully to prevent pain and bleeding. The quick will recede following cutting, thereby enabling trimming to be repeated every few days until the claw is back to its optimum length. The quick is easiest to see in white claws, thus if the dog has at least one white claw start by trimming it and use its length to help guide the amount to trim from any darkly colored claws. If the quick is cut during claw trimming, a styptic pencil containing silver nitrate can be used to cauterize the blood vessels. Instructions and photographs for trimming claws can be viewed at www.vetmed.wsu.edu/ClientED/dog_nails.htm. "Show cut" refers to extremely short claws; caution must be used not to trim the claws too short because the base of the claw encloses the end of the third phalanx (last bone of the toe). Cutting into this bone results in pain and may be followed by a bone infection that may require amputation of the toe.

*Skin and hair coat.* Routine brushing, combing, and bathing help ensure lustrous coats and adequate skin ventilation. Routine grooming minimizes the possibility of a serious external parasite infestation such as fleas, ticks, lice, or mange mites becoming widespread prior to detection. Any cuts or open sores should be cleaned and veterinary care sought if necessary. Care of the hair coat is breed specific. Owners of dogs that will be shown should consult with a mentor who can help teach them breed-specific trimming and grooming techniques. In general, long-coated breeds require linebrushing. The hair coat is sprayed with water and then separated in a line down to the skin starting at the neck. Grooming a completely dry coat often causes the ends of hair to break; spraying the coat with water will prevent this. Linebrushing is accomplished using a metal pin brush or bristle brush. The hair is brushed

**Figure 13.7** Linebrushing should be used in grooming pets with long hair. The hair coat is misted with water and parted in a line down to the skin. The hair is brushed in the opposite direction from which it normally lays. Brushing is continued one layer at a time, making a new part with each layer. *Courtesy Dr. Karen L. Campbell.*

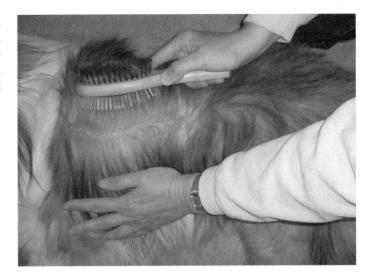

in the opposite direction that it normally lays, one layer at a time. For each layer, a new part is formed and brushed (Figure 13.7). Linebrushing may require an hour or more to complete. Mats are a common problem in long-coated breeds not groomed regularly. Some mats may be "detangled" if saturated with a tangle removal product; others will require "cutting through" and careful combing to remove the mat while minimizing hair loss. Caution must be used when cutting through mats to avoid cutting into the skin.

Terrier breeds are usually groomed by hand stripping (plucking) the old coat to obtain new growth. Without hand stripping, the terrier coat becomes soft and the colors dilute (Chapter 3). Hand stripping does not hurt the dog because the hairs being stripped are dead and ready to be removed. Most terriers tolerate the procedure well. Many terrier breeders believe hand stripping not only maintains a more attractive coat, but also contributes to healthier skin. Hand stripping clears the hair follicles of accumulations of sweat and oil secretions and reduces the incidence of minor bumps and scaliness often associated with clipping or shaving terriers.

A common misconception involves how frequently a dog should be bathed. Dogs with healthy skin and hair coats require bathing only when they become dirty. However, dogs with skin diseases associated with allergies, infections, ectoparasites, or hereditary disorders (Chapter 14) may benefit from being bathed every 2 to 4 days. The choice of a shampoo is important and should be discussed with a veterinarian. Most human shampoos and dish detergents are too harsh to be used on dogs and cats (a human scalp normally has an acidic pH, whereas the skin of dogs and cats has an alkaline pH).

*Teeth.* Dental disease is one of the most common health problems in dogs and cats (Chapter 12). Food particles build up on and between teeth; bacteria grow on these accumulations producing plaque. Toxins from bacteria irritate the gums producing gingivitis (inflammation of the gums). Plaque hardens to form tartar, which may extend downward, under the gum line, resulting in periodontal disease and eventually loss of teeth. Teeth should be checked often for evidence of redness or swelling of gums, tartar, or loose/broken teeth. Feeding dry foods or treats will help minimize the buildup of plaque on teeth. Regular brushing of teeth is the best way to ensure optimum oral health. A variety of commercial dental care products such as canine-specific pastes and brushes can be used between veterinary cleaning appointments (see Figure 12.5). Dogs (and cats) will generally tolerate having their teeth brushed. However, introducing them to this process may require patience. Start by simply handling the pet's mouth several times daily, praising it for letting you rub around the teeth and gums. Next, introduce a soft toothbrush or ribbed finger cot or wet gauze sponge and rub it on the teeth. After a few days, introduce a poultry or malt flavored pet toothpaste (do not use toothpaste for humans because these foam too much and can be irritating when

Chapter 13 Fitting, Grooming, and Showing

swallowed by dogs and cats). Antibacterial gels and rinses containing chlorhexidine (an antiseptic) can also be helpful in reducing the number of bacteria in the mouth. If tartar or redness of the gums is already present, a professional cleaning is indicated.

*Anal sacs.* The anal sacs are paired structures located at approximately the four and eight o'clock positions of the anus. These sacs are lined with glands that secrete an oily substance to help lubricate the anal region and also serve in scent recognition. The sacs should empty when the dog (or cat) defecates. However, this does not always occur and the sacs may become distended and/or infected. The most common early sign of anal sac impaction or infection is the dog licking its perianal region. As the impaction or infection worsens, the dog may "scoot" or drag its anus on the floor or ground. If not treated, the anal sacs may rupture resulting in a draining abscess. Mildly distended anal sacs may be emptied by pressing the anus firmly between a thumb and finger. Impacted anal sacs require insertion of a gloved finger into the rectum. The anal sac is then immobilized between the finger and the exterior thumb and gently squeezed to expel the secretions (Figure 13.8a). If the secretions are difficult to expel, the anal duct should be cannulated (Figure 13.8b) and an antiseptic lubricating solution infused into the sac to soften the impacted debris. A veterinarian should be consulted for management of impacted or infected anal sacs.

## 13.4 SHOWING DOGS AND CATS

Many people of all ages derive a great deal of pleasure and personal fulfillment from showing their favorite pets—animals to which they have given much attention in training and grooming. A large proportion of adult breeders and show participants/supporters received their life-long interest in the public display of beautiful, well-trained, and well-groomed cats and dogs from memorable, pleasant experiences as youths showing their animals (Figure 13.9).

### 13.4.1 CAT SHOWS AND SHOWING

Each year in the United States, more than 400 major cat shows are organized under rules of the major cat associations: the Cat Fanciers' Association (CFA), the American Cat Fanciers' Association (ACFA), the American Cat Association (ACA), the Cat Fanciers' Federation (CFF), the United Cat Federation (UCF), the American Cat Council (ACC), The International

**Figure 13.8** *(a)* Expressing impacted anal sacs. *(b)* Cannulation of an impacted anal sac for infusion of a lubricating agent. *Courtesy Dr. Karen L. Campbell.*

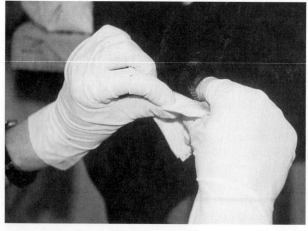

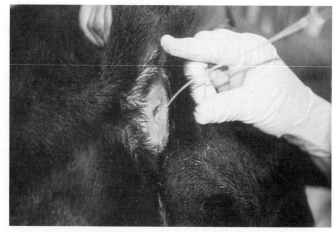

*(a)*                                                                                          *(b)*

**Figure 13.9** Josh Fisher, a junior handler, receiving Best of Breed ribbon with Ch. Karmun Sentinel on the Watch. *Courtesy Jean Fisher, Villa Grove, IL.*

Cat Association (TICA), and Federation Internationale Feline (FIFe). The CFA is the largest with more than 600 affiliated clubs in Canada, Japan, and the United States.

Cat shows are excellent places to learn about breeds and to study peaks of perfection attained by beautiful cats, even if it is not a personal goal to breed or own cats of show caliber. Examples of national events include the Empire Cat Show on Staten Island in New York, Britain's National Cat Show in London's Crystal Palace, and the Cat Club de Paris of France where thousands of beautiful cats can be seen and admired.

Guidelines that govern the basic standards of pedigree cats are those established by cat fancy organizations for the breeds they recognize in each respective country. Breed standards describe the perfect specimen (especially conformation) and acceptable color varieties for show purposes. In addition to breed standard, judges also commonly consider cat temperament, health, and grooming.

In addition to anatomical characteristics, the coat is commonly described in some detail including its color, texture, and quality. Some color varieties require coat markings to be of a particular type and density. Sometimes (e.g., with tortoiseshell-and-white cats) the required proportions of different coat colors and/or positioning of colors are stated. Breed standards also note the eye colors permitted, tail markings, and extent of colorpointing. Some breed standards also note imperfections such as squinting in Siamese and a kinked tail in virtually all breeds.

*Showing nonpedigree cats.* Happily, proud owners of nonpedigree cats are not precluded from exhibiting their special companions because most cat shows include classes for them (commonly called Domestic, Household, or Pet classes). Each cat is judged on its general merits rather than on standards established for a particular breed type.

## 13.4.2 THREE Cs OF SHOWING/JUDGING

Three Cs often cited as basic to breed standards and evaluated by judges of cats are color, coat, and conformation.

Cats described as blue must be blue within the breed standard, although one breed's blue may differ from that of another breed. Because there are no "true blue" cats, one breed's standard might specify bluish-grey whereas another breed might require bluish-white. In the same way, cats of a specific breed have a similar body type (conformation); for example, Siamese cats are not cobby (stocky) and Russian Blue cats are not svelte (slender and sleek).

### 13.4.3   TITLES AWARDED TO CATS

A title is awarded to a show cat that has achieved recognition as an outstanding example of its breed based on rules determined by its breed association. Typically, titles are added to the beginning of the cat's name, before a cattery name prefix. However, some titles are added to the end of the cat's name. Titles may be earned for participation and winning in a specified number of shows or by earning a certain number of points or by producing offspring that have earned titles. A list of titles awarded by cat fanciers is listed in Table 13.1.

### 13.4.4   DOG SHOWS AND SHOWING

Competition with canines is divided into two main branches. The first involves Conformation, or judging a dog's breed characteristics, type, structure, and movement. The second, and a vastly broad area, is considered Performance. Although many people enjoy exhibiting in Conformation, Performance exhibitors outnumber them. The AKC reports more than 2 million exhibitors compete in all types of events annually. Frequently, exhibitors cross over into both areas. Performance is exciting, energetic, demanding, and often utilizes the natural abilities of what the dog was bred to do. These events include Obedience, Agility, Earthdog, Field Hunting, Gun Dog, Herding, Lure Coursing, Coonhound Trials, and Tracking. Associations such as the AKC, UKC, and numerous others award titles to dogs that have demonstrated expertise in a Performance area (usually based on receiving a certain number of points from judges at three or more shows approved by the association) or that have earned a sufficient number of points in Conformation. AKC Performance titles and their corresponding abbreviations are listed in Table 13.4. Conformation-related titles are generally listed before a dog's name while Performance-related titles are listed after the dog's name. Many handlers cross train their dogs and compete in several areas at the same show. Some types of events are breed specific. Handlers find Performance events challenging and a way to utilize the dog's inherent abilities and instincts (Figures 13.10 and 13.11).

Making the decision to show a dog in Conformation or Performance areas should be made only after conducting the appropriate research. Forethought should be given to the specific breed the handler will be exhibiting. Is this breed the right match for the handler in the characteristics of intelligence, attitude, grooming ease, activity level, and lifestyle to ensure a perfect fit? It would be a disservice to both owner and canine to match a very active Border Collie with an owner who lives in a small apartment and works 12-hour days. The owner may come home to find that the dog has taken out its stress on every pair of shoes it could find or the furniture. Potential owners should visit dog shows to see the breed in action and question handlers regarding the specific breed. Web site searches can be done easily by accessing the Internet. The American Kennel Club (AKC) Web site (www.akc.org) is one way to begin searching the 150 different kinds of AKC purebred dogs. It will usually list a parent club's address for further contact about a specific breed. In addition, many breeders will link a homepage to the parent club's Web site; however, breeder reputation usually involves a more thorough search. There are also many good breed-specific books available to a potential owner. We will refer to AKC-approved titles for ease of reference, although there are many types of dog registries, each with their own language of titles.[2]

### 13.4.5   JUDGING CONFORMATION DOG SHOWS

Conformation shows judge a dog's breed characteristics, type, structure, and movement. The goal is to acquire a Championship title or Champion of Record (Ch.). Each breed has what

---

[2]A few of the non-AKC titles are listed in footnote 3. Others can be found by obtaining information from parent breed clubs and performance dog clubs and organizations.

**Table 13.1**
List of Titles Awarded to Cats

| | |
|---|---|
| **Champion (CH)** | In most associations, this is the first title for which a cat can compete in the championship class (and in TICA, also the alter class). The particular requirements depend on the association. |
| **Companion** | In CFF, this title is the rough equivalent of Champion for cats shown in the Household Pet class. |
| **Distinguished Merit (DM)** | A title awarded by CFA to female cats that have produced five CFA Grand Champions or Grand Premiers, and to male cats that have produced 15 Grand Champions or Grand Premiers. This title is also awarded by FIFe to female cats that have produced five FIFe International Champion or International Premier offspring, and to male cats that have produced 10 International Champion/Premier offspring. |
| **Double Grand Champion (DGC or DGCH)** | A TICA title awarded to cats in the championship and alter classes. A cat must first become a Grand Champion before it can begin to compete toward the title of Double Grand Champion. Sometimes the term "Double Grand Champion" is used to refer to a cat that has earned the title of Grand Champion in two associations. |
| **Double Grand Master (DGM)** | In TICA, this title is the rough equivalent of Double Grand Champion for cats shown in the Household Pet class. |
| **European Champion (EC)** | In FIFe, this is the highest title awarded to whole adult cats. A cat must first become a Grand International Champion before it can begin to compete toward the title of European Champion. |
| **European Premier (EP)** | In FIFe, this title is the rough equivalent of European Champion for cats shown in the alter class. |
| **Grand Champion (GC, GRC, GCH, or Gr. Ch.)** | In North American associations, Champions may continue to compete towards this title. A cat must first become a Champion before it can begin to compete toward the title of Grand Champion. The particular requirements depend on the cat association. |
| **Grand Companion** | In CFF, this title is the rough equivalent of Grand Champion for cats shown in the Household Pet class. |
| **Grand International Champion (GIC)** | A FIFe title awarded to whole adult cats. A cat must first become an International Champion before it can begin to compete toward the title of Grand International Champion. |
| **Grand International Premier (GIP)** | In FIFe, this title is the rough equivalent of Grand International Champion for cats shown in the alter class. |
| **Grand Master (GM)** | In TICA, this title is the rough equivalent of Grand Champion for cats shown in the Household Pet class. |
| **Grand Premier (GP, GRP, GPR, or Gr. Pr.)** | In most North American cat associations, this title is the rough equivalent of Grand Champion for cats shown in the Premiership class. |
| **International Champion (IC)** | A FIFe title awarded to whole adult cats. A cat must first become a FIFe Champion before it can begin to compete towards the title of International Champion. |
| **International Premier (IP)** | In FIFe, this title is the rough equivalent of International Champion for cats shown in the alter class. |
| **International Winner (IW)** | In TICA, this title is awarded to cats that have earned an International award. |
| **Master** | In TICA, this title is the rough equivalent of Champion for cats shown in the Household Pet class. |
| **Master Grand Champion (MGC or MGCH)** | In CFF, this is the highest title awarded to cats in the Championship class. A cat must first become a Grand Champion before it can begin to compete towards the title of Master Grand Champion. |
| **Master Grand Companion** | In CFF, this title is the rough equivalent of Master Grand Champion for cats shown in the Household Pet class. |
| **Master Grand Premier** | In CFF, this title is the rough equivalent of Master Grand Champion for cats shown in the Premiership class. |

*Continued*

**Table 13.1**
List of Titles Awarded to Cats (Continued)

| | |
|---|---|
| **National Winner (NW)** | In CFA, this title is awarded to cats that have earned a National award. |
| **Outstanding Dam (OD)** | A title awarded by TICA to female cats that have produced five TICA Grand Champions. |
| **Outstanding Sire (OS)** | A title awarded by TICA to male cats that have produced 15 TICA Grand Champions. |
| **Premier (PR or Prem.)** | In most associations, this title is the rough equivalent of Champion for cats shown in the Premiership class. |
| **Quadruple Grand Champion (QGC or QGCH)** | A TICA title awarded to cats in the championship and alter classes. A cat must first become a Triple Grand Champion before it can begin to compete towards the title of Quadruple Grand Champion. Sometimes the term "Quadruple Grand Champion" is also used to refer to a cat that has earned the title of Grand Champion in four associations. |
| **Quadruple Grand Master (QGM)** | In TICA, this title is the rough equivalent of Quadruple Grand Champion for cats shown in the Household Pet class. |
| **Regional Winner (RW)** | In CFA, this title is awarded to cats that have earned a Regional award. |
| **Supreme Grand Champion (SGC or SGCH)** | In TICA, this is the highest title awarded to cats in the championship and alter classes. A cat must first become a Quadruple Grand Champion before it can begin to compete towards the title of Supreme Grand Champion. |
| **Supreme Grand Master (SGM)** | In TICA, this title is the rough equivalent of Supreme Grand Champion for cats shown in the Household Pet class. |
| **Triple Grand Champion (TGC or TGCH)** | A TICA title awarded to cats in the championship and alter classes. A cat must first become a Double Grand Champion before it can begin to compete towards the title of Triple Grand Champion. Sometimes the term "Triple Grand Champion" is also used to refer to a cat that has earned the title of Grand Champion in three associations. |
| **Triple Grand Master (TGM)** | In TICA, this title is the rough equivalent of Triple Grand Champion for cats shown in the Household Pet class. |
| **World Winner (WW)** | In FIFe, this title is awarded to cats that have been at least best of category at the annual FIFe World Show. This is used after the cat's name with the year it was earned, as follows: EC SF*Rusalka Rodion Raskolnikov, WW96. |

*Source: Cat Fanciers web site: www.fanciers.com/other-faqs/titles.html.*

is known as a standard. It is a constant, written blueprint of what is considered the desirable merits as well as faults and disqualifications for the breed. This standard has been developed over time by breeders and a parent breed-specific club and has been approved by the AKC. Occasionally, a parent club may ask to revise the standards. Sometimes this is to address recent health advances that may be linked to a certain breed characteristic. Other times it is changed to accommodate the characteristics of the particular dogs being exhibited in more recent times. A point system may be assigned for each breed's characteristics to use as a guide in addition to the written description (Table 13.2).

Judges are expected to examine each dog in the ring and not judge them against each other, rather to compare each individually to the breed standard. To obtain approval to judge a specific breed requires breeding/owning a certain number of champions, completing mentoring programs, attending education seminars for judges, passing tests, and satisfactorily completing time as a provisional judge. However, it is important to remember that although each judge is approved to judge a certain breed, the judge will still have his or her own interpretation of the standards. This is why campaigning a dog to a breed Championship can be difficult and frustrating. An interesting note here is that judges come from throughout the United States as well as other countries. They may have preconceived ideas regarding the type of dog most common in their region or what characteristics they particularly like to see. Regional differences may be found in northeastern states, the West Coast, Midwest, and other areas; therefore, a well-informed handler will research a judge's likes or dislikes and make notes following each time they exhibit a dog. A dog becomes a breed Champion of

**Figure 13.10** BISS[3] Ch. Carousel's Sweet Georgia Brown, HXAs, CD, MX, AX, U-AG1, HRD2s, HTDs, TDI, VCX, driving sheep through a Herding Excellent course. *Courtesy Jean Fisher, Villa Grove, IL.*

Record (Ch.) only when achieving the required number of 15 points. A further requirement of the 15 points is having two Majors, or multiple point winnings, consisting of at least three points awarded by two different judges. All the points are then added to make 15 (Table 13.3).

The keys to owning, breeding, and/or handling your own dog to a Conformation Championship is to study, ask questions, observe, take notes, and seek mentoring from well-known, respected breeders. Avoid making quick judgments. Many experienced handlers can help newcomers to dog showing understand the foundations that formed the characteristics of today's breeds. Also, many clubs offer educational seminars on handling, grooming, and training at the show site. For a small fee, local dog clubs usually offer a variety of classes for both novices and those wanting more experience.

---

[3]What do all these letters mean? Each is an abbreviation for a title the dog has received. AKC Performance titles and their corresponding abbreviations are listed in Table 13.4. Numerous other organizations award titles to dogs. Other titles received by this dog include Best in Specialty Show (BISS), United Kennel Club Agility 1 (AG1), American Herding Breeds Association Junior Herding Dog (JHD), Herding Trial Dog Sheep (HTDS), and the American Shetland Sheepdog Association Versatility Certificate Excellent (VCX).

**Figure 13.11** Abbotsford Red Rover, HSAs, CD, OA, OAJ, JHD, with his handler, Josh Fisher, running an Open Agility course. *Courtesy Jean Fisher, Villa Grove, IL.*

**Table 13.2**
Shetland Sheepdog Scale of Points for Judging

| Point Categories | Areas to Evaluate/Score | | Points |
|---|---|---|---|
| General Appearance | Symmetry | 10 | |
| | Temperament | 10 | |
| | Coat | 5 | 25 |
| Head | Skull and stop | 5 | |
| | Muzzle | 5 | |
| | Eyes, ears, and expression | 10 | 20 |
| Body | Neck and back | 5 | |
| | Chest, ribs, and brisket | 10 | |
| | Loin, croup, and tail | 5 | 20 |
| Forequarters | Shoulder | 10 | |
| | Forelegs and feet | 5 | 15 |
| Hindquarters | Hip, thing, and stifle | 10 | |
| | Hocks and feet | 5 | 15 |
| Gait | Gait—smoothness and lack of waste motion when trotting | 5 | 5 |
| | Total | | 100 |
| Disqualifications | Heights below or above the desired range, i.e., 13–16 inches. Brindle color. | | |

*Source: American Kennel Club. Available: www.akc.org/breeds/recbreeds/shetshee.cfm.*


**Table 13.3**
Example of Point System for Earning the AKC Title Champion (Ch.)

| Championship Point Award System |
|---|

To become a breed Champion, each individual dog has to meet the following requirements:
- Older than 6 months of age
- Nonaltered dog or bitch (intact)
- Individually registered with AKC (which also requires the parents to be registered)
- Meet characteristics of specific breed
- Attain 15 points consisting of
  - Two majors (consisting of 3-, 4-, 5-point wins)
  - Each major is awarded by a different judge
  - Subsequent points can be attained by additional majors or single points

| Example Point Schedule[*] | | |
|---|---|---|
| Points Awarded | Minimum # of Dogs in Ring | Minimum # of Bitches in Ring |
| 1 | 3 | 4 |
| 2 | 10 | 10 |
| 3 | 17 | 17 |
| 4 | 21 | 21 |
| 5 | 28 | 28 |

*[*]This is the current Point Schedule for Shetland Sheepdogs in Division 5. Point schedules vary with breed and region of the country.*
*Source: American Kennel Club.*

**Figure 13.12**   Jesse Fisher exhibits a "table" presentation of "Georgia" in Junior Showmanship at Westminster Kennel Club, New York City. *Courtesy Jean Fisher, Villa Grove, IL.*

Of the approximately 3,000 AKC-licensed judges of dogs, fewer than 40 receive invitations from the Westminster Kennel Club to officiate what has been called "the Super Bowl of the Canine Kingdom" at Madison Square Garden in New York City each year (Figures 13.12 and 13.13).

For those select few, judging on the green carpet is an honor wrapped in immense pressure. Throughout the 2-day show that culminates with the Best in Show judging, Westminster judges have about 2 minutes with each dog to make their final decision. Their 2-minute evaluation includes:

- *Viewing the standing dog in direct profile* to evaluate balance and proportion proper for the breed and consider size, color, and grooming;
- *Viewing the standing dog from the front* to evaluate structure proper for the breed from top of head to tip of tail;
- *Viewing the moving dog* to evaluate movement proper for the breed and its function.

**Figure 13.13**   Josh Fisher exhibiting Ch. Karmun Sentinel on the Watch in a "free stack" presentation in Junior Showmanship at Westminster Kennel Club, New York City. *Courtesy Jean Fisher, Villa Grove, IL.*

For the 15 minutes in the show ring, each dog must be at its very best; so must the handler and judge.

In 2 pressure-packed days of judging at Westminster, judges (except for the Best in Show judge) take a backseat to the more than 2,500 dogs entered. But on Tuesday night, eyes of all present for this popular annual event as well as those of the television viewers (4.3 million television viewers tuned in to watch the 2002 show) are on the judge as she/he steps into the center ring and looks at the seven AKC group winners (Sporting, Hound, Working, Terrier, Toy, Non-Sporting, and Herding) for the first time.

Commonly selected 2 years ahead, the judge of the Best in Show competition is sworn to secrecy regarding the assignment and must agree not to judge any Best in Shows during the 5 months immediately prior to Westminster. Further, she/he must not give interviews or watch any of the competition in person or any part of the 6 hours of television coverage prior to judging the Best in Show finale.

Obviously, it is an awesome task to judge the Best in Show at the Garden. Knowing precisely the traits most desired by each of the 150 breeds recognized by the AKC, then evaluating those important breed qualities and elevating to the top of this highly prestigious show the dog believed to best represent the qualities and overall conformation called for in its particular breed standards, requires a very high level of training and experience. Imagine the mounting pressure associated with such an important judging responsibility—picking the most nearly perfect purebred—all staged in front of an anxious, raucous, sold-out Garden crowd and the millions of viewers watching on television.

***Examples of things conformation dog judges look for.*** The AKC recognizes 150 breeds of dogs (see Table 3.2). Each breed has a written standard—a prescribed most-desirable range in height, length of ears, color of coat, clip, gait, and other traits that collectively comprise and determine perfection. The dog coming the closest to the breed ideal wins the class.

The *je ne sais quoi* factor (something that cannot be easily/adequately described or expressed) is what dog show people call "a winning personality" (Figure 13.14). A Best in Show dog not only represents the most nearly perfect physical representative of its breed, but also demonstrates a happy attitude, pleasant demeanor, sparkle, special beauty, and poise; and it is not easily distracted by crowd noise, flowers, and cameras. Now let's consider two breeds having wide differences in conformation.

**Figure 13.14** Ch. Safari's Right Stuff, a breed champion Standard Poodle. *Courtesy Joel E. Haefner and Cynthia A. Huff, Carlock, IL.*

*Greater Swiss Mountain Dog.*  Bred for draft work, Swissies have a calm, steady disposition. Their eyes are dark brown (not too light, not too large or small). Their dense coat is 1 to 1¾ inches long, and its color is jet black with rich rust and white markings. Symmetry of markings is desired. They have a blunt muzzle (not too long or short); long, sloping shoulders; straight and strong forelegs; a wide flat skull; triangular ears with rounded tips that hang close to the head and are level with the top of the skull. The mature male height at highest point on the shoulder is 25½ to 28½ inches, females are 23½ to 27 inches; their body is 10% longer than tall; their straight tail has a white tip; the hindquarters are strong with muscular thighs; they have a powerful gait with good reach; and a level back.

*Miniature Poodle.*  These active, elegant, intelligent, regal dogs outwardly display an air of dignity and distinction. They come in blue, grey, silver, brown, café au lait (color of coffee with milk), apricot, and cream. The coat may show varying shades of the same color. Poodles at Westminster are commonly shown with a Continental clip (pom-poms on hips). Although mature Miniature Poodles can be 10 to 15 inches tall, most show Poodles are 15 inches with height equaling length. Their forequarters display strong, smoothly muscled shoulders; forelegs are parallel and straight when viewed from the front; hindquarters are strong, hind legs are parallel and straight when viewed from the rear; leg bones and muscles are in proportion; eyes are oval-shaped, set far apart, and very dark; they have a long, straight muzzle; their straight tail is set high and up; their gait displays a sound, strong, seemingly effortless movement with head and tail carried high.

## 13.4.6  PERFORMANCE SHOWING

Each Performance event starts with a beginning or introductory level competition and then graduates to an intermediate level and finally to the Masters or Excellent level. A title is earned by achieving a certain number of qualifying scores. Most events require earning three qualifying scores called legs. Once the Masters or Excellent titles have been achieved, a competitor can work toward a Championship title by showing a consistent level of performance for a specified number of performances. Each activity varies in what constitutes a Champion. When competing, it is important to become familiar with the rules for the area in which you are entered. For example, most types of events will not permit dogs younger than 6 months of age to compete, but that varies depending on the specific event. Table 13.4 gives a brief description of the titles that can be earned by the American Kennel Club registry. They are broken into two divisions, prefix titles and suffix titles, in respect to a dog's registered name. Those listed are taken from the AKC Web site as of November 4, 2003.

When training a dog for a specific event, for example, Agility, it is best to work with an experienced handler. Obstacle familiarization and safety issues are ever-present, and knowledgeable instructors are very important (Figures 13.15 and 13.16). Seeking reference from other dog owners or training facilities in your area is usually not difficult. Once you begin training, it is important to be consistent with training methods used. Changing commands or signals frequently can cause confusion, a lack of understanding, or difficulty in forming the bond connection between handler and dog. The dog's trust is placed solely in the handler and is an utmost requirement for a successful team operation. Once that bond is formed, the dog can anticipate a handler's actions or body language and move quickly and fluidly. Many top Agility handler's have said approximately 2 to 4 years are required for a working bond to form between them and their canine teammates. This is after taking the initial classes and then working at least two to three times weekly for several years. It is a major investment of time and effort, classes, and, of course, trial entries.

Exhibitor/dog teams competing in multiple events have a more difficult time reaching their goals. The co-author of this chapter, Jean Fisher, experiences this personally because she exhibits in four or five Performance areas concurrently and also competes in Conformation. Quick progress is often difficult when training in several areas at one time, as consistent exhibiting in one event is impossible. Sometimes it means making a decision as to what

**Table 13.4**
AKC Titles

| Prefix Titles | Title Name | Event |
|---|---|---|
| AFC | Amateur Field Trial Champion | Field Trials |
| Ch. | Champion | Conformation |
| CT | Champion Tracker | Tracking |
| DC | Dual Champion | Conformation |
| | | Field Trials |
| | | Herding |
| FC | Field Trial Champion | Field Trials |
| GDSC | Gun Dog Stake Champion | Field Trials |
| GFC | Grand Field Champion | Coonhound |
| GNCH | Grand Nite Champion | Coonhound |
| GWCH | Grand Water Race Champion | Coonhound |
| HC | Herding Champion | Herding |
| NAFC | National Amateur Field Champion | Field Trials |
| NCH | Nite Champion | Coonhound |
| NFC | National Field Champion | Field Trials |
| NOC | National Obedience Champion | Obedience |
| NGDC | National Gun Dog Champion | Field Trials |
| NOGDC | National Open Gun Dog Champion | Field Trials |
| OTCH | Obedience Trial Champion | Obedience |
| RGDSC | Retrieving Gun Dog Stake Champion | Field Trials |
| SGCH | Senior Grand Champion | Coonhound |
| SGFC | Senior Grand Field Champion | Coonhound |
| SGWCH | Senior Grand Water Race Champion | Coonhound |
| SHNCH | Senior Grand Nite Champion | Coonhound |
| TC | Triple Champion | Multiple Events |
| MACH | Master Agility Champion | Agility |
| VCCH | Versatile Companion Champion | Obedience |
| WCH | Water Race Champion | Coonhound |

| Suffix Titles | Title Name | Event |
|---|---|---|
| AJP | Excellent Agility Jumpers With Weaves "A" Preferred | Agility |
| AX | Agility Excellent | Agility |
| AXJ | Excellent Agility Jumper | Agility |
| AXP | Agility Excellent "A" Preferred | Agility |
| CD | Companion Dog | Obedience |
| CDX | Companion Dog Excellent | Obedience |
| HI | Herding Intermediate | Herding |
| HIAdsc | Herding Intermediate Course A (ducks, sheep, cattle) | Herding |
| HIBdsc | Herding Intermediate Course B (ducks, sheep, cattle) | Herding |
| HICs | Herding Intermediate Course C (sheep) | Herding |
| HS | Herding Started | Herding |
| HSAdsc | Herding Started Course A (ducks, sheep, cattle) | Herding |
| HSBdsc | Herding Started Course B (ducks, sheep, cattle) | Herding |
| HSCs | Herding Started Course C (sheep) | Herding |
| HT | Herding Tested | Herding |
| HX | Herding Excellent | Herding |
| HXAdsc | Herding Advanced Course A (ducks, sheep, cattle) | Herding |
| HXBdsc | Herding Advanced Course B (ducks, sheep, cattle) | Herding |
| HXCs | Herding Advanced Course C (sheep) | Herding |
| JC | Junior Courser | Lure Coursing |
| JE | Junior Earthdog | Earthdog |

| Suffix Titles | Title Name | Event |
|---|---|---|
| JH | Junior Hunter | Hunting |
| LCX | Lure Courser Excellent | Lure Coursing |
| MC | Master Courser | Lure Coursing |
| ME | Master Earthdog | Earthdog |
| MH | Master Hunter | Hunting Test |
| MJP | Master Excellent Jumpers With Weaves "B" Preferred | Agility |
| MX | Master Agility Excellent | Agility |
| MXJ | Master Excellent Jumpers With Weaves | Agility |
| MXP | Master Agility Excellent "B" Preferred | Agility |
| NA | Novice Agility | Agility |
| NAJ | Novice Agility Jumper | Agility |
| NAP | Novice Agility Preferred | Agility |
| NJP | Novice Jumpers With Weaves Preferred | Agility |
| OA | Open Agility | Agility |
| OAJ | Open Agility Jumper | Agility |
| OAP | Open Agility Preferred | Agility |
| OJP | Open Jumpers With Weaves Preferred | Agility |
| PT | Pre-Trial Tested | Herding |
| SC | Senior Courser | Lure Coursing |
| SE | Senior Earthdog | Earthdog |
| SH | Senior Hunter | Hunting |
| TD | Tracking Dog | Tracking |
| TDX | Tracking Dog Excellent | Tracking |
| UD | Utility Dog | Obedience |
| UDVST | Utility Dog Variable Surface Tracking | Obedience |
| UDX | Utility Dog Excellent | Obedience |
| VCD1 | Versatile Companion Dog 1 | Obedience |
| VCD2 | Versatile Companion Dog 2 | Obedience |
| VCD3 | Versatile Companion Dog 3 | Obedience |
| VCD4 | Versatile Companion Dog 4 | Obedience |
| VST | Variable Surface Tracking | Tracking |

*Source: American Kennel Club. Available:* www.akc.org/dic/events/titles.cfm.

**Figure 13.15** Ch. Carousel's Georgia Brown, HXAs, CD, MX, AX, U-AG1, HRD2s, HTDs, TDI, VCX running an agility course. *Courtesy Jean Fisher, Villa Grove, IL.*

**Figure 13.16** Ch. Sentinel Justice Prevails, HSAs, OA, NAJ, TDI, VCX, a "Georgia" son, performing the weave poles on an Excellent Agility course. *Courtesy Jean Fisher, Villa Grove, IL.*

is most important for the immediate time frame or particular show circuit. Show circuits that include more than one type of event on the same show grounds are helpful. For example, many times Conformation shows will also offer Obedience or Agility. Entering several types of events is a challenge many breed enthusiasts find rewarding—to have a dog prove it has correct conformation, intelligence, outgoing temperament, stability, work ethic, and heart to succeed. Finding all of this in one dog is difficult, and owners who have experienced this type of dog know how fortunate they are to have such an animal. They are exceptions! A brilliant example is that of Jean Fisher's Shetland Sheepdog, BISS Ch. Carousel's Sweet Georgia Brown, HXAs, CD, MX, AX, U-AG1, HRD2s, HTDs, TDI, VCX. A true sales representative for Shetland Sheepdogs, "Georgia" combines beauty, brains, and attitude in one fur ball of enthusiasm (Figure 13.17). The path of Jean and "Georgia" has been one of hard work and determination, mistakes and successes, but a pure joy the entire trip. They were rewarded with an Award of Merit at the most prestigious dog show in the United States—the Westminster Kennel Club in Madison Square Garden, New York City (Figure 13.18).

Those aspiring to show dogs in Performance events should begin working with dogs as early as the dogs' level of maturity permits. Working with dogs as puppies not only allows for an easier acceptance to training, but also increases bonding between handler and dog.

**Figure 13.17** Ch. Carousel's Sweet Georgia Brown, HXAs, CD, MX, AX, U-AG1, HRD2s, HTDs, TDI, VCX, running an Excellent Agility course. *Courtesy Jean Fisher, Villa Grove, IL.*

A   Abyssinian. *Courtesy Champion Petfoods, Inc.*

B   American Bobtail. *Courtesy Dr. Allan J. Paul.*

C   American Shorthair. *Courtesy Champion Petfoods, Inc.*

D   Bengal. *Courtesy Cassie Hale.*

E   Birman. *Courtesy Dr. Allan J. Paul.*

F   British Shorthair. *Courtesy Dr. Allan J. Paul.*

G   Korat. *Courtesy Dr. Allan J. Paul.*

H   Maine Coon. *Courtesy Alice Pursell.*

I   Persian. *Courtesy Dr. Allan J. Paul.*

J   Ragdoll. *Courtesy Champion Petfoods, Inc.*

K   Siamese. *Courtesy Mike Nelson and Vern Horning, Graphics: American Nutrition, Inc.*

L   Somali. *Courtesy Sandra Grable.*

# Color Plate 4   Major Breeds of Dogs

A   Alaskan Malamute. *Courtesy Robin Haggard.*

B   American Foxhound. *Courtesy Melissa Maitland.*

C   Basenji. *Courtesy Sandra Grable.*

D   Basset Hound. *Courtesy Mike Nelson and Vern Horning, Graphics: American Nutrition, Inc.*

E   Beagle. *Courtesy Mike Nelson and Vern Horning, Graphics: American Nutrition, Inc.*

F   Belgian Malinois. *Courtesy Jane Rothert, photo by Ann MacKay.*

G   Bloodhound. *Courtesy Melissa Maitland.*

H   Border Collie. *Courtesy Mike Nelson and Vern Horning, Graphics: American Nutrition, Inc.*

I   Boston Terrier. *Courtesy Mike Nelson and Vern Horning, Graphics: American Nutrition, Inc.*

J   Boxer. *Courtesy Sandra Grable.*

K   Bulldog. *Courtesy Mike Nelson and Vern Horning, Graphics: American Nutrition, Inc.*

L   Chinese Crested. *Courtesy Sandra Grable.*

M   Cocker Spaniel. *Courtesy Dr. Allan J. Paul.*

N   Collie. *Courtesy Dr. Karen L. Campbell.*

O   Dachshund. *Courtesy Champion Petfoods, Inc.*

P   Dalmatian. *Courtesy Linda Klippert.*

Q   Doberman Pinscher. *Courtesy Melissa Maitland.*

R   English Pointer. *Courtesy Mike Nelson and Vern Horning, Graphics: American Nutrition, Inc.*

S   English Springer Spaniel. *Courtesy Dr. Dawn Morin.*

T   Flat-Coated Retriever. *Courtesy Pat Norris.*

U   German Shepherd Dog. *Courtesy Kari Bogott.*

V   Golden Retriever. *Courtesy John J. Turnbull.*

W   Great Dane. *Courtesy Champion Petfoods, Inc.*

X   Greyhound. *Courtesy Petfoods, Inc.*

Y   Labrador Retriever. *Courtesy Kim Kensell.*

Z   Mastiff. *Courtesy Sandra Grable.*

AA   Miniature Pinscher. *Courtesy Sandra Grable.*

BB   Miniature Schnauzer. *Courtesy Mike Nelson and Vern Horning, Graphics: American Nutrition, Inc.*

CC   Pekingese. *Courtesy Dr. Thomas Graves.*

DD   Poodle. *Courtesy Joel E. Haefner and Cynthia A. Huff.*

EE   Pug. *Courtesy Mike Nelson and Vern Horning, Graphics: American Nutrition, Inc.*

FF   Shetland Sheepdog. *Courtesy Jason Motsinger.*

GG   Rat Terrier. *Courtesy Mike Nelson and Vern Horning, Graphics: American Nutrition, Inc.*

HH   Rottweiler. *Courtesy Champion Petfoods, Inc.*

II   Shih Tzu. *Courtesy Dr. James E. Corbin.*

JJ   West Highland White Terrier. *Courtesy Sandra Grable.*

A Amazon Parrot. *Courtesy Hartz® Mountain Corporation.*

B Canary. *Courtesy Hartz Mountain Corporation.*

C Triton Cockatoo. *Courtesy Sailfin Pet Shop.*

D Congo African Grey Parrot. *Courtesy Sailfin Pet Shop.*

E Cockatiel. *Courtesy Hartz Mountain Corporation.*

F Finch. *Courtesy Hartz Mountain Corporation.*

G Green Parrotlet. *Courtesy Questhavenpets.com.*

H Green-Winged Macaw. *Courtesy Dr. Michael J. Adkesson.*

I Love Bird. *Courtesy Hartz Mountain Corporation.*

J Parakeet (Budgerigar). *Courtesy Hartz Mountain Corporation.*

K Sun Conure. *Courtesy Jeff Rathmann (photographer) and PetMarket Place pet store, Webster Groves, MO.*

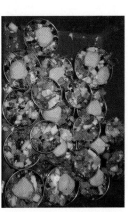

L Nutritious Foods for Birds. *Courtesy Sailfin Pet Shop.*

A    Albino Garter Snake. *Courtesy Dr. Michael J. Adkesson.*

B    Bearded Dragon Babies. *Courtesy Dr. Michael J. Adkesson.*

C    Blue-Tongued Skink. *Courtesy Dr. Michael J. Adkesson.*

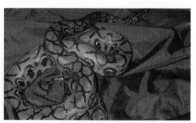

D    Boa Constrictor. *Courtesy Dr. Michael J. Adkesson.*

E    Box Turtles. *Courtesy Dr. Michael J. Adkesson.*

F    Brazilian Rainbow Boa. *Courtesy Dr. Michael J. Adkesson.*

G    Chuckwalla. *Courtesy Dr. Michael J. Adkesson.*

H    Collard Lizard. *Courtesy Jeff Rathmann (photographer) and PetMarket Place pet store, Webster Groves, MO.*

I    Crested Lizard. *Courtesy Jeff Rathmann (photographer) and PetMarket Place pet store, Webster Groves, MO.*

J    Day Gecko. *Courtesy Dr. Michael J. Adkesson.*

K    Fence Swift. *Courtesy Jeff Rathmann (photographer) and PetMarket Place pet store, Webster Groves, MO.*

L    Green Iguana. *Courtesy Dr. Michael J. Adkesson.*

M  Greyband King Snake. *Courtesy Dr. Michael J. Adkesson.*

N  Horned Frog. *Courtesy anonymous contributor.*

O  Leopard Tortoise. *Courtesy Dr. Michael J. Adkesson.*

P  Map and Painted Turtles. *Courtesy Dr. Michael J. Adkesson.*

Q  Poison Dart Frog. *Courtesy Dr. Dorcas P. O'Rourke.*

R  Red-Eared Slider. *Courtesy anonymous contributor.*

S  Red-Footed Tortoise. *Courtesy Dr. Michael J. Adkesson.*

T  Savannah Monitor. *Courtesy Dr. Michael J. Adkesson.*

U  Spiny-Tailed Monitor. *Courtesy Jeff Rathmann (photographer) and PetMarket Place pet store, Webster Groves, MO.*

V  Tegu Lizard. *Courtesy Gonzalo Barquero, Brazil, and Dr. James E. Corbin.*

W  Tokay Gecko. *Courtesy Jeff Rathmann (photographer) and PetMarket Place pet store, Webster Groves, MO.*

X  Waxy Tree Frog. *Courtesy anonymous contributor.*

A   Black   Chinchilla.   *Courtesy Questhavenpets.com.*

B   Black Mouse. *Courtesy Harlan.*

C   Brat Rat (Long Evans). *Courtesy Harlan.*

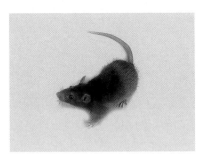

D   Brown Norway Rat. *Courtesy Harlan.*

E   Ferret. *Courtesy Jeff Rathmann (photographer) and PetMarket Place pet store, Webster Groves, MO.*

F   Golden   Syrian   Hamster. *Courtesy Harlan.*

G   Grey   Chinchilla.   *Courtesy Questhavenpets.com.*

H   Grey Mouse. *Courtesy Harlan.*

I   Guinea Pig. *Courtesy Jeff Rathmann (photographer) and PetMarket Place pet store, Webster Groves, MO.*

J   Lop-Eared   Rabbit.   *Courtesy anonymous contributor.*

K   Mongolian Gerbil. *Courtesy Harlan.*

L   New Zealand White Rabbit. *Courtesy Maria Lang (photographer) and Dr. B. Taylor Bennett.*

**Figure 13.18**  Ch. Carousel's Geor-gia Brown, HXAs, CD, MX, AX, U-AG1, HRD2s, HTDs, TDI, VCX, with owner-handler Jean Fisher receiving Award of Merit at Westminster Kennel Club, New York City. *Courtesy Jean Fisher, Villa Grove, IL.*

The dog will learn to be more confident, self-assured, and well adjusted while its behaviors are being modified or taught. Trust on the part of both handler and canine is especially important and is derived through fairness and consistency. As the handler-canine team goes beyond the basics of house manners to competitiveness, they will have a much more enjoyable and rewarding process. Remember, patience is the key. Attempting to rush through critical steps in training may cause problems in the future that may be difficult to correct. Take time to enjoy your puppy! The book *Building Blocks for Performance* by Bobbie Anderson with Tracy Libby is a good reference into training philosophy for puppies.

## 13.5  SUMMARY

In this chapter, we discussed many challenges and opportunities associated with fitting, grooming, and showing dogs and cats. Some are essential on a regular basis (e.g., grooming), whereas others (e.g., showing and certain aspects of fitting) are optional depending on the personal interests and pleasures of owners and breeders.

Dog and cat owners often pay $100 to $300 or more to have professionals train, pamper, and groom their animals in preparation for the keen competition associated with showing.

The American Kennel Club licenses only a few dozen individuals to judge each of the 150 breeds of dogs it recognizes. We noted qualifications of and pressures associated with judging major shows.

Through the owning and showing of well-groomed and well-trained dogs and cats—pets often trained to pose like a mannequin—owners derive untold personal excitement, enjoyment, and pleasure. They enjoy exhibiting examples of how beautiful, admirable, and outstanding a dog or cat can be. Although showing dogs and cats may look like a well-choreographed dance, flawless in presentation and effort, much hard work has occurred behind the scenes. Much consideration has been invested into breeding, grooming, training, and caring for these animals whether they are in the Conformation or Performance ring. For some, it is their livelihood as a professional handler (Chapter 23), but most people involved in showing are doing so as a labor of love and devotion to improving their selected breed(s). They seek the thrill of competition and the close bond the partnership can bring. Owners spare no expense in providing the best training, health, and comfort to their canine and feline friends. The co-author of this chapter, Jean Fisher, is among this latter group of fortunate individuals who recognize the value of such partnership and the joy it brings. She believes her dogs feel this way too.

## 13.6    REFERENCES

1.  American Kennel Club Staff. 1991. *American Kennel Club Dog Care and Training*. New York: Howell Book House.

2.  Anderson, Bobbie, and Tracy Libby. 2002. *Building Blocks for Performance*. Loveland, CO: Alpine Publications.

3.  Saunders, Blanche. 1981. *The Complete Book of Dog Obedience: The Guide for Trainers*. New York: Howell Book House.

4.  Simmons-Moake, Jane. 1991. *Agility Training: The Fun Sport for All Dogs*. New York: Howell Book House.

## 13.7    ABOUT THE AUTHOR

This chapter was co-authored by Jean Fisher, a Certified Veterinary Technician (CVT) employed at the University of Illinois Veterinary Teaching Hospital for 18 years. She is a dog breeder-owner-handler who has exhibited Shetland Sheepdogs and Belgian Sheepdogs in the areas of Conformation, Agility, Herding, and Obedience for more than 10 years. She is the 4-H leader for Obedience and Junior Handling dog training classes for Champaign County and is also an active member of the Dog Training Club of Champaign-Urbana, Illinois. Jean and her dogs have attained many prestigious awards including Best in Specialty Show, High in Trial, Reserve High in Trial, and an Award of Merit at the Westminster Kennel Club show. Many of her dogs are working toward dual Championships. Jean has had articles published in the *Sheltie Pacesetter* and *Sheltie International* magazines.

# 14 COMPANION ANIMAL HEALTH

*When health is absent wisdom cannot reveal itself,*
*art cannot become manifest, strength cannot fight, wealth*
*becomes useless, and intelligence cannot be applied.*

Herophilus (325 BC)
*Physician to Alexander the Great*

# 14.1   INTRODUCTION/OVERVIEW

Whether on the human or companion animal side of the equation, health is precious and priceless. Indeed, the health and quality of life of pets is fundamental to their providing companionship of the highest level possible to owners. Most diseases and ailments that affect humans can also afflict dogs and cats. Additionally, each species has unique infectious diseases and hereditary disorders that may impair their health and quality of life.

Before discussing selected diseases, abnormalities, and other health-related disorders, it is important to have baseline/reference data for healthy dogs and cats as shown in Tables 14.1 and 14.2. Data for other species of companion animals are presented in Chapters 5, 6, and 7.

**Table 14.1**
Selected Vital Statistics for Healthy Dogs by Age Group

|  | Respiration Rate (inhalations/min) | Heart Rate Resting (beats per min, range/avg) | Body Temperature (°F, range/avg) |
|---|---|---|---|
| First 1 to 2 weeks | 15–30 | 200–250/220 | 96–97/96.5 |
| At 4 weeks | 15–30 | 70–220/200 | 100–101/100.5 |
| Adult | 10–30 | 60–160/120 | 101–102.5/101.5 |

**Table 14.2**
Selected Vital Statistics for Healthy Cats by Age Group

|  | Respiration Rate (inhalations/min) | Heart Rate Resting (beats per min, range/avg) | Body Temperature (°F, range/avg) |
|---|---|---|---|
| First 1 to 2 weeks | 15–35 | 200–250/220 | 96–97/96.8 |
| At 4 weeks | 15–35 | 180–220/200 | 100–101/100.5 |
| Adult | 20–30 | 160–240/195 | 101–102.5/101.5 |

In this chapter, we identify and discuss important diseases and disorders of dogs and cats as well as ways of ensuring the best possible health (freedom from physical disease or pain) in companion animals.

# 14.2   INFECTIOUS DISEASES

The field of infectious diseases is among the most rapidly growing areas of human and veterinary medicine. The threat of bioterrorism[1] and the risks associated with zoonotic[2] diseases have resulted in the funding of laboratories and researchers devoted to the study of emerging infectious diseases in addition to those seeking new information pertaining to the diagnosis and treatment of known diseases. This section includes many of the common infectious diseases affecting dogs and cats. Additional information on zoonotic diseases is found in Section 14.3.

## 14.2.1   BACTERIAL DISEASES

*Abscesses.* Abscesses are most commonly associated with fight injuries or penetrating wounds. In cats, most abscesses are caused by *Pasteurella* species, a normal inhabitant of the feline mouth. Anaerobic bacteria (those that flourish under conditions of low oxygen) are commonly found in abscesses associated with puncture wounds in both dogs and cats. Treatment involves draining the abscess and giving oral or parenteral (injectable) antibiotics.

*Bordetellosis.* *Bordetella bronchiseptica* causes upper respiratory tract infections in cats and dogs with clinical signs of sneezing, coughing, and nasal discharge. Diagnosis can be confirmed by culture, and treatment with systemic antibiotics is recommended. Vaccines are available for prevention and are recommended for animals at high risk of exposure—those going to kennels/boarding facilities, cat/dog shows, training classes, and other places with increased risk due to companion animal density.

*Borreliosis (Lyme disease).* *Borrelia burgdorferi* is transmitted by ticks of the genus *Ixodes*. Ticks must be attached to a host for a minimum of 24 hours to transmit the infection. Signs of disease are most common in dogs and humans. The most common sign in dogs is lameness with swelling of one or more joints. Affected humans usually develop a characteristic skin rash; however, rashes are rare in dogs. Other signs include depression, fever, and lack of appetite. Blood tests can be used to detect antibodies to the organism in the serum of an infected animal. Treatment with antibiotics such as doxycycline or amoxicillin is usually followed by rapid improvement. Vaccines containing killed *B. burgdorferi* or selected components (antigens) are

---

[1] Bioterrorism is the use of infectious agents—naturally occurring or genetically engineered to increase their virulence—in warfare or to otherwise harm people. For more information see http://www.bt.cdc.gov/.

[2] Zoonotic diseases are those naturally transmitted between animals and humans (see Section 14.3).

available; however, the effectiveness is controversial and use is generally limited to dogs at high risk of exposure (e.g., hunting dogs in certain geographical areas).

***Brucellosis.*** *Brucella canis* causes infertility, testicular enlargement, late-gestation abortion, ocular inflammation (uveitis), fever, back pain (discospondylitis), lameness (osteomyelitis), and enlarged lymph nodes. Infection in dogs usually occurs during breeding or through contact with aborted tissues. Although infrequent, humans can also become infected through contact with fluids or tissues from an aborting bitch. Treatment is difficult because the organism lives intracellularly. Neutering of affected dogs plus long-term treatment with minocycline or tetracycline plus an aminoglycoside (e.g., gentamicin) may be attempted; however, the prognosis for a cure is guarded. Prevention includes serological testing and mating of only serologically negative dogs. Dogs are rarely infected with other species of *Brucella;* however, infections with *B. melitensis, B. abortus,* and *B. suis* can be acquired by ingesting contaminated milk, aborted fetuses, or fetal membranes of infected livestock. Clinical signs are rarely seen in association with these infections in dogs. Cats are susceptible to *B. suis;* however, this infection is extremely rare and seldom associated with clinical signs of disease in cats.

***Campylobacteriosis.*** Campylobacteriosis is a gastrointestinal infection caused by *Campylobacter jejuni* or *C. coli. Campylobacter* spp can be found in normal dogs and cats; however, in young, debilitated, or stressed animals, these bacteria can cause diarrhea and vomiting. Stained fecal smears will show the presence of small S-shaped rod bacteria. These organisms can be transmitted to humans and other animals. Several antibiotics can be used for treatment with erythromycin considered the drug of choice.

***Cat scratch disease.*** This is a disease of humans caused by *Bartonella henselae.* Infection in humans usually follows a scratch or bite by a kitten. The affected person develops a small bump at the site of the scratch or bite that is followed by enlargement of local lymph nodes, fever, and sometimes muscle pain and nausea. Biopsies of lymph nodes show the presence of the organism. Most cases resolve spontaneously. Up to 60% of cats may be seropositive, indicating infection with *B. henselae;* however, the organism does not cause clinical disease in cats. To minimize the risk of human infection, immunocompromised persons should avoid young cats, all cat scratches or bites should be washed with soap, and cats should not be permitted to lick any wounds on people.

***Chlamydial infections.*** *Chlamydophila felis* is a common cause of conjunctivitis (eyelid infections) and mild upper-respiratory tract disease (sneezing, nasal discharge) in cats. Topical and oral tetracycline are effective treatments. A few cases of suspected transmission to immunosuppressed humans have been reported. Care should be taken to avoid direct contact with discharges from the eyes or nose of infected cats.

***Clostridial disease.*** *Clostridium perfringens* and *C. difficile* can produce diarrhea in some dogs and cats. These organisms are identified on stained fecal smears as large numbers of spore-forming bacteria and are associated with large numbers of white blood cells in the feces. Metronidazole is effective in treatment of both clostridial species.

***Helicobacter infection.*** *Helicobacter* species can be found in the stomachs of 85% to 90% of cats and 70% to 100% of dogs. It is unclear whether these organisms produce disease in animals; however, these species can be important causes of gastric ulcers in humans. It is also uncertain whether animals pose a risk for zoonotic infection of humans. Diagnosis requires a biopsy of the stomach showing the presence of large numbers of these organisms. A combination of two antibiotics plus cimetidine or another gastric acid inhibitor can be used to eliminate the infection.

***Leptospirosis.*** Most infections in dogs are caused by *Leptospira icterohaemorrhagiae* or *L. canicola.* The organisms are acquired from direct contact with an infected animal or from contact with contaminated soil, water, or other substances. Organisms penetrate mucous membranes and replicate in the bloodstream. They may damage blood vessels (vasculitis), eyes (uveitis), kidneys, liver, and the central nervous system. Clinical signs depend on the site of infection and may include fever, lack of appetite, muscle pain, vomiting, icterus, bruising, painful eyes, and coughing. Diagnosis is based on serial antibody titers or finding the organism in the urine or by culture. Treatment includes antibiotics and supportive care. Bacterins containing *L. canicola* and *L. icterohaemorrhagiae* are available but are not 100% effective in preventing infection. Other serovars reported to infect dogs include *L. grippotyphosa*, *L. pomona*, *L. bratislava*, *L. copenhagenii*, *L. autralis*, *L. autumnalis*, *L. ballum*, and *L. bataviae*. Humans are susceptible to infection and should be especially careful to avoid contact with contaminated urine. Infections in cats are rare; reported serovars include *L. canicola, L. grippotyphosa, L. pomona,* and *L. bataviae*. Leptospirosis infections in cats are often asymptomatic (also true in dogs), but may result in kidney and liver disease.

*Mycobacterial infections.* Tuberculous infections are rare in dogs and cats. *Mycobacterium tuberculosis* may be acquired from humans, *M. bovis* from meat products, and *M. avium* from birds. Clinical signs include fever, weight loss, vomiting, diarrhea, coughing, and enlarged lymph nodes. Diagnosis is based on finding nodules containing mycobacterial organisms in the respiratory or gastrointestinal tract. Due to the zoonotic potential and guarded prognosis, most infected pets are euthanized. Nontuberculous infection can be caused by atypical mycobacterial organisms; these infections produce cutaneous or subcutaneous masses and draining tracts. They are differentiated from tuberculosis by culture of the organisms and can be treated by surgical removal followed with long-term antibiotics (usually fluoroquinolones).

*Plague.* Cats are highly susceptible to plague, an infection caused by *Yersinia pestis*. Dogs are rarely affected. The reservoirs are wild rodents, ground squirrels, prairie dogs, rabbits, bobcats, and coyotes. Recent outbreaks in prairie dogs have increased concerns about this typically fatal disease. There are two common forms of plague—bubonic and pneumonic. Bubonic plague involves the skin and lymph nodes, whereas pneumonic plague is an infection of the lungs. Cats acquire bubonic plague by ingesting infected rodents or by being bitten by fleas. The fleas serve as vectors and transmit the bacteria in saliva. The disease is endemic in the southwestern United States. Clinical signs in cats include fever, depression, and enlarged tonsils and lymph nodes. The lymph nodes may rupture and drain fluid containing large numbers of organisms that are highly infectious to humans. Pneumonic plague can be acquired by inhalation of the organism (e.g., via aerosolized droplets) or by spread of the organism via the bloodstream to the lungs. When the infection involves the lungs, the prognosis is grave and the risk of zoonosis is high due to shedding the organism in nasal and respiratory secretions. Aggressive treatment with systemic antibiotics may be effective in early cases of bubonic plague. Cats with pneumonic plague should be euthanized due to the poor prognosis and high threat of zoonotic infections.

*Pyoderma.* Pyodermas are bacterial skin infections. The most commonly isolated organism in pyoderma of both dogs and cats is *Staphylococcus intermedius*. These infections are not contagious (nor are they zoonotic); however, they are often recurrent due to other predisposing factors such as underlying allergies or endocrine disease. Management of skin infections includes the use of topical medications (often in the form of medicated shampoos) and oral antibiotics.

*Salmonellosis.* Many species of *Salmonella* can infect dogs and cats. Infection can be acquired from ingestion of contaminated food or by direct contact with feces of an infected animal. Clinical symptoms include fever, vomiting, diarrhea, and weight loss. Diagnosis is based on culturing feces and treatment involves giving oral antibiotics. Humans can be infected; good hygiene and disinfection of all areas contaminated with feces of an infected animal are important.

*Streptococcal infections.* Many species of the genus *Streptococcus* can infect dogs and cats; some species are considered part of the normal flora. Streptococci may cause infections of the skin, ears, respiratory tract, and urinary tract of dogs and cats. These infections are treated with ampicillin, penicillin, or a cephalosporin. Rarely, dogs and cats can acquire group A[3] streptococci (e.g., *S. pyogenes*) from humans and serve as a source of reinfection for the people in the household. The carrier state can be detected by culturing tonsils and the infection eliminated with antibiotic treatment (may require treatment of all people in the household because humans are more common subclinical carriers than animals).

*Tularemia.* Tularemia is a syndrome caused by *Francisella tularensis*. It may be transmitted by tick vectors (*Dermacentor variabilis, D. andersoni,* and *Amblyomma americanum*) or by ingesting or contacting tissues of infected rabbits or rodents. Outdoor hunting cats are at greatest risk of infection; dogs are rarely affected. Humans can be infected when bitten by an infected cat or from exposure to its body fluids. Clinical signs include fever, oral ulcers, enlarged lymph nodes, and in later stages, liver disease and bone marrow suppression. Mortality rates of infected animals are high.

## 14.2.2   HELMINTH INFECTIONS

The prevention of helminth infections (worms) is important to the health of dogs and cats and is discussed in Chapter 15.

## 14.2.3   RICKETTSIAL INFECTIONS

*Rickettsia.* *Rickettsia* spp are gram-negative, obligate intracellular bacteria that infect mammals and arthropods. Examples of diseases caused by rickettsial organisms include ehrlichiosis, Rocky Mountain spotted fever, salmon poisoning, and typhus (primarily affects humans).

*Ehrlichiosis.* *Ehrlichia* are intracellular organisms transmitted by ticks. The most common vector is the brown dog tick, *Rhipicephalus sanguineus*. Dogs are most commonly affected with *E. canis,* which causes fever, weight loss, bleeding disorders (e.g., nosebleeds), coughing, lameness, and sometimes neurological signs. Serology is helpful in confirming the diagnosis, and antibiotics such as tetracycline are effective in treatment when given in the early stages of the disease. Later in the disease, the bone marrow may be irreversibly damaged and the animal may die from bleeding, anemia, or secondary infections. Cats may be infected with *E. risticii* and show signs of fever, diarrhea, and enlarged lymph nodes. Serological testing facilitates diagnosis. Treating cats with antibiotics such as tetracycline or doxycycline is usually effective in eliminating the infection.

*Rocky Mountain spotted fever.* Rocky Mountain spotted fever is caused by *Rickettsia rickettsii*. It is transmitted by the *Dermacentor* species of ticks. Clinical signs may include fever, a skin rash with bruises, nasal discharge, coughing, enlargement of lymph nodes, joint or muscle pain, and decreased vision. Paired serum samples[4] for antibody titers are useful in confirming the diagnosis. Treatment involves giving tetracycline antibiotics. Tick control is helpful in preventing infection.

---

[3] *Streptococci* spp are classified as belonging to groups A, B, C, D, F, and G. Humans are the natural reservoir hosts of group A streptococci. Group A streptococci can cause infections of the tonsils, oral cavity, ears, skin, joints, lungs, and central nervous system of people.

[4] Paired serum samples are two blood samples taken from an animal at an interval of approximately 3 to 4 weeks for measurement of serum antibody levels against a specific antigen. An increase in the antibody levels (titer) is indicative of the presence of an active infection.

*Salmon poisoning.* Salmon poisoning is a highly fatal disease affecting domestic and wild Canidae in the Pacific Northwest. It is caused by *Neorickettsia helminthoeca,* a rickettsia transmitted by a fluke (*Nanophyetus salmincola,* Chapter 15). The fluke is a parasite of salmon. Dogs (and wild canids) acquire the infection through ingesting an infected salmon. The incubation period is usually 5 to 7 days after ingestion of the parasitized fish; the first sign of infection is a fever of 104.0°F to 107.6°F. Infected dogs stop eating and may die within 7 to 10 days if not promptly treated with tetracyclines or other effective antibiotics.

## 14.2.4   Mycoplasma Infections

Mycoplasmas are the smallest free-living microorganisms found in nature. These organisms obtain nutrients from the surrounding environment. Some are found on the mucous membranes of the respiratory, genitourinary, and gastrointestinal tracts; others affect joints; some are found in the bloodstream. The mycoplasmas of greatest importance in companion animals are *Mycoplasma hemofelis* and *M. hemominutum.*

*Mycoplasma hemofelis (formerly* **Hemobartonella felis***).* *M. hemofelis* and *M. hemominutum* infect red blood cells (RBCs, erythrocytes) of cats causing feline infectious anemia. The organisms may appear as paired cocci, short rods, or small rings on the RBC plasma membrane. Clinical signs are related to the severity of the anemia and include weakness, anorexia with weight loss, and depression. The gums may be very pale. Diagnosis is made by finding the organisms on stained blood smears or by detecting its DNA fragments on PCR[5] tests. Treatment involves giving tetracycline antibiotics. Severely anemic cats may also require blood transfusions.

## 14.2.5   Fungal Infections

*Blastomycosis.* Blastomycosis is caused by *Blastomyces dermatidis,* a fungal organism found in soil. Young, large-breed, male dogs are at highest risk for infection. Cats are rarely affected. The infection is usually acquired by inhalation of spores. Symptoms include fever, chronic coughing, fungal pneumonia, weight loss, draining skin lesions, enlarged lymph nodes, and blindness. Chest radiographs can be highly suggestive of the diagnosis of fungal pneumonia. The organisms may be found on microscopic examination of smears made from skin lesions, lymph node aspirates, or fluid obtained from airways (transtracheal aspirates or bronchoalveolar lavage). Treatment includes the administration of specific antifungal drugs such as ketoconazole, itraconazole, or amphotericin B. Humans are unlikely to acquire infection from a dog or cat except through direct contamination of an open wound; however, like pets, people can become infected by breathing spores in the environment (i.e., from the same source as pets).

*Coccidiomycosis.* Coccidiomycosis is caused by inhalation of spores of the fungus *Coccidioides immitis.* The organisms are primarily found in semiarid areas of the southwestern United States and most commonly affect young, large-breed dogs. Cats are resistant to infection. Symptoms of infection include fever, coughing, fungal pneumonia, weight loss, draining skin lesions, enlarged lymph nodes, lameness, and sometimes neurological signs. Chest radiographs may show signs suggestive of a fungal pneumonia, and the organisms can be found on samples from the skin, lymph nodes, bone, or lungs. Systemic antifungal drugs such as ketoconazole or itraconazole are the treatment of choice.

---

[5]PCR is an acronym for polymerase chain reaction. PCR produces a huge number of copies of a fragment of DNA and allows scientists to determine its nucleotide sequence and specific identity.

***Cryptococcosis.*** *Cryptococcus neoformans* is a yeastlike fungus found in soil contaminated with droppings from pigeons and other birds. Infection occurs via inhalation and is most common in cats. Clinical signs include nodular skin disease, draining tracts, sneezing with nasal discharge, and in severe cases ocular and central nervous system disease. The organism can be found in smears made from skin lesions or nasal swabs. Serological tests can be used for monitoring treatment.[6] Systemic antifungal drugs including itraconazole, ketoconazole, and fluconazole are usually effective but may require a prolonged course of treatment.

***Histoplasmosis.*** Histoplasmosis is caused by the fungus *Histoplasma capsulatum*. This organism is found in soil contaminated with bird or bat droppings. It is most common in the Mississippi and Ohio river valleys. Infection following inhalation results in coughing, fever, and a fungal pneumonia that may spread to involve other organs. Infection following ingestion results in fever, vomiting, diarrhea, and weight loss. Diagnosis is based on finding the organism in samples from the gastrointestinal tract, lymph nodes, liver, bone marrow, or lungs. Radiographs of animals with respiratory disease show a fungal pneumonia and enlargement of lymph nodes near the base of the heart. Systemic antifungal drugs including itraconazole, ketoconazole, and amphotericin B are usually effective in treatment.

***Ringworm (also known as dermatophytosis).*** Dermatophytes are fungi that live in keratinized tissues such as skin, hair, and claws (nails). Three species commonly affect dogs and cats: *Microsporum canis, M. gypseum,* and *Trichophyton mentagrophytes.* Clinical signs of infection include patchy hair loss (often in a concentrically spreading pattern) and deformed claws (Figure 14.1). These organisms can persist for long periods of time on infected hairs and infect many species of mammals, including humans. Diagnosis should be confirmed by culture of the organism. Treatments include topical and systemic antifungal medications, which are administered until the animal is negative on two consecutive fungal cultures. A killed vaccine containing *M. canis* is available for use in cats; however, its efficacy is questionable.

***Sporotrichosis.*** Sporotrichosis is caused by *Sporothrix schenckii,* a fungus found in soil and transmitted primarily by contamination of wounds. Clinical signs include an ulcerated skin nodule and enlargement of draining lymph nodes. Diagnosis is based on finding the organism in smears made from skin lesions. Ketoconazole, itraconazole, and iodides have been used successfully in the treatment of skin lesions.

***Malassezia.*** *Malassezia* spp are yeast organisms found in many dogs (and occasionally cats) with ear infections and skin infections. These are opportunistic organisms that thrive in moist, warm, lipid-rich environments. They are part of the normal flora of the ears and skin of dogs and cats. These organisms proliferate and contribute to disease when the ears or skin is inflamed (often as a result of underlying allergies). *Malassezia* infections are diagnosed by finding budding yeast on cytological examination of ear swabs and acetate tape preps (Chapter 16, Figure 16.5). A variety of topical and systemic antifungal medications can be used to treat *Malassezia* infections; however, long-term control requires identification and management of predisposing factors such as environmental or dietary allergies (see Section 14.4.2).

## 14.2.6 VIRAL DISEASES

***Coronaviral infection.*** Signs of canine coronavirus vary from an inapparent infection to severe vomiting and diarrhea. Puppies are at greatest risk for severe disease. Diagnosis may be

---

[6]Successful treatment results in a decrease in the amount of cryptococcal antigen in the animal's serum.

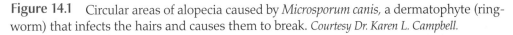

**Figure 14.1**   Circular areas of alopecia caused by *Microsporum canis,* a dermatophyte (ringworm) that infects the hairs and causes them to break. *Courtesy Dr. Karen L. Campbell.*

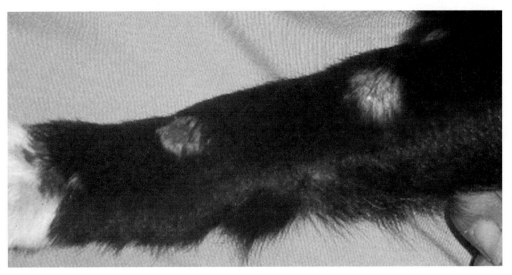

confirmed by virus isolation or finding viral particles on electron microscopy. The disease is usually self-limiting and treatment involves supportive care (may include intravenous fluids and electrolytes). Vaccines are available, but their efficacy is controversial. Feline coronavirus enteritis is similar to that of dogs with symptoms of infection varying from asymptomatic carriers to vomiting and diarrhea. Treatment consists of supportive care.

*Feline infectious peritonitis (FIP).*   FIP is caused by a coronavirus related to the one causing feline coronavirus enteritis. Clinical signs are variable and may include fever, anorexia, weight loss, vomiting, diarrhea, blindness, liver disease, seizures, and in the "wet" form of the disease, fluid accumulations in the abdomen or chest. Diagnosis is confirmed by characteristic findings on biopsies. There is no specific therapy and the disease is usually fatal. A vaccine is available, but its effectiveness is controversial.

*Feline upper respiratory viruses.*   The two most common causes of upper respiratory disease in cats are feline viral rhinotracheitis (FVR, feline herpesvirus–1) and feline calicivirus. Transmission is primarily by direct contact with infected animals. Recovered animals serve as reservoirs and shed the virus intermittently for the rest of their lives. Clinical signs are most severe in kittens and include ocular and nasal discharges, sneezing, fever, anorexia, and in some cases ocular and oral ulcers. Diagnosis is usually based on clinical signs but can be confirmed by virus isolation, PCR, or paired serum antibody titers. Treatment is supportive because the disease is generally self-limiting with a course of 5 to 10 days. Vaccination is helpful in preventing clinical disease but does not eliminate the infection or prevent shedding of the virus.

*Canine distemper.*   Canine distemper is caused by a morbillivirus that is usually acquired by inhalation of viral particles shed in the secretions of infected dogs. It is related to the virus that causes measles in humans. Clinical signs depend on the immune status of the dog. Early symptoms include anorexia, nasal and ocular discharges, fever, coughing, vomiting, and diarrhea. More severe cases affect the central nervous system causing seizures or severe muscle twitching (myoclonus). Chronic cases may result in blindness, progressive weakness, circling, and behavioral changes. Fluorescent antibody testing is used to confirm the presence of the virus in epithelial cells of the conjunctiva. The presence of antidistemper antibodies in cerebral spinal fluid also confirms the diagnosis. Treatment involves supportive care. Vaccines are effective in prevention if started when a puppy is 6 to 8 weeks of age and given every 3 to 4 weeks for a total of three vaccinations followed by boosters every 1 to 3 years.

***Canine infectious tracheobronchitis (ITB, kennel cough).*** The term *tracheobronchitis* refers to inflammation of the upper airways including the trachea (windpipe) and bronchi (airways extending into the lungs). ITB can be caused by several viral and bacterial agents. Mixed infections are common. ITB is highly contagious and is transmitted through aerosolized viral particles and nasal and oral secretions. Viruses causing ITB include canine parainfluenza–3, canine adenovirus type 2 (CAV–2), and canine herpesvirus. *Bordetella bronchiseptica* is the most common bacterial agent in ITB. Signs of ITB include a "honking" cough and low-grade fever. Coughing is stimulated by pressure on the trachea (e.g., pulling on a collar). The coughing usually resolves over 1 to 2 weeks, although it may persist for several weeks in some dogs. Treatment involves supportive care, humidification of the environment, a cough suppressant, and in severe cases, antibiotics to prevent secondary bacterial infections. Vaccines are available to immunize dogs against the causative agents. Immunity is short lasting, thus dogs at high risk of exposure (in boarding kennels or being shown) may require booster vaccinations every 6 months.

***Feline immunodeficiency virus (FIV).*** FIV is caused by a lentivirus that attacks and destroys T lymphocytes in the blood. (T lymphocytes are important for normal functioning of the immune system.) FIV infections usually occur as a result of bite wounds from an infected cat. Infected cats may be asymptomatic for months to years but eventually develop secondary infections with fever, weight loss, coughing, oral ulcers, diarrhea, and other signs. Diagnosis is based on finding antibodies to FIV in the serum. Treatment involves supportive care and management of secondary infections. While similar to the human immunodeficiency virus (HIV); lentiviruses are species-specific—FIV is *not* contagious to humans.

***Feline leukemia virus (FeLV).*** FeLV is caused by an oncovirus. Infection occurs from exposure to secretions, especially saliva, from an infected cat. Some infected cats are able to eliminate the infection within 2 to 4 weeks; others become permanently infected with the virus. Clinical signs associated with infection are variable and include fever, oral ulcers, enlarged lymph nodes, depression, anorexia, vomiting, diarrhea, and secondary infections or neoplasia (cancers including leukemia and lymphosarcoma). Diagnosis is based on finding viral proteins in the blood, tears, or saliva. Test kits for FeLV are widely available for on-site use by veterinarians. Currently, no specific treatment is available beyond symptomatic care and management of secondary infections or neoplasms. Several vaccines are available and are recommended for use in cats at high risk for infection (e.g., outdoor or indoor-outdoor cats).

***Infectious canine hepatitis (ICH; Canine adenovirus type 1, CAV–1).*** ICH is a rare cause of liver disease, fever, vomiting, diarrhea, and corneal edema (cloudy eyes) in dogs. The virus is spread via the oronasal route. Vaccines are based on cross-reactivity (close relationship) between ICH and CAV–2.

***Herpesvirus infections.*** In dogs, herpesviruses are responsible for some cases of infertility, abortion, and upper respiratory tract disease (ocular and nasal discharge, coughing). In cats, herpesviruses are responsible for signs of upper respiratory tract disease, ocular and nasal discharges, and occasional infertility. Diagnosis is based on finding viral inclusion bodies or by virus isolation. Symptomatic treatment may be required on a long-term basis because some animals remain chronically infected.

***Feline panleukopenia virus (Feline parvoviral enteritis).*** This highly infectious virus is transmitted by the fecal-oral route. The virus has a predilection for the rapidly dividing cells of the intestinal tract and bone marrow. Clinical signs include fever, depression, vomiting, and diarrhea. Laboratory findings include a severely depressed white blood cell count that makes the kitten or cat very susceptible to secondary infections, which are often fatal. Treatment involves supportive care and may require plasma or blood transfusions, intravenous fluids and electrolytes, and broad-spectrum antibiotics. Vaccines are highly effective in prevention and should be given as a series of two or three vaccinations starting at 8 weeks of age.

*Canine parvovirus infection.* Parvovirus infections in dogs are similar to feline parvovirus enteritis. In addition to replicating in the intestines and bone marrow, the virus can infect the heart of young puppies producing sudden death. Certain breeds, including Doberman Pinschers and Rottweilers, are particularly susceptible to infection and more likely to die from secondary complications. Vaccines are available but do not provide 100% immunity for all dogs.

*Rabies.* Rabies is caused by a rhabdovirus that is infectious for all warm-blooded animals. The primary reservoirs are bats, skunks, foxes, and raccoons. Infection occurs when saliva from an infected animal enters a wound (e.g., a bite) or the conjunctiva or nasal mucosa. The virus travels along nerve fibers to the central nervous system. Clinical signs include behavioral changes, drooling, difficulty in swallowing, weakness, seizures, and other neurological abnormalities culminating in paralysis and death. Vaccines should be given in accordance with local ordinances (see Section 14.7).

The major viral diseases of dogs and cats are summarized in Tables 14.3 and 14.4.

## 14.3 ZOONOSES AND PUBLIC HEALTH ASPECTS

Zoonotic diseases are those naturally transmitted between animals and humans. Transmission can occur by direct contact with an infected dog or cat; by indirect by contact with secretions, feces, or hairs shed by the animal; or by shared vectors such as ticks and fleas (Table 14.5). Humans are at greatest risk of acquiring a zoonotic disease when their immune system is suppressed by human immunodeficiency virus (HIV) or chemotherapy, or when very young or old. Families with proven or suspected immunodeficient members are often told to avoid pet ownership because of potential health risks. However, the risks of acquiring a zoonotic

**Table 14.3**
Major Viral Diseases of Dogs

| Disease | Common Mode of Transmission | Body Systems Affected | Symptoms |
|---|---|---|---|
| Canine distemper | Contact with body fluids | Respiratory, GI,* and nervous systems; skin and mucous membranes | Fever, nasal discharge diarrhea, convulsions, death |
| Canine hepatitis | Contact with infected animal | Liver, kidney, CNS**, vascular endothelium | Fever, anorexia, liver failure, death |
| Rabies | Bite from or contact with infected animal | CNS, respiratory, salivary glands | Paralysis, temperament change, hypersensitivity, death |
| Canine tracheobronchitis | Contact with secretions of infected animal | Upper respiratory tract | Cough, fever |
| Canine parvovirus | Contact with infected animal | GI tract, lymph nodes, thymus, bone marrow, heart tissue in young puppies | Fever, anorexia, bloody diarrhea; dehydration; respiratory distress, death |
| Canine coronavirus | Contact with feces or fluids of infected animal | GI tract | Anorexia, vomiting, diarrhea |

*Gastrointestinal*
**Central nervous system*

**Table 14.4**
Major Viral Diseases of Cats

| Disease | Common Mode of Transmission | Body Systems Affected | Symptoms |
|---|---|---|---|
| Feline infectious peritonitis (FIP) | Contact with infected animal, highest incidence in multicat facilities | Multisystemic: liver, kidney, intestines, peritoneal cavity, lungs, pleural cavity, CNS, eyes | Fever, loss of appetite, stunted growth in kittens, dull hair coat, icterus, painful eyes, blindness, distended abdomen, respiratory distress, death |
| Feline rhinotracheitis | Contact with infected cat | Respiratory, ocular, reproductive | Rhinitis, sneezing, nasal discharge, conjunctivitis, ocular discharge, ulcerative keratitis, severe infections in kittens born to infected queens |
| Feline calicivirus | Contact with infected cat | Respiratory, ocular, musculoskeletal, GI | Rhinitis, nasal discharge, pneumonia, ulceration of nose, conjunctivitis, keratitis, acute arthritis (lameness), ulcers of tongue and hard palate |
| Feline immunodeficiency virus (FIV) | Cat-to-cat transmission, usually by bite wounds | Blood, lymph nodes, CNS, GI, other systems from secondary infections | Enlarged lymph nodes, gingivitis, stomatitis, rhinitis, conjunctivitis, keratitis, diarrhea, chronic skin and ear infections, fever and wasting, behavioral changes, death |
| Feline leukemia virus (FeLV) | Cat-to-cat transmission, bites, shared dishes or litter pans, transplacental and transmammary | Blood/bone marrow, lymph nodes, GI, respiratory tract, CNS, other systems with tumors or secondary infections | Fever, diarrhea, chronic infections, rhinitis, gingivitis, stomatitis, anemia, tumors, leukemia, wasting, death |
| Feline panleukopenia | Direct contact, infected fomites, virus remains infective in environment for years | Blood/bone marrow, GI, CNS, ocular, reproductive | Severe depression of white blood cell production, vomiting, diarrhea, dehydration, fever, ataxia (incoordination), death |
| Rabies | Bite from or contact with infected animal | CNS, respiratory, salivary glands | Paralysis, temperament change, death |

infection from a healthy animal are minimal. Pet ownership provides many health benefits including greater happiness and decreased depression (Chapter 1). Thus, any decision regarding pet ownership should consider potential risks and potential benefits. Steps can also be taken to minimize the risk of zoonotic transmission of diseases.

Enteric (gastrointestinal) zoonotic agents include bacteria (*Campylobacter* spp, *Escherichia coli, Helicobacter* spp [zoonotic risk is controversial], *Salmonella* spp, *Yersinia* spp), protozoans (*Cryptosporidium parvum, Toxoplasma gondii, Giardia* spp), cestodes (*Echinococcus multilocularis, Dipylidium caninum*), and helminths (*Ancylostoma braziliense, A. tubaeforme, Uncinaria stenocephala, Strongyloides stercoralis, Toxocara cati, T. canis*). The incidence of these infections varies by geographical region, whether pets are kept indoors or

**Table 14.5**
Zoonotic Diseases Associated with Dogs and Cats

| Organism | Associated Syndrome | Role of Pet |
|---|---|---|
| **Bacteria** | | |
| *Campylobacter* | Dogs and cats: vomiting, diarrhea<br>Humans: vomiting, diarrhea | Infectious organisms in feces |
| *Salmonella* | Dogs and cats: fever, diarrhea, vomiting<br>Humans: fever, diarrhea, vomiting | Infectious organisms in feces |
| *Yersinia* | Cats: susceptible to plague<br>Dogs: naturally resistant<br>Humans: susceptible to plague | Exudates and secretions are sources of organisms; can also serve as transport host for infected fleas |
| *Bartonella* | Cats: part of normal flora of mouth<br>Humans: cat scratch disease | Inoculates organism via bites or scratches |
| *Francisella* | Cats and dogs: none reported<br>Humans: fever, localized and systemic infections | Dogs may be reservoirs of ticks |
| *Pasteurella* | Cats: part of normal flora of mouth, can produce abscesses following bite wounds<br>Humans: abscesses following bite wounds | Bite wounds inoculate the organism |
| *Streptococcus* | Dogs and cats: many species found as normal flora, occasionally associated with skin, ear, respiratory or urinary tract infections<br>Humans: are a reservoir and source of group A infections in animals | Group A streptococcal infections are a reverse zoonosis with humans serving as reservoirs and infecting pets |
| *Leptospira* | Dogs: kidney and/or liver disease<br>Humans: kidney and/or liver disease | Infectious organisms shed in urine |
| *Brucella* | Dogs: infertility, abortion<br>Humans: fever, malaise | Infectious organisms in secretions, especially aborted tissues |
| *Chlamydophila* | Cats: ocular and nasal discharges, sneezing<br>Humans: rare conjunctivitis in immunocompromised individuals | Infectious organisms in secretions from eyes and nose |
| *Mycobacterium* | Dogs and cats: atypical forms produce skin lesions, rarely tuberculosis<br>Humans: greatest concern is tuberculosis | Infectious organisms present in exudates from lesions |
| **Protozoa** | | |
| *Giardia* | Dogs, cats, humans: vomiting, diarrhea | Cysts shed in feces are infectious |
| *Cryptosporidium* | Dogs, cats, humans: vomiting, diarrhea | Oocysts shed in feces are infectious |
| *Toxoplasma* | Dogs, cats, humans: signs vary, can include respiratory, gastrointestinal, ocular and/or nervous system disease; cats are the most likely to have ocular signs; in humans transplacental infection of infant is greatest concern | Infected cats shed oocysts, which are infectious after 1 to 5 days of incubation; most human infections are acquired through the ingestion of cysts in undercooked meats |
| **Fungal** | | |
| Dermatophytes (Ringworm) | Dogs and cats: patchy to circular areas of hair loss, deformed growth of claws (nails)<br>Humans: circular areas of skin infection, hair loss, deformed growth of nails | Infectious spores are shed on hairs from infected dogs and cats |
| *Sporothrix* | Dogs and cats: draining skin lesions<br>Humans: draining skin lesions | Transmission is rare; however, lesions from cats contain high numbers of infectious organisms |
| **Viral** | | |
| Rabies | Dogs, cats, humans: progressive neurological disease | Transmission is rare; however, saliva from infected pets contains the virus |

*continued*

311

**Table 14.5**
Zoonotic Diseases Associated with Dogs and Cats (Continued)

| Organism | Associated Syndrome | Role of Pet |
|---|---|---|
| **Cestodes** | | |
| *Echinococcus* | Dogs: tapeworm parasite of intestines<br>Humans: cysts in body tissues | Ova shed in feces are infectious |
| *Dipylidium* | Dogs: tapeworm parasite of intestines<br>Humans: rare tapeworm | Transmission requires ingestion of an infected flea |
| **Helminths (see also Chapter 15)** | | |
| *Ancylostoma* | Dogs and cats: hookworms of intestines<br>Humans: cutaneous larval migrans | Ova shed in feces |
| *Uncinaria* | Dogs: hookworms of intestines<br>Humans: rare cutaneous larval migrans | Ova shed in feces |
| *Toxocara* | Dogs and cats: roundworms of intestines<br>Humans: possible blindness | Ova shed in feces |
| *Strongyloides* | Dogs and cats: roundworms of intestines<br>Humans: rare intestinal and pulmonary parasites | Larvae shed in feces |
| **Ectoparasites (see also Chapter 15)** | | |
| Ticks | Dog, cats, humans: local and systemic reactions, vectors of tick-borne diseases | Dogs and cats are reservoirs for ticks |
| Fleas | Dogs and cats: irritation, skin allergies, transmit tapeworms following ingestion<br>Humans: irritation | Dogs and cats are reservoirs for fleas |
| *Cheyletiella* | Dogs and cats: skin mites, irritation<br>Humans: skin irritation | Dogs and cats are reservoirs for mites |
| *Sarcoptes* | Dogs: skin mites<br>Humans: skin irritation | Dogs are reservoir for mites |

allowed to roam and hunt outdoors, and by preventive healthcare being given to the pets. The major route of transmission of enteric pathogens is by the fecal-oral route through environmental contamination. Screening procedures to identify and treat enteric zoonotic agents include fecal examinations and fecal cultures. Treatment and prevention include good sanitation and specific antibiotics or anthelmintics depending on the type of agent identified.

Cutaneous zoonoses are those that can be transmitted to people by direct contact with pet hair or by contact with exudates from skin lesions of pets. These zoonoses include fungal, parasitic, and bacterial infections. All animals having skin lesions should be examined by a veterinarian for diagnoses and treatment of the condition. The most commonly transmitted fungal infection is ringworm (dermatophytosis) affecting up to 50% of people in a household with an infected pet. To lessen the potential for zoonotic transfer of infection, pets with dermatophyte infections should be treated with both topical and systemic antifungal medications. The house and furnishings should be thoroughly cleaned to remove infected hairs.

Although respiratory infections are common in dogs and cats, only a few are considered zoonotic threats. Group A streptococcal infections are a reverse zoonosis; humans are the primary reservoir and pets are infected by contact with humans. The only zoonotic respiratory infections of life-threatening significance to humans are plague from cats with the pneumonic form of the disease and *M. tuberculosis* in dogs (very rare in dogs). People in contact with a plague-infected cat or a dog with tuberculosis should seek medical advice about prophylactic antibiotics and wear masks, caps, gowns, and gloves. Euthanasia of the pet is recommended due to the high risk associated with handling infected animals.

Bite and scratch wounds are commonly contaminated with bacteria and occasionally with other organisms on the skin or from the oral cavity and saliva. Although the majority of bite and scratch wounds result in only a localized infection, up to 80% of cat bites develop infections requiring medical treatment. Cat bites may be at a higher risk of infection because they create small, deep puncture wounds that are difficult to clean. The most common isolate from cat bite wounds is *Pasteurella multocida,* which can be treated with penicillin derivatives and local drainage. To prevent bites and scratches, dogs and cats should not be teased and people handling them should be trained to use proper restraint techniques. Any wounds should be immediately cleansed and the person referred to a physician who can determine whether prophylactic medications are indicated.

Vector-associated zoonoses occur when an ectoparasite (e.g., flea, tick, mosquito) transmits an infection from an animal to a person. In some cases, the animal serves as a reservoir for the infection as well as a transport host for the vector. Control measures for prevention of vector-associated zoonoses are primarily achieved by prevention of tick and flea infestations of pets.

# 14.4  NONINFECTIOUS DISORDERS OF DOGS AND CATS

A useful memory aid for categorizing health disorders is DAMNIT:

Degenerative, Developmental

Anomalies, Allergies

Metabolic, Miscellaneous

Nutritional, Neoplastic

Infectious, Inflammatory, Immune-mediated

Toxic, Traumatic

## 14.4.1  DEGENERATIVE AND DEVELOPMENTAL DISORDERS

Common examples of disorders in this category include canine hip dysplasia and juvenile cataracts.

*Canine hip dysplasia (CHD).* CHD is a developmental abnormality of the coxofemoral joint (hip) in dogs. In normal hips, the head of the femur (thigh bone) fits smoothly within the acetabulum (hip socket) of the pelvis (Figure 14.2a). Dogs with CHD have abnormalities in the formation of the acetabulum and a subluxation of the head of the femur (Figure 14.2b). This results in abnormal articulation (movements) of the hip joint and progressive abnormalities in the development of the acetabulum (becomes flat and shallow) and head of the femur (becomes irregular in shape). Over time, degenerative arthritis develops in the hip joint. CHD is a polygenic trait—with genes controlling the conformation of the pelvis, pelvic muscle mass, body size, and growth patterns all interacting to influence the development of clinical disease. Although CHD is hereditary, its severity is profoundly influenced by environmental factors, especially nutrition. When puppies are fed for maximum growth, the growth plates in the acetabulum close early resulting in worsening of the malformation. Overweight puppies have additional stresses on the joints and a worsening of hip subluxation. One way to help retard CHD is to keep dogs of breeds highly susceptible to the disease on restricted diets while they are growing. Controlling body weight lessens the severity and even prevents the development of CHD in some dogs. Excessive exercise should be discouraged during the months of most rapid bone growth (2 to 7 months), especially in the larger breeds of dogs.

**Figure 14.2** Radiographs (X-rays) of a dog with normal hips (*a*, left) and one with hip dysplasia (*b*, right). Note the head of the femur of the normal dog has a smooth contour and fits snuggly in the acetabulum (hip socket). In contrast, the head of the femur of the dysplastic dog is misshapen and subluxated (does not articulate with the acetabulum). *Courtesy Dr. Stephen Kneller, University of Illinois.*

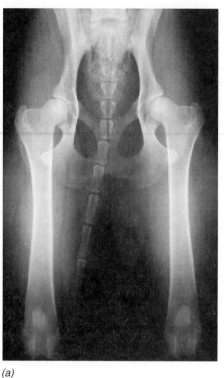
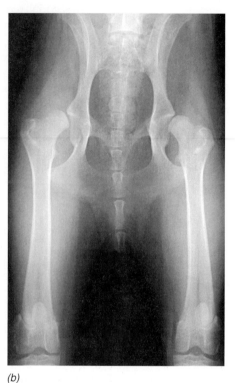

(a)                                                        (b)

CHD accounts for approximately one-third of all orthopedic cases seen by veterinarians. Large-breed dogs are more susceptible than small-breed ones. Overfeeding in the first year of life is a major contributor to the severity of this disease. Radiographs (X-rays) taken when a dog is at least 2 years of age are used to screen breeding stock for the presence of hip dysplasia. Skilled evaluators at the Orthopedic Foundation for Animals (OFA) assign a rating of "excellent, good, fair, or poor" and also the degree of dysplasia (if present) in the dog's hip conformation. Mating only dogs with excellent or good ratings decreases the likelihood of hip dysplasia in the offspring. The ideal femur-to-pelvis angle for most breeds is approximately 30°; evaluation of this angle can be used as a tool in the selection of prospective breeding stock.

Symptoms of hip dysplasia may develop in puppies as young as 4 months or may not be evident until the dog is several years old. Young dogs may show a sudden onset of severe lameness with difficulty jumping and climbing stairs. Older dogs have a more gradual onset of decreased exercise tolerance, stiffness, and lameness that is worsened following exercise. Diagnosis is based on clinical signs, finding pain on manipulation of the hips, and hip radiographs. Mild cases may be managed with exercise restriction and weight reduction. Buffered aspirin and other anti-inflammatory medications may be required to relieve pain. Polysulfated glycosaminoglycan supplements may promote healing of damaged cartilage. Severe cases often require surgical intervention. Total hip replacement surgery replaces the diseased hip with an artificial ball-and-socket joint (prosthetic hip joint). This surgery relieves pain and restores good mobility. An alternative surgery is a femoral head and neck ostectomy (FHO). In the FHO, the head and neck of the femur are removed and a false joint forms between the remaining portion of the femur and the pelvis.

*Juvenile cataracts.* Thousands of dogs develop cataracts every year. Cataracts are opacities of the lens (Chapter 11) or its capsules. The opacities range from barely detectable lesions to complete cataracts that block the transmission of light through the lens causing blindness. Although there are many causes of cataracts in dogs, many are hereditary. The most common mode of inheritance is autosomal recessive. These cataracts may first be evident when the dog is a few months old and progress to cause blindness within 2 to 3 years. Some developmental cataracts are not visible until the dog is middle aged. The Canine Eye Registry Foundation (CERF) (Chapter 4) recommends against breeding any animal with cataracts unless the cause is known to be due to trauma, infection, nutritional deficiencies (Chapter 9), or metabolic diseases (e.g., diabetes mellitus, Section 14.4.3). Surgical removal of cataracts can restore vision for affected animals. However, cataracts are sometimes found in association with another hereditary eye disease—progressive retinal atrophy (PRA). Because cataracts obscure visualization of the retina and thus interfere with the diagnosis of PRA, veterinary ophthalmologists perform an electrodiagnostic test of retinal function (electroretinogram, ERG, Chapter 16) prior to extracting cataracts. (If the dog has PRA, cataract removal will not restore vision.)

## 14.4.2   Anomalies and Allergic Disorders

Anomalies may affect the heart with abnormal configurations of major blood vessels or may affect the eyes with abnormalities in the retina or iris. These disorders are often hereditary. If one puppy or kitten in a litter is affected, the littermates should be examined by veterinary specialists to determine whether special care is needed to minimize subsequent complications resulting from the anomalies.

Allergies are common in dogs and cats. They result from overactivity of the immune system. The allergy-prone animal produces unique antibodies termed IgE against foreign proteins called *allergens.* The IgE antibodies attach to special inflammatory cells called mast cells in the skin and other sites (nasal mucosa, gastrointestinal tract, respiratory tract). IgE antibodies interact with allergens and trigger the mast cells to release inflammatory substances such as histamine, prostaglandins, leukotrienes, and proteolytic enzymes. The most common allergy is to allergens found in flea saliva. Flea-associated allergies can result in severe self-trauma from biting and scratching, extensive hair loss, and secondary skin infections. Environmental allergies are also common (Figure 14.3). These allergens include pollens, molds, danders, and particles from house dust mites and insects. Food allergies can also result in itching and skin lesions and in some cases the additional signs of vomiting and diarrhea. Flea

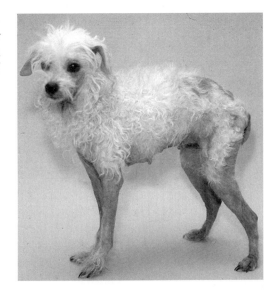

**Figure 14.3**   A 3-year-old dog with environmental allergies. Note the loss of hair around the eyes and on the legs from allergy-induced rubbing and scratching/licking. *Courtesy Dr. Karen L. Campbell.*

and environmental allergies are diagnosed by intradermal skin allergy testing or by finding allergen-specific IgE in the serum. Food allergies are diagnosed by feeding restrictive diets followed by challenge feedings. Flea allergies are best managed by elimination of fleas from the pet and environment. Environmental allergies may be controlled through the use of antihistamines and other anti-inflammatory medications or by desensitization therapy. In desensitization therapy the pet is given injections of dilute concentrations of allergens; over time these injections alter the immune response and the pet becomes tolerant to the allergens. Food allergies are managed by avoiding foods containing the offending allergens.

### 14.4.3    METABOLIC AND MISCELLANEOUS DISEASES

Dogs and cats can be affected by a large number of diseases affecting their metabolism. Some of the most common ones are diabetes mellitus, thyroid disorders (hypothyroidism and hyperthyroidism), adrenal disorders (Addison's disease and Cushing's disease), and chronic renal disease.

*Diabetes mellitus.*  Like humans, dogs and cats can be affected with either type I diabetes mellitus (insulin-dependent) or type II diabetes mellitus (non-insulin-dependent). Type I diabetes mellitus is most common and results from destruction of the insulin-secreting beta cells of the pancreas. It occurs most commonly in middle-aged to older female dogs and middle-aged to older neutered male cats. Type II diabetes mellitus is rare in dogs but is seen in some overweight cats. Clinical symptoms of diabetes include excessive thirst (polydipsia) with increased urination (polyuria), increased appetite (polyphagia), and weight loss. Dogs are susceptible to the development of cataracts, which can result in blindness. Cats sometimes develop such severe rear-limb weakness that they stand with the hocks touching the ground. Measurement of blood glucose concentrations facilitates diagnosis. Treatment involves dietary changes (Chapter 9) and administration of oral hypoglycemic agents for type II diabetes mellitus and insulin injections for type I diabetes mellitus.

*Hypothyroidism.*  One of the most commonly diagnosed metabolic diseases of dogs is hypothyroidism. It is most common in middle-aged dogs with Airedale Terriers, Cocker Spaniels, Dachshunds, Doberman Pinschers, Golden Retrievers, Great Danes, Irish Setters, Miniature Schnauzers, and Old English Sheepdogs being at increased risk. Congenital hypothyroidism is rare but has been reported in Abyssinian cats, American shorthair cats, Scottish Deerhounds, and other breeds. Clinical symptoms include hair loss (alopecia), a dull hair coat, dandruff, lethargy, and infertility. Diagnosis is made by measuring serum thyroid hormone concentrations. Treatment involves daily administration of a thyroid hormone supplement.

*Hyperthyroidism.*  One of the most common metabolic diseases of cats is hyperthyroidism. It is most common in middle-aged to older cats. Clinical symptoms include voracious appetite, weight loss, increases in urination and drinking, restlessness, and an unkempt hair coat. The thyroid glands are enlarged and may be easily felt in the cat's neck. Diagnosis is confirmed by finding high levels of thyroid hormones in the blood. Treatment options include surgical removal of the enlarged glands, giving radioactive iodine, or medications that interfere with secretion of thyroid hormones.

*Addison's disease or hypoadrenocorticism.*  Addison's disease is a metabolic disease caused by inadequate production of hormones from the adrenal gland. Two important hormones produced by the adrenal glands are aldosterone and cortisol. Aldosterone is essential in the regulation of body electrolytes (sodium and potassium levels). Cortisol is essential for the regulation of blood pressure and many other body activities. Clinical signs of hypoadrenocorticism include depression, weakness, muscle tremors, vomiting, and in later stages vascular collapse and death. Finding abnormal concentrations of serum electrolytes and low levels of serum aldosterone and cortisol confirms the diagnosis. Treatment involves giving fluids and electrolytes to correct imbalances and hormone supplements to replace those normally produced by the adrenal glands.

***Cushing's disease or hyperadrenocorticism.*** Cushing's disease is a metabolic disease caused by excessive amounts of corticosteroids. There are three common forms of Cushing's disease. The first is called *iatrogenic* and is due to treating the dog or cat with high doses of a corticosteroid (often given to treat another problem such as destructive inflammation or cancer). The second is *pituitary dependent* and due to a pituitary tumor producing the hormone ACTH (Chapter 11) that causes bilateral adrenal gland hyperplasia and excessive cortisol production. The third form is *adrenal dependent* and due to an adrenal tumor producing excessive amounts of cortisol. Clinical signs include muscle weakness, liver enlargement, redistribution of body fat, hair loss, panting, bruising, a predisposition to secondary infections, and increases in appetite, drinking, and urination (Figure 14.4). Finding elevated blood cortisol concentrations and either an adrenal tumor or bilateral enlargement of the adrenal glands visualized by ultrasound examination of the abdomen will confirm the diagnosis. Treatment of adrenal-dependent hyperadrenocorticism includes removing the neoplastic adrenal gland. Treatment of pituitary-dependent hyperadrenocorticism includes the administration of drugs to decrease corticosteroid production. Treatment of iatrogenic Cushing's disease requires a gradual discontinuation of treatment with corticosteroid medications.

***Chronic renal failure (CRF).*** CRF is common in older dogs and cats. Loss of renal function results in the accumulation of a variety of nitrogenous and nonnitrogenous products of protein catabolism. The failing kidneys are unable to concentrate urine resulting in clinical signs of increased urine production and water consumption. Dietary therapy (with protein and phosphorus restriction) is important in the management of canine and feline chronic renal failure (Chapter 9). Dietary management can benefit dogs and cats with CRF by (1) preventing/ameliorating clinical signs of uremia, (2) minimizing disturbances associated with excesses/losses of electrolytes and minerals, and (3) arresting or retarding progression of CRF. Although additional research is needed to ascertain more precisely the benefits of protein restriction for dogs and cats with CRF, early research indicated that restriced protein and restricted phosphorus diets are beneficial to dogs and cats with renal dysfunction.

***Hairballs.*** Hairballs can be classified as a "miscellaneous" noninfectious disease of cats. The tongue of a cat has tiny barbs that pull out loose hairs; thus, cats commonly swallow hairs while grooming themselves. The swallowed hairs either pass through the body (ending up in the stool) or accumulate in the stomach forming a mass (called a "trichobezoar") that is expelled by vomiting. One of the most effective methods to prevent hairballs is to brush the

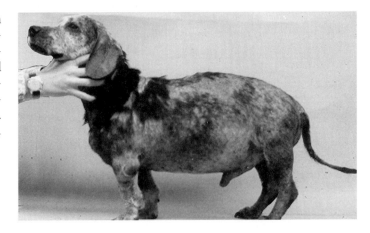

**Figure 14.4** A dog with Cushing's disease (hyperadrenocorticism). Excessive production of cortisol by the adrenal glands results in a pot-bellied appearance and hair loss. *Courtesy Dr. Donna Angarano, Auburn University.*

cat daily; the loose hairs that accumulate on the brush are ones the cat would otherwise have swallowed. Hairballs also have a high lipid content. Special "hairball diets" are available. These diets contain high levels of soy lecithin, which functions as a "dietary detergent" to solubilize the lipids in hairballs, and a high content of fiber to promote movement of hairs through the digestive tract. These diets also have high levels of sulfur-containing amino acids to promote health of skin and coat. Laxatives can also be used to promote movement of the hair through the digestive tract. Overgrooming can be a symptom of ectoparasites, allergies, or psychological problems; thus, cats with frequent hairballs should be evaluated for these conditions by a veterinarian.

## 14.4.4 NUTRITIONAL AND NEOPLASTIC DISORDERS

The most common nutrition-related disorder in dogs and cats is obesity. Other important nutrition-related disorders include food allergies, zinc deficiency in dogs, taurine deficiency in cats, and feline lower urinary tract disease in cats. Dogs and cats can develop neoplasia (cancer) involving each body system and tissue type. Oncology (study of cancer) and oncological surgery are among the most rapidly growing specialties in veterinary medicine.

*Obesity.* Factors contributing to the high incidence of obesity in dogs and cats include their increasingly sedentary lifestyles and the abundance of nutrient-rich, highly palatable pet foods. An animal is technically considered to be obese when its body weight is 20% or more above ideal weight (Figure 14.5). Overweight animals have decreased exercise tolerance, decreased activity of their immune systems, and increased risk of developing diabetes mellitus, cardiovascular disease, degenerative joint disease (arthritis), and cancer. Weight management was discussed in Chapter 9.

*Zinc deficiency.* Two syndromes (types) of zinc deficiency occur in dogs. Type I occurs in rapidly growing puppies, especially those being supplemented with high levels of calcium or being fed diets high in plant fibers (phytates). Calcium competes with zinc for absorption from the intestines and high levels of phytates in the diet can bind zinc and interfere with its absorption. Type II zinc-responsive dermatosis is seen in the arctic dog breeds including Alaskan Malamutes and Siberian Huskies; these dogs are thought to have abnormalities in zinc absorption. Zinc is an important cofactor for many enzymes within the body and has important roles in immune function and maturation of cells of the epidermis (superficial layers of skin). Dogs with zinc deficiency have crusting and fissuring of the skin around their mouth, nose, feet, and pressure points (Figure 14.6). Secondary bacterial skin infections are common. Diagnosis is based on biopsies of the skin and response to zinc supplementation.

**Figure 14.5** An obese cat at high risk of developing diabetes mellitus and other health disorders. *Courtesy Sandra Grable, University of Illinois.*

**Figure 14.6** A Husky with type II zinc-responsive dermatosis. Inadequate levels of zinc result in faulty maturation of skin cells. Crusty lesions are commonly found around the eyes and nose. *Courtesy Dr. Rodney Rosychuk, Colorado State University.*

*Taurine deficiency.* Cats have a unique dietary requirement for the amino acid taurine (Chapter 9). Taurine deficiency can occur when cats are fed low-quality dog foods or "table-food" diets. Taurine deficiency can lead to retinal degeneration and blindness or dilated cardiomyopathy and heart failure. Commercial cat foods have been formulated to contain adequate amounts of taurine, thus taurine supplements are only required when home-cooked foods are being fed.

*Feline lower urinary tract disease (FLUTD).* FLUTD is also known as feline urological syndrome (FUS). Lower urinary tract disease is a common problem in cats. Clinical signs include straining to urinate (dysuria), blood in the urine (hematuria), frequent urination (pollakiuria), inappropriate urination (outside of litter box), and urethral obstruction. Causes of FLUTD include calculi (urolithiasis, urethral or bladder stones), urethral plugs, infections, trauma, and unknown (idiopathic) factors. Although controversial, some people believe dry cat foods may increase the risk of FLUTD, possibly due to decreased total daily water intake, increased fecal water loss, and higher content of magnesium. It is recommended that magnesium intake be limited to less than 20 mg/100 kcal of dietary energy (or 0.04% to 0.10% of dietary dry matter). It is also important to avoid excesses of calcium (ideal is 0.5% to 0.8% of dietary dry matter) and phosphorus (ideal is 0.5% to 0.8% of dietary dry matter). Water consumption should be encouraged as concentrated urine favors supersaturation of minerals and subsequent precipitation. Urine crystals should be evaluated to determine if they are composed of struvite (magnesium-ammonium-phosphate) or calcium oxalate. Struvite crystals are most soluble in acidic urine, and diets favoring acid urine formation are preferred for cats with struvite crystalluria (comprise 48% to 90% of feline uroliths). In contrast, calcium oxalate crystals are most soluble in alkaline urine, and acidifying diets should be avoided for cats with calcium oxalate crystalluria (Chapter 9). Medications are sometimes used to decrease urethral inflammation and spasms. Glycosaminoglycan supplements may help decrease crystal adherence to the mucosa of the urinary tract; the effectiveness of these supplements has not been proven scientifically.

*Neoplasia.* As pets live longer lives, more of them develop neoplasia (cancers). Cancers are diseases in which cells are dividing out of control. Ten common signs of cancer in small animals include (1) abnormal swellings that persist or continue to grow, (2) sores that do not heal, (3) weight loss, (4) loss of appetite, (5) bleeding or discharge from any body opening, (6) offensive odor, (7) difficulty eating or swallowing, (8) hesitation to exercise or loss of stamina, (9) persistent lameness or stiffness, and (10) difficulty breathing, urinating, or defecating. If any of these signs are noticed, the pet owner is encouraged to consult with a veterinarian. When cancer is diagnosed early, many treatment options are available including surgery, radiation therapy, gene therapy, and chemotherapy.

## 14.4.5 INFECTIOUS, INFLAMMATORY, AND IMMUNE-MEDIATED DISORDERS

Infectious diseases were discussed in Section 14.2. Many infectious and immune-mediated diseases are associated with inflammation. Immune-mediated diseases are those in which the immune system is adversely affecting another body system. A normal immune system defends the body against foreign substances and abnormal cell types by producing antibodies and directing cells to destroy any foreign invaders or abnormal cells. In immune-mediated diseases, antibodies and killer cells are "attacking" the animal's own body components. Examples of immune-mediated diseases include autoimmune hemolytic anemia (AIHA), immune-mediated thrombocytopenia (ITP), discoid lupus erythematosus (DLE), systemic lupus erythematosus (SLE), pemphigus foliaceous (PF), myasthenia gravis (MG), and rheumatoid arthritis (RA). In AIHA, antibodies are produced against red blood cells. The red blood cells are destroyed and the dog or cat may die from severe anemia. In ITP, antibodies are produced against blood platelets and the dog or cat may bleed to death. In DLE, the immune system is attacking the cells of the nose resulting in a loss of pigmentation and ulcers on the nose. SLE may involve immune-mediated destruction of blood cells, the kidneys, muscles, and other organs. In PF, autoantibodies disrupt the connections between skin cells resulting in the formation of micro-blisters (vesicles), pustules, and ulcers of the skin (Figure 14.7). In MG, autoantibodies disrupt the receptors for signals between nerves and muscles. This results in profound muscle weakness, easy fatigability, and collapse. RA is characterized by destruction of cells lining the joints. This results in a severe, erosive arthritis with lameness, pain, and joint swelling. Immune-mediated diseases are diagnosed by detecting the presence of auto-reactive antibodies. Management of these diseases requires the use of immunosuppressive drugs and supportive care.

*Otitis externa* is an example of a common, complex, inflammatory disease of the ears. It is associated with inflammation involving the ear pinna and also the vertical and horizontal portions of the ear canal (portions "external" to the tympanic membrane; Chapter 11). Clinical signs of otitis externa may include pain, shaking and scratching at the ears, abnormal odor, and presence of an exudate in the ear canals. A decrease in hearing may also occur. Otitis externa is often multifactorial with a combination of contributory causes including predisposing factors [e.g., pendulous ear pinna, hair in the ears, stenotic (narrow) ear canals, high humidity, moisture from bathing or swimming, and others], primary

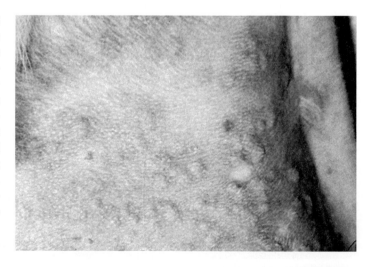

**Figure 14.7** Pustules (small blisters filled with pus) of this dog were caused by the immune-mediated disease pemphigus foliaceous (PF). In PF, antibodies are produced against proteins in the epidermis (superficial layers of skin) resulting in the formation of pustules. *Courtesy Dr. Jennifer Matousek, formerly of the University of Illinois.*

causes [e.g., foreign bodies, tumors, parasites such as *Otodectes cynotis* (ear mites) or ticks, allergies (food, environmental, or contact), or immune-mediated diseases (pemphigus foliaceous), and others], and perpetuating factors [e.g., bacterial and yeast (*Malassezia* spp) overgrowth, progressive pathological changes in the ears, and others]. Treatment must be directed at elimination of bacterial and yeast infections and also identification and management of predisposing factors and primary causes of the disease. *Otitis media* is an inflammatory disease involving the middle ear (Chapter 11) and can be due to extension of infections across the tympanic membrane or through the Eustachian tube. *Otitis interna* refers to an inflammatory disease involving the inner ear resulting in hearing loss and loss of equilibrium (resulting in a head tilt, abnormal eye movements called nystagmus, and circling).

## 14.4.6   TOXINS AND TRAUMA

Free-roaming dogs and cats are at high risk for traumatic injuries from being hit by vehicles or in fights with other animals. A few of the most common toxins causing disease in dogs and cats are those in antifreeze, rodenticides, pesticides, poisonous plants, lead, and excessive zinc. Many medications used for people and even some foods are also toxic to animals.

*Antifreeze toxicity.* Antifreezes contain ethylene glycol, a sweet substance readily consumed by cats and dogs. One teaspoon of antifreeze may contain enough ethylene glycol to cause death in a small cat. Metabolites of ethylene glycol cause a rapid destruction of cells of the renal tubules and result in acute kidney failure. Early signs of intoxication may include inebriation, vomiting, and seizures or coma. Death from kidney failure rapidly follows. Diagnosis can be made by measuring serum ethylene glycol concentrations. If diagnosed early, aggressive fluid therapy and drugs to inhibit the metabolism of ethylene glycol may be effective. Of all poisons, antifreeze causes the highest fatality rate in cats and dogs.

*Rodenticides.* Rodenticides are another common cause of death in dogs and cats. Most rodenticides inhibit the production of vitamin K-dependent clotting factors and result in death from bleeding. Diagnosis is based on a history of exposure, testing of blood or liver to detect the presence of the toxin, and abnormalities in coagulation tests and bleeding time. Treatment involves giving plasma transfusions and high doses of vitamin $K_1$. Other rodenticides contain vitamin D. These rodenticides result in high serum levels of calcium and death from kidney failure. Diagnosis is based on a history of exposure to rodenticides and high levels of serum calcium and vitamin D. Treatment includes administration of fluids and drugs to promote calcium excretion.

*Pesticides.* Cats are particularly susceptible to poisoning from pesticides due to low activity of detoxifying enzymes in the liver. Many products that are safe for use on dogs may result in fatal toxicity if used on a cat (Figure 14.8). Exposure to pesticides can be direct through products applied to or consumed by the pet, or indirect from exposure to products used in the environment. Signs of toxicity vary depending on the product, but often include vomiting, excessive salivation, incoordination (ataxia), depression, and seizures. Treatment varies depending on the product. The ASPCA National Animal Poison Control Center may be contacted for advice.

*Plant toxicity.* Plants in the *Lilium* and *Hemerocallis* genera are very toxic to cats. These plants include Easter lilies, tiger lilies, Japanese show lilies, rubrum lilies, and day lilies. Ingestion of two or three leaves or flowers can be fatal. Early signs include vomiting and depression; death from kidney failure follows within one week. If consumption of the plant is

**Figure 14.8** This topical insecticide used to prevent fleas and ticks on dogs is highly toxic to cats. *Courtesy University of Illinois.*

discovered early, activated charcoal and cathartics can be given to lessen absorption and fluid therapy may preserve kidney function.

*Lead toxicity.* Lead poisoning results from the ingestion of lead-containing paint and paint residues (e.g., paint flakes after sanding), linoleum, plumbing materials, fluid leaking from car batteries, putty, tar paper, lead foil, golf balls, solder, lead weights (e.g., fishing sinkers, drapery weights), shot, or the use of improperly glazed ceramic food or water bowls. Young animals (< 1 year) are at highest risk. Clinical signs of lead toxicity include vomiting, diarrhea, anorexia (poor appetite), abdominal pain, lethargy, blindness, seizures, and hysteria in dogs. Cats may also show vertical nystagmus (abnormal movements of the eyes) and ataxia. Lead interferes with the maturation of RBCs, and nucleated RBCs may be found in circulation. Diagnosis is confirmed by finding concentrations of blood lead > 0.4 ppm. Treatment involves supportive care and the administration of lead chelators.

*Zinc toxicity.* Zinc toxicity may follow the ingestion of zinc-containing nuts from transport cages or plumbing, metal pieces from board games, pennies minted after 1982,[7] or zinc oxide ointment. Excessive serum levels of zinc result in hemolysis (rupturing of red blood cells) and multiple organ failure. Diagnosis may be made by radiographs showing the presence of a metal object in the gastrointestinal tract or by finding high levels of zinc in the blood. Treatment involves removing the zinc-containing material from the gastrointestinal tract and fluid diuresis. Blood transfusions may also be needed.

*Medications.* Many medications used for people are highly toxic for pets. Dogs and cats should never be treated with any medication without the advice of a veterinarian. One common example is acetaminophen (Tylenol™); a single tablet of this common analgesic can be fatal to an adult cat. The topical anesthetic benzocaine can cause fatal disruption of blood cells in dogs and cats. Phosphate-containing enemas can cause fatal kidney damage and should never be used to treat constipation in a cat.

*Chocolate poisoning.* Chocolate contains theobromine, the darker the chocolate the more theobromine it has. Dogs lack the enzymes needed to metabolize theobromine. Four ounces of dark chocolate can kill a toy dog.

---

[7] Pennies minted prior to 1982 were 95% copper and 5% zinc. In 1982 the composition was changed to 96% to 98% zinc with a 2.5% copper coating.

## 14.5   PREVENTIVE MEDICINE

Practicing preventive medicine means being proactive in protecting pets from developing disease. Regular observation is important in assessing an animal's health. It is important to pay attention to such factors as appetite; drinking and elimination habits; body weight and condition; tolerance for exercise and interest in activities such as playing or working; condition of the skin and hair coat; any sneezing, coughing, or difficulty breathing; or any other signs that could be indicative of a developing problem. Preventive healthcare measures include feeding a nutritious diet, providing fresh clean water, paying close attention to sanitation, providing regular veterinary examinations, vaccinating for viral and bacterial diseases, and preventing parasitic infections (including heartworm and flea preventives). Regular grooming is important for the health of skin and hair coat and will also minimize the incidence of hairballs in cats.

Preparing for a pet should include "pet-proofing" the home, just as one would "childproof" the home of a young child. Inspect the house room by room. Look for things a pet could chew or swallow. Hide electrical cords behind furniture or tape them firmly to the walls and floors to keep pets from chewing them. Keep medicines out of reach. Cats will knock opened bottles onto the floor and puppies can chew through childproof caps, so keep medicines in closed drawers or cabinets. Keep windows closed or protected with guards to prevent pets from jumping through an open window (note that lightweight mesh screens may not be strong enough to stop a pet from jumping through them). Cords for curtains and window blinds should be kept out of reach of pets that could become entangled in them and strangle. Close the lids or doors on appliances—cats and dogs have been accidentally shut inside washers, dryers, and even microwave ovens! It is a good idea to check inside appliances before turning them on. Put toilet lids down; kittens like to jump onto things, and more than one has ended up slipping inside a toilet bowl. Make the garage off limits. Some chemicals stored in garages and others that leak from cars—especially antifreeze—are highly toxic to animals that may find them tasty. Cats are also prone to crawl into the vicinity of car engines as they like the heat; injuries from fan belts are not a pretty sight! Provide good footing; slippery floors can be particularly difficult for young puppies and older dogs to walk on without slipping and sliding. Although it can be cute to watch a puppy sliding as it runs down a hall, the sliding may damage bones and joints. Be careful with strings on toys, fishing lines, or from packaging materials. Swallowing a string is a surgical emergency to prevent fatal damage to the intestines.

## 14.6   SANITATION/CLEANLINESS

Most infectious agents are transmitted to susceptible animals in fecal material, respiratory tract secretions, reproductive tract secretions, or urine. Other agents are transmitted by bites or scratches from an infected animal or from a transport host or vector. Many infectious agents are stabile in the environment and can be transmitted by fomites (toys, grooming equipment, leashes, cages, and other inanimate objects). Cleanliness and environmental sanitation are important in preventing the spread of infectious agents.

Persons handling animals suspected of having an infectious disease should thoroughly wash their hands before and after handling the animal. Hand washing should be done using an antiseptic soap with a contact time of at least 30 seconds, being careful to clean under fingernails—which should be cut short. Gloves should be worn when handling animals with possible zoonotic diseases.

Animals with infectious diseases should be housed separately from other animals. Bedding, litter pans, food and water dishes, collars, cages, tables, leashes, and grooming equipment should be disinfected after each use. Equipment including thermometers, stethoscopes, pen lights, bandage scissors, percussion hammers, and clipper blades can all be fomites and should be cleaned and disinfected following each use. In veterinary hospitals and when dealing with a zoonotic infection at home, all persons handling the animal should wear an outer garment such as a smock, gown, or scrub suit, which should be changed between animals and when contaminated with urine, feces, secretions, or exudates.

Cleaning is the physical removal of organic material or soil from objects. Cleaning will remove many microorganisms but alone may not kill them. Sterilization is the destruction of all forms of microbial life. Sterilization requires the use of chemicals (e.g., ethylene oxide gas), steam under pressure, or dry heat. Disinfection is an intermediate measure between physical cleaning and sterilization; it is achieved through the use of germicides. Germicides may be classified as general disinfectants, hospital disinfectants, sanitizers, and sporicides. Disinfectants commonly used for pathogens affecting dogs and cats include hypochlorite (bleach), chlorhexidine, iodophor solutions, glutaraldehyde solutions, quaternary ammonium solutions, phenolic and synthetic phenolic compounds, benzalkonium chloride, and others. Surfaces should be physically cleaned and then in contact with the disinfectant for 10 to 15 minutes for maximum efficacy in killing pathogens.

# 14.7  VACCINATIONS OF DOGS AND CATS

"Vaccination is much like buying casualty insurance. It is an investment to prevent a potential major loss."[8] Vaccinations are given to stimulate the immune system to ward off or prevent infections with specific agents. Vaccines are composed of agents that have been rendered noninfectious but remain capable of inducing an immunological response against the agent. Although no vaccine is always 100% effective, vaccination programs greatly reduce the incidence and severity of infection.

## 14.7.1  TYPES OF IMMUNITY

Immunity refers to the capacity to resist infection against a specific disease. *Natural immunity* is generally due to species differences in susceptibility to an infection. For example, canine distemper virus can cause fatal neurologic disease in dogs but does not affect cats. *Passive immunity* is the transfer of antibodies formed in one animal to another animal. The most important form of passive immunity is that received by newborns from their mothers. Puppies and kittens receive antibodies produced by their mothers through the placenta and also through the ingestion of colostrum (milk containing high levels of antibodies produced the first few days following birth of the litter). Antiserums are another form of passive immunity. Passive antibodies last a relatively short period of time; for example, the maximal length of maternal antibody protection for puppies is about 16 weeks. The third form of immunity is *active immunity*. Active immunity occurs when an animal is exposed to a foreign substance (antigen) and responds by producing antibodies against the antigen. The antibodies neutralize or destroy the antigen so it cannot infect the animal or produce disease.

In vaccines, the antigen is a virus or a bacterium that has been killed or altered (called attenuation) to make it incapable of inducing disease while remaining capable of stimulating the immune system to produce antibodies against it. Once primed by vaccination, the immune system mounts a rapid "memory" response if reexposed to the antigen by another vaccination or the natural form of the pathogen. The rapid memory response destroys an infecting organism before it can replicate and produce disease in the vaccinated animal. Different classes of antibodies are produced depending on the route of exposure to the foreign antigens. Antigens on body surfaces including the surfaces of the respiratory tract, urogenital tract, and gastrointestinal tract stimulate the production of secretory antibodies of the IgA class. IgA antibodies are then secreted into the mucous coverings of the tract and bind to

---

[8] Paul L. Nicoletti, DVM (b. 1932), Professor Emeritus of Pathobiology, College of Veterinary Medicine, University of Florida, Gainesville.

foreign antigens (such as viruses and bacteria) preventing their entrance into the animal's body. Antigens gaining entrance into body tissues (by infection or injection) stimulate the production of antibodies of the classes called IgM and IgG. IgM and IgG antibodies are active in destroying antigens that have gained entry into body tissues or the bloodstream and thus are important in preventing the spread of infections. These antibodies can be measured in diagnostic tests to document whether an animal has been exposed to a specific antigen and has developed an immune response against it.

## 14.7.2 TYPES OF VACCINES FOR DOGS AND CATS

The two major types of vaccines used in veterinary medicine (Table 14.6) are killed vaccines and modified live vaccines. Most vaccines are given by subcutaneous injection, a few vaccines are given by intramuscular injection, and vaccines aimed at stimulating IgA antibodies are usually given intranasally.

Killed vaccines contain dead or inactivated pathogens (e.g., bacteria or viruses). These killed pathogens do not replicate in the host, so there is no possibility of producing an active infection in the vaccinated animal. Killed vaccines contain the foreign antigens associated with the pathogen and thus stimulate the production of antibodies against these antigens. These antibodies are protective against possible future infection with the pathogen. Killed vaccines are stable and have a long shelf life; they are also safe to use in pregnant animals. However, killed pathogens are not as effective as live pathogens in stimulating an immune response. Thus, multiple vaccinations may be required to induce production of adequate levels of antibodies to protect the animal from challenges with pathogenic organisms. To boost the immune system's response to killed pathogens, adjuvants are sometimes added to increase the interactions between the antigens and the cells that produce antibodies. Unfortunately, adjuvants can cause localized inflammation and in some cases this inflammation leads to the formation of invasive cancers known as soft-tissue sarcomas and fibrosarcomas. Vaccine-associated sarcomas and fibrosarcomas are seen in approximately 1 of every 2,000 cats vaccinated with killed rabies and FeLV vaccines. Because of this, newer nonadjuvanted vaccines are being developed for use in cats. Commonly used killed vaccines include those used to protect against rabies, FeLV, canine coronavirus, leptospirosis, Lyme disease, and *Bordetella bronchiseptica*.

Modified live vaccines (MLV) contain a weakened or attenuated form of the pathogen. Following vaccination, the modified pathogen replicates in the host but does not produce clinical disease. The replicating pathogen persists in the body until the animal's immune system destroys it, thus these vaccines are effective at inducing active immunity. If the animal has preexisting antibodies, as do kittens and puppies that have

**Table 14.6**
Types of Vaccines

|  | Description | Advantages | Recommended Use |
|---|---|---|---|
| **Modified live** | Contains a weakened form of the disease agent | Antigens stimulate strong active immunity | Canine distemper; infectious canine hepatitis; canine parvovirus |
| **Inactivated (killed)** | Contains a "killed" form of the infectious agent, which does not cause disease, but stimulates active immunity | Longer shelf life and more stable than modified live vaccines | Rabies, canine coronavirus, leptospirosis, Lyme disease |
| **Intranasal** | Designed to stimulate local IgA* immunity in respiratory tract | Although short-lived, the vaccine is effective against upper respiratory infections | Recommended for dogs that will be exposed to kennel cough; dogs should be revaccinated every 6 months |

*Immunoglobulin A*

passively acquired antibodies from their mothers, the antibodies may inactivate the injected pathogens before they can replicate and there is no boost in active immunity. Because of the variable persistence of maternal antibodies (passive immunity) in young animals, an initial series of two to four vaccinations are recommended to ensure that the animal has been adequately immunized. MLVs commonly used in dogs and cats include the vaccines for canine distemper, infectious canine hepatitis, canine parainfluenza, canine adenovirus, canine parvovirus, feline panleukopenia, feline rhinotracheitis, and feline calicivirus. MLV vaccines should not be used in immunocompromised animals (e.g., cats infected with FIV).

## 14.7.3   Recommended Vaccinations for Dogs

Core vaccines[9] for all dogs include distemper, adenovirus/infectious canine hepatitis, parainfluenza, leptospirosis,[10] parvovirus, coronavirus, and rabies. To decrease the number of injections required for immunization, combination vaccines containing multiple agents are available. Puppy shots typically include canine distemper, canine adenovirus–2 (which also protects against ICH), canine parainfluenza, canine parvovirus, canine coronavirus, and one or two species of *Leptospira*. Initial immunizations begin at 6 to 8 weeks of age and are repeated every 2 to 3 weeks until the puppy is 16 to 18 weeks of age. Annual boosters are usually recommended; under certain circumstances (e.g., when concerned about adverse reactions to the vaccines) serum titers can be measured as an indication of whether a booster vaccination is needed. Depending on local ordinances rabies vaccines are given at 12 to 16 weeks and repeated at 1 year of age. Then a booster is administered every 1 to 3 years.

Noncore vaccines are considered on a case-by-case basis and given to dogs at high risk for exposure to infectious agents such as *Bordetella* or borreliosis (Lyme disease). The immunity from *Bordetella* vaccines is short-lived, requiring revaccination as often as every 6 months. Recommended vaccinations for dogs are summarized in Table 14.7.

## 14.7.4   Recommended Vaccinations for Cats

Core vaccines that should be given to all cats include feline viral rhinotracheitis (FVR), feline calicivirus (C), feline panleukopenia (P) virus, and rabies. FVRCP is available as a combination vaccine requiring a single injection to immunize against all three agents. Passive antibodies acquired from the queen may interfere with the development of active immunity in kittens under 16 weeks of age. It is difficult to predict precisely when maternal antibodies will no longer protect the kitten from disease. Therefore, FVRCP vaccinations should begin at 6 to 8 weeks with boosters given every 2 to 4 weeks until the kitten is 16 weeks of age. FVRCP should be repeated at 1 year of age, and then every 3 years. Depending on local ordinances, rabies vaccine should be given at 12 to 16 weeks of age, at 1 year, and then every 1 to 3 years. Due to the concerns of vaccines leading to the development of soft-tissue sarcomas and fibrosarcomas in cats, it is recommended to use only nonadjuvanted vaccines in cats and to vaccinate adult cats every 3 years (where permitted by law) instead of annually. Standard locations for the administration of vaccines in cats include the following: FVRCP is given subcutaneously over the right forelimb, rabies vaccines are given subcutaneously

---

[9] Core vaccines are those that should be routinely used in preventive health programs.
[10] The effectiveness of leptospirosis vaccination is questioned by some due to the many serotypes of *Leptospira* organisms and the short duration of immunity. In addition, some dogs develop a hypersensitivity reaction to leptospirosis vaccines. For these reasons, many veterinarians believe *Leptospira* should be considered a noncore vaccine and only used when the risk of exposure is high (e.g., hunting dogs).

**Table 14.7**
Recommended Vaccinations for Dogs

| Age | Vaccine | When Recommended |
|---|---|---|
| 2 to 6 weeks | Intranasal vaccine for *Bordetella bronchiseptica* | Recommended for puppies in high-risk environments |
| 6 to 8 weeks | Distemper, hepatitis, parainfluenza, leptospirosis, kennel cough, parvovirus, coronavirus | Core vaccines |
| 10 to 12 weeks | Same as 6 to 8 weeks | Core vaccines |
| 14 to 16 weeks | Same as above | Core vaccines |
| 18 weeks | Parvovirus | High-risk breeds (Doberman Pinschers, Rottweilers) |
| 4 to 5 months (can be given as early as 12 weeks, check local ordinances) | Rabies | 1-year or 3-year core vaccine |
| Every 1 to 3 years for adults | Distemper, hepatitis, leptospirosis, kennel cough, parvovirus, coronavirus, rabies | Core vaccines |
| Prior to exposure | Intranasal vaccine for *Bordetella bronchiseptica* | Animals going to shows, boarding kennels, other at-risk environments |
| Annually | Lyme vaccine | Dogs in high-risk environments, such as hunting dogs in areas with deer ticks |

over the right rear leg (each of these vaccines is administered as low as possible on the leg because limb amputation is required if a vaccine-induced tumor develops).

Noncore vaccines for cats include FeLV, Chlamydia, FIP, *Bordetella*, FIV, and *Microsporum canis.* FeLV is recommended for at-risk FeLV negative cats (e.g., outdoor cats or indoor cats allowed to roam outdoors). The standard administration site for FeLV vaccines is subcutaneously in the left rear leg, as low as possible. The other vaccines are rarely used except for cats in catteries with a high incidence of disease. Recommended vaccinations for cats are summarized in Table 14.8.

**Table 14.8**
Recommended Vaccinations for Cats

| Age | Vaccine | When Recommended |
|---|---|---|
| 6 to 8 weeks | FVRCP | Core vaccine |
| 8 to 12 weeks | FVRCP | Core vaccine |
| 12 to 14 weeks | FVRCP | Core vaccine |
| 14 to 16 weeks | FVRCP and rabies | Core vaccines |
| Every 1 to 3 years for adults | FVRCP and rabies | Core vaccines |
| 8 to 10 weeks and booster at 12 to 14 weeks and annually if at risk | FeLV (test prior to vaccination and only give if negative) | At-risk cats (e.g., outdoors or allowed to roam) |
| Annually on individual basis | FIP, dermatophyte, *Bordetella* | Consider for at-risk catteries |

## 14.7.5    ADVERSE REACTIONS TO VACCINES

Adverse reactions to vaccines are uncommon, although they may occur. Because vaccines activate the immune system, some dogs may develop allergic reactions to vaccine components. Mild allergic reactions are seen as malaise and depression. More significant adverse reactions include vomiting and diarrhea or itching, hives, facial swelling, and respiratory distress. Life-threatening adverse reactions include acute cardiovascular collapse and death. These adverse reactions may develop within 10 to 15 minutes following vaccination. The leptospirosis bacterin component of the vaccine is the most common antigen responsible for adverse vaccine reactions and should be omitted from subsequent vaccines for dogs with a history of a previous adverse reaction to vaccines. In addition, dogs with a history of an adverse vaccine reaction should be premedicated with an antihistamine and carefully observed by a veterinarian for 1 to 2 hours following subsequent vaccinations. Another vaccine-related complication in dogs is the development of hair loss (alopecia) at the vaccination site; this is most commonly seen in small white dogs following rabies vaccinations (Figure 14.9). Various medications may be used in an attempt to stimulate hair regrowth (consult a veterinary dermatologist for advice). The hair loss may be permanent in some dogs. Recently, concern has emerged that administering vaccines too frequently may increase the likelihood the animal will develop an immune-mediated disease; further research is needed to validate this concern. The development of soft-tissue sarcomas at the site of vaccination has become a major concern in feline medicine (Section 14.7.2.)

## 14.7.6    VACCINATION FAILURE

Failure to induce protective immunity is termed vaccination failure. The most common cause of vaccination failure is the persistence of maternal antibodies providing a puppy with passive immunity that inactivates MLV agents and reduces the effectiveness of the vaccine, thereby interfering with development of a longer-lasting, active immunity. The primary reason for giving booster vaccinations to puppies every 2 to 3 weeks until the puppy is 16 to 18 weeks of age is to ensure that the puppy is protected during the waning of maternal passive antibody protection and the development of active immunity. Other reasons for vaccination failure include the use of vaccines that lack potency due to being outdated or mishandled (e.g., stored at improper temperatures), giving the vaccine to an animal already incubating natural disease, or giving vaccines to immunosuppressed animals incapable of developing active immunity.

**Figure 14.9**  Hair loss (alopecia) in this dog was caused by an adverse reaction to a rabies vaccine. Small white dogs are at greatest risk. *Courtesy Dr. Karen L. Campbell.*

# 14.8   ANIMAL EMERGENCIES

Most emergency situations will require veterinary care. Pet owners and handlers should be familiar with first-aid procedures to prevent further injury and stabilize the animal during transport to a veterinary clinic. Knowing how to recognize and treat signs of shock may mean the difference between life and death for severely injured or ill pets.

The first thing to do in all emergencies is to stay calm so you are able to think clearly and help the animal without endangering yourself or others. If the pet is not breathing or if there is not a pulse or heart beat, cardiopulmonary resuscitation (CPR) must be started immediately.

## 14.8.1   RECOGNIZING AN EMERGENCY

Assessment of the severity of the animal's condition requires a knowledge of normal vital statistics. The most rapid way to assess the animal's cardiovascular function is to examine the gums and measure capillary refill time (CRT). Normal gums should be light pink in color. If the gums are very pale or grey, the animal may be in shock or severely anemic. Blue or grey gums can indicate shock, heart failure, lung failure, or poisoning. Bright red gums are seen in some types of poisonings. Capillary refill time is determined by applying pressure to the gums with a thumb. When thumb pressure is removed, the gums will be white but should rapidly refill with blood. The time required to refill with blood is the CRT (normally about 1 second). If the CRT is longer than 2 seconds, the animal has inadequate cardiovascular function and is probably in shock. Normal heart rates are 60 to 160 beats per minute for dogs, 160 to 240 for cats. Elevated heart rates are indicative of fever, heart failure, pain, trauma, and some types of poisonings. Decreased heart rates are seen in some types of poisonings, hypothermia, some types of heart disease, and in the late stages of shock. Normal respiratory rates are 10 to 30 per minute in dogs, 20 to 30 in cats (see Tables 14.1 and 14.2). Elevated respiratory rates are seen with fever, heatstroke, pain, and in the early stages of shock. Decreased respiratory rates are seen with some types of poisoning, hypothermia, and in the late stages of shock. Normal body temperature in dogs and cats is 101°F to 102.5°F. Elevated temperatures are seen in heatstroke, infections, and with pain or trauma. Decreased body temperature is indicative of hypothermia or severe shock.

## 14.8.2   MOVING AN INJURED ANIMAL

When animals are injured or in pain, they may react in unexpected ways and may bite those trying to help them. Thus, it is wise to muzzle injured dogs and cats prior to performing first aid or moving them. Muzzles appropriate for the size of animal should be purchased and kept in first-aid kits. In an emergency, a muzzle may be made using gauze or strips of cloth (Figure 14.10). Animals that have suffered a traumatic injury should be transported on a firm surface to prevent rotation or damage to the spine. If a firm surface cannot be found, a large towel or blanket can be used for a makeshift stretcher or sling.

## 14.8.3   PERFORMING CPR

If an animal is not breathing, and/or if its heart stops beating, immediate CPR should be tried because being without oxygen causes the brain to suffer irreversible damage within 3 to 5 minutes. First establish an airway by opening the pet's mouth to be certain there are no visible objects obstructing the airway (e.g., a golf ball or toy). Gently pull the tongue forward and straighten the neck. Attempt to force air into the chest by holding the pet's mouth closed and blowing into its nose; if successful, the chest will rise. Sometimes giving one or two

**Figure 14.10** Steps in tying a gauze muzzle. (*a*) A loop is formed; (*b*) the loop is tied on top of the animal's nose; (*c*) a second loop is tied under the animal's jaw; and (*d*) a third loop is tied behind the animal's head. (*e*) The finished muzzle. *Courtesy Jason Motsinger.*

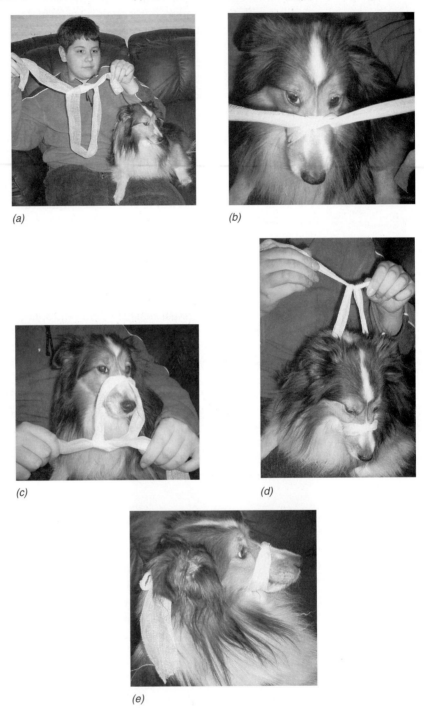

(a)

(b)

(c)

(d)

(e)

breaths will be enough to reestablish normal breathing; if not, continue giving 15 to 20 artificial breaths per minute. If there is no pulse, chest compressions will be needed to circulate blood. The dog or cat should be placed on its side. The chest of a small dog or cat can be compressed by placing the flat of each hand on the chest behind the elbows and pressing simultaneously with both hands to compress the chest approximately 0.5 inch. The inward push should be rapid to increase pressure in the chest and move blood out of it. Repeat the compressions at a rate of 100 per minute. For dogs larger than 25 lb, both hands are placed on the left side of the chest over the heart (which is located where the elbow touches the rib cage when the leg is pulled backward). The chest of medium-sized dogs should be compressed approximately 1 inch with each push; for large dogs compress it 1.5 inches. If the pet is not breathing and does not have a pulse, give 15 chest compressions followed by two breaths and repeat these cycles 4 to 6 times per minute until the pet starts breathing or veterinary assistance is available.

## 14.8.4 SHOCK

Shock occurs when there is insufficient blood circulation to maintain brain and tissue oxygenation. Shock can result from rapid blood loss, severe dehydration, overheating, trauma, or overwhelming infections. Signs of shock include profound weakness, pale-colored gums, slow CRT, cool extremities, a rapid weak heartbeat, and rapid shallow respirations. Emergency care should include keeping the pet warm and quiet, positioning the head a little lower than the body to encourage blood flow to the brain, and transportation to a veterinary clinic for intravenous fluid administration and other supportive care.

## 14.8.5 BLEEDING

Bleeding may be external or internal. A little blood can look like a lot because it spreads quickly. Normal blood volume is approximately 6% to 7% of the body weight of cats and 8% to 9% of the body weight of dogs. Blood loss is generally not life threatening until it exceeds 20% of the blood volume. This means that a 10-lb cat can lose as much as 60 ml (2 oz) of blood and a 50-lb dog can lose 400 ml (13 oz) of blood before the blood loss becomes critical. External bleeding can usually be stopped by covering the wound with a piece of gauze or a clean cloth and applying pressure for 5 minutes. Leave the covering on the wound after bleeding stops to avoid disrupting the clot. If bleeding has not stopped after 5 minutes of direct pressure, it may be necessary to apply pressure to the arteries that supply the area or to use a tourniquet. When applying a tourniquet or collapsing arteries by pushing on a pressure point, pressure should be relaxed for 1 to 2 seconds every 10 seconds to allow enough blood to get through to prevent oxygen deprivation to the tissues.

## 14.8.6 CHOKING

Cats rarely choke due to their slow consumption of food. However, dogs tend to "wolf" down food and may also choke on pieces of toys or a ball. Most dogs will gag and retch and be successful with emesis (vomiting the material from the windpipe). If the object does not come out, the dog will paw frantically at its face and then faint. Carefully open the pet's mouth and see if the object can be located and removed. If the object cannot be seen or dislodged, put the dog on its side and perform a modified Heimlich maneuver. For small dogs (or cats), place one hand over the shoulder blades and place the other hand on the ventral abdomen behind the last rib. Steady the animal with the hand over the shoulder blades and push rapidly upward with the hand on the belly—this should force the diaphragm forward and create pressure inside the airways to force the object out. For larger dogs, use both hands on the abdomen to forcefully move it upward. If the dog does not start breathing on its own, try giving artificial respiration as described in Section 14.8.3 (CPR).

## 14.8.7 HEATSTROKE

Heatstroke is an extremely serious condition in which the body temperature of a dog or cat rises above 104°F. The leading cause of heatstroke is leaving a dog or cat in a hot parked car.

Overexertion, especially on a hot day, can also lead to heatstroke. Brachycephalic breeds are predisposed due to the conformation of the soft palates and airways making it difficult for them to dissipate heat by panting. Young and old animals are also more susceptible to heatstroke than healthy adults. Signs of heatstroke include excessive panting or rapid, frantic breathing, drooling, glassy eyes, deep red gums, and a rapid weak pulse rate. Animals may vomit and rapidly go into shock with loss of consciousness. If the body temperature is above 105°F, immediate action is crucial. The pet should be moved to a cool ventilated area and sprayed or immersed in cool water. Put the pet in front of the car air conditioner vents and transport to a veterinary clinic as soon as possible. Cooling measures should be stopped when the body temperature decreases to 103°F to prevent overcooling and hypothermia. Veterinary care is important as damage to kidneys and other organs may occur. Thus, preventive treatments including intravenous fluids and anti-inflammatory medications may be needed.

## 14.8.8  BURNS

Burns may be thermal, chemical, or electrical. Cold packs should be applied for 20 minutes after a thermal burn. Use cold water if a cold pack is unavailable. For most thermal burns, applying cold will be the only treatment required. Monitor the area closely and if the skin sloughs or is very red, seek veterinary treatment to prevent secondary infections. Chemical burns may be caused by household cleaners or yard chemicals. Flush the area with water and take the pet to a veterinarian for further recommendations. Electrical burns most often occur when a playful puppy or kitten chews into an electrical cord. If the pet is fortunate, the damage will be limited to a nasty burn in the mouth. However, in some cases the electrical shock will stop the heart and the pet will require CPR. Unplug the cord before touching the pet. It is advisable to take pets that have bitten an electrical cord to a veterinary clinic and have them examined for heart or lung damage.

## 14.8.9  FRACTURES

Pets may break a bone in a fall or when injured by a moving object (e.g., car, bicycle, kicking horse). If a single limb is injured, the pet may continue moving by hopping on the other three legs. If the fracture is below the knee or elbow, attempt immobilization by padding with rolls of cotton or a soft towel and splinting with rolls of newspaper, a ruler, or similar materials. Muzzle the pet before moving the limb as it may bite because it is in pain. Keep the pet calm and transport to a veterinary clinic.

## 14.8.10  INSECT STINGS AND SPIDER BITES

Insect stings commonly result in pain, swelling, and itching around the site (Figure 14.11). Pets rarely have severe allergic reactions to insect stings; however, if any signs of shock develop, immediate veterinary care should be sought. For other cases, apply cold compresses for 10 to 15 minutes and call a veterinarian for advice concerning administration of antihistamines or other medications. Brown recluse spider bites cause a painful, fluid-filled blister at the bite site. Within a day the skin around the bite site begins to die and slough; the sloughing may continue for several days. It may be necessary for a veterinarian to surgically remove the dying tissue to prevent severe secondary infections and speed healing of the wound. Black widow spiders have venom that causes neurological signs. Within hours of the bite, the pet will become weak and wobbly, may drool excessively, have difficulty breathing, and have muscles spasms or seizures. Pets bitten by a black widow spider should be taken to a veterinary clinic for supportive treatments.

**Figure 14.11** A swollen nose caused by a bee sting. *Courtesy University of Illinois.*

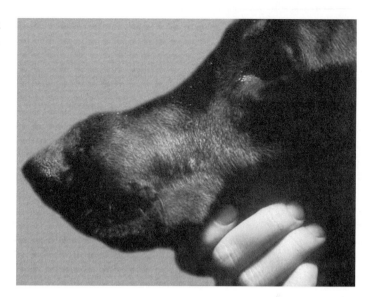

## 14.8.11   SNAKE BITES

It is estimated that more than 15,000 dogs and cats are bitten by poisonous snakes each year.[11] Poisonous snakes of greatest threat to pets in the United States are coral snakes and pit vipers (copperheads, water moccasins, and rattlesnakes). Most pets are bitten on either the face or legs. Immediate signs are swelling, pain, and redness at the site of the bite. Two puncture wounds may be visible corresponding with the two fangs of the snake. Following bites with pit vipers, the pet may show difficulty breathing, vomiting, diarrhea, increased pulse rate, and shock. Later signs are due to tissue damage from enzymes in the snake venom causing bleeding disorders and tissue damage. Veterinary assistance should be sought for administration of antivenom, fluids, and antibiotics. Signs following a coral snake bite may not be apparent for several hours; the venom of this snake causes damage to the nervous system and signs include difficulty swallowing, drooling, a flaccid paralysis, and death from respiratory failure. Immediate veterinary care should be sought for administration of antivenom and systemic antibiotics; do not wait for the onset of clinical signs.

## 14.8.12   POISONING

Pets are inquisitive and frequently ingest poisonous substances. "Pet-proofing" a home is similar to "child-proofing," as prevention of poisoning is much preferred to treatment. First-aid for poisoning is aimed at removing the source of the toxin, decontamination, and seeking veterinary care as soon as possible. See Section 14.4 for additional information on the treatment of poisonings.

## 14.8.13   HEALTH RECORDS

(See also Chapter 8.) Normal neonatal puppies and kittens are pleasingly plump and healthy looking. They sleep and nurse often. Interestingly, newborn puppies and kittens double their birth weight in about 8 to 9 days, which corresponds with the mineral- and protein-rich milks of the bitch and queen (see Table 9.5).

---

[11]http://www.sidyboysfoolin.com/snakebites.html

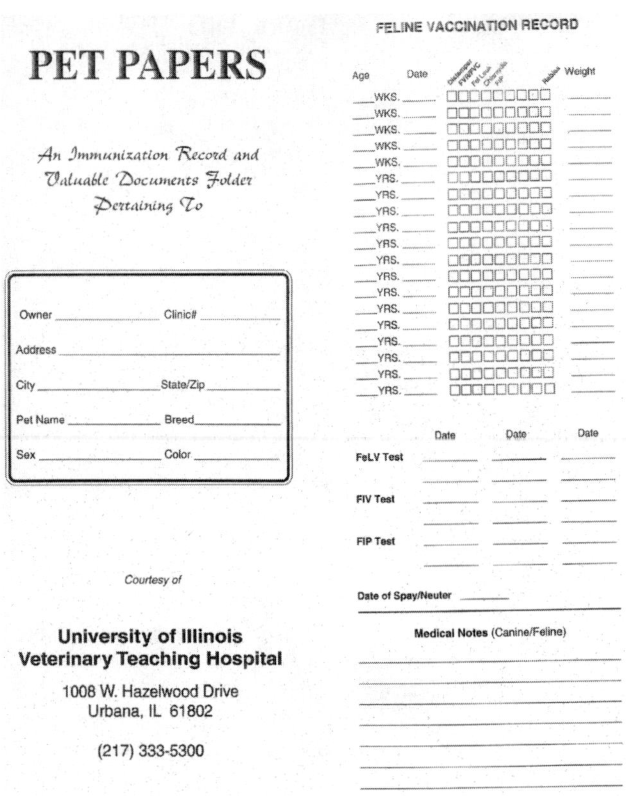

**Figure 14.12** Example of a pet health record. *Courtesy University of Illinois.*

An important early record to keep for puppies is their rate of growth. Record the birth weight, which varies by breed (e.g., toy breeds, 3.5 to 8.8 oz; medium-sized breeds, 7.0 to 10.5 oz; large breeds, 12.3 to 17.6 oz; and giant breeds, 17.6 to 24.7 oz), as does their rate of weight gain (should be 0.03 to 0.06 oz/lb of anticipated adult weight each day for the first 5 months following birth). Other vital signs to record include body temperature (96°F to 97°F for 1 to 2 weeks and 100.5°F by 4 weeks of age), heart rate (about 200 beats per minute), and respiration rate (15 to 30 breaths per minute in puppies and 15 to 35 in kittens 1 to 2 weeks of age) (see Tables 14.1 and 14.2).

Figure 14.12 is a sample form showing basic information desired including what, if any, medication the animal is receiving, age, body weight, previous trip(s) to the veterinary clinic, vaccinations, and other pertinent data.

## 14.9   PET HEALTH INSURANCE

Numerous types of pet health insurance policies are available. The high cost of modern veterinary care can force pet owners to make tough choices related to the health of their pets and their financial resources. Pet health insurance companies work by spreading the cost of veterinary care among a large number of policyholders.

Some pet health insurance companies are like human HMOs or Preferred Provider plans. A hospital or group of hospitals have agreed to provide all the veterinary care the pet needs in return for a fixed monthly fee. These plans usually require policyholders to select veterinarians from a list of participating clinics; specialty care may not be readily available under these plans.

Other pet health insurance companies allow policyholders to choose different levels of service. A basic plan will have a certain per-incident and per-year limit to the coverage and seldom covers wellness and routine pet care. Value plans will have higher per-incident and per-year coverage and may include most routine care. Premium plans will have the highest coverage limits and will cover most routine and specialty clinic examinations and care.

The best plans allow policyholders to choose any licensed veterinarian and will cover a comprehensive list of medical and surgical procedures in addition to wellness care and routine vaccinations. These pet health insurance policies can provide peace of mind to pet owners who know their pet will be covered if illness or accidents occur (see also Section 19.10).

## 14.10 GERIATRIC CARE OF DOGS AND CATS

Aging is inevitable, but with proper care and management, cats and dogs can live healthy lives well into their teens or longer. The aging process varies with species. One year in the life of a cat or dog is equivalent to approximately 7 years of aging in a human. Giant breeds of dogs age faster with 1 year of their life being equivalent to about 10 years of aging in a human. Many veterinarians recommend biannual geriatric examinations for cats and dogs over 7 years of age (5 years for giant breeds).

Good nutrition is critical for good health for animals of all ages (Chapter 9). Diets for older animals ("senior" or "geriatric" formulations) generally have decreased caloric density (older animals tend to be less active and thus require fewer calories), decreased phosphorus (with age kidney function declines and becomes less efficient in excreting excess phosphorus), increased levels of protein (to help maintain muscle mass—although when kidney function declines protein levels may need to be decreased), increased fiber (to prevent constipation), increased levels of antioxidants (to help combat arthritis and other degenerative changes in the body), and increased levels of vitamins (to compensate for increased losses in the urine). Blood chemistry profiles (Chapter 16) should be performed every 6 months in older animals. If these tests detect abnormalities in organ function, additional dietary changes may be warranted (Chapter 9, Section 9.12).

Common problems in geriatric dogs and cats include obesity, arthritis, dental disease, renal (kidney) disease, liver disease, heart disease, cancer, urinary incontinence (urine leaking, especially during sleep), and impairments in taste, vision, and hearing. Older animals are also more susceptible to infections. Geriatric animals may suffer from cognitive dysfunction ("senility" with symptoms of confusion, loss of housetraining, barking for no discernable reason, and other behavioral changes). Fortunately, early recognition and treatment can stabilize many of these disorders and slow the age-associated decline in body functions.

Geriatric examinations typically include a thorough physical examination, complete blood cell counts, biochemical profiles (serum chemistry tests), urinalysis, fecal examinations, heartworm testing, thyroid testing, and evaluation of any lumps or other abnormalities. Additional tests may include an electrocardiogram (ECG or EKG), chest (thoracic) radiographs, abdominal radiographs or ultrasound, blood pressure evaluation, and test for glaucoma. Regular dental care is especially important in older animals (Chapters 12, 13, and 16).

Owners of older animals should watch for any changes in appetite, drinking, urination, body condition, and habits. If changes are detected, they should be reported to the pet's veterinarian. Daily massages or brushing the pet will help to maintain good circulation and a healthy hair coat; additionally, any lumps or signs of disease can be detected in an early (and more easily treated) stage. Older cats may fail to keep their claws short (spend less time using scratching posts) and thus need their claws trimmed every 1 to 2 weeks. Older pets appreciate a soft-padded surface for sleeping or a ramp to make getting onto their favorite chair or bed easier.

## 14.11 EUTHANASIA

*Euthanasia* is derived from the Greek and means "good or gentle death." Modern euthanasia drugs are very fast acting; they are the equivalent of giving an overdose of an anesthetic. The end is peaceful but irreversible.

The decision to end a pet's life is one of the hardest decisions a person ever has to make. It typically involves many emotions. Considerations that may be helpful include: (1) Is the pet's quality of life deteriorated to the point it no longer enjoys what was once a favorite activity? (2) Is the pet in the final stages of a terminal illness? (3) Is the pet suffering pain that cannot be alleviated by drugs? (4) Is the pet's behavior dangerous to others or itself? (5) Has the pet suffered an injury that cannot be repaired? (6) Has the pet reached the point that life is no longer enjoyable with no hope of recovery?

### 14.11.1    BEREAVEMENT

Because the decision to have a closely held family pet euthanized is one of the most difficult decisions owners of companion animals ever face, a feeling of guilt, sadness, and depression may follow. Inasmuch as grief is a common companion of a pet's death, it is often helpful to contact an understanding, caring counselor in the early stage of the bereavement and grief process. Many colleges of veterinary medicine maintain "pet loss support hotlines" to help those dealing with the loss of a beloved pet. The Delta Society can be contacted for a list of referral sources in a given area (see www.deltasociety.org).

### 14.11.2    CARE OF REMAINS

Upon the death of a dog, cat, or other pet, the question arises as to the best method of disposal. Although expensive, it is possible to purchase a marked gravesite in a pet cemetery. This facilitates the opportunity to return periodically and pay tribute to the beloved pet. Personally burying it in the yard or on the farm of a relative or friend is another alternative, although local ordinances should be checked as some prohibit the burying of pets. Many pet owners elect to have their deceased pet cremated. The ashes can be kept in an urn, or may be sprinkled over the pet's favorite outdoor spot, or can be disposed of by the crematory. Other alternatives include having the family veterinarian or local humane society dispose of it. Occasionally, owners will decide to have the pet "stuffed" by a taxidermist; however, it is difficult for a taxidermist to capture the expression of the pet and thus the results may be disappointing. Two additional, although rarely used, methods of preserving a pet are freeze drying (results in a type of mummification) and plasticizing the body.

## 14.12    SUMMARY

A healthy, well cared for animal can be a source of pride for its owner and others who interact with it. In this chapter, we reviewed the most common infectious diseases affecting dogs and cats and provided information related to their prevention and treatment. The interface between diseases affecting animals and humans is becoming blurred by the recognition of similarities in infectious agents affecting different species as well as an increasing number of zoonotic diseases. It is important for those who interact with animals to be familiar with steps that can be taken to minimize the risk of zoonotic transmission of disease.

This chapter reviewed many common infectious and noninfectious diseases affecting the health of dogs and cats. The principles of immunity were reviewed and recommendations provided for vaccination of dogs and cats. Application of the principles of preventive medicine discussed in this chapter will help provide companion animals with maximum potential of living long, healthy lives. An introduction was provided pertaining to the recognition and handling of animal emergencies, which may occur despite efforts to protect animals from illness and injury. Introductions were also provided pertaining to the importance of health records and types of pet health insurance currently available.

We salute readers for being interested in the health and quality of life for their favorite pet(s)!

# 14.13 REFERENCES

1. Ettinger, Stephen J., and Edward C. Feldman, eds. 2002. *Textbook of Veterinary Internal Medicine, Diseases of the Dog and Cat.* 5th ed. Philadelphia: W. B. Saunders.

2. Feldman, Edward C., and Richard W. Nelson. 2004. *Canine and Feline Endocrinology and Reproduction.* 3rd ed. Philadelphia: W. B. Saunders.

3. Greene, Craig E. 1998. *The Infectious Diseases of the Dog and Cat.* Philadelphia: W. B. Saunders.

4. Hand, Michael S., Craig D. Thatcher, Rebecca L. Remillard, and Philip Roudebush. 2002. *Small Animal Clinical Nutrition.* 4th ed. Topeka, KS: Mark Morris Institute.

5. Nelson, Richard W., and C. Guillermo Couto. 1998. *Small Animal Internal Medicine.* 2nd ed. St. Louis, MO: Mosby.

6. Scott, Danny W., William H. Miller, Jr., and Craig E. Griffin. 2001. *Muller & Kirk's Small Animal Dermatology.* 6th ed. Philadelphia: W. B. Saunders.

7. Tilly, Larry P., and Francis W. K. Smith, Jr. 2004. *The 5-Minute Veterinary Consult.* 3rd ed. Philadelphia: Lippincott Williams & Wilkens.

# 15 PARASITES AND PESTS OF COMPANION ANIMALS

*The essence of knowledge is, having it, to apply it.*

Confucius (551–479 BC)
*Chinese philosopher*

# 15.1  INTRODUCTION/OVERVIEW

In conflicts between parasites and hosts, parasites usually win. Why is this true? Because a parasite is an organism living in, with, or on (and at the expense of) another organism known as the host. In most cases, the parasite obtains nutrients from and causes injury to the host; it is ideal for parasite survival if the injury to the host is minimal. Ectoparasites are parasites found on the external skin surface, whereas endoparasites are located within the host (internal; e.g., in the gastrointestinal tract, respiratory tract, urinary tract, circulatory system, muscles, or another internal location). A pest is one that pesters or annoys; many ectoparasites are pests to their hosts, whereas others serve as vectors (carriers) of disease.

Each parasite has a life cycle, which may include several life stages (e.g., eggs, larvae, nymphs, pupae, adults), and involves one or more hosts. The definitive host harbors the adult stage of the parasite. The life cycle of many parasites also includes one or more intermediate hosts, which harbor larval or juvenile stages of parasites.

Parasites of importance in companion animals belong to the phyla Arthropoda, Nematoda, Acanthocephala, Trematoda, Cestoda, and Protozoa. The nematodes, acanthocephalans, trematodes, and cestodes are commonly referred to as helminths or worms. Nematodes are cylindrical worms; cestodes and trematodes are flatworms. In this chapter, we review the major arthropod, helminth, and protozoan parasites of dogs and cats. *The most important information is summarized in Tables 15.1 through 15.6.* Parasites of importance to other companion animals were reviewed in Chapters 5, 6, and 7. Knowing the classification of parasites is helpful in remembering their life cycles—important in understanding ways to prevent infection—and also in selecting parasiticides; medications effective in eliminating members of one phylum are often not effective against parasitic members of other phyla.

# 15.2  PHYLUM ARTHROPODA

Arthropods are a diverse group of invertebrate animals that total over one million species. There are more species of arthropods on earth than all other animals combined! Of this huge number of species, only a few have evolved as parasites of other animals.

Arthropod parasites belong to the classes Insecta and Arachnida. Within the class Insecta are the orders Diptera (flies), Phthiraptera (lice), Siphonaptera (fleas), Hymenoptera (wasps, bees, hornets, ants), and thousands of insects that do not cause harm to companion animals (beetles, butterflies, and others). Within the class Arachnida are the orders[1] Acari (mites and ticks), Araneae (spiders), Scorpionida (true scorpions), and many orders of arachnids that do not directly affect companion animals. The order Acari includes the suborders Astigmata (mites), Prostigmata (mites), Mesostigmata (mites), and Metastigmata (ticks) (Figure 15.1).

## 15.2.1  INSECTS

Ninety percent of the species in the class Arthropoda belong to the order Insecta. Adult insects have bodies with three sections (head, thorax, and abdomen). The thorax typically articulates with three pairs of legs and in many species with two pairs of wings. The head has one pair of antennae. There are 29 orders within the class Insecta; however, only four are of importance to companion animals (Table 15.1): Diptera (flies), Phthiraptera (lice), Siphonaptera (fleas), and Hymenoptera (wasps, bees, hornets, and ants).

Diptera are "true flies"; the adults have a single pair of wings. Their name is derived from the Greek phrase *di pteron*, which means two winged. There are more than 120,000 species of flies; each has a complex life cycle with complete metamorphosis. The life cycle includes (1) *eggs* that hatch into larvae, (2) *larvae* (also called maggots or grubs) that go through one

---

[1] In some classification schemes, Acari, Araneae, and Scorpionida are considered as subclasses of the class Arachnida, whereas other classification schemes consider these to be orders.

**Figure 15.1**   Schematic of arthropod orders parasitizing companion animals.

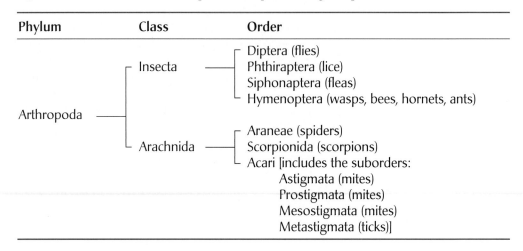

| Phylum | Class | Order |
|---|---|---|
| Arthropoda | Insecta | Diptera (flies)<br>Phthiraptera (lice)<br>Siphonaptera (fleas)<br>Hymenoptera (wasps, bees, hornets, ants) |
| | Arachnida | Araneae (spiders)<br>Scorpionida (scorpions)<br>Acari [includes the suborders:<br>Astigmata (mites)<br>Prostigmata (mites)<br>Mesostigmata (mites)<br>Metastigmata (ticks)] |

**Table 15.1**

Insect Parasites of Dogs and Cats and Their Associated Diseases

| Insect | Disease |
|---|---|
| Mosquito | Vectors for heartworm disease (*Dirofilaria immitis*), viral diseases (West Nile virus)<br>Allergic reactions result in lesions on face of cats and dogs |
| Sand flies<br>    *Phlebotomus, Lutzomyia* | Vectors for *Leishmania* |
| Biting flies<br>    Stable flies (*Stomoxys*)<br>    Black flies (*Simuliidae*) | Fly bite dermatitis, most common on ear tips |
| Fly larvae, especially flesh flies (Sarcophagidae) and blow flies (Calliphoridae) | Myiasis (fly larvae in tissues) |
| *Cuterebra* | Larvae found in a skin nodule, a hole forms in the skin creating a pore for the larvae to breathe through and later emerge through |
| Biting lice<br>    *Trichodectes canis* in dogs,<br>    *Heterodoxus spiniger* in dogs,<br>    *Felicola subrostrata* in cats | Pediculosis; itching and hair loss; intermediate host for tapeworms (*Dipylidium caninum*) |
| Sucking lice<br>    *Linognathus setosus* in dogs | Pediculosis; itching and hair loss |
| Fleas<br>    *Ctenocephalides felis, C. canis,*<br>    *Pulex irritans, Echidnophaga*<br>    *gallinacea, Xenopsylla* | Anemia; itching and hair loss; intermediate hosts of tapeworms (*Dipylidium caninum*) and nematodes (*Acanthocheilonema reconditum*); vectors of *Rickettsia typhi, R. felis, Bartonella* spp; *C. felis* is the cat flea, *C. canis* is the dog flea, *P. irritans* is the human flea, *E. gallinacea* is the poultry sticktight flea, *Xenopsylla* are rodent fleas—however, fleas are not species specific and cross-infections are common |
| Hymenoptera (bees, wasps, hornets, ants) | Venom results in pain, swelling, and allergic reactions that may produce difficult breathing, circulatory collapse, and death |

or more molts before entering a pupal stage, (3) *pupa* that form a protective shell known as the puparium or cocoon and undergo a reorganization and reconstruction (metamorphosis) into (4) the *adult fly.*

Flies may be parasitic as larvae or as adults, but rarely are both life-cycle stages parasitic on animals. Adult flies may feed on blood, sweat, tears, or other secretions of dogs and cats. In addition to causing blood loss, pain, and annoyance, biting flies serve as vectors and transmit a variety of other organisms causing disease in dogs and cats. Larvae of some species (e.g., screwworms, other maggots, *Cuterebra*) parasitize tissues of dogs, cats, and other animals.

The most important family of flies parasitizing dogs and cats is the family Culicidae (mosquitoes). Adult mosquitoes are small, slender flies with long legs. Adult females lay eggs on the surface of water, where the eggs hatch into larvae known as wigglers. Wigglers molt three to four times and enter the pupal stage (tumblers). The adult mosquito emerges from the pupal case and soon flies in search of a mate. Males feed on nectar and plant juices; females require a blood meal prior to egg laying. Diseases associated with mosquitoes include (1) heartworm disease (mosquitoes are vectors of *Dirofilaria immitis;* Chapter 14), (2) viral diseases (mosquitoes are vectors of West Nile virus and other arboviruses; these primarily cause disease in birds, horses, and humans, although dogs and cats can also be infected), and (3) allergies (mosquitoes are sources of antigens; mosquito-bite hypersensitivity can result in severe skin lesions on the face of cats and dogs).

Sand flies (family Psychodidae) include the genus *Phlebotomus* (in the Mediterranean) and the genus *Lutzomyia* (in the Americas). Sand flies are vectors of the protozoa *Leishmania.* Leishmaniasis is a serious infection in dogs and humans and has been reported in cats (Chapter 14). Leishmaniasis is often a chronic disease resulting in enlargement of the spleen and liver, loss of appetite, skin lesions, weight loss, and eventually death. In the United States, leishmaniasis most commonly affects Fox Hounds.

Flies of the genera *Stomoxys* (stable flies) and *Simuliidae* (black flies) have sharp biting mouthparts and feed on the ears, face, legs, and ventral abdomen of cats and dogs. The ear pinnae are most often affected. Bite sites may ooze blood and serum forming a hemorrhagic crust (Figure 15.2). If these lesions are noted, a fly repellant can be applied to the ears to help protect the animal from further fly bites.

Many species of flies lay eggs on the warm, wet skin of an animal with wounds or that is urine soaked. When the eggs hatch, the larvae feed on the diseased skin. The larvae produce proteolytic enzymes that break down the tissue resulting in "punched out" lesions in the skin called fly strike (Figure 15.3). Secondary bacterial infections may result in death of the animal. True screwworms, *Cochliomyia hominivorax,* rarely cause disease in dogs and cats;

**Figure 15.2**  Fly bite dermatitis on the ear (pinna) of a dog. Note hair loss and crusts on the tip of the ear. *Courtesy Dr. Karen L. Campbell.*

**Figure 15.3** Fly larvae (maggots) producing "punched out" lesions (fly strike) in the skin of a dog. *Courtesy Dr. Karen L. Campbell.*

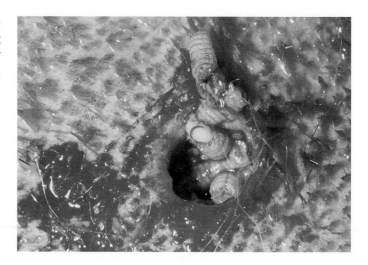

however, larvae of flesh flies (Sarcophagidae) and blowflies (Calliphoridae) can be found in neglected wounds (wounds not promptly treated and protected from flies).

Adult flies of the genera *Cuterebra* are beelike insects that deposit eggs near the burrows of rodents and rabbits. The eggs hatch into parasitic larvae, which may penetrate through the nose, eyelids, mouth, or wound of an animal (e.g., dog, cat, rabbit, or rodent). In dogs and cats, the larvae usually remain within the skin and subcutaneous tissues of the head and neck. A swelling develops around the larva and it forms a fistula (breathing hole). Eventually, the larva drops to the ground, pupates, and emerges as a new adult. Occasionally, the larvae migrate into other tissues such as the brain, and in those cases, may cause death of the host.

The order Phthiraptera (lice) are small, wingless insects that are highly species specific; this means they affect only one host species. They usually spend their entire life on the host, leaving only when transferred by grooming or close contact with a new host. There are two important suborders: Mallophaga (chewing or biting lice) and Anoplura (sucking lice). The life cycle of lice includes eggs (nits) cemented onto the hairs of the host, nymphs that are similar in shape to adults but smaller, and adults. Lice can survive only a few days off a host. Infestation with lice is referred to as pediculosis. Some species of lice serve as vectors for the transmission of disease organisms (e.g., poxvirus in swine) and as intermediate hosts for tapeworms (e.g., *Dipylidium caninum*). Heavy infestations may result in itching, loss of hair, and anemia (from loss of blood).

Mallophaga are 2 to 3 mm long with large, rounded heads and large mandibles fitted with teeth for chewing. The most common biting louse of dogs is *Trichodectes canis.* It is a small yellow louse found on the head, neck, and tail of dogs (Figure 15.4). *Heterodoxus spiniger* is another biting louse and is an ectoparasite of dogs in warm climates. *Felicola subrostrata* is the only biting louse of cats. It is most commonly found on the face, ears, and back. Six families of sucking lice (Anoplura) affect mammals; however, the only one that commonly affects dogs is *Linognathus setosus.* There are also several species of biting, chewing, and sucking lice that affect guinea pigs, rabbits, mice, rats, gerbils, and hamsters (Chapter 7).

Lice are easily diagnosed by finding nits, nymphs, or adults on the hairs of an infested animal. Acetate tape can be used to "trap" the louse for examination and identification using a microscope. Lice can be killed by saturating the animal's coat with an insecticidal shampoo or spray. Ivermectin can also be used to kill lice.

The order Siphonaptera includes over 2,500 species of fleas. Adult fleas are wingless, blood-feeding insects with well-developed hind legs used for jumping onto host animals. They have a complex life cycle with a complete metamorphosis (Figure 15.5). The adult flea lives on its host, where the female lays several eggs daily. The eggs are nonsticky and soon fall from the host. Under ideal conditions (18°C to 27°C and $\geq$ 75% relative humidity), eggs hatch in 4 to 6 days. The flea larvae feed on organic material (such as feces from adult fleas),

**Figure 15.4**  (*a*) Lice on the head of a dog. (*b*) *Trichodectes canis* removed from the dog at 400× magnification. *Courtesy University of Illinois.*

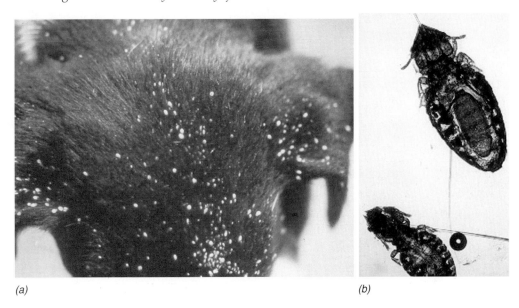

(a)  (b)

and molt twice before spinning an ovoid cocoon and developing into a pupa. Emergence of an adult flea from its cocoon is triggered by increases in temperature, vibration, or physical pressure. The length of this life cycle varies between 14 and 140 days, dependent primarily on temperature.

The most common flea found on dogs and cats is *Ctenocephalides felis. C. felis* parasitizes a wide range of animals including dogs, cats, swine, rodents, birds, and humans. *C. canis* is relatively rare; the majority of fleas found on both dogs and cats are *C. felis*.[2] Other fleas that occasionally parasitize companion animals include *Echidnophaga gallinacea* (the sticktight flea), *Pulex irritans* (the human flea), and *Xenopsylla* spp (the rodent flea).

A flea consumes approximately 0.014 ml of blood per day. A heavy infestation can result in death of the host from loss of blood. Flea saliva contains a variety of proteins that are highly irritating and can also elicit allergic reactions in host animals. Clinical signs of flea infestation include itching, scratching, chewing, hair loss, rashes, and bumps on the skin that may become secondarily infected with bacteria (Figure 15.6). Fleas can serve as intermediate hosts of the tapeworm *Dipylidium caninum* and for the nematode *Acanthocheilonema* (*Dipetalonema*) *reconditum*. Fleas may also serve as vectors for *Rickettsia typhi, R. felis, Bartonella* spp, and other organisms.

The diagnosis of flea infestations is usually easy; fleas are large enough to be seen crawling or jumping on an animal (Figure 15.6b). Finding flea feces is also diagnostic of a flea infestation. Flea feces are comma-shaped particles composed primarily of dried blood (observed by finding a halo of red diffusing out of a flea fecal particle when placed on a dampened cotton ball or piece of gauze). Control of fleas is best achieved by using a program of integrated pest management. This involves the simultaneous use of treatments and control measures designed to interrupt the flea life cycle at multiple stages. Adulticides such as imidacloprid (Advantage®), fipronil[3] (not on rabbits; Frontline®, TopSpot®), pyrethrins, or permethrin[4] (not on cats) are used to kill adult fleas. Insect growth regulators (IGRs) or

---

[2]The differentiation between *C. felis* and *C. canis* is based on differences in the lengths of their teeth (first and second teeth are of equal length in *C. felis,* whereas the first tooth of *C. canis* is about half as long as the second tooth).

[3]**Important note**—do not use fipronil on rabbits; it is highly toxic to them.

[4]**Important note**—do not use permethrins on cats; it is toxic to them. Many cats have been killed by treatment with permethrin-containing insecticides approved only for use on dogs.

**Figure 15.5** Life cycle of *Ctenocephalides felis* (fleas). *Courtesy University of Illinois.*

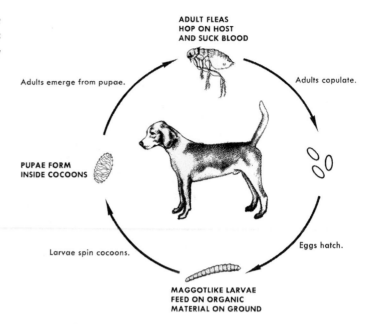

ADULT FLEAS
HOP ON HOST
AND SUCK BLOOD

Adults emerge from pupae.

Adults copulate.

PUPAE FORM
INSIDE COCOONS

Eggs hatch.

Larvae spin cocoons.

MAGGOTLIKE LARVAE
FEED ON ORGANIC
MATERIAL ON GROUND

insect development inhibitors are used to prevent flea larva from completing metamorphosis into adults. Examples of IGRs include methoprene, fenoxycarb, pyriproxyfen, and lufenuron. Lufenuron is unique in being given orally to the host. Lufenuron concentrates in the host's adipose tissue (fat). Fleas feeding on the host for the next month acquire enough lufenuron to sterilize eggs and produce feces that are fatal when consumed by flea larvae. Additional efforts are needed to remove and kill fleas in the environment. Frequent vacuuming plus applications of IGRs and adulticides can be used to kill fleas in the environment.

The order Hymenoptera has over 130,000 species; those of importance as pests to dogs and cats are bees, wasps, hornets, and fire ants. These venomous insects are not parasitic; however, toxins in their venom produce a painful sting or bite and can be fatal in animals with allergies, or those receiving a large number of stings or bites. The face is most often affected (see Figure 14.11), likely a result of the dog or cat sticking its nose into a bush or nest harboring the insects.

**Figure 15.6** (*a*) Hair loss and skin irritation associated with flea infestations in an adult dog. (*b*) Close-up of the fleas feeding on the dog in (*a*). *Courtesy Dr. Karen L. Campbell.*

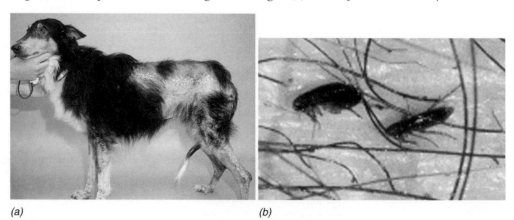

*(a)*                                        *(b)*

## 15.2.2   ARACHNIDS (MITES, TICKS, SCORPIONS, SPIDERS)

Arachnids are arthropods with four pairs of legs and no antennae as adults. Their bodies are usually divided into two parts, the cephalothorax and abdomen; however, these are fused in many ticks. The first pair of appendages is located in front of the mouth. These appendages are used in feeding and are called chelicerae (from Greek *chele keras* meaning claw horn). The second pair of appendages (called pedipalps) is located behind the mouth. In some species, the chelicerae and pedipalps are joined together in a structure called the gnathosoma. The arachnids have simple eyes and no antennae or wings. Orders within this phylum of importance as ectoparasites of companion animals (Table 15.2) include Acari (mites and ticks), Araneae (spiders), and Scorpionida (scorpions).

Arachnids have four basic life-cycle stages: (1) egg, (2) six-legged larva, (3) eight-legged nymph, and (4) eight-legged adult. There may be more than one molt in the larval and nymph stages. In some species, the stages are further divided into prelarva and larva plus protonymph, deutonymph, and tritonymph. Classification of the suborders of Acari (ticks and mites) into Astigmata, Prostigmata, Mesostigmata, and Metastigmata is based on the location of the respiratory opening on the body.[5]

*Astigmata.* This suborder of mites includes the families Sarcoptidae, Psoroptidae, Knemidokoptidae, and Listrophoridae. Infections with mites produce skin irritation, hair loss (feather loss in birds), and thickening of the skin and predispose the animal to bacterial and fungal infections. Mange is the term used for skin disease caused by mites.

Mites in the Sarcoptidae family burrow in the skin and complete their life cycle on a host animal. Three genera are important parasites of companion animals: *Sarcoptes, Notoedres,* and *Trixacarus.*

*Sarcoptes scabiei* causes a highly pruritic (itchy) form of mange. The mites are very small (0.2 to 0.6 mm long) and burrow in superficial layers of skin. They are highly transmissible from one animal to another. The hosts most commonly affected include dogs, swine, sheep, goats, cattle, and humans. Foxes and hedgehogs are believed to serve as reservoirs of these mites. Mites are most numerous in the skin of ears, elbows, legs, and ventral abdomen (Figure 15.7). The host often develops an allergic reaction to mites; this allergic reaction intensifies itching and skin irritation. Diagnosis is made by finding the mites on superficial skin scrapings (Chapter 16) or by response to therapy.[6] Mites can be eliminated from a host using a topical acaricide (selamectin or lime sulfur) or a systemic parasiticide (ivermectin or milbemycin). Treatments are usually repeated weekly or every other week for a month. All animals in contact with an infected individual should be treated. The bedding and grooming tools should also be cleaned.

*Notoedres cati* is a sarcoptiform mite that is smaller than *Sarcoptes scabiei* and has a dorsal anal opening (the anal opening of *Sarcoptes scabiei* mites is posterior). Females are 225 μm in length, males approximately 150 μm. This mite primarily affects cats and is most numerous on the head and ears. Clinical signs include facial rubbing and scratching, hair loss, crusting, and scaling. Lesions may spread to involve the legs and other body areas. The life cycle is similar to that of *Sarcoptes scabiei;* the mites burrow in the superficial layer of the skin and lay eggs within these tunnels (Figure 15.8). The mites are highly contagious to other cats and occasionally infest people, dogs, and rabbits. Other species of *Notoedres* are ectoparasites of bats, mice, and rats. Mites in the genera *Trixacarus* are ectoparasites of rats and guinea pigs (Chapter 7).

Mites in the family Psoroptidae have oval-shaped bodies and are nonburrowing. Some feed on skin scales, others on tissue fluid. The genera of greatest importance as ectoparasites

---

[5]The Astigmata do not have a visible opening (stigmata); the stigmata of the Prostigmata mites are located on the gnathosoma, the stigmata of Mesostigmata mites are located above the legs, and the stigmata of Metastigmata ticks are located behind the last pair of legs.

[6]When treating the animal for mange eliminates the signs of itching and skin disease, a "retrospective" diagnosis of probable mange is made.

**Table 15.2**
Arachnid Parasites of Dogs and Cats and Their Associated Diseases

| Parasite | Host(s) | Disease |
|---|---|---|
| Scabies<br>  *Sarcoptes scabiei* | Dog, fox, pig, sheep, goat, cattle, humans | Highly contagious skin disease associated with severe itching and hair loss |
| Feline scabies<br>  *Notoedres cati* | Cats, occasionally dogs, rabbits, humans | Highly contagious skin disease with itching, hair loss, and crusting of the head and legs |
| Ear mites<br>  *Otodectes cynotis* | Dogs, cats, other carnivores | Dark exudates in ear canals, shaking and scratching at ears, secondary ear infections |
| Fur mite<br>  *Lynxacarus radovsky* | Cats | Hair loss and itching along the back |
| Red mange, demodicosis, follicular mange<br>  *Demodex canis* | Dogs | Hair loss, secondary skin infections (can be severe), passed from bitch to puppies but is not contagious in adult dogs |
| Feline demodicosis, follicular mange<br>  *Demodex felis* | Cats | Hair loss and secondary skin infections, most common in immunosuppressed cats, not contagious to other cats |
| Short, stubby demodex mite<br>  *Demodex gatoi* | Cats | Hair loss and itching, found in epidermis rather than hair follicle, may be contagious to other cats |
| Walking dandruff, cheyletiellosis<br>  *Cheyletiella yasguri*<br>  *Cheyletiella blakei*<br>  *Cheyletiella parasitivorax* | <br>Dogs<br>Cats<br>Rabbits | Scaling and hair loss along the back, itching, cross-infections are common and may transiently infest humans |
| Chiggers<br>  *Trombiculidae* | Many animals | The larvae are parasitic, result in severe itching that persists for days after the larvae falls off or is removed |
| Nasal mite<br>  *Pneumonyssus caninum* | Dogs | Results in sneezing and facial itching |
| Brown dog tick<br>  *Rhipicephalus sanguineus* | Dogs | Intermediate host for *Babesia canis* |
| American dog tick, wood tick<br>  *Dermacentor variabilis* | Mice, rodents, dogs, humans | Vector for Rocky Mountain spotted fever, tularemia, also causes tick paralysis |
| Lone star tick<br>  *Amblyomma americanum* | Many animals, humans | Vector of tularemia and Rocky Mountain spotted fever |
| Gulf Coast tick<br>  *Amblyomma maculatum* | Cattle, horses, sheep, dogs, humans | Feeds in ears, can cause tick paralysis |
| Deer tick<br>  *Ixodes dammini* | Deer, mice, voles, humans, dogs | Larvae transmit *Borrelia burgdorferi* (Lyme disease); other species of *Ixodes* also transmit this infection (*I. scapularis*, *I. pacificus*) |
| Continental rabbit tick<br>  *Haemaphysalis leporispalustris* | Rabbits, birds, dogs, cats, humans | Transmits Rocky Mountain spotted fever, Q fever, and tularemia |
| Spinose ear tick<br>  *Otobius megnini* | Cattle, sheep, goats, horses, dogs | Larvae and nymph forms are parasitic |
| Brown recluse spider<br>  *Loxosceles reclusa* | Not a primary parasite | Bites result in extensive skin necrosis |
| Common brown spider<br>  *Loxosceles unicolor* | Not a primary parasite | Bites result in extensive skin necrosis |
| Black widow spider<br>  *Latrodectus mactans* | Not a primary parasite | Bites may result in muscle cramps, pain, vomiting, diarrhea, paralysis, and death |
| Scorpions<br>  *Scorpionida* | Not a primary parasite | Stings may result in pain, muscle spasms, drooling, vomiting, difficult breathing, paralysis, and death |

**Figure 15.7**   (*a*) Hair loss on the ears of a puppy with sarcoptic mange. (*b*) *Sarcoptes scabiei* mite found on skin scraping viewed at 400×. *Courtesy University of Illinois.*

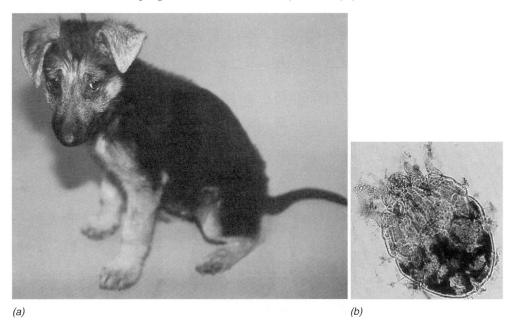

(a)                                                                                              (b)

of companion animals are *Otodectes* and *Psoroptes.* A third genera, *Chorioptes,* includes ectoparasites that affect cattle, sheep, goats, horses, and occasionally rabbits.

*Otodectes cynotis* is a very common mite affecting the ears of dogs, cats, and other carnivores. The mites are most numerous in the deep portion of the ear canal, but are occasionally found on the head, back, and tail. The mites feed on ear secretions. Some infected animals show few clinical signs whereas others shake their head and scratch their ears. The ear canal is often filled with reddish-brown-black debris (Figure 15.9) and may become secondarily infected with bacterial and yeast organisms. The mites are highly contagious to other cats and dogs. Diagnosis is made by visualizing the mites with a magnifying otoscope (Chapter 16) or

**Figure 15.8**   Life cycle of *Notoedres cati. Courtesy University of Illinois.*

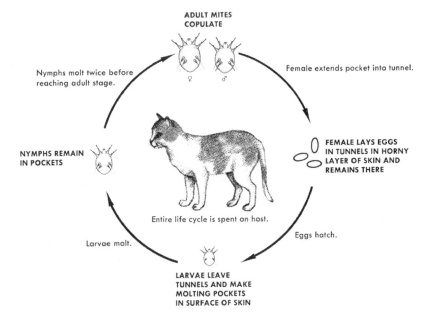

ADULT MITES
COPULATE

Nymphs molt twice before
reaching adult stage.

Female extends pocket into tunnel.

NYMPHS REMAIN
IN POCKETS

FEMALE LAYS EGGS
IN TUNNELS IN HORNY
LAYER OF SKIN AND
REMAINS THERE

Entire life cycle is spent on host.

Larvae molt.

Eggs hatch.

LARVAE LEAVE
TUNNELS AND MAKE
MOLTING POCKETS
IN SURFACE OF SKIN

**Figure 15.9**    (*a*) Dark exudate filling the ear canal of a dog infested with *Otodectes cynotis*. (*b*) *Otodectes cynotis* mites as seen with a microscope. *Courtesy University of Illinois.*

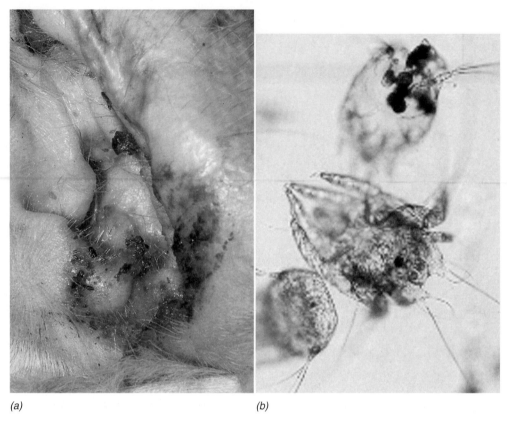

*(a)*                                                            *(b)*

finding the mites on microscopic examination of an ear swab (Chapter 16). Treatment options include topical or systemic acaricides (e.g., pyrethrins, milbemycin, selamectin, ivermectin, and others). All dogs and cats that have been in contact with an infected animal should be treated. Grooming equipment, bedding, and other potential fomites[7] should be cleaned.

*Psoroptes cuniculi* is the ear mite of rabbits (Chapter 7). Other species of *Psoroptes* are mites affecting the ears or skin of sheep, cattle, and horses that rarely affect companion animals. The family Knemidokoptidae includes astigmatid burrowing mites that affect birds (Chapter 5).

Mites in the family Listrophoridae are commonly known as fur mites. These mites have modified legs that grip or clasp hairs. The fur mite of cats is *Lynxacarus radovsky*. These mites have elongated, oval-shaped bodies and are usually located on hairs along the back of the cat. Presence of these mites results in hair loss and itching. Diagnosis is made by finding the mites on hair plucks or skin scrapings. Treatment with topical or systemic acaricides (e.g., lime sulfur or ivermectin) is effective. The fur mite affecting guinea pigs is *Chirodiscoides caviae*. Some fur mites that affect other species also belong to the suborder Prostigmata (e.g., the fur mites of mice and rats, *Myobia musculi, Radfordia affinis,* and *Radfordia ensifera;* Figure 15.10).

***Prostigmata.***    This suborder of mites is highly diverse; it includes the families Demodicidae, Cheyletiellidae, Myobiidae, Trombiculidae, and many others.

Mites of the family Demodicidae are elongated inhabitants of the hair follicles and sebaceous glands (oil glands of skin). A few species are also found as inhabitants of the superficial layers of skin. *Demodex canis* is an important cause of skin disease in dogs. Most dogs harbor a

---

[7] Fomites are objects that are not in themselves harmful but are able to harbor pathogenic organisms and transfer them to an animal.

**Figure 15.10**  A fur mite of rats, *Radfordia ensifera*, has an elongated, oval body and legs that clasp hairs. *Courtesy University of Illinois.*

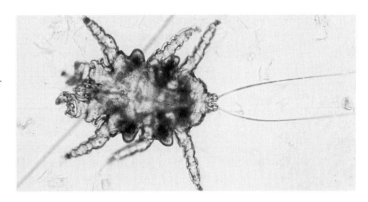

few *Demodex* mites as part of their normal skin flora. The immune system is important in minimizing numbers of mites. Young dogs with immature immune systems may develop circular areas of hair loss caused by large numbers of mites in hair follicles of the face and forelegs. These localized infections often resolve when the dog's immune system matures at approximately 1 year of age. Dogs with defective immune systems (from hereditary immune defects or from other diseases or drugs that impair the function of the immune system) develop a generalized loss of hair and severe skin inflammation with secondary bacterial infections. This generalized form is referred to as demodicosis or "red mange" (Figure 15.11). Demodicosis is not contagious but can be fatal without treatment. The treatment of demodicosis includes topical or oral acaricides plus antibiotics for secondary skin infections. Treatments must often be continued for several months. Dogs that have been affected with demodicosis should not be used for breeding because the immune defects associated with this disease are hereditary.

*Demodex cati* is the follicular mange mite of cats. It is most commonly found in cats with immune deficiencies secondary to infection with feline leukemia virus or feline immunodeficiency virus (Chapter 14). *Demodex gatoi* is a shorter, stubbier demodex mite of cats found in "pits" in the epidermis (skin surface) rather than within hair follicles. *D. gatoi* can result in

**Figure 15.11**   (*a*) Red mange caused by *Demodex canis* in a young bloodhound dog. (*b*) *Demodex canis* mites in a skin scraping viewed at 100× magnification. *Courtesy University of Illinois.*

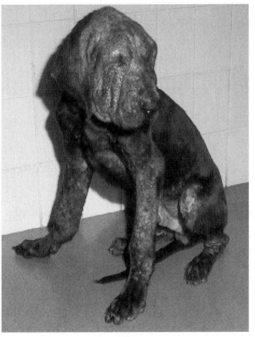

(a)                                                              (b)

excessive hair grooming, itching, and hair loss. Because of its more superficial location, *D. gatoi* can spread to other cats (the only species of demodex mites considered to be contagious). Diagnosis of demodex infections is made by finding large numbers of mites and mite eggs on skin scrapings (Chapter 16).

Parasitic mites of the family Cheyletiellidae have large palpal claws. These mites live on the skin surface and result in itching and large amounts of dandruff. The life cycle is completed on the host. A common descriptive name for the *Cheyletiella* spp is "walking dandruff" as the mites can be observed to move when observed with a magnifying lens. The mites can also be found on skin scrapings or acetate preparations (Chapter 16 and see also Figure 16.5a). *C. yasguri* primarily affects dogs, *C. blakei* primarily affects cats, and *C. parasitivorax* is most frequently found on rabbits; however, cross-infection may occur and humans can be temporary hosts for these mites. The mites can survive in the environment for up to a month, thus treatment must include both animals and their environment. Weekly applications of acaricides should be continued for 6 to 8 weeks.

The family Trombiculidae includes chiggers, which are parasitic as larvae. The nymphs and adults are free living (not parasitic). Chiggers have a bladelike mouth part that penetrates the host's skin. Saliva of the chigger dissolves skin cells. Some dissolved material is ingested by the mite, and the remaining material hardens to form a tube called a stylostome that connects the chigger to its host. Chiggers may feed for several days before disconnecting and falling off the host. Chiggers parasitize a variety of animals including dogs, cats, rodents, and humans. In the Northern Hemisphere, they are most active in late summer and autumn. These mites cause severe itching that can continue for days after the larvae disconnect from the host. The family Myobiidae includes fur mites of rodents (Figure 15.10).

*Mesostigmata.*   Important families in the suborder Mesostigmata include Dermanyssidae, Raillietidae, Halarchnidae, and Rhinonyssidae. Mites in the family Dermanyssidae are bloodsucking parasites of birds and rodents and will transiently affect other animals and humans. *Dermanyssus gallinae* is perhaps the best known. Its common names are "red mite" and "chicken mite." The primary mite of interest in the family Raillietidae is the ear mite of cattle, *Raillietia auris*. The family Halarchnidae includes the dog nasal mite, *Pneumonyssus caninum*. This mite causes sneezing and facial itching in dogs. It can be seen using a magnifying lens or by microscopic examination of a nasal swab. Mites in the family Rhinonyssidae are parasites of the respiratory passages of birds (Chapter 5).

*Metastigmata.*   The suborder Metastigmata includes two major families of parasitic ticks: the Ixodidae (hard-bodied ticks) and the Argasidae (soft-bodied ticks). Ticks are long-lived acarines that directly weaken a host by sucking blood and can also transmit a large variety of diseases (Chapter 14), including borreliosis (Lyme disease), rickettsial diseases (ehrlichiosis, Rocky Mountain spotted fever), piroplasmosis (infection with *Babesia canis*), and tick paralysis. Ticks have four major stages in their life cycle: egg, larva, nymph, and adult. Eggs are laid in the environment. The larvae, nymphs, and adults are all intermittent blood feeders that remain on a host for several days and then drop off. This intermittent feeding pattern results in infestation of several hosts and thus the potential to transmit diseases to many animals. One-host ticks feed on a single host species; two-host ticks have one host species for the larvae and nymph and a second for the adults; three-host ticks feed on a different host species for each life stage (Figure 15.12).

Ixodid ticks have a dorsal plate known as the scutum that covers most of the body of a male tick and the anterior half of a female tick. The family Ixodidae includes 14 genera and over 700 species of ticks. *Rhipicephalus sanguineus* (brown dog tick) feeds almost exclusively on dogs and can be found both outdoors and inside homes and kennels. It is an intermediate host for *Babesia canis* (causes piroplasmosis in dogs). *Dermacentor variabilis* (American dog tick, wood tick) is a three-host tick found outdoors. Its usual hosts are mice and rodents; however, it also feeds on dogs and humans. Diseases spread by this tick include Rocky Mountain spotted fever, tularemia, and tick paralysis. *Amblyomma americanum* (lone star tick) is a three-host tick that feeds on many animals as well as humans.

**Figure 15.12**   Life cycle of a three-host tick. *Courtesy University of Illinois.*

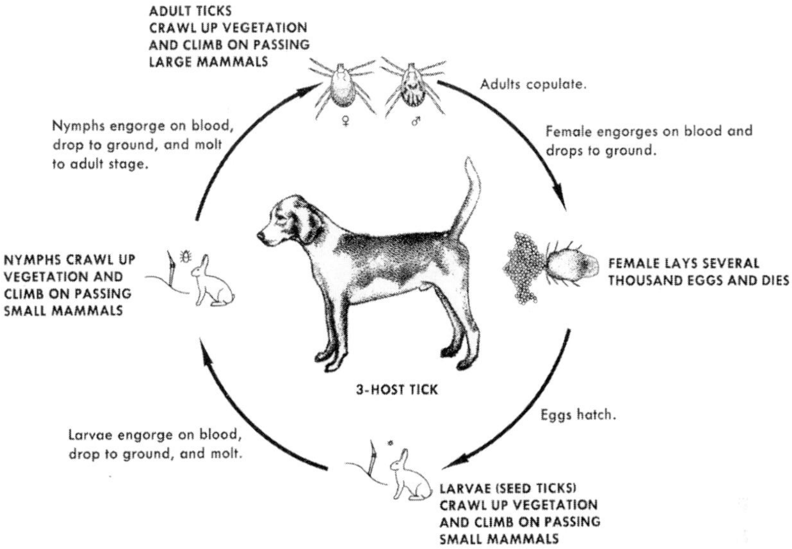

It is a vector of tularemia and Rocky Mountain spotted fever. *Amblyomma maculatum* (Gulf Coast tick) is a three-host tick that preferentially feeds in the ears of cattle, horses, sheep, dogs, and humans. It can result in tick paralysis. *Ixodes dammini* (deer tick) is a three-host tick that feeds on mice and voles as larvae and nymphs and on deer as adults. This tick is often involved in the transmission of *Borrelia burgdorferi,* the causative agent of Lyme disease, to dogs and humans. Other *Ixodes* spp important in the transmission of *B. burgdorferi* include *I. scapularis* and *I. pacificus. Haemaphysalis leporispalustris* (continental rabbit tick) is a three-host tick; adults feed on domestic and wild rabbits whereas larvae and nymphs feed on birds, dogs, cats, and humans. Birds can transport the ticks over long distances. These ticks transmit Rocky Mountain spotted fever, Q fever, and tularemia (Chapter 14).

The family Argasidae (soft ticks) includes four genera and approximately 140 species. Many species of Argasid ticks live in the nests and burrows of their hosts, feeding while the host sleeps. *Argas* spp feed on poultry, grouse, guinea fowl, pigeons, and canaries. They transmit rickettsial organisms and are a cause of tick paralysis. *Ornithodoros* spp are parasites of birds, rodents, and large mammals. They transmit spirochetes (one of which causes relapsing fever in humans). *Antricola* spp are parasites of bats. *Otobius megnini* (spinose ear tick) parasitizes the ears of cattle, sheep, goats, horses, and sometimes dogs. Only the larval and nymph forms are parasitic; these ticks often remain on the host for many months.

The identification of a specific specie of tick is made by examination of its size, shape, color, and special markings (e.g., a single white spot on the scutum of the lone star tick). Amitraz collars, permethrin sprays and spot treatments (toxic to cats), or fipronil sprays and spot treatments (toxic to rabbits) are effective in killing ticks and preventing tick attachment. Environmental control measures include trimming grass and brush and applying pesticides to yards and kennels.

*Araneae.*   Members of this order are commonly known as spiders and are ubiquitous (found virtually everywhere). Spiders of greatest importance in producing disease in companion animals include *Loxosceles reclusa* (brown recluse spider), *Loxosceles unicolor* (common brown spider), and *Latrodectus mactans* (black widow spider). Bites of *Loxosceles* spp initially appear as puncture wounds. The tissue surrounding the bite becomes necrotic and may slough leaving a large ulcer or hole. Bites by *Latrodectus* spp can result in muscle cramps, pain, vomiting, diarrhea, paralysis, and sometimes death. These spiders are most common around woodpiles and in dark corners. Control is aimed at removing hiding spots and spraying insecticides in cracks and crevices.

***Scorpionida.*** Scorpions have large pincers on their mouthparts and a stinging tail. They have been described as resembling miniature lobsters. Most live outdoors under stones or logs. Their sting is painful and can result in muscle spasms, drooling, vomiting, difficult breathing, paralysis, and death in small animals. Applying ice packs to a bite site slows absorption of venom.

## 15.3   PHYLUM NEMATODA

Nematodes are invertebrates with cylindrical bodies and a complete digestive tract. The body wall has three layers: cuticle (outside), hypodermis (middle), and inner muscular layer. The body cavity contains the digestive and reproductive tracts. There are approximately 200 species of nematodes found as parasites of domestic animals. These are commonly referred to as roundworms because of the shape of their cross-sections.

The nematodal life cycle includes several stages: one-cell egg or zygote, morula (embryo appears as a cluster of cells within a shell), vermiform embryo (embryo appears as a tiny worm within a shell), first-stage larva (L1), second-stage larva (L2), third-stage larva (L3), fourth-stage larva (L4), juvenile worm, and adult worm. The L3 larva is generally the infective stage.

There are three variations in larval development: (1) some nematodes complete development of L1 to L3 within the eggshell; (2) in other nematodes, larval development from L1 to L3 occurs in the environment (larvae in the environment are free living and feed on organic material or bacteria); (3) in the third group, the vermiform embryo or L1 is ingested by an intermediate host and develops into an infective L3 within the body of the intermediate host.

The definitive host is the host for the adult nematode. The definitive host can become infected with a nematode in several ways: (1) the definitive host may ingest the L3 (in an eggshell, as a free-living larva, or within an intermediate host); (2) a free-living L3 may penetrate through the skin of the definitive host; (3) the L3 may be injected into the body of the definitive host by the intermediate host (e.g., by a mosquito, flea, or fly); (4) offspring of an infected host may become infected through transplacental or transmammary migration of infective L3.

Six orders of nematodes commonly affect domestic animals (Table 15.3): Rhabditida (threadworms), Strongylida (bursate nematodes), Ascaridida (ascarids), Oxyurida (pinworms), Spirurida (spirurid and filarial worms), and Enoplida.

### 15.3.1   ORDER RHABDITIDA

The order Rhabditida contains a large number of free-living nematodes and species that are parasitic on plants and other invertebrates. Two species are of importance as parasites of dogs.

*Strongyloides stercoralis* is an intestinal threadworm, which can cause diarrhea in young puppies. Only females are parasitic; these females are unique in being parthenogenetic (produce eggs that do not require a male to fertilize them). The eggs hatch in the intestines of dogs. Thus, L1 can be found in dog feces. The subsequent generation is free living in the environment. Dogs and other animals can be infected by skin penetration or by ingestion of milk (transmammary) or contaminated soil containing L3. Following skin penetration, larvae migrate through the body into the lungs where they may cause signs of pneumonia (coughing, difficult breathing). Ingested larvae develop into adults in the intestines. Intestinal threadworms may cause abdominal pain and diarrhea. Threadworms are a zoonotic infection. Larvae penetrating human skin produce a rash called creeping eruption. Control measures include good sanitation with the removal of feces containing infective larvae from the environment, and also treatment of infected dogs with ivermectin.

**Table 15.3**
Nematode Parasites of Dogs and Cats and Their Associated Diseases

| Nematode | Host(s) | Disease |
|---|---|---|
| Threadworm | Dogs | Diarrhea in puppies |
| *Strongyloides stercoralis* | | |
| *Rhabditis (Pelodera) strongyloides* | Dogs | Dermatitis from larvae penetration of skin |
| Hookworm | | Anemia, black tarry feces |
| *Ancylostoma caninum* | Dogs | |
| *A. braziliense* | Dogs and cats | |
| *A. tubaeforme* | Cats | |
| *Uncinaria stenocephala* | Dogs | |
| Lungworm (canine) | Dogs | Coughing, nodules in lungs |
| *Filaroides osleri, F. hirthi, F. milksi* | | |
| Lungworm (feline) | Cats | Coughing, nodules in lungs |
| *Aelurostrongylus abstrusus*/cats | | |
| Roundworm | | Vomiting, diarrhea, constipation, abdominal distension, poor growth; larvae may produce disease in humans |
| *Toxocara canis* | Dogs | |
| *T. cati* | Cats | |
| *Toxascaris leonina* | Dogs and cats | |
| Stomach worm | Dogs and cats | Vomiting, poor appetite, dark tarry feces |
| *Physaloptera* spp | | |
| Stomach worm | Dogs and cats | Gastric ulcers, larvae cause damage to liver and are fatal in cats |
| *Gnathostoma* | | |
| Esophageal worm | Dogs and cats | Nodules in wall of esophagus and stomach, difficulty swallowing |
| *Spirocerca lupi* | | |
| *Dracunculus* | Dogs, cats, other carnivores, humans | Adult worms found in subcutaneous tissues, ulcers form over females |
| Heartworm | Dogs and cats | Torturous pulmonary vessels, coughing, right heart failure; microfilaria are found in blood and transmitted by mosquitoes |
| *Dirofilaria immitis* | | |
| *Acanthocheilonema (Dipetalonema) reconditum* | Dogs | Adults live in subcutaneous tissues, microfilaria are in blood and must be differentiated from those of *Dirofilaria immitis*; these parasites are transmitted by fleas and lice |
| Giant kidney worm | Mink, dogs, pigs, other carnivores, humans | Adult worm may digest the entire parenchyma (tissue) of a kidney, usually affects the right kidney, may be found in the abdominal cavity |
| *Dioctophyme renale* | | |
| *Capillaria plica* | Dogs and cats | Found in the kidneys and urinary bladder, predispose to cystitis and urinary tract infections |
| *Capillaria aerophila* | Dogs and cats | Found in nasal cavity and bronchi, signs include sneezing and wheezing |
| Whipworm | Dogs (rarely cats) | Found in cecum and colon, produce diarrhea with mucus and blood |
| *Trichuris vulpis* | | |

*Rhabditis (Pelodera) strongyloides* is a free-living threadworm that feeds on decaying organic material such as straw. The larvae penetrate the skin of animals and produce skin irritation with itching and hair loss. Diagnosis is made by finding larvae on skin scrapings (Chapter 16). Treatment includes removing decaying organic material from the environment and using a parasiticide to kill the larvae in skin.

**Figure 15.13** Parasitic ova are identified by their characteristic size, shape, and contents. Top row (left to right): *Toxascaris leonina* (zygote), *Isospora canis*, *Toxascaris leonina (morula);* middle row (left to right): *Toxocara canis, Trichuris vulpis, Capillaria sp.;* bottom row (left to right): *Ancylostoma caninum, Uncinaria stenocephala, Spirocerca lupi. Courtesy Dr. James E. Corbin.*

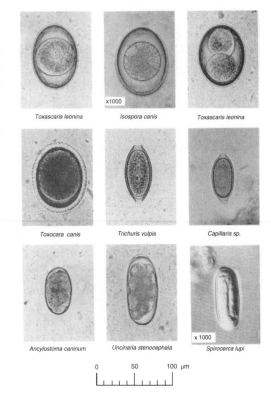

## 15.3.2   Order Strongylida

These are small to medium nematodes characterized by a buccal capsule with teeth used for feeding and a copulatory bursa on males. The eggs are oval with a thin transparent eggshell containing blastomeres. These worms are inhabitants of the digestive and respiratory tracts of mammals. Species of importance in dogs and cats are hookworms (*Ancylostoma caninum, A. braziliense, Uncinaria stenocephala*) and lungworms (*Filaroides osleri, F. milksi, F. hirthi, Aelurostrongylus abstrusus*). Many of these can be diagnosed by identification of their ova on fecal flotations (Figure 15.13; Chapter 16).

Adult hookworms are parasites of the small intestines. The name was derived from the hooklike appearance of the anterior end of adult worms. Adult hookworms feed on blood. The worms secrete an anticoagulant that results in bleeding at the attachment site even after the worm has moved. Bleeding into the intestines produces a black, tarry stool. The blood loss results in anemia (low red blood cell count). Heavy infestations can be fatal. The most common hookworms in dogs are *Ancylostoma caninum, A. braziliense,* and *Uncinaria stenocephala.* The most common hookworms in cats are *A. tubaeforme* and *A. braziliense.* The adult hookworm is 5 to 20 mm long and 0.2 to 0.6 mm in diameter.

Female hookworms produce 16,000 or more eggs daily. The eggs are passed in the feces (see Figure 15.13) and develop into L1 in 1 to 12 days. L1, L2, and L3 stages are free living in the environment. L3 can penetrate skin or be ingested by a mammal. Larvae penetrating the skin migrate into capillaries and are transported to the lungs. Larvae molt in the lungs to L4 and crawl up the airways to the pharynx, are swallowed, and develop into adults in the intestines. Some migrating larvae become arrested (encysted, "stuck") in the muscles of the host. Arrested larvae are reactivated during pregnancy and lactation and are passed to offspring in milk. L3 that are swallowed burrow in the gastric or intestinal mucosa and develop into adults without further migration (Figure 15.14). Humans are susceptible to developing cutaneous larva migrans from skin penetration by infective larvae.

Diagnosis of a hookworm infestation is made by finding ova (eggs) in an animal's feces (see Figure 15.13). Pyrantel pamoate and other anthelmintics (dewormers) are very effective in killing adult hookworms. Prevention of infection consists of good sanitation and regular

**Figure 15.14**   Life cycle of *Ancylostoma caninum* (hookworms). *Courtesy University of Illinois.*

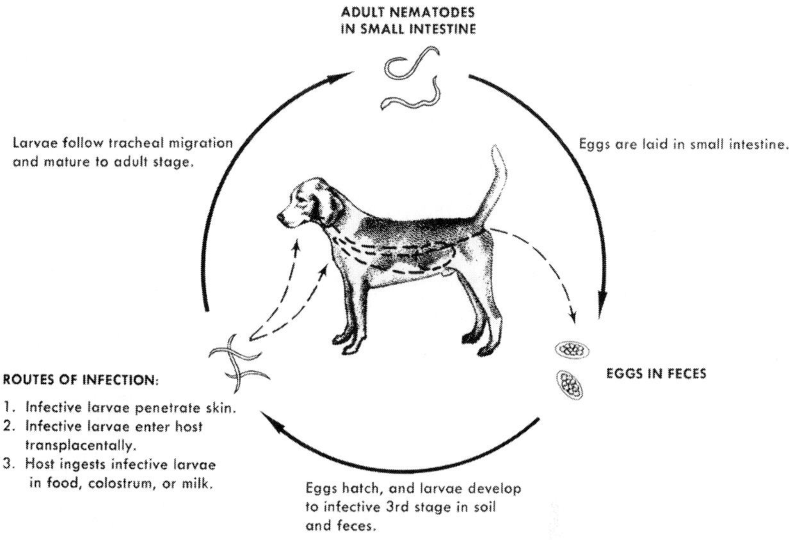

ADULT NEMATODES
IN SMALL INTESTINE

Larvae follow tracheal migration
and mature to adult stage.

Eggs are laid in small intestine.

ROUTES OF INFECTION:

1. Infective larvae penetrate skin.
2. Infective larvae enter host
   transplacentally.
3. Host ingests infective larvae
   in food, colostrum, or milk.

EGGS IN FECES

Eggs hatch, and larvae develop
to infective 3rd stage in soil
and feces.

removal of animal feces. Treating bitches with fenbendazole or ivermectin during late gestation and again in early lactation will significantly decrease transmission of parasites to puppies. Animals that have been heavily parasitized may require blood transfusions.

The nematode lungworms of dogs are *Filaroides osleri, F. hirthi,* and *F. milksi.* These are found in the trachea, lung parenchyma (tissue), and bronchioles of dogs. The lungworm of cats is *Aelurostrongylus abstrusus.* It is found in both lung parenchyma and bronchioles. Adults are 2 to 7 mm long and 0.1 to 0.5 mm in diameter. Eggs of these strongyles hatch before being passed in the host's feces. The L1 is infective if ingested by another dog. The swallowed L1 penetrates the intestines and is transported through blood vessels in the liver and subsequently to the lungs. The larvae feed, molt, and mature within lung tissue.

Clinical signs of lungworm infection include coughing and the finding of nodules in lungs and airways on radiographs or on bronchoscopy (use of a fiber-optic device to visualize inside the airways; Chapter 16). Diagnosis is made by finding larvae in respiratory secretions or in the feces. Bitches and queens can transmit lungworm larvae while licking their offspring. Ivermectin, fenbendazole, and albendazole are variably effective in treating animals with lungworm infections (worms encysted within nodules in the airways and lung tissue are resistant to treatment).

## 15.3.3   ORDER ASCARIDIDA

Ascarids are the large roundworms of dogs and cats. Common species include *Toxocara canis* (dogs), *T. cati* (cats), and *Toxascaris leonina* (dogs and cats). Adult worms are 3 to 18 cm long and 0.2 to 0.3 cm in diameter. Clinical signs associated with ascarid infections include vomiting, diarrhea, constipation, abdominal distension (pot-bellied appearance), and poor growth.

*T. canis* is primarily transmitted transplacentally (across the placenta); almost all puppies acquire these parasites from their dam. *T. cati* is primarily transmitted in the queen's milk. *T. leonina* infections are acquired by ingesting an infective egg or by a carnivore eating a rodent (paratenic host) with arrested larvae in its tissues. Larvae invade walls of the intestines and migrate throughout the body of a host. Some larvae migrate into airways and up to the pharynx to be swallowed; these complete their development into adults in the intestines. Other larvae become encysted (arrested development) in tissues of the host. These encysted larvae are reactivated during pregnancy, and depending on the species, some migrate transplacentally (*T. canis*) (Figure 15.15) and others transmammarily (*T. cati*). When ascarid

**Figure 15.15** Life cycle of *Toxocara canis* (roundworms). *Courtesy University of Illinois.*

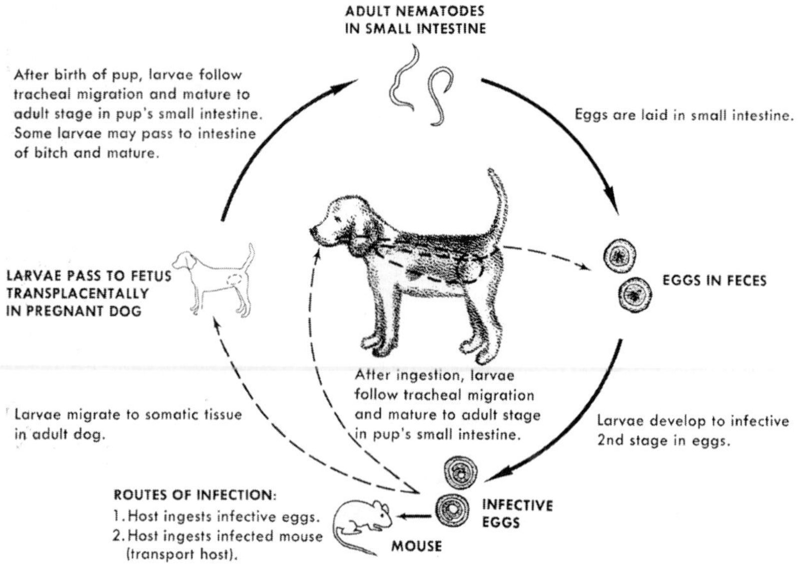

larvae are ingested by humans, the larval migration may cause damage to internal organs (visceral larval migrans) and blindness from migration in the eyes.

Diagnosis of ascarid infections may be made by finding the worms or eggs in an animal's feces (see Figure 15.13). Treatment includes administration of anthelmintics to infected animals and removal and burning of feces. Daily treatment of bitches with 50 mg/kg of fenbendazole from day 40 of pregnancy to day 14 of lactation will prevent infection in puppies.

## 15.3.4   ORDER OXYURIDA

The Oxyurida are pinworms. Although dogs and cats are often erroneously blamed for transmitting pinworms to humans, there are no species of pinworms that affect dogs or cats. Pinworms are important as parasites of rabbits, mice, rats, and other rodents (Chapter 7).

## 15.3.5   ORDER SPIRURIDA

The two major superfamilies within this order are Spiruroidea and Filarioidea. These parasites affect the upper digestive system (esophagus and stomach) and other nonintestinal sites (e.g., circulatory system, connective tissue, or body cavities).

Members of the Spiruroidea superfamily all require an arthropod as an intermediate host. Eggs are passed in feces of the definitive host (e.g., cat or dog) and either the egg or an L1 is ingested by a beetle or another insect (Figure 15.16). Some are then ingested by a transport host (paratenic host) such as a fish, bird, or small mammal. Either the insect or the transport host containing an L3 is ingested by a cat or dog. Completion of the life cycle often involves migration of the L3 within the liver or circulatory system of the cat or dog prior to their reaching maturity.

*Physaloptera* are stomach worms found in dogs and cats. These worms attach to the mucosal surface of the stomach and feed on gastric tissue and blood. Clinical signs of infection with this parasite include vomiting; loss of appetite; and dark, tarry stools containing digested blood. Diagnosis of *Physaloptera* infections can be made by finding the worms in vomitus or eggs in feces. Pyrantel pamoate is effective as an anthelmintic.

*Gnathostoma* is another Spiruroidea worm found in the stomach of carnivores and omnivores. Its larvae develop in aquatic crustaceans such as *Cyclops* (microscopic water crus-

**Figure 15.16** Life cycle of *Physaloptera* (canine stomach worm). *Courtesy University of Illinois.*

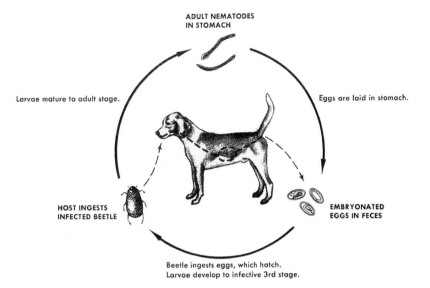

taceans). Amphibians, fish, and snakes may serve as transport hosts that are eaten by the definitive host. The larvae migrate through the liver causing damage to it before moving into the stomach wall. The adult worm can cause perforating ulcers in the stomach. Infections with this parasite are almost always fatal in cats and cause significant morbidity in dogs (weight loss, abdominal pain, liver failure).

*Spirocerca lupi* are esophageal worms of dogs and cats. Adult worms are found in nodules in the wall of the esophagus or stomach. Infection is acquired by ingesting a beetle or a transport host. The larvae migrate through the circulatory system and spend approximately 3 months in the aorta before returning to the esophagus. The nodules have an opening (called a fistula) that allows eggs to be passed into the lumen and through the digestive tract before being expelled in feces. Some animals may exhibit difficult swallowing whereas others show no clinical signs (asymptomatic). Diagnosis is made by finding eggs in the feces (see Figure 15.13) or by visualizing nodules in the esophagus or stomach via endoscopy (esophagoscopy, gastroscopy; Chapter 16).

*Dracunculus* is a parasite of the subcutaneous tissues of carnivores and humans. Skin over the site of a female worm becomes necrotic forming an ulcer. When water contacts the skin over the ulcer, the female worm protrudes out far enough to expel L1 larvae into the water. The life cycle requires ingestion of larvae by a *Cyclops* where it develops into a L3. Dogs and cats (and other carnivores) become infected by drinking *Cyclops*-infested water. Surgical excision is the treatment of choice for removal of *Dracunculus* worms; another technique is to wet the ulcer to lure the worm from its hole then slowly wind it on a stick. The winding may take several days as females can be several inches long and care must be taken not to break the worm (breaking the worms can result in a severe allergic reaction).

Adult females of the superfamily Filarioidea produce L1 rather than eggs. The intermediate hosts for the larvae are blood-feeding arthropods. The larvae develop into infective L3 and are transmitted through the mouthparts of the arthropod during its feeding. The most important members of this family to affect dogs and cats are *Dirofilaria immitis* and *Acanthocheilonema (Dipetalonema) reconditum.*

*Dirofilaria immitis* is the cause of heartworm disease in dogs and cats. Adult worms are 10 to 30 cm long and 1.0 mm in diameter. They live in the pulmonary arteries, right ventricle, and vena cava. L1 are found circulating in the blood and are ingested by mosquitoes feeding on the host. The larvae develop into infective L3 within the mosquito and are transmitted to a new host during the mosquito's next blood meal (Figure 15.17). L3 migrate through the subcutaneous tissues of dogs and cats, molting to L4s, and then into juvenile worms. Juvenile

**Figure 15.17**   Life cycle of *Dirofilaria immitis* (heartworms). *Courtesy University of Illinois.*

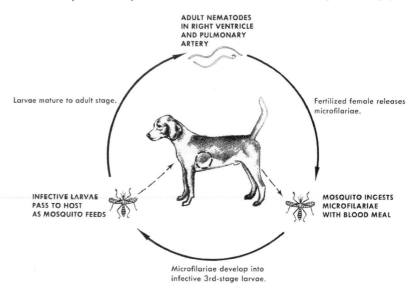

heartworms penetrate into blood vessels. The largest numbers are found in the pulmonary arteries. In addition to causing physical obstruction to blood flow, the larvae cause thickening of the vessel walls resulting in increased blood pressure in the pulmonary vessels, and in severe cases leading to right heart failure.[8] Signs of infection include decreased exercise tolerance, coughing, difficult breathing, an accumulation of fluid in the abdominal cavity (ascites), kidney disease, and death. Mildly affected dogs usually have less than 25 adult heartworms, moderately affected dogs have around 50, and severely affected dogs may have over 100. Cats are not affected as often as dogs (approximately 10% of affected animals are cats). Cats rarely have more than two or three adult heartworms. Signs associated with *D. immitis* infections in cats include coughing, difficult breathing, vomiting, and death.

Many tests can be used to diagnose *D. immitis* infections. The simplest test is to examine a drop of blood under a microscope for the presence of microfilaria. The microfilaria of *D. immitis* are 308 μm long and 7 μm in diameter and wiggle without moving forward. They must be distinguished from the microfilaria of *Acanthocheilonema reconditum* (slightly smaller, 290 × 5.5 μm; tail has a button-hook shape; and show progressive movement). Several techniques can be used to concentrate the microfilaria making them easier to find. One simple technique is to spin blood in a capillary tube and use a microscope to examine the buffy coat above the red blood cells. Microfilaria can be seen wiggling just above the buffy coat layer.[9] The most reliable diagnostic test for finding microfilaria is the Knott concentration technique. In this test, blood is mixed with a 2% formalin solution and then centrifuged. The sediment is mixed with a dye (usually methylene blue) and examined with a microscope. Filter tests are also available that utilize a filter with 5 μm pores. Blood is mixed with a solution causing hemolysis (rupturing of red blood cells) and passed through the filter. The filter is then stained and examined for microfilaria. Heartworm infections with microfilaria in the blood are called patent infections. Approximately 7 months are required for heartworms to mature within animal and produce microfilaria. Ten to 15% of dogs and 80% of cats with heartworm infections do not have microfilaria in their blood; these are termed occult infections. Additional diagnostic tests are required to diagnose occult *D. immitis* infections.

---

[8]In right heart failure, the right ventricle becomes distended and inefficient at pumping blood forward, pressure builds up in the systemic circulation, and fluid leaks into the abdominal cavity. This fluid leakage is called ascites.
[9]The buffy coat layer is the layer containing white blood cells and platelets. It is found between the packed red blood cells and the plasma in a centrifuged microhematocrit tube (Chapter 16).

Immunological tests are available for detecting the antigens of adult *D. immitis* or antibodies against these worms. Antigen tests are very specific in confirming an infection with *D. immitis,* but the tests may be negative in animals infected with fewer than five heartworms. Because cats are typically infected with fewer than five heartworms, antigen tests are negative in more than 40% of cats with heartworm disease. Antibody tests are very sensitive, but a positive test is not conclusive that there is an active infection with *D. immitis.* Animals that were exposed but did not develop active infections and those that harbor a single dead worm may have positive antibody tests, but would not require treatment for heartworm disease.

Radiographs (x-rays) can be used to detect changes in the pulmonary blood vessels associated with heartworm disease (e.g., enlarged, tortuous pulmonary arteries; enlargement of the right ventricle). Echocardiography (ultrasound waves used to evaluate the heart and large blood vessels) can be used to visualize heartworms within the heart and pulmonary vessels (Chapter 16).

Treatment of heartworm disease includes (1) stabilization of heart and lung disease, (2) administration of an adulticide to kill the adult worms, and (3) administration of a microfilaricide to kill circulating microfilaria. These treatments are hazardous because the dying worms may disintegrate and block blood vessels of the lungs causing other complications. In severe cases, the worms may be mechanically removed via vascular surgery. Cats are rarely treated because heartworms have a short life span in cats and treatment complications are often more severe than the disease caused by the worms.

Heartworm prevention is highly recommended for all dogs in environments infested with mosquitoes. Heartworm prevention is also recommended for cats in areas with a high incidence of heartworm disease. There are several types of heartworm preventives available; all are highly effective when given on the appropriate schedule. The mechanism of action is through killing L3 or L4 microfilaria during their migration and developmental stages. Diethylcarbamazine should be given daily starting before the mosquito season and continuing for one month after frosts have killed all mosquitoes. Ivermectin, milbemycin oxime, and selamectin must be given monthly during the mosquito season. Moxidectin is available as an injection given every 6 months. Many of the heartworm preventive products also prevent infections with other parasites (e.g., roundworms and hookworms). For this reason, and because of the variability of mosquito seasons, many veterinarians recommend administration of preventives throughout the year.

*Acanthocheilonema (Dipetalonema) reconditum* adult worms are 1 to 3 cm × 0.1 mm and are located in the subcutaneous tissues of canids (e.g., dogs, wolves, jackals, foxes, coyotes). The L1 microfilaria circulate in blood. The intermediate hosts are fleas and lice. There are no adverse effects on the host, thus the primary concern for dogs infected with *A. reconditum* is that the microfilaria can easily be confused with those of *D. immitis.* Differentiation is made based on the smaller size, button-hook shaped tail, and progressive motility of *A. reconditum* microfilaria. Additionally, the antigen and antibody tests for *D. immitis* will be negative in dogs infected only with *A. reconditum.*

## 15.3.6  ORDER ENOPLIDA

The order Enoplida has two superfamilies with parasites affecting cats and dogs. These are the superfamily Dioctophymatoidea and the superfamily Trichinelloidea.[10]

*Dioctophyme renale* is the giant kidney worm of carnivores and some omnivores. Mink are the principal definitive host; however, dogs, swine, and humans are sometimes affected. It is the largest nematode parasite and may exceed one meter in length. The adult worm is most commonly found in the right kidney. The adult worm frequently ingests the entire parenchyma of the kidney, leaving only the capsule of the kidney surrounding the worm. Eggs are passed in

---

[10]In some classification schemes, Trichinellida and Dioctophymatida are considered as orders of the subclass Adenophorea (formerly known as Aphsmidia).

**Figure 15.18**    Life cycle of *Dioctophyme renale* (giant kidney worm of dogs). *Courtesy University of Illinois.*

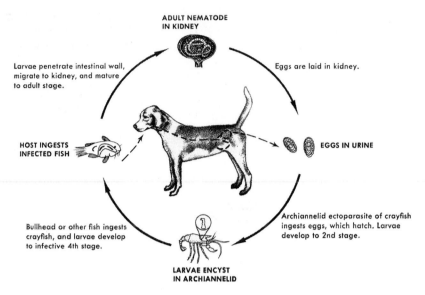

ADULT NEMATODE
IN KIDNEY

Larvae penetrate intestinal wall, migrate to kidney, and mature to adult stage.

Eggs are laid in kidney.

HOST INGESTS INFECTED FISH

EGGS IN URINE

Bullhead or other fish ingests crayfish, and larvae develop to infective 4th stage.

Archiannelid ectoparasite of crayfish ingests eggs, which hatch. Larvae develop to 2nd stage.

LARVAE ENCYST IN ARCHIANNELID

the urine and ingested by annelid worms. The annelid worm may be ingested by a fish or frog that serves as a transport host. Dogs become infected by eating either the annelid worm or a transport host (Figure 15.18). Adult worms may also be found free in the abdominal cavity.

The superfamily Trichinelloidea contains two genera with species of importance to dogs and cats: *Trichuris* and *Capillaria.*

*Trichuris vulpis* is the intestinal whipworm of dogs. The anterior portion of the adult worm is much more narrow than the posterior portion resulting in a shape that resembles a whip. The female produces several thousand eggs daily. The eggs are passed in feces; eggs containing L1 are infective. The L1s hatch in the intestines and burrow in the small intestinal wall for 3 to 10 days, then return to the intestinal lumen and move into the large intestine. The larvae mature slowly, taking 3 months to reach patency (egg production). The adults live about 16 months in dogs. Adults are most commonly found in the cecum but may also be found in the colon and rectum (Figure 15.19). Heavy infections result in diarrhea, often containing blood and mucus. Cats are rarely infected. Diagnosis is based on finding eggs in the stool. The eggs are lemon-shaped and bioperculated (caps cover pores at both ends of the egg; see Figure 15.13). The most

**Figure 15.19**    Life cycle of *Trichuris vulpis* (canine whipworms). *Courtesy University of Illinois.*

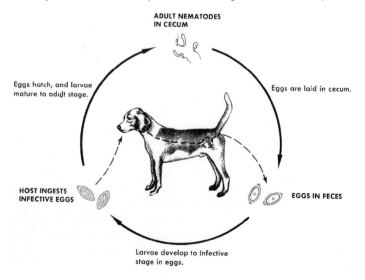

ADULT NEMATODES
IN CECUM

Eggs hatch, and larvae mature to adult stage.

Eggs are laid in cecum.

HOST INGESTS INFECTIVE EGGS

EGGS IN FECES

Larvae develop to infective stage in eggs.

effective anthelmintics for eliminating *T. vulpis* are fenbendazole (given for 3 consecutive days) and milbemycin oxime. Treatments should be repeated three times at monthly intervals.

*Capillaria* spp resemble *Trichuris* but are smaller and found in different tissues. *C. (Eucoleus) aerophila* is found in the nasal cavity and bronchi of cats and dogs. Its life cycle can be direct or involve earthworms as intermediate hosts. Signs of infection include coughing and wheezing. Diagnosis is based on finding eggs in the airways (via transtracheal or bronchiole lavage, Chapter 16) or in feces (see Figure 15.13). *C. (Pearsonema) plica* is a parasite of the urinary bladder and kidneys of dogs and cats. Eggs are passed in the urine and ingested by earthworms in which they hatch and develop into an infective L1. When eaten by a carnivore, the L1s penetrate the intestinal wall, molt to L2s, and invade blood vessels. In kidneys, the L2s penetrate the ureters and move into the bladder where they complete their development into adult worms. The worms can cause inflammation of the bladder (cystitis) and predispose the animal to urinary tract infections. Diagnosis is made by finding eggs in the urine. Ivermectin and fenbendazole (given for 3 consecutive days) are effective treatments.

# 15.4 PHYLUM PLATYHELMINTHES (FLATWORMS)

Platyhelminthes are dorsoventrally flat worms that lack an anus. They contain two important parasitic groups: Trematodes (flukes) and Cestodes (tapeworms).

## 15.4.1 CLASS TREMATODA (FLUKES)

Trematodes are hermaphrodites; adult flukes possess complete male and female reproductive systems. The female portion produces eggs that are passed in feces. The eggs hatch in wet environments, producing a motile miracidium that penetrates the skin of an aquatic snail. Within the snail, the miracidium develops into a sporocyst containing rediae (embryos) and then into cercariae (juvenile flukes with a tail). The cercariae emerge from the snail and swim to penetrate the skin of a definitive host, encyst on vegetation, or penetrate into a second intermediate host. Encysted cercariae develop into metacercariae that are infective when ingested by a definitive host. Once in the definitive host, the juvenile fluke migrates to its preferred life site and develops into an adult fluke. Parasitic flukes may be found in the liver, lungs, and blood vessels of companion animals (Table 15.4).

**Table 15.4**
Flukes of Dogs and Cats and Their Associated Diseases

| Flatworm | Intermediate Host(s) | Definitive Host(s) | Disease |
|---|---|---|---|
| *Alaria* spp | Tadpole of leopard frog ± frog, snake, bird, small mammal | Dogs and cats | Intestinal fluke |
| Salmon poisoning fluke *Nanophyetus salmincola* | Salmon | Dogs | Intestinal fluke: vector for rickettsia producing salmon poisoning |
| Lizard poisoning fluke *Platynosomum fastosum* | Lizard | Cats | Adult found in bile duct, results in vomiting, diarrhea, liver failure, and sometimes death |
| Lung fluke *Paragonimus kellicotti* | Aquatic snails, crabs, crayfish | Cats, dogs, other carnivores, pigs, humans | Coughing and sometimes blood in sputum |
| *Schistosoma* spp | Snails | Dogs | Flukes found in abdominal veins of dogs |
| *Heterobilharzia americana* | Snails | Dogs | Flukes found in abdominal veins of dogs |

***Intestinal flukes.*** *Alaria* spp are intestinal flukes of dogs and cats. These parasites may be passed through the milk to kittens nursing infected queens. The usual intermediate host for *Alaria* spp is a tadpole of the leopard frog. Infected tadpoles are often eaten by frogs, snakes, birds, and small mammals, which serve as paratenic (transport) hosts. Dogs in the Pacific Northwest region of the United States can be infected with another intestinal fluke, *Nanophyetus salmincola.* Dogs acquire this fluke by ingesting salmon containing infective metacercariae. The adult fluke is a vector of the rickettsial organism producing salmon poisoning (Chapter 14). Thus *N. salmincola* is commonly known as the salmon poisoning fluke. Albendazole, niclosamide, and praziquantel are effective in killing intestinal flukes.

***Liver flukes.*** *Platynosomum fastosum* is a small fluke found in the bile duct of cats. Cats become infected by ingesting a lizard containing infective metacercariae. Thus *P. fastosum* is commonly called the lizard poisoning fluke of cats. These flukes can cause vomiting, diarrhea, liver failure, and death. Some cats respond to treatment with praziquantel; however, this drug can be dangerous in cats with liver dysfunction.

***Lung flukes.*** *Paragonimus kellicotti* is the lung fluke of dogs and cats. Eggs can be found in respiratory secretions and feces. The adults live in cystic spaces in the lungs (Figure 15.20). Metacercariae excyst in the intestines, and the juvenile flukes migrate through the peritoneal cavity and penetrate the diaphragm before reaching the lungs. Some infected animals cough and have blood in their sputum. Diagnosis is made by finding the eggs in sputum or feces. Fenbendazole (twice daily for 10 days) or praziquantel can be used to kill these flukes.

***Blood trematodes.*** *Schistosoma* spp and *Heterobilharzia americana* are trematodes found in abdominal veins of dogs. Snails serve as intermediate hosts. Schistosomes are unique in this phylum as adults are either male or female (other trematodes are hermaphrodites). Fenbendazole and praziquantel are effective treatments.

## 15.4.2   CLASS CESTODA (TAPEWORMS)

Cestodes are narrow, long, flat parasites commonly known as tapeworms. The adult has three major parts: scolex, neck, and strobila. The scolex is the anterior part of the tapeworm that anchors the parasite to its host; it is not a "true" head because it has no mouth or sensory organs. The neck connects the scolex to the strobila; it contains cells that produce the

**Figure 15.20**   Life cycle of *Paragonimus kellicotti* (lung fluke of dogs). *Courtesy University of Illinois.*

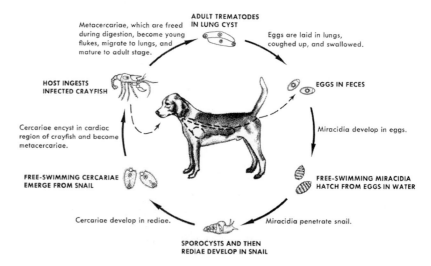

strobila. The strobila is the body or chain of the tapeworm and is formed by segments called proglottids. Cestodes are hermaphrodites; each proglottis develops an ovary and uterus and also one to numerous testes. After fertilization, other organs regress and the proglottis becomes a sac full of eggs. Mature eggs contain a multicellular embryo. Most tapeworms require an intermediate host (some require two) to complete the life cycle. When an infective tapeworm is ingested by its definitive host, most of the larval body disintegrates leaving the scolex and neck. The neck begins to bud off proglottid segments that mature as new buds move the older segments down the chain.

There are 13 orders of cestodes, but only two—the Cyclophyllidea and Pseudophyllidea—are of importance in companion animals (Table 15.5).

**Table 15.5**
Common Tapeworms of the Dog and Cat and Their Associated Diseases

| Tapeworm | Definitive Host(s) | Intermediate Host(s) | Location | Disease |
|---|---|---|---|---|
| *Mesocestoides corti* | Carnivores | 1st Oribatid mite 2nd rodent or reptile | Intestines | May result in anal pruritus in the definitive host |
| *Taenia hydatigena* | Dogs | Sheep, goats, cattle, pigs | Liver, peritoneal cavity | Liver disease in intermediate host |
| *T. multiceps* | Dogs | Sheep, goats, rarely humans | Brain, spinal cord | Neurological disease in intermediate host |
| *T. ovis* | Dogs | Sheep, goats | Muscles | Muscle disease in intermediate host |
| *T. pisiformis* | Dogs | Lagomorphs | Liver and peritoneal cavity | Liver disease in intermediate host |
| *T. serialis* | Dogs | Lagomorphs, rodents, rarely humans | Muscles subcutaneous tissues | Interfere with mobility of intermediate hosts making infected rabbits and rodents easy prey |
| *T. taeniaeformis* | Cats | Rodents | Liver | Liver disease in intermediate host |
| *T. solium* | Humans | Pigs, dogs, humans, others | Muscles, internal organs | Signs dependent on location of cysticercus (intermediate stage) may include blindness and brain damage |
| *Dipylidium caninum* | Dogs and cats | Fleas, biting lice | Intestines | Occasional anal pruritus or diarrhea in definitive host |
| *Echinococcus granulosus* | Dogs | Sheep, ungulates, humans | Liver, lungs | Hydatid cyst disease in intermediate host, may have liver or respiratory disease |
| *E. multilocularis* | Dogs, cats, foxes | Rodents, humans, others | Liver, lungs | Hydatid cyst disease in intermediate host, may have liver or respiratory disease |
| *Diphyllobothrium latum* | Bears, carnivores, humans | 1st *Cyclops* 2nd freshwater fish (pike, trout, salmon, and others) | Intestines | Adult tapeworm may cause abdominal pain and diarrhea |
| *Spirometra mansonoides* | Carnivores | 1st *Cyclops* 2nd amphibians, reptiles, mammals, humans | Subcutaneous cysts | Very rarely cause disease |

***Order Cyclophyllidea.*** There are 13 families within this order; however, only 3 are of importance as parasites of dogs and cats: Mesocestoididae, Taeniidae, and Dilepididae.

The family Mesocestoididae is represented in dogs by the tapeworm *Mesocestoides corti* (North America) and *M. lineatus* (Europe, Asia, Africa). These tapeworms require two intermediate hosts: the first is a beetle or mite and the second is an amphibian, reptile, bird, or small mammal. Dogs and cats are definitive hosts, and *Mesocestoides* attach to the wall of their small intestines.

The family Taeniidae includes several common tapeworms of dogs and cats: *Taenia pisiformis, T. hydatigena, T. ovis, T. taeniaeformis, T. multiceps, T. serialis, Echinococcus granulosus,* and *E. multilocularis.* Gravid (egg-filled) proglottid segments break off the tapeworm and crawl out the anus or exit in feces. As the segments crawl, they expel eggs. Eggs are ingested by an intermediate host and hatch into a hexacanth embryo. The embryo migrates to a developmental site, usually the liver or peritoneal cavity but sometimes skeletal or cardiac muscle or occasionally in the lungs. The hexacanth embryo then grows into an infective second-stage larva. The infective larvae of *Taenia* spp are enclosed in a fluid-filled membrane called a bladder; this stage is called a bladderworm or cysticercus. The infective larvae of members of the genus *Echinococcus* are enclosed in a large fluid-filled membrane termed a hydatid. A single hydatid cyst contains thousands of infective scolices. Humans may become infected through ingestion of tapeworm eggs. Large numbers of hydatid cysts can interfere with liver function and are fatal in severe cases.

The most important tapeworm in the family Dilepididae is *Dipylidium caninum.* This tapeworm is also known as the cucumber seed tapeworm or double-pored tapeworm because newly shed proglottids resemble cucumber seeds and have lateral pores located on each of their long edges. Desiccated (dried) proglottids look like uncooked grains of rice. The intermediate hosts for *D. caninum* are fleas and biting lice (Figure 15.21).

***Order Pseudophyllidea.*** This order contains two genera infective to dogs and cats: *Diphyllobothrium* and *Spirometra.*

*Diphyllobothrium latum* is a very large tapeworm of bears, carnivores, and humans. In humans, it can range from 2 to 10 m long and 2 cm wide. *Diphyllobothrium* requires two intermediate hosts; a freshwater copepod (e.g., *Cyclops*) for the first stage and a freshwater fish for the infective second stage.

*Spirometra* spp is a tapeworm of domestic and wild carnivores. It is smaller than *Diphyllobothrium,* averaging less than 1 m. The first intermediate host is a copepod (e.g., *Cyclops*); the second may be an amphibian, reptile, bird, or mammal.

**Figure 15.21** Life cycle of *Dipylidium caninum* (cucumber seed tapeworm). *Courtesy University of Illinois.*

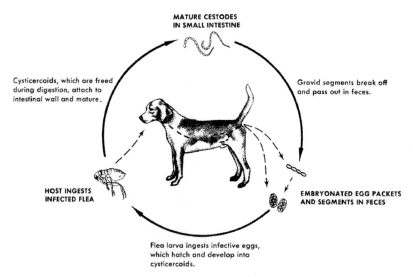

***Diagnosis and treatment of tapeworms.*** The diagnosis of tapeworm infections is made by finding proglottid segments in the feces or around the anus of the infected dog or cat. Identification of species is based on characteristic shapes and features of the proglottids and of eggs within the proglottids. Elimination of tapeworm infections can be difficult and requires use of a cestocide (drug to kill the adult tapeworm) and preventing future ingestion of the intermediate hosts. Praziquantel is the cestocide most commonly used in dogs and cats.

## 15.5 KINGDOM PROTISTA (PROTOZOA)

Protozoans are one-cell organisms. Most are free living, but several members of this kingdom are parasitic and produce serious disease in a host. Three phyla cause disease in dogs and cats: Sarcomastigophora, Ciliophora, and Apicomplexa (Table 15.6).

### 15.5.1 PHYLUM SARCOMASTIGOPHORA

This phylum contains the subphyla Mastigophora (flagellates) and Sarcodina (amoebae).

Flagellates occur in two forms: a motile form known as a trophozoite and a cyst stage. The trophozoite has one or more slender appendages called flagella used for locomotion. Some species live in blood whereas others are parasites of the gastrointestinal or reproductive tracts.

*Trypanosoma cruzi* is a protozoan found in the blood (hemoprotozoan) of dogs in the southern United States and Central and South America. It is banana-shaped, is 3 to 10 times longer than red blood cells, and can be observed "swimming" on microscopic examination of a fresh blood smear. It causes minimal disease in dogs, but humans can also be affected, and cross-reactivity between the trypanosome and human tissues can result in damage to the heart, esophagus, and colon (Chagas' disease). It is transmitted by kissing bugs (a blood-feeding arthropod).

*Leishmania* spp are protozoa with an intracellular stage (called an amastigote) found in phagocytic mononuclear cells of the skin, liver, spleen, lymph nodes, and bone marrow. Clinical signs of infection include scaly skin, hair loss, fever, anemia, enlarged lymph nodes, enlargement of the spleen and liver, weight loss, and eventual death. Humans, dogs, and occasionally cats are hosts for this parasite. *Leishmania* are transmitted by sand flies (*Phlebotomus* and *Lutzomyia*). Diagnosis is made by finding the amastigotes in aspirates or biopsies of the spleen, lymph nodes, or liver. Polymerase chain reaction (PCR) tests[11] can be used to find antigens of *Leishmania.* Several treatments have been used with variable success; these include meglumine antimonite, ketoconazole, and itraconazole (Chapter 14). Prognosis is poor and infected dogs may be a reservoir for human infections. Thus, euthanasia of an affected dog is recommended.

*Giardia intestinalis* is a flagellated protozoan of the upper small intestines of vertebrates. Many animals infected with *Giardia* show no signs of disease. However, other animals have intermittent soft stools or diarrhea. The stools often contain mucus. Diagnosis is made by finding *Giardia* trophozoites in feces by direct examination or utilizing an immunological test (ELISA[12]) or a PCR test to detect *Giardia* antigens in feces. A number of drugs can be used to treat cats and dogs and, thereby, eliminate the infection. Commonly used drugs include albendazole, fenbendazole, and metronidazole. Prompt removal and disposal of feces helps prevent spread of the infection.

---

[11]Polymerase chain reaction (PCR) tests utilize a series of reactions that amplify segments of nucleic acids from DNA gene sequences. These tests can be used to identify infectious organisms that would otherwise be difficult to detect due to their small size or low numbers.
[12]ELISA tests are enzyme-linked immunosorbent assays; these typically utilize an antibody linked to an enzyme. The antibody is specific for the antigen of interest, and the antibody will bind to that antigen if present in the sample being tested. Unbound antibodies are removed by washing. A substrate is added and the enzyme (linked to the antibody, which in turn is now bound to antigen in the sample) changes the substrate into a colored product that is visible to denote a positive test.

**Table 15.6**
Common Protozoa of Dogs and Cats and Their Associated Diseases

| Protozoan | Intermediate Host(s) | Definitive Host(s) | Disease |
|---|---|---|---|
| *Trypanosoma cruzi* | Kissing bug | Dogs, humans | Found free in blood, no disease in dogs but can cause Chagas' disease in humans (a muscle disease) |
| *Leishmania* spp | Sand flies (*Phlebotomus*, *Lutzomyia*) | Dogs, humans, rarely cats | Leishmaniasis: scaly skin, cutaneous nodules, hair loss, fever, anemia, enlargement of liver, spleen and lymph nodes, weight loss, death |
| *Giardia intestinalis* | | Many vertebrates | Giardiasis: diarrhea, soft stools |
| *Entamoeba histolytica* | | Dogs, other animals | Amebic dysentery in humans; dogs usually show no symptoms; fatal in monkeys |
| *Balantidium coli* | | Pigs, dogs | Normal flora in pigs, may cause diarrhea in dogs |
| *Isospora* spp | | Cats, dogs | Coccidiosis, diarrhea especially in kittens and puppies |
| *Cryptosporidium* spp | | Calves, dogs, cats, humans, and many other vertebrates | Diarrhea |
| *Toxoplasma gondii* | Many animals serve as intermediate hosts | Cats | Life cycle is only completed in cats (definitive hosts); tachyzoites (intermediate forms) can cross placenta and cause fetal disease; signs in cats (and intermediate hosts) include blindness, vomiting, diarrhea, fever, seizures, muscle weakness, incoordination, pneumonia, paralysis |
| *Sarcocystis* spp | Ruminants, pigs, horses, | Dogs, cats, humans birds | Diarrhea, decreased appetite, vomiting in humans |
| *Neospora caninum* | | Dogs, ruminants, horses, rarely cats | Nodular skin lesions, pneumonia, liver disease, muscle disease |
| *Hepatozoon* spp | Ticks (*Rhipicephalus sanguineus*) | Dogs, cats | Fever, poor appetite, weight loss, bloody diarrhea, neurological disease |
| *Babesia* spp | Ticks (*Rhipicephalus sanguineus*) | Dogs (*B. canis*, *B. gibsoni*), cats (*B. felis*) | Anemia, depression, poor appetite, enlarged spleen |
| *Cytauxzoon felis* | Ticks | Cats | Fever, depression, poor appetite, dehydration, anemia, icterus, enlargement of liver and spleen, death |

Amoebae also have two forms: trophozoites (motile) and cysts. Only one species is of importance in dogs, *Entamoeba histolytica*. *E. histolytica* parasitizes the large intestine and causes amoebic dysentery in humans. *E. histolytica* may also be found in the liver where it results in necrosis and abscesses. Dogs and other domestic species are occasionally infected. Infections in monkeys and chimpanzees are often fatal. Diagnosis is made by finding the trophozoites or cysts in feces. Metronidazole can be used to treat affected dogs.

## 15.5.2   PHYLUM CILIOPHORA

Only one ciliated protozoan is known to be associated with disease in dogs, *Balantidium coli*. *B. coli* is considered to be part of the normal flora of swine intestines. It can parasitize the cecum and colon of dogs resulting in abdominal pain and diarrhea. Diagnosis is based on finding ciliated trophozoites or cysts in feces. Metronidazole is effective in treating affected animals.

## 15.5.3   PHYLUM APICOMPLEXA

Members of this phylum are also known as sporozoa. These protozoa live within cells and cause destruction of the cells. Those that live in the gastrointestinal tract and other body tissues are called coccidians. Sporozoa living in the blood are called hemosporidians.

The life cycle of the coccidia begins with an oocyst in the feces of an infected animal. The oocyst develops into a sporocyst containing infective sporozoites. When ingested by a new host, the sporozoites invade host cells and multiply into numerous merozoites. The merozoites burst out of the host cell and invade new cells. After several generations, a sexual form develops which produces new oocysts (Figure 15.22).

*Isospora* spp are coccidian parasites of the small intestines of cats and dogs. *Isospora* spp multiplying in intestinal cells cause destruction of the intestinal epithelium; large numbers of these parasites result in diarrhea. Coccidiosis is most commonly seen in young puppies and kittens. Older animals develop immunity to these parasites. Diagnosis is based on identification of the oocysts in fresh feces (see Figure 15.13). Albendazole and sulfadimethoxine can be used to treat infected dogs and cats.

*Cryptosporidium* spp are coccidian parasites that invade the small intestines of numerous animals including cats and dogs. *Cryptosporidium* spp are important as a cause of severe diarrhea in humans and calves. These organisms can produce a watery diarrhea in young kittens and puppies, but are rarely associated with clinical disease in older cats and dogs. Diagnosis is based on finding the very small oocysts in fresh feces. Fluorescent antibody tests are available to help locate the oocysts. There are no drugs currently approved for treatment of cryptosporidiosis in dogs; paromomycin and pyrimethamine have been used experimentally.

*Toxoplasma gondii* is an important coccidian parasite of cat intestines; it causes the disease known as toxoplasmosis. This protozoan is directly infectious to cats, but it can also undergo development in a wide range of intermediate hosts. Intermediate hosts can be

**Figure 15.22**   Life cycle of *Isospora* spp (coccidian parasites of small intestines). *Courtesy University of Illinois.*

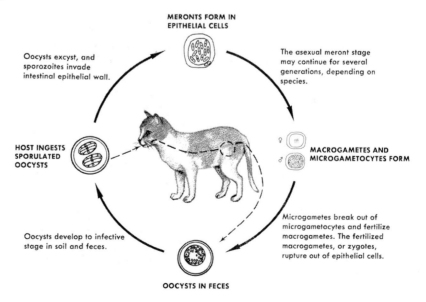

infected by ingesting mature oocysts or tissues of another infected intermediate host. *T. gondii* sporozoites develop into tachyzoites and bradyzoites within the intermediate host. Tachyzoites invade host cells and rapidly multiply resulting in the bursting of host cells. Bradyzoites divide more slowly, distending the host cell to form a tissue cyst. Clusters of cysts containing bradyzoites may form in muscles and cells of the nervous system. Tachyzoites can infect the placenta of pregnant animals and women. When a woman is infected during pregnancy, the brain and eyes of the baby may be affected. To minimize risk of toxoplasmosis, pregnant women should never handle cat litter pans or eat undercooked meats. Dogs and cats can show signs of blindness, vomiting, diarrhea, fever, seizures, muscle weakness, incoordination (stumbling), pneumonia, and/or paralysis.

Infected cats shed oocysts for only a few weeks, thus finding the characteristic oocysts (10 × 12 μm, unsporulated) in cat feces is diagnostic but rare. A number of serological tests are available to detect antibodies to *T. gondii*. Caution must be used in interpreting these test results because they may be indicative of past exposure. Demonstration of a rising titer is needed to prove the presence of an active infection (requires taking two blood samples 3 weeks apart and looking for an increase in the antibody titer in the second sample). Animals with active infections may improve following treatment with clindamycin. To prevent infection, cats and dogs should not be allowed to eat raw prey or undercooked meats. In addition, feces from cat litter boxes should be removed and destroyed daily (oocysts in feces require 24 hours to become infective).

*Neospora caninum* is a coccidian found in dogs, ruminants, and horses. It can be experimentally transmitted to cats and rodents. Tachyzoites multiply in tissues of the nervous system and can produce paralysis in dogs. Nodular skin lesions, pneumonia, liver disease, and muscle disease have also been reported. Puppies can be infected transplacentally. Diagnosis is based on finding antibodies to the parasite in the serum or cerebrospinal fluid (Chapter 16). Clindamycin is the preferred drug for treatment.

*Hammondia hammondi* is a two-host protozoan. Oocysts are infective to mice. Within the mouse, the sporozoites develop into bradyzoites and tachyzoites that are infectious when the mouse is eaten by a cat. Diagnosis is based on finding the oocysts in cat feces.

*Sarcocystis* spp also have an obligatory two-host life cycle. Intermediate hosts include ruminants, swine, horses, and birds. The intermediate hosts become infected through ingestion of infective oocysts. Definitive hosts include dogs, cats, and humans. Definitive hosts are infected by eating undercooked meat or tissues of infected intermediate hosts. Clinical signs are mild in dogs and cats but may include decreased appetite or vomiting. Infected humans show severe diarrhea, vomiting, and breathing problems. Diagnosis is made by finding antibodies to the organism. To prevent infections, dogs and cats should not be allowed to eat raw prey or undercooked meats or tissues.

*Hepatozoon* spp are infectious for dogs; they may also infect cats. Animals become infected by ingesting infected ticks (*Rhipicephalus sanguineus*). Infections may be subclinical (no visible symptoms of disease) or may result in fever, poor appetite, weight loss, bloody diarrhea, and neurological disease. Diagnosis is made by finding the organism in blood smears. A number of treatments have been proposed but none have been proven to be effective.

*Babesia* spp are blood parasites transmitted primarily through tick bites. The most common species affecting dogs is *B. canis*, which is transmitted by the brown dog tick (*Rhipicephalus sanguineus*). Another species affecting dogs is *B. gibsoni* (rare in the United States but common in Africa and Asia). Cats are susceptible to infection with *B. felis* (primarily found in Africa and Asia). Clinical signs of infection include anemia, depression, poor appetite, and enlargement of the spleen. Diagnosis is made by finding the trophozoites in blood smears or by antibody tests demonstrating exposure to this parasite. There are no drugs currently approved for treating *Babesia* infections in dogs and cats.

*Cytauxzoon felis* is a rare but life-threatening protozoan affecting cats. It is transmitted by ticks and causes fever, depression, lack of appetite, dehydration, anemia, icterus, and an enlarged spleen and liver. Diagnosis is made by finding the organism in red blood cells or in aspirates from the spleen and bone marrow. No treatments have proven effective and infected cats usually die within 2 weeks.

# 15.6  SUMMARY

This chapter was devoted to providing basic information pertaining to parasites and pests important to the health and quality of life of companion animals. The most important information for readers to know was summarized in Tables 15.1 through 15.6.

Parasites have no respect for animal size, health, or value. They live on (ectoparasites) or in (endoparasites) their host animal. True to their name, they thrive at the expense of other living organisms, weakening the health, strength, and vigor of their hosts. Pests pester and annoy their hosts; many serve as vectors (carriers) of disease. Major pests of importance to dogs and cats were discussed in this chapter.

Dogs and cats have many internal parasites that cause considerable damage, even death. Four of the most common ones are heartworms, roundworms, hookworms, and tapeworms.

*Heartworms* are caused by a filarial worm found in most countries of the world. Adult worms can infect and reside in the heart and pulmonary arteries where they impede blood flow. With proper veterinary care and treatment, effective methods of controlling heartworm infections in dogs and cats are available.

*Ascarids,* including the common roundworms that parasitize dogs and cats, may leave the alimentary canal and invade the central nervous system of dogs, cats, and humans. Ascarids have been associated with invasion of the eyes causing blindness in humans. Reliable methods of detection are available as well as effective, safe anthelmintics for treating roundworm infestation.

*Hookworms* are tenacious and have the ability to adhere to the lining of the digestive tract where they consume huge quantities of blood. This may be lethal when more blood is consumed than can be generated within the dog's or cat's body. Hookworm eggs identified in the feces are excellent indicators of their parasitizing presence; highly effective anthelmintics are available.

*Tapeworms* can develop in the intestines of dogs or cats when infective larvae are consumed in intermediate hosts, which include fleas, rodents, rabbits, and sheep. Humans can become infected by ingestion of tapeworm eggs produced within dogs, cats, or other definitive hosts. Tapeworms are diagnosed by identification of the eggs within tapeworm segments (proglottids) passed in feces or found crawling in the perineal region. The anthelmintics used for treating nematode infections are rarely effective against tapeworms; however, cestocides such as praziquantel will kill these endoparasites. The important ectoparasites of dogs and cats include fleas, lice, ticks, mites, and many others. The ectoparasites of greatest importance were summarized in Tables 15.1 and 15.2.

Each of the above parasites has a life cycle. These, as well as life cycles of numerous other parasites, were illustrated and discussed in this chapter. Also discussed were important insect pests of companion animals.

Now, let us move to Chapter 16 for information and discussion of common diagnostic and therapeutic procedures and terms important to the care, health, and management of companion animals.

# 15.7  REFERENCES

1. Barriga, Omar O. 1997. *Veterinary Parasitology for Practitioners.* 2nd ed. Edina, MN: Burgess Publishing.

2. Georgi, Jay R., and Marion E. Georgi. 1990. *Parasitology for Veterinarians.* 5th ed. Philadelphia: W. B. Saunders.

3. Hendrix, Charles M. 1998. *Diagnostic Veterinary Parasitology.* 2nd ed. St. Louis, MO: Mosby.

4. Ivens, Virginia R., Daniel L. Mark, and Norman D. Levine. Special Publication 52. *Principal Parasites of Domestic Animals in the United States.* University of Illinois at Urbana-Champaign: Colleges of Agriculture and Veterinary Medicine.

5. Wall, Richard, and David Shearer. 2001. *Veterinary Ectoparasites.* 2nd ed. Oxford, UK: Blackwell Science.

# 16 COMMON DIAGNOSTIC AND THERAPEUTIC PROCEDURES AND TERMS[1]

[1]Please turn to the end of the chapter for complete author affiliation information.

*I have always believed that the best doctor in the world is the veterinarian.*
*He cannot ask his patients "what is the matter?"—he has to know.*

**William Penn Adair "Will" Rogers (1879–1935)**
*American actor and humorist*

# 16.1  INTRODUCTION/OVERVIEW

When Frenchman Julius Herisson developed and introduced a device to measure blood pressure in humans in 1835, physicians sneered. "This is one of the most ridiculous baubles ever foisted on the profession," they said. No one sneers at medical technology today. We are intently interested in alleviating diseases, lengthening life, and ensuring the best quality of life possible in humans and the companion animals with which we bond and share many pleasures of life. The most basic and fundamental component of Herisson's sphygmo-manometer was an idea. Many sound ideas create technologies leading to life-sustaining diagnostic procedures that collectively help ensure an improved life for companion animals and those whose lives they enrich.

The 20th century was a period of continuous high tide for science and technology—an era when great strides were made in expanding the availability and use of diagnostic and therapeutic procedures in the progressive professions of veterinary medicine and affiliated disciplines.

Understanding and utilizing the modern tools and procedures of veterinary medicine will help those working with companion animals provide them with an increasingly high quality of life. Utilizing the common diagnostic and therapeutic procedures discussed in this chapter will enable those interested in, committed to, and associated with companion animal biology and affiliated disciplines achieve the noble goals of preventing and controlling disease as well as treating illnesses and maladies not yet conquered.

Diagnosis involves collecting data about the pet and then analyzing the data to develop a hypothesis regarding the cause(s) of the animal's problem(s). A variety of means are used to collect data including obtaining a history of the problem, doing a physical examination of the patient, performing laboratory tests, using diagnostic imaging, and performing other procedures. We will review several of these procedures in this chapter.

## 16.2   HISTORY AS A DIAGNOSTIC TOOL

Obtaining a complete history can provide important information regarding a patient's problem(s). Letting the pet's owner describe the problem in her/his own words is often helpful. The problem that the owner is most concerned about is termed the chief complaint. The examiner will then want to ask the owner additional questions. Some of the basic information to be included in the history includes signalment (breed, sex, age, and color of the pet), birth history if under 6 months old, medical history of the animal's relatives (sire, dam, littermates), vaccination history, travel history, usual diet and treats, where the pet is housed, contact with other animals, use of the animal, any past medical problems with response to previous treatments, current medical problems with response to any treatments, and a review of the various body systems.

The animal's signalment can provide clues to what is wrong because many diseases have hereditary predispositions, others are color related, some are sex related, and many diseases are most common in specific age groups. For example, a young Chocolate Labrador puppy with hair loss is likely to have a congenital hair follicle disease, a 9-month-old Collie with hair loss is likely to have demodecosis, a 10-year-old Boston Terrier with hair loss and excessive water drinking is likely to have Cushing's disease (hyperadrenocorticism), and a 16-year-old domestic shorthair cat with hair and weight loss is likely to have hyperthyroidism.

The history of an illness can provide clues on the cause of the problem. Infectious, traumatic, and toxic diseases have an acute onset (develop rapidly). Degenerative and neoplastic diseases have an insidious onset and slowly worsen over time. Coughing during excitement is often due to a collapsing trachea, whereas coughing at night is associated with heart failure. Travel history can give clues on exposure to geographically restricted diseases (coccidioidomycosis would only be suspected if an animal has been in the desert of southwestern United States).

A review of the body systems provides information about potential problems the owner may not associate with the chief complaint. Sometimes these associated problems give clues regarding the chief complaint (e.g., the chief complaint may be hair loss; the associated history of excessive water drinking provides a clue that hyperadrenocorticism is a likely cause of the hair loss). Here are some sample questions to use in a review of the systems:

*Integument.* Does the pet scratch or chew itself? Are there any rashes, bumps, or sores? Does anyone in the household have skin problems? Has there been a loss of hair?

*Respiratory.* Is there any sneezing or coughing? Any nasal discharge? Any abnormal sounds associated with breathing?

*Cardiovascular.* Does the pet tire easily? Any changes in activity level? Any coughing at night or associated with exercise?

*Gastrointestinal.* Has there been any vomiting or diarrhea? Any changes in appetite? Any changes in body weight? Any difficulty in swallowing? Any changes in the stools?

*Musculoskeletal.* Is the pet lame? Any changes in activity? Any difficulty getting up?

*Neurological.* Any changes in activity or gait? Is the pet depressed? Have there been any seizures? Any wobbliness or muscle tremors?

*Urinary system.* Any changes in drinking or urination? What color is the urine? Any difficulty in urination? Any changes in urinary habits?

*Reproductive system.* Is the pet neutered? Is there a vaginal or prepucial discharge? If an intact female, when was the last estrous cycle? If an intact male, any difficulty in urinating or defecating?

*Special senses.* Are there any concerns about the pet's ability to see, hear, or smell? Does the pet run into objects? Does the pet respond when its name is called? Is there a response to the ringing of a doorbell? Does the animal shake its head, scratch its ears, paw its nose? Has there been any ocular or nasal discharge? Have there been any major changes in appetite?

Using history sheets or a computer survey form will facilitate the process of obtaining the medical history and also provide a written record or computer file for storing this important information. Review Chapter 8 for examples of history forms (see Figure 8.4).

# 16.3  THE PHYSICAL EXAMINATION AS A DIAGNOSTIC TOOL

A complete physical examination can yield important information regarding the cause of disease in an animal. Components of a physical examination include an overall assessment of the general health of the animal followed by an examination of each body system. Methods used to perform the physical examination include visual observation, palpation (use of fingers to apply light pressure in feeling various parts of the body), auscultation (listening for the sounds within the body), and percussion (use of "tapping" with the fingers to detect the vibrations characteristic of different media such as fluid versus air-filled spaces). Results of physical examination findings should be recorded; these are an important part of an animal's medical record (Chapter 8, see Figure 8.5).

*General impressions.* Is the animal alert? What is its attitude? What is the body condition score (Table 16.1)? What is the general appearance of the hair coat? What is the distribution of any lesions?

*Integumentary system.* Examine every inch of skin to detect areas of hair loss or inflammation. Note the presence of any ectoparasites or debris in the coat. Record the location of any ulcers, crusts, or nodules in the skin (see Figure 8.3 for an example of a topography form). Assess skin thickness; endocrine diseases may be associated with decreased skin thickness (e.g., hyperadrenocorticism) or increased skin thickness (e.g., myxedema in hypothyroidism, Chapter 14).

*Respiratory system.* Begin with examination of the nose and throat. Is there a nasal discharge? Is air moving through both nostrils (hold a glass slide in front of the nares and look for moisture condensation)? Are the tonsils enlarged? Is the soft palate elongated? Palpate the larynx and trachea—does the animal cough? Observe the animal's respirations—any abnormalities on inspiration or expiration? Auscultate the lung fields—any crackles or wheezes? If lung sounds are absent, use percussion to evaluate for a fluid line.

*Cardiovascular system.* What is the color of the mucous membranes? What is the capillary refill time?[2] Palpate the femoral pulse—does it have a regular rhythm and is it strong? Auscultate the heart—are the heartbeats strong and regular? Is the heart rate the same as the pulse rate? Are heart sounds "crisp" or is there a "whishing sound" that would indicate the

---

[2] The capillary refill time (CRT) is a simple method of assessing tissue perfusion. The animal's upper lip is lifted and moderate pressure is applied to the gingival mucosa above the upper teeth resulting in blanching. The finger applying the pressure is removed and the time that it takes for blood to return to the site is the CRT. Normal CRT is less than 2 seconds.

**Table 16.1**
Body Condition Scores (BCS) of Dogs and Cats

| BCS | Ribs | Lumbar Region | Waist, Abdominal Tuck | Overall Impression |
|---|---|---|---|---|
| 1 | Easily visible | Lumbar vertebrae and pelvic bones visible, no palpable fat | Obvious waist and abdominal tuck, no fat in flank folds* | Emaciated |
| 2 | Easily palpable | Backbone and pelvic bones visible and easily felt, minimal fat covering | Waist easily noted when viewing from above, abdominal tuck evident | Thin, underweight |
| 3 | Palpable but not visible | Smooth topline (vertebrae and pelvic bones covered with muscle, minimal fat) | Waist observed behind the ribs when viewed from above, abdomen tucked up when viewed from side | Ideal |
| 4 | Palpable with slight excess of fat over covering | Fat deposits noticeable back and base of tail | Waist discernable when viewed from above but not prominent, abdominal tuck also discernable but not prominent, flank folds hang down with moderate amount of fat and jiggle when walking* | Overweight |
| 5 | Not easily palpable, heavy covering of fat | Massive fat deposits over lumbar areas and tail base | Waist barely visible to absent, no abdominal tuck, abdomen may appear distended, prominent flank folds that sway from side to side when walking* | Obese |

*Assessment of the flank folds pertains only to cats.*

presence of a heart murmur (Table 16.2)? If a murmur is noted, is it continuous or does the intensity vary? Murmurs are further characterized based on where they are heard in relation to the animal's chest—where is the murmur the loudest—right or left side, between which ribs, high or low?

*Gastrointestinal system.* The exam begins with looking at the teeth, tongue, and oral cavity. Auscultation is used to evaluate gut sounds. Palpation is used to evaluate the abdominal organs. Cats and small dogs can be palpated using one hand (thumb on one side and four fingers on the other side of the abdomen). Two hands are used to palpate the abdomen of most dogs. It is important to be gentle and allow the animal time to relax its abdominal muscles. Place the hands (or fingers) in the dorsal cranial abdomen and then gently move them toward each other and slowly down the abdomen; organs will be felt as they slip past the fingers. This procedure is repeated in the central and then caudal regions of the abdomen. Note any areas that are painful to the animal and also the presence of any abnormal structures (this takes practice). Finally, perform a digital examination of the rectum using a lubricated gloved finger. Are there any masses or strictures? What is the consistency of the feces?

*Musculoskeletal system.* Observe the animal's stance (posture). Is the body symmetrical? Is there any evidence of muscle wasting (loss of muscle mass)? Watch the animal move. Is its gait normal? If lameness is present, how severe is it? Flex and extend each joint. Is there any swelling, pain, or crepitus (crackling or grinding) when the joint is moved? Palpate the muscles and bones—is there any swelling, displacement, or pain?

**Table 16.2**
Characterization of Heart Murmurs

| Grade | Description |
|---|---|
| I/VI | Heard only after several minutes of listening in a quiet room |
| II/VI | Very soft but can be heard immediately after placing stethoscope on the chest |
| III/VI | A low to moderately intense murmur, easily heard |
| IV/VI | Moderately loud murmur that can be heard on both sides of the chest, no thrill* |
| V/VI | Loud murmur with a palpable thrill |
| VI/VI | Loud murmur with a palpable thrill that can be heard when the bell of the stethoscope is moved slightly off the chest wall |

| Timing | Duration |
|---|---|
| Systole | Early, mid, late, or holosystolic |
| Diastolic | Early, mid, late |
| Continuous | Continuous |

| Character | Description |
|---|---|
| Plateau | Regurgitant type, same for duration of murmur |
| Decrescendo, crescendo, crescendo-decrescendo | Ejection type, intensity changes |
| Machinery | Continuous, heard throughout systole and diastole |

| Valve Affected | Location on Chest Wall |
|---|---|
| Mitral | Left, low, between 5th and 6th ribs |
| Tricuspid | Right, low, between 3rd and 4th ribs |
| Pulmonary | Left, low, between 3rd and 4th ribs |
| Aortic | Left, mid-thorax, between 4th and 5th ribs |

*The term thrill when used in describing a heart murmur refers to a vibration of the chest wall that can be palpated at the same time the murmur is heard.*

*Neurological system.* Observe the animal's posture, gait, and mental status. Is the animal alert? Is it able to walk normally? Is there any dragging of the toes? The cranial nerves (CN) are evaluated by testing (1) smell (does animal recoil from a volatile substance?) (CN I); (2) pupillary light reflexes (do pupils constrict when a bright light is shined in them?) (CN II, CN III); (3) menace reflex (does the animal blink when a finger is pointed near the eye?) (CN II, CN VII); (4) eye movements (do the eyes follow a moving object?) (CN III); (5) eye position (are the eyes centered?) (CN IV, CN VI); (6) jaw movement (can the dog open and close its mouth?) (CN V); (7) hearing (does the animal respond to having its name called or clippers run near its head?) (CN VIII); (8) gag reflex (does the animal swallow?) (CN IX, CN X); (9) muscle tone of the neck (CN XI); and (10) tongue stretch (does the animal retract its tongue when it is pulled out?) (CN XII). Postural reactions are tested by having the animal walk on front legs and rear legs (wheelbarrowing), hemistand/hemiwalk (left side, right side), hopping (one leg at a time), foot placement (testing proprioception, a foot is placed so the dog/cat is standing on the top of its toes, the animal should flip the foot back to its normal position within a few seconds). A test used to help localize spinal cord lesions is the panniculus reflex. Gentle pinching of the skin along the back should elicit a reflex contraction of the skin. Anal tone should be checked, and the perineal reflex is tested by touching the skin near the anus and looking for a "wink" or contraction of the anal sphincter. Other reflexes that can be tested include the patellar reflex (tapping on the patellar tendon below the kneecap and looking for extension of the stifle), gastrocnemius reflex (tap on the gastrocnemius tendon above the hock and look for slight extension followed by flexion of the hock), and the triceps reflex (tap on the triceps tendon above the elbow and look for a slight extension of the elbow).

*Urinary system.*  Both kidneys can be palpated in cats; usually only the left kidney in dogs (the right kidney is firmly attached to the upper abdominal wall). The kidneys should feel smooth and symmetrical. The urinary bladder can be palpated in the caudal ventral abdomen; it is pear shaped in dogs and spherical in cats. It should have the texture of a water-filled balloon. If the bladder is extremely firm or feels gritty it may contain stones or tumors. In dogs, a rectal examination using a lubricated gloved finger will allow palpation of the urethra, which lies on the ventral floor of the pelvis (should feel like a firm tube).

*Reproductive system.*  The uterus of females may be felt lying dorsal to the bladder. During pregnancy the fetuses may be palpable from approximately days 21 to 28 of gestation (Chapter 10). In male dogs, the prostate gland will be felt as a bilobed, firm, walnut-sized mass lying over the urethra in the cranial part of the pelvic canal. The prostate gland should be symmetrical and nonpainful. An enlarged prostate gland may fall forward over the brim of the pelvis and be palpable as a mass located between the urinary bladder and the pelvis. The testicles should be smooth on the surface and symmetric in size and shape. The penis should be extruded from its sheath and examined for any abnormalities. The mammary glands of females should be palpated and examined for the presence of any swellings, masses, or discharges.

*Eyes.*  Examine the eyes and eyelids for symmetry, inflammation, and discharges. Is the cornea clear? What is the appearance of the lens? Evaluate the pupillary light and menace reflexes reviewed as part of the nervous system examination. Note the placement of the eyelashes. An ophthalmoscope is used to evaluate the retinal vessels and optic nerve.

*Ears.*  Start by examining the ear pinnae for any hair loss, excoriations (scratches), crusts, or scales. Evaluate the animal's hearing by calling its name, ringing a bell, or turning on clippers. Is the external ear canal inflamed? Is there any discharge or exudate in the ears? An otoscope can be used to evaluate the horizontal ear canal and examine the tympanic membrane (Figure 16.1). This may require sedation or general anesthesia if the ear is inflamed and painful. A more complete evaluation of the ear canal can be made using a video-otoscope (Section 16.17, Figure 16.18).

*Lymph nodes.*  Observe the lymph nodes and note any asymmetry or visible enlargements. Each of the superficial lymph nodes should be palpated with notation of any pain or enlargement. Lymph nodes that should always be palpated include the mandibular, prescapular, and popliteal lymph nodes. The axillary and superficial lymph nodes may be difficult to palpate in normal animals. (Review the anatomical locations of lymph nodes in Section 11.7.4.)

## 16.4    BLOOD COLLECTION FOR DIAGNOSTIC TESTS

Blood samples are used for monitoring the health of animals in wellness checks, research, diagnosis and management of diseases, and evaluation for medication-related side effects. Venous blood is easiest to collect and suitable for the majority of diagnostic tests. Veins commonly used for venipuncture include the jugular (neck), cephalic (front leg), saphenous (lateral hind leg), and femoral (medial thigh). Blood may be withdrawn using a needle and syringe or a vacuum tube collection system (Vacutainer™ tubes). Different types of collection tubes with and without additives are used depending on the type of test that will be performed (Table 16.3).

Arterial blood is technically more difficult to obtain and has a higher risk of complications (hemorrhage from the puncture site, thrombosis of the artery, subcutaneous

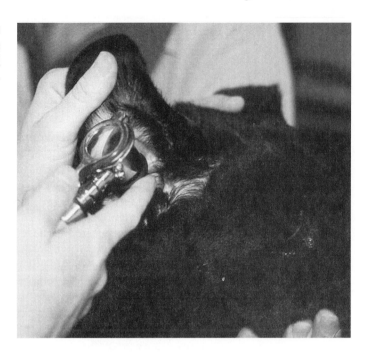

**Figure 16.1** An otoscope is used to examine the horizontal ear canal and tympanic membrane. *Courtesy University of Illinois.*

**Table 16.3**
Types of Blood Collection Tubes and Their Purposes

| Tube Additive(s) | Conventional Stopper Color[*] | Purpose of This Collection Tube |
|---|---|---|
| None | Red | Serum chemistry analyses, serology, blood banking, clotting time 60 minutes, remove serum from tube after centrifuging |
| Clot activator and gel for serum separation | Red/Black | Serum chemistry analysis, suitable for shipping blood to laboratories without requiring separate tube |
| Lithium heparin and gel for plasma separation | Green/Gray | Plasma chemistry analysis, suitable for shipping blood without requiring a separate tube |
| Thrombin | Gray/yellow | For stat[**] serum chemistry analysis, blood clots in less than 5 minutes |
| Sodium heparin, sodium EDTA[***] | Royal blue | For trace-element toxicology and nutritional chemistry analyses |
| Sodium heparin, Lithium heparin | Green | For plasma chemistry analysis |
| Potassium oxalate/ sodium fluoride; sodium fluoride/ sodium EDTA; sodium fluoride | Gray | For glucose measurement (sodium fluoride prevents metabolism of glucose; tubes with potassium oxalate or sodium EDTA do not clot and thus yield plasma) |
| Sodium heparin | Tan | For lead measurement |
| Potassium EDTA | Lavender | For hematology (blood cell counts and analysis), blood typing |
| Sodium citrate | Light blue | For coagulation studies |
| Acid citrate dextrose | Yellow | For blood banking, DNA testing |
| Sodium polyanethol sulfonate | Yellow | For blood cultures |

[*]*These are "traditional" colors used for stoppers containing the additives listed in this table; however, some manufacturers may adopt a different color-code scheme, so users should verify the type of additive present in a tube prior to using the tube.*
[**] *Stat refers to procedures required "as soon as possible" or as emergencies.*
[***]*EDTA is an acronym for ethylenediamminetetraacetate. EDTA forms stable complexes with many metals (chelates) and also prevents blood from clotting.*

hematoma formation). Arterial blood is required for accurate assessment of blood gas and acid/base status of patients. The two sites most often used to obtain arterial blood samples are the femoral artery (medial thigh) and the dorsal pedal artery (metatarsal region).

## 16.4.1 HEMATOLOGY AS A DIAGNOSTIC TOOL

Hematology is the study of blood cells. The most commonly performed tests are measurements of the hematocrit (packed cell volume, PCV, see Figure 11.10), hemoglobin concentration (Hb), erythrocyte count, erythrocyte indices (mean corpuscular volume, MCV; mean corpuscular hemoglobin, MCH; mean corpuscular hemoglobin concentration, MCHC), leukocyte count (white blood cells, WBC), platelet count, and WBC differential count (numbers of neutrophils, basophils, eosinophils, lymphocytes, and monocytes). The morphology of the cells is also noted. Specialized tests on blood cells include the reticulocyte count (numbers of immature erythrocytes), Coomb's test (evaluates for the presence of autoreactive immunoglobulins or complement on the surface of the erythrocytes and is a test for autoimmune hemolytic anemia), and toxicological assays. Coagulation tests measure platelet function and/or the levels of coagulation factors in blood. Relatively simple machines are available to perform these tests in a veterinary clinic or research laboratory. Large commercial and state diagnostic laboratories are often preferred due to their stringent quality assurance programs, ensuring the accuracy of test results. Normal hematology values for cats and dogs are found in Table 16.4.

## 16.4.2 SERUM TESTS AND THE HEALTH OF DOGS AND CATS

Serum chemistry profiles are widely used in the assessment of an animal's health, in the diagnosis of disease, to monitor response to treatment, to evaluate for adverse drug reactions, and to study the impact of various diets and other factors that could affect the well-being of dogs and cats. Tests routinely used in dogs and cats, their purposes, and normal reference ranges are found in Table 16.5. Additional tests performed using serum include hormonal

**Table 16.4**
Normal Hematology Values for Adult Cats and Dogs

| Test | Reference Range* for Adult Cats | Reference Range* for Adult Dogs |
|---|---|---|
| Number of red blood cells (erythrocytes) | $5.0–10.0 \times 10^6/\mu L$ | $5.5–8.5 \times 10^6/\mu L$ |
| Hemoglobin | 8.0–15.0 g/dL | 12.0–18.0 g/dL |
| Hematocrit (packed cell volume, PCV) | 30–45 % | 35–52 % |
| Mean cell volume (MCV) | 39.0–55.0 fl | 60.0–77.0 fl |
| Mean cell hemoglobin (MCH) | 13.0–18.0 pg | 20.0–25.0 pg |
| Mean cell hemoglobin concentration (MCHC) | 30.0–36.0 g/dL | 32.0–36.0 g/dL |
| Platelets | $300–600 \times 10^3/\mu L$ | $200–900 \times 10^3/\mu L$ |
| White blood cell count | $5.5–19.5 \times 10^3/\mu L$ | $6.0–17.0 \times 10^3/\mu L$ |
| Number of neutrophils | $2.5–12.5 \times 10^3/\mu L$ | $3.0–11.5 \times 10^3/\mu L$ |
| Number of neutrophil bands | $0.0–0.3 \times 10^3/\mu L$ | $0.0–0.2 \times 10^3/\mu L$ |
| Number of lymphocytes | $1.7–7.0 \times 10^3/\mu L$ | $1.0–4.8 \times 10^3/\mu L$ |
| Number of eosinophils | $0.0–0.8 \times 10^3/\mu L$ | $0.0–0.8 \times 10^3/\mu L$ |
| Number of basophils | $0.0–2.0 \times 10^3/\mu L$ | $0.0–0.5 \times 10^3/\mu L$ |
| Number of monocytes | $0.0–0.9 \times 10^3/\mu L$ | $0.2–1.4 \times 10^3/\mu L$ |

*Reference ranges may vary among laboratories; a patient's results should always be compared with the reference ranges provided by the laboratory conducting the test.

**Table 16.5**

Routine Serum Chemistry Tests (Biochemical Profile), Their Purpose (System[s] Evaluated), and Reference Ranges for Adult Dogs and Cats

| Name of Test | System(s) Evaluated | Reference Range* for Normal Adult Dogs | Reference Range* for Normal Adult Cats |
|---|---|---|---|
| Albumin | Urinary, liver, gastrointestinal | 2.1–4.3 g/dL | 2.7–3.8 g/dL |
| Alanine aminotransferase (ALT) | Liver | 17–87 U/L | 1–64 U/L |
| Alkaline phosphatase (sometimes referred to as serum alkaline phosphatase, SAP) | Liver, bones, gastrointestinal tract, adrenal gland | 12–110 U/L | 6–93 U/L |
| Amylase | Pancreas (inflammation) | 190–1357 U/L | 200–800 U/L |
| Bicarbonate | Acid/base balance | 15.0–27.0 mmol/L | 13.0–25.0 mmol/L |
| Bile acids | Liver, gall bladder | 0–25 µM/L | 0–25 µM/L |
| Bilirubin | Liver, gall bladder, red blood cells | 0.08–0.50 mg/dL | 0–03 mg/dL |
| Blood (serum) urea nitrogen (BUN) | Urinary system | 7–31 mg/dL | 14–34 mg/dL |
| Calcium | Urinary system, parathyroid gland | 7.9–11.5 mg/dL | 8.4–10.8 mg/dL |
| Chloride | Urinary system, gastrointestinal system (lost in vomit) | 104–125 mEq/L | 112–124 mEq/L |
| Cholesterol | Liver, thyroid function | 109–315 mg/dL | 63–130 mg/dL |
| Creatinine | Urinary system | 0.5–1.6 mg/dL | 0–1.5 mg/dL |
| Gamma-glutamyltransferase (GGT) | Liver, gall bladder | 1–11 U/L | 0–3 U/L |
| Globulins | Liver, immune system, gastrointestinal system | 2.7–4.4 g/dL | 2.6–5.1 g/dL |
| Glucose | Liver, pancreas (insulin production), diet (postprandial) | 65–127 mg/dL | 35–129 mg/dL |
| Lipase | Pancreas (inflammation) | 25–534 U/L | 0–160 U/L |
| Phosphorus | Urinary system, parathyroid gland | 2.4–6.5 mg/dL | 4.0–7.0 mg/dL |
| Potassium | Urinary system, adrenal glands | 3.9–5.7 mEq/L | 3.8–5.4 mEq/L |
| Sodium | Urinary system, adrenal glands | 141–161 mEq/L | 144–156 mEq/L |

*Reference ranges may vary among different laboratories; a patient's results should always be compared with the reference ranges provided by the laboratory conducting the test.

(endocrine) tests to diagnose and monitor the treatment of endocrine diseases; serological tests to measure antibodies to various infectious agents (viruses, fungi, protozoa, rickettsial organisms, bacteria, and others); antigen tests for the detection of heartworms and feline leukemia virus infections; allergen-specific antibody tests for the detection of allergies; antinuclear antibody tests for the diagnosis of systemic lupus erythematosus; toxicological tests for exposure to lead, zinc, antifreeze, warfarin, and many other substances; drug levels to monitor the absorption of medications; and many others.

# 16.5   URINE SAMPLE COLLECTION AND URINALYSIS

Urine samples provide important information regarding the health of an animal. Several methods can be used to obtain urine samples. The simplest is to catch a midstream sample being voided by the animal; this is termed a free catch sample. If the animal has a full bladder and does not cooperate in allowing a container to be put under it while urinating, an attempt can be made to manually express the bladder while holding a container under the urethra (apply firm pressure to the cranial pole of the bladder). A third method—and the one preferred for cultures—is to obtain the urine using cystocentesis. This procedure involves sticking a needle (with syringe attached) through the ventral abdominal wall into the lumen of the bladder and then aspirating urine (by pulling back on the syringe). Abdominal ultrasound can be used to facilitate placement of the needle into the bladder lumen. A fourth method of obtaining urine is by catheterization of the bladder. Aseptic technique (being as sterile as possible) must be used during catheterization to avoid introducing bacteria into the bladder. The final and least desirable method of obtaining urine is to get it off the floor or from a litter box. Urine obtained from these surfaces is likely to be contaminated and is useless for cultures.

The first step in analyzing a urine sample is to look at its physical characteristics. What color is it? Very pale urine is likely very dilute and may indicate a hormonal problem or kidney disease. Red urine is suggestive of trauma or infection. Very dark yellow urine may be due to liver disease. Urine should be clear; if it is cloudy, it may contain crystals or white blood cells. The next step is to measure the specific gravity of the urine, which is usually done using a refractometer. Urine-specific gravities between 1.008 and 1.012 are termed isothenuric (have a concentration of salts similar to plasma) and can be signs of serious kidney disease. Chemical analysis of urine is usually accomplished using a urine dipstick (Figure 16.2). The test strips on a urine dipstick measure different substances in the urine and are color-coded to interpret the results of these measurements. Common substances measured by urine dipsticks include protein, glucose, ketones, blood, and bilirubin. A urine pH indicator is included on the test strip.

The final test commonly performed as part of a routine urinalysis is an examination of the urine sediment. Ten milliliters of urine is placed in a conical-tip centrifuge tube and spun at 1,500 rpm for 5 minutes. The supernatant (fluid) is decanted (poured off), the tube is tapped to resuspend the pellet, and a drop of the sediment is placed on a glass slide and examined using a microscope. The microscopic examination includes looking for epithelial cells, blood cells, casts (a sign of renal tubule disease), crystals, and microorganisms. Other tests that can be performed using urine include a urine culture, urine protein:creatinine levels (used to help determine the severity of protein loss in urine), urine cortisol:creatinine levels (used to screen for hyperadrenocorticism; Chapter 14), and toxicological analyses (screens for poisoning and also drug testing).

**Figure 16.2** Urine dipstick used to measure pH as well as the relative concentrations of protein, glucose, ketones, blood, and bilirubin in urine. *Courtesy University of Illinois.*

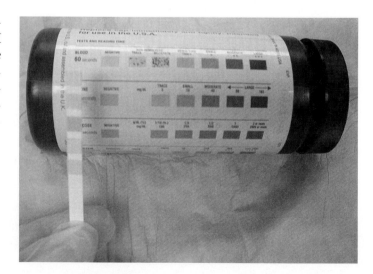

# 16.6   FECAL COLLECTION AND ANALYSES

Feces contain undigested food substances, digestive enzymes, a variety of microorganisms, and may contain eggs of gastrointestinal and respiratory parasites. Fresh feces are preferred for most analyses. Freshly voided feces can be taken from the ground or litter pan. A gloved finger or fecal loop can be inserted into the rectum to remove fresh feces. Feces can be placed in a resealable bag or container and refrigerated until analyzed.

## 16.6.1   FECAL EXAMINATIONS FOR PARASITES

*Direct fecal smear.*  A small amount of feces is suspended into a drop of saline (dilute salt solution) and examined under a microscope. This examination is used to look for trophozoites of protozoa (e.g., *Giardia,* Chapter 15) and may also demonstrate nematode larvae and eggs.

*Fecal flotation.*  Approximately one teaspoon of feces is mixed with a solution[3] of zinc sulfate, sodium nitrate, or sucrose. The solution is filtered through coarse gauze or a strainer to remove large particles and either centrifuged for 5 minutes or allowed to stand for 10 to 20 minutes (Figure 16.3). Parasite ova and cysts will rise to the surface of the solution and can be collected by placing a coverslip over the full tube and then transferring the coverslip to a glass slide for examination using a microscope (see Figure 15.13).

*Fecal sedimentation.*  Sedimentation techniques are used for eggs that do not float. A fecal solution is prepared by mixing feces with water or formalin. The solution is filtered and centrifuged. The supernatant is then discarded and the sediment is examined using a microscope.

*Fecal cytology.*  A direct smear is prepared by mixing fresh feces with a small amount of water or by inserting a cotton-tipped applicator (Q-tips® cotton swab) in the rectum, rolling it along the rectal mucosa, and then putting it onto a glass slide. The slide is allowed to air dry and stained using either a Gram's stain or a Diff-Quick stain. It is then examined using an oil immersion lens (1,000× magnification) for the presence of white blood cells, bacteria (*Clostridium, Campylobacter*), yeast (*Candida, Prototheca*), or fungal organisms (*Histoplasma*).

*Fecal ELISA*[4] for parvovirus.  Fresh feces are tested using the instructions provided with the ELISA kit. A positive test result strongly suggests the presence of parvovirus enteritis.

*Fecal culture.*  Fresh feces or rectal swabs from dogs or cats with persistent diarrhea should be cultured for pathogenic enteric (intestinal) microorganisms including *Clostridium, Campylobacter,* and *Salmonella* spp. The diagnostic laboratory performing the cultures should be contacted to see if special enrichment culture media and transport swabs are recommended (particularly for *Salmonella*).

*Fecal fat.*  Animals with diseases causing interference with the digestion or absorption of fat have large quantities of fat in their stools. This can be demonstrated by mixing feces with Sudan III stain.

*Fecal starch and muscle fibers.*  Animals with diseases causing interference with the digestion or absorption of starch and/or proteins have large quantities of starch and/or muscle

---

[3] The specific gravity of the solution used for fecal flotations should be approximately 1.20.
[4] ELISA is an acronym for enzyme-linked immunosorbent assay. An ELISA links an enzyme to an antibody specific for the antigen (protein, hormone, microorganism, parasite, viral particle, or others) being tested for; if the antigen is present, the enzyme-linked antibody binds to the antigen. A color-producing substrate is then added; the enzyme coverts the substrate to a colored product. If color is produced, the test is positive for the presence of the antigen.

**Figure 16.3**   Fecal flotation using a commercial Ovassay™ kit. Feces are mixed with flotation solution in lower portion of the vial. (*a*) A strainer is inserted into the vial. (*b*) The vial is then filled with the flotation fluid and a coverslip is placed over the top. Parasites and ova will float to the top of the solution and adhere to the coverslip. (*c*) The coverslip is then placed onto a glass slide for examination using a microscope. *Courtesy Sandra Grable, CVT, University of Illinois.*

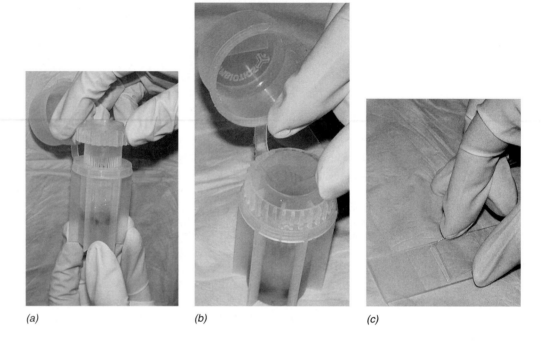

(a)                                                (b)                                                (c)

fibers in their stools. Fecal smears can be stained with 2% Lugol's iodine and examined using a microscope. Starch granules appear as dark blue-black granules and muscle fibers as blue strands in the stained smears.

*Fecal proteolytic activity.*   Inadequate production of digestive enzymes results in chronic diarrhea and weight loss. A simple but controversial test is the film digestion test of fecal proteolytic activity. To perform this test 1 ml of fresh feces is mixed with 9 ml of 5% sodium bicarbonate solution in a test tube. A strip of x-ray film (5 inches long by 0.5 inches wide) is placed in the tube and incubated at 98.6°F (37°C) for 1 hour. If trypsin (a proteolytic enzyme produced by the pancreas) is present in the feces, the gelatin emulsion will be digested from the x-ray film and it then appears clear. False positives and false negatives can be caused by large concentrations of various bacteria in the feces; thus many people consider this test to be undependable. Most veterinarians prefer the more accurate serum tests for trypsin activity (however, serum tests take longer to obtain results and are more expensive).

*Fecal occult blood.*   Bleeding into any part of the digestive system can be detected through the use of fecal occult blood test strips. However, false positives can be seen if the animal's diet contains meat; thus the dog or cat should be fed a meat-free diet for a minimum of 3 days prior to performing this test.

# 16.7   SKIN SCRAPINGS AS A DIAGNOSTIC TOOL

One of the most useful tests in evaluating animals with hair loss and skin disease is skin scraping. Skin scrapings are used to find and identify ectoparasites. Mineral oil is placed on the skin lesion and a #10 scalpel blade is repeatedly scraped across the skin until a small

**Figure 16.4** Performing a skin scraping to examine for ectoparasites. *Courtesy University of Illinois.*

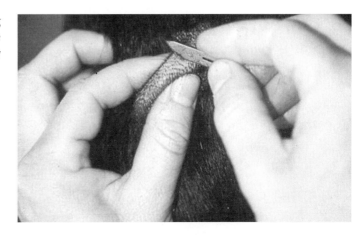

amount of blood oozes from the site (Figure 16.4). The material that has been scraped off of the skin surface is transferred into mineral oil on a glass microscope slide and examined at 100× to 400× magnification. Mites, lice, and nematode larvae are responsible for many skin diseases and most are easy to identify in a skin scraping (Chapters 14 and 15).

# 16.8   IMPRESSION SMEARS (CYTOLOGY) AS A DIAGNOSTIC TOOL

Another useful technique for evaluating skin tumors and other skin diseases is an impression smear (also called cytology because cells are being examined using a microscope). A glass slide or strip of acetate tape is pressed onto the surface of the lesion (Figure 16.5a). Tape is applied sticky side down to the lesion and then sticky side down to a glass slide. The slide is then stained and examined using a microscope at 1,000× magnification. Many bacterial, fungal, yeast, and neoplastic (cancer) diseases can be identified using impression smears (Figure 16.5b).

**Figure 16.5** (*a*) Acetate tape applied to a skin lesion. (*b*) Epithelial cells with bacteria and yeast organisms on a skin impression smear (1,000× ). *Courtesy University of Illinois.*

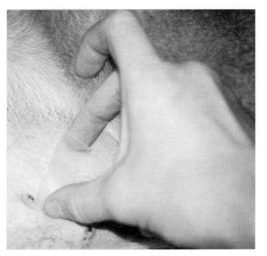

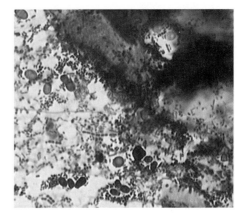

*(a)*                                                    *(b)*

## 16.9    FINE NEEDLE ASPIRATES AS A DIAGNOSTIC TOOL

A small-gauge needle (usually 22 to 25) attached to a 6 cc syringe can be used to obtain a sample of cells from an accessible mass (nodule, cyst, or tumor), joint, body cavity (thorax, abdomen, or subdural space), or organ (spleen, liver, lymph node, bone marrow, or anterior chamber of the eye) for cytological evaluation. The needle is directed into the mass, joint, body cavity, or organ by palpation and visual direction or through the use of ultrasound to guide the placement (Section 16.13). Following placement of the needle, negative pressure is applied by pulling back on the syringe plunger. The pressure is then released and the procedure is repeated two to three times. The needle is then withdrawn and any fluid present in the syringe is ejected onto a glass slide for staining and examination using a microscope. If no fluid is present in the syringe, the syringe is disconnected from the needle, filled with air then reattached to the needle. The needle is then held over a glass slide and the plunger is quickly depressed to "blow" cells out of the needle onto the slide. Special terms are applied when this technique is used in joints (arthrocentesis), the thorax (thoracocentesis), the abdomen (abdominocentesis), and the subdural space [cerebrospinal fluid (CSF) tap].

## 16.10    TRANSTRACHEAL WASH

This procedure is used to obtain samples from the trachea and bronchi for cytology and/or cultures. An 18-gauge, 20-cm through-the-needle catheter is generally used to perform a transtracheal wash. The needle is inserted into the trachea just below the larynx (through the cricothyroid membrane);[5] the catheter is then threaded through the needle and directed down the trachea. A small amount (< 0.5 ml/kg body weight) of sterile saline is quickly injected through the catheter and immediately withdrawn using rapid back and forth movements of the syringe's plunger. Ideally, the dog or cat will cough during this procedure. Coughing facilitates obtaining cells and organisms from deeper in the airways. An alternative technique is to intubate the cat or dog (put a sterile endotracheal tube into the trachea; this requires the use of general anesthesia) and to then pass a sterile urinary catheter down the endotracheal tube for the "washing." This is less desirable due to the requirement for anesthesia (can be dangerous in animals with respiratory tract disease), and also because anesthesia suppresses the cough reflex making it difficult to obtain materials from the lower airways.

## 16.11    CULTURES AS DIAGNOSTIC TESTS

Bacterial and fungal cultures should be performed when either of these types of microorganisms is suspected to be the cause of a disease. A variety of culture collection systems and culture media can be utilized. Specific recommendations should be obtained from the diagnostic laboratory that will be performing the culture. In addition to identifying the causative microorganisms, the laboratory will perform antibiotic susceptibility tests on bacterial isolates. These results are useful in selecting the most appropriate antibiotics to use in treating an infection. Fungal susceptibility testing is not readily available. Many veterinarians and veterinary technicians perform fungal cultures to identify dermatophyte (ringworm) infections (Chapter 14) using a special medium called Dermatophyte Test Media® (DTM) or DermDuet® media. Hairs are plucked from an area with broken hairs and embedded in the

---

[5]The cricothyroid membrane bridges the gap between the cartilages forming the larnyx and those of the trachea; inserting a needle through this membrane facilitates passage of a catheter into the trachea without causing damage to cartilage.

**Figure 16.6**  (*a*) Fungal growth on DermDuet® media. (*b*) Identification of the fungal growth as *Microsporum gypseum* is made by examining a portion of the colony using a microscope (400×). *Courtesy Sandra Grable, CVT, University of Illinois.*

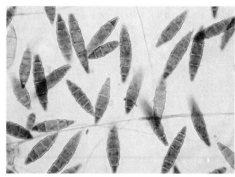

*(a)*                                                                                           *(b)*

fungal culture media. If a dermatophyte is present on the hairs it will grow as a white colony (Figure 16.6a) and produces alkaline metabolites that turn the fungal culture media a red color. Specific identification of the type of dermatophyte is made by examining the colony growth using a microscope (Figure 16.6b).

# 16.12  RADIOGRAPHIC IMAGING AS A DIAGNOSTIC TOOL

Diagnostic imaging enables veterinarians to see a visual representation of body structures not ordinarily visible. Depending on the modality, images of tissues and internal organs are created using x-rays, ultrasound, radioactivity, magnetism, and computer technology.

Diagnostic imaging is used to examine body tissues for abnormalities and to evaluate and monitor those abnormalities. Diagnostic imaging allows detection of internal foreign bodies, bone fractures and dislocations, tumors, abscesses, changes in organ size, and pregnancy (see Figure 10.6). In some instances, diagnostic imaging may allow organ function(s) to be evaluated. This section reviews the basic principles and techniques of radiographic imaging. Other imaging techniques are reviewed in Sections 16.13 through 16.16.

## 16.12.1  RADIOGRAPHY

Radiographs are photographs made using x-rays. Because x-rays can penetrate body tissues, radiographs are images of internal structures (Figure 16.7). Although important advancements in medical science and technology have made new imaging modalities available, radiographs remain the foundation of diagnostic imaging.

Radiographs are useful in diagnosing abnormalities in many body systems. If an animal is limping or reluctant to move, the veterinarian will frequently use radiographs of the limbs or spine as part of the diagnostic evaluations. Radiographs of teeth are used to evaluate dental abnormalities. If an animal is vomiting, losing weight, or has abdominal pain, the veterinarian may want to evaluate radiographs of its abdomen. Animals that tire easily or are coughing may undergo thoracic radiographs so the veterinarian can evaluate their heart and lungs.

Contrast radiography is used to enhance the veterinarian's ability to visualize certain structures or areas, or to evaluate the movement of the contrast agent after it is swallowed. A liquid or gaseous contrast agent is introduced into the animal's body and radiographs are taken. The progress of the substance through the body or appearance of the agent surrounding structures or surfaces allows visualization of details that would not otherwise be visible.

**Figure 16.7** Radiograph of the thorax of a dog. *Courtesy Oklahoma State University.*

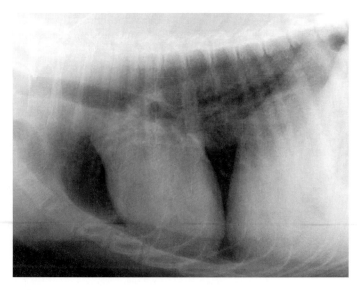

Although radiographic equipment represents a major capital investment, most veterinary hospitals have this very useful imaging modality available. In fact, veterinarians that travel to stables and ranches usually have portable x-ray machines.

## 16.12.2   CHARACTERISTICS OF RADIATION

Radiation is the movement of energy through space. X-rays are a form of radiation similar to visible light but having more energy and penetrating ability. Depending on the energy levels of the x-rays in the beam, and the densities of the tissues being imaged, variable amounts of the energy will penetrate the tissues and strike the cassette containing the x-ray film. Variation in the amounts of energy striking the film causes variation in the degree of darkness on the film when it is processed. It is these variations of darkness on the film that result in an image (picture) of internal structures.

## 16.12.3   RADIATION SAFETY

X-rays are considered ionizing radiation because they have sufficient energy to bring about changes in the atoms they strike. These changes in atoms can damage or kill living cells. Cells most sensitive to radiation are those dividing rapidly (e.g., blood-forming cells in bone marrow). Radiation exposure can also damage genetic material in spermatozoa and ova, possibly resulting in abnormal offspring. Although some irradiated cells may repair their damage, cellular injury due to irradiation tends to be cumulative.

To protect the health of people exposed to radiation in connection with their occupation, state health departments develop and enforce rules pertaining to the safe operation of x-ray equipment. Protective (lead lined) clothing and shielding devices are used to minimize personnel exposure. Monitoring devices are worn to measure exposure and the cumulative dose an occupationally exposed individual receives is tracked. The National Committee on Radiation Protection, a nonprofit organization comprised of experts in the field, has defined a maximum permissible dose of irradiation as being one that does not involve a risk to the health of those occupationally exposed or to their offspring.

For the veterinary patient, the health benefits of a diagnostic radiograph greatly exceed the risks of the small dose of ionizing radiation received during the procedure. Most animals are exposed infrequently if at all during their life. Nonetheless, when intact (not neutered or spayed) animals are examined, the gonads and any fetuses should be shielded.

**Figure 16.8**   An x-ray machine. *Courtesy Oklahoma State University.*

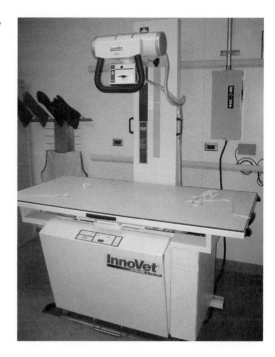

## 16.12.4   PRODUCING X-RAYS

X-rays are produced when electricity is applied to a filament in an x-ray tube. The filament can be compared with the filament in a light bulb. The radiographer controls the number of x-rays produced by the x-ray tube and their penetrating power by changing settings on the x-ray machine (Figure 16.8). A working knowledge of the relationship between these exposure factors and the degree of darkness and contrast on the resulting radiograph is essential in ensuring the production of useful diagnostic images.

## 16.12.5   RECORDING X-RAY IMAGES

Radiographic cassettes are lightproof containers for radiographic film. Intensifying screens are mounted to the inside doors of cassettes (Figure 16.9). Intensifying screens contain crystals that fluoresce when x-rays hit them. Light emitted by the screens exposes the film. It is

**Figure 16.9**   X-ray cassettes. *Courtesy Oklahoma State University.*

important that correct screen and film combinations be used. The most common type of intensifying screens used in veterinary practices are referred to as rare earth screens because the natural compounds used in their production are difficult to extract from the earth. Rare earth screens emit green light when struck by x-rays penetrating the cassette. Film used in cassettes with rare earth screens should be sensitive to green light.

X-ray film is made of an emulsion of silver halide crystals adhered to a clear base. The silver halide crystals are changed when exposed to light emitted by intensifying screens. After being exposed, the film is removed from the cassettes and processed to produce the visual image.

During processing, x-ray film goes through numerous precisely timed steps including submersion in chemical solutions. The first step involves a developing solution containing chemicals that change the light-altered silver halide crystals to black metallic silver. Silver halide crystals not exposed to light are not changed to metallic silver by the developing solution.

The fixer solution contains chemicals that cause the black metallic silver to adhere to the film base and removes silver halide crystals. This results in the areas of exposed film becoming shades of gray and black, depending on how many x-rays penetrated the tissues. The unexposed areas are clear and appear white when viewed.

### 16.12.6 RADIOGRAPH INTERPRETATION

The radiographic image depicts tissues using clear areas, various shades of gray, and black. The color of a particular tissue on the radiograph depends on the number of x-rays able to pass through tissue and hit the cassette. Dense tissues such as bone allow few x-rays to penetrate. As a result, bones appear white on radiographs. Muscles, fluids, and most abdominal organs have similar densities; they appear gray on radiographs. Gas and normal lung tissue block very few x-rays and appear dark or black on radiographs. Tissues can be distinguished from adjacent tissues only if they are of different densities and therefore appear as different colors.

Many abnormalities resulting in changes in size, shape, density, or location of a structure may be diagnosed by examining radiographs. For example, pneumonia results in increased density of lung tissue; a broken bone may result in pieces of dense bone separated by less dense blood or being displaced from their normal location (Figure 16.10).

Numerous abnormalities do not result in radiographic changes (e.g., a tumor within the liver will usually have the same density as other liver tissues and will therefore be the same shade of gray). Other imaging modalities may allow diagnosis of abnormalities and evaluation of structures that cannot be visualized on radiographs.

## 16.13 ULTRASONOGRAPHY AS A DIAGNOSTIC TOOL

An image produced using ultrasound is called a sonogram. Sonograms use a nonionizing form of energy (sound waves) to produce an image. The image is displayed on a monitor in real time. In other words, as the shape and location of tissue changes, the image on the monitor changes. This moving image allows visualization of a beating heart, fetal movement, and the orientation of a biopsy instrument within an organ. Proper clinical use of diagnostic ultrasonography is safe for personnel and the patient because it does not use ionizing radiation. The examination is not painful and most patients do not require sedation.

Ultrasound waves penetrate soft tissues but do not penetrate bone or gas. Although most organs can be visualized using ultrasound, intestinal gas and ribs impose limitations. The heart and chest cavity are imaged by aiming the ultrasound waves between the ribs. Cardiac ultrasonography, also called echocardiography, is used to diagnose and monitor heart disease. It provides information about both structure and function. Echocardiography can be used to diagnose abnormalities in the size of the heart or its chambers and evaluate movement of the heart valves. More sophisticated (and expensive) machines have color monitors that demonstrate the movement of blood between heart chambers and through vessels.

**Figure 16.10**   Radiograph of a fractured bone. *Courtesy Oklahoma State University.*

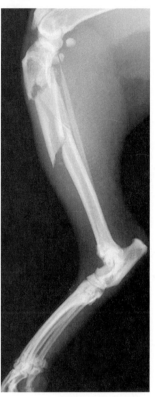

Ultrasonography can be used to diagnose pregnancy in the bitch 16 to 20 days after conception. By day 25, the fetal heartbeat can be visualized as a flickering movement on the monitor. Several abnormalities involving the uterus (e.g., pyometra, an accumulation of pus in the uterus) and ovaries may be diagnosed using ultrasonography. In the male dog, the prostate gland can be evaluated to diagnose enlargement resulting from infection or cancer.

Abdominal ultrasonography is used to identify intestinal obstructions, peritoneal fluid, inflammatory bowel disease, foreign bodies, bladder disease, certain kidney diseases, and vascular anomalies such as congenital portosystemic shunts.[6] Many types of ultrasound machines are available, varying in capabilities and expense. Many veterinary practices currently have ultrasound machines and many more will likely purchase them in the future.

## 16.13.1   PRODUCING THE SONOGRAM

When sound waves hit an object, some bounce back toward the source, creating an echo. The length of time required for the echo to return depends on the distance between source and object. Ultrasound machines direct very high-frequency sound waves into tissues and measure the length of time required for the echo to return. A computer within the ultrasound machine translates the echo into an image on a monitor (Figure 16.11).

The image on the monitor is made of gray and white dots on a black background. Tissues that reflect a majority of the sound waves striking them are displayed on the monitor as bright

---

[6]Portosystemic shunts are abnormal vascular connections between portal veins (vessels that normally carry blood from the gastrointestinal tract to the liver) and the systemic circulation. Blood going through the shunt bypasses the liver and thus is not filtered or detoxified. Animals with portosystemic shunts have a buildup of toxic waste products in their blood and show signs of liver failure.

**Figure 16.11** An ultrasound machine. *Courtesy Oklahoma State University.*

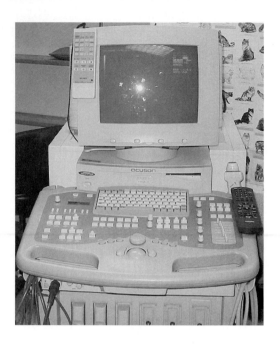

white dots. The greater the number of reflected waves the brighter the dots. If most of the sound waves are transmitted through the tissues and only a few bounce back, the resultant dots will be dark gray. Tissues not reflecting waves will be seen as black areas on the monitor.

The location of dots on the monitor depends on the length of time required for sound waves to return to the machine. Sound waves returning from deep tissues require longer to return and result in dots being nearer the bottom of the image.

### 16.13.2 SONOGRAM INTERPRETATION

The number of waves transmitted through a tissue depends on its physical characteristics, including density and elasticity. As ultrasound waves leave one tissue and enter another, the proportion of waves being transmitted and reflected depends on characteristics of the new tissue. Adjacent tissues are distinguished from each other by differences in the brightness of their image. Because membranes covering organs have different characteristics than organ tissue, most body organs may be visualized. The examination may be recorded on videotape or the image on the monitor may be printed and made part of the patient's medical record.

Ultrasonography is an especially useful diagnostic tool. However, correct interpretations of the images require the training and experience of a skilled professional.

## 16.14 NUCLEAR SCINTIGRAPHY AS A DIAGNOSTIC TOOL

Nuclear scintigraphy, often referred to as nuclear scanning, uses a small dose of a radioactive pharmaceutical to detect abnormalities based on blood flow. Several radiopharmaceuticals having variable affinities for particular tissues or organs are available.

Areas of inflammation, tumors, and other irregularities can result in increased blood flow and an accumulation of radioactivity. These areas are referred to as hot lesions; those showing an absence of radioactivity are referred to as cold lesions. Nuclear scans are helpful in localizing bone abnormalities in cases of subtle lameness. Thyroid scans use a radiopharmaceutical with an affinity for the thyroid gland and can confirm hyperactivity, tumors, or other abnormalities (Figure 16.12).

Most veterinary teaching hospitals and some private hospitals (particularly specialty veterinary hospitals) can perform nuclear scintigraphy (Figure 16.13). The hazards of ra-

**Figure 16.12**   A thyroid scan using nuclear scintigraphy.
*Courtesy Oklahoma State University.*

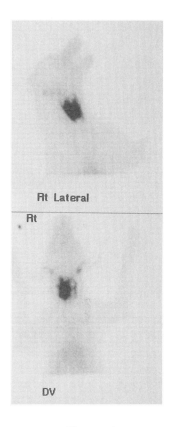

dioactivity pose minimal risk to patients. To help ensure the protection of hospital personnel, protective attire is used, and depending on state law the patient is usually hospitalized 1 to 3 days until the radioactivity has dissipated. During hospitalization, contact with the patient is minimized and precautions are taken in disposing of feces and urine.

**Figure 16.13**   A nuclear scintigraphy unit.
*Courtesy Oklahoma State University.*

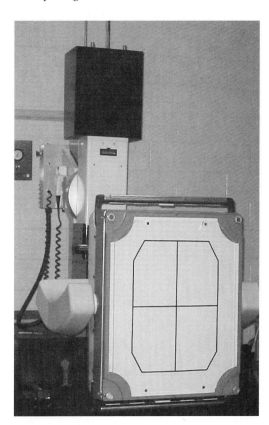

### 16.14.1   PRODUCING THE IMAGE

To perform the scan, a radiopharmaceutical is injected into the patient's bloodstream. The radioactivity is then detected with a special camera that forms an image composed of tiny dots representing sites of radioactivity. Anatomic areas having increased blood flow appear darker on the image. Images do not provide much anatomic detail; instead they are merely an indication of blood flow to an area.

## 16.15   MAGNETIC RESONANCE IMAGING (MRI)

The use of MRI takes diagnostic imaging one stage further. Magnetic resonance imaging uses magnetic fields and the transmission of energy in the form of radio waves to produce an image. The equipment is expensive; except for several teaching hospitals and a few large specialty veterinary practices, MRI is seldom available in veterinary hospitals. Some veterinary hospitals have made arrangements with nearby human facilities or private companies with mobile equipment to image veterinary patients.

MRI is extremely sensitive. Differences in chemical composition result in the ability to differentiate between closely related tissues. The images are particularly useful for visualizing the brain and spinal cord because the bony skull and vertebrae do not interfere with the production of the image and the different types of nervous tissue (gray and white matter) can be differentiated. MRI is generally considered to be risk free except for patients with cardiac pacemakers or certain metal implants.

### 16.15.1   PRODUCING THE IMAGE

The patient is placed in a strong magnetic field and radio waves are used to cause body tissues to emit signals that are translated into an image. The images are cross sections or "slices" of body parts. Because the animal must remain motionless throughout the procedure, general anesthesia is required.

## 16.16   COMPUTED TOMOGRAPHY (CT)

Compared with traditional radiography, computed tomography allows superior imaging of three-dimensional structures. CT uses x-rays and computer technology to produce images that show internal anatomy as thin, cross-sectional slices. Most veterinary teaching hospitals and some private ones (particularly specialty veterinary hospitals) have CT units. CT scans can provide information not available from routine radiographs. The potential health benefits of CT scanning far exceed the risks of ionizing radiation. General anesthesia is required because the patient must remain motionless throughout the procedure.

### 16.16.1   PRODUCING THE IMAGE

The CT unit consists of a moveable bed on which the patient lies. The bed moves slowly through the hole of a doughnut-shaped x-ray machine (Figure 16.14). The x-ray tube and x-ray detectors move around the patient collecting information. Computers translate the information into images. Because the x-rays transverse the patient from many angles, a great deal of detailed anatomical information is available (Figures 16.15 and 16.16).

**Figure 16.14** A computed tomography (CT) unit. *Courtesy Oklahoma State University.*

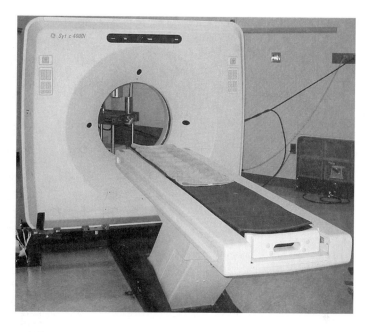

## 16.17  ENDOSCOPY AS A DIAGNOSTIC TECHNIQUE

Endoscopy is the use of long, narrow tubes containing fiber optics. The fibers transmit light through the tube to illuminate objects being examined and transmit images back to a camera or projection screen for viewing (Figure 16.17). Endoscopy is a nonsurgical method of viewing the interior surfaces of the body. Endoscopy is used to diagnose inflammation, bleeding ulcers, strictures, internal parasites, infections, and tumors. A wide variety of endoscopes are available for use in examining different organs of the body. Rigid endoscopes are used for some examinations (ears, joints, bladder, rectum, and distal colon); however, flexible instruments are often preferred as these can go around bends and view deeper regions of the gastrointestinal, respiratory, reproductive, and urinary tracts. Biopsy instruments can

**Figure 16.15** CT image of a herniated intervertebral disk. *Courtesy Oklahoma State University.*

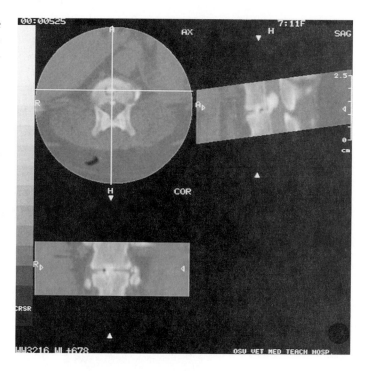

**Figure 16.16** CT image of the nasal sinuses of a cat. *Courtesy Oklahoma State University.*

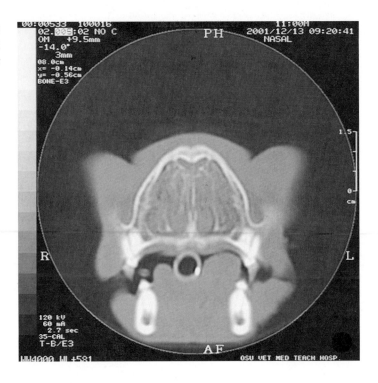

be passed through the scopes to obtain tissue samples for cultures or microscopic evaluation. Brushes can be used to obtain cell samples for cytology. "Grabbing" instruments are passed through the scopes to retrieve foreign bodies, parasites, and calculi (stones). Photographs and digital images taken using endoscopes provide documentation of abnormalities and are useful in monitoring response to therapy.

**Figure 16.17** (*a*) Endoscopy unit for use in dogs and cats. (*b*) Gastroscopy being used to investigate the cause of chronic vomiting in a dog. *Courtesy Kristie Stasi, CVT, University of Illinois.*

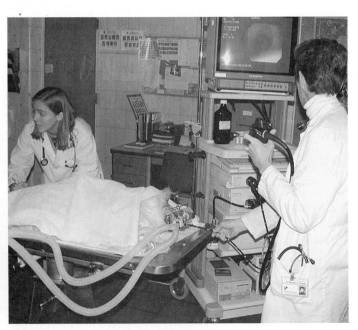

*(a)*         *(b)*

*Esophagoscopy* is the endoscopic examination of the esophagus and is indicated in dogs and cats with suspected esophageal obstructions, masses, or erosions. *Gastroduodenoscopy* is useful in the evaluation of dogs and cats with vomiting, reflux, or small intestinal disease. Stomach and esophageal worms, erosions, ulcers, and tumors may be visualized and biopsies can be taken for the diagnosis of inflammatory and neoplastic diseases (tumors). *Colonoscopy* (examination of the colon), *proctoscopy* (examination of the rectum), and *ileoscopy* (examination of the ileum) are indicated in the evaluation of animals with chronic diarrhea or difficulty in defecation. *Bronchoscopy* (examination of the airways) is used for assessment of airway inflammation or pulmonary hemorrhage, to identify structural abnormalities (e.g., tracheal collapse, strictures, foreign bodies, masses, lung lobe torsions), and to collect samples for microscopic evaluation and culture. *Cystoscopy* is used to examine the urethra and bladder for inflammation, tears, stones (uroliths, calculi), and tumors. *Arthroscopy* is used to evaluate joints for tears in cartilage and ligaments. Arthroscopes are used by surgeons performing arthroscopic joint surgery. *Video-otoscopy* is used to perform thorough cleaning of ears, biopsy masses in the ear canal, retrieve foreign bodies from the ears, and evaluate the patency of the tympanic membrane (Figure 16.18). *Vaginoscopy* is

**Figure 16.18**   (*a*) Video-otoscope for examining the horizontal ear canal and tympanic membrane. (*b*) Normal tympanic membrane visualized using a video-otoscope. Arrow points to a portion of the stapes bone of the inner ear. (*c*) Nasopharyngeal polyp in the horizontal ear canal of a cat visualized using a video-otoscope. Arrow points to the polyp. *Courtesy University of Illinois.*

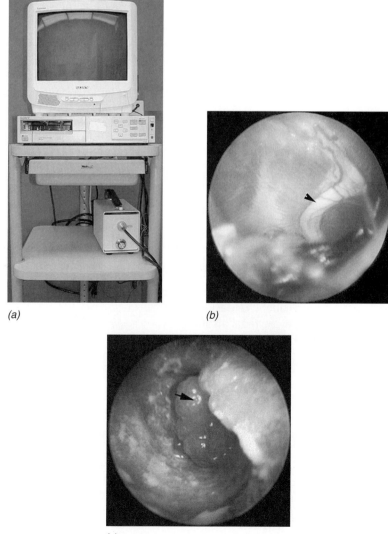

(a)

(b)

(c)

useful in examining for strictures, tumors, or tears in the vagina or cervix. The scope can also be used to direct placement of an insemination pipette or for obtaining uterine biopsies. *Laparoscopy* is the use of a fiber-optic scope (usually rigid) to examine the organs within the peritoneal cavity (abdomen). A small incision is made through the abdominal wall into the peritoneal cavity for insertion of the scope. The abdomen is inflated with carbon dioxide gas to facilitate movement of the laparoscope and visualization of the organs. The laparoscope can be used to guide placement of instruments to obtain biopsies of internal organs (e.g., liver, kidney, lymph nodes) or masses. Some surgical procedures can also be performed using the laparoscope (e.g., tubal ligations).

# 16.18 ELECTRODIAGNOSTICS

Activation of muscle and nerve activity involves electrical impulses. A resting nerve or muscle is polarized with positive charges on the outside of the cell due to extracellular accumulations of sodium ions. When the muscle or nerve is activated, the cell depolarizes, cell membrane permeability increases, and sodium ions rush into the cell generating an electrical impulse. Electrodiagnostic machines measure the electrical impulses associated with depolarization and repolarization of muscles and nerves. Many electrodiagnostic machines have been developed for use in human and veterinary medicine. These machines are also used for research studies in physiology and other sciences.

## 16.18.1 ELECTROCARDIOGRAPHY

One of the most familiar uses of electrodiagnostics is in the use of an electrocardiograph (ECG, EKG) machine for recording electrical activity of the heart. ECG machines are used to document heart rate and rhythm and obtain estimates of the sizes of heart chambers (the latter is based on the determination of the mean electrical axis) and in diagnosing and monitoring many diseases (Table 16.6). For a standard ECG, the dog or cat is posi-

**Table 16.6**
Uses of Electrocardiography

| Evaluation of Cardiac Diseases |
| --- |
| Arrhythmias |
| Chamber enlargement |
| Monitoring therapy: drugs, electrolyte disturbances |
| Documenting disease progression |

| Evaluation of Metabolic Diseases Causing Electrolyte Disturbances |
| --- |
| Adrenal insufficiency (hypoadrenocorticism, Addison's disease) |
| Diabetic ketoacidosis |
| Severe renal insufficiency |
| Eclampsia |

| Monitoring during Anesthesia and Surgery |
| --- |
| Depth of anesthesia |
| Arrhythmias associated with anesthetic drugs |
| Signs of hypoxia |

| Other Uses |
| --- |
| Wellness checks (preventive care, annual physicals) |
| Evaluation of trauma cases |
| Evaluation of animals prior to anesthesia and surgery |

**Figure 16.19** Dog positioned in right lateral recumbency for electrocardiography. *Courtesy Robyn Ostapkowicz, CVT, University of Illinois.*

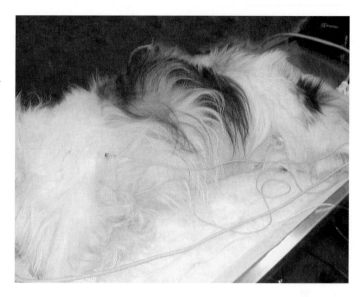

tioned laying on its right side (right lateral recumbency) and held with its front and back legs parallel to each other and at right angles to the body. The electrodes (also known as leads) are attached to the skin of the right and left elbows and the right and left stifles with a 5th lead (called the "V" lead) on the chest (Figure 16.19). The skin is moistened with alcohol or an electrode gel. The most common ECG is taken in what is called the lead II position with the positive electrode on the left hind leg and the negative electrode on the right foreleg. The first impulse during the cardiac cycle is an upward deflection (named the P wave) produced by depolarization of the atria. This is followed by spreading of the impulse in the atrial septum resulting in return of the line to baseline and a straight segment called the PR interval.[7] Next, the ventricles depolarize and contract producing the QRS complex.[8] Repolarization of the ventricles produces a final wave called the T wave (Figure 16.20). Many other lead systems are useful in providing additional information regarding cardiac muscle physiology and the relative sizes and shapes of the cardiac chambers.

*Electromyograms* (EMGs) are used to evaluate muscles. Small electrodes are inserted into muscles, and the electrical activity of the muscle is recorded.

*Electroencephalograms* (EEGs) and *nerve conduction velocity* (NCV) tests monitor the electrical activity of the brain and peripheral nervous system, respectively.

*Electroretinograms* (ERGs) are used in the diagnosis of hereditary eye diseases (e.g., congenital stationary night blindness, progressive retinal atrophy, and others) and in the evaluation of retinal function prior to cataract surgery. Other electrodiagnostic tests used in evaluation of eyesight include the *electrooculogram* (EOG) and *visual evoked potential* (VEP) tests.

*Brainstem auditory-evoked response* (BAER) testing is used to evaluate hearing. This test is particularly useful in screening puppies and kittens of breeds with a high incidence of deafness [e.g., Dalmatian dogs and blue-eyed white cats (Chapter 4)].

---

[7]The PR interval is the time from the beginning of the P wave until the beginning of the QRS complex.
[8]The QRS complex represents activation of the ventricle. The Q wave is a downward deflection resulting from activation of the ventricular septum. The R wave is an upward deflection resulting from activation of the ventricles; it is normally the largest component of the QRS complex. The S wave is a downward deflection following the R wave; it is associated with the end of ventricular activation.

**Figure 16.20** Lead II electrocardiogram showing the P wave (atrial depolarization), QRS complex (ventricular depolarization), and T wave (ventricular repolarization). *Courtesy University of Illinois.*

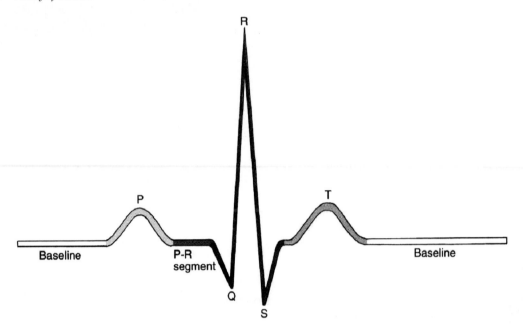

# 16.19   ANESTHESIA

Those working in veterinary hospitals, conducting research with laboratory animals in which anesthesia is administered, and performing related job responsibilities must have a basic knowledge of anesthesia. Anesthesia relies on chemicals or drugs to eliminate sensations and/or provide immobilization. The purposes of anesthesia range from a need to restrain or relax an animal to the complete immobilization required for surgery and certain procedures (e.g., CT and MRI). Anesthetics function by depressing the central nervous system to provide pain relief, sedation, or unconsciousness as required.

Animals must be evaluated prior to the administration of an anesthetic as certain anesthetics are contraindicated in those with preexisting diseases (e.g., heart, kidney, or liver disease). Careful monitoring of the animal is required throughout the anesthetic period. Anesthetic protocols generally include (1) the administration of one or more preanesthetic agents, (2) induction anesthetics(s), (3) maintenance of anesthesia using injectable or inhalation anesthetic(s), and (4) monitoring of vital signs and the depth of anesthesia.

## 16.19.1   PREANESTHETIC AGENTS

Preanesthetic agents are used to accomplish many goals including (1) calming the animal to make it easier to handle, (2) decreasing the amount of the anesthetic agent that will be needed, and (3) minimizing potential side effects associated with anesthetic agents. Preanesthetic agents are usually given 10 to 20 minutes prior to the induction of anesthesia. Anticholinergic agents are used to prevent bradycardia[9] and to minimize salivary and respiratory tract secretions.[10] The two anticholinergic agents used most frequently in dogs and cats are at-

---

[9] Bradycardia is a slow heart rate; many anesthetic agents depress both heart rates and respiratory rates, which may result in inadequate tissue oxygenation. Anticholinergic agents help maintain a normal heart rate during anesthesia.
[10] Many anesthetic agents increase salivation and respiratory tract secretions; this is dangerous because swallowing and coughing reflexes are suppressed by anesthesia and thus these secretions may flow into the lungs impairing respiration.

ropine and glycopyrrolate. Tranquilizers (sedatives) depress the central nervous system and are used to aid in restraint, reduce anxiety, and decrease the amount of the more potent medications that will be required to induce and maintain anesthesia. Commonly used tranquilizers include acepromazine, diazepam (Valium™), xylazine (Rompun™), detomidine (Dormosedan™), and medetomidine (Domitor™). Opioids[11] may also be used to provide sedation and analgesia (pain relief). Examples of opioids commonly used in dogs and cats include hydromorphone, oxymorphone, and butorphanol. A combination of an opioid with a tranquilizer is termed neuroleptanalgesia and results in a synergistic effect yielding sufficient analgesia and sedation for procedures such as casting a broken bone or taking radiographs.

## 16.19.2  INDUCTION AGENTS

Induction anesthetic agents are used to facilitate a smooth and rapid transition from consciousness to unconsciousness. Agents commonly used include barbiturates,[12] propofol, ketamine, tiletamine, and mask delivery of gas anesthetics. Barbiturates commonly used for the induction of anesthesia are the ultra-short-acting ones: thiopental (thiobarbiturate) and methohexital (oxybarbiturate). Barbiturates often cause apnea (animal stops breathing) and thus may require intubation and ventilatory support. Barbiturates are metabolized by the liver and should not be used in animals with liver disease or in sighthounds (these breeds are often unable to metabolize barbiturates). Propofol is a nonbarbiturate, intravenous anesthetic agent that produces a rapid loss of consciousness. It is also rapidly metabolized and is safer than barbiturates for use in animals with liver disease and in sighthound breeds. Ketamine and tiletamine are dissociative agents that produce a cataleptic[13] state. Dissociative agents are usually administered together with a benzodiazepine (e.g., diazepam or zolazepam) for the induction of anesthesia. Inhalation agents can be administered using a face mask or by placing the animal in a Plexiglas chamber that can be filled with the gas (Figure 16.21). Mask induction is primarily used for patients that cannot be safely given injectable induction agents because this technique greatly increases waste gas contamination of the environment.

## 16.19.3  INHALATION ANESTHESIA

Inhalation anesthetics are widely used for the maintenance of anesthesia. These are usually administered through an endotracheal tube, which ensures a patent (open) airway, protects the animal from fluids entering the airway, and facilitates ventilatory support. Advantages of inhalation anesthetics include ability to rapidly change the depth of anesthesia, ability to easily ventilate the animal and administer supplemental oxygen, and rapid recovery when the gas is turned off. However, inhalation anesthetics have a low safety margin and thus require vigilant monitoring of heart rate, respiratory rate, blood pressure, and other parameters. Agents commonly used in dogs and cats include isoflurane, halothane, and sevoflurane.

## 16.19.4  EQUIPMENT NEEDED FOR ANESTHESIA

Prior to the induction of anesthesia, an intravenous catheter should be inserted. This provides an access port for the injection of intravenous anesthetics, fluids to maintain blood volume, and emergency drug administration should complications develop. Immediately

---

[11]The term *opioid* refers to compounds related to opium. These include drugs derived from the opium poppy plant and related compounds (natural and synthetic). These compounds have many effects including inhibition of neurotransmitter release, sedation, and analgesia. Side effects include bradycardia, respiratory depression, impaired thermoregulation, and increased sensitivity to noises.

[12]The term *barbiturate* generally refers to drugs that are derivatives of barbituric acid. Barbiturates have been widely used as tranquilizers, sedatives, and anticonvulsant agents. Over 2,000 different barbiturates have been synthesized, although less than a dozen are in common use.

[13]A cataleptic state is characterized by profound analgesia, unresponsiveness to commands, amnesia, open eyes, involuntary limb movements, salivation, normal to increased heart rate, and spontaneous respiration.

**Figure 16.21** Plexiglas chamber connected to an inhalation anesthetic machine for induction of anesthesia. This is useful for anesthetizing fractious cats. *Courtesy Sandra Grable, CVT, University of Illinois.*

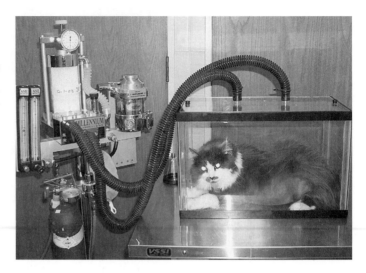

following induction of anesthesia, an endotracheal tube should be inserted to protect the airway, facilitate ventilatory support, and permit subsequent administration of inhalation anesthetic agents. Specialized anesthetic machines are used to deliver inhalation anesthetic agents (see Figure 16.21). These machines have a source of oxygen with a pressure regulator, a flowmeter to control the flow of gases, a vaporizer designed to deliver appropriate concentrations of a specific inhalation agent, a trap (usually a canister containing soda lime) to remove carbon dioxide from exhaled air, and a scavenger system to prevent anesthetic gases from entering the work area.

## 16.19.5 MONITORING OF ANESTHESIA

Vital signs must be monitored throughout the anesthetic period. Irreversible damage occurs within 3 to 5 minutes of inadequate delivery of oxygen to the brain and body tissues. Anesthetic records should include the exact dose of administered anesthetic agents and also list the patient's vital signs at 5-minute intervals. Monitoring the cardiovascular system includes checking the pulse, heart rate and rhythm, capillary refill time, and mucous membrane color. Electrocardiograms (ECG) are helpful in assessing the heart rate and rhythm but should be used in addition to, not instead of, checking the pulse, capillary refill time, and mucous membrane color. (ECGs show the electrical activity of the heart but not its actual functional output.) Pulse oximetry[14] can be used to measure pulse rate and the amount of oxygen being carried by the blood; the percent oxygen saturation of hemoglobin (Hb) should be maintained above 90%. Arterial blood pressure can be measured using oscillometric pulse monitors (similar to blood pressure cuffs used in humans) or with an arterial catheter attached to a manometer or pressure transducer. The respiratory system is monitored by assessing respiratory rates and rhythm plus using a pulse oximeter. Blood gases can be measured to accurately assess the adequacy of ventilation. Ventilatory support is often needed because anesthetic agents depress respiration. This support can be given by manually "bagging" the animal (squeezing extra air into the lungs) or by using a mechanical ventilator. The depth of anesthesia should be carefully monitored. This is assessed by evaluating the position of the eye in the orbit, the presence or absence of reflexes, the degree of muscle relaxation (e.g., jaw tone), and any voluntary movements made by the patient (Table 16.7).

---

[14]A pulse oximeter is a specialized instrument designed to estimate arterial oxyhemoglobin saturation by utilizing selected wavelengths of light. The pulse oximeter consists of a probe attached to the patient's toe, tongue, or ear pinna that is linked to a computerized unit. The unit displays the percentage of Hb saturated with oxygen together with an audible signal for each pulse beat, a calculated heart rate, and in some models, a graphical display of the blood flow past the probe.

**Table 16.7**
Clinical Signs Associated with Depth (Plane) of Anesthesia

| Sign | Too Light | Adequate | Too Deep |
|------|-----------|----------|----------|
| Palpebral reflex (blinking when eyelid is touched) | Present | Slowed | Absent |
| Eyeball position | Centered | Rotated anteromedially | Centered |
| Withdrawal reflex (toe pinch) | Present | Slowed | Absent |
| Anal reflex | Present | Present | Absent |
| Jaw tone | Firm, gagging | Relaxed | Absent |
| Spontaneous movement | Shivering, tense neck/shoulder muscles | Absent | Absent |

# 16.20   LASERS

LASER is an acronym for **l**ight **a**mplification by **s**timulated **e**mission of **r**adiation. Different types of lasers are available, which produce varying wavelengths of light. The lasing media is usually the product the laser is named after. For example, in a $CO_2$ laser, the lasing media is $CO_2$ and it produces a wavelength of 10,600 nm. The energy from the laser light is transferred to tissue where it vaporizes cells.

Lasers cauterize as they cut resulting in less bleeding, less scarring, and more sealing of nerve endings compared to traditional surgery. Lasers are used for removing benign growths (Figure 16.22), spaying, neutering, declawing (Figure 16.23), and other routine procedures. The use of lasers to declaw cats results in less pain and quicker recovery (only 2 days versus approximately a week when claws are surgically removed).

# 16.21   DENTAL EXAMINATION AND CARE

Every pet with teeth requires dental care to maintain optimal health. Home dental care should include the brushing of teeth on a regular basis. Most veterinary clinics and hospitals offer professional cleaning of teeth as part of their preventative healthcare services.

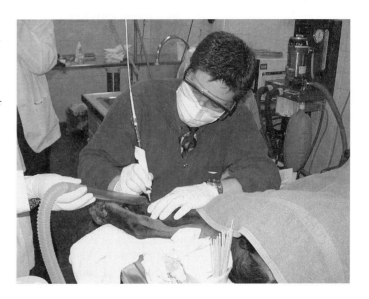

**Figure 16.22** A carbon dioxide ($CO_2$) laser being used to remove a benign granuloma from the rear leg of a dog. *Courtesy Dr. Adam Patterson, University of Illinois.*

**Figure 16.23** A carbon dioxide ($CO_2$) laser being used to declaw a cat. *Courtesy Oklahoma State University.*

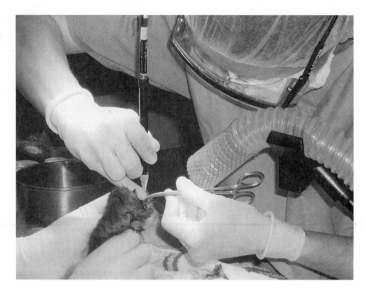

Dental radiographs are used to assess the health of tooth roots and surrounding bones. Animals, like people, develop plaque on their teeth (Chapters 12 and 13). Plaque is a white slippery film composed of food debris, exfoliated cells, and salivary glycoproteins. If allowed to accumulate, plaque becomes mineralized to form dental calculus, which appears as yellowish brown deposits on teeth. Deposits of plaque and calculus support the growth of large numbers of bacteria and other microorganisms that damage the gingival mucosa (gums). This results in gingivitis (inflammation of the gums) and later periodontitis (inflammation of the ligaments and bone around the teeth).

General anesthesia is usually required for thorough teeth cleaning and polishing in dogs and cats. The pet should be positioned with its head slightly lower than its body and a cuffed endotracheal tube in place to prevent fluids from entering the trachea. After the teeth are cleaned, the owners are given instructions on how to brush the teeth at home (using special toothbrushes designed for dogs and cats [see Figure 12.5] or a child's soft-bristled toothbrush and malt- or poultry-flavored toothpastes). Dental rinses containing chlorhexidine or zinc ascorbate are useful in decreasing bacterial growth in the mouth and in promoting the healing of damaged gingiva. Specialists in veterinary dentistry (diplomates of the American Veterinary Dental College, AVDC) offer root canal therapy, orthodontics, and restorative dental services for dogs and cats. Veterinary technicians can become members of the American Society for Veterinary Dental Technicians (ASVDT). The ASVDT was founded in 1994 to promote advances in the knowledge of personnel providing dental health care to animals.

## 16.22 SUMMARY

This chapter highlights common diagnostic and therapeutic procedures. More than 24 centuries ago, Greek physician and father of medicine, Hippocrates (ca 460–377 BC), outlined the principles of modern medicine. His first fundamental tenet was "get the facts," which he further admonished was possible only through accurate diagnoses.

The pioneers of modern medicine and those of related disciplines were farsighted visionaries. They challenged dedicated researchers to push the global frontiers of medical science and technology forward so that humans and their companion animals can benefit from the diagnostic and therapeutic procedures discussed in this chapter—scientific advancements that have aided immeasurably in improving health and quality of life.

There are more secrets of nature to be discovered, more diagnostic and therapeutic procedures to be developed and applied. We must reject the contentment and security associated with status quo. The challenge of changing is great; the responsibility even greater. We encourage students interested in advancing the medical sciences toward new horizons to pursue the challenges and opportunities associated with medical research and technology. The rewards and personal satisfaction of making a positive difference in the lives of people and their companion animals are many and immensely consequential.

## 16.23   REFERENCES

1. Boord, Mona. 2004. Lasers in Dermatology. In *Small Animal Dermatology Secrets,* edited by Karen L. Campbell, 85–90. Philadelphia: Hanley & Belfus.

2. Campbell, John R. 1998. Technology Transfer Initiative. In *Reclaiming a Lost Heritage . . . Land-Grant and Other Higher Education Initiatives for the Twenty-First Century,* 80. East Lansing: Michigan State University Press.

3. Farrow, Charles S. 2003. *Veterinary Diagnostic Imaging: The Dog and Cat.* St. Louis, MO: Mosby.

4. McCurnin, Dennis M., and Joanna M. Bassert. 2002. *Clinical Textbook for Veterinary Technicians.* 5th ed. Philadelphia: W. B. Saunders.

5. McCurnin, Dennis M., and Ellen M. Poffenbarger. 1991. *Small Animal Physical Diagnosis and Clinical Procedures.* Philadelphia: W. B. Saunders.

6. Muir, William III. 2000. *Handbook of Veterinary Anesthesia.* 3rd ed. St. Louis, MO: Mosby.

7. Willard, Michael D., and Grant H. Turnwald. 1999. *Small Animal Clinical Diagnosis by Laboratory Methods.* 3rd ed. Philadelphia: W. B. Saunders.

## 16.24   ABOUT THE AUTHOR

The authors are grateful to Cynthia Coursen Alexander, DVM, for authoring the imaging sections of this chapter. Dr. Alexander graduated with honors from Michigan State University (MSU). She owned and operated Equine Veterinary Service in Potterville, Michigan, for 18 years. While in practice, Dr. Alexander had an adjunct appointment with MSU teaching clinical skills to senior veterinary students and assisting in the instruction of equine anatomy. Dr. Alexander has held many leadership roles in organized veterinary medicine at both state and national levels. In 2000, Dr. Alexander joined the faculty of Oklahoma State University-Oklahoma City. She currently teaches radiology, pharmacology, pathology, and other clinical courses to veterinary technician students.

# 17 PET SITTING, PET MOTELS, AND OTHER BOARDING ARRANGEMENTS

*We can judge the heart of a man by his treatment of animals; he who is cruel to animals becomes hard also in his dealings with people . . . tender feelings toward animals develop humane feelings toward mankind.*

Immanuel Kant (1724–1804)
*German Philosopher*

## 17.1   INTRODUCTION/OVERVIEW

At some point, virtually all owners of dogs and cats must leave their favorite companion animal(s) behind to travel for work, vacation, or spend time caring for a friend or relative. Leaving pets behind can be stressful for both owners and their companion animals. Fortunately, many excellent alternatives are available for providing care for animals while their owners are away. Options include pet sitters, boarding facilities, dog walkers, day care centers, and camps for dogs and cats. This chapter discusses these various options and also some business aspects to be considered in establishing a boarding facility or pet sitting enterprise.

## 17.2   PET SITTERS

Most pets seem to be happiest in their home environment with its familiar sights, smells, and sounds. Staying home means there is no stress from being transported to an unfamiliar kennel or cattery where the animal may be housed adjacent to a disruptive barking or howling dog or a hissing cat. Animals staying at home have lower risks of exposure to ectoparasites and infectious diseases compared with those boarded at kennels or catteries. The pet's diet does not need to change, and depending on arrangements with the caregiver, the pet's normal exercise routine may be available. Pet sitters vary from a neighbor or friend that spends a few minutes each day feeding the pet to a licensed veterinary technician specializing in pet sitting—a person who can monitor glucose levels and make adjustments in diet and insulin dosages for diabetic pets.

Arranging for a friend or neighbor to care for pets during the owner's absence is commonly the least expensive option; however, this can be viewed as an imposition and unfortunately some people prove to be unreliable. A professional pet sitter generally provides excellent care for pets and home while the owner is away.

### 17.2.1   SELECTING A PET SITTER

Opening your home to a stranger and entrusting that person with the care of your pet(s) and possessions is a decision warranting careful consideration. Professional pet sitters should be bonded and insured; ask to see copies of these documents. References should be called and questioned about the reliability of the pet sitter and to learn how happy other clients have been with the care provided to pets and also how any problems were handled. A personal interview should be conducted. The owner should discuss expectations for pet care (feeding and exercise routines, administering medications as needed, how much time will be spent each day interacting with the animals) and home (watering plants, bringing in mail, and crime deterrent measures such as using light and radio timers).

It is important to observe interactions between the prospective sitter and household pets and to determine the competency of the sitter to handle the animal(s) properly. Ask about the sitter's backup plan in case of personal illness or transportation difficulties (vehicle failure or weather related). Discuss expectations for communication during travels—daily reports or only if difficulties arise? Consider who the sitter should contact and who will be authorized to make decisions when the owner is not available. Ask if the sitter feels competent to administer medications. Will the sitter provide or arrange for transportation to a veterinary clinic if the pet(s) become ill during absence of the owner? Does the sitter have a contract that describes services and fees? Other questions may be pertinent to the pet owner's unique situations; the owner should feel comfortably positive about the prospective pet sitter before final arrangements are made.

## 17.2.2 FINAL PREPARATIONS BEFORE LEAVING PETS TO THE CARE OF A PET SITTER

Several important steps should be taken before leaving your pet in the care of another person: (1) Provide the pet sitter with detailed instructions on when and how much to feed. Be certain adequate food (and litter for cats) is available. (2) Outline the play and exercise routines to be followed while you are away. (3) Be sure any outside gates are secured and leave toys and leashes for the sitter to use. Tell the sitter about any hiding places the pet likes to use to help locate a "missing" pet. (4) Leave pictures and descriptions of each pet so the sitter can easily identify them. (5) Provide vaccination and health records as well as information about any chronic or recurrent health problems (e.g., has the animal shown signs of separation anxiety in the past; does the pet frequently vomit or have diarrhea?). (6) Provide instructions for any medications to be given and ensure that the sitter is willing to give these. (7) Provide a travel itinerary and instructions on when and how to contact you for reports or to discuss any problems that may arise. (8) Let the sitter know if a neighbor has extra keys available. (9) Provide timers for lights and a radio or television (if you want these left on). (10) Leave instructions for bringing in mail. (11) Leave instructions for watering plants. (12) Make arrangements to have grass mowed or snow shoveled if either of these is needed while you are away. (13) Provide a list of emergency numbers (e.g., veterinarian, insurance agent, plumber, electrician, and other people to contact) if you cannot be reached. (14) Have cleaning supplies available (mop, bucket, broom, dustpan, vacuum cleaner, trash bags, sponges, rags) and provide instructions on what to do with trash disposal.

## 17.2.3 RESOURCES FOR THOSE WANTING TO START PET SITTING BUSINESSES

Tips can be obtained from others who have been successful in a pet sitting business. A mentor who has experienced the joys and frustrations of starting a new business can provide valuable assistance to others. Several pet sitting organizations provide support to pet sitters through training, certification, marketing, and insurance programs.

The National Association of Professional Pet Sitters (NAPPS; www.petsitters.org) was founded in 1989. The primary goals of NAPPS are to promote excellence among pet sitters and to foster the growth of this industry. NAPPS became a nonprofit membership organization in 1993. Its members must sign a Pledge of Professional Conduct. The organization has a certification program that includes courses on care of animals and establishing and running a business followed by a written examination. Certification by the NAPPS shows that members have met objective criteria of knowledge related to animal care and the business of pet sitting. Benefits of joining the NAPPS include access to a network of other pet sitters, help designing Web sites (www.epetnet.com/napps/hosting.htm), discount registration for the NAPPS Annual Conference, client referrals through the NAPPS nationwide referral service (1–800–296-PETS), informational newsletters and press releases, mediation services to resolve possible complaints from clients, access to NAPPS's credit card processing program (appreciated by clients who prefer paying for services using a credit card), and several insurance programs (liability, bonding, health, dental, life, and disability insurance options).

Pet Sitters International (PSI; www.petsit.com) is another membership association providing support and services to pet sitters. It was founded in 1994 by Patti Moran to promote excellence in pet sitting. Ms. Moran authored the book *Pet Sitting for Profit* (first published in 1987, Howell Book House, Hungry Minds, Inc.) and was the owner of Crazy 'Bout Critters, a pet sitting business in Winston-Salem, North Carolina. In its first decade, PSI grew to over 5,000 members. PSI publishes *The WORLD of Professional Pet Sitting,* a magazine devoted exclusively to professional pet sitters. Through the work of PSI, the first week of March has been designated National Professional Pet Sitters Week. PSI has created four levels of accreditation programs utilizing home study courses: (1) Pet Sitting Service, (2) Pet Sitting Technician, (3) Advanced Pet Sitting Technician, and (4) Master Pet Sitting Professional. Official logos are available for use by those who pass each level of PSI accreditation. Membership benefits include support staff to answer questions about pet sitting, publicity, Web locator listing services, links to group rates for insurance and bonding service providers, the bimonthly *The WORLD of Professional Pet Sitting* magazine, reduced registration fees for PSI's annual Quest for Excellence Conference, and access to legal services at a reduced rate.

Pet Sitters Associates, LLC of Eau Claire, Wisconsin, provides general liability insurance for pet sitting businesses as part of the annual membership dues. It also provides a code of ethics and quarterly newsletters with tips for pet sitters (www.petsitllc.com). The American Pet Sitters Association (TAPSA; www.sitmypet.com) is another organization working to promote and recognize excellence in pet sitting in the United States. It also offers members access to liability and bonding insurance and publishes a bimonthly magazine for pet sitters and dog walkers. It maintains a Web site and member forum, sample brochures and flyers, and an accreditation program to promote excellence among pet sitters.

Vet Tech Pet Sitters Association is an organization for professional veterinary technicians who want to be recognized as professionals in the pet sitting industry (www. vettechpetsit.com). Membership is restricted to licensed veterinary technicians. This organization provides veterinary technicians a way of networking with peers and promoting excellence in pet care and pet sitting. Many pet owners appreciate having a pet sitter with the qualifications of a veterinary technician who can provide nursing care for animals needing extra special care (e.g., neonates, geriatric pets, diabetic pets, and others requiring daily medications).

### 17.2.4    INSURANCE AND BONDING

In today's litigious-prone society, pet sitters are understandably concerned about liability claims should anything happen to the pet or house in the owner's absence. Even if a claim is proven to be without merit, the costs of a defense lawyer can be substantial. If a pet sitter is judged to be negligent, the award may deplete all personal and business assets; therefore, liability protection insurance is recommended for those in the pet sitting business. Because pet sitters have access to client's homes, bonding is also important to provide clients with assurance they will be repaid for any damage or loss of property. A bonding company guarantees that homeowners will be compensated for any losses. If a claim is made, the pet sitter is required to repay the bonding company for any payment made to the homeowner.

## 17.3    PET MOTELS AND OTHER PET BOARDING FACILITIES

Boarding facilities vary from those providing only the basics of clean shelter and food to those offering suites (Figure 17.1) and extensive pampering of pets in their care. Pet owners can select a facility that meets their individual expectations. Some modern pet motels have installed speaker phones to enable owners to communicate with their pets during boarding; this service has been termed "affections connections." Locating a facility is usually easy—boarding kennels advertise in yellow pages, newspapers, pet magazines, flyers, Web sites, veterinary offices, kennel clubs, Welcome Wagons, pet shops, and city councils. Many obtain referrals from satisfied customers. According to the American Boarding Kennel Association (ABKA, www.abka.com), each year over 30 million pet owners utilize the services of a professional boarding kennel. A 2002 industry survey by the ABKA reported the existence of more than 9,000 boarding kennels in the United States and Canada. Boarding kennels are designed specifically to care for pets; many provide add-on services including day care, pet camps, training classes, grooming, and the sale of pet foods, toys, beds, and other supplies.

The basics include a secure pen that is clean, safe, well-ventilated, and an adequate space for the size of the pet. More luxurious facilities provide plush beds and entertainment for the animals (Figure 17.2). The kennel rooms should smell clean. Many pet boarding facilities have optional services including walks, runs, or individual play sessions for pets. Other amenities offered may include grooming sessions, training, bathing, and pet massages.

**Figure 17.1** Two dogs from the same household relaxing on the bed in their luxury suite at the Kennelwood Pet Suites. *Courtesy Kennelwood Village, St. Louis, MO (www.kennelwood.com).*

Although no one can replace the special bond a pet has with its owner, the structure of a boarding facility may occupy the pet's time and help alleviate the anxiety experienced when left unattended at home. Many pet owners prefer boarding kennels, where pets are supervised most of the day, rather than pet sitters who may spend only a few minutes a couple of times a day with the pet.

Dogs and cats requiring specialized care such as diabetic dogs and cats may be best cared for by a medical professional such as a veterinarian or veterinary technician. Many veterinary hospitals offer boarding services for clients. Those utilizing a veterinary hospital for boarding have the assurance that their pet's medical care is being closely supervised. The disadvantages of boarding at a veterinary clinic include a slightly increased risk of exposure to infectious diseases and typically a more austere and sterile environment than in a specialized facility where boarding is the primary business.

The ABKA is a nonprofit international trade association for businesses providing boarding, day care, grooming, training, shipping, and other services for dogs and cats. It was established in 1977 to promote professional standards within the pet boarding industry. The ABKA Voluntary Facilities Accreditation (VFA) Program contains over 200 written standards of operation for boarding kennels. These include standards for the grounds, office and reception areas, record keeping, business practices, personnel, work areas, kennel areas, ani-

**Figure 17.2** Overhead television screens above each pet suite provide entertainment for pets staying at Kennelwood Pet Suites. *Courtesy Kennelwood Village, St. Louis, MO (www.kennelwood.com).*

mal care procedures, environmental control, sanitation, trash and sewage disposal, pest control, fire safety, boarding of animals other than cats and dogs, grooming rooms, kennel vehicles, and community playtime. Kennels meeting ABKA standards are awarded the Blue VFA Ribbon representing their state-of-the-art certification in pet care.

The ABKA has developed training programs for pet care personnel. The first such program is the Pet Care Technician Program. It teaches basic principles of animal care, kennel management, and customer relations. The next level is the Advanced Pet Care Technician Program that provides advanced training in these areas. Advancement to this level requires successful completion of a closed-book examination administered at an approved testing site (ABKA meeting or community college). The third level is the Certified Kennel Operator Program, which requires demonstration of proficiency in kennel management and successful completion of oral and written examinations.

An important role of the ABKA is in providing members with educational resources. It publishes a monthly online newsletter and sponsors regional meetings plus an annual convention. The ABKA Web site has a listing of member kennels sorted by city and state.

## 17.4  SPECIAL CONSIDERATIONS FOR BOARDING CATS

It may be preferable to board cats at a specialty boarding cattery where they will not be disturbed by barking dogs. However, many boarding facilities have separate areas for dogs and cats and this is also acceptable.

The Feline Advisory Bureau (FAB) has established standards of excellence for boarding catteries. This association inspects both the construction and management of a boarding facility. A facility that has passed inspection will display a certificate and/or sign denoting it as a FAB Listed Boarding Cattery. The FAB Web site lists currently approved facilities (www.fabcats.org).

Cats are particularly susceptible to airborne bacteria and viruses, thus "sneeze" barriers should be installed between cats. FAB prefers catteries with outdoor runs; these ensure good ventilation and airflow to help prevent the spread of disease (Chapter 18). However, outdoor runs may be prohibited by local ordinances or impractical in inclement weather, so many facilities offer only indoor housing (Figure 17.3). It is preferable to find a facility that provides each cat with a separate enclosed sleeping area and exercise run with a scratching post, an interesting view, a shelf for resting, and plenty of toys. There should be full height barriers or gaps between runs to prevent direct contact between cats from different households. Group runs and exercise pens should be avoided due to the potential for spread of infectious diseases. Also try to avoid facilities that keep cats in cages; cats need room to play and exercise

**Figure 17.3** Cat condos at the Willow Creek Pet Center, Sandy, UT, have a clear acrylic back facing a window. Cats can climb, bask in the sun, or retreat to the shady area for sleeping. *Courtesy Dr. Rick Campbell.*

**Figure 17.4** Play room for cats at Dad's Creekside Kennels. *Courtesy Dr. James E. Corbin.*

(Figure 17.4). Cats should be transported in a secure carrier to ensure safety traveling to and from the boarding facility. Before selecting the best-fit boarding arrangement, visit the alternatives available.

## 17.5   PREPARING PETS FOR BOARDING

When the decision is made to board the pet, owners should look for a clean, escape-proof, safe, and reliable boarding kennel/cattery properly supervised by experienced, responsible, caring individuals. Owners should visit several facilities and then select the one that best meets their expectations for the level of care they want and can afford. Reservations should be made as soon as travel plans are known. This is particularly important during holidays when the best boarding facilities may be booked months in advance. Because the boarding of dogs and cats (and other pets) is a business, owners should expect to sign a boarding agreement/contract that clearly states rights and responsibilities. Check the hours of operation for leaving and picking up pets.

Owners should be certain all immunizations are up to date prior to boarding their pet. The boarding facility should require proof of these immunizations. Also be sure the pet is free of internal and external parasites. This is for your pet's well-being as well as to protect the health of other animals.

Avoid an emotional farewell; pets are influenced by the emotions of their owners and may become unnecessarily upset and anxious if they sense their owner is distressed. Some dogs and cats may not like the experiences associated with boarding, but certain things tend to help prepare them for such experiences. Dogs/cats that have been well socialized and those that have received obedience training and have been exposed to new situations usually handle boarding better. Apprehensive pets may adapt to longer stays more readily if they are day boarded and/or boarded for a short overnight stay a few times before being left for an extended trip. This will help accustom the pet to the boarding kennel and should decrease its level of anxiety.

Inquire about the kennel's medication policy. If your pet requires medication, be sure you have an adequate supply and the kennel has authorization from you to get a refill should your return be delayed or a mishap occur with the container. Also inquire about the feeding

**Figure 17.5**  Cat napping with a favorite toy. *Courtesy Paul E. Miller.*

policy; some kennels feed a standard food to all boarding animals, others have a variety of foods to offer, whereas others ask owners to bring any special diets with the pets.

Many kennels will allow you to bring a favorite toy or blanket (Figure 17.5); however, some may prohibit this if the animal guards its possessions, which could be hazardous to workers wanting to remove the objects for cleaning. Advise the kennel if a pet has any phobias such as fear of thunder or loud noises. Always provide the kennel with several emergency contact numbers (veterinarian, trusted friend or relative, cell phone or other telephone where you can be reached).

## 17.6   CAT/DOG DAY CARE CENTERS AND CAMPS

Day care pet centers have been established in many areas to provide useful services when companion animals fail to adapt to being left alone in an empty place of residence while family members are at work, school, or other areas outside the home. Dogs suffering from separation anxiety (Chapter 20) often become house wreckers, bark excessively, and/or urinate and defecate inside the house. Although many can be "retrained" using a combination of behavioral modification therapy and medications, others may respond best to involvement in a day care or camp for pets.

Activities at day care centers vary from simple day boarding (a kennel or cage with a blanket and a toy or bone to chew) to elaborate activities. At many facilities, dogs are provided with opportunities to interact with other dogs and people in play activities. Dogs of similar sizes and temperaments may be turned out to romp in exercise yards for group play (Figure 17.6). Balls and other play equipment may be used to encourage active exercising. Individual attention may include providing animals with "lap sitting" time and petting. Many centers also offer grooming services, training, and overnight boarding.

## 17.7   STARTING A DAY CARE/BOARDING KENNEL OR CATTERY

Owning and operating a boarding facility can be a fun and rewarding business for those who enjoy spending time with dogs and cats. Caring for animals belonging to other people is an important responsibility and requires hard work. Those wishing to start a pet boarding or day care business should do as much research as possible, talk with others in the business, and visit as many facilities as possible to decide what arrangements would work best for them. Areas to consider include (1) regulations and licensing (check applicable laws in your state, county, and city); (2) location (zoning permits may be required); (3) building plans (local codes may require county or city approval; be sure to include waste handling plans and isolation facilities); (4) lifestyle requirements—are you prepared to handle numerous calls and inquiries about the companion animals you are caring for? (living on site is advisable for

**Figure 17.6** Group play time at Kennelwood Pet Suites' Doggie Day Camp. *Courtesy Kennelwood Village, St. Louis, MO (www. kennelwood.com).*

security and monitoring of the health of the animals; the busiest times of the year will be holidays and summers, meaning you should expect to be working during those times); (5) a business plan (what type of business do you envision? what services will you provide? will you be able to make a profit? how many employees will you need? who will train the employees?); (6) accreditation by the ABKA or CFA; and (7) insurance (be sure you include this expense in your business plan; Chapter 19).

Managing a pet boarding or day care facility can be rewarding and enjoyable for those who derive personal pleasure from interacting with pets and their owners. It is also an enormous challenge so prospective entrepreneurs should consider all aspects involved in starting a business before embarking on the journey (Chapter 19).

## 17.8   SUMMARY

Pet ownership is a personally rewarding experience. It also carries many important responsibilities including that of making satisfactory provisions for the proper care of pets when an owner is away. Several alternatives are available, including qualified pet sitters, pet motels, and other boarding arrangements. Merits and limitations of these respective alternatives were discussed in this chapter. Caring owners know their pets well and, within their financial means, should select the best-fit care arrangement for their animal(s).

Now let us move to the next two chapters, which address kennel/cattery design and management (Chapter 18) and the business/financial aspects of the companion animal enterprise (Chapter 19).

## 17.9   REFERENCES

1. Bartos, Bob. 1972. *Dogs, Kennels, and Profits.* Los Angeles: Friskies Pet Products.
2. Krack, James. 1990. *Building/Buying and Operating a Boarding Kennel.* 3rd ed. Colorado Springs: American Boarding Kennel Association.

# 18 KENNEL/CATTERY DESIGN AND MANAGEMENT

*The white man must treat the beasts of this land as his brother. What is man without beasts? If all the beasts were gone, man would die of great loneliness of spirit, for whatever happens to beasts also happens to man. All things are connected.*

Chief Seathl (Seattle) (1786–1866)
*Puget Sound Suquamish chief[1]*

---

[1]From speech given at the 1855 signing of the Port Madison Treaty which settled the Suquamish on their reservation across the Sound from Seattle. The city of Seattle bears his name.

## 18.1   INTRODUCTION/OVERVIEW

Kennels and catteries are built for many purposes—personal pets, breeding, training (hunting, protection, and other types of specialized training), boarding, pet day care, animal shelters, animal rescue facilities, and veterinary hospitals. Although modifications may be desirable for some of these purposes (e.g., luxury suites to pamper pets whose owners are willing to pay for extra amenities when boarding, facilities for specialized training equipment, reproductive laboratories, grooming stations, pet supplies sales room, and other special use areas), basic guidelines can be helpful as a starting point for designing facilities that will house dogs and/or cats.

Although kennels and catteries are designed and built primarily for the comfort, health, and safety of animals, these buildings should also be aesthetically pleasing and easy to maintain and keep clean.

## 18.2   EARLY CONSIDERATIONS AND ACTIONS

If the decision is made to build a kennel/cattery and/or pet motel, the first step is to obtain a planning permit. Once it is secured, purchasing land for a building site, holding public hearings, achieving environmental impact approval, and obtaining a license to raise/board animals follow (Chapter 19).

The permit/license obtained indicates the number of animals that may be kept at any given time and must be prominently displayed. Additionally, it specifies details related to food storage, heating, disposal of excreta, fire precautions, maintenance of records, security, and washing/lavatory facilities for employees.

Many counties and states have animal welfare standards and statutes governing the construction of kennels and catteries. Local societies for the protection and care of animals (SPCAs), humane societies, and animal shelters should be contacted for information pertaining to applicable regulations.

Minimum standards include (1) adequate ventilation and temperature regulation, (2) shade from sun and shelter from inclement weather, (3) protection from injury, (4) veterinary treatment when required, (5) isolation facilities, (6) adequate size for the breed/species, (7) a structurally sound building, and (8) good sanitation.

## 18.3   KENNEL/CATTERY DESIGN

Basic considerations in designing kennels and catteries are provisions for safe, clean, escape-proof, properly heated and cooled housing, as well as for food and water. A southern exposure is preferred in the Northern Hemisphere. This enables cats and dogs to enjoy sunlight and the warmth it provides much of the year.

Cats and dogs from different places of residence should not be penned together, share the same air space, or be permitted to come in contact with each other (Figure 18.1).

Adequate exercise and sleeping space must be provided. Each cat weighing over 8.8 lb should have a minimum of 4.0 sq ft of floor space (exclusive of food and water bowls). The minimum amount of floor space per dog is calculated by the United States Department of Agriculture (USDA) Animal Welfare Act as follows: find the mathematical square of the sum of the length of the dog in inches measured from the tip of its nose to the base of its tail plus 6 inches; then divide the product by 144. For example, if the length of the dog from nose to base of tail is 24 inches, the calculation will be $30 \times 30/144 = 6.25$ sq ft of floor space. The minimum height for a primary enclosure should be at least 6 inches higher than the head of the tallest dog in the enclosure when it is in a normal standing position. Dogs should also have access to a covered outdoor run (Figure 18.2) or provided with regular opportunity for exercise. Bitches with litters should not be housed in the same primary enclosure with other adult

**Figure 18.1** These cat condos provide secure, relatively spacious, and attractive housing for the boarding of cats at Kennelwood Pet Suites. *Courtesy Kennelwood Village, St. Louis, MO (www.kennelwood.com).*

animals. Puppies under 4 months of age should not be housed with adult dogs other than their dam. Bitches in heat should not be housed with sexually mature dogs except for breeding. A maximum of 12 compatible adult dogs may be housed in the same primary enclosure.

The kennel/cattery should have a hallway (minimum 4 ft wide, 6 ft is preferable) from which there is access to the pens and runs (Figure 18.3). Many kennels are designed with 4 × 4 ft inside pens connected to 4 × 16 ft outside runs. Pens that are 4 × 6 ft or 6 × 6 ft are more desirable, especially for medium-sized and large-breed dogs. The outside runs for large dogs should be 6 × 20 to 30 ft. Six-foot-wide runs are easier to keep clean as dogs have room to avoid scattering their droppings while running. The entire facility should have a pet-proof exterior guard fence. This prevents outside dogs and cats from transmitting diseases or fighting with kennel/cattery animals and, additionally, prevents escape of animals if a door to a pen or run is accidentally opened or left open. The building should be covered with a waterproof roof that drains away from the facility. Runs should be separated by curbs that prevent urine and feces from passing between pens.

Cats should be provided with a resting surface that is elevated, impervious to moisture, and easily cleaned and sanitized (Figure 18.4). A queen nursing kittens should be provided with additional floor space based on her size, breed, and behavior. Kittens under 4 months of age should not be housed in the same primary enclosure with adult cats other than the

**Figure 18.2** Outdoor runs provide dogs with a secure area for exercise and fresh air. *Courtesy Kennelwood Pet Suites, Kennelwood Village, St. Louis, MO (www.kennelwood.com).*

**Figure 18.3**   Indoor runs with 4-ft hallway in the Willow Creek Pet Center, Sandy, UT. *Courtesy Dr. Rick Campbell.*

dam or foster dam. Cats with a vicious or aggressive disposition must be housed individually. The maximum number of adult cats being housed together should not exceed 12. Intact (unneutered) females should not be housed with sexually mature males except for breeding. Cat pens and runs should be separated by full height "sneeze barriers" to prevent transmission of diseases in respiratory secretions or urine spraying.

Catteries are classified as outdoor (having outdoor runs for cats) or indoor only. Most cats enjoy having an outdoor run. Access to the outdoors helps ensure good ventilation and airflow to help prevent the spread of disease. It is ideal for each cat to have an exercise run attached to an enclosed sleeping area. The sleeping area may be a penthouse (a raised box off the ground accessed by a ramp or ladder) or a full-height house attached to the run. Cats appreciate having a raised, solid shelf for use in resting and sunning.

Wooden buildings should be protected with a wood preservative that does not contain cresol (a derivative of coal tar that is toxic to cats). Because wood absorbs more urine and urine odors and adds to insurance rates and cost of maintenance, concrete or concrete block construction is preferable. Although the initial cost of wood may be less, wood is less durable and more costly to maintain.

**Figure 18.4**   These cat cages include elevated shelves for the cats to rest on. *Courtesy Dr. James E. Corbin.*

All animals admitted must have proper certification of required inoculations. Each cattery and kennel should have an isolation unit located away from the main building complex. Prospective proprietors must ensure through proper design and management that:

- The animals will at all times be kept in accommodations suitable with regard to construction, size and safety of quarters, number of animals to be housed, exercising facilities, temperatures, lighting, ventilation (avoiding cross drafts is highly desirable), sanitation, and overall cleanliness.

- The animals will be supplied with appropriate food, clean/fresh water, and bedding materials.

- All reasonable precautions will be taken to prevent and control the spread of infectious or contagious diseases among animals; this includes provisions for adequate isolation facilities.

- Appropriate provisions are made to protect animals in the event of fire or other emergencies.

- A registration/admissions log includes a description of all animals received, date of their arrival and departure, and name/address/telephone number of the owner(s).

- Food and bedding is stored in a manner that protects these supplies from spoilage, contamination, and infestation with mice, rats, and insects. Food should be stored off the floor and away from walls to allow cleaning underneath and around the containers. Foods requiring refrigeration must be stored accordingly. All open bags of food must be stored in leak-proof containers with tightly fitting lids to prevent contamination, spoilage, and vermin infestation. Only food and bedding currently being used should be stored in rooms containing animals.

- Outdoor shelters are sufficient to protect dogs, cats, and other pets from the elements at all times. The structure must be large enough to allow each animal to sit, stand, and lie in a normal manner and to turn about freely.

## 18.4   SANITATION

Kennels are required to have regular collection and removal of animal excrement, used litter and bedding, waste food, and other garbage and debris. The drainage system must be installed in accordance with local regulations and be properly maintained. One-way valves and airlock traps must prevent the backflow of gases and sewage into the kennel. Disposal and drainage systems should be designed to minimize odors, insects, pests, vermin, and disease hazards. Arrangements must be made for storage and disposal of any animals that die; these cannot be stored in refrigerators or freezers used for food.

The use of impervious materials is ideal for the walls, ceilings, and floors—these surfaces do not absorb fluids and can be thoroughly and repeatedly cleaned and disinfected. Impervious materials and surfaces do not retain odors, and fluids bead on them to be easily removed with pressure spraying or mopping/wiping. Examples of impervious materials include ceramic tile, vinyl, linoleum, and epoxy-coated or sealed concrete.

Outside floor areas in contact with animals may include compacted earth, absorbent bedding, sand, gravel, concrete, or grass. Concrete and gravel are preferred because they are easiest to keep clean. Concrete should be sealed to prevent its harboring parasite eggs and to decrease abrasion to skin and feet. Asphalt flooring is not recommended because it absorbs heat and becomes very hot on sunny days. Moreover, it is difficult to seal and thus harbors parasite eggs and can weep phenolic compounds that cause skin irritation to pets. Runs should have a slope of approximately one-half inch per foot to encourage rapid drainage of rain, melting snow, and water used for cleaning. It can be difficult to maintain slopes in gravel pens because dogs move the gravel around when running and digging. Likewise, earth runs encourage digging and become muddy in wet weather.

All litter and waste should be properly disposed of or incinerated daily. If wire or slatted flooring allows wastes to drop beneath the primary animal enclosures, the area under these floors must be cleaned regularly to prevent accumulations of feces and food wastes and to minimize pests, disease hazards, insects, and objectionable odors. Inside pens may be sloped to drain to a catch basin along the inner hallway or to outside pens. Many kennel designs incorporate both drainage systems with the highest interior floor level in the middle of the pen, thereby ensuring that the area next to the outside door drains to the outside and the inner portion drains to the hallway. Outside runs are usually 1.5 to 2 inches below the inside pen level.

Washing areas including basins, sinks, tubs or showers, and toilet facilities should be readily accessible to kennel workers. Sanitation of primary enclosures and food and water receptacles can be accomplished using (1) pressurized steam, (2) hot water (180°F or higher) with soap or detergent, or (3) detergents and disinfectants followed by rinsing with hot water.

## 18.5 HEATING, COOLING, LIGHTING, AND VENTILATION

USDA Animal Welfare standards state that indoor housing facilities for dogs and cats must be sufficiently heated and cooled to protect them from temperature extremes that would endanger their health and well-being. Ambient temperatures should not fall below 45°F or above 85°F for more than 4 consecutive hours when dogs or cats are present. Dry bedding, solid resting boards, or other methods of conserving body heat must be provided when temperatures are below 50°F. Many kennels utilize floor heating with pipes or heat cables imbedded in the floors. Ceilings should be insulated to moderate temperature fluctuations and also prevent moisture condensation.

Lighting should be sufficient to permit routine inspection of cleanliness of the facility and observation of animals. Animal areas should be lighted in a diurnal cycle using either natural or artificial light.

Indoor housing facilities for dogs and cats should be ventilated in a manner that minimizes odors, drafts, ammonia levels, and moisture condensation. Ventilation can be provided by windows, vents, fans, or air conditioning. Relative humidity should be maintained between 30% and 70%. Auxiliary ventilation such as fans, blowers, or air conditioning should be provided when ambient temperature is 85°F and above. Air filters should be replaced or cleaned and sanitized weekly. Isolation rooms must have a separate ventilation system to prevent spread of disease within the kennel or cattery.

When possible, kennels should be located perpendicular to prevailing winds. This will facilitate cross-ventilation and help keep the buildings cool during hot summer months.

## 18.6 FEEDING, PEST CONTROL, AND SOCIALIZATION

A regular routine of cleaning, feeding, interacting with people, and grooming should be established and followed. Regular meal times should be adhered to and individual animal diets provided. Adult cats and dogs should be fed twice daily; kittens/puppies and geriatric cats and dogs are fed as needed. The foods should be of sufficient quantity and nutritive value to maintain normal condition and body weight of each animal. The diet should be appropriate for the individual animal's age (Chapter 9). Feeding pans should be disposable or made of durable materials that can be easily cleaned and sanitized. If self-feeders are used, they must be cleaned and sanitized regularly. Details of cat and dog diets and special feeding instructions should be provided by the owner(s) and entered on the animal's record card. Provide fresh water daily to every cat and dog. A pest control program is an important aspect of kennel management (Chapter 15). This may involve administering a flea preventive to all animals upon arrival at the kennel with retreatment monthly or as needed. Facilities that had infestations with ticks or mites will need to utilize an acaricide as part of the pest control pro-

**Figure 18.5** Animals being boarded in the Kennelwood Pet Suites benefit from daily handling by a caring person. *Courtesy Kennelwood Village, St. Louis, MO (www. kennelwood.com).*

gram. Infectious diseases such as dermatophytosis (ringworm) can spread rapidly within a kennel or cattery; thus any dog or cat with skin lesions will need to be isolated and examined by a veterinarian.

Animals that are housed individually should be provided daily positive contact with a person (Figure 18.5). Provisions should be made on the animal's admission card with regard to any medications to administer as well as the name of veterinarian(s) to call in case of a health problem or emergency.

# 18.7 ANCILLARY SERVICES

Many boarding kennels provide grooming services and bathe animals prior to their owner's return. Animals that have been boarded for several days will absorb odors from the kennel and thus will smell fresher if given a bath (Figure 18.6). Grooming and bathing provide opportunities to inspect animals and maintain surveillance of any problems with skin infections or ectoparasites in the kennel or cattery. A separate room containing tubs, cage dryers, and grooming tables will facilitate efficient grooming without disturbing other animals in

**Figure 18.6** Dog grooming facilities of the Willow Creek Pet Center, Sandy, UT. (*a*) Elevated tubs of varying sizes are used for bathing small- and medium-sized dogs. This permits safe handling and reduces back strain for the groomers. (*b*) Large- and giant-sized dogs are easily bathed using walk-in showers with flexible spray nozzles. *Courtesy Dr. Rick Campbell.*

(a)                                                          (b)

**Figure 18.7** Grooming room at Kennelwood Pet Suites. *Courtesy Kennelwood Village, St. Louis, MO (www. kennelwood.com).*

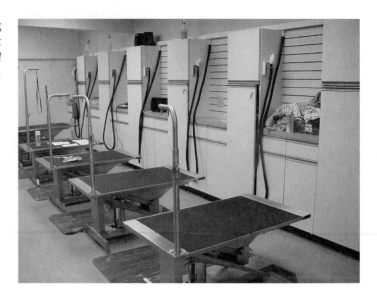

the facility (Figure 18.7). Care must be taken when using cage dryers to avoid overheating the animals.

Many boarding kennels also offer sales of specialty pet foods, toys, and supplies. Owners appreciate the convenience of purchasing a toy or special treat for their pet(s) when leaving them for boarding and also when picking them up (Figure 18.8). When designing plans for a boarding kennel or cattery, space should be allocated for a small shop in the admissions and discharge area (Chapter 19).

Additionally, many boarding kennels include training services as part of their business (Chapter 12). Training may vary from basic dog training (Figure 18.9) and socialization classes to specialized training for field trials, hunting, tracking, and personal or premise protection work. Kennel owners anticipating offering training services should acquire sufficient acreage to accommodate the needs for the type of training being provided.

Kennels and catteries involved in breeding dogs and cats need additional space to accommodate visiting females brought to the kennel/cattery for breeding, a reproductive laboratory (with microscopes for evaluation of sperm motility and vaginal cytology), delivery rooms with nesting boxes, and nurseries for neonates and their mothers. Breeding and raising dogs and cats requires dedication and devotion of innumerable hours in monitoring health, socializing, training, advertising, and finding good homes for each animal.

**Figure 18.8** Pet toys and supplies offered for sale at the admissions and discharge desk of Kennelwood Pet Suites. *Courtesy Kennelwood Village, St. Louis, MO (www. kennelwood.com).*

**Figure 18.9** Training is included as an option at Kennelwood Pet Suites. *Courtesy Kennelwood Village, St. Louis, MO (www. kennelwood.com).*

# 18.8   CERTIFICATION OF BOARDING KENNELS AND CATTERIES

Designing and managing a boarding kennel or cattery in accordance with standards established by leaders in the pet boarding business assures the public that the facility is capable of providing a high level of care to boarders.

### 18.8.1   AMERICAN BOARDING KENNELS ASSOCIATION (ABKA) VOLUNTARY FACILITIES ACCREDITATION (VFA) PROGRAM

The ABKA VFA Program includes over 200 detailed requirements that must be met for a boarding kennel to receive certification as representing the state of the art in animal care and management. The certification program includes assessments of the following areas: (1) grounds, (2) office and reception areas, (3) record keeping, (4) business practices, (5) personnel, (6) work areas, (7) kennel area, (8) animal care procedures, (9) environmental control, (10) sanitation, (11) timely trash and sewage disposal, (12) pest control, (13) fire safety, (14) boarding of animals other than cats and dogs, (15) grooming room, (16) kennel vehicles, and (17) community playtimes. Kennel owners wishing to obtain certification must submit detailed information about their animal care procedures and have an on-site inspection by trained VFA evaluators. More information about this program can be obtained from the ABKA (www.abka.com).

### 18.8.2   CATTERY CERTIFICATION PROGRAMS

Catteries are multi-cat households or facilities maintained for the purpose of perpetuating and protecting the heritage and desirable traits of one or more breeds of pedigreed cats. The Cat Fanciers' Association (CFA, www.cfainc.org) has developed the CFA Approved Cattery Environment Program (ACEP, www.cfainc.org/cattery-environment.html) with two tiers;

veterinarians evaluate catteries using forms developed by CFA and can award either the CFA Approved Cattery Certificate or the CFA Cattery of Excellence Certificate. The Traditional Cat Association, Inc. (TCA, www.traditionalcats.com/index.html) has also established guidelines for certification of catteries to recognize facilities that maintain the welfare of cats. Inspections are conducted by veterinarians using evaluation forms developed by TCA. Catteries scoring 2.0 and above are awarded the TCA Approved Cattery Certificate. Those scoring greater than or equal to 3.1 receive the TCA Cattery of Excellence Certificate.

## 18.9 EMPLOYEES

Taking care of a kennel or cattery is labor intensive. Workers need skills in handling animals, understanding animal behavior, detecting the signs of illness in animals, understanding customer relation skills, understanding and following directions, and working well with others. Associations such as the ABKA offer seminars and training materials for kennel staff. The ABKA has a three-step staff certification program. The first level is the Pet Care Technician Program that teaches the basic principles of animal care, kennel management, and customer relations. It includes home-study materials and a certification examination. The second level is the Advanced Pet Care Technician Program, which includes more advanced training in pet care and an introduction to kennel management. The third level is the Certified Kennel Operator Program that requires managers of boarding facilities to have a minimum of 3 years experience working in a boarding kennel and a demonstrated proficiency in kennel management.

## 18.10 LEGAL CONSIDERATIONS AND INSURANCE

Regulations and licensure of kennels and catteries varies from region to region. Those contemplating starting these businesses should consult with a knowledgeable local attorney and also with applicable county commissioners and city councils. The USDA regulates commercial dog and cat breeders, kennels, and brokers who sell dogs through wholesale channels and those using animals for biomedical research (Chapter 22). Some states have kennel licensing laws that parallel or exceed the standards set by the USDA Animal Welfare Act. Some states have puppy lemon laws that require refunds to buyers if the puppy they acquire is found to have a congenital disease or serious illness. Cities or neighborhood covenants may have statutes limiting the numbers of dogs or cats that can be owned by a household. Many cities have nuisance laws that require animals to be under control at all times and prohibited from barking continuously (Figure 18.10).

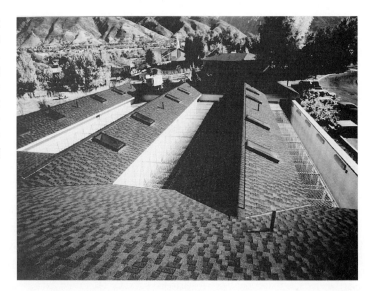

**Figure 18.10** Roof view of indoor/outdoor runs at the Willow Creek Pet Center, Sandy, UT. Three banks of runs house small, medium, and large/giant dogs. Roof height and slope optimize seasonal variations in sunlight for the runs. The entire kennel utilizes radiant heated concrete floors. Positioning the outdoor runs within the complex isolates the dogs from distractions by passing vehicles or people and also decreases the impact of the kennel on the surrounding residential neighborhood. *Courtesy Dr. Rick Campbell.*

Insurance is a necessity for business owners and is also essential for those raising dogs and cats as a hobby. Unfortunately, insurance is becoming increasingly expensive and difficult to obtain. In most states, it is legal for insurance companies to charge higher premiums to pet owners and to even refuse to renew a policy based solely on pet ownership. The AKC provides tips to finding homeowners' insurance on its Web site (www.akc.org/life/homeins/homeowners_inscenter.cfm). The ABKA has an insurance program available to its members (www.abka.com/General_Info.htm).

# 18.11   SUMMARY

In this chapter, we discussed important considerations and alternatives related to planning, designing, constructing, and managing successful service-oriented kennels and catteries. Discussed as well were various environmental aspects including proper heating, cooling, lighting, and ventilation. Important, too, are recommended considerations for proper feeding and watering, socialization, safety, and disease/pest control.

Other important topics discussed include programs to certify acceptable kennel/cattery facilities and management procedures. As with other service-oriented business enterprises, having well-trained employees—persons who derive personal satisfaction in caring for companion animals—is essential to the long-term success of kennel/cattery management and business profitability.

# 18.12   REFERENCES

1. Judy, Will. 1968. *Kennel Building and Plans*. 7th ed. Chicago: Judy Publishing.
2. Science and Education Administration. 1980. *Kennels for Dogs and Other Small Animals*. Washington, DC: U.S. Department of Agriculture, Science and Education Administration.

# 19 BUSINESS/FINANCIAL ASPECTS OF THE COMPANION ANIMAL ENTERPRISE[1]

*If you are selling something, try to make it so good that you would rather be the person who bought it than the person who sold it.*

William Penn Adair "Will" Rogers (1879–1935)
*American actor and humorist*

---

[1]The authors acknowledge with sincere appreciation the contributions to this chapter of Frank C. Robson, President, Robson Properties, Claremore, OK.

## 19.1 INTRODUCTION/OVERVIEW

Many former students—now practicing veterinarians, veterinary technologists/technicians, and others engaged in businesses and industries related to companion animals—have told the authors they would have benefited immeasurably from better preparation for the business and financial aspects associated with their respective career opportunities, assignments, and responsibilities in the companion animal enterprise.

Business principles associated with accounting, financing, risk/insurance, business law, marketing, preparation of a business plan, and other aspects of companion animal services and sales are included in this chapter. Special emphasis is given to veterinary medicine and related enterprises.

## 19.2 GOOD BUSINESS PRACTICES—CODE OF STANDARDS, ETHICS, AND VALUES

Underpinning long-term, successful, service-oriented/results-driven businesses are strong leadership and organizational skills with adherence to a high level of standards, ethics, and values both personally and professionally.

Noble and respected are business owners, members of the management and support team, and other professional associates who demonstrate strong moral principles and values as they interface with clientele and the public. Examples of sound values and virtues include displaying a can-do/will-do attitude, caring about the health and well-being of animals, as well as the economic and personal conditions of business clientele and friends. One client stated this philosophy succinctly to his veterinarians, "I don't care how much you know until I know how much you care."

Deep-rooted intrinsic core values and sound business principles include an impeccable character, unimpaired honesty and integrity, responsibility, service-oriented work ethic, and punctuality. Especially important core values also include a good blend of courage, tenacity, and resilience; plus the capability and commitment to consistently display and practice excellence, fairness, civility, tolerance, and the behavior respected citizens and the public perceive as being correct.

Finally, modern veterinary healthcare and related business professionals are expected to deliver to their clientele the all-important contribution of "peace of mind" in believing their companion animals are receiving the best possible prophylactic and therapeutic healthcare. And, remember to *always thank* your clients and customers for their business!

## 19.3 COMPREHENSIVE PLANNING AND DEVELOPMENT OF STRATEGIES

Planning and strategies are only as good as the data/information on which they are based. Quality data are vital for shaping well-founded decisions that underpin business growth, success, and profitability.

Sources of data based on the latest census are available on the Internet. One example of a business service that has provided timely data for more than 30 years is Environmental Systems Research Institute (ESRI) Business Information Solutions. These ESRI Business Information Solutions' Sourcebooks provide a wealth of information related to demographic data reports and maps including age, education, income, lifestyles, occupation, traffic volume, and other key variables. They also include target marketing analysis, customer profiling and prospecting, equal competitive analysis, site evaluation and selection, response modeling, media and direct mail planning, and other information pertinent to starting a new or expanding an existing business. For additional information, see www.esri.com. Contact information is available in the East and West as follows:

**East**

8620 Westwood Center Drive
Vienna, VA 22182–2214
Tel: 800–292–2224
Fax: 703–917–9061
Email: br@esribis.com
Web: www.esribis.com

**West**

3252 Holiday Court, Suite 200
La Jolla, CA 92037
Tel: 800–394–3690
Fax: 858–677–5420
Email: br@esribis.com
Web: www.esribis.com

## 19.4   DETERMINING BUSINESS FEASIBILITY

Before prospective owners of a new veterinary or other animal-related business move ahead in developing a comprehensive business plan (Section 19.8), they should first outline the need and opportunities for a new business in the area. The nature of the market and competition must be understood. Specific questions should be answered. Who and where are the nearest two or three competitors? Precisely what are the needs and expected benefits of the services to be rendered and products to be sold? What is the projected growth in the clientele base (potential number of people and companion animals in the area to be served)? Are there new market niches not now being serviced? Such niches might include a greater focus on kennel/cattery care, specialty surgeries, general practice, animal inoculations and parasite prevention, specialty petfoods and products, bathing and grooming, dental care, companion animal training, boarding, well-animal health examinations (e.g., routine blood screens), pet nutrition and counseling, dietary recommendations related to supplements or special needs of pets, and other services not now being fully offered.

Consider current and projected economic conditions and stability of employment in the area. What are the present and projected expectations related to the financial outlook for the area and state? What are current and projected interest rates? How may interest rate changes influence the demand and use of animal healthcare services? Review pertinent demographic data per household income level in the area to be serviced, pet population data, and other related factors that will have an impact on the success of the proposed business enterprise.

## 19.5   RESEARCHING THE MARKET AND ANALYZING THE COMPETITION

There are various ways to obtain information related to current and projected market needs and opportunities, as well as methods to analyze the competition before buying an existing, establishing a new, or proposing to expand a current veterinary practice. What is the number of pet animals per veterinarian in the area or city? What is the number of households in the area having income levels sufficient to provide discretionary spending for professional services for pets? Conferring with informed representatives of the area government(s), Chamber(s) of Commerce, marketing specialists, kennel/cattery club members, private financial institution leaders, product and equipment suppliers, and others should be insightful. Additionally, private consultants and consultative groups can assist in evaluating the market and competition and the needs/opportunities for expanded business services.

# 19.6   THE ECONOMICS OF OWNERSHIP: BUYING, BUILDING, OR EXPANDING

There are various ways to pursue entrepreneurship goals in the companion animal enterprise. Because of the initial high costs of building a new or making an outright purchase of an existing veterinary practice or other animal-related business, there are advantages of buying a minority position as a partner in an established business with an option to purchase the balance of the business enterprise in a predetermined period of time at a negotiated fixed price (e.g., two times book value for the first 10%, the balance at book value). A word of caution is appropriate in buying a minority position. Always exercise discretion as it is usually easier to buy than sell a minority position should the partnership have misunderstandings and develop serious conflicts and disagreements that lead to a premature dissolution.

Before buying, building, or expanding a business, determine which direction(s) the city is expanding. Viewing the city or area from a helicopter or small aircraft is a worthwhile way to observe development and expansion of roads, housing, and businesses. What is the traffic pattern and the accessibility (ease of getting off and back on the main road) to the proposed site(s)? If the proposed business is to be located in a new area, review plans for roads and expansion of basic utilities (determine plans for sewer expansion first; water, electricity, and gas are commonly much easier to access).

Critically important, as well, is to review the clientele base, cash flow, and profit/loss status of an existing business over the past 2 years before making a commitment to join or buy into it.

# 19.7   CHOOSING A BUSINESS NAME

Selecting a proposed business name with positive, friendly, and descriptive signage is an early step in preparing the business plan. The business name may include the personal name of the owner(s), the proposed name of the business services offered, and one or more niche services to be provided. Additionally, the street name or general location of the proposed business may be added (e.g., Northwest, Southeast, South Main Street). In choosing a name, consider that the name will be an important part of the highly visible and legible signage in front of the business.

The name and signage should help convey to the public and potential clientele what services/products the business offers, as well as the address and telephone number—all in letters large enough to be easily read and understood. Colors used on signs affect attractiveness and ease in reading. Red, green, or blue on white; white on red, blue, or green; black on orange; and orange on black are time-tested color enhancers on signage. Some of the following words may be useful components of the business logo and/or signage: Alpine Animal Hospital, Blue Ridge Animal Health Center, Complete Care Animal Clinic, Health Guard Animal Clinic, Lakeside Veterinary Hospital, Northwest Petcare Clinic, Northwoods Animal Clinic, Pampered Pets Veterinary Clinic, Preferred Care Veterinary Hospital, Seven Pines Animal Hospital, Westside Animal Clinic.

# 19.8   PREPARING A BUSINESS PLAN

When starting a new or expanding a present business venture, a comprehensive, well-written business plan is essential. It is the entrepreneur's management game plan. The document verbalizes the dreams, hopes, and expectations that motivate the prospective owner to start a new business or purchase outright, buy-in, or expand an existing business. It methodically describes the basic situation of the business, indicates the direction you want to go, and highlights how you propose to get there.

The business plan explains key variables that will affect success or failure and thereby helps prospective owners prepare for different situations that may occur by thinking through each aspect that could go right and what might go wrong. In short, it is the blueprint for creating a new or expanded business venture; it is the bridge linking a visionary idea/proposal, a hoped-for result, and the actual fact of reality.

The written plan provides a clear visualization of each action the prospective business owner plans to implement. The plan includes 3- to 5-year projections about financial, marketing, and operational aspects of the proposed business. It should also address expansion plans (new branches and/or outright expansion). The plan should be thought of as an ongoing process—not merely as the ultimate means-to-an-end goal or outcome. It should be revised/updated annually because it is an important guide for future decisions and actions.

Fundamental functions of the business plan include (1) providing a clearly articulated statement of goals and strategies and (2) serving as a selling document to be shared with prospective financial supporters as well as with potential customers, suppliers, investors, and employees and business partners.

In preparing the business plan, remember that potential investors and lenders have more opportunities to fund business plans than they have capital to invest. The goal of prospective investors and lenders is to minimize risk exposure and maximize the potential return on an investment. This must be met through the cash flow of the business. To help ensure a careful reading of your proposal, the first hurdle is to excite the interest of prospective investors/lenders in the plan.

Each business plan is unique and no format can guarantee success; however, there are specific guidelines to follow in its preparation. Important features and components of the successful business plan, appropriately arranged in a loose-leaf notebook to facilitate revision, include:

*Cover.*  An attractive, labeled, and well-designed cover for the notebook is a nice touch.

*Title page.*  Name(s) and address of the proposed business and key personnel, date prepared, copy number, and contact person(s).

*Table of contents.*  Provide page number of key sections and data/information presented. This page is prepared following completion of the plan.

*Executive summary.*  Provide an overview/summary of the total plan. Although it is at the front of the business plan, it is written after all sections have been completed; it highlights especially important points and ideally creates sufficient interest and excitement to cause readers to continue.

*Vision/goals statement.*  Describe and set forth the strategy, philosophy, and commitment to make the business successful—one with eminent potential and expectations for longevity. Properly researched and articulated, the plan will help calm concerns related to uncertainty involving future client numbers, competition, the economy, and unexpected future events and developments.

*Legal formation.*  Describe the form of ownership and management to be used. The most common forms of small business are sole proprietorships, partnerships, and corporations. Consult with a knowledgeable tax and business attorney and a certified public accountant because the choice of business form affects how the business will be managed, its tax structure, how it will be financed, and personal liability of the owner(s).

*Financial projections.*  It is important to present reasonable financial management projections; pertinent, believable data should be explained and documented. Show how investors/lenders can sell their investment in 3 to 5 years with an attractive rate of return and capital appreciation. Share pro-forma income projections (profit and loss statements) month-by-month for the first year, detailed by quarters for the second and third years, and a 3-year summary as well as the assumptions on which the projections are based.

*Financial needs.* Present a detailed, realistic, and well-documented summary of the total financial needs and contemplated sources of funding and projected revenues, costs, and profits. Share projected cash flow statements, break-even analyses of profits, and planned sources of financing (give a projected timetable to draw on an approved line of credit).

*Market plan.* Focus marketing efforts on those persons/families with the greatest potential to purchase your services and products. Outline the marketing plan and strategy (e.g., the use of newspapers, direct mail, door hangers, yellow pages, radio, television, attractive easy-to-read outdoor billboards and other signage) designed to achieve the business goals in 3 years or less. Identify the projected customer and present competition bases; give examples of targeted clientele/profile of target consumers, proposed methods of identifying and attracting new clientele and customers (e.g., open houses), types of sales promotions and advertising, and credit and pricing policies.

Future marketing and business success will be enhanced by *always thanking* the client/customer at the conclusion of each service given and sale made. Then follow up with a personal note of gratitude within 3 days thanking them for their business. This is a personalized courtesy that will pay handsome dividends in terms of client–customer relations. As one highly successful businessperson who reviewed the manuscript of this chapter said: "That is a low-cost courtesy that gives positive results!"

*Personnel qualifications.* Identify key personnel; share their qualifications such as professional degrees and unique experiences, board certification and licenses (e.g., the veterinarian, veterinary technologist/technician, pet trainers and groomers, and other team professionals), and why the business will be recognized and accepted as a reliable pet healthcare or other service provider. Present a plan for recruiting and providing additional training of personnel on an as-needed basis.

*Competitive prices.* Highlight examples of the services and products to be made available. Where possible, include comparative state, regional, and national average costs of proposed services and the number of pets projected to be seen, treated, trained, groomed, and/or housed.

*Operating plan.* Using charts, diagrams, graphs, and tabular summaries as needed, describe present and/or proposed expanded facilities (include photographs and/or architectural renderings when possible), maps to show the physical location of the business (include nearby businesses), floor plan (proposed space utilization), major items of equipment, kennel/cattery facilities (as appropriate), alternate sources of supplies and medications, proposed specialty products to be offered for sale, purchasing procedures, and accounting-related information (e.g., bookkeeping, billing and credit policies). Include the degree of computerization projected (related to appropriate data/information to be generated) and sample essentials of client/animal records to be maintained.

*Seek counsel of business plan reviewers.* Before contacting potential investors/lenders, have the plan reviewed by successful entrepreneurs, certified public accountants, business-savvy lawyers, suppliers, and others as deemed desirable. Invite them to review your plan and share feedback pertaining to clarity, reasonableness, thoroughness, and overall quality of the plan. Invite constructive criticism regarding other pertinent information/data that can strengthen the business plan.

Seek approval of the plan's reviewers to include their names in your final presentation to potential financing officers. It is important to include the names and backgrounds of the reviewers to help demonstrate the extent to which you attempted to prepare an informative, meaningful, business plan.

*Owner's commitment to work.* Most entrepreneurs are focused, ambitious, self-driven persons who are willing to devote long hours to ensure the success of a proposed business enterprise. One recent study of owner-managers found that 25% worked 70 to 80 hours weekly, 28% worked 60 to 69 hours, and 23% worked 50 to 59 hours per week. Only 24% worked fewer than 50 hours per week. Include in the business plan a statement regarding your work ethic and commitment to the success of the proposed business.

*Pricing and credit strategies.* Extending credit increases business sales but creates the challenge of collecting from the resultant number of customers. It is important to outline credit and payment policies including a credit limit for new clients. Five important "Cs" of credit include (1) *Character*—the fundamental integrity and honesty of the client; (2) *Capital*—the client's net worth (banks have these forms); (3) *Capacity*—the ability of customers to conserve assets and examples of their past commitment pattern in honoring timely payments (there are reliable ways to obtain credit ratings); (4) *Conditions*—projected interest rate and price level, inflation, economic/business cycle, and demographic trends; and (5) *Collateral*—securities that can be designated as a pledge to fulfill the financial obligation.

*Neatness of business plan.* The overall appearance of the business plan will likely be perceived by potential investors/lenders/suppliers and others as reflecting the detail, thoroughness, and professionalism with which you will approach the everyday management and oversight of the proposed business. It may well determine whether you obtain the needed financing and could affect the credit terms. Moreover, it may influence the amount of credit and payment schedule advanced by potential suppliers.

*Business ethics and values code.* As discussed in Section 19.2, business standards, ethics, and values are fundamental underpinnings for long-term, successful businesses. Therefore, it is appropriate to include a short statement by the prospective owner(s) of the proposed new or expanding business regarding personal tenets by which the owner(s) lives.

*Confidentiality.* The business plan represents a great deal of personal time, thought, and effort. It explains important strategies and other aspects of the proposed business, the nature and objectives of client services to be provided, and the current status (new startup, buy-in, buyout, projected business expansion, and legal state of affairs). It contains valuable information and personal ideas that could be useful to present and possible future competitors. Therefore, it should be treated in a confidential manner. Included on the cover and title page should be clear notice that the plan and all information are *proprietary and confidential*. All pages should be numbered; so should each copy. It is important to keep a record of the persons to whom a copy of the business plan is entrusted.

*Length of business plan.* The plan should be long enough to answer relevant questions, but short enough to not bore or dampen the interests of prospective investors and lenders.

*Appendix/supporting documents/references/contacts*

- Short biographies of the professionals and members of the support team may be included as appendices.
- Data, tables, illustrations, charts, photographs, and other items that support and enhance the overall business plan may be included as appendices or as supporting documents within the body of the plan.

*Other possible aids*

- The *United States Small Business Administration (SBA)* offers an extensive selection of information on most business management topics. For information about SBA business development program and services, call the Small Business Answer Desk at 1–800-U-ASK-SBA (800–827–5722).

- A well-written, easy-to-understand book now in its 12th edition that prospective entrepreneurs may find useful is titled *Small Business Management: An Entrepreneurial Emphasis,* Justin G. Longenecker, Carlos W. Moore, and William J. Petty II, South-Western Educational Publishing, 2002 (see http://longenecker.swcollege.com).

- *Service Corps of Retired Executives (SCORE)*—by appealing to an SBA field office, small business owners/managers can obtain free counseling and management workshops and seminar information from SCORE, a group that includes more than 13,000 volunteer business executives. For information, call 1 800-U-ASK-SBA (800–827–5722).

- *Small Business Development Centers (SBDCs)* are sponsored by the SBA in partnership with state and local governments, the educational community, and the private sector to provide assistance, counseling, and training to prospective and existing business people.

- Computer software is available to guide the preparer step-by-step through the preparation of a business plan.

# 19.9 BUILDING A SUCCESSFUL VETERINARY PRACTICE

Human healthcare management firms have studied how people select healthcare providers. Most studies revealed four common factors: (1) *Quality* and totality of services that bear on the ability to satisfy stated or implied needs; (2) *Convenience/Location*—not needing to travel long distances to obtain medical services and ease in parking; (3) *Relationship(s)*—people prefer people who are courteous, caring, friendly—those who sincerely project an "I want to help you" attitude—those who extend compassion in times of bereavement—those who make follow-up calls and/or write thank-you notes; and (4) *Costs/Fees*—some associate higher prices with a higher level of professional experience and expertise, higher-quality services, and better care. Most consumers want options. A fair fee schedule is important; so are professional training, experience, friendliness, the support staff, clean/neat facilities, courtesies extended, convenience, modern equipment, and service quality.

Companion animals may sense some of the aforementioned considerations, but they are unable to converse with veterinary-care professionals. Realistically, the same four factors mentioned in the human healthcare studies above are likely to influence pet owners in choosing veterinary care and services for pets. They notice when the veterinarian and her/his team of professionals are kind and gentle with pets. Properly treated, there is a natural bonding between members of the veterinary team and clients; clients take note of veterinarians and other team members who show respect to them as well as respect and loyalty to co-workers. These traits are noticed and appreciated.

Important is the *image* of the veterinary practice. Image is built over time by high-quality services rendered by friendly professionals and by word of mouth from satisfied clients and customers. Indeed, veterinarians know that the all-important ingredient in successful practices is happy, satisfied clients. Remember to *always thank* your client-customer. Of course, pet owners need veterinarians so opportunity exists for mutual fulfillment.

Numerous surveys have shown that the four primary sources of new clients are (1) new companion animal owners; (2) an established pet owner who moved recently; (3) owners who are dissatisfied with present veterinary services or who have specialty practice needs; and (4) referrals by clients and pet shop recommendations.

Other factors influencing choice of a veterinary practice include reputation, signs (outside signage perpendicular to the street is more effective than signs attached to the building's front), Internet linkages, paid and unpaid mentions by the media, newspapers, yellow

**Figure 19.1** American Animal Hospital Association (AAHA) award-winning, attractive, well-planned 20,000 sq ft Willow Creek Pet Center, Sandy, UT. *Courtesy Dr. Rick Campbell.*

pages in telephone books, contacts made through membership(s) in civic clubs and community activities, scouting, companion animal clubs, clinic open houses, and other opportunities to interact with potential clients and customers.

The physical appearance of the facilities—both exterior and interior—can affect customer selection of a healthcare provider for their pets (Figure 19.1). First impressions are important in this regard. These may be influenced by perceived attitude of the receptionist and other staff, cleanliness and neatness of the facilities, lighting and security, diplomas and licenses, and art work/photographs on the wall.

Most clients appreciate a short wait but also enjoy having publications to read while waiting. Examples include a newsletter that provides information related to pet care and management; a leaflet that gives tips on traveling with your pet; and a brochure describing the scope of the practice, names and photographs of the veterinary clinic's team of professionals, services and products offered, special technical expertise available, hours of operation, street address, telephone number, Web site, e-mail address, and other pertinent information.

The reception-client waiting area is also a good place to have an attractive display of products for sale. Most service businesses have found that the profit side of the profit and loss statement is enhanced by having products for sale (Figure 19.2).

## 19.10   EVALUATING RISKS AND INSURANCE ALTERNATIVES[2]

Risk is a condition in which there is some level of probability that an adverse deviation from a desired outcome may occur. Applied to the veterinary and related business enterprises, risk translates into the level of probability of losses that may occur associated with the assets and earnings potential of the business.

Risk management is reflected in efforts to preserve the assets and earning power of the business. Risk management grew out of insurance management. Risks can be both insurable and uninsurable. Let us consider three categories of risks: (1) some risks could result in bankruptcy, (2) other risks could result in losses that require additional capital to continue the business enterprise, and (3) still other risks could result in minimal losses and be readily covered with current income and/or existing assets.

[2]The authors acknowledge with sincere appreciation the contributions to this section of J. Michael Bale, Director of Risk and Property Management, Oklahoma State University, Stillwater.

**Figure 19.2** Petfoods and supplies in reception-client waiting area of Willow Creek Pet Center, Sandy, UT. *Courtesy Dr. Rick Campbell.*

Fire and/or heavy water damage may be a category-one risk and therefore should be covered by fire/water damage insurance. An example of category-two risk could include expensive equipment needed to perform standard diagnostic tests such as blood profiles, endoscopy, and ultrasonography against damage from a transient sudden burst of current/voltage in an electrical circuit resulting from lightning. Because the risk of lightning damage to equipment can be greatly reduced by installing lightning rods, arresters, and circuit surge protectors, the owner may choose to utilize those preventive measures and assume the lessened risk, or alternatively, to insure at a lower level.

Employee embezzlement might be considered a category-three risk, and because it is a felony covered by the protection of court proceedings, the possible cost of lawyer fees associated with such an unlikely occurrence may be a risk that management is more comfortable in assuming or possibly insuring with higher deductible provisions.

Small business owners must determine the magnitude of loss they could bear without experiencing serious financial difficulty. This helps in deciding to avoid the purchase of unnecessary insurance by covering only those losses exceeding a specified minimum amount. Relating premium costs to probability of loss will be reflected in the cost of insurance, increasing as the probability of insured losses increases.

Customers/clients are the source of business profits. They are also a source of an ever-increasing amount of business risk. Much of this risk is attributable to on-premise injuries and services liability. On-premise risk of liability is an important reason for buying liability insurance. Agents of most insurance companies will write insurance for companion animal owners against liability from their dog/cat biting a person or against other harmful acts by pets.

Although insurance is the traditional method of managing risk, it should be used as the final option in choosing risk management options. Several methods of risk management suggested by the industry should be an integral part of all veterinary-related business enterprises. Properly used, these methods will serve to protect and preserve assets and earning power of the business. These include (1) *risk avoidance*—do not participate in potentially high-risk situations (e.g., locate veterinary service and related business facilities out of flood plain); (2) *risk prevention*—utilize well-documented medical procedures, sanitary facilities, and education (e.g., use of surgical equipment checklists, proper building maintenance and employee training); (3) *risk reduction*—install sprinkler systems, when feasible locate facilities near fire department stations and/or fire hydrants, install smoke detectors, and utilize back-to-work programs; (4) *risk transfer*—use insurance-based contractual agreements among clients, partners, financiers, and others; (5) *risk segregation*—practice the time-tested adage "do not put all your eggs in one basket." Have more than one supplier of mandatory goods and services, prepare back-up computer records stored in a separate location; and (6) use *combinations* of the above methods.

Risk management must be an all-encompassing view of the owner's total assets including property, human resources, liabilities, and net income. Use of the best business practices helps protect assets.

Other considerations of the risk/insurance equation include professional liability, business interruption coverage, workers' compensation, and hiring/firing practices. Many professions and businesses have access to insurance consortiums that provide coverage specific to that profession and business. A comprehensive self-risk analysis of potential risk hazards is a good method of determining what is needed to protect a new or expanding business. Industry lawsuits are examples of what could happen and loss histories of other business enterprises may be helpful as well as examinations of written policies and procedures, contract reviews, hold-harmless agreements, and review of proposed new/expanded facilities.

## 19.10.1   PET HEALTH INSURANCE

Ethical Pet Insurance is a risk-sharing property insurance, the same as car or house insurance. Veterinarians do not commonly sell pet insurance (a state license may be required), but when asked, they will provide clients with an invoice when they pay their bill. This invoice may be submitted for reimbursement, including wellness veterinary healthcare when covered by their policy.

Most pet insurance policies in the United States cover specific services and have limits, co-payments, and deductible provisions much as with human medical insurance policies (Section 14.9). Recent emphasis with pet insurance has been on providing protection against catastrophic costs from unexpected accidents and diseases. Some European countries now have insurance coverage for more than 50% of their pet populations.

Veterinary Pet Insurance (VPI), based in Brea, California, has more than two decades of experience in providing veterinary-specific pet insurance for dogs, cats, birds, and other common pets. The VPI Skeeter Foundation is a nonprofit foundation developed by VPI to help validate and promote the human–animal bond, perpetuate the positive effects that pets have on human health, and educate people about the immense value pets provide to society. For additional information, see www.skeeterfoundation.org, the VPI Web site at www.petinsurance.com, or call 800-USA-PETS (800–872–7387).

## 19.10.2   LEGAL ASPECTS OF ANIMAL COMPANIONSHIP

Legislators of several states are supporting new legislation that would elevate the status of dogs and cats from property to companions or guardianship. Proposed provisions of such a measure would permit people to bring legal action against veterinarians and alleged animal abusers by seeking damages for "loss of companionship" up to $100,000 (or another specified maximal amount).

Now classified as property or chattel, companion animal owners can seek only "fair market value" (replacement) in a lawsuit. This area of public debate and possible legislation will be watched closely by the veterinary and other related professions. It certainly has important risk/insurance implications.

## 19.11  VETERINARY AND SERVICE PROFESSIONALS

Veterinary practices may include special in-house services, such as boarding and dietary consulting and management. Some veterinary clinics have separate business endeavors for dental hygiene and care, bathing and grooming, training pets, and other professional services.

As discussed in Chapter 17, boarding operations can be marketed as a pet hotel, a pet day camp, a pet resort, a pet daycare center, a pet bed-and-breakfast, or a pet vacation facility (Figure 19.3). Some are co-located with bathing/grooming facilities where the groomer serves as boarding manager of the kennel/cattery. The complex staff may include professional trainers of companion animals.

Frequently, an examination room in the veterinary clinic is reserved for routine use by trained professional staff to conduct dental hygiene examinations, nutritional counseling, and other important pet healthcare services.

The wellness center area of the veterinary practice may include information related to specialty dietary products and supplements, treats, dental care, training aids, pet carriers, specialty shampoos, internal and external parasite control products, companion animal-related books, and other publications/materials. The wellness center is often co-located with the reception area and commonly offers products for sale.

## 19.12  RECORDS/ACCOUNTING

Good records are essential to all successful veterinary practices (Chapter 8). These include basic information pertaining to both clientele and their pets. An efficiently organized records storage system is depicted in Figure 19.4. Having readily accessible vaccination records will be useful, for example, in sending friendly reminders to pet owners when the next booster shot or physical examination is due.

The application of accepted good accounting practices is basic in measuring business liquidity, determining the return on investment, accounts receivable/accounts payable, proper accounting of receipts and expenditures, and net income/loss of the business.

Personnel and payroll records are also essential as are certain clientele information (e.g., name, address, telephone number, and proper records of the pet healthcare services provided).

One commonly used accounting program is Quicken®. Whether that program or another user-friendly program is used in-house, or the decision is made to contract with a local CPA/accounting firm for maintaining up-to-date business records, it is very important to

**Figure 19.3** Front entrance to Kennelwood Pet Suites, St. Louis, MO. *Courtesy Kennelwood Village,* www.kennelwood.com.

**Figure 19.4** An efficiently organized records storage system, Willow Creek Pet Center, Sandy, UT. *Courtesy Dr. Rick Campbell.*

determine monthly where you are on the road to business success by having profit and loss statements available for review within 3 to 5 days after the end of each month. Always know where you are businesswise! These profit and loss statements will be useful in preparing tax returns and in keeping your financial institution updated.

## 19.13 LAWS AND REGULATIONS

Numerous state and federal laws/regulations affect practices of veterinary medicine, schools/institutes/programs for the education and certification of veterinary technologists and technicians, and other professionals/staff members of the veterinary clinic support team. Most of these laws/regulations also apply to colleges of veterinary medicine.

Business owners, managers, and other veterinary professionals should be familiar with and conform to state and federal laws/regulations with regard to employee safety and hiring and dismissal matters (e.g., those related to discrimination).

- *Occupational Safety and Health Administration (OSHA).* Since its inception in 1971, the U.S. Department of Labor's OSHA has required companies to maintain a record of workplace illnesses and injuries (i.e., any injury beyond first aid).
- *Environmental Protection Agency Office of Small Business Ombudsman.* For information concerning the assistance and services available to small businesses through this agency, contact Small Business Ombudsman (Mail Code 2131), Room 3423, 401 M Street, SW, Washington, DC 20460, 1–800–368–5888.

## 19.14 SUMMARY

It is essential that veterinarians and others engaged in businesses related to companion animals be well educated and trained as professionals. It is also extremely important that they be knowledgeable about the underpinning principles of owning and managing successful companion animal business enterprises.

Encouraged by former students to include information and discussion of business/financial aspects shared in this chapter, we attempted to prioritize materials and include topics pertinent to good business practices, new and expanded business feasibility considerations, market and professional service opportunities, and choosing a business name. We also discussed the importance of preparing a comprehensive business plan, building a successful veterinary practice and related services, evaluating risk and insurance alternatives, keeping good records, and understanding pertinent laws and regulations. We also included information on persons, groups, agencies, and organizations to contact for answers to specific questions pertaining to establishing and maintaining successful small business enterprises.

We sincerely hope readers found the materials interesting and useful. We encourage readers to share anecdotes and recommendations that may benefit others.

# 19.15   REFERENCES

1. Bower, John, John Gripper, Peter Gripper, and Dixon Gunn. 2001. *Veterinary Practice Management.* 3rd ed. Oxford: Blackwell Science.

2. Catanzaro, Thomas E. 2000. *Veterinary Management in Transition: Preparing for the 21st Century.* Ames: Iowa State Press.

3. Longenecker, Justin G., Carlos W. Moore, and William J. Petty II. 2002. *Small Business Management: An Entrepreneurial Emphasis.* 12th ed. Boston: South-Western Educational Publishing.

4. McCurnin, Dennis M., ed. 1988. *Veterinary Practice Management.* Philadelphia: J. B. Lippincott.

5. Messonnier, Shawn P. 1997. *Marketing Your Veterinary Practice.* Vol 2. St. Louis, MO: Mosby.

6. Opperman, Mark. 1983. *Veterinary Business Management: A Guide to an Efficient and Profitable Practice.* Media, PA: Harwall Publishing.

# 20 COMPANION ANIMAL BEHAVIOR AND SOCIAL STRUCTURE

*Tis sweet to hear the honest watch dog's bark.*

Lord Byron (1788–1824)
*English poet*

## 20.1   INTRODUCTION/OVERVIEW

Behavior is the way an animal reacts to an internal or external stimulus. Natural selection among species played an important role in the innate behavior of companion animals, especially regarding their ability to survive and reproduce in the wild. Behavioral responses are determined by heredity and learning. Cat behavior such as growling, hissing, hair raising, arching of back, showing teeth, ear flattening, and side leaning are innate signals designed to win arguments without injury.

Behavioral problems (Table 20.1) comprise the reason cited for more than a third of the dogs and cats surrendered to animal shelters. Early detection of and providing appropriate intervention with bad behavior (e.g., aggressiveness toward their owner or strangers, separation anxiety, eliminating in unacceptable places in the residence, phobias, or failure of the family pet to get along with people or other pets) is important in keeping pets out of shelters.

Behavioral problems may be due to a variety of factors. One source is a mismatch in owner expectations and normal behavior of the pet. This can be due to inadequate consideration during the selection of a pet (Chapter 4). For example, a Border Collie with the high energy and desire to work, typical of its breed, will be restless if confined to a small apartment with an owner who lacks the time or ability to exercise the dog. Another example would be the cat that kills songbirds in the owner's yard—this is normal behavior for the cat but may be completely unacceptable to the owner. Third are problems due to inadequate socialization of the pet during its first 2 to 3 months of life (Section 20.2). Inadequate early socialization may result in lifelong difficulties in relating to people and new situations. Fourth are problems associated with inadequate training such as the dog that is constantly lunging and pulling on its leash, or the one that jumps on people as a greeting. Fifth are problems

**Table 20.1**
Common Behavioral Problems of Dogs and Cats

| Dogs | Cats |
|---|---|
| Aggression | Aggression |
| Barking | Compulsive grooming |
| Begging for food | Destructive scratching |
| Coprophagy | Fears and phobias |
| Destructive behavior | Inappropriate urination and defecation |
| Disobedient, hard to control | Predatory behavior |
| Excessive submission | Sexual behavior problems |
| Excitability | Stereotypic behavior |
| Fears and phobias | Unfriendliness |
| Inappropriate elimination | Urine marking/spraying |
| Jumping on people | |
| Mounting | |
| Running away | |
| Stereotypic and compulsive disorders | |
| Unruliness | |

due to unintentional owner reinforcement of undesirable behavior; an example being an owner feeding the pet from the table then becoming upset with the animal's constant begging or jumping up to grab food off a plate. A few behavioral problems result from underlying diseases (infections, parasites, trauma, metabolic or neurologic disorders, and others) or from past experiences (being hit by a newspaper, harsh treatment from a person, being attacked by another animal).

Companion animal owners; veterinarians, veterinary technologists, and veterinary technicians; laboratory animal caretakers; owners and managers of pet motels and other boarding facilities; those who fit, groom, show, and train pets; photographers who seek the best pet pose; and other professionals who associate and work with companion animals should be familiar with and understand both normal and abnormal behavioral patterns of pets. They also should recognize and be knowledgeable about the symptoms and ethological (*ethology* is the scientific study of animal behavior) factors associated with normal and abnormal animal behavior. Various aspects of companion animal behavior and social structure are discussed in this chapter.

## 20.2 SOCIALIZATION AND OTHER SOCIAL ASPECTS OF DOMESTIC DOGS AND CATS

Numerous young mammals can be tamed and socialized to humans. This capacity to form social attachments is often confined to a relatively short, sensitive period during early development, after which positive social responses to strangers gradually decline and are replaced by fearful behavior that effectively prevents formation of further attachments. This primary socialization period (PSP) varies from species to species in its time of onset and duration. In dogs and cats, PSP begins at about 3 weeks of age. If puppies or kittens are exposed to individuals of two species during the few weeks of PSP, they will socialize with both species (e.g., to both humans and dogs or cats).

During the first 2 weeks of life, puppies and kittens show little activity beyond sleeping and nursing. During this time, the eyes and ears are closed and the neonate's only response is to tactile stimulation, which may initiate movement toward the mother or littermates. Following opening of the eyes and ears, young puppies and kittens start to notice other animals and begin walking instead of crawling. The period from 3 to 8 weeks is the developmental phase during which dogs and cats learn how to react to others of their species and learn species identification. The critical period of PSP with humans begins around 3 weeks of age and continues until approximately 16 weeks. During the PSP period, the puppy or kitten is responsive to nonmaternal social interactions. A 3- to 5-week-old puppy will actively approach strangers. Investigative behavior is very active between 5 and 7 weeks and should be encouraged to promote an outgoing temperament (disposition). A popular rule of thumb used by dog breeders is the rule of 7s (Table 20.2). Implementing the rule of 7s promotes adaptability and development of outgoing and friendly temperaments in puppies. Social and environmental enrichment should be continued for the next 5 months as puppies and kittens continue to learn how to react to new environments during this time period. Dogs isolated in a kennel environment from 8 weeks to 6 months of age often have difficulty adjusting to more complex environments; this is referred to as "kennelosis" or agoraphobia (fear of new places or experiences).

By 8 to 9 weeks of age, the puppy is beginning to develop stable learning[1] and will remember painful experiences and seek to avoid a similar circumstance. Fear of strangers begins during this period. Fear of strangers was helpful in the wild because it helped young animals venturing outside their dens to avoid predators. Unfortunately, fear of strangers can interfere with the establishment of strong human–animal bonding.

Dogs and cats not socialized with humans prior to 14 weeks of age are commonly forever fearful of humans. Interspecies social acceptance by kittens is highest between 3 and 7

---

[1]Stable learning is the term applied when an animal remembers an experience and will seek to duplicate rewarding experiences and to avoid traumatic ones.

**Table 20.2**
Rule of 7s for Developing a Well-Socialized Dog

| **By the time a puppy is 7 weeks old, he/she should have:** |
| --- |
| 1    Been on 7 different types of surfaces: carpet, concrete, wood, vinyl, grass, dirt, gravel, wood chip |
| 2    Played with 7 different types of objects: big balls, small balls, soft fabric toys, fuzzy toys, squeaky toys, paper or cardboard items, metal items, sticks, or hose pieces |
| 3    Been in 7 different locations: front yard, back yard, basement, kitchen, car, garage, laundry room, bathroom |
| 4    Met and played with 7 new people: children and older adults, someone walking with a cane or stick, someone in a wheelchair or walker |
| 5    Been exposed to 7 challenges: climb on a box, climb off a box, go through a tunnel, climb steps, go down steps, climb over obstacles, play hide and seek, in and out of a doorway with a step up or down, run around a fence |
| 6    Eaten from 7 different containers: metal, plastic, cardboard, paper, china, pie plate, frying pan |
| 7    Eaten in 7 different locations: crate, yard, kitchen, basement, laundry room, living room, bathroom |

weeks of age. Kittens handled by humans for at least 30 minutes each day when they are 4 to 8 weeks of age have the greatest likelihood of being friendly, gregarious, and outgoing toward humans later in life. Animals that are not well-socialized and have not been provided with an interesting and varied environment often develop xenophobia (fear of the unknown). Puppy kindergarten classes are a popular tool to provide puppies with opportunities for socialization with other dogs and people. Traveling to and from classes will familiarize the dog with car travel. Many classes also provide environmental enrichment exercises (e.g., going through tunnels and over bridges) and also an introduction to obedience training (e.g., walking on a loose leash, sitting, lying down, coming on command).

Interestingly, socialization determines the species to which a young mammal will respond in a positive social and sexual manner as an adult. For example, if a wolf cub is hand reared by humans throughout the socialization period, it will subsequently fail to recognize wolves as members of its own species and will not mate with another wolf. However, if obtained by humans after 6 to 8 weeks of age, the wolf cub would have had time to socialize with its parents and littermates and will still be capable of developing a positive relationship with humans. Moreover, as an adult, such an animal would be able to mate successfully while maintaining a relationship with humans.

Mother dogs and cats teach their offspring social behavior skills. Mothers use growls, postures, muzzle bites (dogs), or neck scruffs (cats) to teach submissiveness and to inhibit biting. Owners can use similar techniques to teach proper social order to their pets. The owner should be the dominant individual in the relationship; tools such as Gentle Leaders® work on the same principles as a muzzle bite and neck scruff (Chapter 12, Figure 12.4).

Observational learning is important in kittens. Kittens show greatest interest in observing how their mother behaves/responds but will also learn from watching other cats. Kittens normally learn hunting techniques, food preferences, elimination behavior, and responses to social cues from their mother. Although hand-raised kittens will eventually establish similar behaviors, they learn much quicker when observing other cats and learn most efficiently when observing their mother.

Dogs and cats have a reputation for playfulness. This play is an important feature of the human–companion animal bond because it increases the potential for contact between companion animals and owners. The noncompetitive qualities of pet-oriented play provides a harmless form of recreation for humans of all ages and contributes to the need for enjoyment derived from these companion animals. Types of play behavior in kittens include (1) exploratory play—climbing and jumping; (2) social play—belly-up (kitten lies on back),

**Figure 20.1** Kitten poised for pouncing. *Courtesy Champion Petfood Ltd.*

standing over (second kitten stands over the "belly-up" kitten and the two paw and bite at each other playfully), side-stepping, face-off (kittens paw at each other's faces), pouncing (Figure 20.1), chasing; and (3) object play—manipulation of small objects and toys, batting with paws, pouncing, tossing in air, scooping, grasping, poking, mouthing, and biting (Figures 20.2 and 20.3). Kittens often play in pairs. Many cat owners adopt or purchase kittens in pairs, which enables them to enjoy watching the kittens play together.

Some readers may wonder about the basis for puppy selection testing. There are generally five types of tests: (1) social attraction—will the puppy come when called? (2) following—will the puppy follow a person? (3) restraint—how does the puppy react when held on its back? (4) social dominance—how does the puppy react to touching the top of its shoulders? (5) elevation dominance—how does the puppy respond when held off the ground? Unfortunately, these puppy selection tests have not correlated well with predicting adult personalities and behaviors. As we have noted, many environmental influences can shape a dog's adult personality. However, owners should be cognitive of communicative signs used by their pets. These signs are discussed in Section 20.4.

**Figure 20.2** Kitten trying to manipulate a spot on the floor. *Courtesy Champion Petfood Ltd.*

**Figure 20.3**    Kitten poised to bat at a toy. *Courtesy Texas A&M University.*

## 20.3    SOCIAL STRUCTURE IN THE WILD

The most common social structure of wolves is the pack. Packs consist of small groups of related individuals of different ages. Social ranking is achieved through single-sex hierarchies. The highest-ranking male and female are referred to as the alpha animals. Typically, only the alpha male and female breed. All members of the wolf pack work together to hunt, scavenge, protect the young, and defend their territory. As leaders of the pack, alpha animals exhibit dominant body postures (Section 20.4.1) toward lower-ranking individuals, choose the best sleeping areas, are the first to eat, and initiate pack activities such as hunting and travel. Maintaining a stable social order facilitates cooperation among members of the pack. Fighting among members of a pack is rare.

It is generally accepted that dogs descended from wolves (Chapter 1). An understanding of the social structure of wolves provides insight into the behavior of dogs. However, humans have used selective breeding to drastically modify not only the appearance but also the temperament and behavior of dogs. Thus, dogs show many unique behavioral characteristics. Feral dogs rarely form stable packs. Feral dogs are free-roaming descendants of domestic dogs. Most feral dogs are solitary scavengers that participate in a pack for only a short time—forming a pack to hunt and then disbanding. Feral dogs are attracted by the barks of a lead dog to join in the hunt.

Cats are believed to have descended from the African wildcat (Chapter 1). Perhaps because their normal prey is small animals that are stalked, wildcats are solitary hunters with well-defined territories. In contrast, domestic cats have been bred to live with humans and other companion animals. The majority of domestic cats are flexible and can adapt to a variety of social settings. Many cats develop strong bonds with their owners. An understanding of how cats communicate is useful for those who love and work with them.

## 20.4    COMMUNICATIVE BEHAVIOR

Sound scientific evidence supports the contention of many owners that dogs and cats have emotions and experience sensations similar to those of people—anger, anxiety, depression, fear, grief, guilt, jealousy, joy, pain, regret, sadness, and others. Communication occurs when one animal responds to signals emitted by another. When humans understand the communication patterns of their pets and react appropriately, a strong bond is usually formed. Communicative signals used by dogs and cats include those of visual, auditory, and olfactory impact.

**Figure 20.4** Cat approaching another animal with an offensive posture (vertical tail, confident manner). *Courtesy Champion Petfood Ltd.*

## 20.4.1 VISUAL COMMUNICATION (POSTURAL)

Both the dog and cat employ body postures and various expressive behaviors that owners can interpret. Moreover, both have important repertoires of visual social signals. Dogs and cats utilize facial expressions; changes in position of their body, ears, and tail; and the direction of their gaze to express different emotional states.

The greeting posture includes demonstration of interest in approaching another animal or person. Dogs will commonly wag their tails and jump up while happily barking (auditory communication). Approaches to another dog are often directed toward the inguinal (groin) or perineal (near the anus) region and incorporate sniffing (olfactory communication) or play initiation (Section 20.2). Cats typically greet people or another cat with their tail held vertically and back legs slightly extended (Figure 20.4). They then touch with their nose and begin head rubbing (olfactory communication, Section 20.4.3; head rubbing is known as bunting, see Figure 20.10).

Postural signals to elicit play include the play bow in dogs and rolling or chasing in cats. A play bow consists of lowering the front end while keeping the rear end up. A play bow may be followed by running, circling, or chasing.

Postural signals of an animal on the offensive include those that make the animal appear as large as possible. Hairs on the back are erect (piloerection). The back may be arched (especially in cats). Stance is high on the front toes with body weight shifted forward. A dominant animal will stare at its opponent (Figure 20.5). The pupils of the eyes will be constricted. In cats, the tail is down and the tip will flag slowly. The ears are up and point forward in breeds with erect ears. Dogs on the offensive hold their tails high and show a high frequency wag. Snarling is seen in dogs. Both dogs and cats growl (auditory signaling).

**Figure 20.5** Dog showing a dominant stance with tail held high, ears forward, and staring at opponent. *Illustration by Diana Nicoletti.*

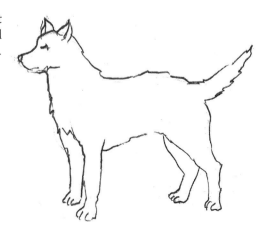

**Figure 20.6**   Submissive posture in a cat demonstrating tucking of the tail and avoidance of eye contact. *Illustration by Diana Nicoletti.*

Postural signals of dominance include those seen in animals on the offensive. In addition, the dominant dog will put its head or forelegs on the neck or shoulder of a lower-ranking dog.

Signs denoting submissive behavior include the avoidance of eye contact and postures to make the animal seem small. Ears and head are lowered, the body is crouched low, and the tail is low or tucked between the hind legs (Figure 20.6). Submissive dogs may grin (lips pulled back) and may lick the face of the dominant dog. A highly submissive dog will lie down and roll partially on its back and may also urinate (Figure 20.7). Submissive dogs are sometimes observed to raise one of their front paws as if trying to ward off an attack.

Dogs may show a combination of body postures when fearful. These animals are often fear biters. The head and tail are lowered as seen in a submissive dog; however, the hairs are raised (piloerection) and the dog snarls and growls (Figure 20.8). This posture is often referred to as one of defensive aggression. It is safest to avoid direct eye contact with a dog showing this posture as staring will be perceived as a threat. Similarly, refrain from standing or reaching over a dog showing signs of defensive aggression.

## 20.4.2   VOCAL COMMUNICATION (AUDITORY)

Five basic sound groups are used in vocal/auditory communication. These include (1) infantile sounds—crying, mews, clicks, whimpering, whining; (2) warning sounds—hissing, barking, growling; (3) eliciting sounds—chirrs, barks, meows, calls, howling; (4) withdrawal sounds—yelping, chatter, screaming; and (5) pleasure sounds—moans, grunts, purring.

Barking is a complex auditory signal having a variety of meanings. Different pitches are used for different situations. Barking is used in greeting; for play solicitation; to sound an alarm; during hunting, tracking, and herding; for defense; to alert others of a threat; to threaten; to

**Figure 20.7**   Submissive posture in a dog lying down and rolling partially on its back. *Illustration by Diana Nicoletti.*

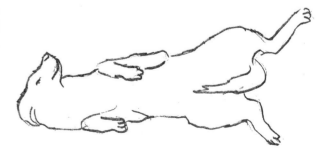

**Figure 20.8** "Fear biter" posture includes lowered head and tail, piloerection of hairs, snarling, and growling. *Illustration by Diana Nicoletti.*

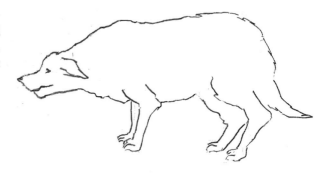

seek contact with another dog or a person; and during times of distress. In general, higher tones are used in greetings and during play, whereas lower tones are used to convey threats.

Dogs occasionally use growls as a greeting. Growling is also used during play, to reinforce dominance, and to show aggression. Dogs may howl when isolated and seeking companionship or as a form of group vocalization reminiscent of the social howling of wolves. Yelping may denote pain, submission, or distress; serve as a greeting; or be emitted during play. Newborn puppies greet their dam by grunting, adult dogs often grunt when being petted by their owners.

The queen uses a chirrup or trill sound to greet her kittens. Adult cats may chirr as a greeting to their owners. Cats use the meow to request food, play, or other interaction with their owners. Purring is widely recognized as denoting pleasure or contentment. Purring is instinctive; it is the first sound a kitten hears because most queens purr loudly and continuously throughout the birthing process. Several theories have been advanced to explain purring in cats. One of the more common explanations is that purring and growling are produced by two so-called false vocal cords (two membranes located above and behind the true vocal cords). Growling is used as a warning by cats on the offensive. Male and female cats have mating calls to attract the attention of members of the opposite sex. Cats sometimes emit a chattering sound when frightened. Hissing is also used when frightened as an attempt to deter threats.

### 20.4.3   OLFACTORY COMMUNICATION

Dogs have the keenest sense of smell of any domestic species. Dogs can detect odors at 1/1,000th of the lowest concentration detected by humans. Scent clues provide species recognition and convey information about gender and reproductive stage of an animal (Figure 20.9). Scents emitted by dogs include odors of urine, feces, and anal sac secretions. Scents emitted by cats include those of urine, feces, and scent glands.

Dogs use urine to mark territory and identify individuals. Dogs exhibit urine overmarking (urinating over the scent of urine deposited by another dog). Female dogs in estrus excrete potent pheromones in their urine that attract male dogs for many miles (Chapter 10).

**Figure 20.9** Dogs meeting each other commonly engage in anogenital sniffing to gather information about the gender and reproductive stage of the other dog. *Illustration by Diana Nicoletti.*

**Figure 20.10** Cats deposit scents on their owners through rubbing the side of their heads (called bunting). *Courtesy Jason and Sarah Motsinger.*

Wolves deposit feces at the periphery of their territories; this pattern is not observed in domestic dogs. Anal sac secretions are normally deposited on feces during defecation; dogs may also excrete anal sac contents when frightened.

Cats use scents to mark their territories. Different postures are used for urine elimination and urine spraying when making scent markings. The posture for urine elimination directs urine onto a horizontal substrate whereas the posture for urine spraying directs urine onto a vertical surface. Unlike dogs, cats do not spray over the urine markings of other cats. There are two amino acids that impart the strongest odors in cat urine: felinine and isovathene. The intensity of the sulfur-like smell of these two amino acids increases during the degradation of urine. Intact (unneutered) adult tomcats excrete approximately 95 mg of felinine daily; intact females excrete 20 mg per day. The role of feces in scent marking is not well understood. Interestingly, during hunting or traveling, cats do not bury their feces whereas feces eliminated at home are buried. Unburied feces provide information to other cats about the cats that deposited them.

Cats have scent glands under the chin, at the corners of the mouth, on the side of the forehead, and between their toes. When a cat rubs its face on an object or person (Figure 20.10), it is transferring scents; this is called bunting. During scratching, cats deposit scents from glands between their toes onto the scratched surface.

# 20.5 INGESTIVE BEHAVIOR

The innate behavior of eating and drinking is critically important to animal survival. The first ingestive behavior trait demonstrated by all neonatal mammals is suckling.

In his classical feeding studies, Russian physiologist Ivan P. Pavlov (1849–1936) discovered an important behavioral fundamental in animals by experimenting with dogs. Pavlov's research showed that the flow of saliva in the dog is a conditioned reflex, not an automatic reaction to the odor of food. Pavlov rang a bell each time he brought food to the test dog. Eventually the dog began to salivate when Pavlov rang the bell—even with no food present. The dog associated ringing of the bell with food, just as it had associated odor with food.

Pavlov appreciated the importance of ensuring experimental control in animal research (Chapter 22) by using a chamber to ensure that the dog attended only to stimuli presented by the experimenter. The wisdom of this was manifested in later studies in which it was demonstrated that dogs accustomed to feeding together will consume more food in the presence of their feeding mates than when fed separately.

## 20.5.1   EATING

There are no reliable ways of measuring the processes involved in an animal's decision to eat. Food preference is commonly evaluated through food intake utilizing the assumption that a greater intake of one food over another indicates higher palatability for that particular food. We commonly define a palatable food as one that achieves a higher intake with respect to the two-pan, free-choice research methodology (Chapter 9). A highly palatable food may not always provide proper nutrition to sustain good health. For example, some cats are so fond of liver they will consume sufficient quantities to develop vitamin A toxicosis and skeletal deformities.

Reflexive behaviors to substances that provide sweet, bitter, or burning sensations evolved early and remain among humans and their companion animals. Appetite is determined in part by things learned from past eating experiences with the same or similar foods. Foods that have previously resulted in nutritional benefits increase appetite and conversely those that resulted in nutritional detriments decrease appetite through a conditioning process. Thus, palatability reflects in part a nutrient-conditioned preference.

Because the olfactory acuity of dogs and cats is much greater than that of humans, it is reasonable to conclude that odor is important in food selection by cats and dogs. Most animals (except the cat) prefer the sweet taste of simple sugars over others. Both cats and dogs tend to reject bitter flavors and, for example, the burning sensation associated with chili.

Cats have an unusually keen sense of smell, and the first thing most cats do when food is offered is smell it. Only if it passes the smell test will it be eaten. It has been estimated that the cat's nasal cavity contains more that 200 million olfactory (scent) cells, compared with fewer than half that number in humans. Cats use scent to detect prey, identify other cats, explore, evaluate their environment, and in their sex lives (e.g., male cats and dogs can detect the scent of female cats and dogs in estrus for many miles). Cats are quite sensitive to the smell of chlorine in water. That is why they prefer to drink from a puddle or pond than from a bowl of clean chlorinated water.

Although flavor is a major motivator in food acceptance by both dogs and cats, the form in which the food is offered is also reflected in their priority of selection. Most dogs prefer a canned meat or semimoist diet to a dry food. Most cats and dogs respond favorably when given a variety of foods, although cats are more finicky than dogs in their feeding habits. Apparently, a few odors, tastes, and flavors are inherently more acceptable or unacceptable to both cats and dogs. Exceptions may include the sweet tooth of dogs and the tendency to reject bitter tastes by both dogs and cats.

Owners often ask why their dog eats grass. In the wild, a favorite part of prey (e.g., rabbits) is the abdominal contents containing partially digested grass. Perhaps because commercial dog foods do not mimic intestinal contents, domestic dogs frequently eat grass. However, dogs lack the enzymes and bacterial flora required to digest cellulose (Chapter 9). Thus, undigested grass, which contains cellulose, may act as an irritant causing vomiting if too much is consumed. Grass eating is considered to be a normal behavior and a problem only if excessive amounts are ingested.

Another common concern of dog owners is that of coprophagy (eating of feces). Dogs may eat the feces of other animals—often preferring that of horses and cats but sometimes eating their own. The attraction of horse feces is likely the presence of partially digested grass, similar to that found in the intestines of several prey animals. Cat feces are eaten due to the tasty diets ingested by cats. The attractiveness to some dogs of eating their own feces may be related to a desire for cleanliness (e.g., bitches eat the feces of their puppies for several weeks after giving birth). Putting meat tenderizer or commercial bitter-tasting powder on the food of the animal whose feces are being consumed may discourage coprophagy through taste aversion. Another strategy is to treat the feces with an extremely bitter substance or with an emetic[2] so the dog learns to associate feces with unpleasant tastes or vomiting. Putting cat litter boxes in an area that is inaccessible to the dog will prevent litter box dining. If highly bothered by the thought of coprophagy, the owner may choose to muzzle the dog when it is in an area providing access to feces.

## 20.6 ELIMINATION BEHAVIOR

Most kittens and puppies begin voluntary elimination at about 3 weeks of age and by 5 to 6 weeks, seek appropriate elimination substrates and sites (Chapter 12). Substrate preferences are developed between 8 and 9 weeks of age. It is best to housetrain puppies and litter box train kittens using the substrate that will be desirable for use when the pet is an adult. For example, a St. Bernard puppy should not be paper trained unless one is willing to accept the large volume of urine it may void on a newspaper lying on the living room floor when it is an adult.

If kittens have access to outdoors, they often eliminate in a place where they can perform earth raking. Indeed, following elimination, kittens perform innate reflective burying of excreta. Because most kittens instinctively prefer eliminating in loose soil, litter box training is relatively simple as long as the kitten is provided an appropriate litter box/tray in an easily accessible location, and the litter is cleaned regularly. Housing the kitten near the litter box facilitates easy access. Moreover, following eating or playing, take the kitten to the litter box until the habit is well established.

The kitten should be placed in front of the litter box with its front paws inside. The owner should move the paws gently forward and backward to move the litter about. Because cats instinctively cover their urine and feces, they usually get the idea quickly. It is desirable to reward success with gentle strokes, kind words of praise, and an occasional treat. The litter should be changed daily and the box/tray cleaned weekly, thereby giving the kitten an expectation of a clean place to eliminate.

If allowed to eliminate outdoors, cats commonly establish their territories by spraying urine (a practice not confined to unneutered cats) with a scent marking that corresponds with scent on their head and flanks in addition to scratching trees and/or posts with their claws.

When allowed to eliminate outdoors in a fenced pen, dogs tend to deposit their urine and feces at particular places (scent posts). In training puppies, it is important to be as regular as possible. Puppies should be taught the proper area in which to eliminate and interrupted immediately when elimination is attempted in an inappropriate place. Putting a newspaper in a corner can be helpful, especially if the entire paper is not removed, so its odor will remind the puppy to eliminate there in the future.

Most puppies eliminate shortly after eating, drinking, sleeping, or playing. The puppy should be taken outdoors to an appropriate elimination area (or to the newspaper in the corner) and rewarded with food, praise, or play when it eliminates. When supervision is impossible, the puppy should be prevented from eliminating in inappropriate locations by keeping it in a crate, a paper-lined room, or an outdoor run. If it is impossible to disrupt indoor elimination during the act and take the puppy to the preferred elimination area, it is useless to punish the puppy after elimination has ceased.

The most common behavioral complaint of cat owners is inappropriate elimination (urinating or defecating outside the litter box or urine spraying inside the residence). There are many possible reasons for inappropriate elimination including substrate preference, substrate aversion, location preference, location aversion, scent marking, and lower urinary tract disease (see discussion on FLUTD in Section 14.4.4). The first step in dealing with a cat exhibiting inappropriate elimination is a veterinary evaluation for FLUTD, and obtain professional treatment if present. Soiled areas should be thoroughly cleaned and covered with

---

[2]Emetics are substances used to induce vomiting.

thick layers of plastic sheeting (plastic serves two purposes: it blocks scent and is unpleasant for the cat to eliminate on). A variety of different size, shape, and depth of litter boxes should be provided in multiple locations and cleaned daily. Substrate aversions may be due to an unpleasant experience such as disliking the fizz of urine mixing with baking soda (contained in the litter) or an association with pain when urinating or defecating during an illness. Location aversions may be due to being frightened by another animal or a loud noise in that area. Startling a cat that is beginning to eliminate outside a litter box and praising it when it uses the box may help reestablish a preference for the litter box. Many cats prefer a very fine litter. Some prefer unscented litter. Trial and error is required to identify litter preferences. Intact male cats may cease urine spraying when neutered; however, this cannot be guaranteed as many neutered male and female cats continue urine marking. Treatment of urine marking/spraying usually requires a combination of social changes and pharmacological treatments. Consulting a veterinarian specializing in feline behavior is advised.[3]

## 20.7    ORIENTATION (NAVIGATION OR HOMING) BEHAVIOR

Domestic cats and dogs have excellent homing instincts as demonstrated by their traveling hundreds of miles to find familiar former territory following a job/house move by their owners. This ability to find their way back home is apparently accomplished by smell and observing landmarks. Similarly, salmon return to the stream where they were hatched to spawn largely through their keen sense of smell. Migratory birds, homing pigeons, and turtles are thought to use celestial bodies in navigating their return to home base.

## 20.8    AGONISTIC (FIGHTING OR AGGRESSIVE) BEHAVIOR

Much of the fighting behavior of companion animals is genetic. The ability to fight in the wild was crucial for survival of individuals and propagation of species. Except for predation, interspecies fighting is not commonly observed among companion animals.

Aggression may be an appropriate or inappropriate threat or challenge that is ultimately resolved by combat or deference (submission). Several types of aggression have been identified (Table 20.3). Aggression is viewed as being appropriate when the animal is defending its owner or home. Aggression directed toward friendly visitors is inappropriate.

Owners of puppies should be encouraged to utilize antiaggressive puppy training. The philosophy is "nothing in life is free" and the puppy is required to show submissive behavior to receive food or attention. For example, food or praise is given when the puppy sits, lies down, or comes on command. Puppies should not be permitted to bite their owner's hand. Tug-of-war type games should be avoided. Puppies should be taught to accept handling of their ears, feet, and mouth (this will be necessary later in life to clean ears, trim claws, and brush their teeth). Puppies should be taught to give toys to the owner on command.

More than a million dog bites are reported annually in the United States.[4] Fifty percent of dog bites result in scars, and 7 to 10 persons are killed each year; the majority of fatalities are children less than 10 years old. Fully 85% of dog bites occur in the owner's home and 62% are inflicted on owners. Dogs exhibiting aggressive behavior should be evaluated by a veterinarian specializing in animal behavior; a combination of behavioral modification training and pharmacological treatments may be prescribed. Castration may help decrease aggressive behavior in intact male dogs.

---

[3]The American College of Veterinary Behaviorists (ACVB) certifies individuals as specialists in animal behavior (http://www.animalbehavior.org/Applied/directory_cert9_97.html). Many other individuals with expertise in animal behavior are members of the Animal Behavior Society (www.animalbehavior.org), the American Veterinary Society of Animal Behavior (www.avma.org/avsab/homepage.html), and other organizations.
[4]http://www.cdc.gov/ncipc/duip/dogbites.htm

**Table 20.3**
Types of Aggression Identified in Dogs and Cats

| | |
|---|---|
| 1 | Dominance aggression |
| 2 | Fear aggression |
| 3 | Interdog/intercat aggression |
| 4 | Protective aggression |
| 5 | Predatory aggression |
| 6 | Territorial aggression |
| 7 | Food-related aggression |
| 8 | Possessive aggression |
| 9 | Play aggression |
| 10 | Maternal aggression |
| 11 | Neurological aggression (brain disease) |
| 12 | Aggression of undetermined cause |

Intercat territorial aggression can be minimized if new cats are introduced in a non-threatening manner. The new cat should initially be caged and handled in a location on the periphery of the established cat's territory. When the established cat is calm and does not react to the owner giving attention to the new cat, the new cat may be gradually moved further into the living areas of the residence.

## 20.9   SEXUAL BEHAVIOR

The courtship and mating of companion animals is controlled largely by hormones. This is one of nature's safeguards in ensuring perpetuation of the species. Male dogs detect estrus (heat) in the bitch via their keen sense of smell. Females in estrus utilize urine marking to attract the attention of males. Additional discussions on normal sexual and maternal behaviors are in Chapter 10.

Objectionable mounting behavior is sometimes directed toward human legs or arms, but may also occur on pillows, towels, blankets, or other objects. This problem is more common among dominant dogs—mounting behavior is one way dogs assert their dominance over subordinate animals. This behavior is reported in 20% to 45% of male dogs and approximately 20% of female dogs. Mounting behavior can generally be controlled by consistent punishment (e.g., scolding the dog and pushing it away); castration may decrease the frequency of this behavior in male dogs.[5]

## 20.10   LEARNING AND PRINCIPLES OF BEHAVIORAL MODIFICATION

Learning in the context of behavioral modification is the establishment of changes in behavior in response to specific stimuli. Positive reinforcement of desired behaviors can be achieved through the use of food rewards, praise, petting, as well as providing attention to and playing favorite games with the pet. Negative reinforcement is provided by collar corrections, verbal reprimands, harsh eye contact, startling noises, electric shocks, swats,

---

[5]Haupt, Katherine. 1997. Sexual Behavior Problems in Dogs and Cats. *Veterinary Clinics of North America: Small Animal Practice* 27(3): 601–615.

offensive sprays, and other unpleasant experiences. A large number of training techniques are used in behavioral modification. Examples include:

- *Habituation or desensitization.* Frequent exposure to a stimulus that does not result in pain eventually results in nonresponsiveness to this stimulus (animal learns to ignore such a stimulus).

- *Sensitization.* Frequent exposure to an unpleasant stimulus results in an increased responsiveness (may include responding to increasing low levels of the stimulus).

- *Classic conditioning.* Consistent pairing of two stimuli results in stimulus–stimulus association. The first signal is often otherwise meaningless but the pairing results in the first signal triggering anticipation of the second signal (as in Pavlov's famous pairing of a ringing bell with food).

- *Operant or instrumental learning.* This is the use of positive (e.g., food, playing, petting) or negative (e.g., sharp word, squirting with water) reinforcement. Eventually the reinforcement may be weaned to intermittent reinforcement (may use a fixed ratio, fixed interval, variable ratio, or variable interval of pairing the reinforcement with the command).

- *Counterconditioning.* When the stimulus causing the undesirable behavior is identified, the pet is given a countercommand (e.g., sit/stay) or food reward to facilitate teaching the pet to replace its undesirable behavior with a desirable one. An example is teaching a dog to sit instead of chasing down the yard fence when a car drives by.

- *Successive approximation (shaping).* The animal is rewarded when it does something approximating what you wanted (e.g., when a puppy lifts a paw in playing, the owner says "shake" and rewards the puppy).

- *Extinction.* Anticipation of a stimulus–response linkage will gradually decrease when the linkage between the stimulus and response is broken. For example, the owner who picks up a puppy when it barks at visitors may be inadvertently reinforcing barking. If the owners are able to totally ignore the puppy's barking, it may cease practicing this behavior.

- *Aversive conditioning.* An animal is taught to avoid an undesirable behavior by coupling that behavior with an aversive result. An example is the use of taste aversion to teach a puppy not to chew on shoes. Shoes can be pretreated with a bitter-tasting substance. Then the puppy is sensitized to this substance by squirting a small amount in its mouth, which results in wincing, excessive salivation, and perhaps vomiting. Subsequently, when the puppy smells or tastes the bitter substance on the shoes, it avoids them.

- *Flooding.* Prolonged exposure to a high level of the stimulus may result in the animal becoming nonreactive to it. The difference between this and habituation (desensitization) is that the level of the stimulus is sufficient to make the animal anxious to the point it gives up. This is dangerous as the animal may injure itself or damage its surroundings while trying to escape from the stimulus. It is usually considered a last resort in behavioral modification training.

## 20.11  APPROACHES TO BEHAVIORAL PROBLEMS

Behavioral problems should be addressed using a scientific approach. To successfully treat behavioral problems requires an understanding of the normal behavior of the species (ethology). Factors that are precipitating an unwanted behavior must then be determined. Next, a strategy to change the animal's behavior should be implemented. This may require a combination of behavioral modification techniques, and on occasion the use of medications. In many instances, the owners will need to be educated regarding their expectations of the time and effort that may be required to change their animal's behavior.

### 20.11.1 IDENTIFY THE PROBLEM

The first step in solving a problem is identifying the problem. Owners should be asked to state the problem in their own words (e.g., "goes crazy during thunderstorms or the shooting of fireworks," "is aggressive toward children," "urinates on furniture").

### 20.11.2 DETERMINE HISTORY OF THE PROBLEM

Question the owner about potential arousing stimuli (e.g., doorbell rings, children yelling, owner putting on coat, owner nervousness), eliciting stimuli (e.g., thunder, stranger entering house, threatening gesture or stare), and potentiating stimuli (e.g., second dog barking, stranger backing away or running). Owners should be questioned regarding their previous responses to the behavior and any effect these have had (e.g., has the owner unintentionally reinforced unwanted barking by giving the dog attention?). Information should be gathered on the family's lifestyle, the amount of time the pet is left alone, feeding and exercise habits, any training the animal has received, where and when the pet was obtained, and also a general medical history.

### 20.11.3 EVALUATE THE PROBLEM

Determine whether the problem is due to normal or abnormal behavior on the part of the pet. If the problem is a manifestation of normal behavior, it will be difficult to change (may be easier to change the owner's expectations). Testing should be performed to rule out an underlying medical disease as the cause of the problem—this is particularly important when dealing with complaints of inappropriate elimination where the animal may have a urinary tract infection or kidney disease as the underlying problem.

### 20.11.4 RECOMMEND TREATMENT ALTERNATIVES AND STRATEGIES

Once a diagnosis has been made for the cause of the behavioral problem, recommendations can be made for appropriate treatment alternatives and strategies. Some problems can only be managed, not solved. For example, if the problem is a cat killing songbirds, management alternatives may include keeping the cat indoors or having the cat wear a noisy bell. A dog that bites handlers may be managed using a muzzle; resolving the problem would involve changing its behavior from one of aggression to friendliness. Obedience training can be helpful in solving some problem behaviors (e.g., dragging the owner on walks, jumping on people, running away), whereas prescription drugs may be required for the management of some obsessive-compulsive disorders [e.g., psychogenic alopecia (pulling out hair), acral lick dermatitis,[6] tail chasing, flank sucking]. Behavioral modification strategies can be complex and may warrant referral to an individual specializing in animal behavior consultation and treatment (e.g., a diplomate of the American College of Veterinary Behavior, a member of the Animal Behavior Society, and other animal behavior specialists).

### 20.11.5 EDUCATE THE OWNER

Many owners are unaware of the importance of social structure in the lives of their pets. Owners should be instructed on methods of establishing themselves as being dominant

---

[6]Acral lick dermatitis, also known as acral lick granuloma, is an ulcerated, firm, thickened lesion on one or more limbs resulting from excessive licking; in many dogs, this excessive licking is an obsessive-compulsive disorder.

over the pet. Information should be given on normal behaviors of dogs and cats and effective methods of restraint, housetraining, claw trimming, basic coat and dental care, nutrition, the importance of toys and exercise, how to discipline the pet, and other basic management and training issues (Chapter 12). Owners must understand the causes of any behavioral problems and the steps that may be successful in managing the problem. The advantages and disadvantages of various approaches should be discussed when applicable. Many times, owner's expectations will need to be changed and myths they have heard will need to be debunked (e.g., rubbing a puppy's nose in its excrement does not help in housetraining).

# 20.12   COMMON BEHAVIORAL PROBLEMS

We have already mentioned several common behavioral problems of dogs and cats: hyperactivity, inappropriate elimination, aggression, xenophobia (fear of unfamiliar people or animals), agoraphobia/kennelosis (fear of new places or situations), coprophagy, excessive barking, mounting, car chasing, and aggression. Many of these can be minimized or eliminated through proper socialization and the establishment of strong human–animal bonding. Unfortunately, a too-strong dependence of a dog on its owner may result in significant separation anxiety when the owner is absent.

Dogs may exhibit a number of behaviors attributable to separation anxiety. Examples include inappropriate elimination (urination and/or defecation), excessive vocalization (barking, howling), destructive chewing and digging, trembling, excessive salivation, vomiting, diarrhea, fearful behavior, excessive licking or chewing on itself (self-mutilation), and overactive greetings when the owner returns. Because separation anxiety is most common among dogs obtained from animal shelters, we can speculate that having lost one owner makes the animal fearful of losing another. Separation anxiety is often triggered by a change in the owner's schedule (e.g., returning to work following a vacation).

The goals of treating a dog with separation anxiety are to decrease its dependency on the owner and increase its security when left alone. A combination of habituation (desensitization) and counterconditioning may prove helpful. The first step is to countercondition the dog by decreasing its anxiety in response to predeparture clues (e.g., the dog is given a special treat while the owner picks up the car keys and puts on a coat). Special "feeding" toys have been designed with hollow centers that can hold a treat or peanut butter and can occupy a dog for a prolonged period of time. Once the dog is comfortable and interested in the treat during the owner's predeparture activities, the next step is moving to graduated departures. Initially the owner may simply move to another part of the house for gradually increasing periods of time. When the dog is content to stay relaxed during these separations, the owner begins leaving the house for short periods, always returning before the pet becomes anxious. In some cases, the dog may benefit from treatment with a tranquilizer or antidepressant medication during this training. The advice of an animal behavior specialist is usually helpful in developing an appropriate behavioral modification training program.

Now let's consider a pair of possible puppy behavior problems: First is destructive chewing. Frequent chewing and destructiveness of nonfood items by puppies may reflect insufficient stimulation as well as curious exploration, investigation, scavenging, teething, and attempting to escape from confinement. To help prevent chewing, puppies should receive regular stimulation from play, training, and exercise sessions, and the opportunity to chew and explore. Rotating a variety of toys made of rawhide, nylon, or durable rubber is recommended.

A second problem associated with puppies is being unruly, overly feisty, and disobedient. Excessive play, jumping up, and nipping are examples of other puppy problems that should be controlled early. Healthy puppies are by nature energetic and should receive regular exercise, attention, and reward-based (e.g., food treats) training. When undesirable behavior is displayed, the owner should walk away, thereby ensuring that no physical punishment is rendered and no reward given.

## 20.13  SELECTED OTHER FELINE BEHAVIORAL TIDBITS OF INTEREST

Cats are born with certain innate instincts. They are by nature nocturnal, predatory, territorial, cautious, and—unlike dogs, which are pack animals—solitary in their hunts. Some natural instincts in cats are modified by domestication, but these instincts are never lost. This explains (at least in part) why stray and feral cats have a higher survival rate than dogs in similar circumstances.

Most domestic cats prefer a quiet life. They are known for their propensity to sleep—often 16 to 18 hours per day. Cats are disturbed by loud noises, which is why clapping one's hands and saying "no" loudly is an effective training tool.

Cats can stare for long periods of time—probably a trait that enables them to watch the movements of their prey closely in the natural state.

A cat showing the belly-up position is generally inviting attention and play, unlike dogs in which the belly-up position is a sign of submission to a more dominant individual.

Unlike dogs, cats are independent thinkers and do not particularly care whether they please their owners. Because of this, cats do not respond well to being taught to perform tricks.

## 20.14  DOG TEMPERAMENTS AND TRAINABILITY

Through selective breeding, humans have developed breeds of dogs for specific purposes. Through this process, traits have been fixed to facilitate performance of specific jobs. Different jobs require different sets of behavioral characteristics. Dogs are easiest to train when the training is relevant to the job for which the breed was developed.

Personal guard dogs need to have a dominant personality and a high degree of territorialism. They are alert and highly reactive but generally are even-tempered and easy to train. They will bond strongly to an owner or family. Examples of breeds developed to guard and protect their owners include the Mastiff, St. Bernard, and Boxer.

Livestock-guarding dogs were developed to protect flocks of sheep from predators. Most are large dogs, which in part deterred predators by their size and presence within the herd. Examples include the Great Pyrenees, Komondor, and Kuvasz. Predatory behavior was selected against in the development of these breeds making it very difficult to teach them to retrieve or even chase a ball. In general, these dogs have low reactivity, low trainability, and moderate aggression.

Herding breeds were developed to round up and move livestock. Herding instinct is related to predatory behavior; however, selection was made against biting or killing behavior. Herding breeds are highly reactive and bond strongly to their owners and trainers. They are generally easy to train, but those selecting these breeds for pets should anticipate that their dog will be programmed to chase and herd moving objects (e.g., cars, bikes), animals, and people (especially children). Examples of herding breeds include the German Shepherd Dog, Border Collie, Collie, Welsh Corgi, Shetland Sheepdog, Australian Shepherd, and others.

Hunting dogs were developed with portions of the predatory sequence including an eagerness to hunt and chase prey. However, as in the herding breeds, the killing bite was selected against. The grabbing bite was maintained in retrieving breeds. These breeds are highly trainable and reactive and generally nonaggressive. Examples include the English Pointer, German Shorthaired Pointer, German Wirehaired Pointer, Brittany Spaniel, English Cocker Spaniel, English Springer Spaniel, English Setter, Gordon Setter, Irish Setter, Golden Retriever, Labrador Retriever, and others.

Terrier breeds were developed to find and kill small rodents. These breeds work independently and kill their prey. They tend to have low to medium trainability, high reactivity, and high levels of aggression. Examples include the Bull Terrier, Norfolk Terrier, Norwich Terrier, Parson Russell Terrier, Scottish Terrier, and Smooth Fox Terrier.

Sighthounds were developed to follow prey using their eyesight and to chase it down. They have powerful hindquarters designed for running. Because they were bred to work independently, they tend to have low to medium trainability and a somewhat aloof personality. They are valued for their quiet dispositions and rarely bark. Examples include the Borzoi, Greyhound, Saluki, and Whippet.

Scenthounds were developed to follow prey using their nose. They have great endurance but lack speed. The developers of these breeds selected dogs that bay or howl when following a scent to enable the hunter to locate the prey. These breeds have a low level of reactivity, low levels of aggression, and are considered to be low to moderately trainable. They have gentle natures and are usually good with children. Examples include the Bloodhound, Basset Hound, and Beagle.

The arctic breeds were developed as sled dogs. Most were kept outdoors and used strictly for work, thus these breeds may not form strong social bonds with their owners. They are low to moderately reactive, generally nonaggressive, and show low to moderate trainability. Examples of these breeds include the Spitz, Norwegian Elkhound, Siberian Husky, and Malamute.

The genetic differences among the different breeds influence the behavior of dogs. However, many environmental factors including the temperaments of the people handling an animal, previous experiences, and training are also extremely important in the behavioral patterns shown by an individual dog.

## 20.15  SUMMARY

In this chapter, we discussed numerous important behavioral aspects of companion animals including primary socialization, which is the process whereby young puppies and kittens form social attachments and relationships with humans and other animals (e.g., their mothers and littermates).

Ethology—the scientific study of animal behavior—is an interesting but complex topic. Most behavioral practices of dogs and cats can be traced to the ability of their ancestors to survive in the wild. An understanding and appreciation of animal behavior is fundamental to a close bonding and an enjoyable relationship between pets and their owners.

Behavioral responses in companion animals are determined by heredity and learning. Hereditary responses are innate and are demonstrated by animals behaving in certain ways without needing to learn the behavior. This is often referred to as instinctive behavior. Other behaviors are learned through training and experience. Animals tend to act and respond in accordance with past experiences. For this reason, it is important to help them have good experiences early and often. The rapidity with which animals learn is determined in large part by intelligence. By consistently exposing them to the habits and experiences the owner wants them to display throughout their lives, most companion animals learn quickly especially when praise and treats are provided in connection with desirable learned and practiced behaviors. Animals communicate with each other by sight, sound, smell, and touch. Communication plays an important role in companion animal behavior and interactions with other animals.

Typical categories of behavior include ingestive, eliminative, orientation, agonistic, sexual, investigative, and gregarious. These were discussed in this chapter as were numerous behavioral problems and ways to prevent, minimize, and resolve them in cats and dogs.

Now let's move to Chapter 21 for a review of therapeutic and service uses of companion animals—topics of special interest to both pet owners and the general public.

## 20.16  REFERENCES

1.  Askew, Henry R. 2003. *Treatment of Behavior Problems in Dogs and Cats.* 2nd ed. Berlin, Germany: Blackwell Publishers.

2.  Beaver, Bonnie V. 1999. *Canine Behavior: A Guide for Veterinarians.* Philadelphia: W. B. Saunders.

3.  Hart, Benjamin L. 1978. *Feline Behavior*. Santa Barbara, CA: Veterinary Practice Publishing.

4.  Hetts, Suzzane. 1999. *Pet Behavior Protocols*. Lakewood, CO: American Animal Hospital Association (AAHA) Press.

5.  Landsberg, Gary M., Wayne Hunthausen, and Lowell Ackermann. 2003. *Handbook of Behaviour Problems of the Dog and Cat*. 2nd ed. New York: Saunders.

6.  Lindsay, Steven R. 2000. *Handbook of Applied Dog Behavior and Training* (1). Ames: Iowa State Press.

7.  Lindsay, Steven R. 2001. *Handbook of Applied Dog Behavior and Training* (2). Ames: Iowa State Press.

8.  Overall, Karen L. 1997. *Clinical Behavioral Medicine for Small Animals*. St. Louis, MO: Mosby.

9.  Swartz, Stephanie. 1997. *Instructions for Veterinary Clients Canine and Feline Behavior Problems*. 2nd ed. St. Louis, MO: Mosby.

10. Thorne, C., ed. 1992. *The Waltham Book of Dog and Cat Behaviour*. Oxford, England: Pergamon Press.

# 21 THERAPEUTIC AND SERVICE USES OF COMPANION ANIMALS

*The relationships between people and animals are important
and beautiful. We link ourselves with the world of nature
through the animals that share our lives.*

Leo K. Bustad, DVM, PhD (1920–1998)
*Former dean, College of Veterinary Medicine
Washington State University*

## 21.1   INTRODUCTION/OVERVIEW

Pets serve many roles in the lives of people: friend, surrogate child, surrogate parent, a partner in work and leisure activities, and others. In all these roles, strong bonds are commonly formed between pets and their human companions. The enjoyment of living with a pet as well as the therapeutic qualities of their presence has been recognized for many years (Chapter 1). Numerous studies have shown that being in the same room with an animal has a calming effect on people, often reducing blood pressure and heart rate. Pets serve as social catalysts, increasing communication and interactions among humans. Many people enjoy dog-related sports and activities (Chapter 13). This chapter focuses on the therapeutic and service roles of companion animals in modern society.

## 21.2   HUMAN–COMPANION ANIMAL BOND

In today's predominantly urban society, pets live with their owners in houses, apartments, and other residences where close contact leads to bonding and mutual dependency. Companion animals provide their owners with an outlet for nurturing and loving. Single adults and childless couples often treat their pets as surrogate children. Children with working parents are often greeted when they return home from school by a family pet that makes the home seem safe and welcoming. Pets provide the elderly with companionship, a stimulus for social interactions with other people, protection, and motivation to remain active and independent. Many pet owners view their pets as reflections of themselves, their interests, and personal lifestyles.

The 2003/2004 National Pet Owner Survey conducted by the American Pet Products Manufacturers Association (APPMA) shows that 62% of all U.S. households own one or more pets. Approximately half of all companion animal owners consider their pets as children or family members, sharing love, companionship, and friendship with their owners. Virtually all pet owners talk to their pets. Over 60% of all cat, dog, bird, and small animal owners purchase gifts for their pets and celebrate their birthdays; 65% buy them treats on a regular basis. The National Pet Owners Association reports that 41% of U.S. pet owners display the pet's picture in their home. Two-thirds of America's cats and dogs are allowed to sleep on their owner's bed or anywhere they want. Sales of pet products in the United States now total over $31 billion annually, outpacing the human toy industry ($20.3 billion) and the candy industry ($24 billion). For additional information and data, visit the APPMA Web site at www.appma.org.

## 21.3   SERVICE DOGS AND CATS

Dogs, and in some instances cats, can be trained to assist people with a variety of physical impairments, disabilities, and medical disorders. Service animals facilitate independent living and increase both the confidence and self-worth feelings of physically challenged

**Figure 21.1** Malinda Carlson, Companion Animal Biology graduate of the University of Illinois, training a guide dog to assist a vision-impaired individual. *Courtesy Guide Dogs for the Blind, San Rafael, CA.*

humans. Service dogs provide companionship and enhancement of confidence, independence, and freedom for their human partners. Service dogs have been trained individually to work or perform tasks for the benefit of a person having a disability. They are recognized under the Americans with Disabilities Act (1990), and their owners are entitled to have these animals accompany them in public places and facilities (including airplanes) (www.usdoj.gov/crt/ada/animal.htm and www.access-board.gov/about/ADA%20Text.htm).

## 21.3.1   GUIDE DOGS

Guide dogs are trained to assist people with impaired vision (Figure 21.1). More than a dozen schools in the United States train guide dogs. These schools have rigorous, effective training programs; only emotionally stable, attentive dogs are selected for placement as guide dogs. Typically, persons obtaining their first guide dog spend 4 weeks in a residential facility and are taught how to care for and work with their dog. This on-site training may be shortened to 2 to 3 weeks for those obtaining a new dog to replace a retired one. Many of these schools were founded as charitable organizations and do not charge for dogs provided to persons who are legally blind. Information on guide dogs and schools can be obtained from the National Federation of the Blind (www.nfb.org), American Council of the Blind (www.acb.org), and Guide Dogs of America (www. guidedogsofamerica.org).

## 21.3.2   HEARING ASSISTANCE DOGS

Hearing dogs are trained to alert people to specific sounds in the environment. For example, a hearing dog may run back and forth between its owner and the door when the doorbell rings to alert the owner that someone is there. The dog may use nosing and pawing to wake its owner when an alarm clock rings. Hearing dogs may be trained to alert their owners for other sounds including an oven buzzer, telephone ring, baby cry, name call, or smoke alarm. They are trained to make physical contact and lead their deaf partners to the source of the sound. As with vision-impaired people, the hearing-impaired person generally spends several weeks in training to learn how to work most effectively with the hearing dog. There are many organizations that train dogs to serve as hearing dogs. Examples of organizations involved with training and placing hearing dogs include Paws With A Cause (www.pawswithacause.org), Assistance Dogs International, Inc. (www.adionline.org/hearing.html), National Education for Assistance Dog Services (www.neads.org), and Dogs for the Deaf, Inc. (www.dogsforthedeaf.org). Many of these organizations obtain dogs from animal shelters and provide them with basic obedience training followed with specific training to serve as a hearing dog.

### 21.3.3  MOBILITY ASSISTANCE DOGS

Trained to help those with physical disabilities or impairments, mobility assistance dogs carry objects, pick up dropped items, open and close doors, turn lights on and off, retrieve items such as telephones, assist their handler in getting in/out of bed and in/out of wheelchairs, carry items in a dog backpack, put items in/out of washers and dryers, push life-line buttons to call for help, and assist their owners in dressing and undressing (Figure 21.2). Mobility dogs can pull wheelchairs and help stabilize walkers (Figure 21.3). They may circle around their owner to prevent bumping from/with other people, and they provide help if their owner falls. Mobility dogs assist people having spinal cord injuries, brain injuries, muscular dystrophy, multiple sclerosis, spina bifida, cerebral palsy, balance problems, arthritis, and other disorders (http://4pawsforability.org/mobilitydogs.htm).

### 21.3.4  MEDICAL ALERT DOGS

Medical alert dogs are trained to obtain assistance for owners with medical diseases such as diabetes, migraine headaches, asthma, panic attacks, and seizures disorders. These professionally trained dogs may be able to respond to subtle changes in their owner's scent or to subtle behavioral clues. In some cases, the owner can then take medications to ward off the

**Figure 21.2** Mobility assistance dog pushing an elevator call button for its owner. *Courtesy Paws With A Cause,* www.pawswithacause.org.

**Figure 21.3** Mobility assistance dog pulling its owner in a wheelchair. *Courtesy Paws With A Cause,* www.pawswithacause.org.

impending health problem. Should preventive measures fail to stop a problem, these dogs are trained to push buttons that result in calling 911.

## 21.3.5 PSYCHIATRIC SERVICE DOGS

Psychiatric service dogs assist owners in waking and bring medications to them. They help their owner in getting out of the house and into public or social settings. These dogs provide emotional support to owners during times of stress. Dogs trained to alert their owner to an impending mood attack provide the owner with advance warning so they can take medications, utilize conditioned behavioral skills, or contact a physician for assistance. Psychiatric service dogs help schizophrenic owners differentiate between sounds that are real or imaginary and can take their owner home when he/she becomes disoriented. Psychiatric service dogs assist people suffering from posttraumatic stress disorder by providing emotional support and by helping them differentiate between real threats and unfounded fears.

## 21.3.6 MINIMAL PROTECTION SERVICE DOGS

Minimal protection service dogs are trained to protect disabled handlers by using low levels or minor displays of territorial-type aggression.

## 21.3.7 CATS

Cats are less amenable to working as service animals, however, they can be trained to respond to hand signals (e.g., from owners with speech impairments) or to sounds such as rattles or bells. Cats and kittens can be selected according to personalities that best fit the needs of a handicapped person (http://flamepointcat.com/therapy.htm).

## 21.4 ANIMAL-ASSISTED THERAPY

Therapy dogs are often privately owned by individuals who wish to make a difference in the lives of others by sharing their pets with patients in nursing homes, assisted-living facilities, hospitals, prisons, and other institutions (Figure 21.4). Providing regular pet visits increases the quality of life for the people being visited and those who care for these people (Chapter 1). Dogs bring a sparkle into what might otherwise be a boring, depressing day. Such visits provide a subject for conversation and rekindle fond memories of previously owned pets. Dogs are happy to see the residents regardless of the person's physical appearance or state of health. Almost everyone responds to the wagging of a dog's tail and its eagerness in seeking attention and petting. Contact with animals lowers blood pressure, relieves stress, and alleviates depression. Organizations such as the Delta Society's Pet Partners® Program (www.deltasociety.org/dsa000.htm) and Therapy Dogs International, Inc. (www.tdi-dog.org) provide certification of a dog's aptitude for interactions with a wide range of people (young to elderly, active to physically impaired and confined to wheelchairs). These organizations may provide insurance to those volunteering their time and pets for work in animal-assisted activities. Some long-term care facilities keep one or more companion animals (dogs, cats, birds, fish, and others) on site where these animals can be shared and enjoyed by many persons each day.

Animal-assisted activities (AAA) provide opportunities for motivational, educational, recreational, and/or therapeutic benefits to enhance human quality of life. These are basically meet-and-greet activities that involve pets visiting people. There are no specific treatment goals planned for the visits. Rather, the visits are spontaneous with no set time requirements and volunteers are not required to record detailed notes (www.deltasociety.org/aboutaaat.htm).

Benefits derived from AAA include outward focus (individuals watch and talk to and about the animals instead of dwelling on their own problems), nurturing (individuals can provide care and give attention to the pet), rapport (the animal breaks the ice and facilitates interactions between the patient and the pet's owner or a therapist), acceptance (the patient feels accepted by a nonjudgmental pet), entertainment (even people who do not particularly like animals can be entertained by watching others react with pets), socialization (residents readily exit their rooms to interact with the animals and also increase their interactions with each other and the staff), mental stimulation (talking about the animals brings back fond memories and brightens the atmosphere), physical contact and touch

**Figure 21.4** Having a dog visit brightens the day for workers and residents in a nursing home. *Courtesy Paul E. Miller, Sullivan, MO.*

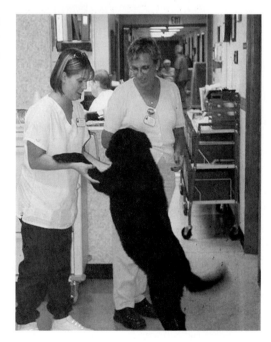

(provides positive physical contact to those who might otherwise have little or no opportunity for nonclinical touching), physiological benefits (many studies have documented that petting an animal or even observing one lowers blood pressure, calms, and relaxes people), and empathy (children can learn to empathize with others through interactions with animals).

Animal-assisted therapy (AAT) (also called animal-facilitated therapy, Chapter 1) is a goal-directed intervention in which an animal meets specific criteria and is an integral part of the therapeutic process. AAT is directed and/or delivered by health/human service professionals with specialized expertise and within the scope of practice of their profession. Key features distinguishing AAT from AAA include: (1) AAT establishes specified goals and objectives for each individual, (2) AAT documents progress toward meeting goals, and (3) AAT is directed by a health or human service professional (www.deltasociety.org/aboutaaat.htm). Many types of goals can be achieved through AAT programs (Table 21.1).

## 21.5    PET-ORIENTED CHILD PSYCHOTHERAPY

The phrase *pet therapy* was first conceptualized and formalized as a mode of child psychotherapy in 1964 by Boris Levinson, an American child psychiatrist. Levinson noted that using a dog as a focal point in therapy sessions greatly helped establish rapport with children. Soon other child psychiatrists were using pets in therapy sessions. Children engaged in pet therapy soon demonstrate improvements in communication, self-esteem, and interest in other activities. One of the first pet-facility psychiatric programs was initiated in 1977

**Table 21.1**
Goals of Animal-Assisted Therapy Programs

| Physical |
|---|
| • Improve fine motor skills |
| • Improve wheelchair skills |
| • Improve balance when standing |

| Mental Health |
|---|
| • Increase verbal interactions among group members |
| • Increase attention skills (e.g., paying attention, staying on task) |
| • Develop leisure/recreation skills |
| • Increase self-esteem |
| • Reduce anxiety |
| • Reduce loneliness |

| Educational |
|---|
| • Increase vocabulary |
| • Aid in long- or short-term memory |
| • Improve knowledge of concepts such as size and color |

| Motivational |
|---|
| • Improve willingness to be involved in group activity |
| • Improve interactions with others |
| • Improve interactions with staff |
| • Increase exercise |

*Source: www.deltasociety.org/aboutaaat.htm.*

**Figure 21.5**   Proud and contentment-filled child with her pet. *Courtesy Mike Nelson and Vern Horning, Graphics, American Nutrition Inc.*

at The Ohio State University. Patients selected a dog from a nearby kennel and had daily interactions with it. Dogs were found to act as social catalysts, helping to form positive links between patients and staff. Patients report an increased self-respect, independence, and confidence. Emotionally disturbed and learning-disabled children as well as troubled inner-city youth benefit from interactions with animals. They become more responsive and optimistic, more communicative and responsible, and more caring and compassionate. Pets help bolster morale and self-esteem, thereby improving a child's quality of life (Figure 21.5). Many researchers believe having the opportunity to touch pets, which seek physical contact with people, results in a form of therapeutic intimacy. Pets do not offer opinions or criticisms; they are attentive but silent observers with an empathetic gaze. In contrast to human therapists, dogs can snuggle up and lick a child, and the child in turn can kiss and hug it. Loving and being loved by an attentive creature helps heal disappointments and other hurts in life.

## 21.6   POLICE DOGS

Police dogs are trained for specific tasks depending on the goals of the unit. Specially trained police dogs serve police departments and a variety of other government and law enforcement agencies. Police dog handlers are trained professionals who understand how to work most effectively with their dog in serving as a team (Figure 21.6).

### 21.6.1   TRACKING POLICE DOGS

Tracking police dogs are trained to swiftly follow the scent of a person to aid in catching a fleeing criminal suspect or locate a missing person (Figure 21.7). Breeds commonly used as tracking police dogs include German Shepherd Dog and Belgian Malinois. These special dogs are trained to attack the person only if their handler gives them a verbal command to do so.

**Figure 21.6** Members of the Stamford hospital patrol force with their dogs. *Courtesy Christopher Byrne, Dogs Unlimited L.L.C., www.K9one.com.*

### 21.6.2   SEARCH AND RESCUE POLICE DOGS

Search and rescue police dogs are also highly motivated and fast-paced trackers. Their training also emphasizes agility and ability to work under a variety of settings (Section 21.7). Only friendly, outgoing dogs are selected for this work; their response to finding a person is one of exuberant enthusiasm rather than the threatening stance of the criminal-tracking police dog. The German Shepherd Dog, Belgian Malinois, Golden Retriever, and other sporting breeds are used in search and rescue work.

### 21.6.3   PATROL AND TRACKING POLICE DOGS

Trained to assist in apprehending criminals (Figure 21.8), patrol and tracking police dogs will chase down and corner a suspect, even under gunfire. They are also trained to protect their handler. The German Shepherd Dog is the most common breed used in police patrol work.

### 21.6.4   NARCOTICS DETECTION DOGS

Narcotics detection dogs utilize their highly sensitive sense of smell to locate the source of odors associated with various narcotics (Figure 21.9). These dogs can be trained to detect a

**Figure 21.7** A police dog trained by Dondi Hydrick at Cross Creek Training Academy Inc. tracking a suspect who crossed a paved road. *Courtesy Dondi Hydrick, www.cctapolick9.com.*

**Figure 21.8** Patrol police dog being trained to apprehend a criminal. *Courtesy Chistopher Byrne, Dogs Unlimited L.L.C.,* www.K9one.com.

wide variety of substances including marijuana, cocaine, crack, hashish, methamphetamine, ecstasy, heroin, LSD, oxicotin,[1] special K, and other illicit or recreational drugs.

### 21.6.5 CADAVER DETECTION DOGS

Trained to detect the odor of bodies in various stages of decomposition, cadaver detection dogs are used in criminal investigations and also at disaster sites such as the 1995 Oklahoma City bombing and the 2001 terrorist attacks of the New York City World Trade Center. These dogs are trained to give a clear signal without disturbing possible crime scene evidence (e.g., to bark or lay down to indicate a find rather than digging or pawing).

### 21.6.6 EXPLOSIVE DETECTION DOGS

Explosive detection dogs are trained to detect varying quantities of explosive agents as well as locate explosive residues. These dogs are capable of locating concealed firearms and ammunition in suitcases, lockers, vehicles, and other locations (Figure 21.10). They are used by police forces, customs agents, border patrol units, and the armed forces.

## 21.7 USDA BEAGLE BRIGADE

The USDA (United States Department of Agriculture) Beagle Brigade is a group of dogs that detects prohibited fruits, plants, and meat at United States international airports and mail facilities. Beagle Brigade dogs sniff the baggage of international travelers as they proceed through Federal Inspection Service areas. When they sniff prohibited agricultural items in

---

[1]Oxicotin is a prescription medication used to treat patients that have a narcotic addiction. It is sometimes used illegally as a recreational drug.

**Figure 21.9** A well-trained narcotics detection dog searching bags for illegal drugs. *Courtesy Christopher Byrne, Dogs Unlimited L.L.C.,* www.K9one. com.

passenger luggage or packages, the dogs sit to alert their handlers (Plant Protection and Quarantine officers). The dogs are taught to sit when they smell the scents of citrus fruit, mango, beef, or pork. The officers then check the passengers' bags, confiscate any prohibited items, and give the dogs a food reward. Beagles were selected for these searches because of their acute sense of smell and their gentle nature with people. Their natural love of food makes them effective detectives and happy to work for treats. Beagles remain calm in crowded, noisy locations such as busy airport baggage claim areas. Beagle Brigade dogs wear distinctive green jackets with "Protecting American Agriculture" on both sides (Figure 21.11). After 1 year of experience, beagles sniff out prohibited material correctly 80% of the time. Their success rate rises to about 90% after 2 years of experience.

**Figure 21.10** Explosive detection dog searching a Federal Express truck near the World Trade Center, September 2001. *Courtesy Christopher Byrne, Dogs Unlimited L.L.C.,* www.K9one.com.

**Figure 21.11** A member of the USDA Beagle Brigade searches luggage at the Orlando International Airport. *Courtesy Department of Agriculture National Detector Dog Training Center, Orlando, FL.*

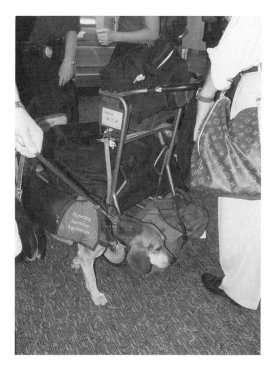

## 21.8   SEARCH AND RESCUE DOGS

Search and rescue dogs have been utilized in disasters since World War I. Some are trained to look only for survivors, others to find cadavers and body parts. In 1989, the Federal Emergency Management Agency (FEMA) established criteria for Urban Search and Rescue dog teams. Each team must pass rigorous tests for certification. The dogs must demonstrate response to commands, agility skills, barking alert skills to notify rescuers of a victim, and a willingness to overcome fears of wobbly surfaces and tunnels. Many years of training are required to achieve certification, and teams must then be recertified every 2 years. In addition to their efforts as members of search and rescue squads, these dogs serve as unofficial team mascots and morale boosters. Dogs working at the 2001 New York City World Trade Center disaster site were important "therapy animals" as exhausted rescue workers offered water to the dogs, gave them a loving pat, and asked their handlers where they were from (Figure 21.12). Search and rescue dogs are happiest when working and become discouraged when no survivors are

**Figure 21.12** Canine "Ray" with handler Christopher Byrne and federal police officer during the recovery efforts at the World Trade Center disaster site, New York City, September 2001. *Courtesy Christopher Byrne, Dogs Unlimited L.L.C.,* www.K9one.com.

found; to help bolster their dog's enthusiasm, handlers hide in debris and wait for the dog's enthusiastic barks of discovery. Search and rescue dogs at the World Trade Center disaster site worked shifts of up to 12 hours; they were plagued by dust, irritants, and damaged footpads but worked diligently through these discomforts. These dogs have tremendously admired work ethics. During work at Ground Zero,[2] a 28-year veteran firefighter and paramedic from the Miami Valley Urban Search and Rescue Task Force, Terry Trepanier, said, "To know what they've been through and to see how they performed makes you proud. And they never complain. They just wag their tails and say 'Let's go.'"

## 21.9  PERSONAL PROTECTION DOGS

Personal protection dogs are trained to show aggression on command from their owner or in situations where the owner is being threatened by another person. Dogs selected for use in personal protection should be highly obedient and trustworthy in the presence of other people. These dogs are taught to tolerate being petted and handled by strangers and to display aggression only when given the protect command. Breeds commonly trained for use in personal protection include German Shepherd Dog, Rottweiler, Doberman Pinscher, and Boxer.

- *Deterrent or Level 1 personal protection dogs* are taught to lunge, show teeth, and bark aggressively when given the protect command. They will stay on task despite threats from an aggressor. These dogs typically work on a leash.

- *Defense or Level 2 personal protection dogs* have been trained not only to lunge and bark at aggressors but also to proceed with biting and holding the offender using their powerful jaws. These dogs increase their grip if the aggressor strikes at them. The dogs are trained to release their bite on command.

- *Offense or Level 3 personal protection dogs* are trained to chase and knock down a fleeing person. They will hold the person down until commanded to release their bite (Figure 21.13).

## 21.10  PREMISE SECURITY DOGS

Premise security dogs often work in teams with handlers in areas that would be dangerous for security guards to work alone (e.g., high crime/gang areas). These dogs are used to protect large areas and those with restricted views (e.g., construction sites, warehouses, parking garages). The high visibility and aggressive barking of security dogs provide an effective deterrent to intruders (Figure 21.14). These dogs are trained to threaten intruders and to chase and knock them down on command. In nonthreatening situations, the dogs are well-mannered and can be handled by strangers. They respond immediately and aggressively to commands given by their handler. Breeds commonly used for security work include the German Shepherd Dog, Belgian Malinois, Dutch Shepherd, Rottweiler, Bouvier, Doberman Pinscher, and American Bulldog. Many of these dogs are imported from Europe. European dog breeders focus primarily on work ethic and disposition when selecting dogs for breeding, whereas some dog breeders in the United States tend to place their highest priority on conformation, or "beauty before brains."

The cost of training any specialized working dog is high; trainers require temperament and health testing as part of the selection process. Certification of dogs for use in security and other specialized work teams involves proficiency testing and staged real-life scenarios to better ensure that handlers and dogs are fully prepared for situations they may encounter in everyday work.

---

[2]Live, Love & Laugh with Golden Retrievers Web page, Disaster Urban Search & Rescue report on dogs working at World Trade Center site in September 2001, http://landofpuregold.com/disaster.htm.

**Figure 21.13** Training a dog to chase and apprehend a fleeing person. *Courtesy Dondi Hydrick (trainer), Cross Creek Training Academy, Inc.,* www.cctapolicek9.com.

## 21.11 SUMMARY

When humans began the domestication and development process of companion animals some 15,000 years ago, it is doubtful they envisioned a future time when these highly intelligent, trainable, precious pets would serve such important roles in the mental and physical health and well-being of humans as they now do. In this chapter, we discussed the

**Figure 21.14** Chris and canine "Ray" patrolled the Federal Building following the 2001 World Trade Center disaster. *Courtesy Christopher Byrne, Dogs Unlimited L.L.C.,* www.K9one.com.

all-important bonding of people with pets. We emphasized the special services and companionship provided by guide dogs, hearing dogs, mobility assistance dogs, medical alert dogs, psychiatric service dogs, and minimal protection service dogs. We discussed the contributions of companion animals in animal-assisted therapy as well as pet-oriented child psychotherapy. We discussed the important services rendered by police dogs, search and rescue dogs, personal protection dogs, and premise security dogs. We cited numerous organizations that help facilitate many important services provided to physically impaired humans. Additional photographs of dogs at work are shown in Color Plate 12. We commend and salute the women and men of these important, caring/sharing organizations. Simply knowing that these examples of compassion and service exist make us proud of being part of today's caring society.

In Chapter 22, we discuss the important and interesting contributions of companion animals in the advancement of biomedical research—advancements that protect and enhance the lives of millions of children and adults globally.

## 21.12   REFERENCES

1. Bustad, Leo K. 1990. *Compassion: Our Last Great Hope.* Renton, WA: Delta Society.

2. Chandler, Cynthia. 2001. *Animal-Assisted Therapy in Counseling and School Settings* (electronic resource). Greensboro, NC: ERIC Clearinghouse on Counseling and Student Services.

3. Fine, Aubrey H., ed. 2000. *Handbook on Animal-Assisted Therapy: Theoretical Foundations and Guidelines for Practice.* San Diego, CA: Academic Press.

4. Jackson, Donna M. 2003. *Hero Dogs: Courageous Canines in Action.* New York: Little, Brown.

5. Pfaffenberger, Clarence J. 1976. *Guide Dogs for the Blind, Their Selection, Development, and Training.* Amsterdam; New York: Elsevier Science.

6. Schroeder, Howard, ed. 1985. *Working Dogs.* Mankato, MN: Crestwood House.

## 21.13   WEB SITES

1. www.usdoj.gov/crt/ada/animal.htm
2. www.access-board.gov/about/ADA%20Text.htm
3. www.nfb.org
4. www.acb.org
5. www.guidedogsofamerica.org
6. www.pawswithacause.org
7. www.adionline.org/hearing.html
8. www.neads.org
9. www.dogsforthedeaf.org
10. http://4pawsforability.org/mobilitydogs.htm
11. http://flamepointcat.com/therapy.htm
12. www.deltasociety.org/dsa000.htm
13. www.tdi-dog.org
14. www.deltasociety.org/aboutaaat.htm
15. www.K9one.com
16. www.cctapolicek9.com
17. http://landofpuregold.com/disaster.htm

# 22 ANIMALS IN BIOMEDICAL RESEARCH[1]

[1]The authors acknowledge with appreciation the contributions to this chapter of Dr. Stanley E. Curtis, a professor with the Department of Animal Sciences, University of Illinois at Urbana-Champaign. Dr. Curtis has been actively involved in animal care and use issues nationwide for more than 3 decades.

473

*I maintain this as an incontestable fact. The results of a properly conducted and properly appreciated experiment can never be annulled, whereas a theory can change with the progress of science.*

Carl von Voit (1831–1908)
*German chemist and physiologist*

## 22.1   INTRODUCTION/OVERVIEW

Animals have been used in scientific experimentation for more than 2,000 years. In the third century BC in Alexandria, Egypt, the philosopher and scientist Erasistratus used animals to study body functions.

Five centuries later, the Greek physician and medical physiologist Claudius Galen (129–ca. 201) used dogs, monkeys, and pigs to prove his theory that veins carry blood, not air. In succeeding centuries, numerous animals were studied to further determine how the body functions. Knowledge gained through the experiment conducted in 1622 by English physician and anatomist William Harvey (1578–1657) demonstrated the circulation of blood. This was followed by research showing the effect of anesthesia on the body in 1846 and the relationship between bacteria and disease in 1878.

Of the nearly 100 Nobel Prizes awarded in Medicine and Physiology during the 20th century, more than two-thirds involved the use of animals. An example was awarding the 1923 Nobel Prize to two medical scientists who discovered insulin. This discovery has helped extend the productive lives of millions of human diabetics. It was made possible by studying dogs following the removal of the pancreas. In the absence of insulin secretion by the pancreas, neither dogs nor humans can utilize sugar without insulin injections or some other medication.

Other important discoveries by 20th century Nobel Laureates enabled major medical advances in diagnosis, treatment, and prevention of many health-related conditions including heart disease, cancer, poliomyelitis, measles, smallpox, and massive burns. New 20th-century procedures—including open heart surgery, blood transfusions, organ transplantations, laser surgery, and others—were developed in whole or in part by experiments that involved animals. Animal models are useful in determining safety and efficacy of new drugs, vaccines, and other compounds (Figure 22.1) used by humans. Small mammals are often used in developing sources of antigens or attenuated agents that can be used for vaccination of domestic animals (e.g., mouse-passaged bovine lymphoid cells, which when injected into cattle elicit protective immunity to a lethal dose of East Coast Fever).

**Figure 22.1**   Monkeys participating in research studies to evaluate the efficacy of cholesterol-lowering compounds. *Courtesy Dr. Steele F. Mattingly.*

**Figure 22.2** Group nutrition studies with dogs in China. *Courtesy Dr. James E. Corbin.*

"Man's best friend" has provided important insights into nutritional considerations and problems in humans throughout recorded history. Twentieth-century examples include Ivan Pavlov's work with digestive glands (1910). The Russians erected a monument to honor the dog used in Pavlov's studies of digestion. Early on, dogs were used to determine if foods were safe for humans. This was based on the assumption that if a food poisoned a dog it would also poison a human.

Other consequential 20th-century research with dogs included Joseph Goldberger's (1874–1929) work with experimental black tongue of dogs and its relationship to pellagra in humans (1928), Schaumann's studies on the value of heating foods (1913), G. R. Cowgill's research on the role of vitamin B in the nutrition of dogs (1921), E. Mellanby's research on canine rickets and osteomalacia (1921), and numerous studies on vitamins A and D deficiencies and bone disease.

Much of the early research related to individual nutrient requirements of humans was conducted first with dogs (Figure 22.2). Indeed, dogs probably have contributed more to the fundamental findings of human nutrition than any other species. Their contributions to specific vitamin requirements in humans are excellent examples. Dogs were used extensively at the Mayo Clinic in developing low-residue diets for humans suffering from enteric diseases.

## 22.2 ANIMAL MODELS FOR HUMAN RESEARCH

The use of animal models holds numerous advantages over human subjects in scientific research including increased control over genetics (Figure 22.3) and environment, improved protocol compliance, and availability of invasive techniques. Creating animal models through the use of surgical procedures is a common strategy employed in biomedical research (Figure 22.4).

At the turn of the 20th century, rats, mice, birds, and fish represented about 90% of the animals used in biomedical research and numbered approximately 30 million. Because the above animals are unregulated species in the United States (Section 22.9.2), their numbers cannot be determined with precise accuracy. Approximately 1.1 million animals representing species frequently considered to be companion animals contributed to medical research, experiments, testing, and teaching in 2002. The approximate numbers were rabbits, 244,000; guinea pigs, 245,000; hamsters, 180,000; dogs, 68,000; nonhuman primates (mostly chimpanzees and monkeys), 52,000; cats, 24,000; and numerous amphibians (especially frogs). The number of farm animals (143,000) and other animals (180,000) used in animal research totaled 323,000 in 2002.[2]

---

[2]Animal Welfare Report. FY 2002. Report of the Secretary of Agriculture to the President of the Senate and the Speaker of the House of Representatives. USDA Animal and Plant Health Inspection Service.

**Figure 22.3** Depicted here is American Association for Laboratory Animal Science (AALAS) Laboratory Animal Technician, Lindsay Barnett, who manages a breeding colony of genetically engineered mice. Research mice in the Notre Dame Freimann Life Science Center are identified by ear tagging. Ear biopsies are taken for use in genotyping when the animals are tagged. *Courtesy Kay Stewart and the University of Notre Dame.*

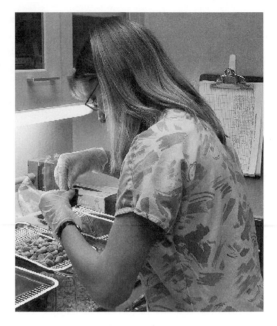

The number of companion animals and other regulated animal species used in medical research has decreased by approximately one-half during the past three decades. And although the number of unregulated animals is not reported (Section 22.9), it is believed that their use has decreased significantly, as well. Chapters 6 and 7 include discussions pertaining to companion animal use in laboratory animal research.

## 22.2.1 PREFERRED ANIMAL MODELS

The dog is, in many cases, the best research model for studying human physiology and diseases. Both canines and humans are monogastric and have many gastrointestinal structural similarities. Rats and swine utilize a significant amount of cecal fermentation, whereas the ceca of dogs and humans are underdeveloped.

**Figure 22.4** Depicted here, Deborah Donahue performs microsurgery on mice using a dissecting microscope. Ms. Donahue is a certified Laboratory Animal Technician who specializes in microsurgery on mice. She earned the B. S. degree in Animal Science from Cornell University. *Courtesy Kay Stewart and the University of Notre Dame.*

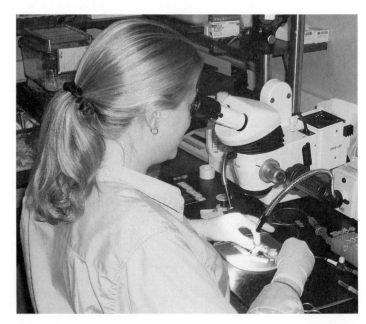

Canine and feline colons are proportionate in length to human colons. Both dogs and humans have bacteria in the colon that contribute a significant amount of colonic fermentation. Dog and human colons have bacteroides, bifidobacteria, and lactobacilli as predominant bacterial species.

Dogs and humans also have similar dietary habits. Although dogs evolved from carnivores, they have adapted to an omnivorous diet. With the exception of vitamin C, dogs generally require the same essential nutrients as humans (Chapter 9). Moreover, dogs are susceptible to many of the same diseases as humans, including cancers, diabetes, kidney disease, and obesity.

Although dog and cat nutrition may overlap, cats are generally a less-desirable model for human nutrition studies than dogs. Cats are strictly carnivores, whereas dogs and humans are omnivores. Cats require the amino acid taurine for conjugation of bile acid and are sensitive to a deficiency in the amino acid arginine (Chapter 9). Cats are unable to convert beta carotene to vitamin A or to convert tryptophan to niacin, making preformed vitamin A and niacin essential nutrients for cats. Still another unique dietary need of cats is their requirement for the omega-6 fatty acid arachidonic acid, as cats are unable to synthesize a sufficient amount from linoleic acid. Additionally, the small body of cats makes it difficult to obtain adequate amounts of blood, fecal, tissue, or urine specimens for research purposes.

Because dogs and humans have many similarities, it is, of course, also useful to apply many human health- and nutrition-related findings to similar conditions in dogs. For example, because few data are available on the nutritional requirements of unhealthy animals, recommendations for feeding sick dogs often have been empirically derived from human studies. An example is with hepatobiliary disease, which is characterized by diminished liver function leading to an accumulation of metabolic waste products in the blood, anorexia, and weight loss.

Both dogs and cats develop spontaneous tumors, many of which have histopathological and biological traits similar to human tumors. This makes certain forms of animal cancer good research models for studying cancer in humans.

# 22.3   ENSURING EXPERIMENTAL CONTROL IN ANIMAL RESEARCH

Because minimizing unregulated variables is so critical to ensure statistical precision in animal research, we share the following anecdote of how a leading petfood manufacturer cooperated in supporting the monumental research efforts of Dr. Jonas E. Salk in developing a prophylactic vaccine to ward off poliomyelitis, a much dreaded disease as recently as the 1950s.

When co-author James E. Corbin, an animal scientist specializing in nutrition, was Director of Pet Foods Research and Laboratory Animal Research for the Ralston Purina Company in St. Louis, Jonas Salk (1914–1995), American physician and research scientist, developed the first effective vaccine for preventing poliomyelitis. In 1953, Dr. Salk announced the development of a trial vaccine. Among the first to receive the vaccine were Dr. Salk, his wife, and their three sons. The vaccine soon was found safe and early evidence indicated it was effective. It was tested further during a mass trial in 1954 when 1,830,000 schoolchildren were vaccinated. This trial was sponsored by the National Foundation for Infantile Paralysis. The vaccine was pronounced safe and highly effective in April 1955. Dr. Salk received the Congressional Gold Medal for his "great achievement in the field of medicine."[3]

An important development in the support needed to advance Dr. Salk's research in preparing a consistently reliable vaccine occurred when he approached his medical colleague and friend, surgeon William H. Danforth, M.D., son of W. H. Danforth, founder of the Ralston Purina Company. Dr. Salk asked Dr. Danforth to confer with his father seeking his

[3]*The World Book Encyclopedia, Vol. 17.* 1965. Chicago: Field Enterprises Educational Corp, 62.

corporate assistance in formulating and manufacturing a nutritionally complete monkey diet. Dr. Salk was obtaining rhesus monkeys for research in vaccine development from India, where they were being fed a nutritionally incomplete diet and displaying significant variability. Consequently, Dr. Salk's work with the inadequately fed monkeys was yielding unsatisfactory, inconsistent results when he used their kidney tissues in preparing polio vaccine. Company founder Danforth embraced the idea and right away called Dr. Corbin to his office, asking him and his research associates to develop a "monkey chow" and test it to determine its value for growing and supporting healthy monkeys.

Dr. Corbin and his research colleagues accepted the challenge, and in only a few months they had formulated a new Monkey Chow™. Young weanling monkeys fed the diet grew quickly and reproduced fast-growing offspring, confirming that the new Monkey Chow formulation was nutritionally complete—just what Dr. Salk had hoped for. Ralston Purina then manufactured and shipped the world's first commercial primate chow to India.

Dr. Salk was highly pleased with the results, and told Dr. Corbin he believed that having the benefits of the more uniform and healthy monkeys reduced by 4 to 5 years the time required to bring his safe, highly successful polio vaccine to the global public. Figure 22.5 depicts Dr. Jonas E. Salk with a monkey fed the new Monkey Chow in India and brought to his research laboratory at the University of Pittsburgh.

## 22.4 EXPERIMENTAL CATEGORIES SUPPORTED BY COMPANION ANIMALS

Animals undergird three important areas of biomedical science: (1) biomedical and behavioral research, (2) teaching, and (3) drug toxicity and product testing.

Biomedical research advances our understanding of how biological systems function and enhances the knowledge base of human and veterinary medicine. Biochemical experimentation is conducted in accordance with the principles of the scientific method. This method established two basic requirements of conducting a valid experiment: (1) control of variables that might affect the response of the animal so only one factor (or set of factors) is changed at a time, and (2) replication of results confirmed by other scientists/research laboratories.

Behavioral research is directed toward determining the factors affecting animal behavior and how the animal responds to different specific stimuli. Much behavioral research is environmental in nature but some involves the study of responses to physical stimuli or manipulation of systems such as the brain and nervous system (Chapter 20).

**Figure 22.5** Laboratory technician, Francis Yurochko (*l*) holds the rhesus monkey as Dr. Jonas Salk (*r*) examines an important contributor to the research development of the Salk polio vaccine. *Courtesy The Jonas Salk Trust and the Family of Jonas Salk.*

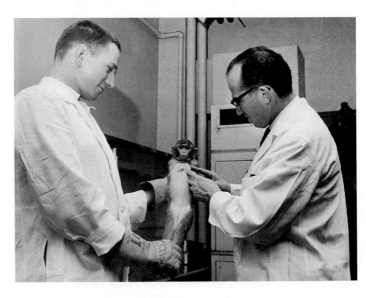

Demonstrations involving live animals are conducted to educate and train students in curricula including biology, medicine, veterinary medicine, animal science, physiology, psychology, and affiliated sciences. However, approximately two-thirds of medical schools in the United States have abandoned the use of live animals in laboratory classes for alternatives such as molded rubber and plastic replicas, computer simulators, and videos. Numerous colleges of veterinary medicine have followed suit with similar alternatives (see Section 22.8).

Animals also are employed to determine efficacy and safety of new medicines as well as the possible toxicity of chemicals—such as those that may be found in cosmetics and bath soaps—to which humans or animals may be exposed. Most of these experiments are conducted by research scientists of commercial companies to fulfill requirements of the Food and Drug Administration (FDA) or other governmental agencies.

The toxicity of a chemical compound is commonly expressed by means of an $LD_{50}$ (lethal dose) value. This is a statistical estimate of the dose necessary to kill 50% of a population of the test species (commonly 10 or more mice or rats) under specific conditions. Although not absolutely required by law, these tests have become a standard measuring tool in applications to the FDA for approval of drugs and for meeting certain requirements of the Environmental Protection Agency (EPA) and the Consumer Product Safety Commission (CPSC).

As with other kinds of scientific research, biochemical experiments are of two types: basic and applied. Basic biomedical research is conducted to increase the knowledge base and understanding of the chemical, physical, and functional mechanisms of life and disease processes. The primary purpose of applied research is to attain specific targeted objectives such as development of a new drug, therapy, or surgical procedure. Applied research commonly involves the application of existing knowledge, much of which has been obtained through basic research. Applied research can be either experimental or clinical. Experimental research is conducted with animal or nonanimal alternatives (e.g., see Section 22.8). Clinical trials are the final step in the biomedical research process. These trials are performed only after the procedure, drug, or device being investigated has been determined to be effective and then thoroughly tested in experimental research conducted under regulatory guidelines designed to ensure safety, efficacy, and protection of human users.

## 22.5   USE OF ANIMALS IN BIOMEDICAL RESEARCH

The basic assumption of biomedical research is that scientists should develop ways of relieving and minimizing human and animal suffering. The primary purpose in using animals in biomedical research is that animals serve as research surrogates for humans. Animal rights advocates argue that if the tests are for the benefit of humans, then people should serve as the experimental subjects. There are certain limitations, however, in using humans as test subjects. The process of aging, for example, can be observed through experiments with rats, which have an average life expectancy of only 3 years, or with certain types of monkeys, which live 15 to 20 years. Many nutritional and physiological experiments require either numerous subjects of the same weight and genetic makeup or special diets or specific external environmental conditions. These types of conditions and requirements make employment of human subjects difficult or impossible. By using animals in such studies, scientists can observe subjects of uniform age and background and in sufficient numbers to determine if the findings are consistent and applicable to a larger population.

Animals are important in research because they have body systems and functions that react and interact with stimuli similar to those of humans. The more true this is with a particular animal, the more valuable that animal is for use in a particular experimental application. For example, dogs are well-suited for biomedical research because of the relative size of their organs compared with those of humans. The first successful kidney transplant was

**Figure 22.6** Surgery room with ample fluorescent lighting and double overhead surgery lights in a well-equipped veterinary medicine surgery suite, Willow Creek Pet Center, Sandy, UT. *Courtesy Dr. Rick Campbell.*

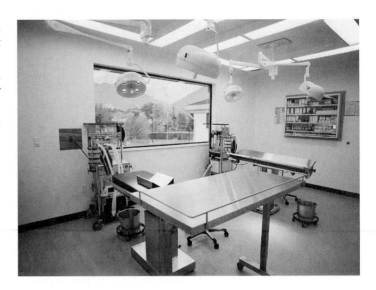

performed in a dog, and the techniques used to save the lives of "blue babies" and other human babies with structural heart defects were first developed with dogs. Open-heart surgery techniques, coronary-bypass surgery, and heart-transplantation procedures all were developed using dogs, Shetland ponies, and calves. Numerous surgical procedures now utilized in human medicine were first developed and tested in well-equipped veterinary surgery facilities much like the one depicted in Figure 22.6.

Another important factor in biomedical research is the amount of information available about a particular animal. Mice and rats play an especially important role in research and testing, in part because repeated experiments and regulated breeding (purpose bred; Section 22.9.7) have created an enormous pool of scientific data with which the findings from a new experiment can be compared and given meaning. The rapid rate of reproduction by mice and rats (Chapter 7) also makes them important in studies of genetic and other experiments that require observations over several generations. Additionally, humans cannot be selectively, intentionally mated to produce inbred strains as can be done with purpose bred animals. Therefore, humans could not be substituted for animals in studies in which an inbred strain is essential (Figure 22.7).

Virtually all scientists concur that research is necessary to reduce disease and suffering in humans as well as in animals. Because biomedical advances depend on research with animals, it can be argued that not using them would be unethical because it would deprive humans and animals of the benefits of biomedical research findings and applications.

**Figure 22.7** Purpose bred (inbred strain) mice are housed in this near germ-free, environmentally controlled research facility to study anti-cancer drugs. *Courtesy Dr. Steele F. Mattingly.*

## 22.6  ANIMAL WELFARE AND ANIMAL RIGHTS DEBATES AND CHALLENGES

Controversy related to the involvement of animals in biomedical research is not new. Such use has been opposed for philosophical and emotional reasons in Europe for more than 300 years and in the United States for over 100 years. The animal cause movement is complex and diverse. It is rooted in a broad spectrum of attitudes, concerns, and philosophies. Historically, most groups within the movement have tended to be identified primarily as anticruelty or antivivisectionist. Now, the distinction is mainly between animal rights and animal welfare.

Those concerned with *animal welfare* do not necessarily propose to ban the use of research animals. Instead, their primary efforts in recent years have been to influence the passage of legislation to ban experimentation with animals obtained from animal shelters/public pounds and to improve the care and housing of animals kept in research facilities. In many cases, they have been joined in these efforts by animal rights organizations.

The *animal rights* movement is more complex and embraces a broader range of attitudes and activities. This movement received its most noteworthy impetus from three books published in the 1970s. These books created the atmosphere and introduced the concepts, terms, and rationales for what became the animal rights philosophy and movement.

The first of these books, *Animals, Men and Morals* (edited by S. Godlovitch, R. Godlovitch, and J. Harris. 1971. New York: Taplinger, Inc.), was an anthology that revived and presented many long-held, dormant thoughts and views regarding the relationships between humans and animals. One was the view that animals have rights.

The 1971 book led to the 1975 publication of two books that form the basic animal rights philosophy currently espoused. One, *Victims of Science* (R. Ryder. 1975. London: Davis-Poynter), introduced the concept of *speciesism*. The other, *Animal Liberation: A New Ethic for Our Treatment of Animals* (by Australian philosopher Peter Singer. 1975. New York: Random House), placed the concepts of animal rights and speciesism in a philosophical framework that questioned humans using animals in several contexts including food and clothing.

The debate revived by Singer had been opened by René Descartes (French mathematician and philosopher, 1596–1650), who defended the use of animals in experiments by insisting that because animals could respond to stimuli in only one way—according to the arrangement of their organs—they lacked the ability to reason and think. This, he opined, made them similar in nature and function to machines. Humans, on the other hand, through their ability to think and talk, were capable of responding to stimuli in a variety of ways. Descartes argued that these differences made animals inferior to humans, and given their machine-like nature, people may use them as they would use a machine, including as objects of experiments.[4] He also argued that animals can learn only by experience and not by the teaching-learning method used by humans (e.g., humans do not have to experience everything, such as touching a hot stove, to learn it is harmful, whereas animals do).

Singer's book was not merely a philosophical treatise, it also was a polemical call to action. He deplored the attitude of humans toward nonhumans as a "form of prejudice no less objectionable than racism and sexism" and urged that liberation of animals become the next cause after the civil rights and women's rights movements.[5] His book enjoyed considerable popularity that brought to the animal rights cause the same commitment, zeal, and tactics that had been employed in support of the civil rights, women's rights, and antiwar movements of the 1960s and 1970s.

Dozens of state and national animal rights organizations were successful in recruiting thousands of new members who devoted themselves to conducting demonstrations, lobbying legislators, circulating petitions, raising funds, and communicating with and through the media. Many members of numerous groups have embraced the philosophy espoused by

---

[4]Descartes, René. http://www.philosophypages.com/hy/4b.htm
[5]Singer, Peter. 1975. *Animal Liberation: A New Ethic for our Treatment of Animals.* New York: Random House.

Singer (1975) that all human relationships with nonhuman animals should be altered so as to end all forms of exploitation—whether for business, companionship, food, fur, leather, research, pleasure, sport, or transportation. The concentration of effort on animal experimentation and modern farms—another target—was a strategic decision based on the belief that these would be the easiest causes for which public support could be generated and funds raised.

The most active and visible organization nationally has been People for the Ethical Treatment of Animals (PETA). Its basic philosophy is that all mammals have been created equal—a belief expressed by a PETA leader in the aphorism "a rat is a pig is a dog is a boy"—and that people's treatment of animals represents superracism and fascism. PETA advocates the release of all animals, including companion animals, back to nature. Its activities have included raids on research laboratories in which animals were being used in experimentation and the distribution of edited videotapes and other materials stolen by terrorist organizations.

The radical Animal Liberation Front (ALF) has been identified by Scotland Yard as an international underground terrorist organization that is active in the United Kingdom, France, Canada, and the United States. Unlike some terrorist groups, which have focused their activities on entering laboratories to free animals, ALF has destroyed records, equipment, and facilities, including the burning of an animal disease diagnostic laboratory under construction at the University of California-Davis College of Veterinary Medicine in 1987.

Efforts of animal rights activists to interfere with the conduct of research are of great concern to scientists. So are the harassment of and death threats to researchers and their families. These "raids of liberation" have caused millions of dollars in damage to laboratories, theft of hundreds of laboratory animals, and destruction of valuable research records. These break-ins have interrupted, delayed, and forced the abandonment of some research. They have increased the cost of conducting research by requiring more security at laboratories, the replacement of stolen animals and equipment, and the repetition of experiments. Inevitably, the loss of valuable data and records has slowed and hindered efforts to prevent and treat human diseases.

Legislation to ban the use of animals in biomedical research and to establish the belief that animals have rights similar to those of humans has been introduced in numerous state legislative assemblies as well as in the U.S. Congress.

## 22.6.1   Validity and Conduct of Animal Research

Animal rights activists insist that animal experiments are unnecessary for scientific and medical progress. Many scientists insist animal models are required to provide the information needed to advance health and quality of life for humans and animals. Furthermore, many scientists argue there are no alternative scientific methods for obtaining the same quality of information as that gained through using animals in research. Scientists point out that global life expectancy of humans more than doubled (from 30 to 65 years) during the 20th century—a century during which the fruits of animal research contributed immensely to decreased infant mortality, effective treatments for many diseases, the virtual elimination of some life-threatening diseases, and an enhanced quality of life for humans and animals in general.

Noteworthy is the fact that virtually every advance in medical science in the 20th century—from antibiotics and vaccines to antidepressant drugs and organ transplants—were achieved directly or indirectly through the use of animals in laboratory experiments (Table 22.1). The result of those experiments has been the control or elimination of many infectious diseases, including smallpox, poliomyelitis, and measles, and the development of numerous life-saving techniques, including blood transfusions, burn therapy, and open-heart and brain surgery. Overall, this has meant a longer, healthier, better-quality life for humans. For many, it has indeed meant life itself. Moreover, many humans do participate in biomedical research by way of clinical trials as well as in testing new treatments of cancer and other diseases. Thus, humans have shared with nonhuman animals the experiences of pain and inconvenience for the sake of science and the advancement of medicine.

**Table 22.1**
Selected Medical Advances Made Using Investigational Animals

| | |
|---|---|
| Pre-1900 | Treatment of anthrax, beriberi (thiamine deficiency), rabies, and smallpox |
| Early 1900s | Treatment of pellagra (niacin deficiency) and rickets (vitamin D deficiency), electrocardiography, and cardiac catheterization |
| 1920s | Blood transfusions and intravenous feeding, discovery of insulin (greatly improved diabetes control), and discovery of thyroxine |
| 1930s | Therapeutic use of sulfa drugs, prevention of tetanus, development of anticoagulants, modern anesthesia, and neuromuscular blocking agents |
| 1940s | Treatment of rheumatoid arthritis and whooping cough; therapeutic use of antibiotics including penicillin, Aureomycin, and streptomycin; discovery of the Rh factor; treatment of leprosy; and prevention of diphtheria |
| 1950s | Prevention of poliomyelitis, development of cancer chemotherapy, open-heart surgery, influenza vaccine, and cardiac pacemaker |
| 1960s | Prevention of rubella; corneal transplant and coronary bypass surgery; therapeutic use of cortisone; development of radioimmunoassay for measuring minute quantities of antibodies, hormones, and other bodily substances |
| 1970s | Prevention of measles, modern treatment of coronary insufficiency, heart transplant, development of nonaddictive pain killers |
| 1980s | Use of cyclosporin and other antirejection drugs to facilitate organ transplantation; artificial heart transplantation; identification of psychophysiological factors in depression, anxiety, and phobias; development of monoclonal antibodies for treating disease; development of anesthetics that do not trigger malignant hyperthermia in humans |
| 1990s | Development of fast-clotting powder and impregnated bandages; application of magnetic resonance imaging (MRI) and computed tomography (CT) scans, which facilitate superior imaging of three-dimensional structures (e.g., these scientific techniques allow medical professionals to distinguish between cancerous and healthy tissue and to detect aneurysms in the brain, heart, and other body organs); the 2003 Nobel Prize in medicine was awarded for the highly important advancements made in MRI |

Scientists believe it is important for the public to understand that had scientific research using animals in the first two decades of the 20th century been restrained, as antivivisectionists and other activists were then—and still are—advocating, millions of Americans alive and healthy now would never have been born or would have suffered an early death. Indeed, their parents or grandparents would have died from diphtheria, scarlet fever, tuberculosis, diabetes, appendicitis, or one of numerous other maladies.

Animal rights activists attribute advances in human longevity and health to public health measures and improved nutrition. Scientists concur that such measures have been important. However, for most infectious diseases, improved public health and nutrition have played only a minor role. This is made clear when one considers the marked reduction in the incidence of infectious diseases such as whooping cough, rubella, measles, and poliomyelitis. Virtual eradication of these and most other infectious diseases was not achieved until vaccines and drugs discovered through research using animals were developed. And development of an effective vaccine against AIDS (acquired immunodeficiency syndrome) and SARS (severe acute respiratory syndrome) depends greatly on continued research studies using animals.

Use of animals in biomedical research also has directly benefited animals. Obvious examples have been advances in veterinary medicine and surgery such as vaccines for rabies and distemper and the development of products to protect against the often fatal affliction of heartworm infestations in dogs and cats (Chapters 14 and 15). In farm animals, the control of hog cholera and tuberculosis and brucellosis in cattle was achieved through research

**Table 22.2**
Selected Veterinary Medicine Advances Made Using Investigational Animals

• Immunization against anthrax, distemper, infectious hepatitis, parvovirus, rabies, and tetanus
• Treatment for animal parasites (e.g., heartworm)
• Orthopedic surgery for dogs and horses
• Surgery to correct hip dysplasia in dogs
• Experimental radiation techniques and immunotherapy for cancer in dogs
• Identification and prevention of brucellosis and tuberculosis in cattle and hog cholera in swine
• Treatment of feline leukemia
• Improved nutrition for pets
• Experimental laser techniques and surgical applications
• Improved adhesives for surgical use and to glue horseshoes to damaged/weak hooves
• Vaccine to protect horses against the West Nile virus
• Nuclear-transfer cloning of pigs whose organs (minus the two genes that cause pig organs to be rejected by humans) may avert immune reaction (rejection) when transplanted to humans (xenotransplantation)

with animals. Selected breakthroughs and developments in veterinary medicine attributable to research with animals are noted in Table 22.2.

## 22.6.2   ADVANCING BEHAVIORAL RESEARCH

Behavioral research with animals in a laboratory setting has been a special target of animal rights activists who assert that it subjects animals to stress without producing meaningful results. Psychologists and other behavioral scientists believe discoveries made through behavioral research may not only save the lives of individual animals but also provide information important in saving entire species from extinction.

Cited as an example of information gained through behavioral experiments on imprinting—the tendency of an animal to identify with and relate to the first species it comes into contact with—has been useful in training captive-born animals to relate to members of their own species and has been important, for example, in helping condors (a large vulture of the high Andes and California coast) survive and propagate in the wild. Animal studies are helping save the musk ox from extinction in Alaska. Research conducted on the sexual behavior of animals threatened with extinction has led to successful reproduction in captivity, an essential step in restoring their numbers in the wild. Some of this research is being conducted in the wild but much of it can be conducted only using controlled conditions in a laboratory. (Cloning is a recent scientific development that also holds promise of helping to perpetuate threatened species, Chapter 25.)

Behavioral research with animals has benefited humans as well (Chapter 20). For example, fundamental information on how people learn was discovered with experiments on animals in laboratories. The learning principles and behavioral modification therapies discovered or developed through animal experiments are being used to treat conditions such as enuresis (bedwetting), addictive behaviors (tobacco, drugs, alcohol), and compulsive behaviors such as anorexia nervosa.

Biofeedback techniques that have become a useful means of treatment for a number of conditions were developed through behavioral research with animals. The use of biofeedback enables people to control what are normally automatic body functions such as blood pressure, heart rate, and muscle tension. Biofeedback helps cardiac patients reduce the risk of heart attack by controlling blood pressure; assists persons paralyzed with spinal injuries to raise their blood pressure and permit them to sit up; and often relieves the discomfort of migraine headaches, insomnia, and lower back pain. Many of these afflictions had no reliable treatment before biofeedback, which was developed through studies of the rat's nervous system.

Experiments with cats have enhanced the understanding of the corpus callosum, a band of fibers connecting the left and right sides of the brain needed for transferring information from one side to the other. This finding, for which a Nobel Prize was awarded in 1981, led to the development of new treatments for patients with strokes, language disorders, brain damage, intractable epilepsy, and other neurological conditions.

Studies conducted with mice and monkeys have helped establish and explore the relationship between stress and conditions including heart disease, hypertension, and breakdowns in the immune system that leave individuals vulnerable to disease. Such studies are now leading to a better understanding of the nature of psychosomatic illnesses in humans.

# 22.7   UNEXPECTED DIVIDENDS OF RESEARCH

Experiments often produce findings that have no apparent relevance to their original purpose but which contribute to another line of inquiry or lead to important new discoveries. This factor is called research serendipity (the phenomenon of discovering valuable information or things not being sought). Examples of serendipity include important medical advances such as the discovery of penicillin. Another example is a natural antibiotic from the skin of frogs that was discovered during an experiment involving the removal of eggs from a frog through an incision in the skin. The alert investigators noted that the frog's skin was not infected with bacteria under conditions highly favorable for bacterial growth and conjectured that some antibacterial substance must be present.

# 22.8   ALTERNATIVES TO USING ANIMALS IN RESEARCH AND TESTING

The concept of alternatives was first introduced in 1959 by two British scientists. It was defined in terms of what is commonly known as the three Rs—*refinement, reduction,* and *replacement.* In their 1959 book, *The Principles of Humane Experimental Techniques,* W. M. S. Russell and R. L. Burch contended that a moral approach to scientific investigation requires scientists to work toward *refinement* of techniques to reduce potential suffering, toward *reduction* in the number of animals needed, and when possible toward *replacement* of animals by nonanimal techniques.

Animal rights activists have attempted to make the word *alternatives* virtually synonymous with the word *substitutes.* They advocate two: (1) *in vitro* research (cell, tissue, and organ cultures), and (2) computer simulation of biological systems in the form of mathematical models as well as the use of videos and films. Both methods play important roles in biomedical research and have permitted the performance of certain experiments not possible with animals. Additionally, both have avoided the need to use animals in some stages of research. However, they cannot serve as total replacement of live animals in experiments because they cannot reproduce the intact biological system provided by live animals. Moreover, tissue cultures, test tubes, and computer simulations cannot show the full impact of AIDS, Alzheimer's disease, cancer, heart attacks, SARS, or strokes.

The Health Research Extension Act passed by Congress in 1985 (Section 22.9.4) requires the National Institutes of Health (NIH) to establish a plan for conducting, evaluating, and disseminating research to train scientists in methods of biomedical and behavioral experimentation that (1) do not require the use of live animals, (2) reduce the number of animals, and (3) produce less pain and distress than those currently in use.

Computer simulations—promoted as the great hope of future research by animal rights activists—have aided in developing new mechanisms and techniques for scientific inquiry. However, both computers and computer models and simulations have sufficient inherent limitations that it is unlikely they will ever totally replace animal experiments. One limitation is the very nature of simulation. The validity of any model depends on how closely it resembles the original. Much remains unknown about the body and the various biological systems of humans and animals. For example, how the body metabolizes many chemicals or drugs or the manner in which brain cells transmit the sensory signals that create vision are unknown; therefore, they cannot be accounted for in a computer model. Until complete knowledge of a particular biological system is developed (via study of live animals), it will be impossible to construct a model that will predict or accurately represent the reaction of the system to various stimuli. Also, to construct a computer simulation that would fully replace the use of a live animal in behavioral or biomedical research would require a complete understanding of the subject behavior or biomedical unknown, which in turn would eliminate the need for computer simulation.

No known method of study can exactly predict or reproduce the characteristics and qualities of a living, intact biological system. Therefore, to understand how such a system functions in a particular set of circumstances or how it will react to a given external or internal stimulus, it becomes necessary at some point to conduct experiments with animals to ascertain the reaction(s). There is simply no alternative to this approach. To make advances in modern medicine for improved health and quality of life in animals and humans, the use of animals is an essential element of health-related research.

### 22.8.1  GERIATRIC MEDICAL CHALLENGE

As the knowledge base and skills in the basic sciences continue to improve, it is likely that human and companion animal life spans will be lengthened still further. This will bring new challenges. Alzheimer's dementia is an example (nearly 50% of our people show symptoms by age 85). Unless scientific research provides new prophylactic measures and/or effective therapeutic applications, lengthened life span in humans could be viewed as a curse rather than a blessing.

Without satisfactory alternatives to using animal models in research methodology needed to answer basic questions about the pathogenesis of mammalian diseases, we can expect laboratory research animals to be involved in discovering new procedures and products to improve the quality of human life as well as that of companion and other animals.

## 22.9  LEGISLATION AND REGULATORY ACTIVITIES AFFECTING THE USE OF ANIMALS IN RESEARCH

The first legislation concerning animal experimentation was enacted in the United Kingdom in 1876 via the Cruelty to Animals Act. Animal welfare and animal rights organizations have been successful in persuading the U.S. Congress to enact legislation that establishes standards for and governs the treatment of animals used in research, teaching, and testing. These same groups have convinced some state, county, and municipal governing bodies to pass laws prohibiting the release of animals (primarily dogs and cats) from public pounds to researchers.

### 22.9.1  ANIMAL CARE LAWS

When humans assume the custody of animals, they incur a responsibility and a moral obligation to avoid when possible—and when impossible to reduce to an absolute minimum—

any pain or distress that animals might experience before, during, and following experimentation, demonstration, or testing.

Scientists support setting high standards for the proper care and treatment of laboratory animals. Indeed, it was the research community that first developed the concept and principles of humane treatment of animals early in the 20th century and established the first formal guidelines for humane care and use of research animals through the NIH in 1963. Further evolution along this line led to the present NIH Office of Laboratory Animal Welfare (OLAW) and the independent, voluntary accrediting organization, Association for Assessment and Accreditation of Laboratory Animal Care International (AAALAC, www.aaalac.org/about.htm).

## 22.9.2   LABORATORY ANIMAL WELFARE ACT (AWA) OF 1966

This law regulated dealers who handle dogs and cats as well as laboratories that use dogs, cats, hamsters, guinea pigs, rabbits, or nonhuman primates. Individuals or organizations buying or selling dogs and cats for laboratory use are required to be licensed and are held to certain standards of care and housing for the animals as promulgated by the Secretary of the U.S. Department of Agriculture (USDA), the federal agency responsible for enforcing the law. Laboratories using these animals were required to register with the USDA and to identify and keep records on each animal used. The law was amended in 1970 to change the name to AWA and broaden the definition of *animal* to include any warm-blooded animal used in research, exhibition, or the wholesale pet trade. In 1972, horses and farm animals not used for biomedical research (also birds, mice, and rats) were excluded (unregulated).

With public pressure mounting to revise the 1966 act, the U.S. Congress passed the Food Security Act of 1985 containing an amendment called the Improved Standards for Laboratory Animals Act. This amendment as well as the Health Research Extension Act of 1985, which amended the Public Health Service Act of 1978, codified the NIH voluntary guidelines. They directed the NIH to establish a plan for research in biomedical experimentation that does not use animals, reduces the number of animals used, and minimizes animal pain and suffering.

Moreover, these regulations established specific and stricter requirements for the care and treatment of all animals used in federally funded research. These regulations require appropriate use of pain-relieving drugs, veterinary care, and euthanasia on all animals used in experiments. They set detailed and specific requirements regarding cage sizes, feeding and watering, lighting, and sanitation for animals residing in laboratories. To ensure adherence to the guidelines, research institutions are required to establish an animal oversight committee (typically called the Institutional Animal Care and Use Committee, IACUC) to approve and supervise all procedures involving live animals including regular inspections of animal holding areas and to review ongoing procedures to ensure compliance with approved protocols. The committee must include at least one veterinarian, one nonscientist, and one person not affiliated with the institution. An institutional veterinarian, formally appointed by the institution, has oversight over animal care and use and advises the IACUC on a variety of matters.

The Animal Welfare Act amendments of 1985 addressed the psychological well-being of nonhuman primates and the exercising of dogs only. However, the guide, which is the set of guidelines that all facilities accredicted by AAALAC International must follow, addresses the social and physical needs of all species. Many types of social and environmental enrichments are employed for animals in both the biomedical research field and in the zoo arena, with pair or group housing being the starting point for all social species. Physical structures such as PVC tubing, paper tubes, and molded plastics are added to enclosures giving the animals choices in their environment and also adding complexity to their environment (Chapters 6 and 7). The animal's ability to display normal behaviors such as grooming, exploration, foraging, and burrowing is one of the basic goals of an enrichment program. Modifying mal-adaptive behaviors such as self-mutilation, stereotypic pacing, and aggression can also be accomplished through the provision of species-appropriate enrichment.[6]

## 22.9.3    THE GOOD LABORATORY PRACTICES ACT OF 1978

This federal legislation included a list of scientific procedures that a toxicological researcher should follow, as does the EPA. Included are detailed procedures to follow for animal care, feeding, handling, separation of species, disease control and treatment, sanitation, food and water inspection, bedding, and pest control.

## 22.9.4    THE HEALTH RESEARCH EXTENSION ACT OF 1985

This federal legislation is often referred to as the National Institutes of Health Authorization Bill, which establishes provisions for the care and use of animals in research funded by the Public Health Service (PHS) under whose aegis the NIH (the major public source of bio-medical funding in the United States) operates. Important provisions of the bill established NIH guidelines (previously a matter of contractual agreement between NIH and the re-searcher/institution) as a legal agreement. Violation of these rules now permits seizure of federal funding from an offending institution.

In response to the Health Research Extension Act of 1985, the U.S. Department of Health and Human Services (USDHHS) contracted for the publication of two important documents: (1) *Guide for Grants and Contracts,* and (2) an updated *Guide for the Care and Use of Laboratory Animals* (first published under another title by another organization in the 1960s), as mentioned above. The second publication included provisions to follow pertain-ing to institutional policies, husbandry, veterinary care, and physical plant.

## 22.9.5    NATIONAL INSTITUTES OF HEALTH REVITALIZATION ACT OF 1992

In addition to stipulations concerning acts of violence against research facilities receiving federal research funds, the act included other sections relevant to animal research including instructions for the director of NIH to prepare a plan dealing with methods of the three Rs already mentioned—*reduction, refinement,* and *replacement* of live animals in biomedical re-search—and with the reliability and validity of these methods, and to submit the plan to Congress. The director of NIH was also instructed to establish a committee known as the In-teragency Coordinating Committee on the Use of Animals in Research to advise the direc-tor on the above plan with inputs from the EPA, FDA, and National Science Foundation (NSF), among other agencies.

---

[6]We are grateful to Kay Stewart, RVT, LATG, CMAR, for authoring this paragraph. Ms. Stewart graduated with distinction from the Purdue University School of Veterinary Medicine Veterinary Technology Program. She joined the Freimann Life Science Center (FLSC) at the University of Notre Dame as a technician when it opened in 1985. She currently serves as Associate Director of the FLSC. Through AALAS, Ms. Stewart earned technologist certification in 1987 and Certified Manager of Animal Resources (CMAR) in 2000. She teaches laboratory animal science classes and serves as faculty advisor to the Notre Dame Pre-Veterinary Medicine Club. Because of her keen interest in the welfare/state-of-being of animals used in biomedical research, she organized and provides leadership/oversight in the environmental enrichment program for research animals at Notre Dame. She has presented numerous invited papers at national and international meetings on Notre Dame's environmental enrichment program.

### 22.9.6 THE ANIMAL ENTERPRISE PROTECTION ACT OF 1992

This federal law includes provisions for fines and imprisonment of those found guilty of causing disruption to the functioning of an animal enterprise resulting in economic damage exceeding $10,000. The act also imposes sentences of up to 10 years or life imprisonment on persons causing serious bodily injury or death of another person during the course of an offense. It also directs the U.S. Attorney General and Secretary of Agriculture to study the extent and effects of both domestic and international animals rights terrorism on activities using animals for food, fiber, or agricultural research or testing, and to issue a report to Congress. The resulting document titled *The Report to Congress on Animal Enterprise Terrorism* was issued in August 1993.

### 22.9.7 POUND LAWS

Most states, cities, and communities have some form of law(s) requiring the seizure, detention, and humane destruction of stray or unclaimed animals. The stated purpose of these laws is to protect human health and safety. Millions of animals arrive at public pounds or private animal shelters each year—some delivered by owners, others collected by animal control officers (Chapter 24). Through most of the 20th century, scientists could obtain these animals from most public pounds and some private shelters after a specified period, which gave owners an opportunity to claim their lost pets. The use of pound animals reduced the expense of conducting experiments and also meant that homeless animals could contribute to science rather than be directly euthanized.

Many experiments require the use of purpose bred animals—those bred for a particular trait or whose genealogy or physiology must be known for experimental results to be valid. Such animals are available from breeders and professional dealers. Other experiments can be conducted with random-source animals whose ancestry and physiologic history are unknown. Such animals are preferable in some experiments because their unknown and varied backgrounds approximate the diversity characteristic of the human population. Until recently, one supply of random-source animals has been public pounds. Banning the use of public pound animals in many localities has added millions of dollars annually to the cost of biomedical research using dogs.

## 22.10 SUMMARY AND CODA

In this chapter, we discussed the use of animals in biomedical and behavioral research. We noted many important contributions of such research to the health and quality of life of humans and animals alike. We pointed out, as well, differing philosophies and views of animal welfare and animal rights activists. There are major differences of opinion regarding philosophic and moral issues related to the use of animals in biomedical research, behavioral research, education and training, and toxicological testing in the United States. Many believe being opposed to the use of animal experimentation seems inconsistent among animal rights activists who authorize for their sick child or parent the use of medications derived through experiments or testing that utilized animals.

The arguments and activities of animal welfare and animal rights activists and organizations present people of the United States with an ongoing public issue entailing some fundamental decisions that must be faced regarding the extent to which animals may be used in biomedical research. The basic issue raised by the philosophy of the animal rights movement is whether humans have the right to use animals in ways that may cause them to suffer and/or die. To accept the moral viewpoint of the animal rights movement would require a total ban on the use of animals in any scientific research, education, and testing.

Implementation of this option would effectively arrest most basic biomedical and behavioral research, teaching, and product testing.

This may inevitably shift the debate from one of animal rights to one of moral responsibilities to be met by humans. How humane or moral would it be to remove hope for human victims of AIDS, Alzheimer's disease (it has been estimated that if a prevention or cure for this extremely depressing, degenerating, brain-ravaging illness that confuses the mind and contorts the body is not discovered, 14 million people in the United States will suffer from it by 2040), aging and congenital defects, cancer (kills more than 550,000 Americans annually), disabling injury, heart disease (more than 700,000 die from heart disease in the United States annually), SARS, strokes (strike 600,000, and over 160,000 die from cerebrovascular diseases annually),[7] diabetes, kidney failure, and many other diseases and disorders knowing that solutions to these dreaded maladies afflicting a large and growing proportion of an ever-older population can be lessened or solved only through biomedical research with animals? Sometimes it is necessary to inflict pain or sacrifice the life of an animal in the laboratory in order to avoid or eliminate pain, or even save the lives of humans and animals.

Many primates suffered and died in the process of developing vaccines for poliomyelitis. But because they did the number of poliomyelitis cases in the United States decreased from 58,000 in 1952 to 4 in 1984. (Worldwide cases of polio decreased from 350,000 in 1988—when the World Health Organization's Global Polio Eradication Program began—to less than 700 in 2003).[8] That means in each of the intervening years, and hopefully for every year in the future, tens of thousands of humans have not, do not, and will not spend their lives in wheelchairs or walking in braces or lie trapped and suffering in iron lungs. Given these results, defenders of animal use in biomedical research might ask whether the use of those primates served a moral purpose. The answer by most citizens presumably would be a resounding "yes."

The American Medical Association (AMA) has been an outspoken proponent of biomedical research for more than 100 years. The AMA believes research involving animals is essential to maintaining and improving human health and quality of life. Through polls and other means, the vast majority of people in the United States have indicated their support of the humanely responsible use of animals in biomedical research and testing. At the same time they have expressed a strong wish that animals be protected against unnecessary pain and suffering.

Hopefully, when the whole of the debates related to the use of animals in the biomedical sciences is sorted out, majority judgment will prevail. It is convential wisdom among political scientists that people respond to truthful information and that citizens' ability to make sound decisions—provided they have the facts at hand—should never be underestimated.

# 22.11 REFERENCES

1. Allen, T., and K. Clingerman. 1992. *Animal Care and Use Committees Bibliography.* Beltsville, MD: USDA, National Agricultural Library (Pub. No. SRB92–16), 38.

2. Godlovitch, S., R. Godlovitch, and J. Harris, eds. 1971. *Animals, Men and Morals.* New York: Taplinger, Inc.

3. Kitchell, R. L., H. H. Erickson, E. Carstens, and L. E. Davis. 1983. *Animal Pain Perception and Alleviation.* Bethesda, MD: American Phy. Soc., 231.

4. Milken, Michael. 2003, July 14. American Science, American Lives. *Wall Street Journal.*

---

[7]National Alzheimer's Association, www.alz.org; USDHHS, Centers for Disease Control and Prevention, National Center for Health Statistics, Hyattsville, MD 20782, (301) 458–4000;
http://www.cdc.gov/nchs/fastats/lcod.htm
[8]http://www.who.int/features/2004/polio/en/

5. Mitruka, B. M., H. M. Rawnsley, and D. V. Vadehra. 1976. *Animals for Medical Research: Models for the Study of Human Disease.* New York: John Wiley and Sons, 591.

6. Morton, D. B. and P. H. M. Griffiths. 1985. Guidelines on the Recognition of Pain, Distress, and Discomfort in Experimental Animals and an Hypothesis for Assessment. *Vet. Rec.* 116: 431–436.

7. Orlans, F. B., R. C. Simmonds, and W. J. Dodds, eds. 1987, January. In *Laboratory Animal Science.* Special issue. Published in collaboration with the Scientists Center for Animal Welfare.

8. Russell, W. M. S. and R. L. Burch. 1959. *The Principles of Humane Experimental Techniques.* London: Methuen & Co., 238. (Reprinted as a special edition in 1992 by the Universities Federation for Animal Welfare.)

9. Ryder, R. 1975. *Victims of Science.* London: Davis-Poynter.

10. Singer, Peter. 1975. *Animal Liberation: A New Ethic for Our Treatment of Animals.* New York: Random House.

11. Smith, C. P. 1991. *Animal Models of Disease Bibliography.* Beltsville, MD. USDA, National Agricultural Library, 31.

12. Stephens, M. L. 1986. *Alternatives to Current Uses of Animals in Research, Safety Testing, and Education.* Washington, DC: Humane Society of the United States, 86.

13. Williams, C. H., ed. 1988. *Experimental Malignant Hyperthermia.* New York: Springer-Verlag.

14. Zurlo, J., D. Rudacile, and A. M. Goldberg. 1994. New York: Mary Ann Liebert Publishers, 86.

## 22.12  WEB SITES

1. www.aaalac.org/about.htm

2. http://www.who.int/features/2004/polio/en/

3. Descartes, René http://www.philosophypages.com/hy/4b.htm (click on animals)

# 23 CAREER OPPORTUNITIES ASSOCIATED WITH COMPANION ANIMALS

*Our association with animals cannot be taken for granted, but rather must be tempered with virtuous care and compassion. In the bonding of humans and animals in beneficial ways, not only are the lives of people enriched, but those of animals we enjoy and whose training and services we appreciate can be accorded bona fide care, love, and compassion.*

Susan B. Anthony (1820–1906)
*American suffragist*

# 23.1 INTRODUCTION/OVERVIEW

Throughout life we make many important decisions—choosing a spouse, pursuing an education, selecting a residence. Few are more consequential than that of choosing a career. And few factors, if any, are more fundamental in selecting a career than one that promises true job satisfaction—being able to perform respected work that provides opportunities to attain emotional, personal, and professional fulfillment. Although financial aspects should be considered, most people want more out of employment than a paycheck.

Moreover, job satisfaction can best be achieved when we possess and convey a positive attitude and outlook on life and our chosen profession. This was emphasized by W. Clement Stone, a business executive in Chicago who, in celebrating his 100th birthday in April 2002, reflected on his experiences in life and said, "There is very little difference in people, but that little difference makes a big difference. And, that little difference is attitude and the big difference is whether it is positive or negative."[1]

For persons having an innate love of animals—those who derive great pleasure and personal satisfaction in *making a positive difference* in the health and lives of animals and their owners—we describe and suggest numerous career opportunities in this chapter that can help facilitate and fulfill those respected goals.

# 23.2 VETERINARY MEDICINE

Pursuing the study of veterinary medicine is an educational challenge that gives one the opportunity to obtain a rather broad-based yet professional education. Career options are many: private practice (small and/or large animal) and then possibly one or more specialties (Section 23.3); teaching, research, extension/outreach/public service; public health, industry, governmental agencies; ecosystem health and conservation biology; laboratory animal medicine and biomedical research (Chapters 7 and 22); military and homeland security; zoo medicine. Mention of numerous others follows in this chapter.

Veterinarians have an important role in food production, understanding the transmission of food-borne illnesses, and evaluating food production and processing practices. Additionally, veterinarians have a keen understanding of zoonotic diseases (i.e., animal diseases that can be transmitted to humans). As a result, veterinarians play an important public health role ranging from developing life-sustaining strategies involving zoonoses and bioterrorism to preventing injuries such as companion animal bites (Chapters 14 and 24).

The veterinarian, as with many other health and allied professionals, is linked with companion animals through an important triangular relationship. Because of the unique opportunity to witness firsthand the interactions of companion animals with their owners and families, the veterinarian is placed in a role that has a direct effect on human mental health and physical well-being (Figure 23.1). The practicing veterinarian is often in the position to observe the reactions of owners to the loss of companion animals, especially if the decision is made to euthanize their closely held animal friend that has reached a condition in which it clearly cannot regain its health and live a normal life with the absence of pain (Chapter 14). This can be especially traumatic in cases when the companion animal aided deaf and/or blind persons (Chapters 12 and 21).

Veterinarians are the only doctors trained to protect the health of both animals and humans. Not only are they educated to meet the health needs of numerous animal species, they also contribute immensely in ensuring a high level of food safety, public health, and environmental protection (Chapters 14 and 22).

[1]Stone, W. Clement. 2002. Investing in Illinois. *Newsletter of the University of Illinois Foundation and Private Giving* 39: 28–29.

**Figure 23.1** Dr. Rick Campbell, Willow Creek Pet Center, Sandy, Utah, discusses the findings of an x-ray taken of a dog's stifle (knee) with the pet's owner. *Courtesy Dr. Rick Campbell.*

## 23.2.1   CARING PROFESSIONALS

According to numerous consumer surveys, veterinarians consistently rank among the most respected professionals in society. American Veterinary Medical Association (AVMA) data indicate that about 80,000 veterinarians were actively practicing in the United States in 2003. The profession is growing at a rate of approximately 3% annually.

In taking the Veterinarian's Oath, graduates solemnly swear to use their "scientific knowledge and skills for the benefit of society through the protection of animal health, the relief of animal suffering, the conservation of animal resources, the promotion of public health, and the advancement of medical knowledge."

## 23.2.2   PROTECTING THE HEALTH OF ANIMALS AND SOCIETY

Employment opportunities are virtually endless under this career umbrella and include private or corporate clinical practice, teaching and research in both veterinary medicine and human medicine, regulatory medicine (state and federal), public health, and military service.

*Private or corporate clinical practice.*   Approximately three-fourths of U.S. veterinarians are engaged in private or corporate clinical practice. Of those, more than three-fourths practice small animal medicine (including companion animals, exotic pets, and laboratory animals). Other veterinarians limit their practice to the healthcare of production animals (farm/ranch/commercial operations) and advise owners regarding the best approaches to the medical aspects of animal production and services. Some treat horses exclusively, others treat a combination of several species including birds, rodents, reptiles, and amphibians.

*Teaching and research.*   Veterinarians teach veterinary students, other medical professionals, and scientists. Veterinary college faculty members conduct research, instruct, and develop continuing education programs that assist practicing veterinarians in acquiring new knowledge and skills. They are employed in colleges and universities, governmental agencies, zoos, aquariums, museums, private/public research institutions, and industry and are dedicated to finding new and improved ways to prevent diseases and treat animal and human health disorders. Veterinarians have made many significant contributions to human health. Examples include their helping to conquer malaria and yellow fever, solving the mystery of botulism, producing an anticoagulant used to treat people with certain forms of heart disease, and helping define/develop surgical procedures applicable to humans such as open heart surgery, hip and knee joint replacements, and limb and organ transplants.

Veterinarians who work in pharmaceutical and biomedical research firms develop, test, and supervise production of medications and biological products (e.g., antibiotics and vaccines) for human and animal use (Chapter 22). These veterinary researchers often have specialized training in fields such as bacteriology, immunology, laboratory animal medicine, pathology, pharmacology, and virology.

Numerous veterinarians are employed in management, technical sales and services, and other positions in agribusiness, petfood, and pharmaceutical companies. Additionally, they are recruited by the agricultural chemical industry, private research testing laboratories, zoos, and the feed, livestock, and poultry industries.

*Regulatory medicine.* Veterinarians employed by the U.S. Department of Agriculture's Food Safety and Inspection Service (FSIS) help protect the public from unsafe meat and poultry. They ensure that food products are safe and wholesome through carefully monitored nationwide inspection programs.

To help prevent the introduction of foreign diseases into the United States, veterinarians are employed by state and federal regulatory agencies to quarantine and inspect animals brought into the country. They supervise interstate shipments of animals, test for certain diseases (Chapter 16), and oversee campaigns to prevent and eradicate diseases such as tuberculosis, brucellosis, and rabies that pose threats to animal and human health (Chapter 14).

U.S. Department of Agriculture (USDA) veterinarians in the Animal and Plant Health Inspection Service (APHIS) monitor the development and testing by private companies of new vaccines to ensure their safety and efficacy. APHIS veterinarians are also responsible for enforcing laws for the humane treatment of animals used in research, testing, and teaching (Chapters 6, 7, 14, 22, and 24).

*Public health.* Veterinarians serve as epidemiologists in city, county, state, and federal agencies investigating animal and human disease outbreaks such as food-borne illnesses, influenza, rabies, Lyme disease, and West Nile viral encephalitis (Chapter 14). Through inspections, they help ensure the safety of food processing plants, restaurants, and public water supplies.

Veterinarians working in environmental health programs study the effects of pesticides, industrial pollutants, and other contaminants on animals and humans. At the U.S. Food and Drug Administration (FDA), veterinarians evaluate the safety and efficacy of medicines and food additives. Veterinarians also work at the Agricultural Research Service, Fish and Wildlife Service, Environmental Protection Agency, Centers for Disease Control and Prevention, National Library of Medicine, and National Institutes of Health.

There is a shortage of public health veterinarians in the United States. In response to this need the American Association of Veterinary Medical Colleges launched the "5X7" initiative in 2002. This initiative seeks to train 500 additional public health veterinarians by 2007. The University of Minnesota College of Veterinary Medicine (UMCVM) was one of the first to respond to this need and opportunity by increasing its class enrollment by 10 veterinary students beginning in the fall of 2003. The UMCVM also offers a DVM/MPH program. For more information, visit http://160.94.9.156/ahc_content/colleges/vetmed/.

*Military service.* Veterinarians in the U.S. Army Veterinary Corps are among the front ranks in protecting the United States against bioterrorism. They are responsible for food safety, veterinary care of government-owned animals, and biomedical research and development (Chapter 22). Officers with special training in laboratory animal medicine (Chapter 7), pathology, microbiology, and related disciplines conduct research in military and other governmental agencies (Figure 23.2).

**Figure 23.2** Major Kim Lawlor, DVM, riding an elephant during an exercise called Cobra Gold in Thailand. *Courtesy United States Veterinary Corps.*

In the U.S. Air Force, veterinarians serve in the Biomedical Science Corps as public health officers. They manage communicable disease control programs at air force bases throughout the world and work toward limiting the spread of HIV, influenza, hepatitis, and other infectious diseases through education, surveillance, and vaccination.

*Other professional activities.* Zoological medicine, aquatic animal medicine, aerospace medicine (shuttle astronauts), animal shelter medicine (Chapter 24), sports medicine (race horses, Greyhounds), and wildlife management also employ veterinarians.

## 23.3   COLLEGES OF VETERINARY MEDICINE AND VETERINARY SPECIALTIES

There are 28 AVMA-accredited colleges of veterinary medicine in the United States, four in Canada, and five in other countries. Each college is evaluated regularly by the Council on Education of the AVMA and must maintain a high level of quality in faculty, programs, and facilities to remain accredited.

Following completion of the required veterinary medical curriculum, many graduates choose to pursue additional education and training in one or more of the following 20 AVMA-recognized veterinary specialties: American Board of Veterinary Practitioners, American Board of Veterinary Toxicology, American College of Laboratory Animal Medicine, American College of Poultry Veterinarians, American College of Theriogenologists, American College of Veterinary Anesthesiologists, American College of Veterinary Behaviorists, American College of Veterinary Pharmacology, American College of Veterinary Dermatology, American College of Veterinary Emergency and Critical Care, American College of Veterinary Internal Medicine, American College of Veterinary Microbiologists, American College of Veterinary Nutrition, American College of Veterinary Ophthalmologists, American College of Veterinary Pathologists, American College of Veterinary Preventive Medicine, American College of Veterinary Radiology, American College of Veterinary Surgeons, American College of Zoological Medicine, and the American Veterinary Dental College (Figure 23.3).

**Figure 23.3** Sandra Manfra, diplomate of the American College of Veterinary Surgeons and the American Veterinary Dental College, performing root canal surgery on a Cocker Spaniel dog. *Courtesy Dr. Manfra, University of Illinois.*

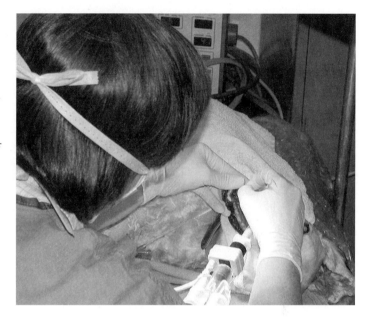

### 23.3.1   Postgraduate Veterinary Education

Positions exist for which postgraduate education is preferred or required in molecular biology, laboratory animal medicine, toxicology, immunology, diagnostic pathology, environmental medicine, or other specialties. The benefits of using scientific methods to breed and raise livestock, poultry, and fish, together with a growing need for effective public health and disease control programs, will continue to demand the professional expertise of veterinarians.

Career opportunities in veterinary medicine also include leadership and service opportunities in state and national veterinary associations as well as in many private business enterprises and consultancies. For additional information related to careers in veterinary medicine, visit www.avma.org/communications/brochures/veterinarian/veterinarian_faq.asp.

## 23.4   CAREERS AS VETERINARY TECHNICIANS AND VETERINARY TECHNOLOGISTS

Veterinary technicians and veterinary technologists provide important technical support to veterinarians, biomedical researchers, and other scientists. Through the 1950s, veterinarians trained their own employees, assigning numerous tasks and procedures as they deemed appropriate and in accordance with applicable Veterinary Practice Acts. These on-the-job trained individuals were referred to as animal assistants, animal attendants, and veterinary assistants or nurses. They were trained to meet specific needs of an individual practice.

To meet technical/technological needs and opportunities of an expanding veterinary profession and a more mobile population, formal new academic programs began in the 1960s. Today, there are approximately 100 veterinary technician programs (2- to 3-year Associate of Science degree) and veterinary technology programs (4-year Bachelor of Science degree) in the United States (plus others in Canada). To ensure and maintain a high standard of excellence, these programs are accredited by the AVMA. Following completion of the training program, technicians must pass a state board licensing examination. Depending on the state or area, those passing the licensing examination are designated as a Certified Veterinary Technician (CVT), Registered Veterinary Technician (RVT), or Licensed Veterinary Technician (LVT).

Professional veterinary technicians and veterinary technologists are educated to assist with physical examination and patient history (Figure 23.4), client education (to include nutritional counseling), dental hygiene examinations and prophylaxis, caring for hospitalized

**Figure 23.4** Angela Acker, RVT, assisting James Baker, DVM, with an ophthalmologic (ocular) examination of a patient at the Baker Animal Clinic, Stillwater, OK. *Courtesy Cynthia Coursen Alexander, DVM.*

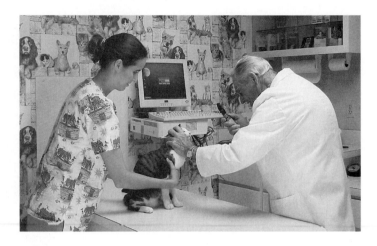

patients, administration of medications and vaccines, clinical laboratory procedures, radiology, anesthesiology, surgical assisting, office/hospital management, pet day care, grooming and training, and biomedical research.

In addition to many of the above areas of responsibility, veterinary technicians and veterinary technologists in research may also supervise the operation of research animal colonies and facilities (Figure 23.5) and assist in the design and implementation of research projects (Chapter 22).

## 23.4.1 SPECIFIC CAREER OPPORTUNITIES

The first postgraduate job for about 85% of graduate veterinary technicians and veterinary technologists is in a private veterinary practice. Companion animal practices lead the list in

**Figure 23.5** June A. Peer, trained technician, performs thorough physical examinations of guinea pigs at least once monthly at the Notre Dame Freimann Life Science Center. Checking the teeth for overgrowth is one component of the examination. June is a graduate of the Purdue University Veterinary Technology program. She is an AALAS-certified Laboratory Animal Technician and is currently studying for the Laboratory Animal Technologist examination. *Courtesy Kay Stewart and University of Notre Dame.*

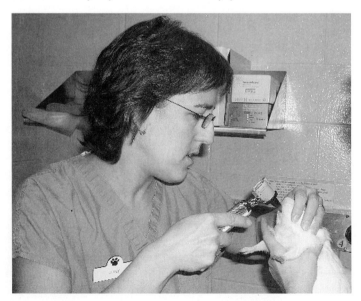

**Figure 23.6** Kristie Stasi, CVT, sets up equipment for an endoscopic procedure (Chapter 16) in the University of Illinois Internal Medicine special procedures room. *Courtesy Kristie Stasi.*

the number of employment opportunities. Professionally trained veterinary technicians and veterinary technologists are very important associates in a veterinary practice (Figures 23.6, 23.7, and 23.8). They perform many functions and routine services including scheduling and recording data and, thereby, allow the veterinarian(s) to stay in the examination room or surgical suite longer, which enables the generation of additional income needed to make the practice a business success (Chapter 19).

In the field of biomedical research, veterinary technicians have an opportunity to specialize through the certification program offered by the American Association for Laboratory Animal Science (AALAS). There are three levels of certification based on education, experience in the field, and passing a written examination. The first level, Assistant Laboratory Animal Technician (ALAT), focuses on the basic biology and husbandry of commonly used animal species in the biomedical field. The second level, Laboratory Animal Technician (LAT), addresses more of the technical aspects and the comparative physiology and anatomy of commonly used animals, whereas the third level, Laboratory Animal Technologist (LATG), encompasses the regulatory requirements and facility operations. AALAS also offers managerial training that is specific to the laboratory animal facility setting through the Institute for Laboratory Animal Managers (ILAM) and the Certified Manager of Animal Resources (CMAR) programs. ILAM is a hands-on 2-week program, whereas the CMAR

**Figure 23.7** Robyn Ostapkowicz, B.S. Biology, CVT, recording an electrocardiogram (ECG, EKG, Chapter 16) of a dog. Robyn is the clinic coordinator of the veterinary cardiology service at the University of Illinois. *Courtesy Robyn Ostapkowicz.*

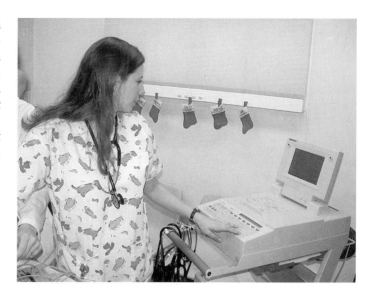

**Figure 23.8** Dylan McAllister, CVT, assisted by Susan Hewitt, CVT, collecting blood from a donor cat for the University of Illinois's Small Animal Clinic Blood Bank. *Courtesy Dylan McAllister.*

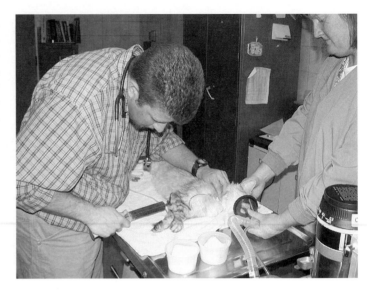

program is a self-study program offered through a joint effort between AALAS and the Institute of Certified Professional Managers.[2]

Numerous other opportunities include teaching (e.g., the University of Illinois College of Veterinary Medicine employs more than 30 veterinary technicians/technologists), humane societies, animal shelters, boarding facilities, kennels and catteries, biomedical research, diagnostic laboratories (Figure 23.9), zoo/wildlife medicine (Figures 23.10 and 23.11), industry, military service, herd health managers, and veterinary supplies sales.

The National Association of Veterinary Technicians in America (NAVTA) has a Committee for Veterinary Technician Specialists (CVTS) that governs groups of technicians who excel at their professional discipline of special interest. As of 2004, the CVTS had given full recognition to two academies—the Academy of Veterinary Emergency and Critical Care Technicians and the Academy of Veterinary Technician Anesthetists—and provisional recognition to the Academy of Veterinary Dental Technicians (Figure 23.12). For information about the NAVTA and related topics, visit www.navta.net and www.avma.org/careforanimals/CFAsiteindex.asp.

Additionally, a wide variety of veterinary technician specialty organizations are available for those interested in pursuing advanced training and expertise in a certain area. Examples of these organizations include Veterinary Technician Anesthetist Society, Veterinary Technician Animal Behavior Society, Society of Veterinary Behavior Technicians, American Society of Veterinary Dental Technicians, Association of Zoo Veterinary Technicians, and numerous individual state veterinary technician associations. Additionally, many veterinary technicians become members (or associate members) of various veterinary societies and academies that include both veterinarians and veterinary technicians—these include the American Academy of Veterinary Dermatology (Figure 23.13), International Veterinary Acupuncture Society, American Holistic Veterinary Medical Association, and many others.

---

[2]We are grateful to Kay Stewart, RVT, LATG, CMAR, for authoring this paragraph. Ms. Stewart graduated with distinction from the Purdue University School of Veterinary Medicine Veterinary Technology Program. She joined the Freimann Life Science Center (FLSC) at the University of Notre Dame as a technician when it opened in 1985. She currently serves as Associate Director of the FLSC. Through AALAS, Ms. Stewart earned technologist certification in 1987 and CMAR in 2000. She teaches laboratory animal science classes and serves as faculty advisor to the Notre Dame Pre-Veterinary Medicine Club. Because of her keen interest in the welfare/state-of-being of animals used in biomedical research, she organized and provides leadership/oversight in the environmental enrichment program for research animals at the University of Notre Dame. She has presented numerous invited papers on Notre Dame's environmental enrichment program at national and international meetings.

**Figure 23.9** Christy ElAmma, CVT, trimming tissues for histopathology (microscopic examination) in a veterinary diagnostic laboratory. *Courtesy Christy ElAmma, University of Illinois.*

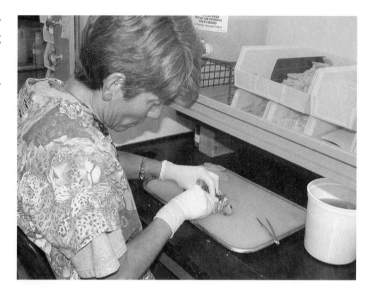

## 23.5   BIOTERRORISM-RELATED EMPLOYMENT OPPORTUNITIES

Following the September 11, 2001, bioterrorist attacks, there has been a recognized need to change the culture and emphasis governmental agencies—city, state, and federal—give to potential biological terrorism against animals and humans. Moreover, the heightened concerns and realities of the overall responsibilities of homeland security must be shared and actions coordinated if major disasters are to be averted.

Traditional discussion and planning regarding bioterrorism have focused largely on humans as the primary target. Yet equally as vulnerable are animal and crop resources. Indeed, a terrorist attack(s) on the nation's food supply and food-related economy could be equally

**Figure 23.10** Zoos offer a unique opportunity for veterinary technicians. Having the opportunity to handle and care for exotic animals is a dream of many animal lovers. Kay Stewart, RVT, is a graduate of the Purdue University Veterinary Technology program. She earned Laboratory Animal Technologist certification through AALAS. She is also a Certified Manager of Animal Resources. Although the Associate Director of animal facilities at the University of Notre Dame is her current position, volunteering time at the local zoo is a favorite pastime. *Courtesy Angel Ott, Kalamazoo, Michigan.*

**Figure 23.11** Depicted here is Laura Arriga, RVT and graduate of Purdue University Veterinary Technology Program, assisting with an annual examination of a colobus monkey. She is employed at the Potawatomi Zoo in South Bend, Indiana. *Courtesy Laura Arriga, Kay Stewart, and the Potawatomi Zoo.*

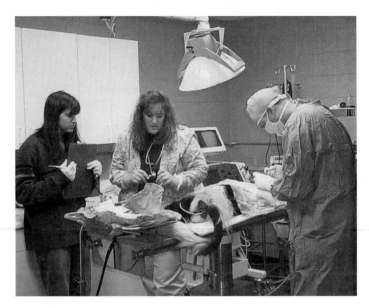

as attractive to terrorists because of the secondary effects on humans and the economically devastating impact on agricultural commodities, livestock, companion animals, wildlife, and related industries and organizations.

Many of these new and expanded efforts will require additional professionals who are trained to care for and manage animals. For example, dogs trained to function at Ground Zero and others trained to aid in the inspection of imports will be an increasingly important component of the national effort to heighten the level of preparedness against bioterrorism. Companion animal biology graduates will find rewarding career opportunities in breeding, training, and caring for these specially trained dogs.

Other companion animal biology and veterinary technician graduates will be needed to care for and manage purpose-bred rodents and other laboratory animals (Chapter 22) utilized in studies of pathogens and biologics that could be used by bioterrorists.

## 23.5.1 CENTERS FOR DISEASE CONTROL AND PREVENTION (CDC)

Operating under the aegis of the U.S. Department of Health and Human Services, the CDC functions to protect human health and safety by preventing and controlling diseases and injuries, and by promoting healthy living through strong partnerships with local, state, national, and international groups and organizations.

**Figure 23.12** Jeanne Vitoux, CVT, is the clinic coordinator for the University of Illinois Veterinary Dental Clinic; she was a founder and first president of the Academy of Veterinary Dental Technicians. *Courtesy Jeanne Vitoux, University of Illinois.*

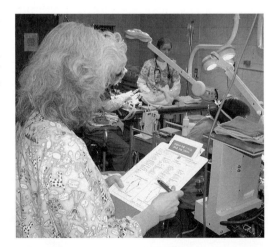

**Figure 23.13** Sandra Grable, CVT, an associate member of the American Academy of Veterinary Dermatology and the clinic coordinator of the University of Illinois Veterinary Dermatology Clinic, preparing allergens for use in intradermal allergy testing and immunotherapy. *Courtesy Sandra Grable, University of Illinois.*

The CDC has outlined three categories of biological agents that could be used by bioterrorist groups: (1) *Category A* poses a high risk to national security because they can easily be disseminated or transmitted person to person, cause high mortality with potential for major public health impact, and cause public panic and social disruption. These agents include variola major (smallpox), *Bacillus anthracis* (anthrax), *Yersinia pestis* (bubonic plague), *Francisella tularensis* (tularemia), filoviruses (Ebola hemorrhagic fever), and *Clostridium botulinum* toxin (botulism); (2) *Category B* agents are moderately easy to disseminate, cause moderate morbidity and low mortality rates, and require specific enhancements of diagnostic capacity. They include *Coxiella burnetti* (Q fever), *Brucella* spp (brucellosis), *Burkholderia mallei* (glanders), alphaviruses (Venezuelan/eastern/western encephalitides), ricin toxin (*Ricinus communis*, castor beans), and pathogens that are food- or waterborne (*Salmonella* spp and *Escherichia coli* O157:H7); and (3) *Category C* agents include emerging pathogens that could be engineered for mass dissemination and include Nipah virus, tickborne hemorrhagic fever viruses, yellow fever, and multi-drug-resistant tuberculosis.

To better prepare for the potential threats of bioterrorism on public health and safety, the CDC is expanding manyfold the number of professionals employed in these important efforts.

## 23.5.2 UNITED STATES DEPARTMENT OF AGRICULTURE

The USDA is the second largest governmental agency. Among many other responsibilities, it is charged with preserving food safety and animal health. Ensuring a comprehensive mechanism to deal with potential zoonotic disease outbreaks is a monumental task. Employees of the veterinary corps (veterinarians, veterinary technologists, and veterinary technicians) are critical stakeholders in efforts to strengthen and ensure the nation's overall public health preparedness. This calls for an expanded group of these professionals. (See Section 23.2.2.)

## 23.5.3 FOREIGN ANIMAL DISEASE (FAD) SURVEILLANCE

The Veterinary Services Division (VSD) of the APHIS (Animal and Plant Health Inspection Service) of the USDA conducts nationwide state/federal cooperative programs for control and eradication of animal diseases, suppresses spread of disease through control of interstate and international movements of livestock, keeps informed of the overall animal disease situation nationally and internationally, administers regulations to ensure humane treatment of transported livestock and laboratory animals, oversees the licensing of biologics (vaccines, diagnostic reagents, serums), collects and disseminates information on morbidity and mortality, and provides training for USDA-APHIS employees and others in some related governmental agencies.

Inasmuch as veterinarians have been asked to take a lead role in bioterrorism preparedness initiatives, a stronger national network of veterinary diagnostic laboratories across the United States is being developed. This will strengthen the nation's biosecurity by developing an early detection mechanism of a biological attack or a foreign animal disease outbreak.

An important adjunct to this nationwide diagnostic network, whose critical purpose is to prevent the spread of disease, is quick and accurate testing of bioterrorism agents as well as foreign animal disease agents. The use of equipment based on the polymerase chain reaction can provide a diagnosis within an hour as opposed to routine culture tests that can require 2 or more days.

The challenge of protecting pets, food-producing animals, and wildlife populations from diseases that could be associated with bioterrorism attacks is important and monumental. The role of the veterinary workers (including veterinarians, veterinary technologists, and veterinary technicians) in the whole of bioterrorism is being expanded enormously. City and state health departments are assessing their capacity to respond to bioterrorism, outbreaks of infectious diseases, and other public health threats and emergencies.

The FDA is recruiting persons to work in the agency's Center for Biologics Evaluation and Research aimed at countering bioterrorism.

Considering the increased number of professionals needed to meet the challenging responsibilities of local, state, and national governmental agencies and groups mentioned above, it is readily apparent that thousands of career opportunities await graduates of companion animal biology, animal science, veterinary medicine, veterinary technology, veterinary technicians, and related disciplines in ensuring a high level of preparedness in national bioterrorism matters.

Fortunately, federal funding has been authorized to recruit and train additional professionals to help prevent and respond to public health emergencies as well as to upgrade infectious disease surveillance and investigation, enhance the readiness of hospitals and communications capacities, and to improve communications among hospitals, cities, and local/state health departments.

## 23.6   ADDITIONAL CAREER OPPORTUNITIES RELATED TO COMPANION ANIMALS

Depending on individual interests, level of entrepreneurial spirit, and numerous other career-related considerations, there is an abundant number of other opportunities associated with companion animals. The following are examples by categories:

■   *Animal sciences.* This multifaceted discipline has numerous opportunities in teaching companion animal biology, care, and management courses and researching in specific areas such as animal nutrition (Chapter 9), animal behavior and psychology—a growing discipline (Chapter 20), environmental sciences, animal state-of-being (welfare), diseases (Chapter 14), genetics, and the biology/physiology of reproduction (Chapter 10). The latter includes artificial insemination, cloning, embryo transfer, molecular biology, and nonsurgical sterilization. Graduates with advanced degrees find challenging leadership and research opportunities with feed, petfood, and pharmaceutical companies; state and federal governmental agencies; as well as in laboratory animal research (Chapter 22) and consultancies with the private sector.

■   *Animal behaviorists.* One of the major causes of pet relinquishment is due to a mismatch between owner expectations and animal behavior (Chapter 20). Animal behaviorists can assist pet owners with the initial selection of a compatible pet, care and training of pets, and correction of any behavioral problems that develop. Programs of study in animal behavior generally begin with a Bachelor of Arts (B.A.) or Bachelor of Science (B.S.) degree and continue with advanced training in psychology. In addition to private behavioral consultation services, animal behaviorists work in college research and teaching, other research institutions, zoos, aquariums, conservation groups, and as animal trainers. For more information, contact the Animal Behavior Society (http://www.animalbehavior.org).

- *Authorship/editing.* Career opportunities exist to write and edit articles and prepare materials for catalogs; dog, cat, and other companion animal magazines; newspapers; technical magazines; specialty books; textbooks; and other print/electronic means of public communication.

- *Companion animal breeder.* The challenge of breeding superior animals is of interest to many, particularly when linked with fitting, grooming, and showing. Raising and selling puppies, kittens, and other companion animals to pet stores or to prospective owners as well as raising and selling purpose-bred animals for research purposes can provide profitable employment and job satisfaction.

- *Conservation biology/ecosystem health.* Numerous job opportunities are available as field biologists in research and field studies aimed at determining ways to ensure the perpetuation of wildlife species and to improve the health and quality of life of these animals.

- *Foreign service.* Career opportunities are available to persons in veterinary medicine and the animal sciences with organizations such as Food and Agricultural Organization (FAO), United States Agency for International Development (USAID), International Livestock Research Centers, the World Bank, and other governmental and private organizations.

- *Groomers.* Professional pet and dog groomer shortages are chronic and career opportunities are almost limitless. Animal groomers provide bathing, trimming, brushing, combing, and related services (Chapters 12 and 13). The pet grooming industry is generally recession proof as many dedicated pet owners want the very best grooming service available to assist them in caring for their prized animals (Figure 23.14). There are about 6,000 dogs and cats for every grooming business in the United States. For more information on careers in dog and cat grooming, contact the National Dog Groomers Association of America, Inc. (http://www.nationaldoggroomers.com) or visit http://www.petgroomer.com.

- *Judges.* Although limited in number, there is a need for qualified judges of breed shows, field trials, and other dog and cat competitions and exhibits (Chapter 13).

- *Kennel and cattery management.* Although limited in number needed, opportunities exist for companion animal graduates in this important area of employment (Chapter 18). Additional employment opportunities include working as an animal care attendant at a

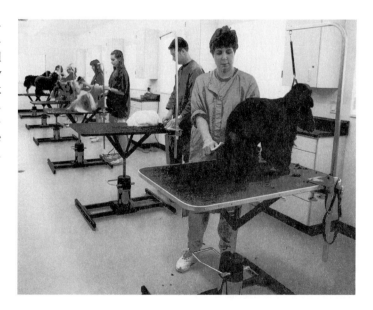

**Figure 23.14** The demand for grooming services is high as evidenced by this row of busy groomers at Willow Creek Pet Care Center, Sandy, UT. These groomers provide service for an average of 1,000 pets each month. *Courtesy Dr. Rick Campbell.*

kennel, cattery, or with colonies of research animals. Animal care attendants feed, water, clean, and monitor the health of animals.

- *Management/ownership of pet motels, other boarding facilities, and pet sitting businesses.* The increasing number of these businesses gives rise to profitable and challenging career opportunities (Chapter 17). Pet motels and boarding facilities are rapidly growing areas of employment with many of these businesses expanding to offer day camps, grooming, training, and many other services in addition to traditional pet boarding. Pet sitters are also in great demand to provide in-home care and exercise of pets while owners work or travel.

- *Ownership/entrepreneurship/management of businesses manufacturing or selling petfoods, pet supplies, and pets.* For those with a special affinity for companion animals and exotic pets, career opportunities exist to own (Figure 23.15) and/or manage businesses manufacturing or selling petfoods and treats/snacks, pet supplies (e.g., toys and accessories, cushions and beds, training equipment, cages and environment enrichment accessories, and other items to promote the health and well-being of pets), and as a means of providing other pet services and sales (Chapters 2 and 19). Pet stores specializing in selling birds may hand rear chicks (Chapter 5). Hand rearing requires specialized equipment and constant attention by trained individuals (Figure 23.16). Benefits of hand rearing include producing chicks that are well-socialized to people, become better pets, and can be sold for premium prices.

- *Professional handler.* Showing dogs (Figure 23.17) and other companion animals for a fee is another personally satisfying career opportunity.

- *Professional trainer.* Substantial personal pleasure and satisfaction can be derived from training dogs to serve the physically impaired (e.g., blind, deaf; Chapter 21) as well as for obedience (Chapter 12), field trials, herding (Figure 23.18), police (guard dogs), tracking, hunting, and performance (movies, television).

**Figure 23.15** (*a*) This attractive petfoods and supplies business is owned by Mark Fisher, B.S. Animal Sciences, University of Illinois. (*b*) Owners frequently bring their pets in the store with them; Mark provides consultation on diets and proper nutrition. *Courtesy Mark Fisher.*

(a)

(b)

**Figure 23.16** This 7-week-old African Grey parrot is being syringe fed by Michelle Hinson, B.S. Zoology, Western Illinois University. Michelle is an avian nursery caretaker for Sailfin Pet Shop, Inc., Champaign, IL. *Courtesy Michelle Hinson and Sailfin Pet Shop.*

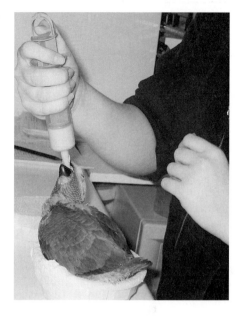

■ *Animal-assisted therapists.* This growing area links companion animals with improving the quality of life for the physically and mentally impaired, especially among an aging human population, and provides increasing career opportunities for those who derive personal satisfaction from working with these members of society (Chapters 1 and 21).

■ *Regulatory/animal control/animal shelters.* As discussed in Chapter 24, there is a great need and expanding opportunities for those who derive satisfaction in organized ways and means of managing unwanted companion animals. Job opportunities in animal control include animal control officers, humane officers, cruelty investigators, field supervisors, rabies control/bite investigators, kennel managers, shelter managers, operations managers, animal care technicians, animal care attendants, animal groomers, clerks, secretaries, dispatchers, receptionists, adoption counselors, executive directors, public relations staff, photographers, newswriters, humane educators, and fundraisers.

■ *Private environmental consulting firms and groups.* Increasing public interest in preserving and enhancing the environment in which animals—both domesticated and nondomesticated—live bodes well for those seeking career opportunities in environmental-related disciplines.

**Figure 23.17** Professional handlers showing dogs at an American Kennel Club conformation show. *Courtesy Melissa Maitland, Great Lakes Nutrition, Inc.*

**Figure 23.18** Border Collie rounding up cattle (yearling heifers) to move into a corral for vaccination and deworming. *Courtesy Ronald W. Rogers, DVM, and John Tate (photographer).*

■ *Other career opportunities.* Limited in number but important in professional services rendered include those specializing in pet photography, pet transportation, zoo animal caretakers, and other part- or full-time jobs available to those seeking employment that permits them to associate with and serve companion animals and exotic pets.

## 23.7   SUMMARY

The enduring enjoyment and personal satisfaction derived from bonding with animals by those dedicated to the quality of life among companion animals cause these individuals to explore the many professions and employment endeavors that provide lifelong career opportunities to interact with their favorite animal species. The thought and possibility of helping improve and preserve the health and life of animals provide sufficient motivation to consider carefully such career opportunities.

In this chapter, we outlined and discussed numerous career alternatives, which can provide many choices for those committed to quality of life for companion animals to achieve those goals and concurrently have respectable employment that promises to provide a high level of enjoyment and personal satisfaction.

Discussion also included the expanding employment opportunities associated with the nation's commitment to achieve a high level of preparedness in matters related to bioterrorism against companion and food-producing animals as well as wildlife.

We hope we have been successful in reinforcing an already instilled desire among students and others to investigate further the many possibilities associated with a career in companion animal biology and related disciplines. "In the long run you hit only what you aim at. Therefore, though you should miss in the short run, it is best to aim at something high."[3]

## 23.8   REFERENCES

1. American Society of Zoologists. 1982. *Careers in Animal Biology.* Chicago, IL: American Society of Zoologists.

2. National Research Council of the National Academies of Science. 2004. *National Need and Priorities for Veterinarians in Biomedical Research.* Washington, DC: National Academies Press.

[3]Henry David Thoreau (1817–1862), American writer.

# 23.9   WEB SITES

1.   www.phppo.cdc.gov
2.   www.navta.net/
3.   www.uwrf.edu/animal-science/careers.html
4.   www.kids4research.org
5.   www.ans.iastate.edu/ugrad/careers.html
6.   www.cals.ncsu.edu/an_sci/undergrad/careers.html
7.   www.ansci.uiuc.edu/labs/companion/index.htm
8.   www.hsus.org/ace/352

# 24 MANAGING UNWANTED COMPANION ANIMALS

*The greatness of a nation and its moral progress
can be judged by the way its animals are treated.*

Mohandas K. Gandhi (1869–1948)
*Indian nationalist leader*

510

## 24.1   INTRODUCTION/OVERVIEW

History has recorded that companion animal control problems are not merely a modern development. In the Middle Ages, a French monarch decreed that all dogs belonging to peasants be required to drag heavy blocks of wood from their collars. This practice essentially prohibited such dogs from hunting on royal game preserves or impregnating purebred hounds of the nobles.

By mid-nineteenth century, records show that animal control workers of New York City caught approximately 1,000 stray dogs weekly. The thousands of unsupervised small animals, particularly in the poorer sections of New York City, coupled with the fact that many were not receiving needed medical treatment, prompted the Women's Auxiliary to the American Society for the Prevention of Cruelty to Animals (ASPCA) to set up a series of temporary shelters for small homeless animals in 1908. Months later, they launched a program to teach children about proper animal care and responsibility. In 1909, the Auxiliary Leaders established a dispensary and outpatient clinic for animals whose owners could not afford to pay for medical treatment, as had been done in London and other European cities.

By 1910, the farsighted, benevolent women started the Animal Medical Center (AMC) of New York City. This nonprofit veterinary hospital is now the world's largest for the medical care of small animals. It employs more than 130 veterinarians representing 27 specialty areas for small animals and is a leading center for veterinary medicine and clinical advances. The AMC is dedicated to improving the quality of life for all animals. This commitment is evident in its motto: *Pro Bono Animalium Hominumque . . . for the benefit of animals and humanity.* For additional information about this world-renowned animal hospital, visit www.amcny.org/asp/homepage/default.asp.

A major problem faced by most large cities and numerous small towns and communities in the United States is an animal surplus as well as a surplus of poorly supervised pets. Precious public funds are being diverted from social and other programs to monitor dog and cat bites, maintain rabies surveillance programs, capture and remove unwanted and vicious straying animals, replant foliage "dug up" from city flowerbeds, remove dead animals from streets and highways, and respond to calls and complaints against irresponsible animal owners. These and related topics are discussed in this chapter.

## 24.2   UNWANTED COMPANION ANIMALS

The term *unwanted* most commonly refers to once-owned dogs and cats that have been abandoned or taken to a shelter, whereas *uncontrolled* commonly refers to owned animals permitted to run at large. The term *stray* refers to rejected or lost animals running at large, and *feral* refers to wild animals and/or those that have escaped from domestication and become wild. *Free-roaming* pets include strays (ownerless and feral) as well as uncontrolled pets.

Interestingly, there are proportionately many more unwanted adult dogs and cats than there are unwanted puppies and kittens. Indeed, most unwanted cats and dogs *were wanted* at one time in their lives. Unfortunately, once the cuteness of the baby stage has been outgrown and the realities and responsibilities associated with pet ownership are evident, many animals that were acquired on an impulse are abandoned.

## 24.3   RESPONSIBLE OWNERSHIP OF COMPANION ANIMALS

Let us consider an important tenet of companion animal ownership: namely, that ownership is a privilege, and as with other privileges in life, animal owners incur moral, financial, legal, and other responsibilities. The first and foremost responsibility is to provide a good home, which includes sufficient space and environmental conditions that are comfortable and conducive to an acceptable quality of life (Chapter 12). These include providing safe, health-sustaining foods (Chapter 9), clean air and water, an ample display of attention, and demonstrating that the animal is wanted and welcomed. Important as well is to demonstrate respect for and compliance with the requirements of thorough animal control.

Major reasons/purposes of owning a dog or cat in the first place are companionship, protection, hunting and sporting purposes, and exhibition as a hobby or from a recreational standpoint (Chapters 3, 4, 13, and 21). Based on the above purposes of ownership, it is illogical that animals be permitted to wander unattended or be left to fend for themselves. How can the owner of free-roaming animals receive the rewards and personal pleasures associated with responsible ownership?

Dogs, like children, can serve as glue or as a solvent to family and/or neighborhood relations. When dogs are permitted to roam unleashed throughout the neighborhood—sometimes leaving unwelcome tracks and/or deposits—frictions, disagreements, and unkind verbal exchanges frequently follow between the animal owner and neighbors.

Numerous segments of society have a stake in and share an inherent responsibility for animal control. These include (1) humane associations and societies; (2) law enforcement and judicial agencies; (3) legislative and local governing bodies; (4) merchandisers of pets including those raising, brokering, dealing with, and retailing pets; (5) registries of purebred companion animals (a significant number of animals surrendered to shelters are purebreds); (6) regulatory forces of state and federal animal and public health agencies; (7) the veterinary medical profession; and (8) responsible citizens.

# 24.4   REASONS COMPANION ANIMALS BECOME HOMELESS

Numerous reasons can be cited why companion animals become rejected, disposed of, and/or homeless. These include death of an owner/guardian; change in lifestyle resulting in an insufficient amount of time for the pet; transfer of employment location (e.g., a military or other job-related transfer); change in owner's work schedule; new baby in the family; moving to housing where animals are not allowed; discovery of serious health problems with the owner or pet; changes in the owner's economic, family, or social situation; lack of compatibility with the animal; unanticipated or disliked pet behavior (e.g., barking, difficulty in housetraining, inappropriate elimination, destructive behavior, aggression, separation anxiety, and others, see Chapter 20); children leaving for college and/or moving out; allergies; prospective spouse does not like pets; and owner apathy or irresponsibility.

Another major factor is simply that too many dogs and cats are born each year. Many litters are unplanned or produced by stray animals, and others are planned but without sufficient consideration being given to the implications of contributing to the oversupply of dogs and cats. A national, nonprofit organization, In Defense of Animals (IDA), based in Mill Valley, California, dedicated to ending the exploitation and abuse of animals, estimates that 50% of all *licensed* companion animals are intact (i.e., not spayed or neutered). The percentage of intact animals is likely to be even higher in unlicensed pets. Intact animals reproduce at a remarkable rate. In 8 years, one female cat could give rise to over 2.4 million offspring (Table 24.1)! Dogs can reproduce at an even greater rate due to the larger number of puppies produced per litter. Each day, approximately 10,000 human babies are born in the United States and around 70,000 puppies and kittens are born. This imbalance between potential homes and an oversupply of puppies and kittens is worsening day by day and will continue to do so until improved animal population control is achieved. The most effective way to achieve animal population control is to spay and neuter more pets. There are many health advantages for the spayed and neutered pet in addition to helping curb the problem of unwanted/free-roaming dogs and cats (Chapters 10 and 14).

**Table 24.1**
Reproductive Potential of Sexually Mature Healthy Cats under Optimal Environmental Conditions

| Length of Time | Number of Cats |
|---|---|
| Day 1 | 2 |
| One year later | 12 |
| Two years later | 66 |
| Three years later | 382 |
| Four years later | 2,201 |
| Five years later | 12,680 |
| Six years later | 73,041 |
| Seven years later | 420,715 |
| Eight years later | 2,423,316 |

*Source: www.spay.org. Intact cats have the potential to reproduce at a phenomenal rate. This table shows the potential numbers of cats resulting from uncontrolled breeding starting with a single pair of cats. Population figures in this table are based on an average of two litters of kittens per year, an average of 2.8 surviving kittens per litter, and 50% females.*

## 24.5 PROBLEMS AND HAZARDS ASSOCIATED WITH UNWANTED/FREE-ROAMING DOGS AND CATS

Some dog and cat owners fail to demonstrate proper ownership responsibility by not honoring local leash laws. This leads to problems and hazards with free-roaming dogs and cats. These include adding to environmental pollution; creating public health and safety problems associated with dogs and cats that attack/bite children and adults; disseminating diseases; adding to fly control problems (e.g., many flies feed on dog feces, which exacerbates problems associated with flies); disturbing the peace; free-roaming and/or runaway animals that maim and/or kill wildlife and livestock as well as other dogs and cats; causing damage to property; serving as nuisances to neighbors; contributing to canine and feline overpopulation and abandonment; creating traffic hazards that often result in accidents; and undermining the overall public appreciation for companion animals by instilling fears in the minds of people of all ages. Four of the above are discussed in Subsections 24.5.1 through 24.5.4.

### 24.5.1 DOG AND CAT ATTACKS AND BITES

Researchers estimate that more than 5 million children and adults in the United States are attacked and/or bitten annually by dogs and cats (less than 10% by cats). Over a million such injuries require medical treatment. Medical bills that follow these bites result in more than $100 million of insurance claims being filed annually. According to the U.S. Postal Service, mail carriers experience more than 29,000 dog bites annually. This results in millions of dollars in sick pay and medical expenses. Placing a monetary value on the personal anxiety and pain experienced by being attacked or bitten by a dog or cat is impossible. Anxiety levels are heightened by not knowing if the biting animal was rabid or the carrier of another feared zoonotic disease (Chapter 14). Biting among dogs is a behavior transmitted from their wolf ancestors. As predators, wolves chase, catch, and chomp a fleeing prey. A child running away from a dog may prompt the dog to give chase and bite the child. Hence, the sage recommendation that a child remain motionless when approached by a strange dog. This, of course, is not the normal behavior associated with fear.

Before many insurance companies write property insurance policies, they determine whether the client has any pets. Because of possible liability exposures associated with pets, some companies will not issue popular homeowner policies to those owning certain types of pets (e.g., a pet lion or tiger, and increasingly refusing to insure owners of some breeds of dogs). Moreover, when property insurance is written, the rates are commonly higher when certain pets are kept.

## 24.5.2 Potential Health Hazards Associated with Stray/Feral Dogs and Cats

Zoonotic diseases of cats and dogs were discussed in Chapters 14 and 15. In addition to the possible dog/cat attacks and biting mentioned above, stray dogs and cats can transmit a variety of diseases and parasitic infections to livestock and domestic pets in rural areas. These include rabies, roundworms, tapeworms, fleas, ticks, toxoplasmosis, and leptospirosis (a spirochete passed in the urine of infected dogs). Cats can spread toxoplasmosis by defecating in areas such as children's sandboxes.

## 24.5.3 The Overpopulation of Feral Cats

Ferals are cats, or descendants of cats, that at one time had been domesticated. The feral population includes strays, abandoned cats, and their progeny. There are an estimated 6.0 million feral cats in the United States and 1.5 million in Great Britain. The interbreeding of feral cats yields what is often referred to as alley cats. The American Humane Association and the Humane Society of the United States (www.hsus.org) report that, given the opportunity to breed and the avoidance of a serious disease outbreak, one fertile feral female and her offspring can produce 420,000 cats in 7 years (see Table 24.1). This is ample evidence of the need to control stray and feral cat populations.

There are diverse opinions regarding the best method of controlling feral cat numbers. Several organizations have been formed to promote and fund trap-neuter-release (TNR) programs. Feral cats are trapped, vaccinated, neutered, and then released back at their original location. One beneficial advantage of TNR programs is that they result in a relatively stable population of cats in an area. Some studies have shown if feral cats are euthanized, other feral cats move in to replace those that were eliminated. When a colony includes all neutered animals, the colony is stable and the cats defend their territory preventing new cats from moving in and establishing a new breeding colony. Ideally, the feral cats would be tamed and adopted, but this is not always possible and there are already millions of tame cats being euthanized in shelters due to a shortage of homes. (An estimated 8 to 10 million homeless cats and dogs turn up in shelters annually. Approximately half are euthanized because no one adopts them.) The long-term solution is prevention of cat abandonment through public education on responsible pet ownership and the neutering of pet cats.

## 24.5.4 The Counterbalancing Effects of Cat and Bird Populations

Numerous wildlife conservation groups return periodically to fundamental concepts associated with *triage*, which involves controlling one animal species, such as cats, to protect another, such as birds. A study conducted by University of Wisconsin wildlife ecologist Dr. Stanley Temple reported that rural cats kill an estimated 39 million birds annually in Wisconsin. It is not a simple matter to reconcile such a loss of birds among those who seek to preserve the beauty and utility of birds with the firm views of benevolent persons who advocate less confinement of cats, which results in their having the opportunity to utilize and develop their innate hunting skills and quickness in catching birds.[1]

## 24.5.5 IMPACT OF RELEASED PETS ON BIOLOGICAL NICHES

The accidental or intentional release of nonindigenous (nonnative) pets into the wild has become one of the most important issues in ecological health and conservation biology. Broadly considered, biological niches are the address, profession, and social circle of an organism. In nature, different species have evolved to form a community of organisms comprising and occupying the available biological niches. When nonindigenous animals (and plants) are released into an area, these invaders may either die or survive by displacing and/or disrupting the community of organisms native to the area. The release of exotic fish into streams and lakes can destroy native fish populations. An example of this is the fear that Asian carp (originally imported into the United States by fish farmers) will destroy the ecosystem of the Great Lakes. There is concern that the Disney movie *Finding Nemo* may encourage children to set their aquarium fish "free" into streams and lakes. There is great concern that pet rodents and carnivores escaping from homes (or being intentionally released by owners no longer wanting the animals) could decimate natural populations of animals. This has led to bans prohibiting the keeping of many exotic animals as pets. For example, keeping ferrets as pets is illegal in California, Hawaii, and numerous other locations. Flocks of parrots released or escaping from homes has become a significant problem in Florida. Responsible pet ownership includes the requirement to protect native animals from decimation by abandoned and released pets. Owners no longer able or wanting to keep a pet have the responsibility to transfer the animal to a new home or animal shelter or to have it humanely destroyed (Section 24.8).

## 24.6 CONTROLLING COMPANION ANIMAL POPULATIONS

The number of companion animals in the United States is increasing much faster than the human population. This has resulted in an abundance of dogs and cats relative to the number of responsible prospective owners who are committed to providing these companion animals with comfortable homes and proper feeding and care as well as good health, quality of life, and access to veterinary services.

The unwanted dog and cat challenge can be approached in numerous ways. Examples are discussed in Subsections 24.6.1 through 24.6.5.

### 24.6.1 PUBLIC EDUCATION

The most important key to controlling the problem of unwanted companion animals is educating the public on both the responsibilities of pet ownership and problems associated with abandoned animals and overpopulation. Numerous organizations have accepted this challenge. One such program is the *Pet Food Institute Initiative*. The Pet Food Institute was organized in 1958 as the National Trade Association of petfood manufacturers (Chapter 2). An important activity of this organization has been a consumer education program directed toward helping people become better pet owners—making persons more aware of personal responsibilities to their pets and to their communities.

---

[1]Sterba, James P. 2002, October 11. Kill Kitty? Question Has Fur Flying in Critter Crowd. *Wall Street Journal*, A1, A4.

### 24.6.2   GOVERNMENTAL COMPANION ANIMAL CONTROL

To aid in the oversight of municipal regulations (e.g., leash laws) and enforcement related to unwanted companion animals, cities and municipalities should have an animal control program with adequate supervision and implementation. Leash laws can help reduce the random matings that result in unwanted kittens and puppies.

### 24.6.3   ANIMAL SHELTERS

These important nonprofit units/organizations have a constant inflow of cute puppies and kittens, which all too often become adult dogs and cats that fail to fit their owner's family lifestyle, are too difficult to train, or are too expensive to keep.

It is much preferred and less costly to do the homework necessary to locate a compatible breed and individual dog or cat prior to adoption or purchase (Chapters 3 and 4) than to subject the family and pet to the problems, anxieties, and unpleasant experiences resulting from unanticipated incompatibilities.

Animal shelters play an important, commendable role in efforts to provide a temporary home in conjunction with efforts to place pets in a new home. Animal shelters vary in their policies associated with adoption and length of time an animal will be kept. Some are "no kill" animal shelters, meaning they never euthanize a healthy animal (an extensive list of these can be found at www.saveourstrays.com/no-kill.htm). Most shelters do not permit animals to be removed for use in research. Procedures to be followed in obtaining animals for use in research are regulated by the Animal and Plant Health Inspection Service of the USDA (Chapter 22). Many shelters require that animals be neutered or spayed prior to being released for adoption. Procedures have been developed to safely spay and neuter dogs and cats as young as 6 weeks of age.

### 24.6.4   ANIMAL ADOPTION PROGRAMS

Because so many unwanted dogs and cats are unceremoniously dumped or abandoned, the supply currently exceeds the demand. Therefore, only animals well-suited for adoption should be placed and then only with responsible owners. Guidelines for adoption include (1) a cat or dog deemed adoptable should be at least 6 weeks of age and weaned (younger animals may be cared for by foster homes until old enough for placement); an example of the successful raising of an abandoned newborn is depicted in Figure 24.1; (2) it should be in excellent physical condition and free of disease or injury; (3) it should be even tempered (not vicious) and free of undesirable habits; and (4) in most cases, it is best if the animal is spayed or neutered prior to adoption.

Adoption recommendations for suitable owners include having (1) proof of an age of legal responsibility; (2) no substance abuse problem; (3) a permanent residence; (4) the ability to provide an environment appropriate to the animal's needs; and (5) proof of adequate financial resources and a willingness to pay for proper pet care. If the pet is for children, their parents should meet the above criteria.

There are several organizations and groups whose primary purpose is to assist individuals wanting to adopt a pet. An example is www.petfinder.com. This nonprofit service has numerous sponsors that include Merial, Nestlé-Purina, PetCare, and Petco and helps orchestrate the adoption of pets. Since its inception in 1995, the Web site has grown to represent more than 5,000 animal shelters and rescue groups in the United States. It has been instrumental in finding homes for more than a million previously unwanted pets. At any given time, Petfinder.com has more than 100,000 adoptable pets on the site. Dogs and cats represent more than one-half of the total pets available through this adoption service. Petfinder.com has a resource library and provides an ongoing search for adoptable pets as well as opportunities for volunteers to assist.

**Figure 24.1** This cat was found as a newborn kitten near a city dumpster where it had been left to die. The fortunate finders (Miranda and Elliott Wall) quickly carried the seemingly destined-to-die neonate to their grandmother's house where warmed evaporated cow's milk was administered via an eye dropper. A call to the family veterinarian resulted in the recommendation to purchase milk replacer from the local pet store and follow directions on the label. This, in addition to other recommendations enhancing its probability of survival, were followed and the kitten survived. The cat is now 18 months of age and is a healthy, happy cat named "Lucky." *Courtesy Miranda and Elliott Wall.*

Second Chance is a no-kill adoption agency committed to finding homes for neglected and/or unwanted companion animals. PETsMART, Inc. provides adoption centers at each location (see www.petsmart.com/adoptions/index.shtml for local store information). Online service organizations dedicated to the matching of potential adopters play an important role in decreasing the unwanted companion animal problem.

Hundreds of radio stations in the United States participate through public service time in a *Pet Patrol Radio Directory* established by the American Humane Association to broadcast missing pet notices. Numerous television stations also provide public service time in showing dogs and cats available for adoption at the local animal shelter or city pound.

Another organization helping provide unwanted animals with lifesaving care and loving homes is North Shore Animal League America (www.nsalamerica.org). Since its founding in 1944, this nonprofit organization proudly boasts of having saved the lives of nearly a million pets. An important component of its services is the Alex Lewyt Veterinary Medical Center, delivering timely medical treatment that includes spaying and neutering more than 400 dogs and cats each week.

## 24.6.5 Animal Rescue Organizations

There are numerous breed-specific rescue organizations and also organizations that facilitate the rescuing of non-purebred animals. For a listing, visit www.ecn.purdue.edu/~laird/animal_rescue/breed_rescue_organizations.

The majority of rescue organizations are breed specific and take unwanted purebred animals from owners who can no longer keep the pets and also purebred animals from local animal control shelters. The animals are often placed in foster homes until a new home can be found. If an owner is seeking rescue help for a breed different from the one an organization specializes in rescuing, the organization will aid in locating someone else to provide assistance.

## 24.7 STERILIZATION/PREGNANCY-PREVENTION PROCEDURES

There are several ways to help control unwanted animals. Compulsory neutering/spaying is practiced in many cities. More recently, the use of endocrine-related agents/compounds and mechanical devices have gained increased acceptance as tools in companion animal population control.

In 2003, the Federal Food and Drug Administration approved Neutersol®, an injectable drug that causes chemical castration in puppies between the ages of 3 and 10 months. Research studies at the University of Missouri-Columbia showed the drug to be 99.6% effective. Few adverse side effects were reported. This provides a noninvasive, lower-cost alternative to surgical castration of puppies.

In addition to preventing increases to the already high numbers of unwanted pets, there are many benefits to spaying and neutering cats and dogs (Chapter 10). The spayed female has no heat cycles (messy in dogs, which drip blood around the house; disruptive in cats where the female vocalizes and may be hyperactive in seeking a mate). Spaying reduces or eliminates the risk for mammary, ovarian, and uterine cancer in females. The neutered male has less desire to roam, is less likely to develop prostatic disease, will not develop testicular cancer (if surgically neutered), and is less likely to engage in territorial marking and spraying (Chapter 20). Neutered cats are also less likely to fight with other cats.

Many organizations promote low-cost spay and neuter clinics; one such program is SPAY/USA (visit www.spayusa.org, or 1-800-248-SPAY, for referrals to low-cost spay/neuter programs). Many states have special "dog and cat sterilization" license plate programs that fund dog and cat sterilization programs.

## 24.8 EUTHANASIA

*Euthanasia* is derived from Greek words *eu-* (easy or good) and *thanatos* (death) and therefore means an easy or painless death.

Although controversial and debatable, euthanasia is often the kindest alternative in the control of unwanted animals when the pet is too young, too old, unhealthy, or otherwise generally unsuited for adoption.

Any method utilized in euthanasia must be dignified and humane (Chapter 14), and those who administer the procedure must be professionally competent, properly trained, and responsible.

The thought of putting a dog or cat to sleep is revolting and unacceptable to most children and many adults. Yet it is important for children to realize that all living plants and animals will eventually die. It is good for them to know that animals cannot be expected to live as long as people.

## 24.9 SUMMARY

The unwanted companion animal problem is, in effect, a complex web of interrelated challenges and issues. These can best be addressed and resolved through the collective and cooperative efforts of members of humane organizations, animal shelters, rescue groups, veterinary medical associations, governmental officials, and other interested parties and individuals throughout the nation working together with singleness of purpose to find satisfactory and effective solutions to unwanted animal problems.

In this chapter, we noted numerous ways to approach and manage the once wanted, later unwanted, companion animal challenge. These include public education programs that emphasize and encourage responsible pet ownership, improved animal control rules and regulations, additional funding for nonsurgical methods of sterilization, surgical sterilization programs for low-income pet owners, support of animal shelters and other adoption agencies, and enactment of new and enforcement of existing leash laws that prevent random animal mating.

## 24.10  REFERENCES

1. Beck, Alan M. 1973. *The Ecology of Stray Dogs: A Study of Free-Ranging Urban Animals.* Baltimore, MD: York Press.

2. Hughes, K. L., and M. R. Slater. 2002. Implementation of a Feral Cat Management Program on a University Campus. *J. Appl. Anim. Welf. Sci.* 5(1): 15–28.

3. Hughes, K. L., M. R. Slater, and L. Haller. 2002. The Effects of Implementing a Feral Cat Spay/Neuter Program in a Florida County Animal Control Service. *J. Appl. Anim. Welf. Sci.* 4(5): 285–298.

4. Hull, S. D. 2002, November 15. Declining Songbird Population Continues to Raise Issues. *J. Am. Vet. Med. Assoc.* 221(10): 1382.

5. Levy, J. K., D. W. Gale, and L. A. Gale. 2003, January 1. Evaluation of the Effect of a Long-Term Trap-Neuter-Return and Adoption Program on a Free-Roaming Cat Population. *J. Am. Vet. Med. Assoc.* 222(1): 42–46.

6. Olson, P. N., M. V. Kustritz, and S. D. Johnston. 2001. Early-Age Neutering of Dogs and Cats in the United States (a Review). *J. Reprod. Fertil. Suppl.* 57: 223–232.

7. Olson, P. N., and C. Moulton. 1993. Pet (Dog and Cat) Overpopulation in the United States. *J. Reprod. Fertil. Suppl.* 47: 433–438.

8. Patronek, G. J. 1998, January 15. Free-Roaming and Feral Cats—Their Impact on Wildlife and Human Beings. *J. Am. Vet. Med. Assoc.* 212(2): 218–226.

9. Phillips, Rutherford T., chair. 1976. *National Conference on Dog and Cat Control.* Denver, CO: The American Humane Association.

10. Selby, L. A., J. D. Rhoades, J. E. Hewett, and J. A. Irvin. 1979, July–August. A Survey of Attitudes toward Responsible Pet Ownership. *Public Health Rep.* 94(4): 380–386.

## 24.11  WEB SITES

1. www.amcny.org/asp/homepage/default.asp

2. www.saveourstrays.com/no-kill.htm

3. www.petfinder.com

4. www.petsmart.com/adoptions/index.shtml

5. www.nsalamerica.org

6. www.ecn.purdue.edu/~laird/animal_rescue/breed_rescue_organizations

7. www.spayusa.org

# 25 TRENDS/FUTURE OF COMPANION ANIMALS AND RELATED FUNCTIONS

*No one can guarantee the future. The best we can do is size up the chances, calculate the risks, estimate our ability to deal with them, and then make our plans with confidence.*

Henry Ford (1863–1947)
*American automobile manufacturer*

## 25.1 INTRODUCTION/OVERVIEW

Companion animals heighten public interests and acquire special significance as the level of their contributions to the companionship and well-being of humans improve. This is especially true as they enhance the quality of life—and in some cases preserve life—for their owners (Figure 25.1) and the physically and mentally impaired.

**Figure 25.1** Depicted here is the April 2003 presentation of the "Heroes of Hartz Award," which celebrates the powerful human–companion animal bond given annually to honor animals, organizations, and people who have experienced and/or participated in the extraordinary. The 2003 honorees exemplified courage, dedication, and the epitome of bonding between a dog and her favorite human friend. Indeed, on a bone-chilling (6°F) night in January 2003, Joan Maguire of Fort Salonga, New York, prepared to take her faithful companion, Foxy, for a walk. On their way out, the 82-year-old owner slipped on the outside steps and broke her hip. For 90 minutes, Foxy lay on top of Mrs. Maguire to keep her warm and barked frantically. Neighbors eventually responded to the barking and sought help in getting Mrs. Maguire to the hospital. Foxy was able to save her owner's life because caring volunteers at Huntington, New York's Little Shelter Animal Adoption Center had first saved Foxy's life. Volunteers found Foxy at a municipal animal pound where she was scheduled to be euthanized. The pretty, part Pit Bull, had been hit by a car and left with a badly damaged, broken leg. Little Shelter's veterinarian performed surgery, installing a steel plate to save Foxy's leg. After 16 months of rehabilitation at the shelter, Foxy was adopted by Mrs. Maguire. By instinct, and perhaps gratitude, Foxy showed how a rescued pet may in turn save the lives of others. We commend Hartz Mountain for its near 80-year commitment to enhancing the human–animal bond. *Photograph courtesy Hartz Mountain Corporation.*

Inasmuch as companion animals are kept and appreciated globally, the petfood industry is expanding worldwide, the ingredients of petfoods are broadly available, and diseases and parasites affecting the health and quality of life of companion animals respect no boundaries (Chapters 14 and 15), it is prudent to consider major factors—those expected to have a global impact—as we look and plan ahead in matters that will likely affect future developments impacting companion animals.

One of the most universally tested and accepted precepts is that research and education lead the way in advancing the quality of life of people and companion animals. The search for knowledge is as old as humanity. Modern equipment, techniques, and procedures greatly aid scientists in their unending search for new knowledge (Chapters 16 and 22). The decades to follow will bring even more tools and techniques to research in the animal and veterinary sciences. Many of those scientific advances will be utilized to provide improvements in the health, management, and quality of life of companion animals.

Science is never static. As new truths are discovered, ideas are modified or discarded and the body of knowledge continues to grow. Humanity is at a time when history is ahead of schedule—when events people once thought might happen in the decades to follow have already occurred. This leads us to a discussion of exciting, far-reaching developments in research and education. "All men by nature desire to know."[1]

# 25.2 RESEARCH AND EDUCATION TO FACILITATE FURTHER ADVANCEMENTS

In his book, *The Rise and Fall of Great Powers,* Paul Kennedy pointed out that over the past 500 years global winners have risen to power because of first, education and research; second, technology transfer and application; and third, the economic development that follows.

Research is the process of gaining new knowledge and discovering new relationships and truths. Research is as essential to progress in animal and veterinary medical sciences applicable to companion animals as oxygen is to the marathon runner. Research often results in change; however, deliberate change usually means greater opportunity and an improved quality of life for people and the companion animals that share the lives of hundreds of millions of humans globally.

## 25.2.1 TYPES OF RESEARCH

Research is the use of systematic methods to evaluate and test ideas and/or to discover new knowledge. There are two types of research: (1) basic or fundamental, and (2) applied or directed. Basic research is important because it supplies the fundamental knowledge needed for applied research. Fundamental research is often a challenge to plan and direct, and its results are often unpredictable. It consists of exploring the unknown, moving forward the frontiers of knowledge.

Animal and veterinary medical research (including biochemistry, applied and molecular biology, dermatology, entomology, genetics, hematology, immunology, microbiology, nematology, neurology, nutrition, oncology, parasitology, pathology, pharmacology, physiology, radiobiology, toxicology, and virology) employ a systematic method of efficiently obtaining and applying knowledge in their perpetuation of companion animals and their services to humans.

Since the inception of research colleges and universities, scientific investigators have been engaged in the spawning and transfer of science and technology. Examples of these

[1]Aristotle (384–322 BC), Greek philosopher.

monumental scientific breakthroughs and technology transfers are those underpinning the therapeutic procedures discussed in Chapter 16. As a result of additional scientific discoveries and improved equipment and instrumentation, we can expect more medical procedures worthy of note and expansion to be introduced in the decades ahead.

## 25.3 BIOLOGY AND GENETIC ENGINEERING

The epoch-making scientific advances being made in the life sciences are attracting greater ethical analyses than ever before. The brilliant scientific research of James Watson of the United States and Francis Crick of the United Kingdom in the early 1950s, which unraveled the nature of the genetic code (the gene is a portion of the DNA, deoxyribonucleic acid molecule), was heralded with universal acclaim. In the 1970s, the development of gene-splicing techniques for transporting foreign genes into bacteria provided the impetus for a 2-year self-imposed moratorium by the scientists. In 1980, the Recombinant DNA Research and Development Notification Act was passed and the idea of local, state, national, and international participation in the regulation of genetic engineering was born. In 1997, the National Bioethics Advisory Commission (Rockville, Maryland) was instructed to review legal and ethical issues associated with cloning and its findings. This resulted in the 1997 Cloning Prohibition Act aimed to prohibit the cloning of humans.

### 25.3.1 BIOTECHNOLOGY IN THE PET INDUSTRY

Biotechnology research is not a high priority with many people, at least not until they or close family members become a victim of a life-threatening disease (Chapter 22). With an aging society, biotechnology will likely be embraced by more people in the future.

It will be interesting to observe how strongly the tools of biotechnology are embraced by the pet industry. Can we expect to see more cloned pets? Will we see more health benefits for pets in terms of improved nutrition, less disease, and extended life spans? Much of technology, per se, is neither good nor bad; it is how it will be used and how well it is explained to customers and the public that will determine the level of its acceptance in the pet industry. Considering the rapidity with which genetic engineering is advancing, our ability to make worthwhile improvements in petfood products, nutrition, and health are immense but often not easily measured. The challenges and opportunities are great; so are the abilities and prospects for major improvements in the quality of life for humans and companion animals.

### 25.3.2 BIOTECHNOLOGY AND FORENSICS

Criminals take note: It is virtually impossible to enter a residence in which a dog or cat lives without being contaminated by its hair. Indeed, dog and cat hairs are a common source of evidence found at crime scenes. DNA assays confirm linkages with household pets.

## 25.4 CLONING AND BIOGENETICS

A clone is a genetic identical. Before 1997, development of twins naturally or through embryo splitting was the only way to have clones of farm animals.

Mammals have been cloned from embryonic cells for years. Indeed, natural cloning occurs because identical twins are clones of each other. Clones were developed through embryo splitting by dividing the embryo when it was still early in development. When differentiation[2] of tissues begins, clones cannot be developed through embryo splitting.

Since the elucidation of DNA by Watson and Crick five decades ago, researchers have hypothesized that adult cells contain all the genetic information necessary to duplicate an individual. In the spring of 1997, Dr. Ian Wilmut of the Roslin Institute in Scotland announced the successful cloning of an adult ewe. His research associates joined cells from the mammary gland of an adult ewe with an egg cell from which the contents of the nucleus had been removed. This cell developed and the resulting lamb became the first reported successful birth of a live mammal as a clone of an adult. She was named "Dolly." Dolly possessed the same genetic material as the adult that donated the mammary gland cell. The most challenging and monumental scientific breakthrough was making the mammary gland cell revert to an undifferentiated cell.

One logical use of cloning may be in the development and production of pharmaceuticals. For example, a scientist may insert a human gene (through transgenesis[3]) into a pig or sheep so that it produces a pharmaceutical product in its milk. People could obtain the drug by simply drinking the milk. Normal reproduction of such an animal would yield some offspring with the gene, others without it. However, if clones were developed, each one would have the gene and would produce the product. Another example is in developing lines of pigs for organ transplants in humans. Genes could be inserted into pigs that would provide certain characteristics of a human immune system, thereby reducing the frequency of organ rejection. Clones of these pigs could be developed as organ sources rather than waiting for a human donor.

Another possible application of transgenic technology could be as follows: Some humans have a form of hemophilia called hemophilia B, which results from a deficiency of the blood-clotting factor IX. Treatment with blood-clotting factor IX is needed to prevent death from bleeding. Human blood-clotting factor IX is in very limited supply. It may be possible to create a transgenic cow by inserting the human gene for the production of human blood-clotting factor IX. When this cow lactates, her milk may contain large quantities of blood-clotting factor IX that could then be separated from the milk by chromatography procedures and used in human medicine. Importantly, once the ideal transgenic for a specific substance is found, that animal can be cloned for production on a larger, infinite basis.

Genetic engineering may also be used to produce new varieties and breeds of animals. The "Night Pearl" zebra fish is the first gene-altered pet to go on public sale, and it is the first ornamental fish to be genetically engineered.

## 25.4.1 CLONING A CAT

In February 2002, scientists at Texas A&M University[4] reported cloning the first house pet—a female domestic shorthair cat named "cc" (Figure 25.2).

One interesting observation was that although DNA of the cloned cat was identical to the donor source, the cloned cat's fur differed from that of the donor. This is not a total surprise. Indeed, it is well known that pigmentation in multicolored animals results not only from genetic factors, but also from developmental factors not controlled by genotype. Researchers at Duke University[5] demonstrated that common nutrients can influence which

---

[2] Differentiation is the process of acquiring individual characteristics such as occurs in the progressive differentiation of cells and tissues of the embryo. The process of transformation of mother cells into different kinds of daughter cells (brain, kidney, liver) is irreversible.

[3] A transgenic animal is one into which cloned genetic material has been transferred; it is created artificially from two or more sources and incorporated into a single recombinant molecule.

[4] Texas A&M University press release: "Texas A&M Clones First Cat," February 14, 2002. www.tamu.edu/aggiedaily/press/020214cat_pics.html.

[5] Waterland, R. A., and R. L. Jirtle. 2003. Transposable Elements: Targets for Early Nutritional Effects on Epigenetic Gene Regulation. *Molecular and Cellular Biology,* 23(15): 5293–5300.

**Figure 25.2**   (*a*) This female cat was the world's first feline to be cloned. (*b*) This photograph depicts "cc," the first feline clone, at 7 weeks of age. DNA testing verified that "cc" is genetically identical to the cat shown in (*a*). *Courtesy College of Veterinary Medicine, Texas A&M University.*

(a)

(b)

genes turn on and off in a developing fetus. They changed the color of baby mouse fur by feeding pregnant mice four supplements: vitamin $B_{12}$, folic acid, choline, and betaine. Mice given the four supplements gave birth to babies predominantly with brown coats. Pregnant mice not fed the supplements gave birth mostly to babies with yellow coats. The supplemental nutrients turned down expression of a gene called "agouti," which affects fur color, and it is now clear that nutritional supplements given the mother can permanently alter gene expression in her offspring without altering the genes themselves.

The unhappy consequences of the death of companion animals may be reduced by the possibility of cloned replacements. On the other hand, because the vast majority of humans reject the unnatural cloning of humans, they may reject it for their closest animal companions as well.

Perhaps cloning will be used to help save threatened exotic species of animals. The technology is probably available because cats, cattle, horses, mice, pigs, and sheep have been cloned.

## 25.4.2   Cloning a Horse

The first cloned horse, a foal named "Prometea," was born in May 2003. The Halflinger mare that gave birth to Prometea was also the source of her DNA, meaning she and her foal are identical twins genetically. She was created in the Laboratory of Reproductive Technologies in Cremona, Italy.[6] An adult skin cell was fused to an empty egg and after a few days the resulting embryo was returned to the horse's womb. The foal appeared healthy and genetically identical to her mother. Although clones may look similar to their DNA doubles, it is not yet known whether they will retain the same temperament and athletic ability.

---

[6] Galli, C., I. Lagutina, G. Crotti, S. Colleoni, P. Turini, N. Ponderato, R. Duchi, and G. Lazzari. 2003. Pregnancy: A Cloned Horse Born to its Dam Twin. *Nature.* 424: 635.

### 25.4.3 THERAPEUTIC CLONING

Therapeutic cloning is known to scientists as nuclear transplantation to produce stem cells.

The development of powerful technologies that have made cloning possible has forced scientists, public policymakers, and citizens to grapple with issues never confronted heretofore. There is almost universal agreement that cloning humans is so unsafe—and to many so morally repugnant—that it has been banned. But many scientists, including Shirley Tilghman (molecular biologist and president of Princeton) and David Baltimore (Nobel Laureate biologist and president of California Institute of Technology), have said it would be unfortunate if this ban on human cloning were extended to nuclear transplantation to produce stem cells.[7] Here a nucleus from an adult cell is placed in an unfertilized egg without a nucleus and the reconstituted cell is allowed to divide multiple times to produce stem cells. This has enormous potential for treating incurable diseases. Deriving stem cells in this way permits the replacement of diseased tissues with healthy ones that are unlikely to be rejected by the body as foreign.

A genuine dialogue is needed among members of the public, scientists, and governments around the world to explore the ramifications of these issues and to generate a favorable environment to conduct further research in controversial endeavors.

## 25.5 EDUCATION COMPONENT OF THE HUMAN–COMPANION ANIMAL EQUATION

The human–companion animal bond is so important and powerful it should be included in educational settings beginning with children's earliest classroom experiences and be continued throughout their formal and informal lifelong learning. Topics discussed should include the personal pleasures and benefits associated with ownership of companion animals, the human–animal bond, and responsibilities associated with human–animal relationships for maximal benefits of people and society (Chapter 21).

In 1852, English Cardinal and writer John Henry Newman[8] (1801–1890) advocated in *The Idea of the University* teaching science or philosophy, which imparted liberal knowledge with no practical, commercial, or professional end. He wrote about "the principle that all Knowledge is a whole and the separate Sciences parts of one."

From Newman's perspective, liberal education seeks to "open the mind, to correct it, to refine it, to enable it to know; and to digest, master, rule, and use its knowledge, to give it power over its own faculties, application, flexibility, method, critical exactness, sagacity, resource, address, eloquent expression." In such an educational setting, knowledge-hungry people from different disciplines "learn to respect, to consult, and to aid each other." Even though students are unable to pursue all branches of learning, they are better able to appreciate the vast body of knowledge and the principles on which it is built and rests. Such an experience educates and stimulates the mind and builds character and independent thinking.

From the educational and real-life perspectives of this book's authors, teachings related to the human–companion animal bond should be included in a liberal education. Indeed, it is important in the course of a student's educational experiences to acquire an understanding of both human and animal behavior and the interdependence and interrelationships between humans and animals. Courses in anthropology, history, philosophy, psychology,

[7]Tilghman, Shirley, and David Baltimore. 2003. Therapeutic Cloning is Good for America. *Wall Street Journal,* February 26: A16.

[8]Newman, John Henry. *The Idea of a University.* Revised edition, 2001. Accessed at http://www.newmanreader.org.

sociology, and the humanities are the backdrop against which this bond can be explored by the best minds and supported by the most respected research.

Persons having had the learning experiences mentioned above are better prepared to respond and assist with serious life-threatening challenges of companion animals as exemplified in the following narrative of a dog named Ruffy.

This story of Ruffy began when, as a starving flea-infested puppy, he was found in a Wal-Mart parking lot. A Good Samaritan took the abandoned dog to the office of Dr. Dennis Meenen at the Prairie Creek Veterinary Hospital in Rogers, Arkansas. Dr. Meenen had a reputation for placing stray dogs in good homes. After examining and treating Ruffy, he contacted acknowledged animal lovers, Nancy and Robert Clarke, who had recently lost two dogs. When Nancy saw the small, shy dog cringing in its cage, she adopted him that day. Over the next 10 years, Ruffy became a special member of their family.

Then a life-threatening accident occurred in which the lower half of Ruffy's body was crushed by the wheel of a pickup truck and the small, mixed-breed dog was dragged briefly over the asphalt street. The unconscious dog was loaded into the pickup for a fast trip to the family veterinarian. Dr. Meenen examined the dog and told the family that Ruffy's injuries were massive, so extremely severe that his only hope was the emergency and critical care section of the University of Missouri (MU) College of Veterinary Medicine (Dr. Meenen's insight into the capability of the hospital, its clinicians, and professional staff came from first hand experience—he graduated from the College in 1978). Dr. Meenen knew the MU Teaching Hospital's professionals possessed an intangible and difficult-to-define esprit de corps and concern for animals that transcends the norm. This included a well-earned reputation for going above and beyond the usual service to animals—not only through excellent veterinary medical procedures and treatments, but by special bonding with their patients and for helping them heal through emotional, as well as medical, support. Dr. Meenen knew half-crushed Ruffy would need that kind of special care.

Dr. Meenen stabilized Ruffy to help ensure that he would survive the chartered flight to MU. A series of six major surgeries, numerous blood transfusions (several veterinary students and clinicians brought in their own dogs as volunteer blood donors), an abundance of TLC, and the best in veterinary medicine aided his recovery. As Ruffy became more mobile, he was started on motion exercises and fitted with a canine version of a wheelchair—a cart that substituted two wheels for his hind legs (Figure 25.3).

A month after being admitted to the ICU (Intensive Care Unit) of the MU Veterinary Teaching Hospital, receiving multiple medical procedures and heavy-duty bonding with veterinary students and clinicians, Ruffy, with the aid of his little cart, walked out of the hospital en route to Rogers, Arkansas, for further recuperation. The story of Ruffy became well-known when his photo graced the calendar cover of the Arkansas Gazette's Pet of the Year.

**Figure 25.3** "Ruffy" in his cart. *Courtesy University of Missouri College of Veterinary Medicine, and Nancy and Robert Clarke.*

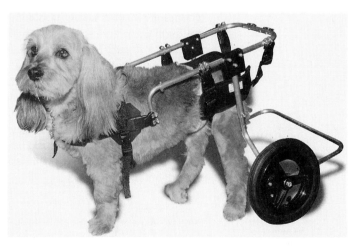

## 25.6   TRENDS RELATED TO THE COMPANION ANIMAL ENTERPRISE

With 77 million cats and 60 million dogs in 2002, the U.S. petfood market was, according to the Pet Food Institute, a $12.4 billion business (sales of cat and dog food increased 4.4% over the previous year).[9] Continued growth in the petfood market can be expected for several reasons: (1) a continuing increase in the number of pets; (2) change in the type of pets owned (e.g., a shift toward more cats and smaller breeds of dogs); (3) an increasing consumer willingness to spend money on their pets (includes increased spending per pet); (4) an increasingly isolated consumer lifestyle; (5) increasing crime rates (more dogs owned for protection); (6) improved position of pets in society as economic wealth and disposable incomes increase; (7) demographic changes (an increase in single-person households, increase in average age of consumers, the tendency to marry and have children later in life); and (8) increased pet pampering and consumers looking for higher levels of care for their pets.

Worldwide there were an estimated 250 million cats and 235 million dogs in 2002. The largest petfood markets that year were North America, 38%; Western Europe, 32%; Asia-Pacific, 16%; and Latin America, 11%. Dog/cat populations (in millions) of selected countries were United States, 60/77; China, 23/53; Brazil, 30/12.5; Russia, 10/8; Japan, 10/7.8; and Great Britain, 6.1/7.5. Two petfood companies had a combined total of 50% of the market (Nestlé-Purina, 26%; Mars, 24%). Of the cat/dog food marketed, 77% was dry, 23% wet.[10]

### 25.6.1   TRENDS IN CONSUMER PREFERENCES

Trends in human foods usually affect trends in petfoods (e.g., convenience, low fat, no additives). Consumers are increasingly more interested in their pet's health. This has contributed to the growth of premium petfoods whose major selling point is their nutritional content. Marketing trends of petfoods that follow trends in human foods include an increase in vegetarian foods (especially for dogs) and foods free of preservatives and artificial colors.

With new products in the premium sector has come redesigned packaging with special emphasis on displaying the nutritional qualities of foods using premium-quality packaging and highlighting environmental concerns. This has resulted in a greater demand for reusable or low-waste packaging.

Snacks and treats are increasing in quality and quantity. Examples include vegetarian snacks based on cheese, pasta, and cereals; meat-based snacks with added fiber such as wheat germ; healthy snacks such as those based on "natural" ingredients (e.g., with no coloring or additives); and snacks that improve oral hygiene.

*Dry versus wet* petfoods are increasing because pets can be fed with less expense on dry food (Chapter 9), dry petfood is more convenient and easier to handle (by both consumers and retailers), and dry petfood is commonly more of a premium product. Additionally, it has traditionally been produced in varieties tailored to the needs of specific types of pets.

*Freeze-dried* petfoods designed to feed pets during travel or emergency situations are increasing.

*Manufacturer's marketing strategy* is aimed to sell the owner—not the pet; appeal to health positioning (healthy skin in pets, look and feel young), appeal to convenience in a time-pressed environmental lifestyle.

*Environmental demands* continue to be stressed. Packaging supplies that use less energy and fewer natural resources are being emphasized as is the use of more recycling/recovering materials. Greater use is being made of more technically advanced, environmentally friendly products.

---

[9]Crossley, Adrienne. 2003, March 31–April 2. The Future of the Petfood Industry. *Petfood Forum 2003.* Chicago, IL.
[10]Russell, Inge. 2003. Solving Pet Problems with Biotechnology. *Petfood Forum Europe Proceedings.* Amsterdam, The Netherlands.

## 25.6.2   NUTRITION–RELATED TRENDS/STUDIES/ DEVELOPMENTS

Increasing consumer awareness of the pet health benefits of functional foods is reflected in an increasing interest in omega fatty acids, probiotics,[11] prebiotics, and botanicals (Chapter 9). This has been reflected in new products promoting functional benefits for cats and dogs related to:

- joint/bone/cartilage benefits (anti-arthritis, anti-inflammatory)
- oral care benefits (dental/breath)
- immune system benefits (antioxidants)
- skin/coat benefits (fatty acids, protein)
- digestive benefits (pre- and probiotics)
- odor control benefits (yucca schidigera and alfalfa)
- hairball control benefits (high fiber formulations to help facilitate quick exit of cat hair ingested through grooming)

*Functional fibers* include soluble fiber, insoluble fiber, fermentable fiber, and dietary fiber.[12] Cellulose is a purified extracted fiber from woody plants. It is the principle carbohydrate constituent of plant cell membranes and is used to lower the energy density of petfood diets. An increasing percentage of dogs and cats are obese. For normal healthy dogs, a diet containing between 2.5% and 4.5% fiber is adequate. For obese dogs, dietary fiber may be increased to 9% to 10% using food ingredients such as wheat bran and barley products. To decrease weight in cats, they should be given less to eat and fed less frequently. A list of recommended daily allowances for calories and nutrients is available online at http://dels.nas.edu/banr/petdoor.html.

*Nutraceuticals* are dietary supplements in a purified or extracted form that are safe and have health and well-being benefits. Examples of nutraceuticals applied in dog foods are beet pulp, fructooligosaccharides, mannanoligosaccharides (MOS), mussel extract, glucosamine, chondroitin sulfate, carnitine, omega-3 fatty acids, and antioxidants. Mannanoligosaccharides may be derived from the cell walls of yeast.

Numerous food nutraceuticals and other compounds are being studied to determine whether they can be used to enhance the beneficial functioning of the immune system. Interest in this field of scientific research is escalating as the proportion of older citizens continues to increase.

Supplementation with fructooligosaccharides has been shown experimentally to enhance the colonic microbial population. Fructooligosaccharide supplementation was also found to decrease fecal concentration of several putrefactive compounds. This is of special interest to pet owners through possibly less flatulence in dogs and less odor in cat litter boxes. Research advances in the area of MOS are benefiting pet intestinal health and immunity.

When added to petfood, an extract of yucca has been shown to reduce the potency of feline fecal odors by up to 49% and that of dogs by 56%.

*Palatability* of petfoods is again under study. Heretofore petfood palatability trials have relied on feeding and observing. Recent biotechnology research has, through mapping the human and mouse genomes, discovered specific taste receptors that are expected to clarify

---

[11]It can be anticipated that in the years ahead scientists will custom-design probiotics to improve digestive functions and the overall health of companion animals.

[12]The American Association of Cereal Chemists defines dietary fiber as "the edible parts of plants or analogous carbohydrates that are resistant to digestion and absorption in the human small intestine with complete or partial fermentation in the large intestine. Dietary fiber includes polysaccharides, oligosaccharides, lignin, and associated plant substances. Dietary fiber promotes beneficial physiological effects including laxation and/or blood cholesterol attenuation and/or blood glucose attenuation." For additional information, go to http://www.aaccnet.org/definitions.

the intricacies of bitter, sweet, and umami (the "fifth" taste, which resembles that of monosodium glutamate). The first human receptor activated by bitter tastes was identified in 2002 using biotechnological techniques.

In 2001, researchers identified a gene for sweetness. Based on the above and other recent developments related to palatability, petfood manufacturers are rethinking the uses of palatants and flavors, fermentation and cheese products, and custom premixes.

*Minerals and their interaction(s)* with other food components are being investigated. Selenium, for example, has been shown to have positive effects on the feathers of chickens. Still to be tested is whether these findings can be applied to pet birds. The bioavailability of minerals in pets deserves further study.

## 25.6.3   PET HEALTH–RELATED CONSIDERATIONS

Lycopene is a common dietary carotenoid that gives tomatoes their red color. Unlike numerous vitamins, lycopene is heat-stable. Researchers are using genetic biotechnological techniques to increase lycopene levels in crops such as tomatoes. It has been shown experimentally that lycopene is effective against prostate cancer, breast cancer, and heart disease. Numerous manufacturers are adding tomato products to dog food believing that dogs can share in these health benefits of lycopene.

*Allergen-free pets.*   Not only can new proteins be introduced into an animal's genetic makeup, but proteins can also be removed providing they are not critical to the animal's survival. Because millions of people experience more allergic reactions to cats than dogs, research efforts are being directed to produce allergen-free cats first, then dogs. Because the cat allergen is not believed to be essential to the health of cats, it is believed that its removal is possible. The cat allergen that causes the allergic reaction in humans is not due to pet fur. Instead the allergen is found in cat saliva, urine, or dander (dead skin flakes containing the protein cling to shedded pet hair, carpeting, or furniture). For additional information, see www.transgenicpets.com.

## 25.6.4   PET INDUSTRY SALES OF NEW PRODUCTS

The pet industry in the United States is large and growing. Indeed, according to the American Pet Products Manufacturers Association, Americans spent $31 billion on their pets in 2003.[13] These expenditures will likely increase as new devices and toys are introduced. An example is the possible purchase of a new device called "Bow-Lingual" if it proves accurate in translating dog barks into one of nearly 200 English phrases. The battery-powered device consists of two parts: a cordless microphone that attaches to a dog's collar, and a wireless receiver that displays the translations and other data on a tiny screen. It will record a dog's barks for up to 12 hours while its owner is away. It is based on analysis of thousands of barks from more than 80 breeds of dogs.

The Japan Acoustic Laboratory (JAL) grouped the barks into one of six moods according to the bark's digital voiceprints or sound wave shape: happy, sad, frustrated, needy, on-guard, and assertive. An independent evaluator of the device found it most reliable in gauging a dog's general mood (happy, sad, and others noted above). The Japanese toy manufacturer Takara has sold over 250,000 units in Japan and is working on a "Meow-Lingual" for cats.

*Product life cycles.*   Overall, product life cycles are becoming shorter. New, improved products sell well. Manufacturers and marketers like the "just-in-time" sales approach to new products.

*Other market opportunities.*   New and extended opportunities exist for private label manufacturers; franchised specialty chains are also growing.

---

[13]http://www.appma.org/press/press_releases/2003/nr.

## 25.6.5   EXPANDING THE NUMBER OF COMPANION ANIMALS IN DEVELOPING COUNTRIES

On a global scale, the primary reservation and concern with increasing the number of companion animals in developing countries is that companion animals would in many cases compete with humans for already scarce food supplies. Additionally, the average per capita living area is considerably greater in developed countries than in the less developed ones.

From the standpoint of pets competing with humans for food, the species of pet kept could be a factor. For example, rabbits feed largely on leafy plants and vegetables that are commonly readily available in developing countries.

# 25.7   FUTURE OF COMPANION ANIMALS AND RELATED ENTERPRISES

"It is difficult to say what is impossible, for the dream of yesterday is the hope of today and the reality of tomorrow."[14] We possess neither genie nor crystal ball. Any predictions will be based in large part on the philosophy of Patrick Henry (American statesman and orator, 1736–1799) who once said, "I have but one lamp by which my feet are guided, and that is the lamp of experience. I know of no way of judging the future but by the past."[15]

To predict the future of companion animals it is necessary to predict the demographics and future of people, because humans will in large part determine the future of companion animals. We need to understand why people acquire pet animals. In many single-person households, companion animals are kept for security purposes as much as for companionship. The companionship provided by animals is not merely an alternative to human companionship; rather it offers something additional, different, and special (Chapter 1). In our aging society, numerous opportunities exist for animal-assisted therapy among senior citizens (Chapter 1) and pet services for the physically impaired (Chapter 21). An increasingly financially capable populace; readily available petfoods, products/supplies; as well as superbly trained healthcare professionals in veterinary medicine and affiliated disciplines add to the sound basis for believing that companion animals and related enterprises will flourish in the decade ahead and well beyond.

We believe present trends emphasize the desire and need for humans to enjoy life to the fullest extent possible. For an increasingly large proportion of people of many countries, youth and adults alike seek a greater return to nature in order to realize a greater sense of unity and fulfillment. This means a closer association with companion animals. Appreciating the fact that life expectancy is increasing, the opportunity for greater numbers of the elderly, sick, and impaired to enjoy the benefits of regular, close association with dogs and cats also bodes well for the future of companion animals and those who tend to their health, food, and other important service needs.

## 25.7.1   GLOBAL CHALLENGES AND OPPORTUNITIES FOR EXPANDING THE COMPANION ANIMAL ENTERPRISE

For several years, China has been a world leader in terms of gross domestic product (GDP) growth. An important question related to the growth of exported petfoods is "will China provide significant commercial opportunities for the petfood industry?"

Important to note is that more than a billion of China's people live in rural areas. And consider here the fact that only among the urban population of China are dogs and cats kept

---

[14]Robert H. Goddard (1882–1945), American physicist.
[15]Platt, Suzy, ed. 1989. *Respectfully Quoted: A Dictionary of Quotations Requested from the Congressional Research Service*. Washington, DC: Library of Congress.

as pets. In rural areas, dogs and cats are kept primarily as working animals—dogs as guards and cats to hunt mice, rats, and other rodents.

Moreover, only a small percentage of China's 1.3 billion inhabitants live within financial reach of "consumer heaven." Less than half of them own cars, telephones, central heating, and air conditioning. Notwithstanding the fact that only about 135 million people (~10%) live in urban areas, the potential petfood market in China is large. There are some 25 cities with a population between 1 and 2 million and 13 with a population of more than 2 million (10 with a population of over 6 million). There are approximately 225 million households in China.[16]

Worldwide there are a large number of other emerging markets. These include developing countries in Africa, Asia, Latin America, as well as the former socialist nations of Eastern Europe.

# 25.8 AREAS RELATED TO COMPANION ANIMALS NEEDING ADDITIONAL RESEARCH

To help the companion animal enterprise achieve its growth potential, additional research is needed in several areas. Let us consider selected ones.

## 25.8.1 COMPANION ANIMAL NUTRITION RESEARCH NEEDED

Notwithstanding the tremendously important research advances made in recent decades in dog and cat nutrition, the need remains for further studies to ascertain more precisely the relationships between nutritional requirements and deficiencies and the overall health and quality of life for companion animals. For example, additional research studies are needed to determine the precise intakes of nutrients required to prevent deficiency diseases in dogs and cats at varying ages and levels of work. Most of the essential nutrients of dogs and cats are known but some have yet to be established and refined. An obvious example is the exact nature of the essential fatty acid requirements, particularly the question of whether the omega-3 fatty acids are required (Chapter 9) and, if so, at what levels? Another area needing additional research pertains to the nature of trace mineral requirements in dogs and cats.

Although non–nutrition-related diseases can affect the nutritional needs and nutritional status of companion animals, little is known in practical terms of the impact of various diseases of dogs and cats on their nutrient needs or metabolism. Additional research is needed to provide pertinent new data and information on this important practical matter.

## 25.8.2 ANIMAL HEALTH–RELATED RESEARCH NEEDED

The use of scientific advancements such as those associated with monoclonal antibodies hold promise for expanded use in diagnosis, control, and eradication of diseases detrimental to companion animals. Additional research is needed.

*Psychosocial benefits.* The psychosocial benefits of companion animals need further researching. Unfortunately, much of the literature about companion animals is encumbered by generalizations and observations without a scientifically sound data base. This calls for new and additional research regarding the health-related benefits that accrue in the later years to owners and associates of companion animals. Additionally, research is needed to determine the "carrying capacity" of pet populations (e.g., the kinds and numbers of pets that can be ecologically and socially maintained without adverse consequences).

---

[16]Marsh, Frederick O. 2001, October. Unleashing Opportunities in Global Petfood: Strategies to 2010. *Petfood Forum Europe.* Amsterdam, Netherlands.

Other areas worthy of further research relate to effective zoonoses control programs and acceptable animal control guidelines that make it possible for pet ownership to continue in a fashion that benefits both animals and people.

***Practical parasite kits.*** Much is known about parasite prevention and treatment related to companion animals (Chapter 15). Additional research could enable parasitologists to make greater use of techniques associated with molecular biology in developing diagnostic kits that lay people could utilize in detecting the parasite(s) involved and applying the best treatment and prevention techniques.

***Bioinstrumentation.*** Practical applications of advances in bioinstrumentation and micro-electronics are needed to better use single and multichannel systems for transmitting selected physiological signals from surgically implanted units. Such techniques could aid in better understanding the normal and changed physiological and psychological state of companion animals under normal and test conditions including heavy work.

***Genetic engineering technologies.*** These technologies hold promise in removing undesirable animal traits and to create breed types better suited to specific needs of owners and the elderly.

***Research animals.*** As discussed in Chapter 22, companion animals can contribute to additional research needed in studying SARS and other emerging diseases important to humans.

# 25.9 SUMMARY

"We must sail sometimes with the wind and sometimes against it, but we must sail, not drift nor lie at anchor."[17] At no time in the history of humanity have we enjoyed a more healthy and productive companion animal enterprise. Much of that progress may be attributed to the dedicated men and women who enhance the mental and physical lives of humans in general through their inquisitive minds and an intense desire to improve the status and lives of animals and those who have chosen to dedicate their considerable efforts and ingenuity to specifically help ensure healthy companion animals. We thank and salute those unselfish, service-oriented individuals who have made available the science-based procedures discussed in Chapter 16, therapeutic and service uses of companion animals discussed in Chapter 21, and biomedical research advances discussed in Chapter 22. May we be ever vigilant in expanding that scientific base for the benefit of future generations.

The future of companion animals and their diverse support and service base will be shaped by economics, demographics, science and technology, global markets, the strength of the human spirit, the values and honorable passions we hold dear, and the ideals we preserve and defend tenaciously.

High-quality biology, care, health, and management of companion animals begins with people—those who are curious and anxious to conduct the research needed to push the frontiers of knowledge forward—those who teach and communicate the current state of knowledge to students; the dedicated core of veterinarians, veterinary technologists and technicians, and others who provide healthcare services to companion animals—those engaged in the modern, service-oriented petfood, pet products and supplies industry—the all-important animal owners, and others who desire to provide an ever-increasing quality of life for companion animals. Indeed, unlike the great pyramids, the health and quality of life of companion animals will not stand untended.

We have deep respect for leaders of the companion animal–related disciplines and businesses. As they develop and expand their capabilities, they extend further the expectations of those who enjoy the association and companionship of and services rendered by pet animals—great and small.

---

[17]Oliver Wendell Holmes (1809–1894), American physician and author.

As we approach the end of this book we are loathe to close because we believe the topics discussed are a vital prerequisite to the health and quality of life of animals that mean so much to their grateful owners and others who benefit immeasurably from these special animals. Our purpose in writing this book and chapter was not so much to predict the future as to help enable it.

If through studying the book, readers have become more deeply committed to knowing and understanding more about the biology, care, health, and management of companion animals and the important role they play in providing companionship and services to their owners and admirers, then we have achieved our foremost objective in writing it. We thank the readers for taking the journey with us!

Finally, we know who holds the future. We have a deep and abiding faith in the future of those who make up the core of the enormous and expanding support team of the companion animal family. They comprise a dedicated group of owners, teachers, researchers, businesses, and industries—professionals all, who display the spirit and challenge advanced 24 centuries ago in the following quotation: "But the bravest are surely those who have the clearest vision of what is before them, glory and anger alike, and yet, notwithstanding, go out to meet it."[18]

## 25.10   REFERENCES

1.  Adams, J. A. 2002. Managing Product Development for Effectiveness. *Petfood Forum Europe 2002*, 177–187.

2.  Bickel, Lennard. 1974. *Facing Starvation: Norman Borlaug and the Fight against Hunger.* New York: Readers Digest Press. Distributed by E. P. Dutton & Co., Inc.

3.  Campbell, John R. 1998. *Reclaiming a Lost Heritage: Land-Grant and Other Higher Education Initiatives for the Twenty-First Century.* East Lansing: Michigan State University Press.

4.  Fahey, George, C., Jr. 2003. Research Needs in Companion Animal Nutrition. In *Petfood Technology*, 57–61. Mt. Morris, IL: Watt Publishing Co.

5.  Harmison, Lowell T., ed. 1973. *Research Animals in Medicine.* National Conference on Research Animals in Medicine sponsored by the National Heart and Lung Institute. Washington, DC: U.S. Government Printing Office.

6.  Kennedy, Paul. 1989. *The Rise and Fall of Great Powers.* New York: Vintage Books-Random House.

7.  Lyons, T. Pearce, and Kate A. Jacques, eds. 2002. *Nutritional Biotechnology in the Feed and Food Industries.* Nottingham, United Kingdom: Nottingham University Press.

8.  Naisbitt, John, and Patricia Aburdene. 1990. *Megatrends 2000.* New York: William Morrow Co., Inc.

9.  Peters, Tom J., and Nancy K. Austin. 1985. *A Passion for Excellence.* New York: Random House, Inc.

10. Roberts, A. L., M. E. O'Brien, and G. Subak-Sharpe. 2001. *Nutraceuticals: The Complete Encyclopedia of Supplements, Herbs, Vitamins and Healing Foods.* New York: Berkley Publishing Group.

## 25.11   WEB SITES

1.  www.tamu.edu/aggiedaily/press/020214cat_pics.html

2.  http://www.appma.org/press/press_releases/2003/NR

---

[18]Thueydides (ca. 400 BC), Greek historian.

# APPENDIX A

## PREFIXES

**a-, an-**   absence, without, separated from (anaerobic)

**ab-**   away, from (abnormal)

**ad-**   toward (adduction)

**ante-**   before (antenatal)

**anti-**   against (antibiotic)

**bi-**   double, two (bilateral)

**bio-**   life (biology)

**co-, con-**   together, along with (congenital)

**di-**   two parts (dipeptide)

**ec-, ecto-, ex-**   outside (ectoparasite, exogenous)

**endo-**   inside (endoparasite, endogenous)

**epi-**   upon, over (epithelium)

**hemi-**   half (hemisphere)

**hydro-**   pertaining to water (hydrocephalus)

**hyper-**   excess, greater than (hyperglycemia)

**hypo-**   lower, less than (hypoglycemia)

**infra-**   beneath, below (infrascapular)

**inter-**   between (intercellular)

**intra-**   within (intracellular)

**iso-**   equal (isotonic)

**kilo-**   thousand (kilogram)

**leuko-**   whiteness (leukocyte)

**macro-**   large (macroorganism)

**mal-**   bad (malnutrition)

**meso-**   middle (mesoderm)

**micro-**   small (microorganism)

**milli-**   thousandth (milligram)

**mono-**   one, single (monogamous)

**morpho-**   shape, form (morphology)

**multi-**   many, number (multicellular)

**myo-**   pertaining to muscle (myocarditis)

**neo-**   new (neoplasm)

**pseudo-**   false (pseudopregnancy)

**sub-**   beneath, deficiency of (subnormal)

**supra-**   above (suprarenal)

**sym-**   with, together (symbiosis)

**zoo-**   animal (zoology)

# ROOT WORDS AND PREFIXES

**adeno-**   gland (adenitis)

**branch-, broncho-**   pertaining to the trachea branches (bronchitis)

**cardi-, cardio-**   pertaining to the heart (cardiology)

**cyto-**   pertaining to the cell (cytology)

**dermo-**   skin (dermatitis)

**entero-**   the intestine (enteritis)

**gastro-**   stomach (gastritis)

**glosso-**   tongue (glossitis)

**hemo-, hemato-**   blood (hematocrit)

**hepat-**   liver (hepatectomy)

**lac-, lact-, lacto-**   pertaining to milk (lactation)

**neuro-**   pertaining to nerve or neurology (neurologist)

**odonto-**   teeth (odontoblast)

**osteo-**   bone (osteomalacia)

**oto-**   ear (otology)

**patho-**   disease (pathological)

**pod-**   foot (podiatry)

**rhi-, rhino-**   nose (rhinoscope)

# SUFFIXES

**-algia**   pain in a part (neuralgia)

**-ase**   enzyme ending (amylase)

**-blast**   cell that builds (osteoblast)

**-cide**   causing death (bactericide)

**-cyte**   mature cell (monocyte)

**-ectomy**   excision, cut out (splenectomy)

**-fuge**   driving out (vermifuge)

**-itis**   inflammation (dermatitis)

**-logy**   study of (physiology)

**-lysis**   breakdown (hemolysis)

**-malacia**   abnormal softness (osteomalacia)

**-oid**   similar in shape (spheroid)

**-oma**   tumor (sarcoma)

**-opia**   pertaining to the eye (myopia)

**-osis**   state, condition of (neurosis)

**-rhea**   excessive discharge or excretion (diarrhea)

**-stat**   stop (hemostat)

**-tomy**   incision, cut into (laparotomy)

**-uria**   abnormalities of urine or of urination (albuminuria)

# APPENDIX B

## CONVENIENT CONVERSION DATA AND ABBREVIATIONS

teaspoon = t
tablespoon = T
cup = c
pint = pt
quart = qt
gallon = gal
ounce = oz
pound = lb
inch = in
foot = ft
square foot = sq ft
yard = yd

milliliter = ml or cc
gram = g
kilogram = kg
centimeter = cm
meter = m
kilometer = km

1 milliliter (ml or cc) = 12 drops
1 teaspoon = 60 drops (5 cc or ml)
1 tablespoon = 3 teaspoons
1 pint = 32 tablespoons
1 cup = 0.5 pint
1 quart = 2 pints (946.4 cc) and 1/8 peck
1 gallon = 4 quarts (3.785 liters, 8.345 pounds)
1 kilogram (kg) = 2.2046 pounds
1 liter = 1.0567 quarts (1 kg water)
1 foot = 0.3048 meter

1 meter = 3.2809 feet (39.37 inches)
1 kilometer = 0.6214 mile

## TEMPERATURE CONVERSION

$$C = \frac{°F - 32}{1.8}$$

$$°F = (°C \times 1.8) + 32$$

## U.S./METRIC CONVERSION

| When you know | Multiply by | To find |
| --- | --- | --- |
| ounces | 28.35 | grams |
| pounds | 0.45 | kilograms |
| teaspoons | 4.93 | milliliters |
| tablespoons | 14.78 | milliliters |
| fluid ounces | 29.57 | milliliters |
| cups | 0.236 | liters |
| pints | 0.473 | liters |
| quarts | 0.946 | liters |
| gallons | 3.785 | liters |
| inches | 2.54 | centimeters |
| feet | 0.3048 | meters |
| yards | 0.9144 | meters |
| miles | 0.6214 | kilometers |

## METRIC/U.S. CONVERSION

| When you know | Divide by | To find |
| --- | --- | --- |
| grams | 28.35 | ounces |
| kilograms | 0.45 | pounds |
| milliliters | 4.93 | teaspoons |
| milliliters | 14.78 | tablespoons |
| milliliters | 29.57 | fluid ounces |
| liters | 0.236 | cups |
| liters | 0.473 | pints |
| liters | 0.946 | quarts |
| liters | 3.785 | gallons |
| centimeters | 2.54 | inches |
| meters | 0.3048 | feet |
| meters | 0.9144 | yards |
| kilometer | 0.6214 | miles |

| Measure of Mass | | | | |
|---|---|---|---|---|
| Grains | Drams | Ounces | Pounds | Metric Equivalents (grams) |
| 1 | 0.0366 | 0.0023 | 0.00014 | 0.0647989 |
| 27.34 | 1 | 0.0625 | 0.0039 | 1.772 |
| 437.50 | 16 | 1 | 0.0625 | 28.350 |
| 7000 | 256 | 16 | 1 | 453.5924277 |

# APPENDIX C

## TABLES OF WEIGHTS AND MEASURES

| Metric Weight | | | | | | | | |
|---|---|---|---|---|---|---|---|---|
| Microgram | Milligram | Centigram | Decigram | Gram | Dekagram | Hectogram | Kilogram | Metric Ton |
| 1 | | | | | | | | |
| $10^3$ | 1 | | | | | | | |
| $10^4$ | 10 | 1 | | | | | | |
| $10^5$ | 100 | 10 | 1 | | | | | |
| $10^6$ | 1000 | 100 | 10 | 1 | | | | |
| $10^7$ | $10^4$ | 1000 | 100 | 10 | 1 | | | |
| $10^8$ | $10^5$ | $10^4$ | 1000 | 100 | 10 | 1 | | |
| $10^9$ | $10^6$ | $10^5$ | $10^4$ | 1000 | 100 | 10 | 1 | |
| $10^{12}$ | $10^9$ | $10^8$ | $10^7$ | $10^6$ | $10^5$ | $10^4$ | 1000 | 1 |

| Metric Measure (Volume) | | | | | | | | |
|---|---|---|---|---|---|---|---|---|
| Microliter | Milliliter | Centiliter | Deciliter | Liter | Dekaliter | Hectoliter | Kiloliter | Myrialiter |
| 1 | | | | | | | | |
| $10^3$ | 1 | | | | | | | |
| $10^4$ | 10 | 1 | | | | | | |
| $10^5$ | 100 | 10 | 1 | | | | | |
| $10^6$ | $10^3$ | 100 | 10 | 1 | | | | |
| $10^7$ | $10^4$ | $10^3$ | 100 | 10 | 1 | | | |
| $10^8$ | $10^5$ | $10^4$ | $10^3$ | 100 | 10 | 1 | | |
| $10^9$ | $10^6$ | $10^5$ | $10^4$ | $10^3$ | 100 | 10 | 1 | |
| $10^{10}$ | $10^7$ | $10^6$ | $10^5$ | $10^4$ | $10^3$ | 100 | 10 | 1 |

## Metric Measure (Length)

| Micrometer* | Millimeter | Centimeter | Decimeter | Meter | Dekameter | Hectometer | Kilometer | Myriameter | Megameter | Equivalents |
|---|---|---|---|---|---|---|---|---|---|---|
| 1 | 0.001 | $10^{-5}$ | | | | | | | | 0.000039 inches |
| $10^3$ | 1 | $10^{-1}$ | | | | | | | | 0.03937 inches |
| $10^4$ | 10 | 1 | | | | | | | | 0.3937 inches |
| $10^5$ | 100 | 10 | 1 | | | | | | | 3.937 inches |
| $10^6$ | 1000 | 100 | 10 | 1 | | | | | | 39.37 inches |
| $10^7$ | $10^4$ | 1000 | 100 | 10 | 1 | | | | | 10.9361 yards |
| $10^8$ | $10^5$ | $10^4$ | 1000 | 100 | 10 | 1 | | | | 109.3612 yards |
| $10^9$ | $10^6$ | $10^5$ | $10^4$ | 1000 | 100 | 10 | 1 | | | 1093.6121 yards |
| $10^{10}$ | $10^7$ | $10^6$ | $10^5$ | $10^4$ | 1000 | 100 | 10 | 1 | | 6.2137 miles |
| $10^{12}$ | $10^9$ | $10^8$ | $10^7$ | $10^6$ | $10^5$ | $10^4$ | 1000 | 100 | 1 | 621.370 miles |

*Formerly called the micron and abbreviated μ.*

# GLOSSARY

**Abdomen**  The anatomical region between the chest and hindquarters.

**Abdominal**  Pertaining to the **abdomen** (belly).

**Absorption**  Transfer of the products of digestion from the intestine to blood or lymph or the transfer of interstitial fluid into cells.

**Acariasis**  An infestation with mites or ticks.

**Acetonemia (ketosis)**  A condition characterized by an abnormally elevated concentration of ketone (acetone) bodies in the body tissues and fluids.

**Acoustic**  Pertaining to sound.

**Acral lick dermatitis**  An ulcerated, firm, thickened lesion on one or more limbs resulting from excessive licking; in many dogs, this excessive licking is an obsessive-compulsive disorder. Also known as acral lick granuloma.

**Acute**  Severe, rapid onset, usually short duration.

**Adipose tissue**  Fatty tissue.

*Ad libitum* **(ad lib)**  At pleasure; commonly used to express the availability of feed/food on a **free-choice** basis.

**Adrenal**  Pertains to near the kidney; the adrenal glands are located near the kidneys.

**Afebrile**  Without fever.

**Afferent**  To carry to/toward; afferent neurons carry impulses from the **peripheral nervous system** toward the **central nervous system**.

**Agglutinate**  To adhere, unite, or combine into a group or mass (as with clumps of certain body cells). The *clumping* is often caused by antibodies attaching to cells and cross-linking them together.

**Agglutination titer**  The highest dilution of a serum that causes clumping together of cells, especially bacteria or red blood corpuscles.

**Agoraphobia**  An abnormal fear of new places, especially open spaces.

**Agouti**  Agouti or ticked hairs have bands of light and dark pigmentation.

**AIDS**  Acquired Immune Deficiency Syndrome. A disease of the human immune system that is caused by infection with HIV and characterized by a severe reduction in the number of helper T cells. See **HIV**.

**AKC**  American Kennel Club.

**Albino**  Animal deficient in pigmentation.

**Alimentary**  Pertaining to food.

**Allele**  Pair of genes that produce alternative physical characteristics (e.g., the genes for longhair or shorthair).

**Allergen**   Any substance that gives rise to the formation of antibodies and the resultant allergic reaction. Also called an **antigen**.

**Allergic**   Hypersensitivity; inflammatory reaction of the immune system, usually directed against a foreign protein. A hypersensitive state acquired through exposure to a particular **allergen**.

**Alley cat**   A homeless or stray cat.

**Alopecia**   Loss of hair (baldness).

**Alter**   To neuter or spay a cat or dog.

**Alveoli**   Tiny air sacs within the lungs.

**Amastigote**   One of the life stages of a protozoan parasite.

*Amblyomma americanum*   The lone star tick.

*Amblyomma maculatum*   The gulf coast tick.

**Amoebae**   Protozoa **parasites**.

**Amylase**   An enzyme produced by the pancreas, breaks down starches.

**Amyloid**   A glycoprotein related to **immunoglobulins**.

**Amyloidosis**   An accumulation of **amyloid** in various body tissues; clinical signs may include those of renal failure and diabetes mellitus due to destruction of the functional cells of the kidney and pancreas, respectively.

**Anagen**   Growth phase of the hair cycle; active hair follicle.

**Analgesia**   A state without pain.

**Analgesic**   Substance that relieves pain.

**Anaphylactic shock (anaphylaxis)**   An **immunological** reaction with increased sensitivity to a normally nontoxic protein or other antigen when injected with it for the second time. It may cause a severe or even fatal reaction with respiratory failure or cardiovascular collapse.

*Ancylostoma braziliense*   Hookworm of cats and dogs.

*Ancylostoma caninum*   Hookworm of dogs.

*Ancylostoma tubaeforme*   Hookworm of cats.

**Androgen**   Male sex hormone.

**Anemia**   A condition in which the blood is deficient in hemoglobin or red blood corpuscles or both. It is characterized by paleness of the skin and mucous membranes, loss of energy, and palpitation of the heart (unduly rapid action of the heart, which is felt by the individual).

**Anesthesia**   Loss of the feeling of pain, touch, cold, or other sensation, produced by ether, chloroform, morphine, and other compounds; by hypnotism; or as the result of hysteria, paralysis, or disease.

**Anestrous period**   That time when the female is not in estrus; the nonbreeding season.

**Anestrus**   Absence of estrus; clinically refers to the period of the reproductive cycle in which the female animal is in sexual quiescence.

**Angora**   Originally, the name given to the first cats with long fur seen in Europe, so called after Angora (now Ankara) in Turkey from where they were thought to have come. Now a recognized breed.

**Animal husbandry**   A branch of agriculture concerned with the care and production of domestic animals.

**Animal shelter**   An organization that cares for stray and other homeless pets and seeks new homes for them.

**Animal welfare**   Political issue or philosophic discussion concerning animal **state-of-being**.

**Anisocoria**   Unequal sized pupils of the eyes.

**Annelid**   An animal of the phylum Annelida.

**Anogenital**   Denotes both the anal and **genital** regions.

**Anoplura**   Sucking lice.

**Anorexia**   Lack or loss of the appetite for food.

**Anterior**   Denotes the front or forward part. It means the same as the **ventral** surface of the body in human anatomy.

**Anthelmintic**   Against worms, deworming agents.

**Antibiotic**   Antibacterial substance produced by microorganisms that inhibits other microorganisms.

**Antibody**   A protein substance (modified type of blood-serum globulin) developed or synthesized by lymphocytes in response to an antigenic stimulus. Each antigen elicits production of a specific antibody. In disease defense, the animal must have had an encounter with the pathogen (**antigen**) or have received colostrum milk before a specific antibody can be found in its blood.

**Anticholinergic**   Blocking the action of acetylcholine.

**Anticoagulant**   Substance added to blood to prevent clotting.

**Antigen**   A high-molecular-weight substance (usually protein) that, when foreign to the body of an animal, stimulates the formation of a specific antibody and reacts specifically in vivo or in vitro with its corresponding antibody.

**Antioxidant**   A substance that prevents oxidation of other molecules.

**Antiseptic**   From the Latin *anti* meaning "against" and *sepsis* meaning "putrefaction." A substance that prevents growth and development of microorganisms either by destroying them (**bactericidal** action) or inhibiting their growth (bacteriostatic action).

**Antiserum**   A serum that contains an antibody or antibodies. It gives temporary protection against certain specific infectious diseases.

**Antitoxin**   Antibody formed against poisonous toxins, such as bacterial exotoxins, which specifically neutralizes (counteracts) the effects of the toxin. Diphtheria antitoxin, obtained from the blood of horses infected with diphtheria, is injected into persons to make them immune to diphtheria or to treat them if they are already infected.

*Antricola*   A genus of soft ticks.

**Aorta**   The major arterial trunk that carries blood from the heart to be distributed by branch arteries throughout the body.

**APF**   Animal protein factor. The original label given to vitamin $B_{12}$.

**APHIS**   Animal and Plant Health Inspection Service (of **USDA**).

**Apnea**   Absence of breathing (occurs following injection of some forms of anesthetics).

**Arbovirus**   Any one of several groups of small viruses transmitted by arthropods such as mosquitoes and ticks. Yellow fever, dengue, and equine encephalomyelitis are caused by arboviruses.

*Argas*   A genus of ticks parasitic on birds.

**Argasidae**   A family of ticks that lack a **scutum**.

**Arm**   The anatomical region between the shoulder and elbow joints consisting of the humerus and associated muscles. Sometimes referred to as "upper arm."

**Arrhythmia**   A condition without rhythm, abnormal heart beat/cycle.

**Arteriosclerosis**   A progressive thickening and hardening of the walls of the arteries, often associated with high blood pressure or with chronic disease of the kidneys.

**Ascarids**   Roundworms.

**Ascites**   In right heart failure, the right ventricle becomes distended and inefficient at pumping blood forward, pressure builds up in the systemic circulation, and fluid leaks into the abdominal cavity. This fluid leakage is called ascites.

**Aseptic**   Free from living germs that cause disease, putrefaction, or fermentation.

**Aspiration**   Refers to the inhalation of foreign materials into the lungs.

**Assay**   The determination of the purity of a substance or the amount of any particular constituent of a mixture.

**Asthenia**   Lack or loss of strength.

**Asymmetry**   Lacking balance or proportion.

**Asymptomatic**   Presenting no symptoms of disease; may be a silent carrier that is infectious to other animals or the disease may be so mild that no symptoms are apparent.

**Ataxia**   Muscular incoordination; stumbling.

**Atheroma**   Fatty degeneration of the walls of arteries.

**Atherosclerosis**   A fatty degeneration of the connective tissue of the arterial walls. A form of **arteriosclerosis** characterized by **atheroma**. Lesions within the arteries with plaques containing cholesterol and other lipid materials.

**Atony**   Lack of physiological tone especially of a contractile organ, without movement.

**Atrophy**   A defect or failure of nutrition or physiological function manifested as a wasting away or diminution in the size of a cell, tissue, organ, or part.

**Attenuate**   To make (microorganisms or viruses) less virulent.

**Aural**   Pertaining to the ears.

**Auscultation**   Listening to sounds arising within organs (as the lungs or heart) as an aid to diagnosis and treatment.

**Autonomic nervous system**   Refers to motor neurons that control functions such as breathing, heart rate, digestion, and other vital functions of the body.

**Autopsy**   Examination, including dissection, of a dead animal to learn the cause and nature of a disease or cause of death. Also called **postmortem** examination or **necropsy**.

**Avian**   Pertaining to all species of birds, including domestic fowls.

**Avian pox**   A viral infection of birds that results in skin lesions and feather loss; secondary bacterial infections may be fatal.

**Aviary**   A place for keeping birds confined, a fenced area or building housing birds.

**Axilla**   Equivalent to the human armpit, the area where the forelimb joins the chest.

**Axillary**   Pertaining to the **axilla** (armpit).

**Azotemia**   A condition with an excess of nitrogenous wastes in blood.

*Babesia*   A genus of protozoal parasites that infect red blood corpuscles (cells).

**Back**   Dorsum, specific area varies depending on the applicable breed standard. In some standards, it is defined as the vertebrae between **withers** and **loin**.

**Bactericidal**   From the Latin *caedere* meaning "to kill." Capable of destroying bacteria.

**Bacterin**   A suspension of killed bacteria (**vaccine**) used to increase disease resistance.

**Balanced**   A consistent whole; symmetrical, typically proportioned as a whole or as regards its separate parts (e.g., balance of head, balance of body, or balance of head and body).

*Balantidium coli*   A large protozoal parasite that primarily affects humans and swine.

**Barbiturate**   Generally refers to drugs that are derivatives of barbituric acid. Barbiturates have been widely used as tranquilizers, sedatives, and anticonvulsant agents. Over 2,000 different barbiturates have been synthesized, although less than a dozen are in common use.

**Barred**   A term used to describe striped markings on fowl.

**Basal metabolism (BM)**   The chemical changes that occur in the cells of an animal in the fasting and resting state, when it uses just enough energy to maintain vital cellular activity, respiration, and circulation as measured by the basal metabolic rate (**BMR**). Basal conditions include thermoneutral environment, resting, postabsorptive state (digestive processes are quiescent), consciousness, quiescence, and sexual repose. It is determined in humans 14 to 18 hours after eating and when at absolute rest. It is measured by means of a calorimeter and is expressed in calories per square meter of body surface.

**Basophil**   A white blood cell with basophilic cytoplasmic granules.

**Basophilic**   Staining with basic stains.

**Basophilic stippling**   **Erythrocytes** containing blue-staining aggregates, often associated with lead poisoning but can also be seen when a large number of erythrocytes are being produced by the bone marrow.

**Battery**   A series of cages or pens.

**Bay**   The prolonged bark or voice of the hunting hound.

**Beak**   The horny tissue covering the jaws of a turtle or bird, the bill of a bird; any of various rigid projecting mouth structures (as of a turtle).

**Bench show**   A dog show at which the dogs competing for prizes are "benched" or leashed on benches.

**Benign**   Harmless, not malignant.

**Best in show**   A dog-show award to the dog judged best of all breeds.

**Bilateral**   On both sides.

**Bile**   Fluid secreted by liver and stored in gall bladder, which aids digestion of fat by emulsification.

**Biliary**   Pertaining to bile.

**Bilirubin**   A breakdown product of hemoglobin, a bile pigment.

**Bioassay**   The use of animals to determine the active power of a compound as compared with the effect of a standard preparation.

**Biologicals**   Medicinal preparations made from living organisms and their products, used for the prevention or detection of disease; they include **serums**, **vaccines**, **antigens**, and **antitoxins**.

**Biological value (BV)**   The percent utilization of protein within the animal body, expressed by the formula ($N$ = nitrogen):

$$\% \text{ BV} = \frac{N \text{ intake} \left[ (\text{fecal } N - \text{metabolic } N) + (\text{urinary } N - \text{endogenous } N) \right]}{N \text{ intake} - (\text{fecal } N - \text{metabolic } N)} \times 100$$

**Biology**   The study of life.

**Bioperculate**   Having two opercula (circular lidlike structure at the end of the egg; bioperculate eggs have these at both ends).

**Biopsy**   Removal of tissue from a living animal for study (microscopic or chemical).

**Bioterrorism**   The use of infectious agents—naturally occurring or genetically engineered to increase their **virulence**—in warfare or to otherwise harm people.

**Biotic**   Pertaining to life or living matter.

**Bird dog**   A sporting dog trained to hunt birds.

**Bitch**   A female dog.

**Bite**   The relative position of the upper and lower teeth when the mouth is closed. See **level bite**, **scissors bite**, **undershot**, **overshot**.

**Blastomere**   A cell produced during cleavage, an early embryonic stage.

**Blaze**   A white stripe running up the center of an animal's face, usually between the eyes.

**Bloat**   A digestive disturbance with abdominal distension.

**Bloom**    The sheen of a coat in prime condition.

**BMR**    Basal metabolic rate.

**Board**    To feed, house, and care for a dog, cat, or other animal for a fee.

**Body**    The anatomical section between the forequarters and the hindquarters.

**Body length**    Distance from the point of the shoulder to the rearmost projection of the upper thigh (point of the buttocks).

**Bradycardia**    A slow heart rate.

**Break**    Term used to describe changing of coat color from puppies to adult stages.

**Breakaway collar**    A collar designed to break if the animal wearing it becomes entangled, thus protecting the animal from strangulation.

**Breastbone**    Bone in forepart of chest, the sternum.

**Breed**    Animals having a common origin and characteristics that distinguish them from other groups within the same species.

**Breeder**    A person who breeds dogs, cats, or other animals. Under **AKC** rules, the breeder of a dog is the owner (or if the dam was leased, the lessee) of the **dam** when she was bred.

**Breed true**    To produce a kitten or puppy very much like the parents.

**Brindle**    A fine even mixture of black hairs with hairs of a lighter color; usually tan, brown, or gray.

**Brisket**    The forepart of the body below the chest, between the forelegs, closest to the ribs.

**Broken color**    Self color broken by white or another color.

**Broken-up face**    A receding nose together with a deep stop, **wrinkle**, and **undershot** jaw (e.g., Bulldog, Pekingese).

**Bronchiole**    Tiny air passages that lead to the **alveoli**.

**Brush**    A bushy or plume-like tail of a longhaired cat.

**Buccal capsule**    Lining of the mouth, within the mouth cavity.

**Buffer solution**    A substance in a solution that makes the degree of acidity (hydrogen-ion concentration) resistant to change when an acid or base is added. See **pH**.

**Buffy coat layer**    The layer containing white blood cells and **platelets**. It is found between the packed red blood cells and the **plasma** in a centrifuged microhematocrit tube.

**Bunting**    To strike or push with the horns or head; an example is cats rubbing their heads on their owners' legs or on furniture.

**Bursa**    A pouch-shaped body cavity.

**Buttocks**    The rump or hips.

**Calciferol**    Commonly known as vitamin $D_2$.

**Calcify (calcification)**    To deposit or secrete calcium salts that harden. An injured **cartilage** sometimes calcifies. The process by which organic tissue becomes hardened by a deposit of calcium salts.

**Calcium oxalate**    A crystalline salt ($CaC_2O_4$) normally deposited in many plant cells and in animals sometimes excreted in urine or retained in the form of urinary **calculi**.

**Calculus**    (pl. **calculi**) An abnormal concretion (hard substance) occurring within the body, usually composed of mineral salts; dental calculus is formed when tartar calcifies; urinary calculi are stones (**uroliths**) of varying mineral compositions.

**Calling**    The noise made by a female cat when in **estrus**.

**Calorie**    Measure of energy equivalent to heat required to raise temperature of 1 gm of water from 14.5°C to 15.5°C. Kilocalorie (kcal) normally used in nutrition equals 1,000 small calories.

**Canidae**   A family of carnivorous mammals (superfamily Arctoidea), most with long coarse fur, comparatively long limbs with strong nonretractile claws, head rounded to elongated with well-developed often somewhat pointed muzzle and jaws, ears erect or drooping, and eyes with rounded pupils; includes dogs, wolves, jackals, foxes, and related extinct animals.

**Canine**   Pertaining to the dog family; includes dogs, wolves, jackals, and others; from *canis* (Latin word for dog).

**Canines**   The two upper and two lower sharp-pointed teeth next to the incisors; fangs.

**Cannibalism**   The practice of some humans and animals of eating the flesh of their own kind.

**Cannula**   A small tube made for insertion into a body cavity (as for drainage) or into a duct or vessel.

**Cannulate**   To insert a **cannula** into.

*Capillaria*   A genus of nematodes parasitic in birds and mammals.

**Capillary refill time (CRT)**   A simple method of assessing tissue **perfusion**. The animal's upper lip is lifted and moderate pressure is applied to the gingival mucosa above the upper teeth resulting in blanching. The finger applying the pressure is removed and the time required for blood to return to the site is the CRT. Normal CRT is less than 2 seconds.

**Carapace**   The upper shell of a turtle or tortoise.

**Carbohydrates**   Materials consisting chemically of carbon, hydrogen, and oxygen. The most important carbohydrates are the starches, sugars, celluloses, and gums. They are so named because the hydrogen and oxygen are usually in the proportion to form water $(CH_20)_n$. Formed in plants by photosynthesis, carbohydrates constitute a large part of animal food.

**Carcinogen**   Any cancer-producing substance.

**Cardiac**   Pertaining to the heart.

**Cardiomyopathy**   Disease affecting the heart; results in ineffective pumping of blood to the body.

**Cardiovascular**   Pertaining to the heart and blood vessels.

**Carnassial teeth**   The shearing teeth, generally the fourth upper premolar and the first lower molar.

**Carnivorous (carnivores)**   Meat eating. Carnivorous animals include dogs, cats, bears, weasels, and other mammals.

**Carpals**   Bones of the pastern joints equivalent to the human wrist.

**Carpus**   Joint in the distal forelimb equivalent to the human wrist.

**Carrier**   A bearer and transmitter of a causative agent of an infectious disease; *especially* one who carries the causative agent of a disease (as typhoid fever) systemically but is immune to it; an individual (as one **heterozygous** for a recessive) having a specified gene that is not expressed or only weakly expressed in its **phenotype**.

**Cartilage**   A firm but pliant type of tissue forming portions of the skeleton of vertebrates. The proportion of cartilage in the skeleton of young animals is greater than in mature animals. Also called *gristle*.

**Casein**   A protein in milk; purified casein useful as a source of protein for purified diets.

**Castrate**   To remove the testicles of a male animal.

**Cataleptic state**   Characterized by profound **analgesia**, unresponsiveness to commands, amnesia, open eyes, involuntary limb movements, salivation, normal to increased heart rate, and spontaneous respiration.

**Cat fancy**   Refers to pedigree cats and their breeding, cat clubs and societies, and things associated with them.

**Cattery**   Building or enclosure where cats are kept.

**Caudal**   Denotes a position toward the tail (rump) or **posterior** end (same as *inferior* in human anatomy).

**CD (Companion Dog)**   A suffix used with the name of a dog that has been recorded a Companion Dog by AKC as a result of having won certain minimum scores in Novice Classes at a specified number of AKC-licensed obedience trials (Chapter 13).

**CDX (Companion Dog Excellent)**   A suffix used with the name of a dog that has been recorded a Companion Dog Excellent by AKC as a result of having won certain minimum scores in Open Classes at a specified number of AKC-licensed obedience trials.

**Cecotrophs**   Small, soft fecal pellets resembling a cluster of grapes produced at night; rabbits ingest their cecotrophs directly from the anus.

**Cecum**   The blind pouch in which the large intestine begins and into which the ileum opens from one side.

**Cellulose**   A polysaccharide making up a large portion of the fibrous parts of plants.

**Celsius**   A centigrade temperature scale. Can be converted into Fahrenheit by using the formula $(°C \times 9/5) + 32 = °F$ (Appendix B).

**Central nervous system**   Refers to the brain and spinal cord.

**Cephalic vein**   The vein located on the craniomedial aspect of the forelimb, it is frequently used for blood collections and the intravenous administration of medications.

**Cercariae**   Tadpole-shaped larval trematode worms produced in the molluscan host by rediae and later freed into water to encyst as a metacercaria or to actively penetrate a suitable definitive host.

**Cerumen**   Ear wax.

**Cervical**   Referring to the neck.

**Cestodes**   Taxonomic group comprising tapeworms.

**Champion (Ch.)**   A prefix used with the name of a dog that has been recorded a Champion by AKC as a result of defeating a specified number of dogs in specified competition at a series of AKC-licensed dog shows.

**Character**   Expression, individuality, and general appearance and deportment (attitude) as considered typical of a breed.

**Chatter**   To click repeatedly or uncontrollably.

**Cheilosis**   Condition characterized by lesions of the lips and corners of the mouth.

**Chemotherapy**   Treatment of disease with chemicals, term commonly used in reference to anticancer drugs.

**Chest**   The part of the body or trunk that is enclosed by the ribs.

**Chigger**   A larva (mite) infesting humans, domestic animals, some birds, snakes, turtles, and rodents. Its bite results in inflamed spots accompanied by intense itching.

**Chippendale front**   Named after the Chippendale chair. Forelegs out at elbows, **pasterns** close, and feet turned out.

**Chitin**   An amorphous horny substance that forms part of the hard outer covering of insects, crustaceans, and certain other invertebrates and occurs also in fungi; it is a polysaccharide structurally similar to cellulose except that the repeating unit is derived from acetylglucosamine instead of glucose.

**Chlorine**   A disinfectant containing sodium hypochlorite, commonly known as bleach. Bleach is inexpensive and depending on the concentration used will kill most bacteria, viruses, fungi, and mycloplasma organisms. It is very caustic to tissues and equipment and does not work well in the presence of organic debris. The most commonly used dilution is one-half cup to a gallon of water.

**Choke collar**   A leather, chain, or cloth collar fitted to the dog's neck in such a manner that the degree of tension exerted on an attached leash tightens or loosens it.

**Cholesterol**   A white, fat-soluble substance found in animal fats and oils, bile, blood, brain tissue, nervous tissue, the liver, kidneys, and **adrenal** glands. It is important in metabolism and is a precursor of certain **hormones**.

**Choriomeningitis**   Inflammation of the cerebral meninges (membranes that envelop the brain).

**Chromosomes**   Dark-staining rodlike or rounded bodies visible under the microscope in the nucleus of the cell in the metaphase of cell division. Chromosomes occur in pairs in body cells, and the number is constant for a species. Chromosomes carry genes arranged linearly along their length and control heredity. The backbone of a chromosome is the **DNA** molecule.

**Chronic**   Of long duration; opposite of acute.

**Cilia**   A hairlike process found on many cells that is capable of vibratory or lashing movement and that serves in free-swimming unicellular organisms and in some small multicellular forms as an organ of locomotion or in the higher animal as a producer of a current of fluid.

**Ciliated**   Provided with **cilia**.

**Cirrhosis**   A chronic disease of the liver characterized by a destruction of liver cells and an increase of **connective tissue**.

**Claws**   A sharp usually slender and curved hard covering of the toe of an animal (analogous to human nails).

**Clinical**   Visible, readily observed externally.

**Clip**   The method of trimming the coat in some breeds, notably the Poodle.

**Cloaca**   Chamber where **urogenital** and digestive systems empty.

**Cloning**   (1) Growing a colony of genetically identical cells or organisms in vitro; (2) transplantation of the nucleus of an adult somatic cell into an ovum, which then develops into an embryo genetically identical to the original adult.

**Coat**   Hair on the body of an animal. Some breeds of animals possess two coats, an outer coat and an undercoat.

**Coccidian**   In the order of protozoans.

**Colibacillosis**   Infection with *Escherichia coli*.

**Colitis**   Inflammation of the **colon**.

**Collar**   The marking around the neck, usually white. Also, leather, chain, or cloth worn around the neck for restraining or leading an animal when a leash is attached.

**Colon**   The part of the large intestines that extends between the **cecum** and the **rectum**.

**Colonocytes**   Colonic epithelial cells.

**Colorpointing**   Having a deeper shade of the body color on the ears, nose, feet, and tail.

**Colostrum**   The first milk secreted pre- and **postpartum**.

**Comfort zone**   The temperature interval during which no demands are made on the temperature-regulating mechanisms. The temperature at which humans and animals feel most comfortable. Also called **thermoneutral zone**.

**Commensal**   Organisms that work together.

**Condition**   Health as shown by the **coat**, general appearance, weight, health, and fitness.

**Conformation**   The physical form or physical traits of an animal; its shape and arrangement of parts, form and structure, body **type**.

**Congenital**   Acquired during fetal development or existing at birth.

**Conjunctiva**   The delicate membrane that lines the eyelids and the exposed part of the **sclera**.

**Conjunctivitis**   Inflammation of the inner eyelids.

**Connective tissue**   Tissue that binds together (collagen, cartilage, tendons, ligaments, bone matrix).

**Contralateral**   Pertaining to the opposite side.

**Convection**   Either cooling or warming of an animal by wind (breezes) according to whether the wind is cooler or warmer than the surface temperature of the animal.

**Copepod**   A subclass of Crustacea that includes minute aquatic forms found in fresh or salt waters.

**Coprophagy**   The ingestion of **feces**.

**Copulation**   The act of mating.

**Copulatory (plug)**   In some species (e.g., hamsters) a gelatinous material forms in the vagina following **copulation.**

**Core vaccines**   Those **vaccines** that should be routinely used in preventive health programs.

**Cornea**   The transparent part of the coat of the eyeball which covers the **iris** and **pupil** and admits light to the interior.

**Correlation**   A measure of how two traits vary together. A correlation of $+1.00$ means that as one trait increases the other also increases—a perfect *positive* relation. A correlation of $-1.00$ means that as one trait increases the other decreases—a perfect *negative,* or *inverse,* relation. A correlation of 0.00 means that as one trait increases, the other may increase or decrease—no relation. Thus, a correlation coefficient may lie between $+1.00$ and $-1.00$.

**Cow-hocked**   When the **hocks** turn toward each other.

**Cranial**   Of or pertaining to the head or the anterior (front) or superior end of the body.

**Crepitus**   A grating or crackling sound or sensation; often associated with joint disease when the ends of bones and/or **cartilages** rub against each other.

**Crepuscular**   Active in twilight.

**Crest**   The upper, arched portion of the neck.

**Crop**   The pouched enlargement of the esophagus (gullet) of many birds that serves as a receptacle for food and for its preliminary maceration (softening for digestion).

**Cropping**   The cutting or trimming of the ear **pinna** to induce the ears to stand erect.

**Crossbreds**   Animals whose parentage is known. A term used, for example, when one **purebred** cat or dog is mated with a purebred of another breed.

**Cross-breed**   The mating of two different **purebred** animals.

**Croup**   The back part of the back above the hind legs.

**Crown**   The highest part of the head, the topskull.

**CRT**   See **capillary refill time**.

**Crude fiber (CF)**   That portion of feedstuffs composed of cellulose, hemicellulose, lignin, and other polysaccharides that serve as the structural and protective parts of plants. It is high in forages and low in grains. **Poultry** and swine are limited in their ability to digest fiber, whereas ruminants (cattle and sheep) can benefit from it through **rumen** bacterial activity. It is the least digestible part of a feed/food.

**Crude protein (CP)**   The total protein in a feed/food. In calculating the protein percentage, the feed/food is first chemically analyzed for its nitrogen content. Because proteins average about 16% (1/6.25) nitrogen, the amount of nitrogen in the analysis is multiplied by 6.25 to give the CP percentage.

**Cryoepilation**   Using extreme cold (freezing) to remove hair (or eyelash) from a **follicle**.

**Cryptorchid**    A male in which the testes have not descended into the **scrotum**, hidden (**testicles** are retained in the **abdominal** cavity).

*Cryptosporidium*    A protozoan intestinal **parasite**.

*Ctenocephalides felis*    Cat flea. *C. felis* parasitizes a wide range of animals including dogs, cats, swine, rodents, birds, and humans.

**Cuttlebone**    The shell of cuttlefishes that is sometimes used for supplying cage birds with lime and salts and a surface to wear down their **beaks**.

**Cyanosis**    Refers to blue mucous membranes caused by inadequate oxygenation of arterial blood.

**Cynology**    The study of canines.

**Cyst**    A sac lacking an opening but having a distinct membrane; a capsule or round sheath formed about certain cells such as **protozoa**.

**Cystocentesis**    Puncture of the bladder; clinically refers to the procedure in which urine is withdrawn from the bladder using a syringe and needle by puncture through the abdominal wall.

*Cytauxzoon felis*    A protozoal **parasite** of feline blood cells.

**Cytology**    The science relating to the study of cells and their origin, structure, and function.

**Cytoplasmic**    Pertaining to cell matter (cytoplasm).

**Dalton**    A unit of mass; 1 Dalton is 1/16 of the mass of the oxygen atom.

**Dam**    The female parent.

**Daylength**    The length of daylight in hours during a 24-hour period.

**Deamination**    Removal of the amino group ($-NH_2$) from an amino acid.

**Decarboxylation**    The removal of carboxyl group ($-COOH$) from an organic acid.

**Deciduous**    Not permanent but cast off at maturity. Used in reference to milk (baby) teeth.

**Definitive host**    The final **host**, the animal in which a **parasite** reaches sexual maturity and reproduces.

**Deglutition**    The act of swallowing.

**Dehydration**    An abnormal depletion of body fluids.

**Dementia**    A general designation for mental deterioration.

**Demodectic mites**    Mites of the family Demodicidae; elongated inhabitants of hair **follicles** and/or the skin.

**Dentition**    Reference to the number of teeth characteristic of a species and to their arrangement in the jaws.

**Depth of chest**    Measured from the **withers** to the lowest point of the **sternum**.

*Dermacentor variabilis*    American dog **tick**.

**Dermatitis**    Inflammation of the skin.

**Dermatophytosis**    Any skin infection caused by a **fungus** (e.g., ringworm).

**Dermis**    The layer of skin below the epidermis, consisting of a dense bed of vascular connective tissue (also known as the corium).

**Desquamate**    To peel or come off in layers or scales, as the **epidermis** in certain diseases. The shedding of epithelial cells.

**Dewclaw**    An extra **claw** or functionless digit on the inside of the leg; a rudimentary fifth toe on dogs.

**Dewlap**    Loose, pendulous skin under the throat.

**Diabetes mellitus**    A metabolic disease in which the ability to utilize carbohydrates for energy is lost, usually due to inadequate production of **insulin** by the **pancreas**; results in **hyperglycemia**, **glycosuria**, **polyphagia**, **polydipsia**, **polyuria**, and in some cases **ketoacidosis.**

**Diagnosis**   Recognition or identification of a disease or abnormality.

**Diastolic**   Pertaining to expansion; the phase of the **cardiac** cycle when the heart muscles are relaxed allowing the heart chambers to expand and fill with blood.

**Dietary fiber**   As defined by the American Association of Cereal Chemists, "the edible parts of plants or analogous **carbohydrates** that are resistant to digestion and **absorption** in the human small intestine with complete or partial fermentation in the large intestine. Dietary fiber includes polysaccharides, oligosaccharides, **lignin**, and associated plant substances. Dietary fiber promotes beneficial physiological effects including laxation and/or blood **cholesterol** attenuation and/or blood glucose attenuation" (http://www.aaccnet.org/definitions).

**Differential count**   Numbers of the different types of white blood cells.

**Differentiation**   The process of acquiring individual characteristics, such as occurs in the progressive diversification of cells and tissues of the embryo. The transformation of mother cells into different kinds of daughter cells (brain, kidney, liver). This process is irreversible.

**Digest**   (1) A flavor enhancer produced by the enzymatic degradation of animal tissues. Proteolytic enzymes are used to partially digest ground **poultry** viscera, fish, liver, and beef lungs. Degradation is stopped by the addition of an acid (usually phosphoric acid). The resulting liquid solution is called digest. The digest is sprayed on the exterior of many dry petfoods at a level of 3% to 5%. Digest may also be dried and dusted onto the surface of a food following the application of fat. (2) To break down or transform, as in the **digestion** of food.

**Digestible energy (DE)**   That portion of the energy in a feed/food that can be digested or absorbed into the body by an animal.

**Digestible protein (DP)**   That portion of the protein in a feed/food that can be digested or absorbed into the body by an animal.

**Digestion**   The process of making food absorbable by dissolving and breaking it down into simpler chemical compounds through the action of **enzymes** and other secretions of the **alimentary** tract.

**Digestion coefficient (coefficient of digestibility)**   The difference between the nutrients consumed and the nutrients excreted expressed as a percentage.

**Digitalis**   A valuable drug having diuretic properties that increase the contractility of the heart, made from the dried leaves of the foxglove. It is commonly used in the treatment of heart diseases.

*Dioctophyme renale*   Giant kidney worm.

*Diphyllobothrium latum*   The broad tapeworm.

**Diptera**   Flies.

*Dipylidium caninum*   The cucumber seed tapeworm.

*Dirofilaria immitis*   Heartworms.

**Discospondylitis**   An infection involving the vertebral column.

**Disease**   Any deviation from a normal state of health that impairs vital functions of animals.

**Disinfect**   To destroy or render inert disease-producing germs (**pathogens**) and harmful microorganisms and to destroy **parasites**.

**Disinfectant**   Substance that destroys disease-causing microorganisms (germs).

**Disposition**   The temperament, or spirit, of an animal.

**Disqualification**   A decision made by a judge or by a **bench show** committee following a determination that a dog has a condition that makes it ineligible for any further competition under the dog show rules or under the standard for its **breed**.

**Distal**    Remote, as opposed to close or **proximal**; away from the main part of the body.

**Distemper teeth**    Teeth discolored or pitted as a result of distemper (similar discolorations can be caused by other diseases or by certain medications).

**Distichiasis**    A double row of eyelashes.

**Diuresis**    Increased production of urine.

**Diurnal**    Active chiefly in the daytime; having a daily cycle.

**DNA**    Deoxyribonucleic acid. Contained in chromosomes and is responsible for the transmission of hereditary characteristics from generation to generation.

**Dock**    To shorten the tail by cutting.

**Dog**    A male dog; also used collectively to designate both male and female.

**Dog show**    A competitive exhibition for dogs at which the dogs are judged in accordance with an established standard of perfection for each breed.

**Dog Show, Conformation (Licensed)**    An event held under **AKC** rules at which championship points are awarded. May be for *all breeds* or for a single breed (Specialty Show).

**Domesticate (domestication)**    To bring a wild animal or fowl under control and to improve it through careful selection, mating, and handling so that its products or services become more useful to humans. Domesticated animals are bred under the control of humans.

**Dominant gene**    A gene that transmits a physical characteristic even if it is in the **genotype** of only one parent.

**Dorsal**    Pertaining to the back, or more toward the back portion; opposite to **ventral**. It means the same as **posterior** in human anatomy.

**Down in pastern**    Weak or faulty pastern (metacarpus) set at a pronounced angle from the vertical.

**Drop ear**    The ends of the ear folded or drooping forward as contrasted with erect or prick ears.

**Dual champion**    A dog that has won both a **bench show** and a **field trial** championship.

**Duct**    A canal (tube) that conveys fluids or secretions from a gland.

**Dysbiosis**    Disruption of intestinal **flora**.

**Dysecdysis**    Abnormal shedding.

**Dysphagia**    Difficulty swallowing.

**Dysplasia**    Abnormal growth or development (as of organs, tissues, or cells); anatomic structure that presents some abnormality as a result of such growth; in **hip dysplasia**, abnormal growth and development of the acetabulum and head of the **femur** results in misarticulation of the hip **joint**.

**Dyspnea**    Difficult or labored breathing.

**Dystocia**    Abnormal or difficult labor (**parturition**) causing difficulty in delivering the fetus and placenta.

**e.g.**    "For example," from the Latin *exempli gratia*.

**Easy keeper**    An animal that does well on a minimum amount of food.

**Eccrine**    Glands or tissues that secrete a substance without a breakdown of their own cells (e.g., sweat glands). See **exocrine**.

**Ecdysis**    Shedding or **molting** of skin; **desquamation** or sloughing.

**Ecology**    The study of the relation of organisms to their environment, habits, and modes of life.

**Ectoblast**    An embryonic cell layer.

**Ectoderm**   The outermost of the three primary germ layers of the embryo; gives rise, for example, to the skin, hair, and nervous system.

**Ectoparasites**   **Parasites** inhabiting external body surfaces.

**Ectotherm**   A cold-blooded animal; **poikilotherm**.

**Ectropion**   Eversion, an outward rolling of the eyelids.

**Eczema**   An inflammatory skin disease of humans and animals characterized by redness, itching, loss of hair, and the formation of scales or crusts.

**Edema**   The presence of abnormally large amounts of fluid in the intercellular tissue spaces of the body; swelling.

**Edematous**   Accumulations of abnormally large amounts of fluid in the intercellular tissue spaces of the body; affected areas have a swollen appearance.

**Efferent**   To carry away, out; efferent neurons carry impulses from the **central nervous system** to the **peripheral nervous system** (motor pathways leading to muscle activation).

**Efficacy**   Effectiveness. For example, the effectiveness of an antibiotic in the therapy of a certain disease.

**Egg binding**   Failure of the normal progression of an egg through the oviducts and **uterus**.

**Ehrlichiosis**   Disease resulting from infection with *Ehrlichia* organisms (intracellular organisms transmitted by **ticks**).

**Elastin**   The yellow proteinaceous **connective tissues** of the skin that provide structural support for the blood vessels and thermostat mechanism. It is obtained when elastic tissue is boiled in water.

**Elbow**   The joint between the **upper arm** and the **forearm**.

**Electrocardiogram**   A recording of electrical activity of the heart, also called an ECG or EKG.

**Electroepilation**   Using an electric current to remove hair (or eyelash) from a **follicle**.

**Electrolyte**   Any solution that conducts electricity by means of its ions.

**Electromagnetic radiation**   Radiation consisting of electric and magnetic waves that travel at the speed of light (e.g., light waves, radio waves, gamma rays, **x-rays**).

**ELISA**   Enzyme-linked **immuno**sorbent **a**ssay. An ELISA links an **enzyme** to an **antibody** specific for the **antigen** (**protein**, **hormone**, microorganism, **parasite**, viral particle, or others) being tested for; if the antigen is present, the enzyme-linked antibody binds to the antigen. A color-producing substrate is then added; the enzyme converts the substrate to a colored product. If color is produced, the test is positive for the presence of the antigen.

**Emaciated**   Very thin.

**Emaciation**   Wasting away physically, losing flesh so as to become very thin.

**Embryology**   The science relating to the formation and development of the embryo.

**Emetic**   Substance that induces vomiting.

**Encephalitis**   An inflammation of the brain that results in various **central nervous system** disorders.

**Encephalitozoonosis**   **Encephalitis** affecting animals.

**Encyst**   To form or become enclosed in a cyst or capsule.

**Endemic (enzootic)**   Pertaining to a disease commonly found with regularity in a particular locality; an endemic disease.

**Endocrine**   Pertaining to glands that produce secretions that pass directly into the blood or lymph instead of into a duct (secreting internally). **Hormones** are secreted by endocrine glands.

**Endogenous**   Internally produced in the body (e.g., **hormones** and **enzymes**).

**Endoparasites**   **Parasites** that inhabit the body tissues or cavities.

**Endoscope**   An instrument for visualizing the interior of a hollow organ or part.

**Endoscopy**   Examination of a body part using an **endoscope**.

**Endotoxins**   Toxic substances (such as those causing typhoid) retained inside the bacterial cells until the cells disintegrate.

**Endotracheal**   Pertaining to within the **trachea** (endotracheal tubes are commonly used to administer inhalation anesthetics).

**Energy balance (EB)**   The relation of the gross energy consumed to the energy output (energy retention). It is calculated as follows: EB = GE − FE − UE − GPD − HP.

**Enteric**   Pertaining to the **intestines**.

**Enteritis**   Inflammation of the **intestine**.

**Enteropathy**   A disease of the intestinal tract.

**Enterotoxemia**   A condition characterized by the presence in the blood of **toxins** produced in the **intestines**; systemic illness caused by toxins produced in the **intestines**.

**Entozoa**   Internal parasites (e.g., stomach worms).

**Entropion**   Inversion; inward rolling of the eyelids.

**Envenomization**   The injection of a poison into an animal (as by a wasp).

**Environment (environmental)**   The sum total of all external conditions affecting the life, performance, and **state-of-being** of an animal.

**Enzootic**   Occurring endemically among animals (i.e., continuously prevalent among animals in a certain region).

**Enzyme**   A complex **protein** produced in living cells that causes changes in other substances within the body without being changed itself. A biological catalyst. Body enzymes are giant molecules.

**Eosinophil**   A type of **leukocyte** (white blood cell) that stains red with commonly used stains; these cells are involved in **allergic** reactions and in defense against **parasites**.

**Eosinophilia**   Increased numbers of **eosinophils** (often associated with parasitism).

**Eosinophilic gastroenteritis**   Inflammation of the stomach and **intestines** with an infiltration of **eosinophils**; may be associated with underlying parasitic infections, allergies, or mycotic (fungal) organisms.

**Epidemic**   Rapid spread of disease with many animals or people affected concurrently, referred to as **epizootic** in animals.

**Epidermal**   Of, relating to, or arising from the epidermis.

**Epidermis**   Epiderm (the outer layer of skin or tissue).

**Epiglottis**   The cartilaginous flap that covers the opening to the glottis (larynx).

**Epinephrine**   A drug used to arrest **hemorrhage** and to stimulate heart action. It is a slaughterhouse by-product obtained from the **adrenal** glands. Also called *adrenaline*.

**Epiphora**   Outward flowing of tears onto the face.

**Epiphysial (epiphyseal)**   Pertaining to or of the nature of an **epiphysis**. The end of a long bone that has ossified.

**Epiphysis**   A portion of bone separated from a long bone by **cartilage** in early life but later becoming a part of the larger bone. It is at this cartilaginous **joint** that growth in length of bone occurs. It is also called the *head* of a long bone.

**Epithelium**   Cells covering cutaneous and mucous surfaces (epithelial tissues—lining of intestinal and respiratory tracts, skin).

**Epizootic**   Designating a widely diffused disease of animals that spreads rapidly and affects many individuals of a kind concurrently in a region, thus corresponding to an epidemic in humans.

**Erosion**   The superficial destruction of a surface area of tissue.

**Eructation**   The act of belching, or casting up gas from the stomach.

**Erythema**   A severe redness of the skin associated with local inflammation.

**Erythrocyte**   A red cell; clinically refers to red blood cells (corpuscles).

**Erythrocyte** or **red blood cell count**   Number of red cells per cubic millimeter of blood.

**Erythropoiesis**   The production of **erythrocytes** (the red blood cells that transport oxygen). Occurs in the red bone marrow.

**Essential amino acid**   An amino acid that cannot be synthesized in the body from other amino acids or substances or that cannot be made in sufficient quantities for the body's needs.

**Essential fatty acid**   A fatty acid that cannot be synthesized in the body or that cannot be made in sufficient quantities for the body's needs. Linoleic and alpha-linolenic acids are essential for humans.

**Estivation**   A physiological state of stupor that occurs when reptiles are exposed to temperatures above their preferred optimal temperature, and their preferred body temperature is exceeded. Sometimes called *summer* hibernation.

**Estrogen**   Female sex **hormone**.

**Estrous**   Having characteristics of **estrus**; the **estrous cycle**.

**Estrous cycle (estrous period)**   The period of sexual receptivity (**heat**) in female mammals. Also called **heat period**.

**Estrus**   Stage of the estrous cycle in which the female is receptive to the male. Also known as being in **heat**.

**Etiology**   The cause of a disease.

**Ethology**   The scientific study of animal behavior.

**Euthanasia**   Painless or humane killing.

**Even bite**   Meeting of front teeth at edges with no overlap of upper or lower teeth.

**Eviscerate**   To remove the entrails, lungs, heart, and certain other organs from an animal or fowl when preparing the carcass for human consumption.

**Excise**   To cut out or off.

**Excoriation**   Scratches, abrasion.

**Excyst**   To emerge from a cyst.

**Excystation**   The escape from a cyst; especially a stage in the life cycle of parasites occurring after the cystic form has been swallowed by the host.

**Exocrine (eccrine)**   Secreting outwardly, into, or through a duct.

**Exotoxins**   Soluble **toxins** that diffuse out of living bacteria into the environment (the culture medium of the living host). Toxins secreted by living microorganisms.

**Expiration**   The act of breathing air out.

**Expression**   The general appearance of all features of the head as viewed from the front and as typical of the **breed**.

**Exsanguination**   The withdrawal of blood, as by the bloodsucking insects.

**Extracellular**   Pertaining to something outside a cell.

**Extraocular**   Outside the eye.

**Exudate**   An abnormal seeping of fluid through the walls of **vessels** into adjacent tissue or space.

**Eyeteeth**   The upper **canines**; also called **fangs**.

**Fahrenheit**   The temperature expressed in degrees, where 32° and 212° are the freezing and boiling temperatures, respectively, of water at sea level. A more universal and scientific graduation of temperature is the centigrade Celsius scale. Fahrenheit is converted into Celsius by using the formula (°F − 32) 5/9 = °C (Appendix B).

**Fallow**   Pale cream to light fawn color; pale; pale yellow; yellow-red.

**Fancier**   A person especially interested and usually active in some phase of the sport of **purebred** dogs and/or cats.

**Fangs**   See **canines**.

**FAO**   Food and Agriculture Organization. A specialized international agency of the United Nations that collects and disseminates information on the production, consumption, and distribution of food throughout the world. It was organized in 1945 and has some 120 member nations; it is headquartered in Rome.

**Fascia**   A thin sheet or band of fibrous **(connective) tissue** covering, supporting, or binding together a muscle, part, or organ.

**Fastidious**   Having complex nutritional requirements, organisms that are difficult to grow in cultures. Also describes a picky eater.

**Fawn**   A brown, red-yellow with hue of medium brilliance.

**FDA**   Federal Food and Drug Administration.

**Febrile**   Relating to fever; elevated body temperature.

**Feces**   Bodily waste discharged through the anus, excrement.

*Felicola subrostrata*   Lice affecting cats.

**Feline**   Of a cat or the cat family.

**Fermentation**   A chemical change with effervescence; an enzymatically controlled anaerobic breakdown of an energy-rich compound (as a **carbohydrate** to carbon dioxide and alcohol or to an organic acid); an enzymatically controlled transformation of an organic compound.

**Femoral**   Pertaining to the **femur** or thigh.

**Femur**   Thigh bone. Extends from hip to **stifle**.

**Feral**   Living wild but descended from domesticated dogs and cats.

**Fetch**   The retrieval of game by the dog; also the command to do so.

**Fibrinogen**   A soluble **protein** present in the blood and body fluids of animals that is essential to the coagulation of blood.

**Fibroblasts**   **Connective tissue** cells whose function is one of repair. Also called *fibrocytes* and *desmocytes*.

**Fibroma**   A **benign** tumor consisting mainly of fibrous tissue.

**Fibrous carbohydrates**   Those feed/food components (e.g., cellulose and hemicellulose) not readily digested by animals.

**Fibula**   The outer and smaller of the two bones of the lower rear leg.

**Field Champion (Field Ch.)**   A prefix used with the name of a dog that has been recorded a Field Champion by AKC as a result of defeating a specified number of dogs in specified competition at a series of AKC-licensed **field trials** (Chapter 13).

**Field trial**   A competition for certain Hound or Sporting Breeds in which dogs are judged on ability and style in finding or retrieving game or following a game trail.

**Finch**   Any of numerous songbirds (especially families Fringillidae, Estrildidae, and Emberizidae) having a short, stout, usually conical bill adapted for crushing seeds.

**Fistula**   A congenital or acquired passage leading from an abscess or hollow organ to the body surface or from one hollow organ to another and permitting passage of fluids or secretions.

**Flaccid**   Limp; weak; flabby.

**Flag**   A long tail carried high, usually referring to one of the Pointing Breeds of dogs. Also used in reference to an **estrous** bitch; denotes lifting and deviation of the tail to the side to facilitate **copulation**.

**Flagellates**   Microorganisms with a filamentous appendage (tail, whip).

**Flank**   The side of the body between the last rib and the hip.

**Flat-sided**   Ribs insufficiently rounded as they approach the **sternum** or breastbone.

**Flat withers**   A fault that is the result of short upright shoulder blades that unattractively join the **withers** abruptly.

**Floating rib**   The last, or 13th rib, which is unattached to other ribs.

**Flora**   The bacteria in or on an animal. Usually referred to as *bacterial flora*, as in the digestive tract.

**Flush**   To drive birds from cover, to force them to take flight; to spring.

**Foci**   Localized regions of a disease or the primary sites of generalized **disease**.

**Follicle**   A small cavity or deep narrow-mouthed depression (e.g., a hair follicle).

**Fomites**   Objects that are not in themselves harmful but are able to harbor pathogenic organisms and transfer them to an animal.

**Food hopper**   A container with an upright, usually clear, reservoir to hold food that is dispensed through a trough-like opening at the bottom.

**Forceps**   A pliers-like instrument used in medicine and certain research studies for grasping, pulling, and compressing.

**Forearm**   The bones of the forelegs between the elbow and the **pastern**; includes the radius and ulna.

**Foreface**   The front part of the head below the eyes; **muzzle**.

**Forequarters**   The combined front assembly from its uppermost component, the shoulder blade, down to the feet.

**Formalin**   A solution of formaldehyde that usually contains about 37% formaldehyde by weight; it is used to preserve tissues for microscopic examination.

**Foster mother**   A bitch or other animal used to nurse offspring not her own.

**Free choice**   Providing animals free access to diets and thereby allowing them to eat at will (**ad lib**).

**Frill or ruff**   Fur around the head that is brushed up to form a frame for the face.

**Front**   The forepart of the body as viewed head on (i.e., forelegs, chest, brisket, and shoulder line).

**Frontal bone**   The skull bone above the eyes.

**Fuller's earth**   A naturally occurring sedimentary clay composed mainly of alumina, silica, iron oxides, lime, magnesia, and water, in extremely variable proportions.

**Fumigant**   A liquid or solid substance that forms vapors that destroy **pathogens**, insects, and **rodents**.

**Fungicide**   An agent that destroys **fungi**.

**Fungus**   (pl. **fungi**) Plant that contains no chlorophyll, flowers, or leaves. They obtain their nourishment from dead or living organic matter.

**Fur ball**   Fur that may be swallowed by an animal when cleaning itself and may form a feltlike mass in the stomach. Also called a **hairball** or **trichobezoar**.

**Gait**   The pattern of footsteps at various rates of speed, each pattern distinguished by a particular rhythm and footfall. Examples include walking, pacing, and trotting.

**Gastric**   Pertaining to the stomach.

**Gastric hairballs**   See **trichobezoar**.

**Gastritis**    An inflammation of the stomach, especially of the lining or mucous membrane.

**Gastroenteritis**    Inflammation of the stomach and **intestines**.

**Gastrointestinal**    Pertaining to the stomach and **intestines**.

**GE**    See **gross energy**.

**Gene**    The unit of heredity that determines the physical characteristics, growth, and function of animals.

**Genealogy**    Recorded family ancestors.

**Genes**    One of the elements of the nucleus, genes consist of portions of deoxyribonucleic acids linearly arranged in fixed positions; they control the hereditary characteristics of an organism and control the synthesis of specific polypeptide chains (proteins).

**Genetics**    A branch of biology that deals with the heredity and variation of organisms based on their **DNA** (**genes**).

**Genital (genitalia)**    Pertaining to reproductive organs, especially external portions.

**Genome**    The total genetic composition of an individual or population that is inherited with the **chromosomes**. A haploid set of chromosomes with their genes.

**Genotype**    The set of genes inherited from both parents.

**Germ**    A small organism, microbe, or bacterium that can cause disease in humans and/or animals. Early embryo; seed.

**Germicide**    A substance that kills disease-causing microorganisms (pathogens).

**Gestation**    Pregnancy (**gravidity**). The period from conception to birth of young.

*Giardia intestinalis*    **Protozoal parasite** of the **intestinal** tract.

**Gingival**    Pertaining to the gingiva (gums).

**Gizzard**    Muscular stomach of birds; the muscular enlargement of the **alimentary** canal of birds that has usually thick muscular walls and a tough horny lining for grinding the food and when the **crop** is present follows it and the **proventriculus**.

**Gland**    An organ that produces a specific secretion to be used in, or eliminated from, the body.

**Glossal**    Pertaining to the tongue.

**Gluconeogenesis**    The process of producing new glucose from **noncarbohydrate** sources.

**Glycosuria**    Presence of glucose in the urine.

**Gnat**    Any **dipterous**, biting insect.

**Gnathotheca**    Lower jaw of birds.

**Gnotobiotic**    "Known biota"; often used to indicate germ-free animals or others with known microorganisms.

**Gonad**    The gland of a male or female that produces the reproductive cells; the **testicle** or ovary.

**Gout**    A **metabolic disease** marked by a painful inflammation of the joints, deposits of urates in and around the joints, and usually an excessive amount of uric acid in the blood.

**Gram-negative**    Bacteria not holding the purple dye when stained by Gram's method.

**Gram-positive**    Bacteria holding the purple dye when stained by Gram's method.

**Gram stain**    A method for the differential staining of bacteria by which they are treated with Gram's solution after being stained with gentian violet, and are then treated with alcohol and washed in water with the result that certain species retain the dye while others are decolorized; see **Gram-negative** and **Gram-positive**.

**Granivorous**    Eating seeds and grains.

**Granuloma**    A mass or nodule composed of fibrous tissue and inflammatory cells (**macrophages**, **neutrophils**, **eosinophils**, and other **leukocytes**).

**GRAS**    Generally recognized as safe. This term is commonly used by the **FDA** and others when referring to food and feed additives.

**Gravid**    Pregnant or full of eggs.

**Gray matter**    Refers to cell bodies of the brain.

**Grit**    Sedimentary rock that consists of angular, sand-sized grains and small pebbles; ingested by birds to assist in the grinding of foods.

**Groom**    To brush, comb, trim, or otherwise make an animal's **coat** neat.

**Gross energy (GE)**    The amount of heat, measured in **calories**, released when a substance is completely oxidized in a bomb calorimeter containing 25 to 30 atm of oxygen (heat of combustion).

**Groups**    Dog breeds as grouped in seven divisions by the **AKC** to facilitate judging.

**Grub**    Also called warble fly. The **larvae** may be found in the **subcutaneous** tissues of animals. An example affecting cats, rabbits, and other species is *Cuterebra*.

**Gun dog**    A dog trained to work with its master in finding live game and retrieving game that has been shot.

**Hackles**    Hair on neck and back raised involuntarily, usually in fright or anger.

**Hairball**    A mass of compressed hair that has been ingested. Also known as a **trichobezoar**.

**Hallmark**    A distinguishing characteristic such as the spectacles of the Keeshond.

**Handler**    A person who handles a dog in the show ring or at a **field trial**.

**Harness**    A leather or cloth strap shaped around the shoulders and chest with a ring at its top over the **withers**, used especially in service dogs.

**Heartworm disease**    Presence of *Dirofilaria immitis* **parasites** in the heart and pulmonary arteries.

**Heartworm prevention**    Medications given to kill immature stages of *Dirofilaria immitis*.

**Heat (estrus)**    Term used to describe the behavior of females during the fertile portion of the **estrous cycle**. Females are receptive to breeding during this time and may actively seek a mate.

**Heat increment (HI)**    The increase in heat production following consumption of food when an animal is in a **thermoneutral** environment. It consists of increased heats of fermentation and of nutrient **metabolism**.

**Heat period**    That period of time when the female will accept the male in the act of mating; in **heat**; **estrous period**.

**Heatstroke**    Physiological stress induced by excessive heat impairing functioning and causing injury or death.

**Heat tolerance**    The ability of an animal to endure the impact of a hot environment without suffering ill effects.

**Heel**    See **hock**; also a command to the dog to keep close beside its handler.

**Height**    Vertical measurement from the withers to the ground; referred to usually as shoulder height. See **withers**.

**Helminth**    An intestinal worm or wormlike **parasite**.

**Hematocrit**    Measure of the ratio or percentage of blood cells to total blood; packed cell volume (PCV).

**Hematology**    The study of blood.

**Hematoma**    A swelling containing blood.

**Hematopoietic**    Pertaining to blood production.

**Hematuria**    Blood in the urine.

**Hemoglobin**   The red pigment in red blood cells of animals and humans that carries oxygen from the lungs to other parts of the body. It is made of iron, carbon, hydrogen, and oxygen and is essential to animal life.

**Hemolysis**   Breaking up of red cells with liberation of **hemoglobin**. The freezing of blood will cause hemolysis. Many microorganisms are able to hemolyze red blood cells by production of hemolysins.

**Hemophilia**   An inherited condition in which blood does not clot normally. An animal so affected may be called a *bleeder*.

**Hemorrhage**   Bleeding.

**Hemostat**   An instrument for compressing a bleeding vessel.

**Hemostasis**   The process of blood clotting; arrest of bleeding.

**Hepatic**   Pertaining to the liver.

**Hepatitis**   Inflammation of the liver, may be caused by infectious and noninfectious agents.

*Hepatozoon*   A genus of **parasites** found in red blood cells or liver cells.

**Herbivorous (herbivore)**   Pertaining to animals that subsist on grasses and plants.

**Hermaphrodite**   An individual possessing both male and female reproductive organs. May be capable of producing both ova and sperm; however, this seldom is true.

**Hermaphroditism**   Having both male and female reproductive organs.

**Herpetology**   A branch of zoology dealing with reptiles and amphibians.

*Heterodoxus spiniger*   **Lice** affecting dogs.

**Heterozygous**   Having two different **alleles** for a particular characteristic—one from each parent.

**Hexacanth**   Having six hooks; the earliest stage of a tapeworm embryo.

**Hibernation (hibernate)**   The ability of an animal to pass the winter in a dormant state in which the body temperature drops to slightly above freezing and metabolic activity is reduced nearly to zero. Hibernation is probably an adaptation to prevent starvation during periods of food scarcity.

**Hibernator**   An animal that **hibernates**.

**High standing**   Tall and upstanding with plenty of leg.

**Hilar lymph nodes**   Located near the bifurcation of the **trachea** at the base of the heart.

**Hindquarters**   Rear assembly (pelvis, thighs, **hocks**, and paws).

**Hinge**   Of a turtle's shell is a mobile **suture** in the **plastron** (e.g., box turtle) or **caudal carapace** (e.g., hingeback turtle).

**Hip dysplasia**   An orthopedic disease with abnormalities in the hip socket (acetabulum or pelvis) and the head of the femur (thigh bone).

**Histology**   Microscopic study of tissue structure.

**HIV**   Human immunodeficiency virus. Any of a group of retroviruses and especially HIV-1 that infect and destroy helper T cells of the immune system causing the marked reduction in their numbers that is characteristic of AIDS. Also called AIDS virus. See **AIDS**.

**Hock**   The **tarsus**, collection of bones of the hind leg forming the joint between the second thigh and the metatarsus; the dog's true heel.

**Homeostasis (homeokinesis)**   The physiological regulatory mechanisms that maintain a state of balance or equilibrium within the body.

**Homeostatic**   (adj. of **homeostasis**) A relatively stable state of equilibrium; a tendency to uniformity or stability in the normal body states.

**Homeothermic (homoiothermic)**   Having a relatively uniform body temperature maintained nearly independent of the environmental temperature. Warm-blooded animals (mammals and birds) are homeothermic (endotherms).

**Homozygous**   Having identical **alleles** from each parent for a particular characteristic.

**Hormone**   A chemical substance secreted by an **endocrine** gland that has a specific effect on the activities of other organs at a location away from point of secretion.

**Host**   Any animal or plant on or in which another organism lives as a **parasite**. Infected **invertebrates** (which are actually hosts in the true sense) are usually referred to as *biological vectors*.

**Hot rocks**   Artificially heated synthetic rocks used to provide warmth for **ectotherms** and other animals.

**Hound**   A dog commonly used for hunting by scent or sight.

**Humerus**   The bone that extends from the shoulder to the elbow.

**Husbandry**   The cultivation or production of plants and animals; the scientific control and management of a branch of farming and especially of domestic animals; methods or apparatus for the propagation, rearing, training, exercising, amusing, feeding, milking, grooming, housing, controlling, handling, or general care of a living animal.

**Hydatid**   A larval tapeworm comprised of a fluid-filled sac from which daughter cysts and scolices form; occasionally forms a proliferating spongy mass that invades into tissues of the host.

**Hydrogenate**   To combine with hydrogen (to reduce).

**Hydrolysis**   Chemical decomposition in which a compound is broken down and changed into other compounds by taking up the elements of water. The two resulting products divide the water, the hydroxyl group (OH) being attached to one and the hydrogen atom (H) to the other.

**Hydrolyze**   Splitting a compound into smaller fragments by the addition of water (the hydroxyl group, OH, is added to one fragment and hydrogen, H, is added to the other).

**Hymenoptera**   Order of insects with four membranous wings (bees, wasps, hornets).

**Hyper-**   A prefix signifying above, beyond, or excessive.

**Hyperestrogenism**   Syndrome associated with unusually high amounts or duration of estrogen levels in the blood, results in swelling of female genitalia and suppression of the production of cells by the bone marrow.

**Hyperglycemia**   Elevated levels of glucose in the blood.

**Hyperplasia**   The abnormal multiplication or increase in the *number* of normal cells in normal arrangement in a tissue.

**Hyperpnea**   Panting (as in dogs); deep breathing.

**Hypertonic**   Greater osmotic pressure.

**Hypertrophy**   Increase in size; abnormal enlargement due to an increase in the *size* of its constituent cells.

**Hypo-**   A prefix signifying under, beneath, or deficient.

**Hypocalcemia**   A significant decrease in the concentration of ionic calcium, which results in convulsions, as in tetany or parturient paresis.

**Hypoglycemia**   Below normal levels of glucose in blood.

**Hypophysectomy**   Surgical removal of the pituitary gland (hypophysis).

**Hypothermia**   The lowering of body temperature.

**Hypothyroidism**   A state of low thyroid; clinically refers to below normal levels of thyroid hormones in blood.

**Hypotonic**   Lower osmotic pressure.

**Hypotrophy**   Degeneration or loss of vitality of an organ.

**Hypoxia**   Below normal levels of oxygen in tissues.

**Hysterectomy**   Partial or total removal of the **uterus**.

**Hystricomorph rodent** A member of the suborder Hystricomorpha (e.g., New and Old World porcupines, guinea pig, Proechimys); is characterized by having an enlarged infraorbital foramen (opening associated with the eye socket).

**Iatrogenic** Disorder resulting from the actions of medical personnel (e.g., disease resulting from administration of drugs).

**Icteric** Jaundiced, a condition resulting from increased levels of bilirubin; results in a yellowish color to mucous membranes.

**Icterus** Yellow-stained tissue from bile pigments.

**Idiopathic** Diseased state of spontaneous or unknown origin.

**i.e.** "That is," from the Latin *id est*.

**IGRs** Insect growth regulators. Products or materials that interrupt or inhibit the life cycle of an insect.

**Ileitis** Inflammation of the **ileum** (lower portion of the small **intestines**).

**Ileocecal valve** Valve at the junction of the lower end of the small and large **intestines** (junction of ileum and cecum).

**Ileum** The last division of the small **intestine** constituting the part between the jejunum and large intestine.

**Ill-being** Condition or state of an animal experiencing illness.

**Imbibe** To drink or inhale; to absorb moisture.

**Immunity** The power that an animal has to resist and/or overcome an infection to which most or many of its species are susceptible. Active immunity is attributable to the presence of **antibodies** formed by an animal in response to an **antigenic** stimulus. Passive immunity is produced by the administration of preformed antibodies.

**Immunize** To make an animal immune to a disease by **vaccination** or inoculation (Chapter 14).

**Immunocompromised** Having the immune system impaired or weakened (as by drugs or illness).

**Immunoglobulins** A family of proteins in body fluids that have the property of combining with **antigen** and, in the situation in which the antigen is pathogenic, sometimes inactivating it and producing a state of immunity. Also called **antibodies**.

**Immunologic** Producing **immunity**.

**Immunosuppression** Inhibition of the immune response as with antimetabolites, corticosteroids, chemotherapy agents, viruses, poor nutrition, or other diseases that destroy or inhibit the activity of **lymphocytes**, **immunoglobulins**, and other components of the immune defense system.

**Implantation** The attachment or embedding of the fertilized ovum in the uterine wall.

**Inappetence** Lack of appetite; no apparent desire to eat.

**Inbreeding** The mating of closely related animals of the same standard breed.

**Incisors** The six upper and six lower front teeth between the **canines**. Their point of contact forms the **bite**.

**Incus** The middle ossicle (bone) of the inner ear, also called the anvil.

**Infection** The invasion and presence of viable bacteria, viruses, and/or **parasites** in a host that result in disease.

**Infestation** An invasion of the body by arthropods, including insects, mites, and **ticks**.

**Inflammation** The reaction of a tissue to injury, which tends to destroy (through increased white blood cells) or limit the spread of the injurious agent. It is characterized by pain, heat, redness, and swelling.

**Ingest** To eat or take in through the mouth.

**Inguinal** Pertaining to the groin.

**Inhalants**    Something (as an **allergen** or medication) that is inhaled (breathed in).

**Innocuity**    Harmlessness; some **vaccines** have more innocuity and **efficacy** than others.

**Insidious**    More dangerous than is apparent.

**In situ**    In the natural or normal place; the normal site of origin.

**Inspiration**    The act of breathing air in.

**Insulin**    A hormone produced by the **pancreas** that regulates the metabolism and utilization of **carbohydrates**, **lipids**, and amino acids.

**Intact**    Not spayed or **neutered**.

**Integument**    A covering layer, as the skin of animals or the body wall of an insect.

**Integumentary**    Pertaining to covering over (skin).

**Inter-**    A prefix meaning between.

**Interdigital**    Between the toes.

**Intermediate hosts**    An animal harboring the immature stage(s) of a **parasite**.

**Intermuscular**    Situated between muscles.

**Interspecies**    Between species.

**Interstitial nephritis**    Inflammation involving the renal tubules (kidneys).

**Intestine**    The lower portion of the **alimentary** canal from the stomach to the anus. Also called the *gut* or *bowels*.

**Intra-**    A prefix meaning within.

**Intracaudal**    Situated or applied within the tissues of the tail.

**Intracellular**    Situated or occurring within a cell or cells.

**Intracranial**    Within the cranium.

**Intracutaneous**    Into or within the layers of skin.

**Intradermal**    Within the **dermis**. An injection into the layers of skin.

**Intramuscular**    Within a muscle (e.g., an injection into muscle).

**Intraocular**    Inside the eye.

**Intraperitoneal**    Within the cavity of the body that contains the stomach and intestines. Administration through the body wall into the **peritoneal cavity**.

**Intrasternal**    Within the sternum.

**Intravenous**    Pertaining to within a vein, a route for administering medications directly into a vein.

**In utero**    Within the **uterus** (intrauterine).

**Invertebrates**    Animal species without a backbone. Aquaculture examples include shrimp and oysters.

**In vitro**    Outside of body; within an artificial environment, as within a glass or test tube.

**In vivo**    Within the living body.

**Ionizing radiation**    Any radiation displacing electrons from atoms or molecules, thereby producing ions (e.g., alpha, beta, and gamma radiation and shortwave ultraviolet light). Ionizing radiation may produce severe skin or tissue damage.

**Iris**    The colored membrane surrounding the **pupil** of the eye.

**Iso-**    From the Greek word *isos* meaning "equal." A prefix or combining form meaning "equal," "alike," or "the same" (e.g., isocaloric refers to the *same* caloric value).

*Isospora*    A genus of sporozoan **parasites**; order Coccidia.

**Isothenuric**    Urine with a specific gravity of 1.010, the same concentration of solutes as blood plasma.

**Isotonic**    Characterized by equal osmotic pressure (e.g., a solution containing just enough salt to prevent the destruction of the red corpuscles when added to the blood would be considered isotonic with blood). Describing a solution having the same concentration as the system or solution with which it is compared.

**-itis**  A word termination (suffix) denoting *inflammation* of a part indicated by the word stem to which it is attached.

**IU**  International unit. A unit of measurement of a biological (e.g., a vitamin, hormone, antibiotic, antitoxin) as defined and adopted by the International Conference for Unification of Formulas. The potency is based on the bioassay that produces a particular biological effect agreed on internationally. See **USP**.

*Ixodes dammini*  The deer **tick**.

**Ixodides**  The hard **ticks**.

**Judge**  The arbiter in the dog show ring, **obedience trial**, or **field trial**.

**Jugular vein**  The large vein located in the ventrolateral neck, lateral to the **trachea**; it is frequently used for the collection of large volumes of blood.

**Keel bone**  The breastbone of birds; the **sternum**.

**Kennel**  Building or enclosure where dogs are kept.

**Kennelosis**  Exhibiting fear of places outside the kennel in which a dog was raised.

**Keratin**  An insoluble complex protein that constitutes hair, horn, claws, and feathers.

**Keratinization**  The development of or conversion into keratin or keratinous tissue; cornification. See **keratosis**.

**Keratoconjunctivitis**  Inflammation of the **cornea** and **conjunctiva** (inner eyelids).

**Keratoconjunctivitis sicca**  Dry eye, inadequate tear production resulting in inflammation of the **cornea** and **conjunctiva** (inner eyelids).

**Keratosis**  Any horny growth, such as a wart, causing the cornification, or hardening, of the epithelial skin layers.

**Ketoacidosis**  A metabolic disease caused by catabolism (breaking down) of body fat stores; is often associated with untreated **diabetes mellitus**.

**Ketosis**  See **acetonemia**.

**Killed virus**  A virus whose infectious capabilities have been destroyed by chemical or physical treatment.

**Kilo-**  A prefix that multiplies a basic unit by 1,000.

**Kilocalorie (kcal)**  Equivalent to 1,000 small **calories**.

**Kindling**  To bring into being; term used for **parturition** in rabbits.

**Kitten**  A cat up to the age of 9 months in Britain or 8 months in the United States.

**Knee joint**  **Stifle** joint.

**Kwashiorkor**  A syndrome produced by a severe **protein** deficiency, with characteristic changes in pigmentation of the skin and hair, edema, anemia, and apathy.

**Labile**  Changeable or unstable. Readily or continuously undergoing chemical, physical, or biological change or breakdown.

**Lactate (lactating)**  To secrete or produce milk.

**Lactose**  Milk sugar. A disaccharide composed of one molecule of glucose and one molecule of galactose.

**Lagomorphs**  Any of an order (Lagomorpha) of gnawing herbivorous mammals having two pairs of **incisors** in the upper jaw one behind the other and comprising the rabbits, hares, and pikas.

**Larva**  (pl. **larvae**) The immature form of insects and other small animals; is unlike the parent or parents and must undergo considerable change of form and growth before reaching the adult stage (e.g., white grubs in soil or decayed wood are larvae of beetles). Caterpillars, maggots, and screwworms are also larvae.

**Larvicide**  A chemical used to kill the larval or preadult stages of **parasites**.

**Larynx**   The modified upper part of the respiratory passage of air-breathing vertebrates bounded above by the glottis and continuous below with the **trachea**; it contains the vocal cords.

**Latent**   Hidden or not readily apparent.

**Lateral**   Of, at, from, or toward the side or flank.

**LD$_{50}$**   Amount of a substance that will kill one-half of a test group of similar organisms.

**Lead**   (1) To guide or cause to follow; (2) a strap, cord, or chain attached to the collar or harness to restrain or lead the dog or cat, leash; (3) the conductors for an electrocardiogram; these are attached to specific locations on the body; (4) a soft, grayish-blue metal with poisonous salts.

**Leg**   Used in reference to dog shows, the term denotes one of several (often three) competitions or events to be won or qualified in to earn a special award; as in earning a *leg* towards the Companion Dog award by receiving a score above 170 out of the 200 points possible in a Novice-level dog obedience trial.

*Leishmania*   A genus of protozoal **parasites**.

**Lens**   A highly transparent biconvex body in the eye that focuses light rays entering the eye onto the **retina**.

**Lesion**   Alteration in tissue due to injury, infection, or another disease.

**Lethal**   Deadly; causing death.

**Lethargy**   Abnormal drowsiness; the quality or state of being lazy, sluggish, or indifferent.

**Leukocyte** or **white cell count**   Number of white cells per cubic millimeter of blood.

**Leukocytes**   The white blood cells. They have amoeboid movements and include the lymphocytes, monocytes, **neutrophils, eosinophils**, and basophils.

**Leukocytosis**   Increased numbers of white blood cells.

**Leukopenia**   A deficiency of white blood cells.

**Level bite**   When the front teeth (incisors) of the upper and lower jaws meet exactly edge to edge. Pincer bite.

**Libido**   Sexual desire or instinct.

**Lice**   Small nonflying biting or sucking insects that are true **parasites** of humans, animals, and birds.

**Life cycle**   The changes in form and mode of life that an organism goes through between recurrences of the same primary stage; life history.

**Ligaments**   Tissues connecting bones and/or supporting organs.

**Ligate**   To tie up, or bind, with a ligature.

**Lignin**   A compound that in connection with **cellulose** forms the cell walls of plants and thus of wood. It is practically indigestible.

**Limit-fed**   See **limit feeding**. Refers to any method of feeding that restricts free access to diets and sets limits on quantity of feed/food consumed each day.

**Limit feeding**   Feeding animals to maintain weight and growth but not enough to fatten or increase production. Feeding animals less than they would like to eat.

**Limiting amino acid**   The essential amino acid of a protein that shows the greatest percentage deficit in comparison with the amino acids contained in the same quantity of another protein selected as a standard.

**Line breeding**   The mating of related dogs of the same breed, within the line or family, to a common ancestor, as, for example, a dog to his granddam or a bitch to her grandsire.

*Linognathus setosus*   A sucking louse.

**Lipase**   An enzyme produced by the **pancreas**; breaks down fats.

**Lipid**   Any one of a group of organic substances that are insoluble in water but are soluble in alcohol, ether, chloroform, and other fat solvents and have a greasy feel. They include fatty acids and soaps, neutral fats, waxes, steroids, and phosphatides (by U.S. terminology).

**Lipolysis**   The hydrolysis of fats by **enzymes**, acids, alkalis, or other means to yield glycerol and fatty acids.

**Litter**   Group of kittens or puppies born at one delivery period. Such individuals are called **littermates**. Also, the accumulation of materials used for bedding farm animals.

**Littermates**   A group of siblings in a **litter**-bearing species that are born at the same time.

**Local infection**   An infection restricted to a small area, as an ear infection.

**Loin**   Region of the body on either side of the vertebral column between the last ribs and the hindquarters.

**Lordosis**   The crouched position in which the **queen** assumes indicating she is ready to **copulate**.

**Lumbar**   Pertaining to the loins.

**Lumen**   The cavity on the inside of a tubular organ (e.g., the lumen of the stomach or intestine).

**Lymph**   A transparent, slightly yellow liquid found in the **lymphatic** vessels.

**Lymphatic**   Pertaining to **lymph**.

**Lymphocyte**   A variety of white blood corpuscles that originate from the **lymph nodes** and thymus or bursa (birds).

***Lynxacarus radovsky***   Cat fur mite.

**Lysin**   Antibody that causes the death and dissolution of bacteria, blood corpuscles, and other cellular elements.

**Lysozyme**   A substance present in human nasal secretions, tears, and certain mucus. It is also present in egg whites, in which it hydrolyzes polysaccharidic acids. It is bactericidal for only a few saprophytic bacteria and is inactive against pathogens and organisms of the normal flora.

**Macro-**   Large or major. Abnormal size or length. Usually a prefix.

**Macrophages**   Large **phagocytes**. Large scavenger cells, the function of which is primarily **phagocytosis** and the destruction of many kinds of foreign particles. These cells are strategically located in the spleen, liver, bone marrow, small blood vessels, and **connective tissue**. Collectively, they compose the monocyte-macrophage system (formerly known as the reticuloendothelial system).

**Malady**   A disease or disorder of the animal body.

**Malignant**   Harmful, threatening life, having the properties of invasion and metastasis (spread).

**Mallophaga**   Biting lice.

**Malocclusion**   Improper occlusion (positioning of the teeth); abnormality in the coming together of teeth.

**Mandible**   Bone of the lower jaw.

**Mandibular**   Pertaining to the **mandible** (lower jaw bone).

**Mandibular lymph nodes**   Located under the **mandible** (lower jawbone).

**Mane**   Long and profuse hair on top and sides of the neck.

**Manometer**   An instrument for measuring blood pressure or the pressure of gases and vapors by balancing the pressure against a column of liquid or against the elastic force of a spring or an elastic diaphragm.

**Marasmus**   Progressive wasting and emaciation.

**Mask**   Dark coloring seen on the face of some cats such as the Siamese, Colorpoints, and Himalayans, and also the foreface of some dogs (e.g., Mastiff, Boxer, Pekingese).

**Mastication**   The chewing of food.

**Mastitis**   Inflammation of mammary tissue.

**Match show**   Usually an informal dog show at which no championship points are awarded.

**Mate**   To breed a female and male.

**Maxilla**   The upper jawbone in vertebrates.

**Meconium**   The first excreta of a newborn animal.

**Medial**   Pertaining to the middle.

**Median**   Situated in the middle; mesial.

**Median plane**   A line or plane (from **cranial** to **caudal**) dividing an animal into two equal halves.

**Mediastinal lymph nodes**   Located in connective tissue between the right and left lungs.

**Megacalorie (Mcal)**   Equivalent to 1,000 kcal or 1,000,000 cal. A megacalorie is equivalent to a therm.

**Megaesophagus**   Distension of the esophagus, often associated with a lack of normal peristalsis; foods are often **regurgitated** rather than being passed into the stomach for digestion.

**Melanin**   A dark-brown or black pigment found in hair and/or skin.

**Melanoma**   A tumor of melanocytes, usually containing dark pigment (**melanin**).

**Melena**   Blood in **feces**.

**Menadione**   Synthetic vitamin K.

**Merle**   Coloration of blue-grey often flecked with black.

**Mesenchyme**   Embryonic connective tissue that gives rise to the **connective tissues** of the body and blood vessels.

**Mesenteric lymph nodes**   Located in the **connective tissue** membranes between segments of the intestines.

**Mesentery**   A membrane that supports a **visceral** organ, particularly the **intestine**, and contains the vessels and nerves that supply that organ.

**Mesocestoididae**   A genus of tapeworms.

**Meta-**   A prefix meaning "between" or "among"; indicating change, transformation, or exchange; after or next.

**Metabolic**   Relating or pertaining to the nature of **metabolism**.

**Metabolism**   The sum total of the chemical changes in the body, including the building up (anabolic, assimilation) and the breaking down (catabolic, dissimilation) processes. The transformation by which energy is made available for body uses.

**Metabolizable energy (ME)**   The food-intake gross energy minus fecal energy minus energy in the gaseous products of digestion (largely methane) minus urinary energy. For birds and monogastric mammals, the gaseous products of digestion need not be considered.

**Metatarsus**   Rear **pastern**.

**Metestrus**   Beyond/after **estrus**; clinically refers to the period of the reproductive cycle after sexual receptivity.

**Metritis**   Inflammation of the **uterus**.

**Micro-**   Small or minor. Usually a prefix, designating *tiny* or *microscopic in size*. Also, a prefix that divides a basic unit by 1 million.

**Microflora**   The flora consisting of microorganisms. Commonly used in reference to the bacteria populating an area of the body.

**Microscopic**   Invisible to the unaided eye; clinically refers to something that requires visualization by use of a microscope.

**Microvilli**   Fingerlike projections on the surface of intestinal cells; these increase surface area for absorbing nutrients.

**Middle ear**   Also know as the tympanic bulla, the cavity between the tympanic membrane and the inner ear; the Eustachian tube (Chapter 11) connects the tympanic bulla and the pharynx.

**Milk teeth**   First (**deciduous**) teeth.

**Milli-**   A prefix that divides a basic unit by 1,000 (Appendix B).

**Miracidium**   The free-swimming ciliated first larva of a trematode; it seeks out and penetrates a snail that serves as the intermediate host in which it develops into a sporocyst.

**Miscellaneous Class**   A competitive class at dog shows for dogs of certain specified breeds for which no regular dog show classification is provided.

**Miticide**   A compound that is destructive to mites. An acaricide.

**Mitosis**   Clinically refers to cellular reproduction.

**MLD**   Minimum lethal dose. Smallest amount of a drug or poison that will cause death.

**Modified live virus**   A virus that has been changed by passage through an unnatural host, such as hog cholera virus passed through rabbits, so that it no longer possesses **pathogenic** characteristics but will stimulate **antibody** production and immunity when injected into susceptible animals; an **attenuated** virus.

**Molar**   A tooth adapted for grinding; the back teeth.

**Molecular**   Pertaining to a little mass (molecule).

**Molt (molting)**   The shedding and replacing of feathers (usually in the fall). Snakes and certain arthropods also shed their outer covering and develop a new one.

**Monestrous (monoestrous)**   One **estrous cycle** per year.

**Mongrel**   Dog of mixed ancestry (commonly no one breed is clearly recognized).

**Monoclonal antibodies**   These are **antibodies** with specificity against only one set of **antigenic** determinants. They are produced in large quantities by hybridomas and have virtually revolutionized immunology. Specific *monoclonal antibody* has been used successfully for immunotherapy of cancer patients. Such antibody can also be used to develop immunodiagnostic techniques.

**Monocyte**   A large phagocytic leukocyte with an oval or horseshoe-shaped nucleus.

**Monogamous**   The condition or practice of having a single mate during a period of time.

**Monogastric**   Having only one stomach or stomach compartment.

**Monorchid**   A male with only one **testicle** descended. A unilateral **cryptorchid**.

**Monoxenous parasite**   A **parasite** that requires only one host for its complete development.

**Morbid**   Diseased or unhealthy.

**Morbidity**   The state of being diseased. The ratio of the number of sick individuals to the total susceptible population.

**Moribund**   Near death.

**Morphologically**   (adv. of morphologic) Pertains to the form and structure of animals and their body parts.

**Morphology**   The science of the forms and structures of animal and plant life without regard to function.

**Mortality**   Death; death rate.

**Morula**    A developing embryo at day 5 or 6 postfertilization characterized by a solid cluster of about 32 cells.

**Multiparous**    Producing many (more than one) offspring at one time. Also having experienced one or more **parturitions**.

**Murine**    Pertaining to the family Muridae including rats and mice.

**Muscle catabolism**    Breaking down of muscle tissues. The proteins from the muscles may be deanimated and used for energy. Destructive **metabolism** involves the release of energy and results in breakdown of muscle.

**Musculature**    The muscular system of any body part.

**Musculoskeletal**    Pertaining to the muscles and skeleton.

**Mustelid**    Any member of the family Mustelidae, fur-bearing carnivores including the weasels, skunks, badgers, and others.

**Mutation**    Change in a **gene** often resulting in a different **phenotype** that results in a change in hereditary characteristics between two generations; also known as a rogue gene.

**Muzzle**    The head in front of the eyes—nasal bone, nostrils, and jaws. **Foreface**. Also, a strap or wire cage attached to the **foreface** to prevent the dog from biting or from picking up food.

**Muzzle band**    White marking around the **muzzle**.

**Mycoses**    Infection with or disease caused by a fungus.

**Mycotoxins**    Toxic metabolites produced by molds during growth on a suitable substrate.

**Myelin**    The lipid-containing sheath that surrounds nerve fibers.

**Myiasis**    A disease due to the presence of fly larvae in warm-blooded animals.

**Myxofibroma**    Tumor involving **connective tissue** and containing mucin plus fibrous material.

**Nails**    The horny sheathes protecting the upper end of each finger and toe of humans and most other primates (are flat; other animals have **claws**—which are curved).

**Nano-**    A prefix that divides a basic unit by 1 billion ($10^9$); 1 nanometer is 1 billionth of a meter.

**Nape**    Back of the neck of an animal.

**Narcosis**    State of deep unconsciousness produced by a drug.

**Nares**    Nasal openings, nostrils.

**Nasolacrimal**    Pertaining to the nose and lacrima (tears); relating to the duct that drains tears from the eyes to the nose.

**Nasopharynx**    Pertaining to both the nose and **pharynx** (throat).

**National Research Council**    See **NRC**.

**Natural immunity**    Immunity to a disease or infestation that results from qualities inherent in an animal. See **immunity**.

**Necropsy**    An examination of the internal organs of a dead body to determine the apparent cause of death. Also called **autopsy**, **postmortem**.

**Necrosis**    Death of tissue, usually in individual cells, groups of cells, or small localized areas.

**Necrotic**    Dead.

**Nematodes**    Members of the phylum (Nematoda or Nemata) of elongated cylindrical worms parasitic in animals or plants or free living in soil or water.

**Neonatal**    The period immediately following birth; the offspring is called a **neonate**. Relating to or affecting the newborn human infant or animal.

**Neonate**    Newborn animal.

**Neoplasia**    The formation of tumors; a tumorous condition.

*Neospora caninum*    A coccidial **parasite** that can cause paralysis in dogs.

**Nephritis** Inflammation of the kidneys.

**Nephropathy** An abnormal state of the kidney; one associated with or secondary to some other pathological process (disease).

**Nephrotoxic** A substance harmful to the kidneys (renal poison).

**Net energy (NE)** The difference between **metabolizable energy** and the **heat increment**; includes the amount of energy used either for maintenance only or for maintenance plus production.

**Neuroleptanalgesia** Joint administration of a tranquilizing drug and an **analgesic** especially for relief of surgical pain.

**Neurological** Pertaining to nerves and the nervous system.

**Neuter** To **castrate** or **spay**.

**Neutered** Castrated or spayed.

**Neutropenia** A deficiency of **neutrophils**.

**Neutrophils** Also known as polymorphonuclear cells (PMNs); white blood cells involved in **phagocytosis** and killing of microorganisms.

**NFE** Nitrogen-free extract. Component of feed/food consisting of the soluble carbohydrates.

**Nick** A breeding that produces desirable offspring.

**Nit** The egg of a louse or similar insect.

**Nitrogen balance** The nitrogen in the food intake minus the nitrogen in the feces minus the nitrogen in the urine. Nitrogen retention.

**Nocturnal** Pertaining to night; active or awake at night.

**Nodes** Small, distinct masses of one kind of tissue enclosed in a different kind of tissue; lymph nodes are small masses of lymphatic tissue and serve as the main sources of **lymphocytes**, are a site of antibody production by lymphocytes, and also serve as a defense system for the removal of noxious agents such as bacteria and fungal organisms carried into the nodes by lymph vessels.

**Nonpedigree** One or both parents unregistered.

**Nose** Organ of smell; also, the ability to detect by means of scent.

*Notoedres cati* A **sarcoptiform** mite affecting cats.

**NRC** National Research Council. A division of the National Academy of Sciences established in 1916 to promote effective utilization of the scientific and technical resources available. This private, nonprofit organization of scientists periodically publishes bulletins giving the nutrient requirements of domestic animals. Copies are available through the National Academy of Sciences—NRC, 2101 Constitution Avenue NW, Washington, DC 20418, http://www.nas.edu/nrc/.

**Nuclear** Pertaining to the **nucleus**.

**Nuclear scans** Imaging of a body part or organ using a radioactive nucleotide.

**Nucleus** A deep-staining body within a cell, usually near the center; the heart and brain of the cell, containing the chromosomes and genes. Also, the small, positively charged core of an atom. All nuclei contain both protons and neutrons, except the nucleus of ordinary hydrogen, which consists of a single proton.

**Nulliparous** Having never given birth to viable young.

**Nutraceutical** A nutrient that produces a healthful effect beyond its normal nutritional effect.

**Nutrient** A substance (element or ingredient) that nourishes the metabolic processes of the body. It is one of the many end products of digestion; element or compound, mineral, vitamin, amino acid.

**Nutrient-to-calorie ratio**   An expression of nutrients in weight per unit of energy needed. For example, the protein-to-calorie ratio is expressed as the grams of protein ($N \times 6.25$) per 1,000 kcal **metabolizable energy** (grams of protein per 1,000 kcal ME). This same dimension may be extended to other nutrients such as grams of calcium per 1,000 kcal.

**Nutriment**   That which is required by an animal for growth and replenishment of tissues and fuel (nourishment).

**Nutrition**   The science encompassing the sum total of processes that have as a terminal objective the provision of nutrients to the component cells of an animal.

**Nymph**   The immature stage of insects having only three stages (egg, nymph, and adult) in their development. Nymphs resemble adults in form and appearance (as contrasted with larvae, which do not resemble their adults) but do not have wings.

**Obedience Trial (Licensed)**   An event held under **AKC** rules at which a "**leg**" toward an obedience degree can be earned.

**Obedience Trial Champion (OTCH)**   A prefix used with the name of a dog that has been recorded an Obedience Trial Champion by the **AKC** as the result of having won the number of points and First Place wins specified in the current Obedience Regulations (Chapter 13).

**Obese**   Fat; overly fat.

**Obligate parasite**   A **parasite** incapable of living without a **host** (these are only found in or on animals because they cannot survive as free-living organisms in the environment).

**Occipital protuberance**   A prominently raised **occiput** characteristic of some dog breeds.

**Occiput**   Upper back point of the skull.

**Ocular**   Of, relating to, or connected with the eye.

**Oestrus**   See **estrus**.

**Oil gland**   Gland at the base of the tail in chickens, ducks, turkeys (and most wild birds) that secretes an oil used by birds in preening their feathers. Also called *preen gland*.

**Olfactory**   Pertaining to smell and the sense of smell.

**Omnivore**   An animal that subsists on feed/food of every kind (plant and animal), as with humans.

**Omnivorous**   Subsisting on all types of food—plant and animal.

**Oocysts**   Sporozoan zygote undergoing sporogenous development.

**Opaque**   Not letting light through; neither transparent nor translucent.

**Open bitch**   A bitch that can be bred; not pregnant.

**Open Class**   A class at dog shows in which all dogs of a breed, champions and imported dogs included, may compete; also an intermediate level class in AKC Obedience Trials.

**Ophthalmologist**   Doctor or veterinarian specializing in the study of eyes.

**Ophthalmology**   Study of the eye.

**Opioid**   Refers to compounds related to opium. These include drugs derived from the opium poppy plant and related compounds (natural and synthetic). These compounds have many effects including inhibition of neurotransmitter release, sedation, and **analgesia**. Side effects include **bradycardia**, respiratory depression, impaired thermoregulation, and increased sensitivity to noises.

**Opisthotonos**   A form of tetanic spasm (**seizure**, convulsion) in which the head and heels are bent backward and the body bowed forward.

**Optic**   Pertaining to vision, sight.

**OR**   Operating room.

**Orchidectomy**  Surgical removal of the testes.

**Orchiectomy**  Excision of the testes, castration, **orchidectomy**.

**Orchitis**  An inflammation of a testis.

**Organ**  Group of tissues organized to perform a specific function.

**Oriental**  Some specific shorthair varieties; also the shape of some cats' eyes, for example, the Siamese.

*Ornithodoros*  A species of soft-bodied **ticks**; some species transmit diseases.

**Oropharynx**  The mouth and throat.

**Orthopnea**  Inability to breathe except in the upright position. This state is common during dehydration exhaustion and with congestive heart failure.

**Osseous tissue**  Bone tissue.

**Osteogenesis**  Formation of bone.

**Osteolysis**  A loss of calcium salts from the bones that causes them to become fragile.

**Osteomalacia**  A condition marked by softening of the bones, pain, tenderness, muscular weakness, and loss of weight. It commonly results from a deficiency of vitamin D or of calcium and phosphorus. May also be caused by an overactive parathyroid gland.

**Osteomyelitis**  An infection involving bone.

**Osteoporosis**  A reduction in total bone mass. This disorder of bone metabolism occurs in middle life and older age in both men and women. The bone becomes porous and thin due to a failure of the osteoblasts (bone-forming cells) to lay down bone matrix. This disorder may result from (1) a dietary deficiency of calcium and/or protein, (2) a lowered calcium absorption, or (3) a hormonal disturbance. In companion animals, osteoporosis is usually associated with metabolic or endocrine diseases.

**Osteosclerosis**  Abnormal hardening of bone.

**Otic**  Pertaining to the ears.

**Otitis**  Inflammation of the ear.

*Otobius megnini*  Species of ear **ticks**, affects rabbits and cattle.

*Otodectes cynotis*  Ear mites of dogs and cats.

**Otoscope**  An instrument used to examine the ears (Chapter 16).

**Ototoxic**  An agent that is toxic or damaging to components of the inner ear or auditory nerve; causes deafness.

**Outcrossing**  The mating of unrelated individuals of the same **breed**.

**Overshot**  The front teeth (**incisors**) of the upper jaw overlap and do not touch the front teeth of the lower jaw when the mouth is closed.

**Ovicide**  Substance that kills **parasites** in the egg stage.

**Oviparous**  Producing offspring from eggs that hatch outside the body.

**Ovulation**  Shedding of the **ovum** (egg) from the follicle of the ovary.

**Ovum**  (pl. **ova**) Female reproductive cell; egg.

**Oxidase**  Enzyme that activates oxygen.

**Pack**  Group of dogs.

**Pads**  Tough, shock-absorbing projections on the underside of the feet; soles.

**Paired feeding (food equalizing)**  A method of comparing nutritional effects at an arbitrary low level set by the animal that consumes the least food. **Littermates** or twins (especially monozygous ones) are considered best for paired-feeding studies.

**Paired serum samples**  Two blood samples taken from an animal at an interval of approximately 3 to 4 weeks for measurement of serum antibody levels against a specific antigen. An increase in the **antibody** levels (**titer**) is indicative of the presence of an active infection.

**Palatability**   Relative acceptance or relish of feeds/foods by animals.

**Palate**   The roof of the mouth consisting of the structures that separate the mouth from the nasal cavity.

**Palmar**   Pertaining to the palm; in animals refers to the soles of the forefeet.

**Palpation**   The act of feeling; palpation of female to determine pregnancy.

**Palpebral**   Pertaining to the palpebra (eyelid).

**Pancreas**   A gland having both **endocrine** and **exocrine** functions. Produces **digestive enzymes** that function in the breakdown of **proteins**, **fats**, and **carbohydrates** and the **hormone insulin**.

**Pancytopenia**   A deficiency of all cells; clinically often used to refer to a deficiency of white blood cells (**leukopenia**).

**Pandemic**   Prevalent (as a disease) throughout an entire country or continent or the world.

**Papanicolaou stain**   A method of staining smears of various body secretions from the respiratory, digestive, or genitourinary tracts. It is used to diagnose cancer or the presence of a malignant process. Exfoliated cells of organs, such as the stomach or uterus, are obtained, smeared on a glass slide, and stained for microscopic examination. It was named for the Greek-born physician George Papanicolaou who developed it. The slides are also known as Pap smear.

**Papillomatosis**   Development of multiple papillomas ("warts," **benign** tumors of epithelium).

**Parakeratosis**   An abnormality of the **stratum corneum** (horny layer of epidermis) of the skin (especially a condition where nuclei are retained in the upper layers showing defective keratinization); is often associated with a deficiency of zinc.

**Paralysis**   A loss of motor function; immobility.

**Parasite**   An organism that lives at least for a time on or in and at the expense of living animals.

**Parasiticide**   An agent or drug destructive to **parasites**.

**Parasitologist**   Individual specializing in the study of **parasites** and parasitism.

**Paratenic host**   A **host** in which the parasite does not undergo any development; a transport host.

**Parenchyma**   The essential and distinctive tissue of an organ or of a portion of a plant.

**Parenteral**   Pertaining to administration by injection, not through the digestive (food) tract, such as **subcutaneous**, **intramuscular**, intramedullary, **intravenous**.

**Paresis**   Partial **paralysis** that affects the ability to move but not the ability to feel.

**Parotid glands**   Salivary or other glands located near the ear of an animal. In amphibians (e.g., toads) the parotid glands are specialized skin glands clustered behind the head that secrete toxins.

**Parthenogenesis (parthogenesis)**   Reproduction by the development of an egg without its being fertilized by a spermatozoon (e.g., drone bees). It occurs in certain lower animals and has been observed in turkeys. It does not occur in mammals.

**Parturition**   The act or process of giving birth to young.

**Passerine**   Belonging to or having to do with the very large group of perching birds; includes more than half of all birds, such as the warblers, sparrows, chickadees, wrens, thrushes, and swallows.

**Passive immunity**   Disease immunity given to an animal by injecting the blood serum from an individual already immune to that disease. Newborns also receive passive immunity by **absorption** of antibodies from **colostrum**. See **immunity**.

**Pastern**   Commonly recognized as the region of the foreleg between the carpus or wrist and the digits.

**Patella**   Kneecap.

**Patency**   The condition of being open or unobstructed.

**Patent**   Affording free passage, open, unobstructed; period when adult **parasites** produce eggs.

**Pathogen**   Disease-producing virus or microorganism.

**Pathology**   Science dealing with the study of disease particularly relating to structural or functional changes in organs or tissues.

**PCR**   See **polymerase chain reaction**.

**Pecking order**   The system of social order in **poultry** exhibited by animals higher on the dominance scale physically pecking at animals lower on the scale. The term may also be applied to other social groups of animals.

**Pectoral**   Pertaining to the breast.

**Pediculosis**   Infestation with lice.

**Pedigree**   A list of an animal's ancestors, usually only those of the five closest generations.

**Pelvis**   Hip bone.

**Perch**   A bar or peg on which a bird sits.

**Percussion**   The act of tapping or striking the surface of a body part (as chest or abdomen) to learn the condition of the parts beneath by the resultant sound.

**Percutaneous**   Performed or introduced through the skin, as an injection.

**Perfusion**   The act of pouring through or immersing in a physiological fluid (e.g., blood or saline).

**Perianal**   Around the anus.

**Pericardium**   The membrane that encloses the cavity containing the heart.

**Perineal hernia**   A deviation and protrusion of the rectum into the perianal tissues due to weakening and stretching of the connective tissues supporting the rectum. This results in fecal impaction within the rectum and difficulty in defecation.

**Perineum**   The anatomical region of the body between the thighs, especially between the anus and the genitals.

**Periocular**   Around the eye.

**Periodontal**   Pertaining to around the teeth.

**Periosteum**   The membrane that covers bone.

**Peripheral nervous system**   Refers to all nerves outside of the brain and spinal cord; includes cranial nerves, spinal nerves, nerves of the limbs, sensory nerves, and others.

**Peristalsis**   The rhythmic contractions and movements of the alimentary canal.

**Peritoneal cavity**   Inside the abdomen, the abdominal cavity is lined by a smooth, transparent serous membrane known as the **peritoneum**.

**Peritoneum**   The membrane that lines the abdominal cavity and invests the contained viscera (digestive organs).

**Peritonitis**   Inflammation of the **peritoneum**.

**Per oral**   Administration through the mouth.

**Per os**   Oral administration (by the mouth) (PO).

**Per se**   By, of, or in itself; as such.

**Persians**   Breed of cats with long fur; referred to as Longhairs in Britain.

**Pest**   One that pesters or annoys.

**Pesticide**   A compound used to control any plant or animal considered to be a **pest**.

**pH**   A symbol used (with a number) to express acidity or alkalinity in analyzing various body secretions, chemicals, and other compounds. It represents the logarithm of the reciprocal (or negative logarithm) of the hydrogen-ion concentration (in gram atoms per liter) in a given solution, usually determined by the use of a substance (indicator) known to change color at a certain concentration. The pH scale in common use ranges from 0 to 14, pH 7 (the hydrogen-ion concentration, $10^{-7}$ or 0.0000001, in pure water) being considered neutral; 6 to 0, increasing acid; and 8 to 14, increasing alkali.

**Phagocytes**   From the Greek *phago* meaning "eat" and *kytos* meaning "cell." Defensive cells (**leukocytes**, or white blood cells) of the body that ingest and destroy bacteria and other infectious agents. See **macrophages**.

**Phagocytosis (phagocytizing)**   The engulfing of microorganisms, cells, or foreign particles by **phagocytes** (certain forms of **leukocytes**).

**Phallus**   Penis.

**Pharyngeal**   Of or pertaining to the **pharynx**.

**Pharynx**   The tube, or cavity, that connects the mouth and nasal passages with the esophagus (throat).

**Phenolic compounds**   Products used for disinfecting purposes derived from the distillation of coal tar. Examples of brand names include Lysol, One Stroke Environ, and O-Syl. These products kill many microorganisms and viruses; however, they are highly toxic to birds, reptiles, amphibians, and cats.

**Phenotype**   Physical appearance of an animal.

**Pheromone**   A substance secreted externally by certain animal species (especially insects) to affect the behavior (especially sexual) or development of other members of the species.

**Phospholipid**   A lipid-containing phosphorus that on hydrolysis yields fatty acids, glycerin, and a nitrogenous compound. Lecithin, cephalin, and sphingomyelin are examples. Also called *phosphatide*.

**Physiological saline**   A salt solution (0.9% NaCl) having the same osmotic pressure as blood plasma, also called *normal saline*.

**Physiology**   The science that pertains to studying functions of organs, systems, and the whole living body (Chapter 11).

**Pica**   A craving for unnatural articles of food, such as is seen in hysteria, pregnancy, and phosphorus deficiency. A depraved appetite.

**Pico-**   Prefix; divides a basic unit by 1 trillion ($10^{12}$). Same as *micromicro*.

**Pied**   Comparatively large patches of two or more colors. Piebald, parti-colored.

**Pigeon-toed**   Toes pointing in.

**Pile**   Dense undercoat of soft hair.

**Piloerection**   Hair erection.

**Pinna**   The largely cartilaginous projecting portion of the external ear (plural *pinnae*).

**Pinocytosis**   The absorption of liquids by cells.

**Pithing**   A method of animal slaughter in which the spinal cord is severed to cause death and/or to destroy feeling.

**Placebo**   In Latin, means "I shall please." An inactive substance or preparation given to please or gratify a patient. Also used in controlled studies to determine the **efficacy** (virtue) of medicinal substances.

**Plantar**   Pertaining to the sole; in animals refers to the soles of the hindfeet.

**Plaque**   A bacterial-laden film that coats teeth and the mucous membranes of the oral cavity (mouth).

**Plasma**   Fluid separated from blood that has been prevented from clotting; the liquid portion of blood or lymph in which the corpuscles or blood cells float.

**Plastron**   The lower shell of a turtle or tortoise.

**Platelet**   A blood cell involved in clot formation and in the plugging of holes in blood vessels (**hemostasis**).

**Pleural**   Pertaining to the pleura (tissue lining the **thoracic** cavity).

**Plexiglas**   An acrylic resin or plastic.

**Plumage**   The feathers of a bird.

**Plume**   A long fringe of hair hanging from the tail as in Setters.

**Pneumatic bones**   Hollow bones; bones of birds containing air sacs.

**Pneumonia**   A condition of the lungs; clinically refers to inflammation of the lungs with consolidation.

**Pocket pets**   A term used to describe small pets such as hamsters, gerbils, rats, mice, guinea pigs, and rabbits. They are cute, usually inexpensive, and are commonly kept as pets. Pocket pets are especially popular with children.

**Poikilotherms (ectotherms)**   Cold-blooded animals; animals having a body temperature that varies with the environment. Ocean fish exemplify cold-blooded species.

**Point**   The immovable stance of the hunting dog taken to indicate the presence and position of game; term also used to refer to areas of the body differing in color from the trunk or back color.

**Police dog**   Any dog trained for police work (Chapter 21).

**Pollakiuria**   Increased frequency of urination.

**Polychromasia**   A condition of many colors; clinically refers to erythrocytes with variable staining (usually indicative of a mixture of immature and mature **erythrocytes**).

**Polycystic**   Disease characterized by the formation of multiple cysts (e.g., within the kidneys).

**Polycythemia**   Excess of red cells.

**Polydipsia**   An excessive thirst.

**Polyestrus**   More than one **estrous cycle** per year.

**Polygamous**   Having more than one mate at one time.

**Polygenes**   Groups of **genes** that act together to produce hereditary characteristics.

**Polymerase chain reaction (PCR) tests**   Utilizes a series of reactions that amplify segments of nucleic acids from **DNA** gene sequences. These tests can be used to identify infectious organisms that would otherwise be difficult to detect due to their small size or low numbers.

**Polyneuritis**   Inflammation of many nerves concurrently.

**Polyp**   A smooth, stalked, or projecting growth.

**Polyphagia**   Increased appetite, excessive hunger, overeating.

**Polypnea**   A condition in which the respiration rate is increased; rapid, shallow breathing.

**Polytocous**   Pertains to animals that normally produce more than one young per gestation.

**Polyunsaturated**   Fatty acids having multiple double bonds within the carbon chain.

**Polyuria**   Increased urine production.

**Popliteal**   Pertaining to the **caudal** thigh (behind the **stifle** or knee).

**Portal system**   The system of blood vessels conveying blood from the digestive organs and spleen to the liver.

**Portosystemic shunts**   Abnormal vascular connections between portal veins (vessels that normally carry blood from the gastrointestinal tract to the liver) and the systemic circulation. Blood going through the shunt bypasses the liver and thus is not filtered or detoxified. Animals with portosystemic shunts have a buildup of toxic waste products in their blood and show signs of liver failure.

**Posterior**   Pertaining to the rear.

**Postmortem**   An examination of an animal or human body after death. **Necropsy, autopsy**.

**Postnatal**   Occurring after birth.

**Postpartum**   Postparturient; following birth.

**Poultry**   Domesticated birds kept for eggs or meat.

**ppm**   Parts per million (1 mg/liter).

**Prebiotic**   A nondigestible food ingredient that beneficially affects the host by selectively stimulating the growth and/or activity of one or a limited number of bacteria in the colon that can improve the host's health.

**Precipitin**   **Antibody** that forms a precipitate with its soluble **antigen**. An antibody formed in blood **serum** as a result of inoculating with a foreign protein.

**Precocious**   Exhibiting mature qualities at an unusually early age.

**Predatism**   Intermittent parasitism, such as the attacks of mosquitoes and bedbugs on humans.

**Predator**   Any animal, including an insect, that preys on and devours other animals (e.g., a coyote or dog preying on sheep). Some predators, such as ladybugs, may be beneficial in that they kill and eat **parasites**.

**Preen gland**   See **oil gland**.

**Prehension**   The seizing (grasping) and conveying of food to the mouth.

**Premortal**   Existing or occurring immediately before death.

**Prepartum**   Occurring before birth of the offspring; before **parturition**.

**Prescapular**   Located before the scapula (shoulder blade).

**Prescapular lymph nodes**   Located in front of the scapula (shoulder blade).

**Prick ear**   Carried erect, usually a slight point at the tip.

**Primates**   Humans, monkeys, and the great apes.

**Primiparous**   Bearing or having borne only one young or set of young.

**Probiotics**   Live (viable) microorganisms found in the gastrointestinal tract of healthy animals. They are intended to influence gut **microflora** by preferentially populating it with nonpathogenic organisms to the exclusion of potential **pathogens** (Chapter 9).

**Prodome**   A symptom indicating the onset of a disease.

**Proestrus**   Before **estrus**; clinically refers to the period of the reproductive cycle before sexual receptivity.

**Professional handler**   A person who shows dogs for a fee (Chapter 23).

**Progeny**   The offspring of animals.

**Progesterone**   A **hormone** produced by the corpus luteum of the ovary important in the maintenance of pregnancy; it is also produced by the adrenal cortex and the placenta.

**Proglottids**   Segments of a tapeworm formed by a process of strobilation in the neck region of the worm; contain both male and female reproductive organs.

**Prognosis**   Prediction of course or outcome of a disease.

**Prolapse**   Abnormal protrusion of a part or organ; displacement of an organ from its normal location.

**Prolapse (rectal)**   In a rectal prolapse, a portion of the caudal rectum protrudes through the anal orifice.

**Prophylactic**   A preventive, preservative, or precautionary measure that tends to ward off disease.

**Prophylaxis**   The prevention of disease.

**Proprioception**   The reception of stimuli produced within the organism; the ability to know where different parts of the body are positioned.

**Prostate gland**   A glandular body that produces fluids that comprise much of the fluid in semen, it surrounds the base of the male urethra and has ducts that open into the floor of the urethra.

**Protective antibodies**   Antibodies that when combined with pathogenic organisms render them noninfectious.

**Protein**   A substance composed of amino acids, containing about 16% (molecular weight) nitrogen. Thus, protein content is computed by multiplying the chemically determined value for nitrogen by the factor 6.25 ($N \times 6.25$).

**Protein equivalent**   A term indicating the total nitrogenous contribution of a substance in comparison with the nitrogen content of protein (usually plant protein). For example, the nonprotein nitrogen (NPN) compound urea contains approximately 45% nitrogen and has a protein equivalent of 281% ($6.25 \times 45\%$).

**Proteinuria**   Presence of protein in the urine.

**Protozoa**   Single-cell microscopic animals.

**Proventriculus**   Glandular stomach of birds.

**Proximal**   Nearest; closer to any point of reference; opposite to **distal**.

**Pruritus**   A sensation producing the desire to itch; the state of itching.

**Pseudocyesis**   False pregnancy.

**Pseudopregnancy**   See **pseudocyesis**.

**Psittaciformes**   Any member of the group of more than 300 species of generally brightly colored, noisy, tropical birds, to which the general name parrot may be applied.

**Psychobiology**   The branch of biology that considers the interactions between body and mind in the formation and functioning of personality; the scientific study of the personality function.

**Psychogenic**   Originating in the mind (brain); controlling emotions.

**Psychro-**   From the Greek *psychros* meaning "cold." The prefix denoting relations to cold.

**Psychroenergetics**   Science dealing with the effect of ambient temperature and humidity on conversion of feed/food into bodily heat and energy.

**Puberty**   The age at which the reproductive organs become functionally operative and secondary sex characteristics develop.

**Pubic**   Pertaining to the pubes (hair growing over pubic area) or pubic bones. The lower part of the hypogastric region.

**Pubic symphysis**   The junction between the two halves of the pubes (ventral portion of the pelvic bones).

**Public health**   An organized effort to prevent disease, prolong life, and promote physical and mental efficiency. Also, the health of the community taken as a whole.

**Pudic**   Pertaining to the external genital parts, especially of the female.

**Pulse oximeter**   A specialized instrument designed to estimate arterial oxyhemoglobin saturation by utilizing selected wavelengths of light. The pulse oximeter consists of a probe attached to the patient's toe, tongue, or ear **pinna** that is linked to a computerized unit. The unit displays the percentage of hemoglobin (Hb) saturated with oxygen together with an audible signal for each pulse beat, a calculated heart rate, and in some models, a graphical display of the blood flow past the probe.

**Pupa**   The quiescent or inactive stage during which an immature insect or larva transforms into an adult.

**Pupal stage**   Period in the life history of insects between the caterpillar, or grub, stage and the mature, or adult, insect.

**Pupate**   To change from an active immature insect into the inactive **pupal stage**.

**Pupil**   The contractile aperture in the iris of the eye, it is round in most vertebrates and elliptical in cats.

**Puppy**   A young dog, generally under 12 months of age.

**Pure culture**   A population of microorganisms that contains only a single species. Cultures are useful in the manufacture of many animal products (e.g., cheeses and yogurt).

**Purebred**   Animal whose parents are of the same **breed**.

**Purified diet**   A mixture of the known essential dietary nutrients in a pure form that is fed to experimental (test) animals in nutrition studies.

**Purpose bred**   Laboratory animals (often inbred) bred specifically for use in research (Chapters 7 and 22).

**Pus**   A liquid inflammatory product consisting of **leukocytes**, **lymph**, bacteria, dead tissue cells, and fluid derived from their disintegration.

**PVC**   Polyvinyl chloride; widely used in pipes for plumbing and other fixtures.

**Pyknotic nucleus**   A degeneration of the nucleus of a cell; the nucleus shrinks in size and the chromatin condenses to a solid amorphous mass.

**Pyoderma**   Purulent infection of the skin.

**Pyometra**   Pus in the **uterus**.

**Pyrexia**   A fever or febrile condition. An abnormal elevation of body temperature.

**Quality**   Refinement, fineness.

**Quarantine**   Commonly thought of as the segregation of the active case of an infectious disease, but more technically, it includes compulsory segregation of exposed susceptible animals or individuals for a period of time equal to the longest usual incubation period of the disease to which they have been exposed. A regulation under police power for the exclusion or isolation of an animal to prevent the spread of an infectious disease.

**Queen**   A mature female cat.

**Quick**   The inner, non-keratinized, living portion of the **claw**, **nail**, or hoof of an animal; cutting into the quick results in pain and bleeding.

**Rabies**   An infectious viral disease of the central nervous system invariably fatal in mammals. Early symptoms include fever and hyperexcitability followed by paralysis of the muscles used in swallowing, progresses to convulsions or **paralysis** and death.

**Radiant energy**   Energy that is being transferred through space by electromagnetic waves.

**Radiant heat**   Heat transmitted by radiation (such as that of the sun) as contrasted with that transmitted by conduction or **convection**.

**Radiograph**   A record or photograph produced by x-rays or other rays on a photographic plate, commonly called an *x-ray picture* (Chapter 16).

**Radiology**   The science that deals with the use of all forms of ionizing radiation in the diagnosis and therapy of disease.

**Rales**   Abnormal respiratory sound; rattling in chest.

**Random mating**   System of breeding in which each male has equal chance to mate with each female of a group.

**Rangy**   Long-bodied, usually lacking depth in chest.

**Ration**   The food allowed an animal for 24 hours. A *balanced* ration provides all the nutrients required to nourish an animal for 24 hours.

**Recessive gene**   A **gene** that will transmit a characteristic only if another identical gene appears in the allele.

**Rectum**   The terminal portion of the colon, proximal to the anal opening.

**Redia**   A larval stage in the development of flukes. Redia of liver flukes of cattle, sheep, and goats occurs in snails.

**Rediae**   A larva produced within the sporocyst of many trematodes that in turn either produces another generation of rediae or develops into a cercariae.

**Redleg**   Disease of frogs characterized by **hemorrhages** in the legs and on the abdomen associated with **immunosuppression**, colonization with *Aeromonas* bacteria, and **septicemia**.

**Register**   To record with the **AKC**, CFA, or a breed association a dog's or cat's sire, dam, color, date of birth, and other details required by the registry (e.g., name of breeder and owner).

**Regurgitation**   (1) The casting up of incompletely digested food (as by some birds in feeding their young), in medical terminology is usually used to refer to food that has not yet reached the stomach; (2) the backward flow of blood through a defective heart valve.

**Relative humidity (RH)**   The ratio of the weight of water vapor contained in a given volume of air to the weight that the same volume of air would contain when saturated. The quantity of water vapor that air can hold when saturated increases with temperature. The RH is expressed as a percentage. For example, if a sample of air at a given temperature contains 30% of the water vapor that it is possible for it to contain at that temperature, it is 30% saturated and therefore has a relative humidity of 30%.

**Remiges**   Flight feathers.

**Renal**   Pertaining to the kidneys.

**Reproductive**   Pertaining to producing offspring.

**Reptiles**   Class of the subphylum Vertebrata; cold blooded, creeping; snakes, turtles, lizards (Chapter 6).

**Reservoir host (reservoir)**   An animal that harbors the same species of **parasite** as humans. Also, an animal that becomes infected and serves as a source from which other animals can be infected.

**Respiratory**   Pertaining to breathing.

**Respiratory quotient (RQ)**   The RQ is used to indicate the *type* of food being metabolized. This is possible because carbohydrates, fats, and proteins differ in the relative amounts of oxygen and carbon contained in their molecules. Also, the relative volumes of oxygen consumed and carbon dioxide produced during **metabolism** of each type of food vary. Respiratory quotient is calculated as follows:

$$RQ = \frac{\text{volume CO}_2 \text{ produced}}{\text{volume O}_2 \text{ consumed}}$$

**Reticulocyte**   An immature red blood cell containing remnants of ribosomes and endoplasmic reticulum.

**Reticuloendothelial system**   A widely spread network of body cells concerned with blood cell formation, bile formation, and engulfing or trapping of foreign materials, which includes cells of bone marrow, lymph, spleen, and liver. Currently the preferred terminology is the monocyte-macrophage system.

**Retrices**   Tail feathers.

**Retrieve**   A hunting term. The act of bringing back shot game to the handler.

*Rhabditis (Pelodera) strongyloides*   A free-living nematode; larval forms can invade the hair **follicles** of animals.

**Rhamphotheca**   Keratinized covering of the **beak** of birds.

**Rhinitis**   Inflammation of mucous membranes of the nose; snuffles in rabbits.

**Rhinopneumonitis**   Inflammation of the nose and lungs.

**Rhinotheca**   Upper jaw of birds.

*Rhipicephalus sanguineus*   Brown dog **tick**.

**Rickettsia**   Pleomorphic rod-shaped microorganisms that live intracellularly in biting arthropods and when transmitted to animals or humans, by the bite of an arthropod **host**, cause a number of serious diseases (e.g., Rocky Mountain spotted fever and typhus).

**Rickettsiae**   Intracellular **parasites** (i.e., ones that multiply inside the living cells of other larger organisms). In size, they are intermediate between bacteria and viruses.

**Ridgling**   Any male animal whose **testicles** fail to descend normally into the scrotum. Also called **cryptorchid**.

**Rigor mortis**   The stiffness of body muscles that is observed shortly after the death of an animal. It is caused by an accumulation of **metabolic** products, especially lactic acid, in the muscles.

**Rodent**   Classification of mammals (order Rodentia) characterized by chisel-shaped **incisor** teeth; mostly vegetarian, gnawing animals—mice, rats, gerbils, hamsters, and others (Chapter 7).

**Rodenticide**   Poison that is lethal to rodents.

**Roman nose**   A nose whose bridge is so comparatively high as to form a slightly convex line from forehead to nose tip. Ram's nose.

**Rostral**   Pertaining to the nose.

**Rudder**   The tail.

**Ruff**   Thick, longer hair growth around the neck.

**Rumen**   The first (large) compartment of the stomach of a **ruminant**.

**Ruminant**   Cud-chewing animal with four-compartment stomach such as cattle, sheep, and goats.

**Rupture**   The forcible tearing or breaking of a body part.

**Sable**   A lacing of black hairs over a lighter ground color. In Collies and Shetland Sheepdogs, a brown color ranging from golden to mahogany.

**Sagittal**   Anteroposterior plane or section parallel to the long axis of the body.

*Sarcocystis*   **Protozoal parasites** of dogs, cats, and humans.

**Sarcoptiform**   Mites of the family Sarcoptiforme that generally burrow within the superficial skin layers producing intense **pruritus**. This infestation is called scabies.

**Satiety**   Full satisfaction of desire; may refer to satisfaction of sexual arousal, appetite.

**Saturated fat**   A completely hydrogenated fat; that is, each carbon atom is associated with the maximum number of hydrogens.

**Saturated fatty acid**   A carboxylic acid in which all of the carbons in the chain are separated by a single bond. Fatty acids are completely hydrogenated (i.e., each carbon atom is associated with the maximum number of hydrogens). Palmitic and stearic acids are examples.

**Saturates**   Molecules that contain no double bonds within their carbon chains.

**Schistosomiasis**   Infestation with a schistosome, or blood fluke.

**Scissors bite**   A bite in which the outer side of the lower **incisors** touches the inner side of the upper incisors.

**Sclera**   The tough, white, supporting covering of the eyeball, which encompasses all of the eyeball except the **cornea**.

**Scrotum**   Pouch of skin that contains the testes.

**Scutes**   The horny plates of the shell of a turtle or tortoise.

**Scutum**   A bony, horny, or chitinous plate that forms the upper surface of the body of certain insects.

**Seasonally polyestrous**   Describes the tendency of some species, or some breeds within species, to have multiple **estrous cycles** primarily during only one season of the year.

**Sebaceous glands**   Small multilobulated glands located in the **dermis** that usually open into the hair **follicles**; secrete an oily or greasy material composed of fats and other lipids that soften and lubricate the hair and skin.

**Sebum**   The thick, semifluid substance composed of **lipids** and epithelial debris secreted by the sebaceous glands.

**Secondary infection**   Infection following an infection already established by other organisms.

**Seeing Eye dog**   A dog trained as a guide for the blind (Chapter 21).

**Seizure**   A sudden attack (as of disease); convulsions.

**Self-feeding**   Any feeding device by means of which animals can eat at will. See **ad libitum**.

**Senescence**   The process or condition of growing old; aging.

**Sensory**   Pertaining to sensation. The eyes and ears are *sensory organs*. *Sensory nerves* convey impulses from the sense organs to a nerve center. Thus, some nerves are *sensory* and pick up sensations from sense organs and carry them to main cords and the brain, whereas others are *motor* and carry impulses from the brain and main nerves to the muscles, which respond to the stimulation.

**Septicemia**   Systemic disease associated with the presence of pathogenic microorganisms or their toxins in the blood, also called *blood poisoning*.

**Serological**   Pertaining to the use of blood **serum** of animals in various tests, which aids in detecting and treating certain diseases.

**Serotype**   The type of microorganism as determined by the kind and combination of constituent **antigens** associated with the cell.

**Serum**   The clear portion of animal fluids, separated from its cellular elements. Blood serum is the clear, pale-yellow, watery portion of blood that separates from the clot when blood coagulates.

**Serum therapy**   The treatment of clinical cases of disease with serum of immunized animals.

**Sex linked**   A characteristic such as the tortoiseshell coat that is associated more with one sex (usually female) than the other.

**Sexual dimorphism**   Differences between males and females in physical characteristics such as size and coloration.

**Shoulder height**   Height of dog's body as measured from the **withers** to the ground. See **withers**.

**Sibling**   In genetics, a brother or sister.

**Sickle hocked**   Inability to straighten the hock joint on the back reach of the hind leg.

**Signalment**   The systematic description of an animal for purposes of identification, including its breed, age, sex, and color.

**Silver sand**   A sharp, fine sand of a silvery appearance often used for grinding.

**Sinistral**   Of or pertaining to the left side; left; or left-handed.

**Sinusitis**   Inflammation of a sinus.

*Siphonaptera*   Fleas.

**Sire**   The male parent.

**Sled dogs**   Dogs worked usually in teams to draw sleds.

**Social insect**   Any insect that lives with others of its kind in a somewhat organized colony, as ants, bees, and wasps.

**Soluble**   Designating a substance that is capable of being dissolved in another.

**Soundness**   The state of mental and physical health when all organs and faculties are complete and functioning normally, each in its correct relation to the other.

**Spay**   To perform a surgical operation on a female's reproductive organs to prevent conception; to neuter or alter.

**Spayed**   Neutered female.

**Spaying**   The neutering of a female.

**Spectacles**   Transparent scales covering the eyes of reptiles, normally shed with the rest of the skin during ecdysis (molting); if retained, the eyes will appear to have an opaque haze over them.

**Sperm**   (sing. **spermatozoon**, pl. **spermatozoa**) A mature male germ cell.

**Spermatogenesis**   The formation and development of **spermatozoa**.

**Spermatozoa**   Mature male germ or reproductive cells.

**SPF**   Specific pathogen-free. Free of certain pathogenic organisms but not necessarily free of others.

*Spirometra*   Tapeworm affecting carnivores.

**Splayfoot**   A flat foot with toes spreading. Open foot, open-toed.

**Splenomegaly**   Enlargement of the spleen, may be associated with tumors, hematomas (blood-filled cavities), or **hematopoietic** disorders (abnormalities in formation of red blood cells and/or excessive destruction of red blood cells).

**Sporocyst**   A saccular body that is the first reproductive form of a trematode in a molluscan host; cells bud off from their inner surface and develop into rediae within the cavity of the sporocyst.

**Sporozoans**   Any of a large class (Sporozoa) of strictly parasitic protozoans that have a complicated life cycle usually involving both asexual and sexual generations often in different hosts and include important pathogens (as malaria parasites and babesias).

**Sporozoites**   Usually motile infective forms of some **sporozoans** that is a product of sporogony and initiates an asexual cycle in the new host.

**Spraying**   A male habit of directing their urine onto objects; this "marks" the site with the individual's scent and is a way male animals establish their territory.

**Sputum**   Something expectorated and usually consists of saliva with or without mucus or other materials from the respiratory passages.

**Stable learning**   Term applied when an animal remembers an experience and will seek to duplicate rewarding experiences and to avoid traumatic ones.

**Stance**   Manner of standing.

**Standard**   A description of the ideal dog or cat of each recognized **breed** to serve as a word pattern by which dogs and cats are judged at shows.

**Standard metabolic rate (SMR)**   Reflects an animal's basic maintenance energy requirement. It is useful and important in studies of thermal physiology and productive efficiency to have such a reference metabolic rate. Because metabolic rate increases during thermal stress, an animal's reference metabolic rate should be measured in the **thermoneutral zone** of effective environmental temperature. Moreover, because metabolic rate increases postfeeding (due to the **heat increment** of feeding), the reference value should be determined sometime after the animal has absorbed its last meal. Additionally, because physical activity increases metabolic rate, to be meaningful the reference value must reflect metabolic rate when the animal is resting. Thus, the SMR takes these three conditions into account and is said to occur in a fasting, resting animal held in thermoneutral surroundings. Standard metabolic rate is based on the 0.75 power of body weight, the value commonly called metabolic body size. By means of SMR, comparisons can be made among animals of different sizes and species. In human physiology, SMR is called basal metabolic rate (BMR). In animal science literature, standard metabolic rate is synonymous with *fasting metabolic rate* and with *resting metabolic rate*.

**Stapes bone**   The innermost bone of the ear of mammals, shaped like a stirrup, located between the **tympanum** and the **incus**.

**State-of-being**   The state of an animal's conditions of life.

**Stenosis**   A condition of narrowing.

**Sterilize**   To remove or kill all living organisms. To render an animal infertile.

**Stern**   Tail of a sporting dog or hound.

**Sternum**   Breastbone.

**Sterol**   Any of a group of high-molecular-weight alcohols, such as ergosterol and **cholesterol**.

**Stifle**   The joint of the hind leg between the thigh and the second thigh; knee.

**Stomatitis**   Inflammation of the mouth.

**Stool**   Fecal material; evacuation from the digestive tract.

**Stratum corneum**   The outermost layer of the epidermis in most species of animals; it is fully keratinized and consists of "horny cells" composed of keratin (an insoluble protein) within an intercellular lipid matrix that protect the skin from desiccation and invasion by microorganisms.

***Streptococci* spp**   Classified as belonging to groups A, B, C, D, F, and G. Humans are the natural reservoir hosts of group A streptococci. Group A streptococci can cause infections of the tonsils, oral cavity, ears, skin, joints, lungs, and central nervous system of people.

**Stress**   The sum of all nonspecific biological phenomena caused by adverse conditions or influences. It includes physical, chemical, and/or emotional factors to which an individual fails to make satisfactory adaptation and that cause physiological tensions that may contribute to disease.

**Stricture**   An abnormal narrowing of the lumen of a tubular organ.

**Strobilation**   Asexual reproduction by transverse division of the body of a tapeworm into segments that develop into proglottids.

***Strongyloides stercoralis***   Intestinal threadworm.

**Studbook**   Book in which breeders register the pedigrees of animals.

**Stud dog**   A male dog used for breeding purposes.

**Subclinical**   A disease condition without clinical manifestations; inapparent.

**Subcutaneous (subcutaneously)**   Situated or occurring beneath the skin, a common route of administering **vaccines** and medications.

**Subcutis**   Beneath the skin, **subcutaneous**.

**Sublumbar lymph nodes**   Located on the dorsal surface of the abdominal cavity, underneath the lumbar muscles.

**Suckling**   A young, unweaned animal.

**Supplement (supplemental)**   Refers to the addition of minerals, vitamins, or other minor ingredients (bulkwise) of a diet.

**Supra-**   A prefix meaning on, above, over, or beyond.

**Suture**   The seam between horny plates (**scutes**) of the shell of a turtle or tortoise; material used to oppose and hold layers of tissues for healing.

**Symbiotic**   (adj. of symbiosis) Associated in symbiosis; living together in a mutually beneficial relationship.

**Symmetry**   Pleasing balance among all parts of the body.

**Symptom**   Sign or evidence of a disease or other abnormality.

**Syndrome**   A group of signs and symptoms that occur together and characterize a disease; a disturbance or abnormality.

**Synergism**   The state of working together, **commensal**.

**Synovia (synovial fluid)**  A viscid fluid containing synovin, or mucin, and a small proportion of mineral salts. It is secreted by the synovial membrane and resembles the white of an egg. It is contained in joint cavities, bursae, and tendon sheaths.

**Syrinx**  The vocal organ of birds that is a special modification of the lower part of the trachea or of the bronchi or of both.

**Systemic**  Pertaining to or affecting the body as a whole.

**Systolic**  Pertaining to contraction, the phase of the cardiac cycle when the heart muscle is contracting.

**Tabby**  A striped, blotched, or spotted coat.

**Tachycardia**  A rapid heart rate.

**Tactile**  Pertaining to the touch.

**Taeniidae**  Tapeworms.

**Tail set**  How the base of the tail sets on the rump.

**Tarsus**  **Joint** in the **distal** hind limb, equivalent to human ankle.

**Tartar**  **Plaque** that has hardened into a more solid substance.

**Tasselfoot**  Scaly condition of the feet of canaries produced by infection with *Cnemidocoptes* spp mites.

**TD (Tracking Dog)**  A suffix used with the name of a dog that has been recorded a Tracking Dog as a result of having passed at least two **AKC**-licensed tracking tests. The title may be combined with the UD title and shown as UDT (Utility Dog Tracking).

**TDX (Tracking Dog Excellent)**  A suffix used with the name of a dog that has been recorded a Tracking Dog Excellent as a result of having passed at least two AKC-licensed tracking dog excellent tests. The title may be combined with the UDT title and shown as UDTX.

**Telogen**  Resting phase of the hair cycle; club hairs (not growing).

**Temperament**  Disposition of an animal.

**Tendon**  The tissue connecting muscle to bone.

**Teratogenic**  A substance that causes physical defects in a developing embryo/fetus.

**Terrier**  A group of dogs used originally for hunting vermin (pest rodents).

**Testicles**  The male gonad; gland that produces spermatozoa.

**Tetraplegia (quadriplegia)**  Refers to **paralysis** of four limbs.

**Therapeutically**  Used for a curative purpose (as in the treatment of a medical disorder).

**Therapy**  Treatment of disease.

**Thermolysis**  The loss or dissipation of body heat.

**Thermoneutral zone**  The relatively narrow zone of effective environmental temperature in which heat production at the animal's minimal or thermoneutral rate is offset by net heat loss to the environment without the aid of special heat-conserving or heat-dissipating mechanisms. Thus, the animal is under neither cold nor heat stress. See **comfort zone**.

**Thermoneutrality**  The state of thermal balance between an organism and its environment so that the body thermoregulatory mechanisms are inactive. The **thermoneutral zone** is also referred to as the **comfort zone**.

**Thigh**  The hindquarter from hip to stifle.

**Thoracic**  Pertaining to the thorax (chest).

**Thorax**  The chest.

**Thrombocyte**  A blood **platelet**.

**Thrombocytopenic**  Deficiency of blood platelets.

**Thrombosis**   The formation or presence of a blood clot within a blood vessel.

**Thyroid**   Gland in the neck that helps to regulate many processes of growth and development.

**Thyroxine**   A hormone produced by the thyroid gland; regulates **metabolism** and the activity of many other cells, also called T4 because it contains four molecules of iodine.

**Tick**   Any of the various bloodsucking arachnids that fasten themselves to warm-blooded animals. Some are important **vectors** of diseases.

**Ticking**   One, two, or three bands of contrasting color on each hair of the fur as in the Abyssinian cat.

**Tissue**   Group of similar cells forming a distinct structure.

**Titer**   The quantity of a substance required to produce a reaction with a given volume of another substance, or the amount of one substance required to correspond to a given amount of another substance. *Agglutination titer* is the highest dilution of a serum that causes clumping of bacteria.

**Tocopherols**   Compounds with vitamin E activity.

**Tonicity**   The state of tension or partial contraction of muscle fibers while at rest; normal condition of muscle tone.

**Tonsils**   Prominent masses of lymphoid tissue that lie one on each side of the throat.

**Topline**   The dog's outline from just behind the withers to the tail set.

**Torticollis**   Wry neck, twisting of the neck.

**Tortoiseshell**   Patched coat pattern; commonly comprised of black and orange but also seen in other color mixtures.

***Toxascaris leonina***   Roundworms affecting cats.

**Toxemia**   Generalized blood poisoning, especially a form in which the toxins produced by pathogenic bacteria enter the bloodstream from a local lesion and are distributed throughout the body.

**Toxicants**   Toxic agents, poisons.

**Toxicology**   The branch of science that deals with poisons and their effect on living organisms.

**Toxins**   Poisons produced by certain microorganisms. They are products of cell metabolism. The symptoms of diseases caused by bacteria, such as diphtheria and tetanus, are due to toxins.

***Toxocara canis***   Roundworms affecting dogs.

***Toxocara cati***   Roundworms affecting cats.

**Toxoid**   A detoxified toxin. It retains the ability to stimulate formation of antitoxin in an animal's body. The discovery that toxin treated with formalin loses its toxicity is the basis for preventive immunization against such diseases as diphtheria, poliomyelitis, and tetanus.

***Toxoplasma gondii***   A protozoan; causes the disease toxoplasmosis.

**Toy dog**   One of a group of dogs characterized by very small size.

**Trachea**   The windpipe; in mammals it extends from the throat to the bronchi.

**Tracheobronchitis**   Inflammation of the **trachea** and bronchi.

**Transgenic animal**   An animal into which cloned genetic material has been transferred; it is created artificially from two or more sources and incorporated into a single recombinant molecule.

**Transitory**   Brief; momentary; lasting only a short time; fleeing; transient.

**Trauma**   Injury; wound.

**Trematoda**   A class of the Platyhelminthes, which includes the flukes.

**Tremor**   An involuntary trembling or quivering.

**Trichobezoar**   A hairball; concretion within the stomach or intestines formed of hairs.

*Trichodectes canis*   Biting lice affecting dogs.

*Trichuris vulpis*   Whipworms.

**Tricolor**   Three coat colors together; black, white, and tan.

**Trim**   To groom the coat by plucking or clipping; to cut off the distal tips of **claws**.

**Triple Champion**   A dog that has won bench show, field trial, and obedience trial championships.

**Trophozite**   A protozoal form that replicates asexually.

*Trypanosoma cruzi*   A protozoan found in the blood of dogs; causes Chagas' disease in humans.

**Tuberculin**   A biological agent derived from the growth and further processing of the tubercle bacilli that is used for detection or diagnosis of tuberculosis in animals and humans.

**Tufts**   Hair growing from the ears or between toes.

**Tumor**   A mass of abnormal tissue growing in or on the plant or animal body.

**Tympanic membrane**   A thin membrane that closes the cavity of the middle ear like the head of a drum and separates the middle ear cavity from the external ear canal.

**Type**   The characteristic qualities distinguishing a breed; the embodiment of a standard's essentials. See **conformation**.

**Typhus**   Disease caused by a rickettsial organism; it is characterized by high fever, depression, and a dark red rash.

**UD (Utility Dog)**   A suffix used with the name of a dog that has been recorded a Utility Dog by AKC as a result of having won certain minimum scores in Utility Classes at a specified number of AKC-licensed obedience trials. The title may be combined with TD or TDX title and shown as UDT or UDTX.

**Ulcer**   A break in skin or mucous membrane that is characterized by loss of surface tissue.

**Umbilicus**   Navel.

*Uncinaria stenocephala*   Hookworms affecting dogs.

**Underline**   The combined contours of the brisket and the abdominal floor.

**Undershot**   The front teeth (**incisors**) of the lower jaw overlapping or projecting beyond the front teeth of the upper jaw when the mouth is closed.

**Uniparous**   Producing only one egg or one offspring at a time.

**Unsaturated fat**   A fat having one or more double bonds; not completely **hydrogenated**.

**Unsound**   An animal incapable of performing the functions for which it was designed; often refers to lameness.

**Unthrifty**   Not thriving, in poor general body condition, underweight with a dull hair coat.

**Upper arm**   The humerus or bone of the foreleg between the shoulder blade and the forearm.

**Uremia**   A condition where the kidneys fail to excrete nitrogenous wastes and other substances; these build up in the blood producing toxic signs associated with kidney failure.

**Urethra**   The canal that in most mammals carries urine from the bladder and, in mature males, serves to transport semen.

**Urinary**   Pertaining to urine.

**Urogenital (genitourinary)**   Pertaining to the urinary and genital tracts (including the kidneys and sex organs).

**Urolith**   A calculus (stone) in the urinary tract.

**Urolithiasis**   A condition with stones/calculi in the urinary tract.

**Uroliths**   A urinary calculus or stone.

**Uropygial gland**   The preen gland (used by birds to waterproof their feathers). See **oil gland**.

**USDA**   United States Department of Agriculture.

**USDHHS**   United States Department of Health and Human Services.

**USP**   United States Pharmacopoeia. A unit of measurement or potency of biologicals that usually coincides with an international unit. See **IU**.

**USPHS**   United States Public Health Service.

**Uterus**   The womb; organ in which the fertilized egg implants and the embryo develops.

**UV-B (Ultraviolet-B)**   A section of the UV spectrum with wavelengths between 270 and 320 nm.

**Uvea**   Includes the **iris**, ciliary body, and choroid of the eye (see Chapter 11); these are pigmented vascular structures of the eye.

**Uveitis**   Inflammation of the **uvea**.

**Vaccination**   From the Latin *vacca* meaning "cow." Artificial **immunization**. To inoculate with a mildly toxic preparation of bacteria or a virus of a specific disease to prevent or lessen the effects of that disease.

**Vaccine**   A suspension of attenuated or killed microorganisms (bacteria, viruses, or rickettsiae) administered for the prevention, amelioration (improvement), or treatment of infectious diseases.

**Vascular**   Relating to blood vessels.

**Vasculitis**   Inflammation of a blood or lymph vessel.

**Vas deferens**   The excretory duct of the testis; provides a conduit for sperm to travel to the urethra.

**Vasectomy**   Surgical removal of all or part of the **vas deferens**. The male becomes sterile but retains **libido**.

**Vasoconstriction**   Constriction of blood vessels.

**Vasodilation**   The dilation of blood vessels resulting from stimulation by a nerve or drug or **hormone**.

**Vector**   From the Latin *vector* meaning "carrier." An organism, such as a mosquito or **tick**, that transmits microorganisms that cause disease.

**Venom**   Poisonous secretion of bees, scorpions, snakes, and certain other animals.

**Vena cava**   One of the large veins by which blood is returned to the right atrium of the heart of a vertebrate animal.

**Ventilation rate**   The volume of air exhaled per unit time.

**Ventral**   Denoting a position toward the abdomen or belly (lower) surface. It means the same as **anterior** in human anatomy.

**Ventricular fibrillation**   Very rapid, uncoordinated contractions of the ventricles of the heart, resulting in the loss of synchronization between heartbeat and pulse beat. Ventricular fibrillation often results in death.

**Vermicide**   Substance that kills internal parasitic worms.

**Vertebral column**   Spinal column; comprised of vertebrae and intervertebral discs.

**Vessels**   Tubes or canals (e.g., arteries) in which body fluids are contained and conveyed or circulated, such as the blood or lymph.

**Viral vaccine**   A preparation of killed microorganisms, living attenuated organisms, or living fully virulent organisms that is administrated to produce or artificially increase immunity to a particular disease. An example of a viral vaccine is a preparation containing the virus of cowpox in a form used to vaccinate humans against smallpox.

**Viremia**    An infection of the bloodstream caused by a virus.

**Virucide**    A chemical or physical agent that kills or inactivates viruses; a disinfectant.

**Virulence**    The degree of pathogenicity (ability to produce disease) of a microorganism as indicated by case fatality rates and/or its ability to invade the tissues of a host.

**Virus**    One of a group of minute infectious agents. They are characterized by a lack of independent metabolism and by the ability to replicate only within living host cells. They include any of a group of disease-producing agents composed of protein and nucleic acid. Viruses are filterable and cause such diseases in people as rabies, poliomyelitis, chicken pox, and the common cold.

**Viscera**    The internal organs of the body, particularly in the chest and abdominal cavities, such as the heart, lungs, liver, **intestines**, and kidneys.

**Visceral**    Pertaining to the **viscera** (organs).

**Vitamin D$_3$**    A fat-soluble alcohol important in calcium metabolism in animals; $D_3$ is the metabolically active form and is produced by the action of ultraviolet irradiation on ergosterol-derived sterols (e.g., $D_2$) present in the skin.

**Vivarium**    A place, especially an indoor enclosure, for keeping and raising living animals and plants under natural conditions for observation or research.

**Viviparous (viviparously)**    Producing living young (as opposed to eggs) from within the body in the manner of nearly all mammals, many reptiles, and a few fishes.

**VO$_2$**    VO$_2$ max is the maximum volume (V) of oxygen (O$_2$) in milliliters that can be used by mammals per kilogram of body weight in one minute while breathing air at sea level. Because O$_2$ consumption is related linearly to energy expenditure, measuring O$_2$ consumption is indirectly measuring an animal's maximal capacity to perform work aerobically.

**Walk**    Gaiting pattern in which three legs are in support of the body at all times, each foot lifting from the ground one at a time in regular sequence.

**Water bottle**    Pet water bottles have a reservoir connected to a stainless steel tube and are designed to hang inside animal cages or pens to provide a sanitary source of drinking water.

**Water hopper**    Pet water dispenser, generally has a see-through top reservoir connected to a small troughlike base from which the pet drinks.

**Webbed**    Connected by a membrane. Webbed feet are important for water-retrieving breeds.

**Wheal**    A flat, usually circular, hard elevation of the skin, commonly accompanied by burning or itching. Its formation follows an irritation or other means of increasing the permeability of the vascular walls of the skin.

**Whelp**    An unweaned puppy; to give birth to (term used in various carnivores, such as dogs).

**Whiskers**    The long stiff hairs (bristles) protruding from an animal's face, such as cats.

**White matter**    Refers to myelinated nerve fibers in the brain.

**WHO**    World Health Organization. An agency of the United Nations founded in 1948. It seeks to promote worldwide health and prevent outbreak of disease. It assists countries in strengthening public health services. It plans and coordinates international efforts to solve health problems, with special attention to malaria; tuberculosis; and venereal, virus, and parasitic diseases. It works with member countries and other health organizations to collect information on epidemics; to develop international quarantine regulations; and to standardize medical drugs, vaccines, and treatment. More than 120 countries belong to WHO, which is headquartered in Geneva, Switzerland.

**Winners**    An award given at dog shows to the best dog (Winners Dog) and best bitch (Winners Bitch) competing in regular classes.

**Wirehair**    A coat of hard, crisp, wiry texture.

**Withers**    The highest point of the shoulders immediately behind the neck.

**Worms**    Soft-bodied, elongated invertebrates.

**Wrinkle**    Loose, folding skin on forehead and foreface.

**Wry mouth**    Lower jaw does not line up with upper jaw.

**Xenophobia**    Fear of strangers or of anything new or strange.

**X-rays**    Radiation produced when electrons in a vacuum tube are projected at very high tension and velocity to strike a metallic target. These are electromagnetic waves, but their wavelength is only about one-thousandth of that of visible light. X-rays are sometimes called roentgen rays, after their discoverer, Wilhelm Roentgen. See **radiograph**.

**Yolk peritonitis**    The breaking of an egg inside a bird resulting in an inflammatory reaction within the animal's abdominal cavity.

**Zo-**    The prefix *zo-* implies *animal*.

**Zoologist**    An individual that studies the branch of biology concerned with the classification and the properties and vital phenomena of animals.

**Zoonosis**    (pl. **zoonoses**) Those diseases and infections that can be transmitted between vertebrate animals and humans.

**Zoonotic diseases**    Those diseases naturally transmitted between animals and humans.

# INDEX

Page numbers followed by *f* denote figures; *t,* tables; and *n,* footnotes. The abbreviation "CP" precedes color-plate designations.

A  Courtesy Jeff Rathmann (photographer) and PetMarket Place pet store, Webster Groves, MO.

B  Courtesy Questhavenpets.com.

C  Courtesy Champion Petfoods, Inc.

D  Courtesy Tom Schaefges.

E  Courtesy Jeff Rathmann (photographer) and PetMarket Place pet store, Webster Groves, MO.

F  Courtesy Dr. Michael J. Adkesson.

G  Courtesy Dr. Michael J. Adkesson.

H  Courtesy Dr. Michael J. Adkesson.

I  Courtesy Champion Petfoods, Inc.

J  Courtesy Champion Petfoods, Inc.

K  Courtesy Jane Rothert.

L  Courtesy Maria Lang (photographer) and Dr. B. Taylor Bennett.

M  Courtesy Paul E. Miller, Sullivan, MO.

N  Courtesy Melissa Maitland.